Materials Handbook for Hybrid Microelectronics

621.38173
M425k
(pb)
Materials handbook for hybrid microelectronics.
c.2

Materials Handbook for Hybrid Microelectronics

Joseph Alison King, Editor
Teledyne Microelectronics
Los Angeles, California

Associate Editors
Julia Freer
Karen Kuty
Ho Nam
Karl Wee
Robert Woodard

Artech House
Boston • London

Library of Congress Cataloging-in-Publication Data

Materials handbook for hybrid microelectronics.
Bibliography: p.
1. Hybrid integrated circuits — Materials — Handbooks, manual, etc. I. King, Joseph A. (Joseph Alison)
TK7874.M343 1988 621.381'73 88-19333
ISBN 0-89006-325-7

Published by
ARTECH HOUSE, INC.
685 Canton Street
Norwood, MA 02062

International Standard Book Number: 0-89006-325-7
Library of Congress Catalog Card Number: 88-19333

10 9 8 7 6 5 4 3 2 1

CONTENTS

PREFACE xv
ACKNOWLEDGMENTS xvii
INTRODUCTION 1
Units 1
Thermophysical Properties 2
Density 2
Specific Heat 2
Thermal Conductivity 3
Thermal Expansion Coefficient 4
Elastic Constants 5
Electrical Properties 7
Resistivity 7
Temperature Coefficient of Resistance 8
Dielectric Constant, Dielectric Strength, and Dissipation Factor 9
Band Gap 10
Mechanical Properties 10
Strength 10
Hardness 13
References for Introduction 14
ELEMENTS 16
Aluminum 17
Specific Heat vs. Temperature 18
Thermal Conductivity vs. Temperature 19
Expansion Coefficient vs. Temperature 20
Electrical Resistivity vs. Temperature 21
Young's Modulus vs. Temperature 22
Poisson's Ratio vs. Temperature 23
Yield Strength vs. Temperature 24
Ultimate Strength vs. Temperature 25
Elongation vs. Temperature 26
Hardness vs. Temperature 27
Chromium 28
Specific Heat vs. Temperature 29
Thermal Conductivity vs. Temperature 30
Expansion Coefficient vs. Temperature 31
Electrical Resistivity vs. Temperature 32
Young's Modulus vs. Temperature 33
Poisson's Ratio vs. Temperature 34
Yield Strength vs. Temperature 35
Ultimate Strength vs. Temperature 36
Elongation vs. Temperature 37
Hardness vs. Temperature 38

Copper 39
Specific Heat vs. Temperature 40
Thermal Conductivity vs. Temperature 41
Expansion Coefficient vs. Temperature 42
Electrical Resistivity vs. Temperature 43
Young's Modulus vs. Temperature 44
Poisson's Ratio vs. Temperature 45
Yield Strength vs. Temperature 46
Ultimate Strength vs. Temperature 47
Elongation vs. Temperature 48
Hardness vs. Temperature 49
Germanium 50
Specific Heat vs. Temperature 51
Thermal Conductivity vs. Temperature 52
Expansion Coefficient vs. Temperature 53
Electrical Resistivity vs. Temperature 54
Young's Modulus vs. Temperature 55
Young's Modulus for Single Crystal 56
Poisson's Ratio vs. Temperature 57
Yield Strength vs. Temperature 58
Ultimate Strength vs. Temperature 59
Elongation vs. Temperature 60
Hardness vs. Temperature 61
Gold 62
Specific Heat vs. Temperature 63
Thermal Conductivity vs. Temperature 64
Expansion Coefficient vs. Temperature 65
Electrical Resistivity vs. Temperature 66
Young's Modulus vs. Temperature 67
Poisson's Ratio vs. Temperature 68
Yield Strength vs. Temperature 69
Ultimate Strength vs. Temperature 70
Elongation vs. Temperature 71
Hardness vs. Temperature 72
Iron 73
Specific Heat vs. Temperature 74
Thermal Conductivity vs. Temperature 75
Expansion Coefficient vs. Temperature 76
Electrical Resistivity vs. Temperature 77
Young's Modulus vs. Temperature 78
Poisson's Ratio vs. Temperature 79
Yield Strength vs. Temperature 80
Ultimate Strength vs. Temperature 81
Elongation vs. Temperature 82
Hardness vs. Temperature 83
Molybdenum 84
Specific Heat vs. Temperature 85
Thermal Conductivity vs. Temperature 86
Expansion Coefficient vs. Temperature 87
Electrical Resistivity vs. Temperature 88
Young's Modulus vs. Temperature 89
Poisson's Ratio vs. Temperature 90
Yield Strength vs. Temperature 91
Ultimate Strength vs. Temperature 92
Elongation vs. Temperature 93
Hardness vs. Temperature 94

Nickel 95
- Specific Heat vs. Temperature 96
- Thermal Conductivity vs. Temperature 97
- Expansion Coefficient vs. Temperature 98
- Electrical Resistivity vs. Temperature 99
- Young's Modulus vs. Temperature 100
- Poisson's Ratio vs. Temperature 101
- Yield Strength vs. Temperature 102
- Ultimate Strength vs. Temperature 103
- Elongation vs. Temperature 104
- Hardness vs. Temperature 105

Silicon 106
- Specific Heat vs. Temperature 107
- Thermal Conductivity vs. Temperature 108
- Expansion Coefficient vs. Temperature 109
- Electrical Resistivity vs. Temperature 110
- Young's Modulus vs. Temperature 113
- Young's Modulus for Single Crystal 114
- Poisson's Ratio vs. Temperature 115
- Yield Strength vs. Temperature 116
- Ultimate Strength vs. Temperature 117
- Elongation vs. Temperature 118
- Hardness vs. Temperature 119

Silver 120
- Specific Heat vs. Temperature 121
- Thermal Conductivity vs. Temperature 122
- Expansion Coefficient vs. Temperature 123
- Electrical Resistivity vs. Temperature 124
- Young's Modulus vs. Temperature 125
- Poisson's Ratio vs. Temperature 126
- Yield Strength vs. Temperature 127
- Ultimate Strength vs. Temperature 128
- Elongation vs. Temperature 129
- Hardness vs. Temperature 130

Tantalum 131
- Specific Heat vs. Temperature 132
- Thermal Conductivity vs. Temperature 133
- Expansion Coefficient vs. Temperature 134
- Electrical Resistivity vs. Temperature 135
- Young's Modulus vs. Temperature 136
- Poisson's Ratio vs. Temperature 137
- Yield Strength vs. Temperature 138
- Ultimate Strength vs. Temperature 139
- Elongation vs. Temperature 140
- Hardness vs. Temperature 141

Tin 142
- Specific Heat vs. Temperature for White Tin 143
- Specific Heat vs. Temperature for Grey Tin 144
- Thermal Conductivity vs. Temperature 145
- Expansion Coefficient vs. Temperature 146
- Electrical Resistivity vs. Temperature 147
- Young's Modulus vs. Temperature 148
- Poisson's Ratio vs. Temperature 149
- Yield Strength vs. Temperature 150
- Ultimate Strength vs. Temperature 151
- Elongation vs. Temperature 152
- Hardness vs. Temperature 153

Tungsten ... 154
Specific Heat vs. Temperature ... 155
Thermal Conductivity vs. Temperature ... 156
Expansion Coefficient vs. Temperature ... 157
Electrical Resistivity vs. Temperature ... 158
Young's Modulus vs. Temperature ... 159
Poisson's Ratio vs. Temperature ... 160
Yield Strength vs. Temperature (Recryst.) ... 161
Yield Strength vs. Temperature (Wrought) ... 162
Ultimate Tensile Strength vs. Temperature (Wrought) ... 162
Ultimate Strength vs. Temperature (Recryst.) ... 163
Elongation vs. Temperature (Recryst.) ... 164
Elongation vs. Temperature (Wrought) ... 165
Hardness vs. Temperature ... 166
Typical Strength Values ... 167
References for Elements ... 168

ALLOYS ... 196

96.5Sn 3.5Ag ... 197
Specific Heat vs. Temperature ... 198
Thermal Conductivity vs. Temperature ... 199
Expansion Coefficient vs. Temperature ... 200
Electrical Resistivity vs. Temperature ... 201
Young's Modulus vs. Temperature ... 202
Poisson's Ratio vs. Temperature ... 203
Yield Strength vs. Temperature ... 204
Ultimate Strength vs. Temperature ... 205
Shear Strength ... 206
Elongation vs. Temperature ... 207
Hardness vs. Temperature ... 208
Sn-Ag Phase Diagram ... 209
95Sn 5Ag ... 210
Specific Heat vs. Temperature ... 211
Thermal Conductivity vs. Temperature ... 212
Expansion Coefficient vs. Temperature ... 213
Electrical Resistivity vs. Temperature ... 214
Young's Modulus vs. Temperature ... 215
Poisson's Ratio vs. Temperature ... 216
Yield Strength vs. Temperature ... 217
Ultimate Strength vs. Temperature ... 218
Elongation vs. Temperature ... 219
Hardness vs. Temperature ... 220
88Au 12Ge ... 221
Specific Heat vs. Temperature ... 222
Thermal Conductivity vs. Temperature ... 223
Expansion Coefficient vs. Temperature ... 224
Electrical Resistivity vs. Temperature ... 225
Young's Modulus vs. Temperature ... 226
Poisson's Ratio vs. Temperature ... 227
Yield Strength vs. Temperature ... 228
Ultimate Strength vs. Temperature ... 229
Elongation vs. Temperature ... 230
Hardness vs. Temperature ... 231
Au-Ge Phase Diagram ... 232

94Au 6Si 233
Specific Heat vs. Temperature 234
Thermal Conductivity vs. Temperature 235
Expansion Coefficient vs. Temperature 236
Electrical Resistivity vs. Temperature 237
Young's Modulus vs. Temperature 238
Poisson's Ratio vs. Temperature 239
Yield Strength vs. Temperature 240
Ultimate Strength vs. Temperature 241
Elongation vs. Temperature 242
Hardness vs. Temperature 243
Au-Si Phase Diagram 244
80Au 20Sn 245
Specific Heat vs. Temperature 246
Thermal Conductivity vs. Temperature 247
Expansion Coefficient vs. Temperature 248
Electrical Resistivity vs. Temperature 249
Young's Modulus vs. Temperature 250
Poisson's Ratio vs. Temperature 251
Yield Strength vs. Temperature 252
Ultimate Strength vs. Temperature 253
Elongation vs. Temperature 254
Hardness vs. Temperature 255
Au-Sn Phase Diagram 256
99.9Fe 0.1C 257
Specific Heat vs. Temperature 258
Thermal Conductivity vs. Temperature 259
Expansion Coefficient vs. Temperature 260
Electrical Resistivity vs. Temperature 261
Young's Modulus vs. Temperature 262
Poisson's Ratio vs. Temperature 263
Yield Strength vs. Temperature 264
Ultimate Strength vs. Temperature 265
Elongation vs. Temperature 266
Hardness vs. Temperature 267
Fe-C Phase Diagram 268
55Cu 45Ni (Cupron) 269
Specific Heat vs. Temperature 270
Thermal Conductivity vs. Temperature 271
Expansion Coefficient vs. Temperature 272
Electrical Resistivity vs. Temperature 273
Young's Modulus vs. Temperature 274
Poisson's Ratio vs. Temperature 275
Yield Strength vs. Temperature 276
Ultimate Strength vs. Temperature 277
Elongation vs. Temperature 278
Hardness vs. Temperature 279
Cu-Ni Phase Diagram 280
80Ni 20Cr (Nichrome) 281
Specific Heat vs. Temperature 282
Thermal Conductivity vs. Temperature 283
Expansion Coefficient vs. Temperature 284
Electrical Resistivity vs. Temperature 285
Young's Modulus vs. Temperature 286
Poisson's Ratio vs. Temperature 287
Yield Strength vs. Temperature 288

Ultimate Strength vs. Temperature 290
Elongation vs. Temperature 291
Hardness vs. Temperature 292
Ni-Cr Phase Diagram 000
58Fe 42Ni (Alloy 42) 293
Specific Heat vs. Temperature 294
Thermal Conductivity vs. Temperature 295
Expansion Coefficient vs. Temperature 296
Electrical Resistivity vs. Temperature 297
Young's Modulus vs. Temperature 298
Poisson's Ratio vs. Temperature 299
Yield Strength vs. Temperature 300
Ultimate Strength vs. Temperature 301
Elongation vs. Temperature 302
Hardness vs. Temperature 303
Fe-Ni Phase Diagram 304
50Pb 50In 305
Specific Heat vs. Temperature 306
Thermal Conductivity vs. Temperature 307
Expansion Coefficient vs. Temperature 308
Electrical Resistivity vs. Temperature 309
Young's Modulus vs. Temperature 310
Poisson's Ratio vs. Temperature 311
Yield Strength vs. Temperature 312
Ultimate Strength vs. Temperature 313
Elongation vs. Temperature 314
Hardness vs. Temperature 315
80Pb 20In 316
Specific Heat vs. Temperature 317
Thermal Conductivity vs. Temperature 318
Expansion Coefficient vs. Temperature 319
Electrical Resistivity vs. Temperature 320
Young's Modulus vs. Temperature 321
Poisson's Ratio vs. Temperature 322
Yield Strength vs. Temperature 323
Ultimate Strength vs. Temperature 324
Elongation vs. Temperature 325
Hardness vs. Temperature 326
Pb-In Phase Diagram 327
89Ni 11P 328
Specific Heat vs. Temperature 329
Thermal Conductivity vs. Temperature 330
Expansion Coefficient vs. Temperature 331
Electrical Resistivity vs. Temperature 332
Young's Modulus vs. Temperature 333
Poisson's Ratio vs. Temperature 334
Yield Strength vs. Temperature 335
Ultimate Strength vs. Temperature 336
Elongation vs. Temperature 337
Hardness vs. Temperature 338
Ni-P Phase Diagram 339
63Sn 37Pb 340
Specific Heat vs. Temperature 341
Thermal Conductivity vs. Temperature 342
Expansion Coefficient vs. Temperature 343
Electrical Resistivity vs. Temperature 344

Young's Modulus vs. Temperature ... 345
Poisson's Ratio vs. Temperature ... 346
Shear Strength ... 347
Yield Strength vs. Temperature ... 348
Ultimate Strength vs. Temperature ... 349
Elongation vs. Temperature ... 350
Hardness vs. Temperature ... 351
Sn-Pb Phase Diagram ... 352
54Fe 29Ni 17Co (Kovar) ... 353
Specific Heat vs. Temperature ... 354
Thermal Conductivity vs. Temperature ... 355
Expansion Coefficient vs. Temperature ... 356
Electrical Resistivity vs. Temperature ... 357
Young's Modulus vs. Temperature ... 358
Poisson's Ratio vs. Temperature ... 359
Yield Strength vs. Temperature ... 360
Ultimate Strength vs. Temperature ... 361
Elongation vs. Temperature ... 362
Hardness vs. Temperature ... 363
Fe-Ni-Co Phase Diagram ... 364
88Pb 10Sn 2Ag ... 365
Specific Heat vs. Temperature ... 366
Thermal Conductivity vs. Temperature ... 367
Expansion Coefficient vs. Temperature ... 368
Electrical Resistivity vs. Temperature ... 369
Young's Modulus vs. Temperature ... 370
Poisson's Ratio vs. Temperature ... 371
Yield Strength vs. Temperature ... 372
Ultimate Strength vs. Temperature ... 373
Elongation vs. Temperature ... 374
Hardness vs. Temperature ... 375
Pb-Ag-Sn Phase Diagram ... 376
92.5Pb 5Sn 2.5Ag ... 377
Specific Heat vs. Temperature ... 378
Thermal Conductivity vs. Temperature ... 379
Expansion Coefficient vs. Temperature ... 380
Electrical Resistivity vs. Temperature ... 381
Young's Modulus vs. Temperature ... 382
Poisson's Ratio vs. Temperature ... 383
Yield Strength vs. Temperature ... 384
Ultimate Strength vs. Temperature ... 385
Elongation vs. Temperature ... 386
Hardness vs. Temperature ... 387
66Sn 36Pb 2Ag ... 388
Specific Heat vs. Temperature ... 389
Thermal Conductivity vs. Temperature ... 390
Expansion Coefficient vs. Temperature ... 391
Electrical Resistivity vs. Temperature ... 392
Young's Modulus vs. Temperature ... 393
Poisson's Ratio vs. Temperature ... 394
Yield Strength vs. Temperature ... 395
Ultimate Strength vs. Temperature ... 396
Elongation vs. Temperature ... 397
Hardness vs. Temperature ... 398
Al-Au Phase Diagram ... 399
Au-Pb-Sn Phase Diagram ... 400
References for Alloys ... 401

INSULATORS 437
AlN 438
Specific Heat vs. Temperature 439
Thermal Conductivity vs. Temperature 440
Expansion Coefficient vs. Temperature 441
Dielectric Constant vs. Temperature 442
Dielectric Strength vs. Temperature 443
Dissipation Factor vs. Temperature 444
Young's Modulus vs. Temperature 445
Poisson's Ratio vs. Temperature 446
Compressive Strength vs. Temperature 447
Ultimate Strength vs. Temperature 448
Flexural Strength vs. Temperature 449
Hardness vs. Temperature 450
Al_2O_3 (96% Ceramic) 451
Specific Heat vs. Temperature 452
Thermal Conductivity vs. Temperature 453
Expansion Coefficient vs. Temperature 454
Dielectric Constant vs. Temperature 455
Dielectric Strength vs. Temperature 456
Dissipation Factor vs. Temperature 457
Young's Modulus vs. Temperature 458
Poisson's Ratio vs. Temperature 459
Compressive Strength vs. Temperature 460
Ultimate Strength vs. Temperature 461
Flexural Strength vs. Temperature 462
Hardness vs. Temperature 463
Al_2O_3 (99% Ceramic) 464
Specific Heat vs. Temperature 465
Thermal Conductivity vs. Temperature 466
Expansion Coefficient vs. Temperature 467
Dielectric Constant vs. Temperature 468
Dielectric Strength vs. Temperature 469
Dissipation Factor vs. Temperature 470
Young's Modulus vs. Temperature 471
Poisson's Ratio vs. Temperature 472
Compressive Strength vs. Temperature 473
Ultimate Strength vs. Temperature 474
Flexural Strength vs. Temperature 475
Hardness vs. Temperature 476
Al_2O_3 (Sapphire) 477
Specific Heat vs. Temperature 478
Thermal Conductivity vs. Temperature 479
Expansion Coefficient vs. Temperature 480
Dielectric Constant vs. Temperature 481
Dielectric Strength vs. Temperature 482
Dissipation Factor vs. Temperature 483
Young's Modulus vs. Temperature 484
Poisson's Ratio vs. Temperature 485
Compressive Strength vs. Temperature 486
Ultimate Strength vs. Temperature 487
Flexural Strength vs. Temperature 488
Hardness vs. Temperature 489
BeO 490
Specific Heat vs. Temperature 491
Thermal Conductivity vs. Temperature 492
Expansion Coefficient vs. Temperature 493

Dielectric Constant vs. Temperature ... 494
Dielectric Strength vs. Temperature ... 495
Dissipation Factor vs. Temperature ... 496
Young's Modulus vs. Temperature ... 497
Poisson's Ratio vs. Temperature ... 498
Compressive Strength vs. Temperature ... 499
Ultimate Strength vs. Temperature ... 500
Flexural Strength vs. Temperature ... 501
Hardness vs. Temperature ... 502
Carbon (Diamond) ... 503
Specific Heat vs. Temperature ... 504
Thermal Conductivity vs. Temperature ... 505
Expansion Coefficient vs. Temperature ... 506
Dielectric Constant vs. Temperature ... 507
Dielectric Strength vs. Temperature ... 508
Dissipation Factor vs. Temperature ... 509
Young's Modulus vs. Temperature ... 510
Poisson's Ratio vs. Temperature ... 511
Compressive Strength vs. Temperature ... 512
Ultimate Strength vs. Temperature ... 513
Flexural Strength vs. Temperature ... 514
Hardness vs. Temperature ... 515
SiO_2 (Fused Quartz) ... 516
Specific Heat vs. Temperature ... 517
Thermal Conductivity vs. Temperature ... 518
Expansion Coefficient vs. Temperature ... 519
Dielectric Constant vs. Temperature ... 520
Dielectric Strength vs. Temperature ... 521
Dissipation Factor vs. Temperature ... 522
Young's Modulus vs. Temperature ... 523
Poisson's Ratio vs. Temperature ... 524
Compressive Strength vs. Temperature ... 525
Ultimate Strength vs. Temperature ... 526
Flexural Strength vs. Temperature ... 527
Hardness vs. Temperature ... 528
SiO_x (Silicon Monoxide) ... 529
Specific Heat vs. Temperature ... 531
Thermal Conductivity vs. Temperature ... 532
Thermal Conductance ... 533
Expansion Coefficient vs. Temperature ... 534
Dielectric Constant vs. Temperature ... 535
Dielectric Constant vs. Deposition Rate ... 536
Percent Change in Capacitance vs. Temperature ... 537
Dielectric Strength vs. Temperature ... 538
Dissipation Factor vs. Temperature ... 539
Reduced Young's Modulus vs. Temperature ... 540
Poisson's Ratio vs. Temperature ... 541
Yield Strength vs. Temperature ... 542
Ultimate Strength vs. Temperature ... 543
Elongation vs. Temperature ... 544
Hardness vs. Temperature ... 545
Hardness vs. Deposition Pressure ... 546
Ta_2O_5 ... 547
Specific Heat vs. Temperature ... 548
Thermal Conductivity vs. Temperature ... 549
Expansion Coefficient vs. Temperature ... 550
Dielectric Constant vs. Temperature ... 551

Percent Change in Capacitance vs. Temperature 552
Dielectric Strength vs. Temperature 553
Dissipation Factor vs. Temperature 554
Reduced Young's Modulus vs. Temperature 555
Poisson's Ratio vs. Temperature 556
Yield Strength vs. Temperature 557
Ultimate Strength vs. Temperature 558
Elongation vs. Temperature 559
Hardness vs. Temperature 560
References for Insulators 561

COMPOUND SEMICONDUCTORS 578

GaAs 579
Specific Heat vs. Temperature 580
Thermal Conductivity vs. Temperature 581
Expansion Coefficient vs. Temperature 582
Electrical Resistivity vs. Temperature 583
Band Gap vs. Temperature 584
SiC 585
Specific Heat vs. Temperature 586
Thermal Conductivity vs. Temperature 587
Expansion Coefficient vs. Temperature 588
Electrical Resistivity vs. Temperature 589
Band Gap vs. Temperature 590
Ta_2N 591
Specific Heat vs. Temperature 592
Thermal Conductivity vs. Temperature 593
Expansion Coefficient vs. Temperature 594
Electrical Resistivity vs. Temperature 595
Band Gap vs. Temperature 596
$CrSi_2$ 597
Specific Heat vs. Temperature 598
Thermal Conductivity vs. Temperature 599
Expansion Coefficient vs. Temperature 600
Electrical Resistivity vs. Temperature 601
Band Gap vs. Temperature 602
References for Compound Semiconductors 603

GLOSSARY 610

PREFACE

A systematic method was used to reduce the mountain of available data to the information listed in this book. The method adopted was to find three independent original references for each property that demonstrated reasonable agreement. The final value listed was chosen from these three as an average or, in some cases, as what is believed to be the most appropriate value. The references associated with the tables of properties list the citing and the value of the property at the indicated temperature.

Extensive use was made of review articles. Particularly heavy use was made of the information provided in *Thermophysical Properties of Matter*, published by the Thermophysical Properties Research Center (TPRC). The property values listed by the TPRC are the synthesis of numerous studies sometimes reduced to a polynomial expressing the property value as a function of temperature. The material property value given by the polynomial expression is typically the value listed in the property table. However, the property values listed in the references are from the original work used to develop the polynomial expression. The accuracy of the polynomial expression varies with temperature. Near room temperature the accuracy is typically 5%. The values listed in the references may deviate from the values listed in the tables by this amount.

The information listed in this book is believed to be the most accurate available. The data faithfully represent material properties of the materials tested. The user of this book should take some care to assure the processing history of the material in question does not deviate wildly from the processing history of the materials tested.

Finding three independent references for some of the solder alloys was not always possible. For these materials, the published property data are somewhat scarce. In these cases, heavy reliance is made on industrial sources of the materials to provide room temperature data. None of the industrial sources have a sufficient data base to provide material properties as a function of temperature. Additional experimental work is required to further profile these materials. Suggestions for improvement of this handbook are welcome and appreciated. Any errors that may be discovered should be brought to the attention of the publisher.

ACKNOWLEDGMENTS

The project that produced this book is the result of the creative interaction between industry (Teledyne Microelectronics) and higher education (Harvey Mudd College). It represents the symbiotic relationship that can exist between these diverse institutions. It is hoped that other projects like the Harvey Mudd Clinic Program can be initiated to strengthen the relationship between industry and higher education.

INTRODUCTION

"Physical properties are determined by functional relationships between measurable quantities." [I1] Thermal linear expansion, for example, is defined by the relationship that describes changes in length associated with temperature changes. A relationship between mass and volume determines density. Note that mass and volume can be measured without regard to crystallographic orientation whereas direction is implicit in the measurement of thermal linear expansion. If the linear expansion varies with direction, the complete behavior of the crystal cannot be described by a single number. For such materials, behavioral description requires the use of tensors of various ranks.

Examples of physical properties that are generally anisotropic are electrical conductivity (current density produced by an applied electric field), thermal conductivity (heat flow produced by a thermal gradient), piezoelectric susceptibility (polarization resulting from mechanical stress) and elasticity (deformation caused by elastic stress). Some properties are independent of direction for all crystallographic systems. Examples of these properties include heat capacity and density. Special circumstances arise for materials that solidify into cubic arrays. These materials are isotropic in electrical and thermal conductivity. This should not be taken to imply that cubic crystals are isotropic for all properties. All crystals are anisotropic for some of their properties.

The material properties presented in this book are listed primarily as scalar quantities even though the properties may in fact depend on crystallographic direction. The use of scalar values is justified on the basis of polycrystalline aggregates with sufficient randomness to obscure directional dependence. In most cases, the properties are measured on polycrystalline materials. For the elastic constants, however, the isotropic values are calculated using anisotropic tensor coefficients. For the materials that are used as single crystals, an attempt is made to provide properties for the minimum number of crystallographic directions required to completely describe the material.

UNITS

The units used in this databook are from the International System of Units (SI), centimeter-gram-second (CGS), meter-kilogram-second (MKS), and foot-pound-second (FPS) systems of units. In the following property definitions and explanations, the most general form of units is supplied. The general forms of the units are expressed in terms of the defined units M(mass), L(length), t(time), T(temperature) and A(current). The specific units listed are for the most convenient form in the hybrid microelectronic industry. Conversion factors are provided for expressing the data in various unit systems. Other abbreviations are listed in the Glossary.

THERMOPHYSICAL PROPERTIES

Density

Density is the concentration of matter. It is expressed as mass per unit volume. The units of density are M/L^3. Density is a scalar and is independent of crystallographic direction. For the CGS system, density is expressed in g/cm^3. For conversion, one g/cm^3 = 10^3 kg/m^3 = 0.03613 lb/in^3. Except where noted, the densities listed are bulk values. The bulk density will generally be higher than the corresponding thin film density due to the high frequency of indigenous imperfections present in thin films. Annealed thin films have densities that approach the bulk values.

The density of a solid material is only a weak function of temperature. As the temperature increases, the distance between atoms gets larger and the number of vacant lattice sites increases leading to a decrease in density. However, the increase in interatomic separation is small and the increase in lattice vacancies is negligible up to about one-half the melting point. For these reasons, the variation of density with temperature is not given. The densities listed are for room temperature except where noted. The decrease in density with increasing temperature can be calculated using the thermal expansion coefficient, accurate up to one-half the melting point.

Specific Heat

Specific heat is a measure of a material's ability to store heat energy. The general units of specific heat are L^2/t^2-T. In the CGS system, the units of specific heat are cal/g-K. For conversion, 1 cal/g-K = 4.184 J/g-K = 1 Btu/lb-°F. Specific heat is a scalar and is independent of direction for all crystal systems.

The terms specific heat, heat capacity, and thermal capacity are sometimes used synonymously although their definitions differ. The CRC *Handbook of Chemistry and Physics* (53rd Ed.) defines *heat capacity* as the quantity of heat required to increase the temperature of a substance one degree. The *thermal capacity* is the amount of heat necessary to produce a unit temperature change of a unit mass of material. Thermal capacity is therefore heat capacity per unit mass. The same handbook defines the *specific heat* of a material as the ratio of its thermal capacity to that of water at 15°C. The thermal capacity and specific heat are sometimes considered numerically equivalent even though the thermal capacity of water at 15°C is 0.99976 cal/g-°C. Some sources give specific heat the definition the CRC handbook reserves for thermal capacity. In this instance, the specific heat and heat capacity per unit mass are truly numerically equivalent. The property listed in this book is heat capacity per unit mass and is given the name *specific heat*.

When a quantity of heat, Q, is added to a material to create an increase in temperature from T_1 to T_2, the *mean* specific heat of a unit mass of material is defined by

$$c_m = Q/(T_2 - T_1) \tag{1}$$

The limiting value of the above ratio as the temperature changes by $\mathrm{d}T$ is defined as the true specific heat, c_t, or

$$c_t = \delta Q/\mathrm{d}T \tag{2}$$

where δQ is used to indicate that the process is path dependent. Raising the temperature one degree under constant pressure requires a different amount of heat than under conditions of constant volume. There are, in fact, infinitely many different specific heats. The most common specific heats are under conditions of constant pressure, c_p, and constant volume, c_v. Values for the true specific heat at constant pressure, c_{tp}, are listed in this book.

Thermal energy is stored in solids primarily as lattice vibrations (phonons) and electron kinetic energy. Ferromagnetic and ferrimagnetic materials also have a magnetic contribution to the specific heat. Lattice vibrations are the largest component. Phonon contribution to the specific heat varies according to T^3. The electronic contribution to the specific heat varies linearly with T and is dominant only at very low (cryogenic) temperatures. The magnetic contribution to the specific heat varies according to $T^{3/2}$. Like the electronic component of specific heat, the magnetic component is small except at low temperatures. For a rigorous presentation of specific heat, the reader is referred to Kittel [I2] or Ashcroft and Mermin [I3].

Thermal Conductivity

The *thermal conductivity* of a material is the time rate of heat conduction per unit area per unit temperature gradient. The thermal conductivity coefficient, k, is defined using the steady-state heat flow along a rod with a temperature gradient $\mathrm{d}T/\mathrm{d}x$ using the expression

$$j = -k(\mathrm{d}T/\mathrm{d}x) \tag{3}$$

where j is the flux of thermal energy or the energy transmitted across a unit area during a unit of time. The dependence on the thermal gradient, $\mathrm{d}T/\mathrm{d}x$, implies the transfer process is random. If the heat was transferred directly down the rod without interruption, the thermal flux would depend only on the temperature difference between the two ends of the rod and not on the temperature gradient.

The general units of thermal conductivity are $M \cdot L/t^3 \cdot T$. The values listed are expressed in cal/cm-K-sec. For conversion, 1.0 cal/cm-K-sec = 4.184 × 10^2 W/m-°C = 0.001927 Btu-in/ft^2-s-°F. Thermal conductivity is in general not a scalar quantity. However, for the special case of cubic materials, it is independent of direction [I4]. For tetragonal and hexagonal materials, the conductivity must be determined parallel and perpendicular to the nonsymmetric axis.

Heat in solids is transferred by electrons, lattice waves (phonons), magnetic excitation, and electromagnetic radiation. The total thermal conductivity is the sum of the conductivity of the individual carriers. Energy is transferred by particle collision or wave interaction, which accounts for the random nature of the process.

The principal heat carriers in metals are electrons or phonons. The total thermal conductivity is the sum of the electronic component and the phonon component. As a general rule, in highly conductive (pure) metals the electronic component is dominant over all temperatures. For less pure metals and poorly conducting metals and alloys, phonons become increasingly important at and below room temperature. For alloys, electronic conduction is decreased at room temperature and below. For semimetals and degenerate semiconductors, electron and phonon contribution are comparable except at low temperatures where the number of electrons in the conduction band is small, decreasing the electronic conduction of heat. For semiconductors, the electronic contribution is generally negligible except at high temperatures. For insulators, the dominant carriers are phonons except at very high temperatures. An excellent review of thermal conductivity is provided in Volumes 1 and 2 of *Thermophysical Properties of Matter* [I5].

Thermal conductivity is dependent on the mean free path of the dominant carrier. *The dependence of the thermal conductivity on the mean free path makes it difficult to uniquely define the thermal conductivity for some materials.* This is the case when the smallest dimension of the sample approaches the mean free path. This is true for insulators at low temperatures and for transparent, single crystalline solids at high temperatures where phonons dominate the transport mechanism. In metals this should be considered for very thin wires and films [I6].

Thermal Expansion Coefficient

The linear thermal expansion coefficient is a measure of the amount a material expands or contracts as its temperature changes. The unit for the expansion coefficient is inverse temperature. For CGS, MKS, and SI, 1/K (or 1/°C) is used. For conversion, 1/K = 5/9 1/°F. For cubic materials, only one coefficient need be specified. For hexagonal and tetragonal single crystals, two coefficients are needed for unique description of the expansion behavior.

There are various definitions for the thermal coefficient of linear expansion as shown by Touloukian, *et al.* [I7]. The *mean* coefficient of linear expansion is given by the formula

$$\alpha_m = (L_2 - L_1)/L_1(T_2 - T_1) \tag{4}$$

where L_1, L_2, T_2, and T_1 represent the length and the temperature at two conditions. The limiting value of this ratio at constant pressure as the temperature changes by a small amount dT is defined as the *true* coefficient of linear expansion. This can be written as follows:

$$\alpha = (1/L)(\partial L/\partial T)_p \tag{5}$$

If the functional dependence of the linear expansion can be fitted to a polynomial, then α can be found by differentiation. For example, if the experimental measurements provide the expression:

$$(L_T - L_{293})/L_{293} = a_0 + a_1T + a_2T^2 + a_3T^3 + \ldots \tag{6}$$

then α is given by the expression

$$\alpha = (L_{293}/L_T)(a_1 + 2a_2T + 3a_3T^2 + \ldots) \tag{7}$$

Since the overall linear growth is small in comparison to L_{293}, $L_{293}/L_T \simeq 1$. This approximation is accurate over a large temperature range. The true linear expansion coefficients are provided in this compilation.

Thermal expansion of solids is the result of increased vibrational displacement about the mean position in the solid lattice. Like specific heat and thermal conductivity, the thermal expansion has phonon, electronic, and magnetic contributions. The total expansion of a solid is the sum of these three contributions. The relative strength of the individual contributions depends on the material and temperature. A theoretical review of the thermal expansion of solids is given in Volume 12 of *Thermophysical Properties of Matter* [I8].

Elastic Constants

All materials respond to an applied stress by changing their shape. In the elastic region, the change in shape is described by various elastic constants. The most commonly used constants are Poisson's ratio, Young's modulus, shear modulus, and bulk modulus. Any two of these constants are necessary and sufficient to describe isotropic solids. The values listed in the accompanying tables are Young's modulus and Poisson's ratio. The general units of the moduli are the same as the units for stress. The general form for the units of stress is $M/L\text{-}t^2$. For conversion, 1 $\text{lb}_f/\text{in}^2 = 6.895 \times 10^3$ newtons/m^2 $= 7.031 \times 10^{-2}$ kg_f/cm^2. Poisson's ratio is unitless.

Hooke's law states that stresses and strains are linearly related. This law holds for most solids to a very high degree of accuracy. In the general case a fourth order tensor containing 3^4 or 81 elastic constants is required to correlate stress and strain in an elastic solid. Thermodynamic arguments and crystal symmetry are used to prove that the maximum number of coefficients needed to describe the behavior of an isotropic material reduces to two. The number of constants required for a cubic single crystal is reduced to three. As the symmetry of the crystal decreases, the number of constants required to completely characterize a material increases [I11].

Young's modulus, E, can be used to correlate stress and strain using the simplified form of Hooke's law

$$\sigma = E \epsilon \tag{8}$$

In this expression σ is stress in lb_f/in^2 or psi and ϵ is a unitless ratio of the change in length divided by the original length. Under an applied load, the relationship between transverse and longitudinal strain provides the definition for Poisson's ratio

$$\epsilon_{trans}/\epsilon_{long} = -\nu \tag{9}$$

Young's modulus and Poisson's ratio can be measured directly using stress-strain and physical measurements with reasonable precision. Alternately, the elastic stiffness or compliance can be measured with extreme precision using ultrasonic techniques and this data can be averaged to place bounds on Young's modulus and Poisson's ratio. It should be noted, however, that the problem of averaging the elastic stiffness coefficients to obtain the isotropic elastic constants cannot be solved uniquely [I9]. Since the elastic stiffness and compliance coefficients are the most accurate data available in the literature, we have chosen to use this data with the method of Hashin and Shtrikman [I10] to determine the least upper bound and greatest lower bound in cubic materials for Young's modulus and Poisson's ratio. The graphs demonstrating the effect of temperature on Young's modulus and Poisson's ratio rely heavily on this data. The data tables are supplemented with information obtained by mechanical testing.

The elastic constants are determined by the interatomic potential. As a result, they are not a strong function of composition until the degree of impurity reaches levels associated with alloys (> 1 weight percent). A more dramatic variation of the elastic constants is caused by the degree of anisotropy and inhomogeneity that exists. In some cases, severe anisotropic behavior resulting from preferred textures developed during manufacture may cause the elastic properties to fall outside the values shown. In these cases, for the most accurate data, the direction of the preferred orientation should be determined. Elastic constants can then be determined for particular crystallographic orientations using the method described by Lekhnitskii [I11]. A brief but thorough description of elastic coefficients is provided by Ledbetter and Reed [I12] and Simmons and Wang [I13].

ELECTRICAL PROPERTIES

Resistivity

Electrical conductivity is determined by the quantity of electricity crossing a unit area per unit potential gradient per unit time. *Electrical resistivity* is the inverse of electrical conductivity. The general units of electrical resistivity are $M\text{-}L/A^2\text{-}t^3$. The common units of measurement are ohm-cm or μΩ-cm. Over the broad spectrum of materials, the electrical resistivity varies by 10^{32}. This is by far the largest variation of any material property.

Resistivity changes significantly with temperature. On approaching absolute zero, the resistivity reaches a plateau value known as the *residual resistivity* (barring superconductive behavior). The residual resistivity is caused by the interaction of conduction electrons with indigenous imperfections and impurity atoms present in the material. Subtraction of the residual resistivity from the total resistivity measured at any temperature gives the resistivity for a "pure, perfect" material with a strain-free lattice. The temperature dependent portion of the resistivity arises from the interaction of conduction electrons with phonons or thermal vibrations in the metal lattice. As the temperature increases, the thermal vibrations increase, which in turn hinders the movement of electrons. The residual resistivity has not been subtracted from the measured resistivities provided.

The resistivity of a metal is a strong function of the purity and perfection of the crystal lattice. Small amounts of impurity atoms can have a profound effect on the interactions of electrons with the lattice ions. An even stronger dependence on purity is found in the case of semiconductors where the intentional addition of less than 1 ppm impurities is used to produce materials with the proper electronic behavior. For semiconductors, resistivities or conductivities are provided for typical doping levels. Intrinsic (undoped) resistivities are provided where appropriate.

The resistivities provided are for bulk materials except where noted. The resistivities of thin films are, in general, higher than the bulk values listed. There are two reasons for the increased resistance in thin film conductors. For films with thicknesses between 2 and 300 angstroms, the mean free path of the electron is reduced due to interaction with the surface. As the film thickness increases, the resistance of the film decreases and eventually reaches a plateau value that is higher than the properties of the bulk material. Second, thin films invariably have more imperfections and lattice strain than the bulk material. This is inherent to the thin film deposition process. If the film is annealed at a temperature sufficient to reduce the number of lattice imperfections, the resistivity of the film can approach the bulk resistivity.

The resistance, R, of a thin film with width, w, length, L, and thickness, d, is given by the expression

$$R = \rho L/wd \tag{10}$$

If the width and length of the film are equal, the resistance of the film is given simply by

$$R_s = \rho/d \tag{11}$$

The resistance of one square of film is therefore independent of the size of the square. The quantity is the sheet resistance and is expressed in ohms/square. This is the most common method for expressing the electrical performance of thin and thick film resistors. Table I provides a compilation of typical resistance values for some film resistor materials. For a more complete discussion of resistivity, the reader is referred to Kittel [I2] or Ashcroft and Mermin [I3].

TABLE I
Electronic Properties of Selected Materials [I14]
R_s in ohms/square and TCR in ppm/K

	R_s As deposited	R_s Annealed	TCR Film	TCR Bulk
Au	1.6–22.2	0.45–4.95	1600–2800	3670
Ni	11.6–28.5	1.8–15.6	3000–5000	5550
Cr	19.5–172.5	9.8–62.0	$<$100–600	2300
Mo	13.9–99.5	17.2–49.0	$<$100–200	4450
Al(a)	0.41	0.36	2800	4190
Ta	25.6–768.0	(b)	$<$100	3540
W	64.9–4390	63.9–4225	−200–100	4290

(a) Only one film examined.
(b) Resistance did not change after heating in vacuum.

Temperature Coefficient of Resistance

The temperature coefficient of resistance (TCR), like the thermal coefficient of expansion, can be defined arbitrarily in many ways. The TCR provided in this book is defined by the expression

$$\mathrm{TCR} = (R_2 - R_1)/R_{300}(T_2 - T_1), \tag{12}$$

where

R_2 = resistivity at temperature, T_2;

R_1 = resistivity at temperature, T_1;

R_{300} = resistivity at 300 K.

The TCR listed in Table I is for the specific temperature range $-55°C$ to $+155°C$. The units for TCR are ppm/K or ppm/°C. The conversion relationship is 1 ppm/°C = 1 ppm/K = 9/5 ppm/°F.

The explicit formulation for the change in resistance with change in temperature is derived by taking the appropriate partial derivatives of Equation (10) with respect to temperature. It is customary to ignore effects arising from geometric changes. This implies the change in resistance with temperature is solely a function of the change in resistivity of the material. This approximation is usually very accurate.

The TCR of thin film materials is generally lower than the TCR for bulk materials, as shown in Table I. Using thicker films and heat treatment subsequent to deposition can lead to a film with TCR approaching the bulk value. The contributions to the resistivity by lattice imperfections and impurities is much larger in thin films. Therefore the temperature dependent part of the resistivity is less in a thin film than in the bulk material. Annealing reduces the amount of lattice imperfections, which leads to a TCR closer to the bulk value.

Dielectric Constant, Dielectric Strength, and Dissipation Factor

The basic component for storing electric charge is the capacitor. Capacitors are typically parallel plate arrangements with a dielectric material sandwiched between conducting plates. When an electric bias is placed across the capacitor, no current flows. Instead, polarization of the dielectric material occurs. Electrical charge is converted into mechanical energy by the displacement of electrons and/or ions from their equilibrium positions within the dielectric material's solid lattice. The magnitude of the capacitance, C, is determined by the overlapping electrode area, a, the thickness of the dielectric, d, and the *dielectric constant* or *permittivity* of the dielectric material, ϵ. The resulting capacitance is given by the expression

$$C = \epsilon_0 \, \epsilon \, a/d \qquad (13)$$

where ϵ_0 is the permittivity of free space. The *dielectric constant* is a measure of a material's ability to store charge. As expressed in the above equation, ϵ is unitless. The dielectric constant can depend on the crystallographic direction.

As the bias on a capacitor is increased, a maximum is reached in a material's ability to store charge. Above this limit the capacitor breaks down and electric current is transferred between the conductor plates. The maximum possible voltage that can be placed across the capacitor is called the *dielectric strength* of the capacitor. Electrical breakdown is generally a consequence of impurities, cracks, flaws, and other imperfections present in the material rather than the inherent electrical characteristics of the material. Dielectric strength is therefore not a true material property but rather a measure of the capacitor's perfection. The dielectric strength is measured in volts per distance. It does not depend directly on the crystallographic direction of measurement.

In an alternating electric field the total potential obtainable depends on the ability of the ions and electrons to reorient themselves as the polarity of the electric field changes. There is invariably a delay in the polarization in an alternating electric field. In a loss-free capacitor, if the voltage varies sinusoidally, the current leads the voltage by 90°. For capacitors produced from real materials, the current leads the voltage by $90° - \delta$, where δ is called the *dielectric loss angle*. The *dissipation factor* is usually defined as $\tan\delta$. The product of ϵ and $\tan\delta$ is designated the *loss factor* and is a measure of the energy lost (as heat) by a capacitor in an ac circuit. In the 10^2 to 10^8 kHz frequency range, the ionic solids used in thin film capacitors store energy by the displacement of electrons and ions from their equilibrium positions. Electrons can respond to changing electric fields at frequencies less than 10^{12} kHz. Ionic movement is only slightly slower. As a first approximation, the dielectric constant and loss factor are therefore independent of the frequency in the 10^2 to 10^8 kHz range.

Band Gap

The quantized nature of energy requires that electrons fill discrete energy levels. In condensed solids, the interactions of the atomic nuclei transform these discrete energy levels into bands of allowed energy levels. Within each band are discrete levels so close together that the difference between any two adjacent levels is immeasurably small. Separating the bands of allowed energies are forbidden gaps. In order for an electron to change energy bands, it must gain or lose the amount of energy that separates the two bands. The energy that separates two adjacent bands is called the *band gap*. The size of the band gap is typically expressed in electron volts. For conversion, 1 eV = 1.602×10^{-19} Joules.

MECHANICAL PROPERTIES

Strength

The ability of a material to sustain an applied load is determined by its *strength*. Strength is usually made an intensive quantity by dividing the applied load by the cross-sectional area on which it acts. The units of strength are therefore the units of stress. The general units of stress are $M/L\text{-}t^2$. For conversion, 1 psi = 6.895×10^3 N/m^2 = 7.031×10^{-2} kg$_f$/cm^2.

The ability of a material to sustain a load depends, to a large extent, on how the load is applied. As a result, the measured strength of a material is largely dependent on the type of test used. The most common test for measuring strength is the uniaxial tensile test [I15]. Typical stress-strain output from a tensile test is shown in Figure 1. Three types of behavior are shown. Various strength terms are also shown. The maximum load sustained divided by the original cross-sectional

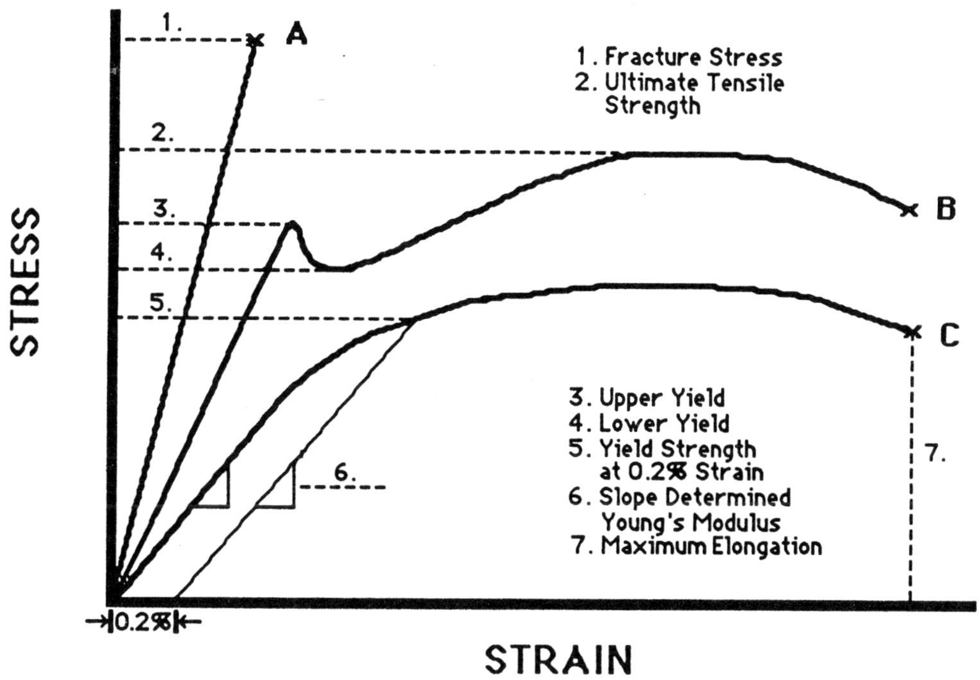

Figure 1 Property definitions.

area is the *ultimate tensile strength*. The *yield strength* is variously defined as the "strength at which the strain increases with no additional stress" or the "strength characterized by a specified level of permanent strain." Both these definitions for yield strength are used in this book. Note that the strength values are the engineering definitions and not the true stress. The *maximum elongation* (the change in length divided by the original length) occurs at break. The maximum elongation is a measure of a material's ability to sustain plastic deformation, and also a measure of the ductility of the material.

The popularity of the yield and ultimate strength do not stem from their wide applicability but rather from the ease of testing. These properties are quite good for describing the mechanical behavior of materials subjected to uniaxial loads. However, most materials are subjected to more complex loading patterns. The uniaxial tensile properties can still be used as a general ranking of materials subjected to complex loading patterns even though the applicability is somewhat restricted. Performance under complex loading conditions can be readily handled using elasticity theory and an assumed failure criteria.

Brittle solids need not be tested in uniaxial tension. Their low capacity for plastic deformation makes a bend test [I16] or compression test [I17] more appropriate. In such tests, the measured parameters are still force and displacement. These parameters are used with the original length and original cross-sectional area to determine the *compressive strength* and *flexural strength*. Both these quantities are defined by the maximum loads sustained prior to catastrophic failure.

Plastic deformation is determined by the movement and generation of *dislocations* within the solid lattice. It is the interaction of dislocations with impurity atoms, grain boundaries, precipitate particles, and other dislocations that determines material strength. As the dislocation density increases, the strength of a material generally increases while the ductility generally decreases. Well-annealed crystals have dislocation densities on the order of 10^4 to 10^6 cm/cm^3. Cold-worked materials and as-deposited metallic films have dislocation densities on the order of 10^{10} to 10^{12} cm/cm^3. The strength and plastic response of thin films is therefore approximated more closely by cold-worked materials. The presence of residual stresses in thin films complicates the comparison.

The yield strength of thin metallic films is found to be as much as 200 times higher than the strength of the corresponding bulk material. Part of this higher strength is associated with increased dislocation density. The increased strength is also attributable to the hindrance of dislocation motion and generation by the film surface. Thicker films that have received thermal treatments subsequent to deposition have mechanical properties that approach the bulk values. Hoffman has provided an extensive review of the residual stresses and strengths of thin films [I18].

Hardness

The *hardness* of a material is its resistance to penetration. The resistance to penetration is measured by the size of the indentation (length, diameter, or depth) left by a standard penetrator using a specific load for a predetermined application time. There are a large number of hardness tests currently in use. No one hardness test is useful for all materials. In spite of the underlying simplicity of a hardness test, the theoretical treatment of the penetration is quite complex. Consequently, there is only limited correlation between hardness tests. The correlations that do exist are for a limited number of materials and of questionable accuracy. The individual hardness tests cited in this text are given below.

The Rockwell test determines hardness by the depth of penetration. There are 21 different Rockwell tests which vary in the indenter used and the load applied. The most common types are a 1.6 mm diameter ball and a 100 kg load (Rockwell B) and a diamond pyramid indentor with a 150 kg load (Rockwell C). The Brinell test determines hardness by the diameter of the indentation. The standard test uses a 10 mm ball and a 3000 kg force. The Brinell hardness number is equal to the applied force divided by the area of the impression. The size of the indenter and the force varies widely. The Vickers hardness test uses a square based diamond pyramid. Hardness is determined by the applied force divided by the actual area of the impression. The diamond pyramid number sometimes reported in the literature is identical to the Vickers number. The Knoop hardness is a microhardness technique that measures hardness by the length of an impression from a rhombic-based pyramidal diamond with included angles of 172 and 130 degrees. The Knoop hardness number is given by the applied force divided by the projected cross-sectional area. Vickers hardness can also be used as a microhardness technique.

REFERENCES FOR INTRODUCTION

I1. Nye, J.F., *Physical Properties of Crystals*, Oxford: Clarendon Press, 1957.

I2. Kittel, Charles, *Introduction to Solid State Physics*, 6th Ed., New York: John Wiley and Sons, 1986.

I3. Ashcroft, N.F., and N.D. Mermin, *Solid State Physics*, New York: Holt, Rinehart and Winston, 1976.

I4. Nye, J.F., *Physical Properties of Crystals*, Oxford: Clarendon Press, 1957.

I5. Touloukian, Y.S., *et al., Thermophysical Properties of Matter*, TPRC Data Series, Vols. 1 and 2, 1970.

I6. *Ibid.*, 3a.

I7. Touloukian, Y.S., *et al., Thermophysical Properties of Matter*, Vol. 12, Thermal Expansion, Metallic Elements and Alloys, 4a, 1975.

I8. *Ibid.*, 1a–25a.

I9. Landau, L.D., and E.M. Lifshitz, *Theory of Elasticity*, London: Pergamon, 1959, 40.

I10. Hashin, Z., and S. Shtrikman, *J. Mech. and Phys. Solids*, Vol. 10, 1962, 343–352.

I11. Lekhnitskii, S.G., *Theory of Elasticity of an Anisotropic Body*, San Francisco: Holden-Day, 1963, 41–67.

I12. Ledbetter, H.M., and R.P. Reed, *J. Phys. and Chem. Ref. Data*, Vol. 2, No. 3, 1974, 531–618.

I13. Simmons, G., and H. Wang, *Single Crystal Elastic Constants and Calculated Aggregate Properties: A Handbook*, 2d Ed., Cambridge: The MIT Press, 1971.

I14. Holland, L., ed., *Thin Film Microelectronics*, Chapman and Hall, 1965 12.

I15. ASTM E8-85b, Standard Method of Tension Testing of Metallic Materials, *Annual Book of ASTM Standards*, Section 3.01, 1986, 124–169.

I16. ASTM F394-78, Standard Test Method for Biaxial Flexural Strength (Modulus of Rupture) of Ceramic Substrates, *Annual Book of ASTM Standards*, Section 10.04, 1986, 433–439 and ASTM F417-78, Standard Test Method for Biaxial Flexural Strength . . . Electronic Grade Ceramics, *ibid.*, 440–443.

I17. ASTM C773-82, Standard Test Method for Compressive (Crushing) Strength of Fired White Ware Materials. *Annual Book of ASTM Standards*, Section 10.04, 1986, 15–18.

I18. Hoffman, R., "The Mechanical Properties of Thin Condensed Films," in *Physics of Thin Films*, G. Hass and R.E. Thun, eds., Vol. 3, New York: Academic Press, 1966, 211.

ELEMENTS

ALUMINUM

PROPERTY	@ 300K	UNITS	REFERENCES
Density	2.71	g/cm^3	Al1 Al2 Al3
Specific Heat	0.211	cal/g-K	Al4 Al5 Al6
Thermal Conductivity	0.566	cal/cm-s-K	Al7 Al8 Al9
Thermal Exp Coef 2	$23.2x10^{-6}$	1/K	Al10 Al11 Al12
Melting Point	933.52	K	Al13
Resistivity 1	$2.73x10^{-6}$	Ohm-cm	Al14 Al15 Al16
TCR (218K to 398K)	4190	ppm/K	Al14 Al15 Al16
Young's Modulus 2	$1.02x10^{7}$	psi	AL17 Al18 Al19
Poisson's Ratio	0.346	-----	Al17 Al18 Al19
Yield Strength	1500	psi	Al20 Al21 Al23
Ult Tensile Strength	6500	psi	Al20 Al21 Al22
Elongation at Break	50	%	Al20 Al22 Al23
Hardness	17	Brinell	Al24 Al25 Al26

1 Aluminum deposited on substrates held at 150C-200C in vacuums in the range of 10^{-5} to 10^{-6} torr produces thin films with resistivities in the range of 3 to 4 x 10^{-6} ohm-cm. A sheet resistance of 0.1 ohms/sq can be achieved with 3000 to 4000 angstom films. [Al27]

2 Recent work on aluminum thin films on quartz suggests the modulus of elasticity for 99.999% pure material deposited in vacuum of $8x10^{-4}$ Pa is about one-half the bulk value and the expansion coefficient is about twice the bulk value. Heating the film subsequent to deposition will cause the properties of the film to approach the bulk values.[Al28]

Mechanical properties listed are for fully annealed, bulk material. 75% cold work will increase the strength of 0.08 inch sheet to 25,000 psi, increase hardness to 47 and reduce the elongation(3 inch gauge length) to 5.4%.[Al29] Thin polycrystaline aluminum films have a reported maximum tensile strength of 57,000 psi with a total strain at fracture of 0.5 to 0.8%.[Al30] Heating cold worked material to temperatures in the range of 400 to 600K will cause recovery and recrystallization erasing the effects of cold work.

ALUMINUM

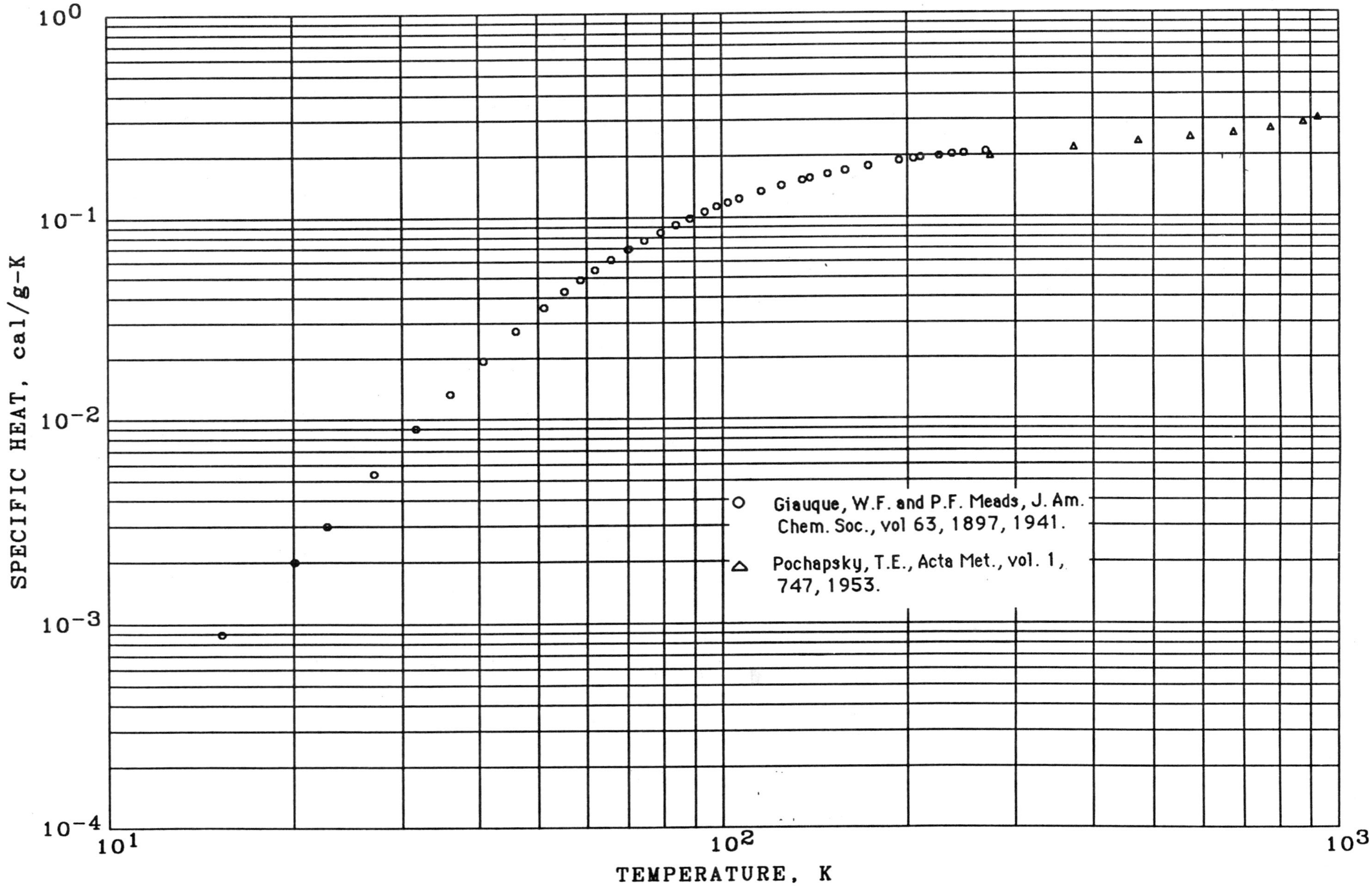

SPECIFIC HEAT, cal/g-K
TEMPERATURE, K
Giauque, W.F. and P.F. Meads, J. Am. Chem. Soc., vol 63, 1897, 1941.
Pochapsky, T.E., Acta Met., vol. 1, 747, 1953.

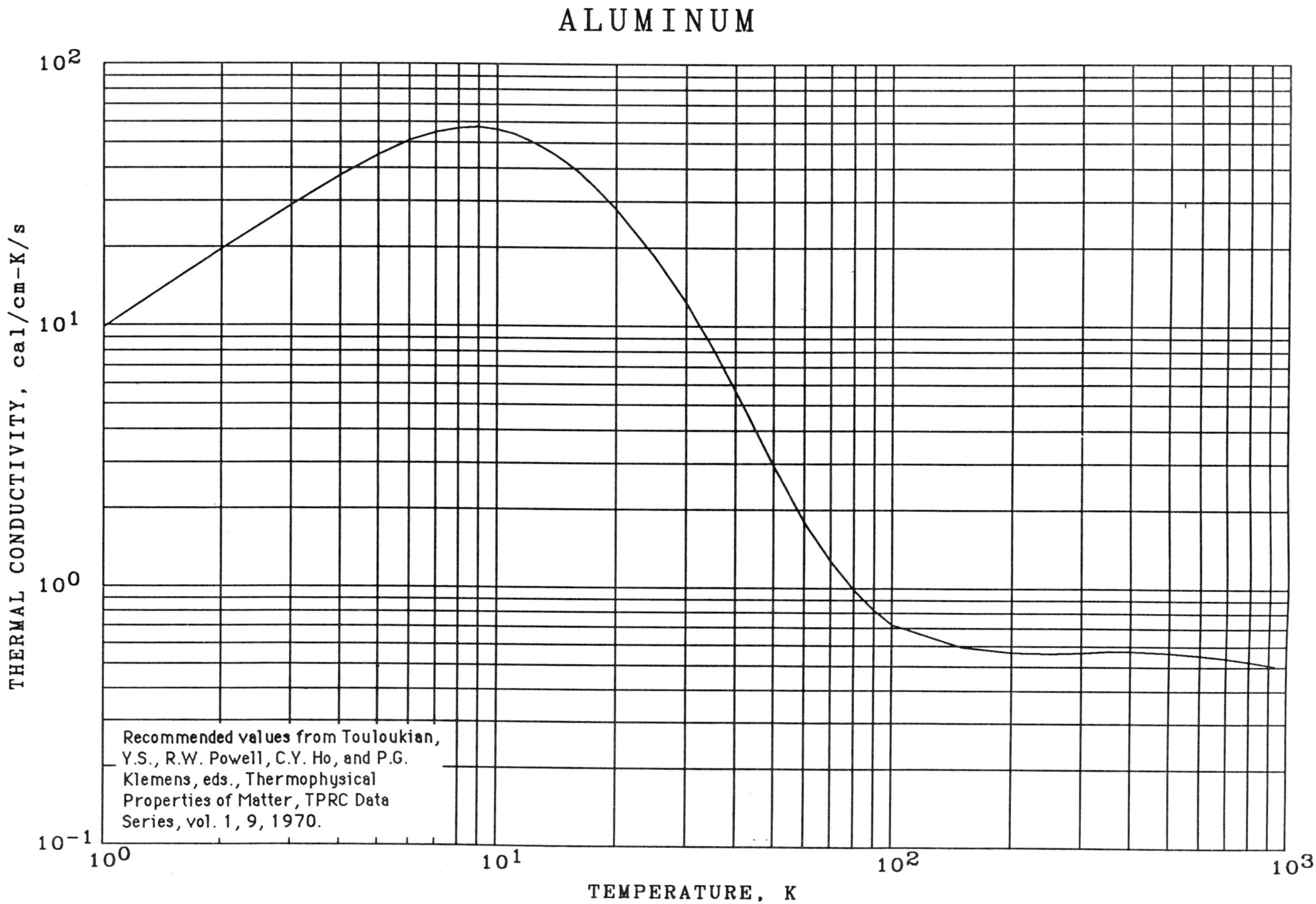
ALUMINUM
10^2
10^1
10^0
10^-1
THERMAL CONDUCTIVITY, cal/cm-K/s
10^0
10^1
10^2
10^3
TEMPERATURE, K
Recommended values from Touloukian, Y.S., R.W. Powell, C.Y. Ho, and P.G. Klemens, eds., Thermophysical Properties of Matter, TPRC Data Series, vol. 1, 9, 1970.

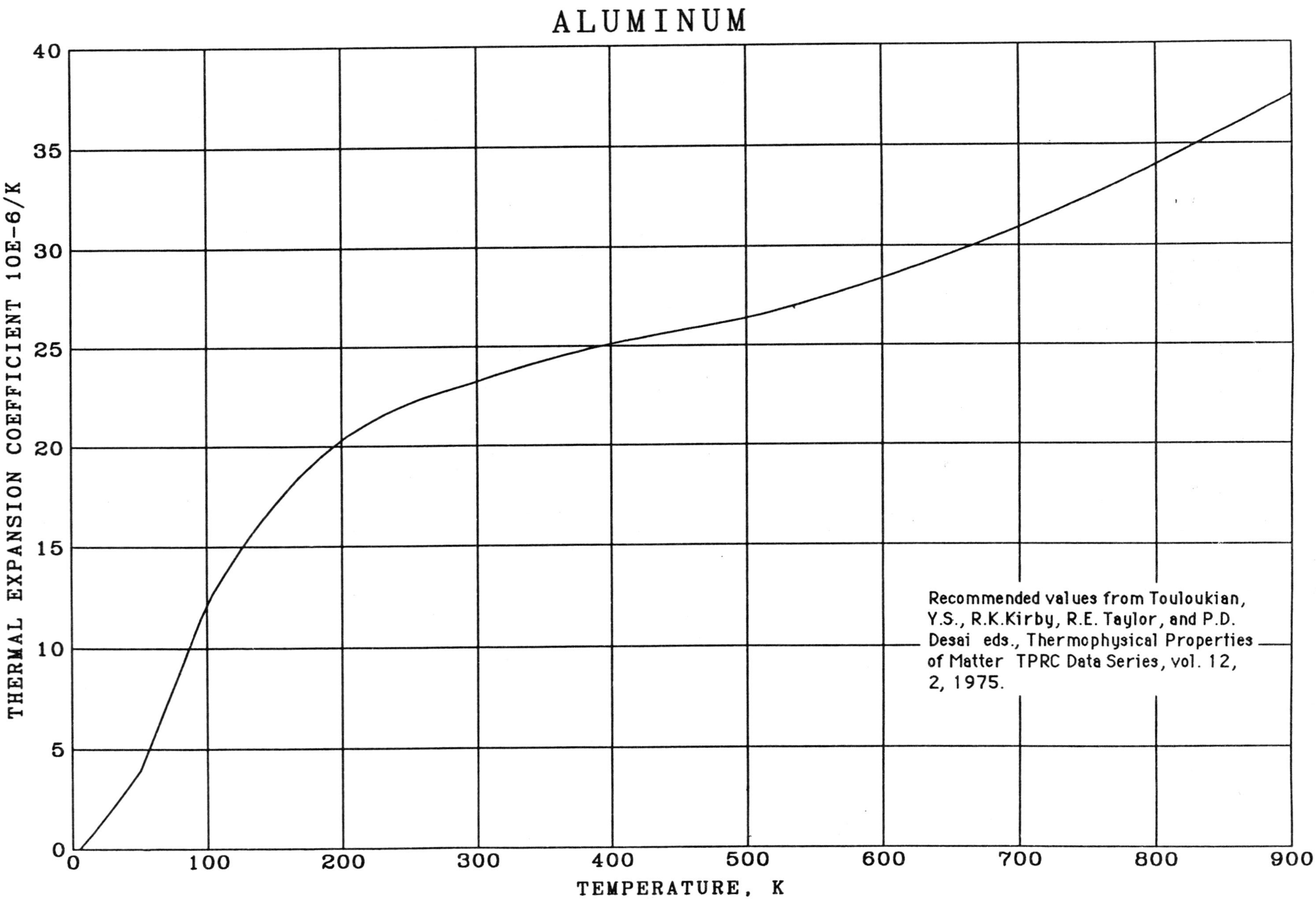
ALUMINUM
THERMAL EXPANSION COEFFICIENT 10E-6/K
TEMPERATURE, K
0
5
10
15
20
25
30
35
40
0
100
200
300
400
500
600
700
800
900
Recommended values from Touloukian,
Y.S., R.K.Kirby, R.E. Taylor, and P.D.
Desai eds., Thermophysical Properties
of Matter TPRC Data Series, vol. 12,
2, 1975.

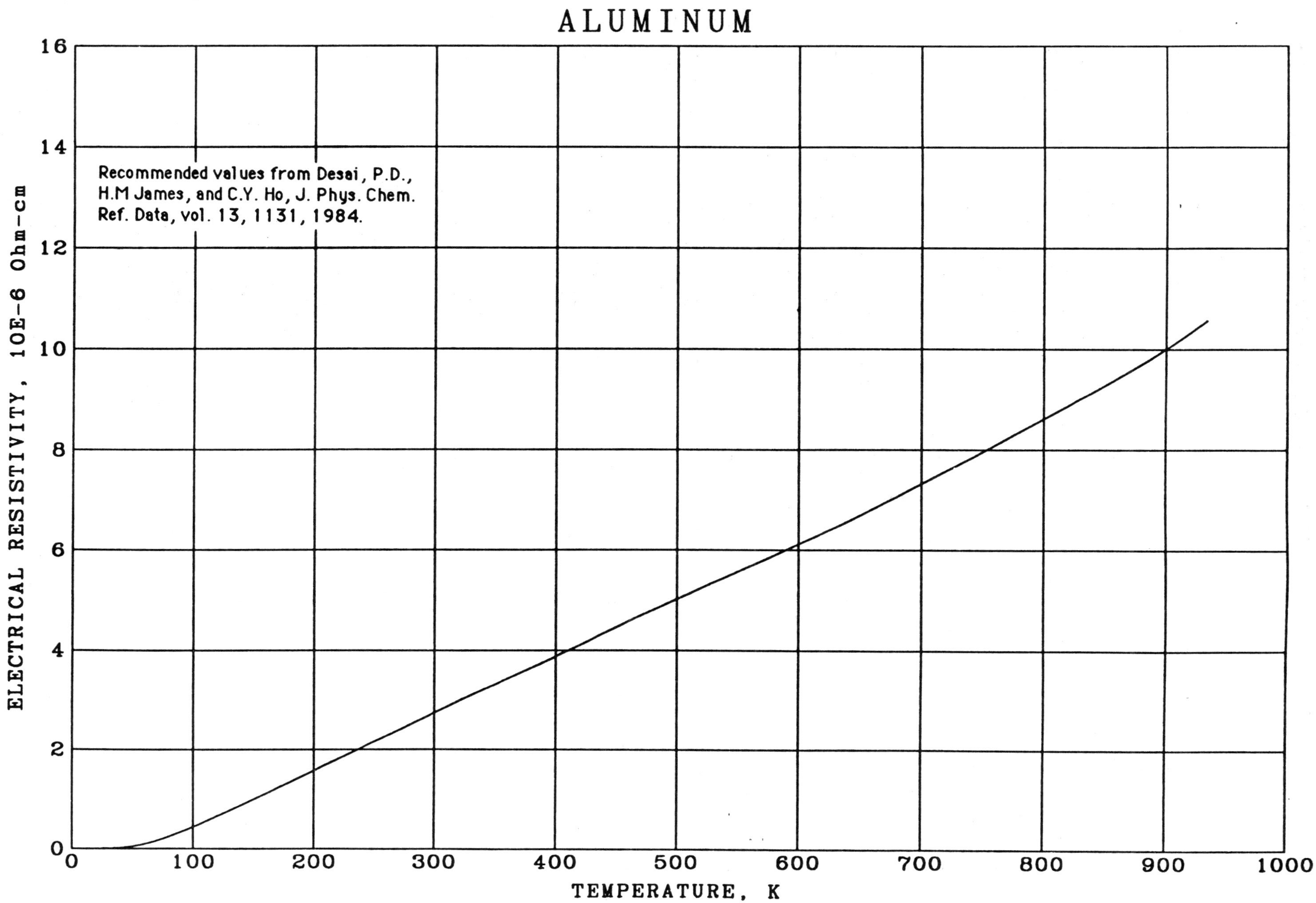
ALUMINUM
Recommended values from Desai, P.D., H.M James, and C.Y. Ho, J. Phys. Chem. Ref. Data, vol. 13, 1131, 1984.
ELECTRICAL RESISTIVITY, 10E-6 Ohm-cm
TEMPERATURE, K
0
2
4
6
8
10
12
14
16
100
200
300
400
500
600
700
800
900
1000

ALUMINUM

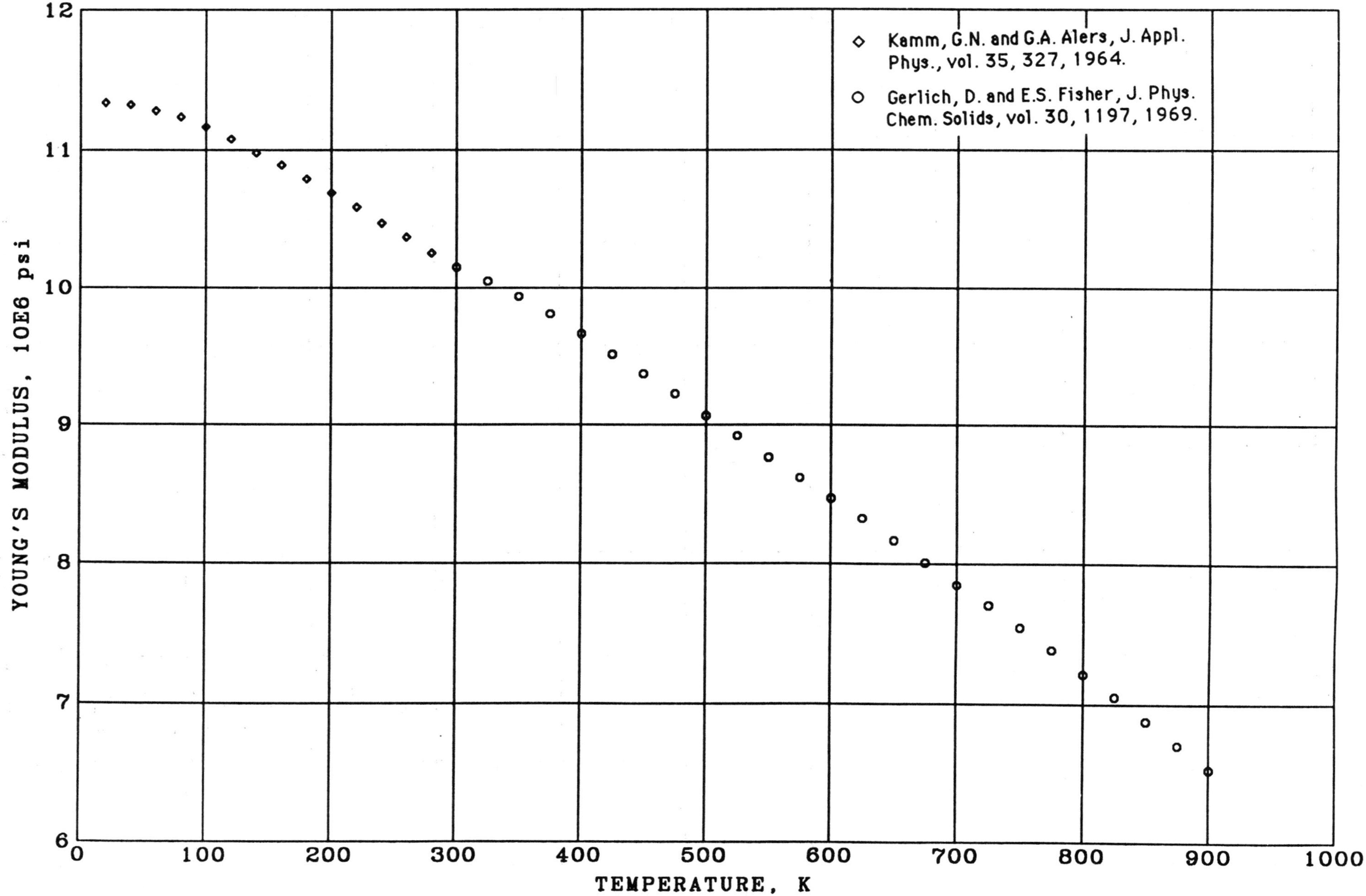
Kamm, G.N. and G.A. Alers, J. Appl. Phys., vol. 35, 327, 1964.
Gerlich, D. and E.S. Fisher, J. Phys. Chem. Solids, vol. 30, 1197, 1969.
YOUNG'S MODULUS, 10E6 psi
TEMPERATURE, K
12
11
10
9
8
7
6
0
100
200
300
400
500
600
700
800
900
1000

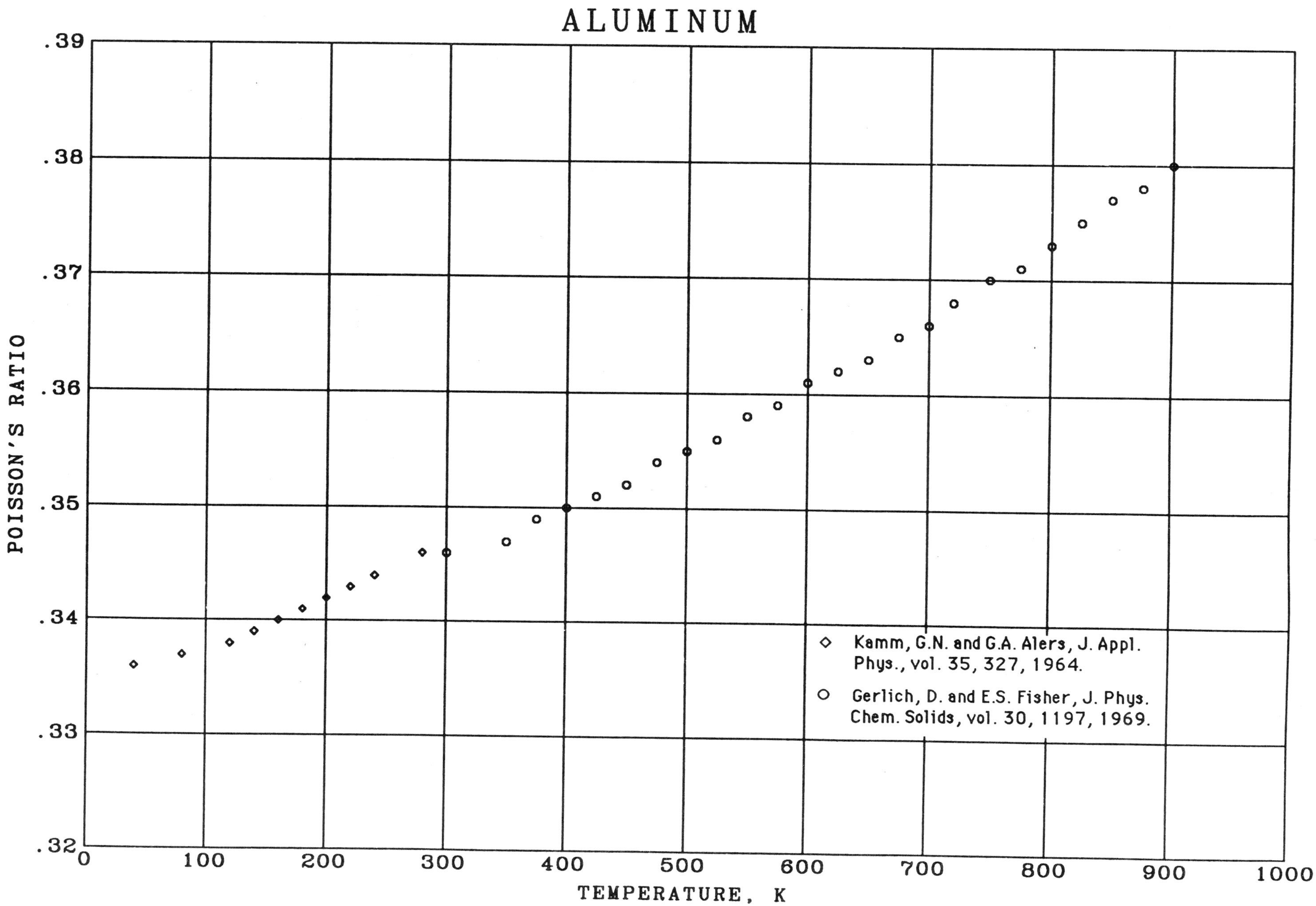
ALUMINUM
POISSON'S RATIO
.39
.38
.37
.36
.35
.34
.33
.32
0
100
200
300
400
500
600
700
800
900
1000
TEMPERATURE, K
Kamm, G.N. and G.A. Alers, J. Appl. Phys., vol. 35, 327, 1964.
Gerlich, D. and E.S. Fisher, J. Phys. Chem. Solids, vol. 30, 1197, 1969.

ALUMINUM

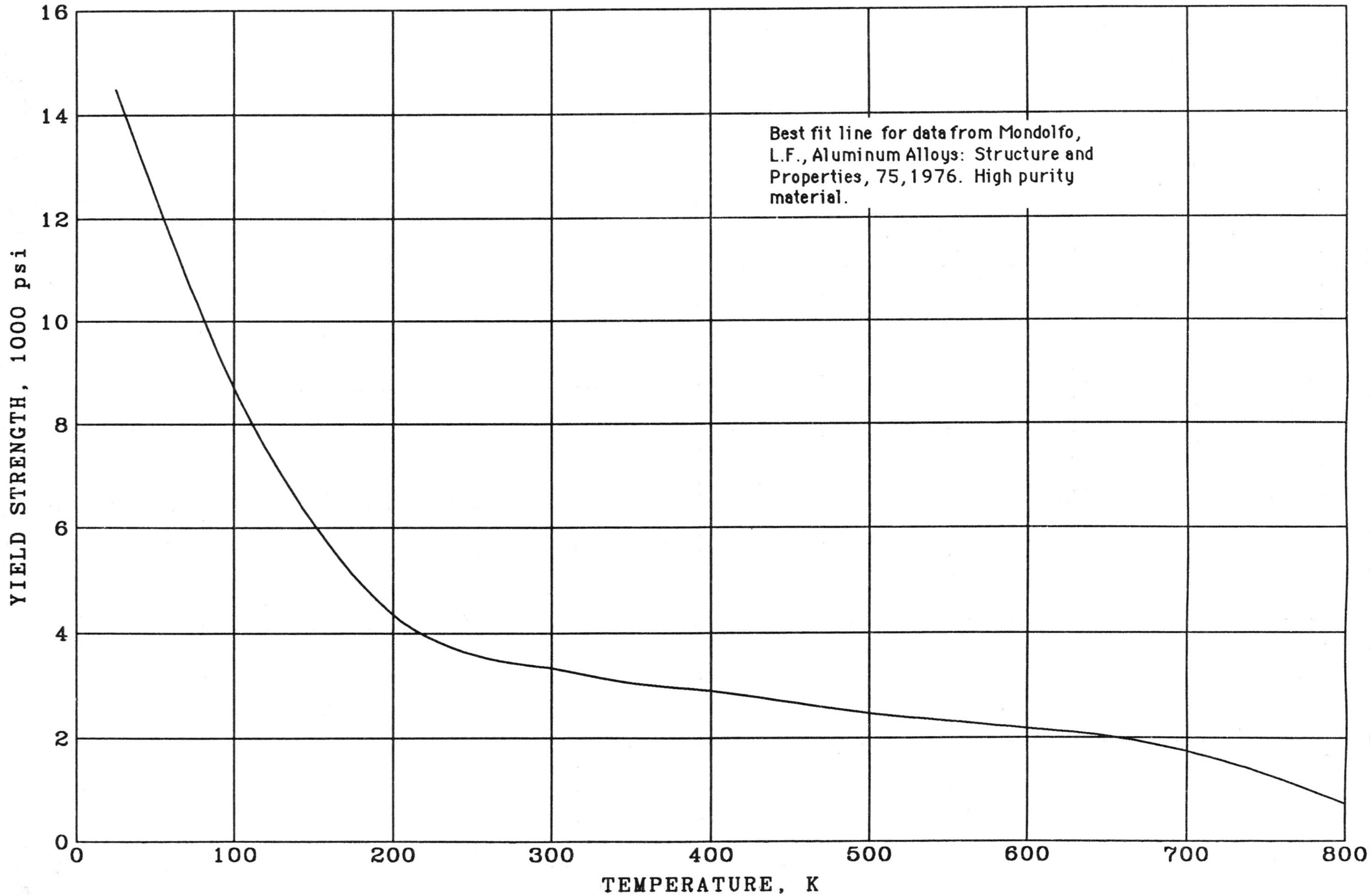

16
14
12
10
8
6
4
2
0
YIELD STRENGTH, 1000 psi
0
100
200
300
400
500
600
700
800
TEMPERATURE, K
Best fit line for data from Mondolfo, L.F., Aluminum Alloys: Structure and Properties, 75, 1976. High purity material.

ALUMINUM

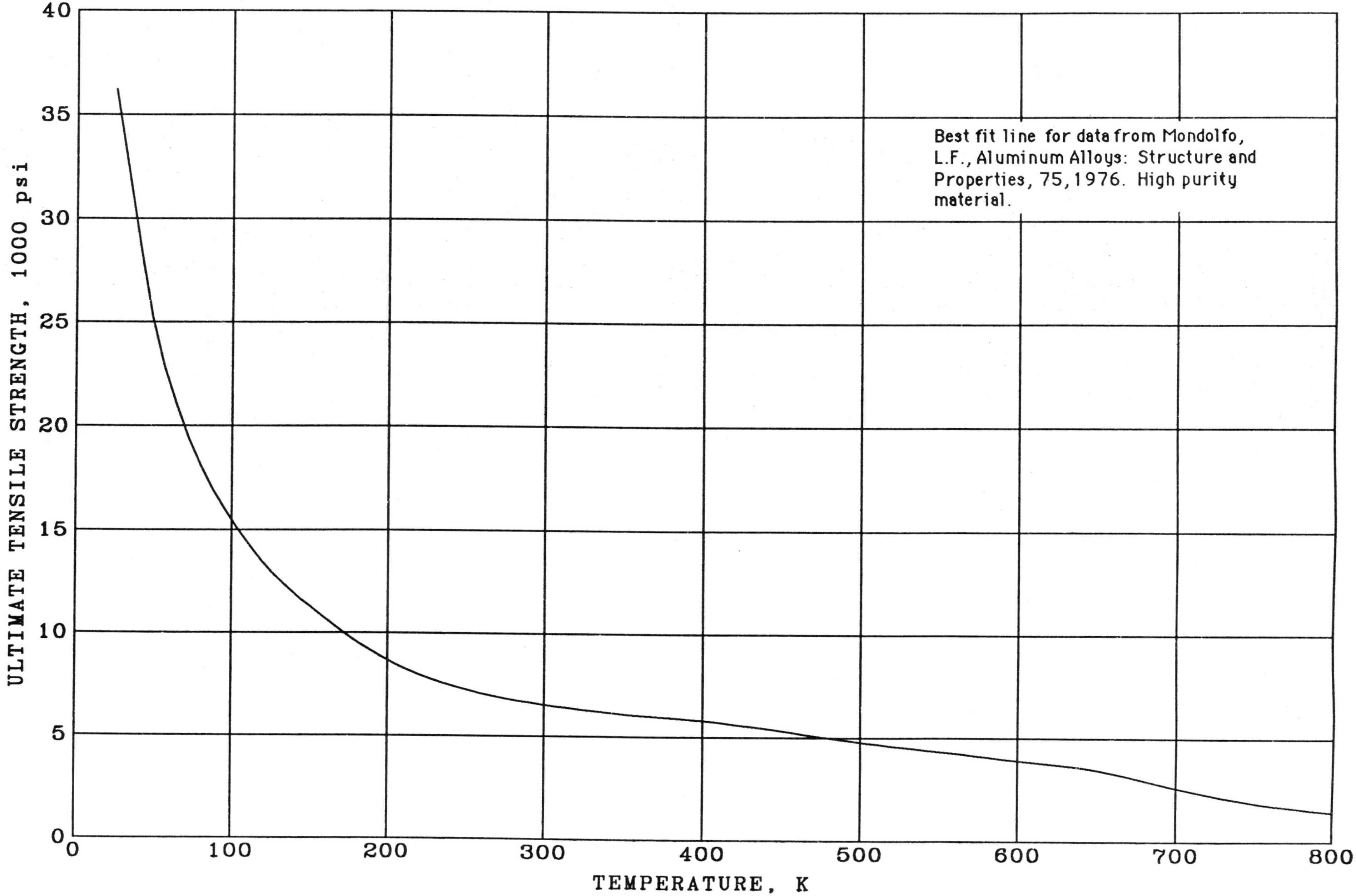

40
35
30
25
20
15
10
5
0
0
100
200
300
400
500
600
700
800
ULTIMATE TENSILE STRENGTH, 1000 psi
TEMPERATURE, K
Best fit line for data from Mondolfo, L.F., Aluminum Alloys: Structure and Properties, 75, 1976. High purity material.

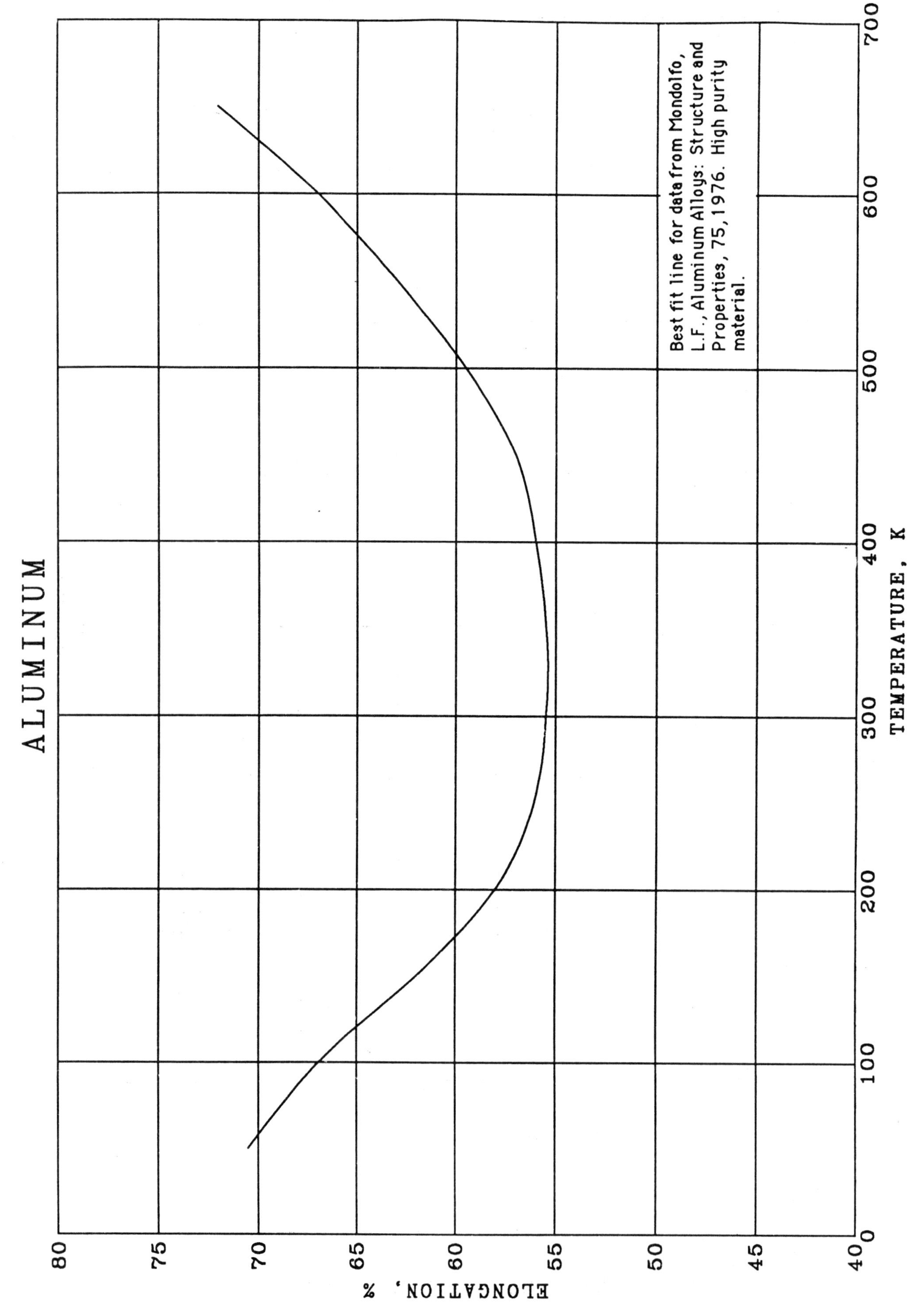
ALUMINUM
Best fit line for data from Mondolfo, L.F., Aluminum Alloys: Structure and Properties, 75, 1976. High purity material.
TEMPERATURE, K
ELONGATION, %
0
100
200
300
400
500
600
700
40
45
50
55
60
65
70
75
80

ALUMINUM

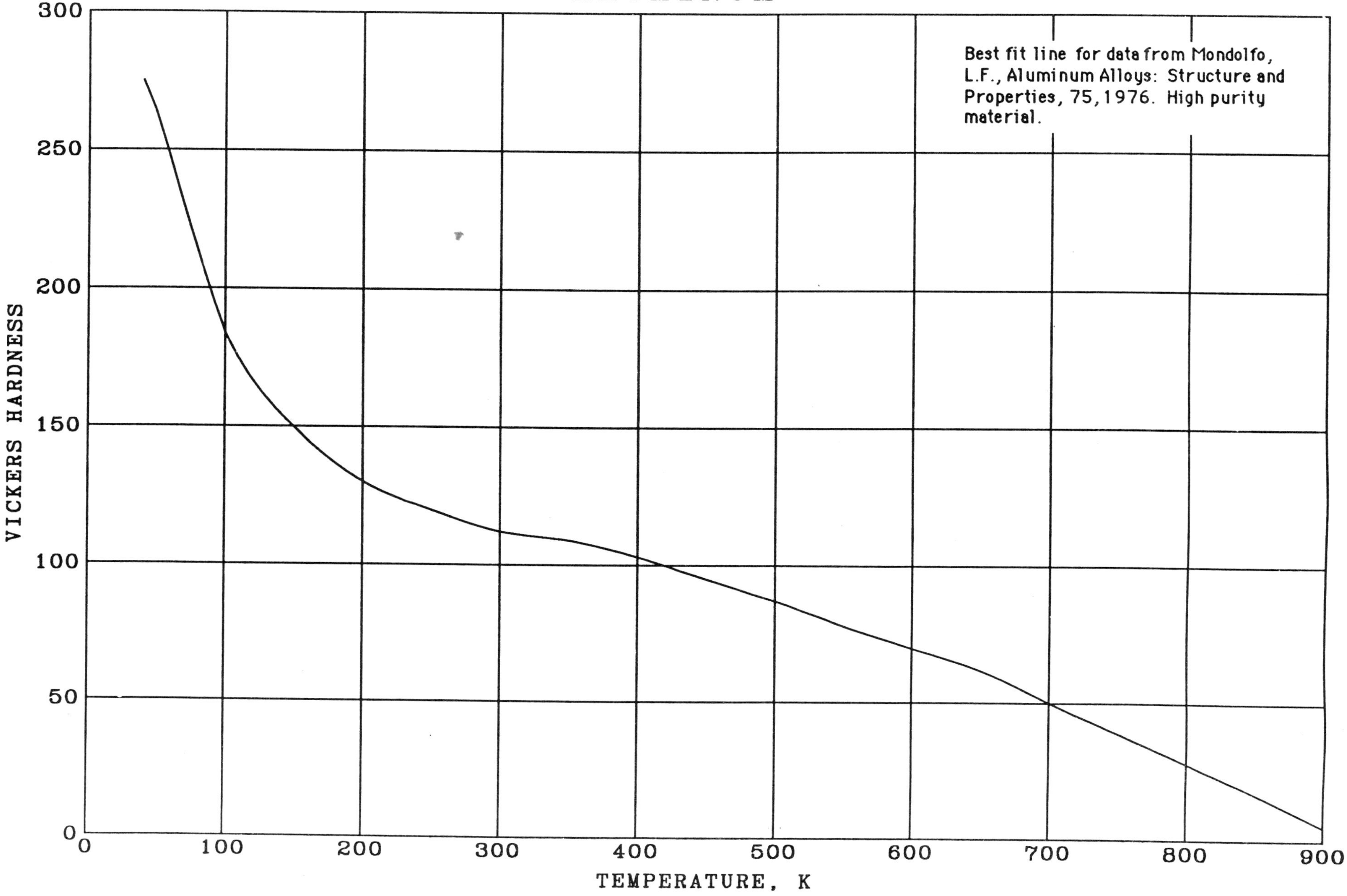

300
250
200
150
100
50
0
VICKERS HARDNESS
0
100
200
300
400
500
600
700
800
900
TEMPERATURE, K
Best fit line for data from Mondolfo, L.F., Aluminum Alloys: Structure and Properties, 75, 1976. High purity material.

CHROMIUM

PROPERTY	@ 300K	UNITS	REFERENCES
Density	7.16	g/cm^3	Cr1 Cr2 Cr3
Specific Heat	0.1077	cal/g-K	Cr4 Cr5 Cr6
Thermal Conductivity	0.216	cal/cm-s-K	Cr7 Cr8 Cr9
Thermal Exp Coef	$5.1x10^{-6}$	1/K	Cr10 Cr11 Cr12
Melting Point	2130	K	Cr13
Resistivity [1]	$13.5x10^{-6}$	Ohm-cm	Cr14 Cr15 Cr16
TCR (218K to 398K)	2300	ppm/K	Cr14 Cr15 Cr16
Young's Modulus	$4.03x10^7$	psi	Cr17 Cr18 Cr19
Poisson's Ratio	0.215	-----	Cr17 Cr18 Cr19
Yield Strength [2]	------	psi	Cr20 Cr21 Cr22
Ult Tensile Strength	20000-40000	psi	Cr20 Cr21 Cr23
Elongation at Break	0	%	Cr20 Cr21 Cr23
Hardness	89	Brinell	Cr24 Cr25 Cr26

[1] Deposition of chromium in 10^{-5} to 10^{-6} torr onto substrates in the temperature range of 100 to 400C produces films with resistivities in the range 15-90 micro-ohm-cm.[Cr27]

[2] Chromium, like most body centered cubic materials, exhibits a dramatic ductile to brittle transition as the temperature is lowered. The transition temperature for chromium varies from -15C to 450C depending primarily on the interstitial impurity content (300 wppm nitrogen increases transition to 450C) and grain size. In the brittle condition, chromium exhibits intergranular or cleavage fracture with virtually no elongation. In this condition, there is no yield strength.

Mechanical properties are for fully annealed, bulk samples. Wrought material stress relieved at 800C can have tensile strengths to 60,000 psi and elongations (0.5 inch gauge length) of 50%.[Cr28] Heating cold worked material to 1000 to 1100C results in recovery and recrystallization erasing the effects of cold work.

The Neel temperature for chromium is 311K. The associated magnetic transition has dramatic effects on the physical properties as shown in the accompanying graphs.

CHROMIUM

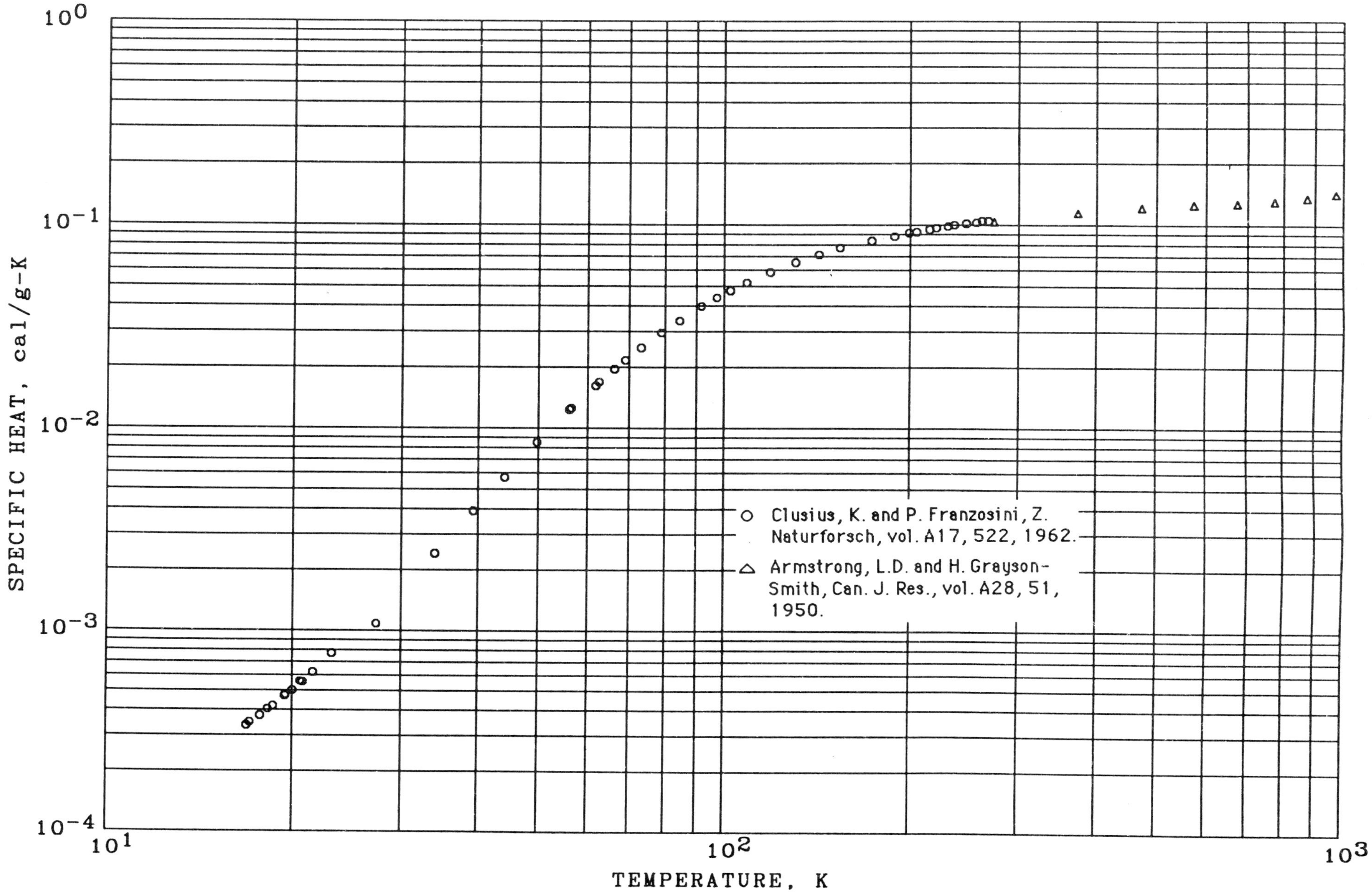

SPECIFIC HEAT, cal/g-K
TEMPERATURE, K
10^0
10^{-1}
10^{-2}
10^{-3}
10^{-4}
10^1
10^2
10^3
○ Clusius, K. and P. Franzosini, Z. Naturforsch, vol. A17, 522, 1962.
△ Armstrong, L.D. and H. Grayson-Smith, Can. J. Res., vol. A28, 51, 1950.

CHROMIUM

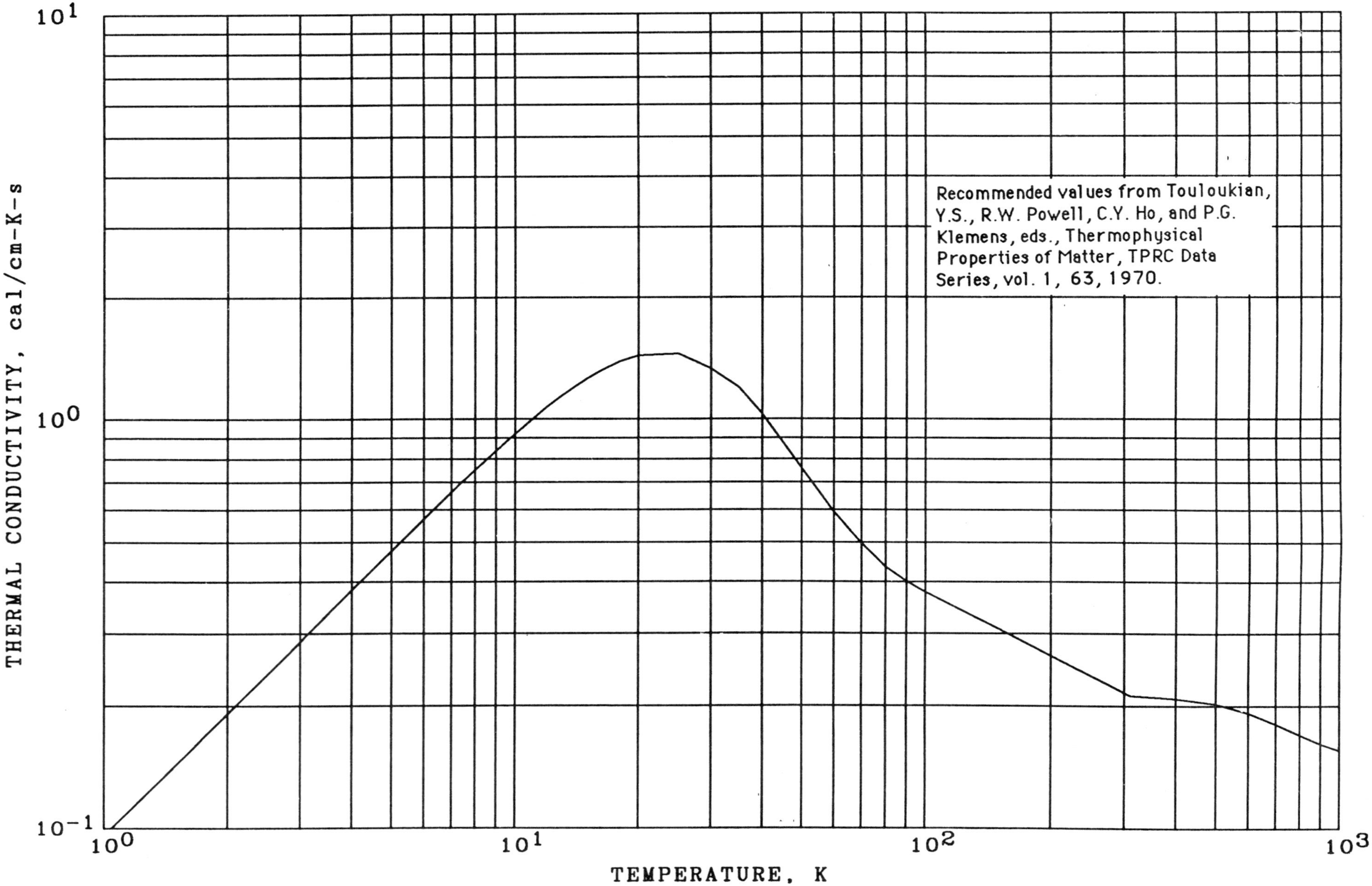

Recommended values from Touloukian, Y.S., R.W. Powell, C.Y. Ho, and P.G. Klemens, eds., Thermophysical Properties of Matter, TPRC Data Series, vol. 1, 63, 1970.
THERMAL CONDUCTIVITY, cal/cm-K-s
10^1
10^0
10^{-1}
10^0
10^1
10^2
10^3
TEMPERATURE, K

CHROMIUM

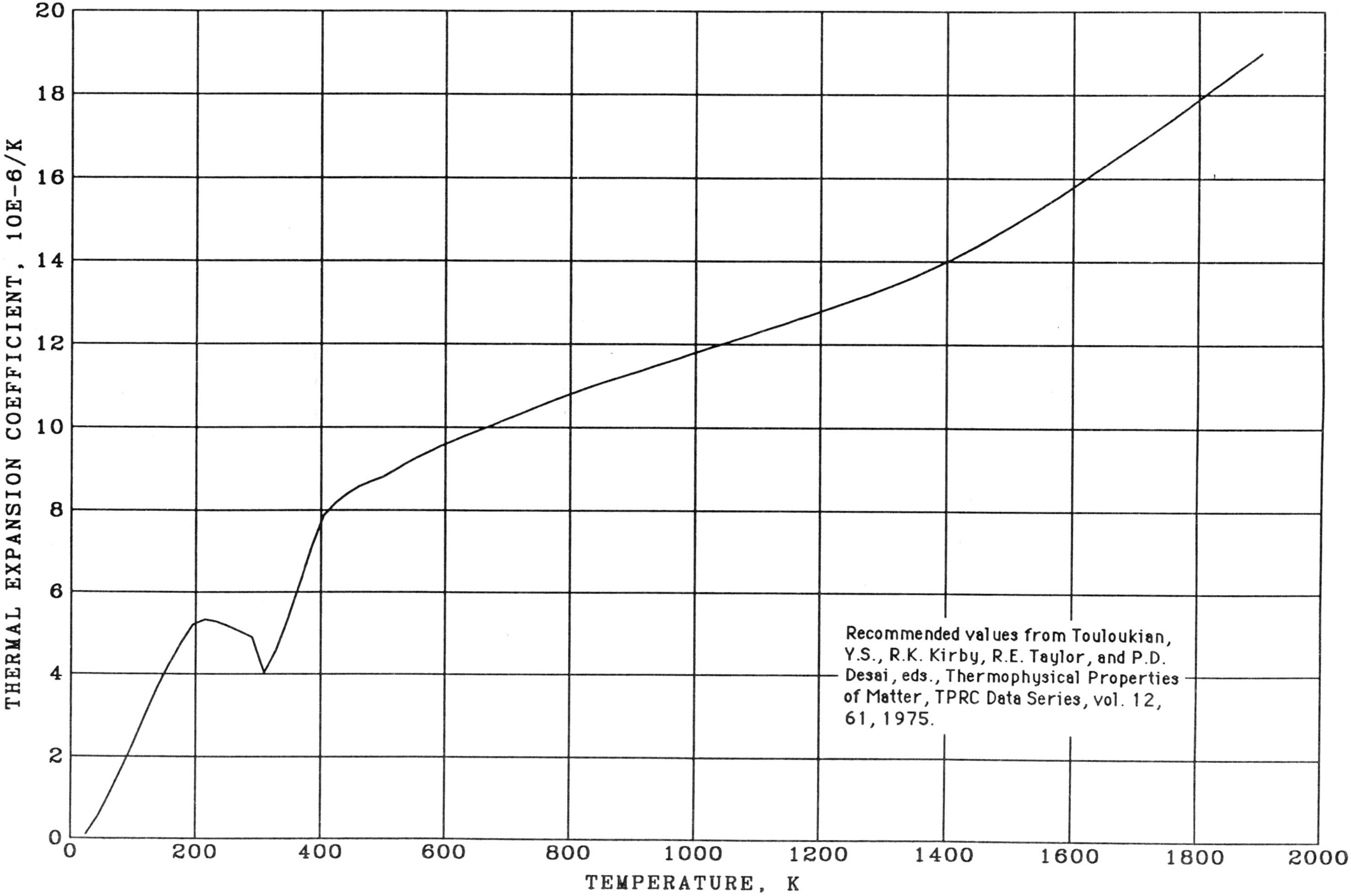

THERMAL EXPANSION COEFFICIENT, 10E-6/K
TEMPERATURE, K
0
2
4
6
8
10
12
14
16
18
20
0
200
400
600
800
1000
1200
1400
1600
1800
2000
Recommended values from Touloukian, Y.S., R.K. Kirby, R.E. Taylor, and P.D. Desai, eds., Thermophysical Properties of Matter, TPRC Data Series, vol. 12, 61, 1975.

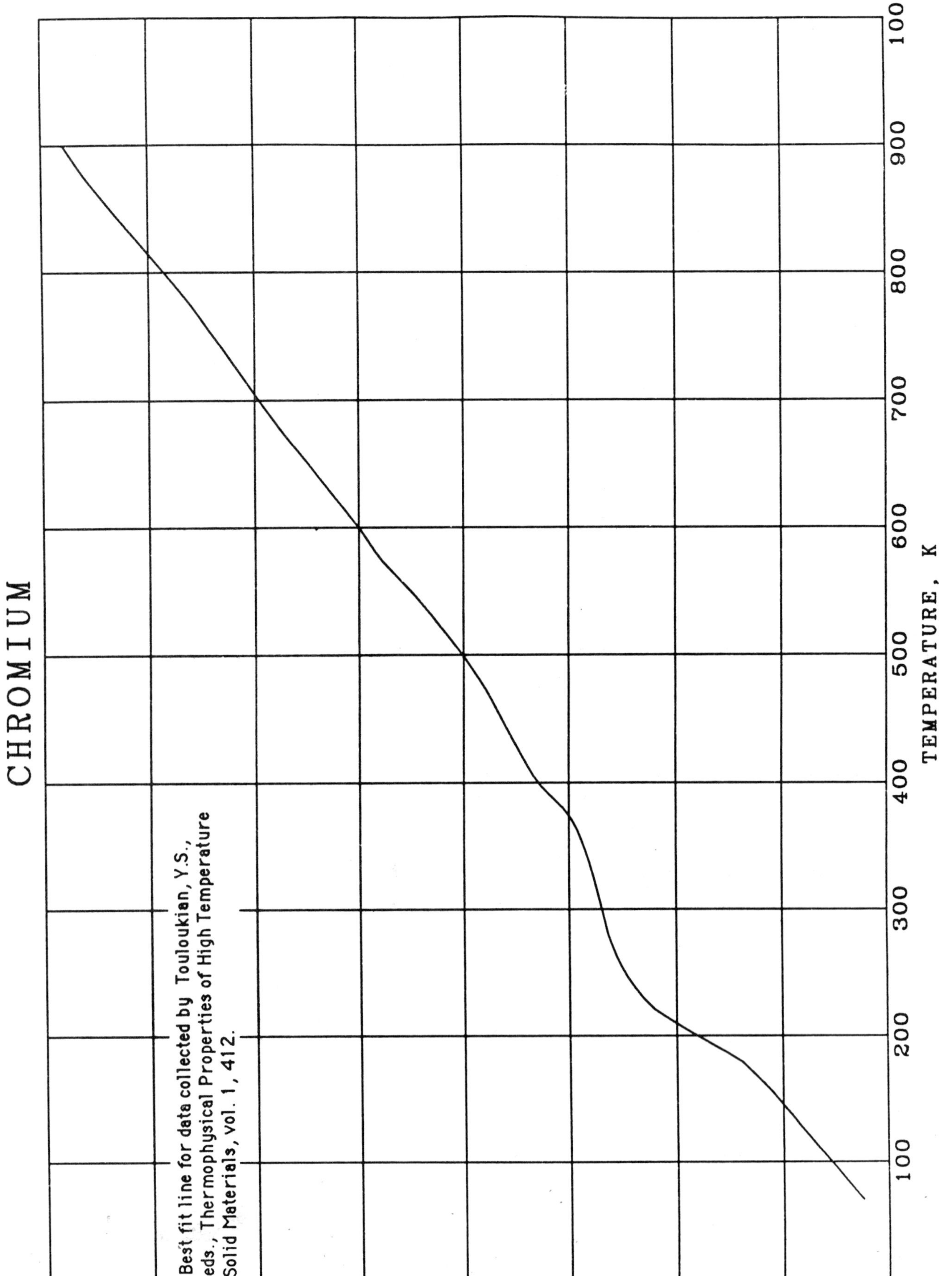
CHROMIUM
Best fit line for data collected by Touloukian, Y.S., eds., Thermophysical Properties of High Temperature Solid Materials, vol. 1, 412.
TEMPERATURE, K
0
100
200
300
400
500
600
700
800
900
1000
ELECTRICAL RESISTIVITY, 10E-6 Ohm-cm
0
5
10
15
20
25
30
35
40

CHROMIUM

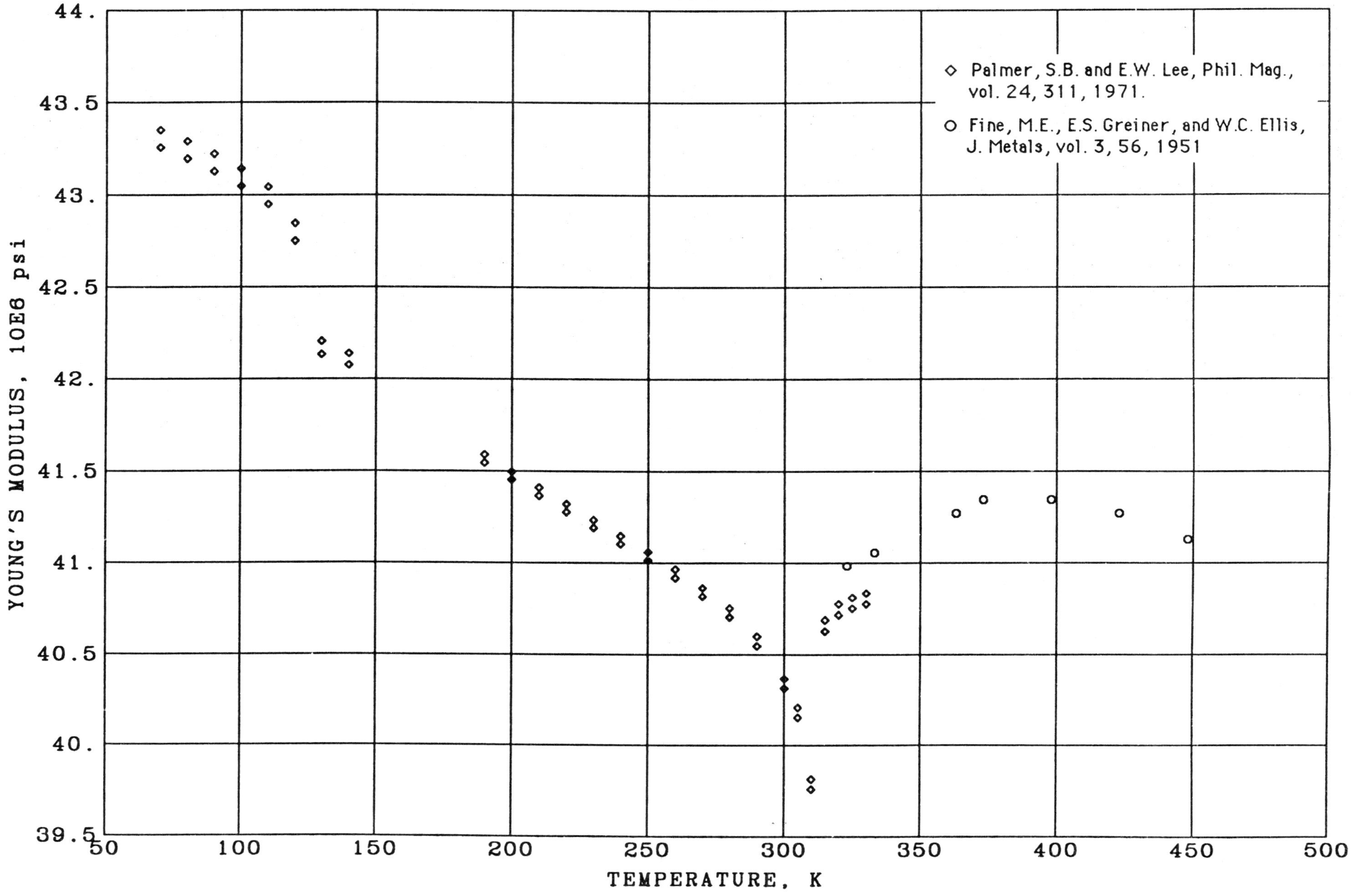
YOUNG'S MODULUS, 10E6 psi
44.
43.5
43.
42.5
42.
41.5
41.
40.5
40.
39.5
50
100
150
200
250
300
350
400
450
500
TEMPERATURE, K
◇ Palmer, S.B. and E.W. Lee, Phil. Mag., vol. 24, 311, 1971.
○ Fine, M.E., E.S. Greiner, and W.C. Ellis, J. Metals, vol. 3, 56, 1951

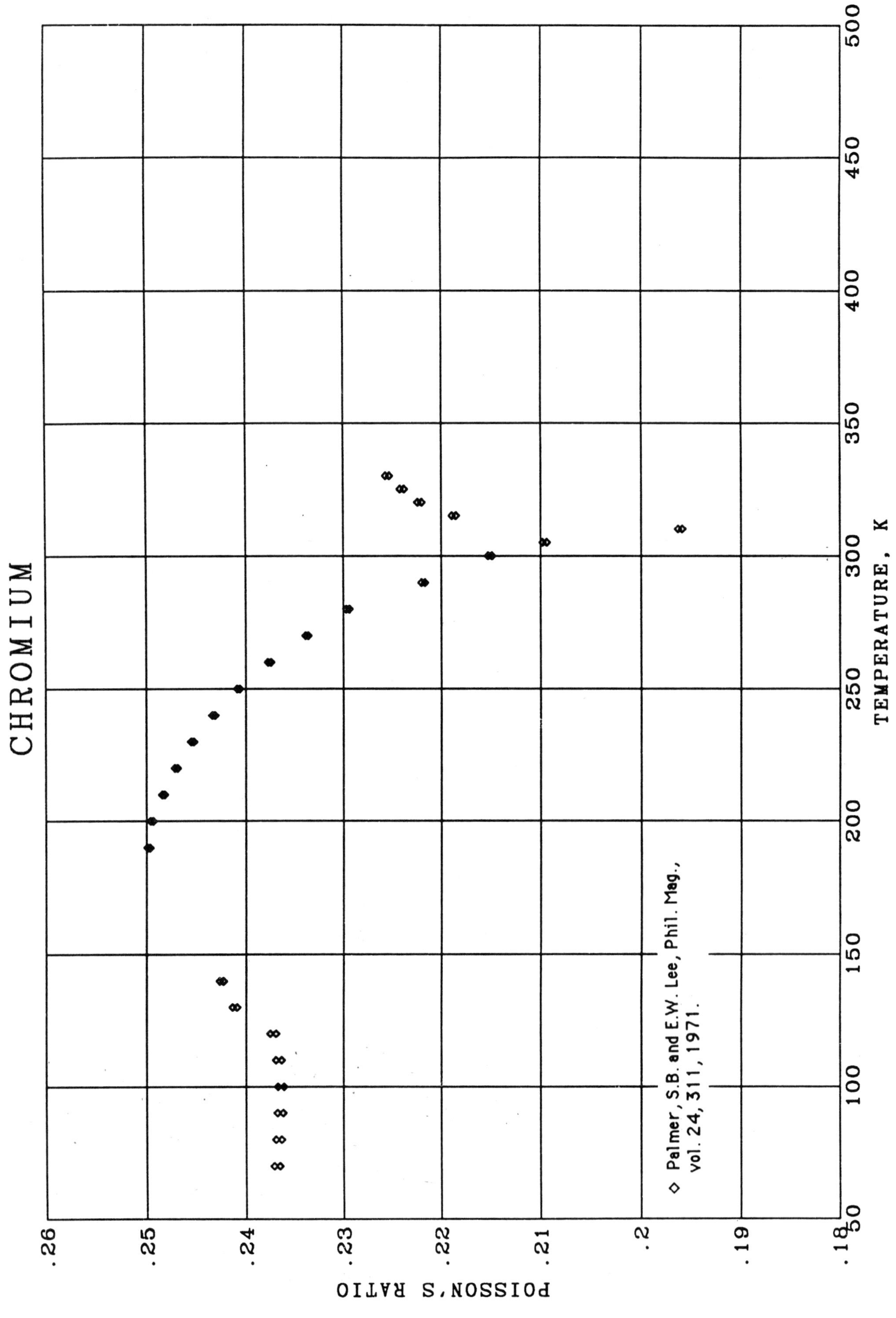
CHROMIUM
POISSON'S RATIO
.26
.25
.24
.23
.22
.21
.2
.19
.18
TEMPERATURE, K
50
100
150
200
250
300
350
400
450
500
◇ Palmer, S.B. and E.W. Lee, Phil. Mag.,
vol. 24, 311, 1971.

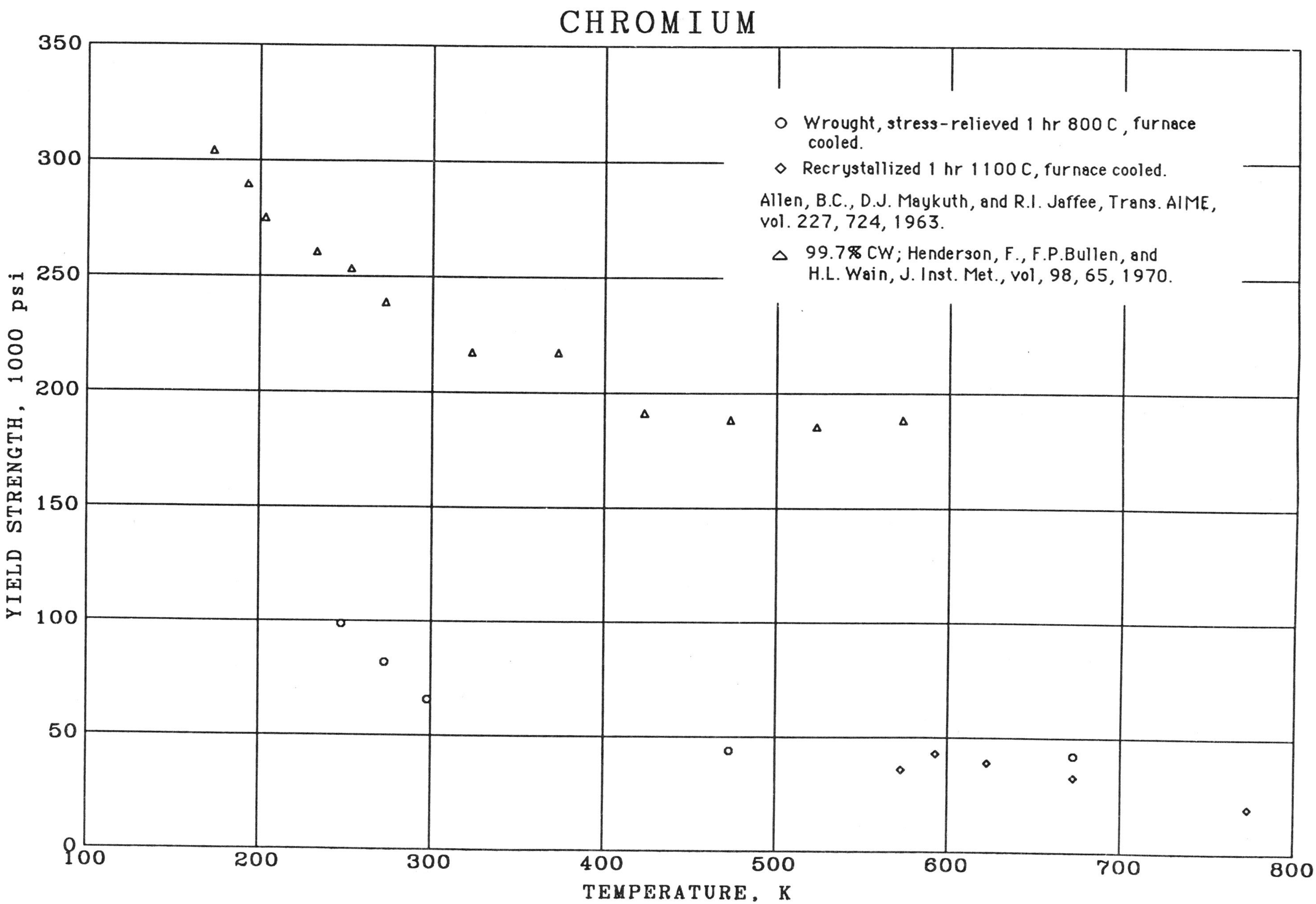
CHROMIUM
YIELD STRENGTH, 1000 psi
TEMPERATURE, K
350
300
250
200
150
100
50
0
100
200
300
400
500
600
700
800
○ Wrought, stress-relieved 1 hr 800 C, furnace cooled.
◇ Recrystallized 1 hr 1100 C, furnace cooled.
Allen, B.C., D.J. Maykuth, and R.I. Jaffee, Trans. AIME, vol. 227, 724, 1963.
△ 99.7% CW; Henderson, F., F.P. Bullen, and H.L. Wain, J. Inst. Met., vol, 98, 65, 1970.

CHROMIUM

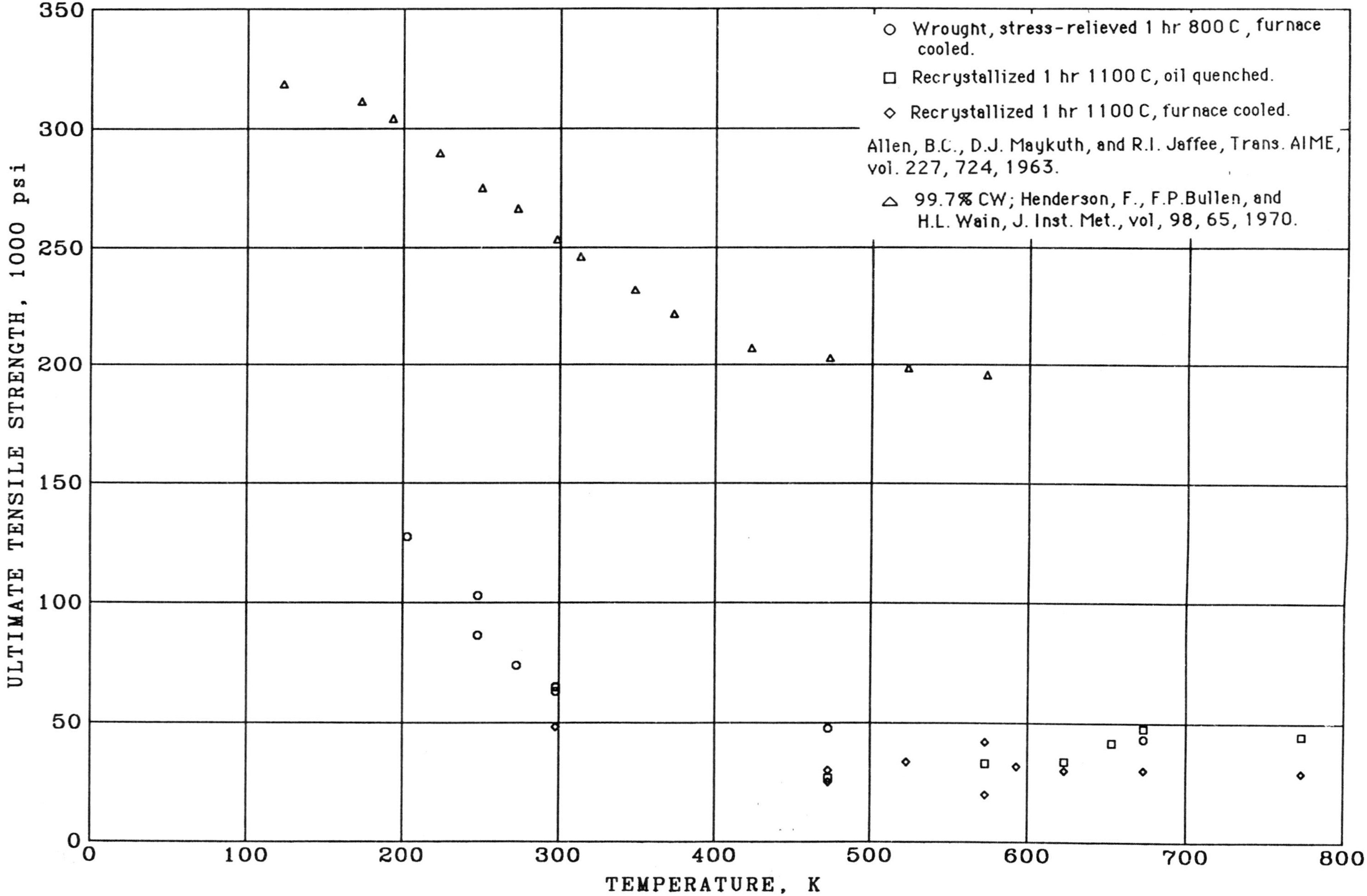

ULTIMATE TENSILE STRENGTH, 1000 psi
350
300
250
200
150
100
50
0
0
100
200
300
400
500
600
700
800
TEMPERATURE, K
○ Wrought, stress-relieved 1 hr 800 C, furnace cooled.
□ Recrystallized 1 hr 1100 C, oil quenched.
◇ Recrystallized 1 hr 1100 C, furnace cooled.
Allen, B.C., D.J. Maykuth, and R.I. Jaffee, Trans. AIME, vol. 227, 724, 1963.
△ 99.7% CW; Henderson, F., F.P. Bullen, and H.L. Wain, J. Inst. Met., vol, 98, 65, 1970.

CHROMIUM

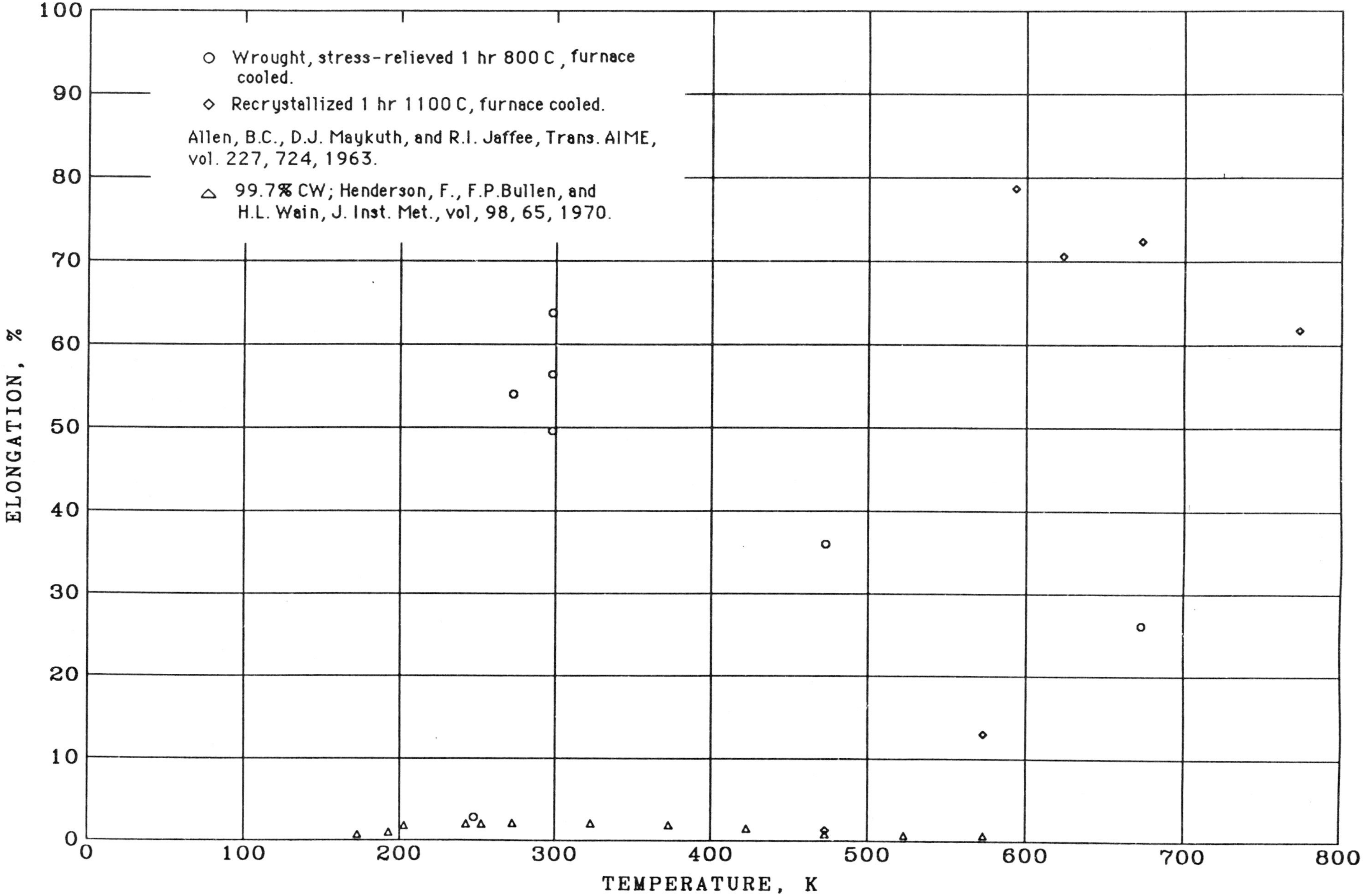
○ Wrought, stress-relieved 1 hr 800 C, furnace cooled.
◇ Recrystallized 1 hr 1100 C, furnace cooled.
Allen, B.C., D.J. Maykuth, and R.I. Jaffee, Trans. AIME, vol. 227, 724, 1963.
△ 99.7% CW; Henderson, F., F.P.Bullen, and H.L. Wain, J. Inst. Met., vol, 98, 65, 1970.
ELONGATION, %
TEMPERATURE, K
0
10
20
30
40
50
60
70
80
90
100
0
100
200
300
400
500
600
700
800

CHROMIUM

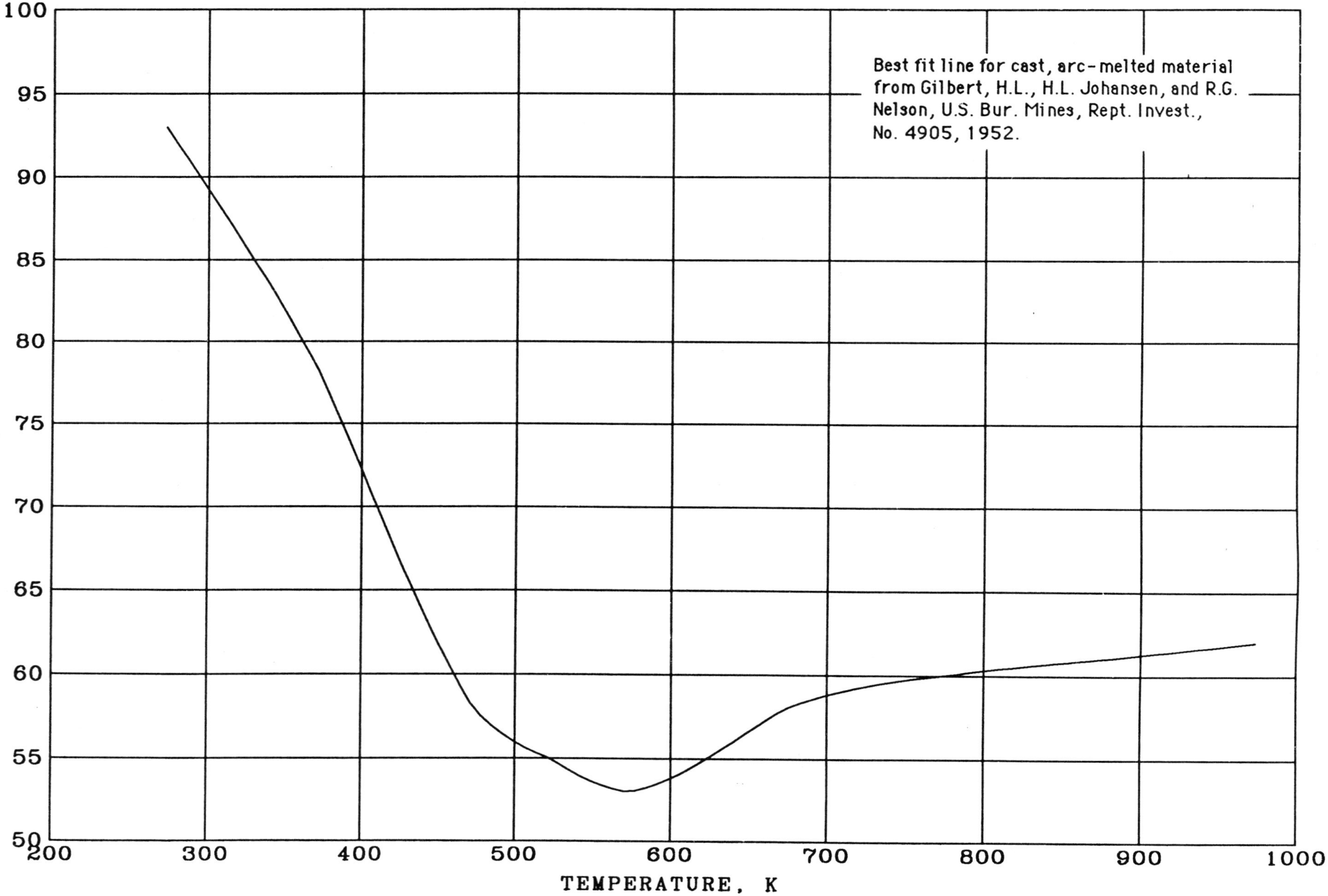

Best fit line for cast, arc-melted material from Gilbert, H.L., H.L. Johansen, and R.G. Nelson, U.S. Bur. Mines, Rept. Invest., No. 4905, 1952.
BRINELL HARDNESS
100
95
90
85
80
75
70
65
60
55
50
200
300
400
500
600
700
800
900
1000
TEMPERATURE, K

COPPER

PROPERTY	@ 300K	UNITS	REFERENCES
Density	8.94	g/cm^3	Cu1 Cu2 Cu3
Specific Heat	0.09118	cal/g-K	Cu4 Cu5 Cu6
Thermal Conductivity	0.9508	cal/cm-s-K	Cu7 Cu8 Cu9
Thermal Exp Coef	16.12×10^{-6}	1/K	Cu10 Cu11 Cu12
Melting Point	1356.6	K	Cu13
Resistivity	1.73×10^{-6}	Ohm-cm	Cu14 Cu15 Cu16
TCR (218K to 398K)	3930	ppm/K	Cu14 Cu15 Cu16
Young's Modulus	1.92×10^{7}	psi	Cu17 Cu18 Cu19
Poisson's Ratio	0.339	-----	Cu17 Cu18 Cu19
Yield Strength	10,000	psi	Cu20 Cu21 Cu22
Ult Tensile Strength	32,000	psi	Cu20 Cu21 Cu22
Elongation at Break	51	%	Cu20 Cu21 Cu23
Hardness	37	Brinell	Cu21 Cu24 Cu25

Mechanical properties listed are for fully annealed, bulk material. 95.6% cold work producing 0.083 inch diameter wire increases the strength to 90,000 psi.[Cu26] 88.7% reduction at 160C producing 0.025 inch diameter wire will reduce elongation to <5%.[Cu27] Polycrystalline copper thin films have a reported tensile strengths varying from 50,000 to 120,000 psi [Cu28] with one report as high as 140,000 psi and a corresponding strain at failure of 0.5 to 0.8%[Cu29]. Heating cold worked material to temperatures in the range 100 to 300C results in recovery and recrystallization erasing the effects of cold work. For more information on the effects of cold work, see graphs of mechanical properties -v- temperature.

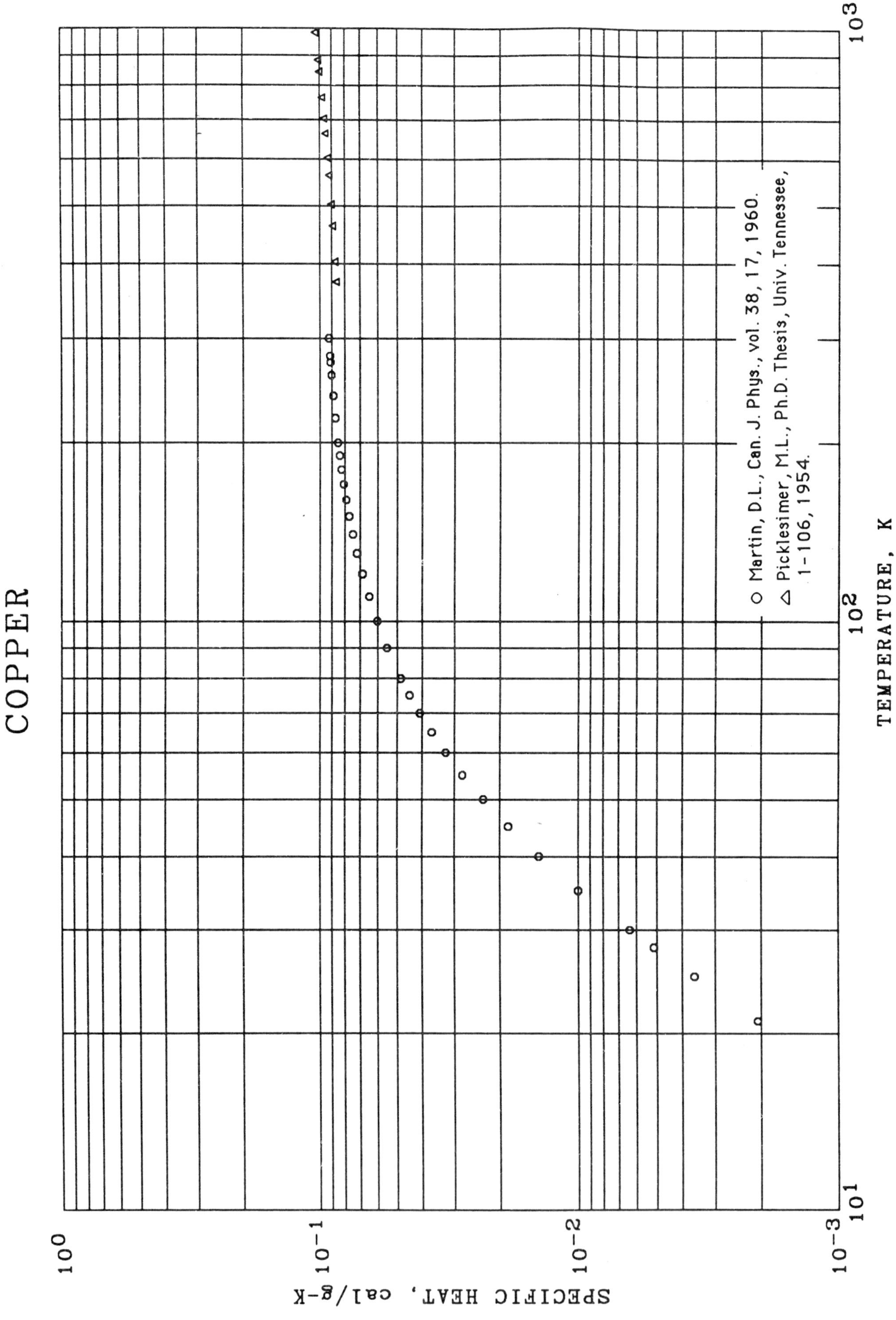
COPPER
SPECIFIC HEAT, cal/g-K
10^0
10^-1
10^-2
10^-3
TEMPERATURE, K
10^1
10^2
10^3
○ Martin, D.L., Can. J. Phys., vol. 38, 17, 1960.
△ Picklesimer, M.L., Ph.D. Thesis, Univ. Tennessee, 1-106, 1954.

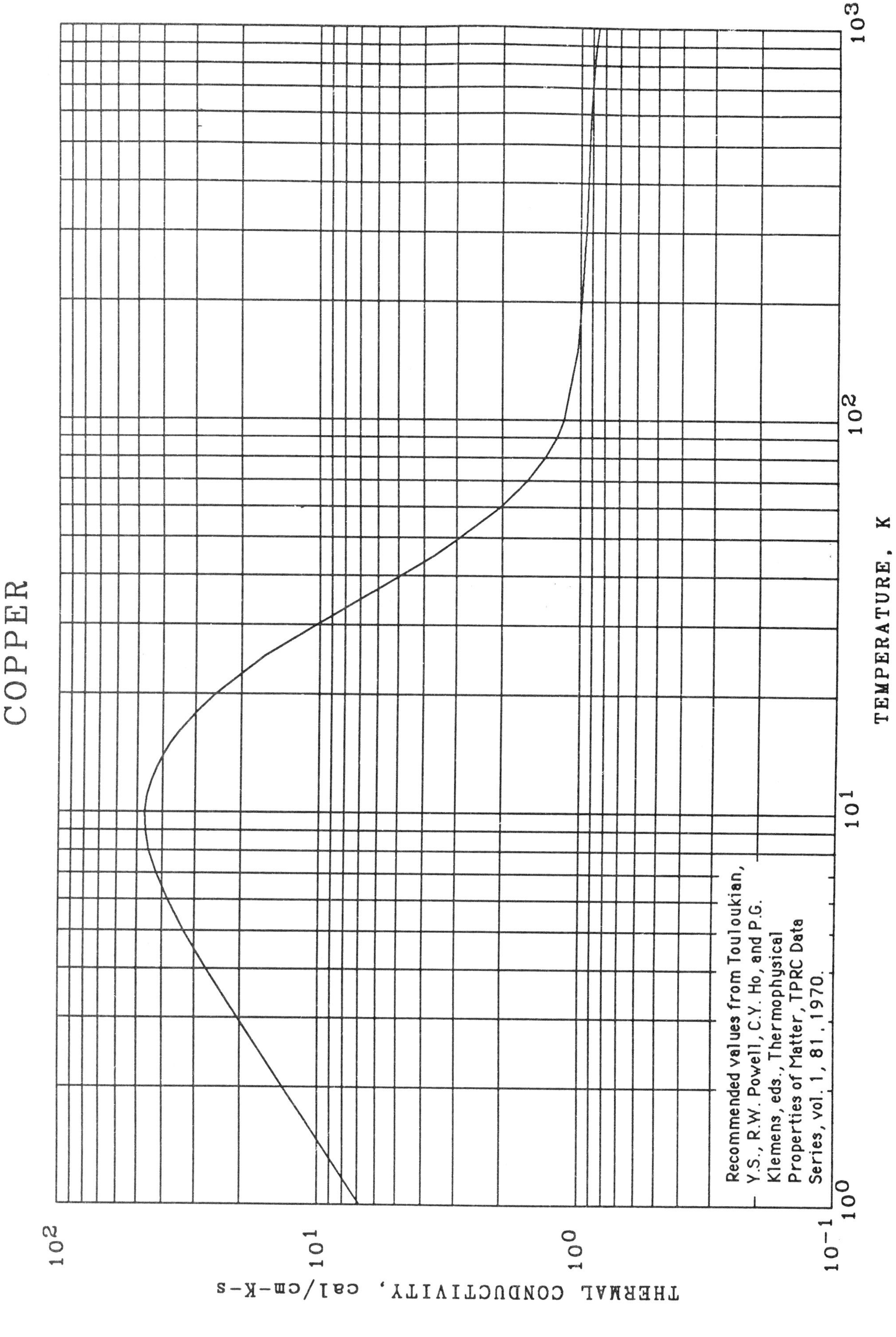
COPPER
THERMAL CONDUCTIVITY, cal/cm-K-s
TEMPERATURE, K
10^{-1}
10^{0}
10^{1}
10^{2}
10^{0}
10^{1}
10^{2}
10^{3}
Recommended values from Touloukian, Y.S., R.W. Powell, C.Y. Ho, and P.G. Klemens, eds., Thermophysical Properties of Matter, TPRC Data Series, vol. 1, 81, 1970.

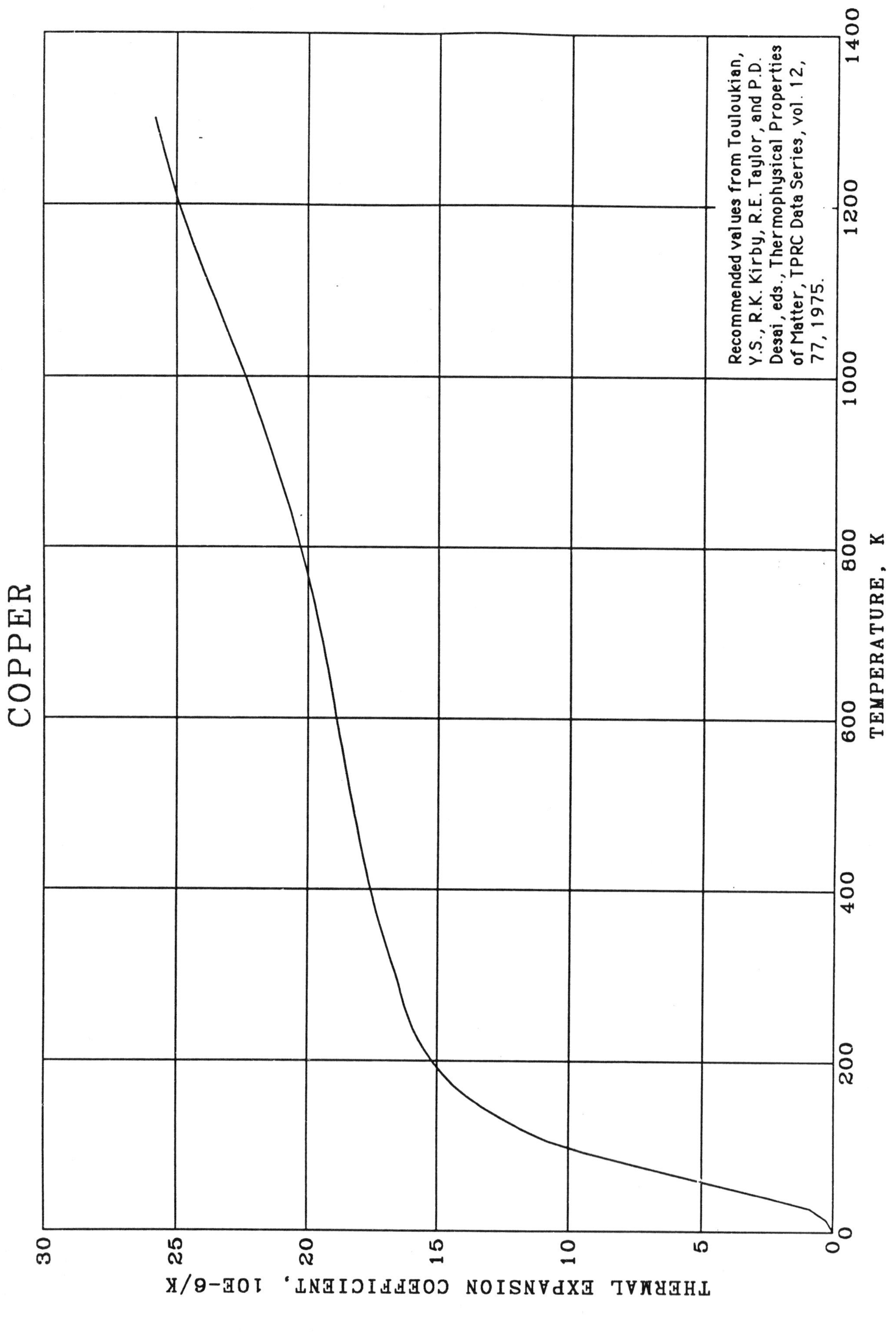

COPPER
Recommended values from Touloukian, Y.S., R.K. Kirby, R.E. Taylor, and P.D. Desai, eds., Thermophysical Properties of Matter, TPRC Data Series, vol. 12, 77, 1975.
TEMPERATURE, K
0
200
400
600
800
1000
1200
1400
THERMAL EXPANSION COEFFICIENT, 10E-6/K
0
5
10
15
20
25
30

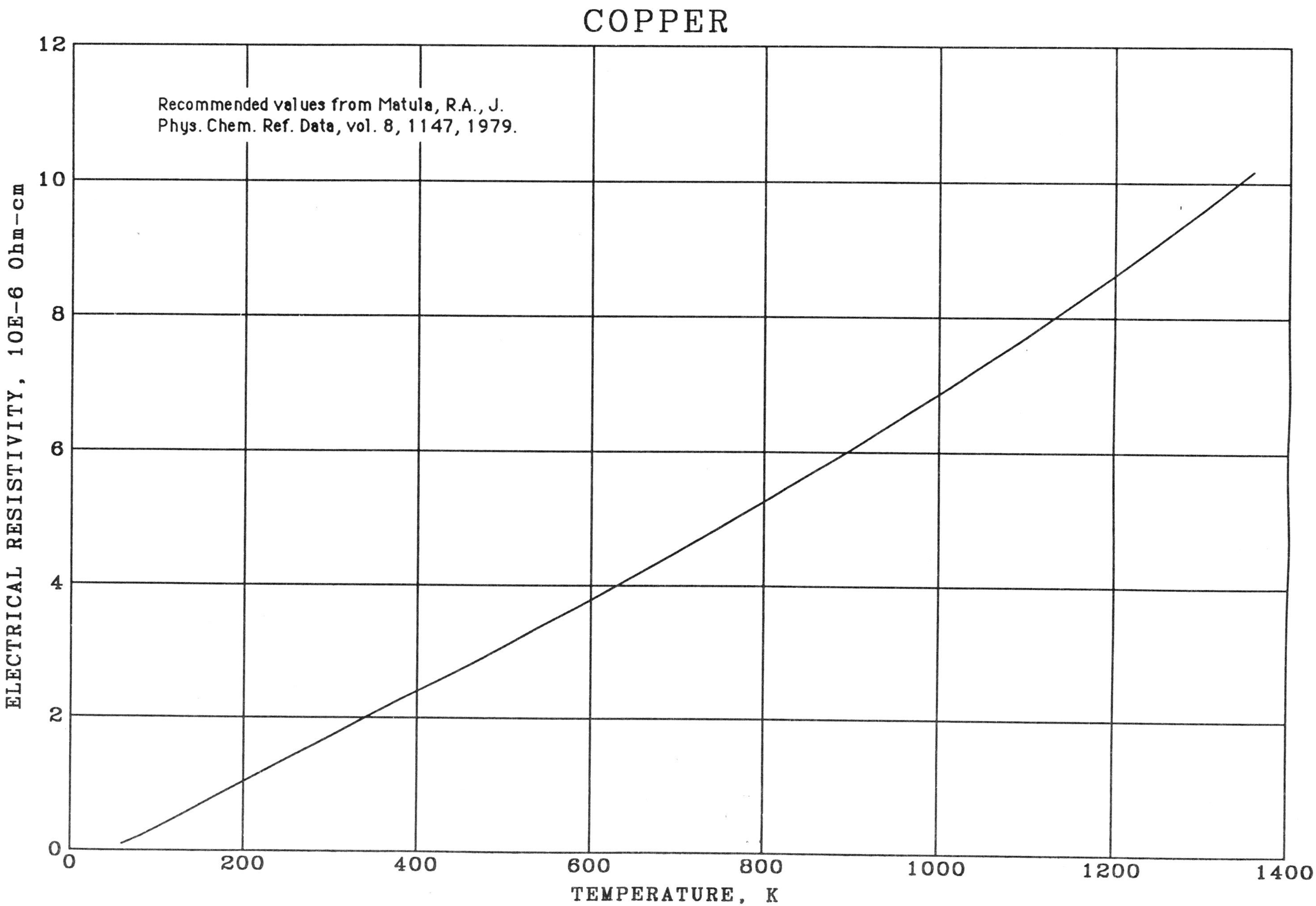
COPPER
Recommended values from Matula, R.A., J.
Phys. Chem. Ref. Data, vol. 8, 1147, 1979.
ELECTRICAL RESISTIVITY, 10E-6 Ohm-cm
TEMPERATURE, K
0
2
4
6
8
10
12
0
200
400
600
800
1000
1200
1400

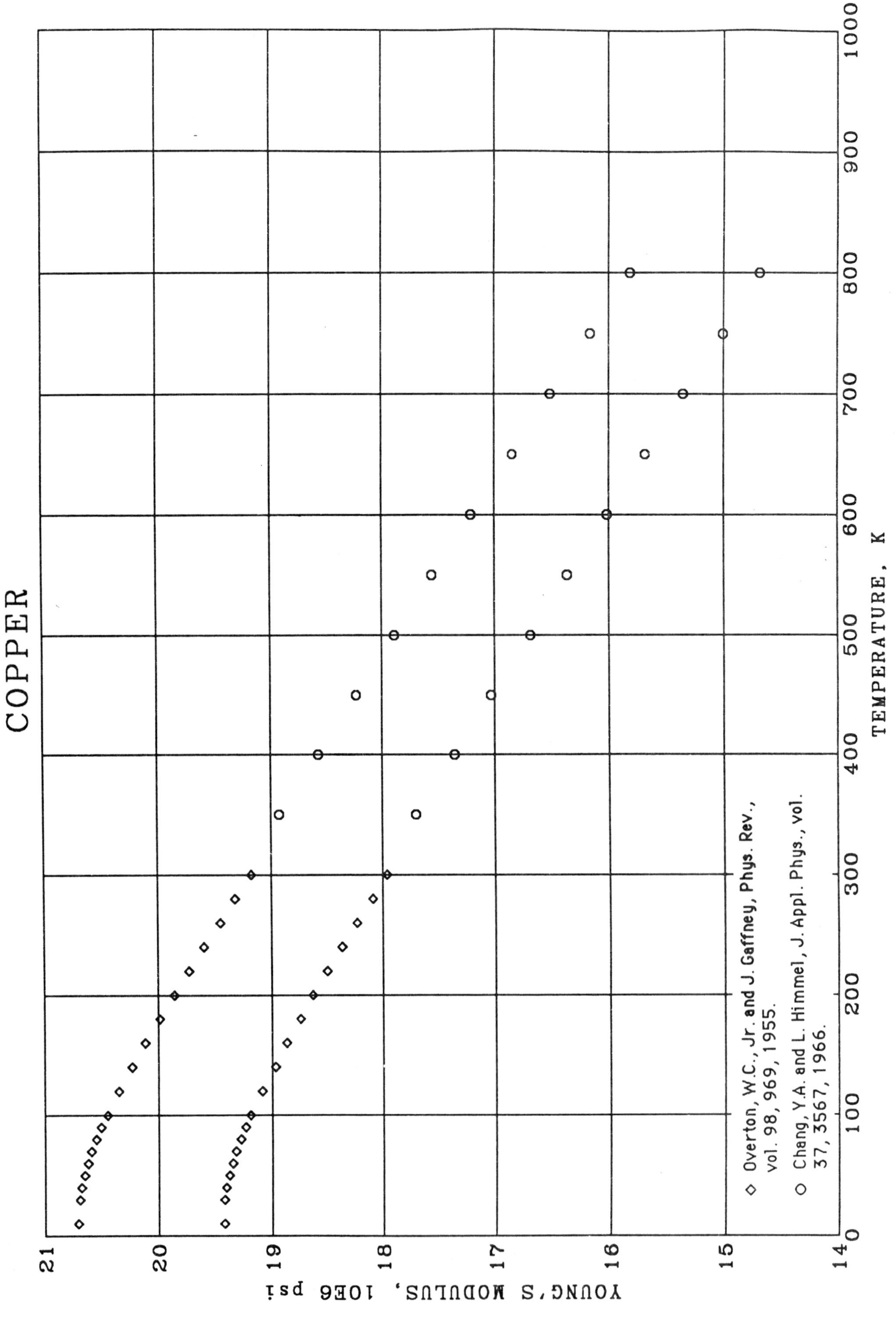
COPPER
TEMPERATURE, K
YOUNG'S MODULUS, 10E6 psi
0
100
200
300
400
500
600
700
800
900
1000
14
15
16
17
18
19
20
21
◇ Overton, W.C., Jr. and J. Gaffney, Phys. Rev., vol. 98, 969, 1955.
○ Chang, Y.A. and L. Himmel, J. Appl. Phys., vol. 37, 3567, 1966.

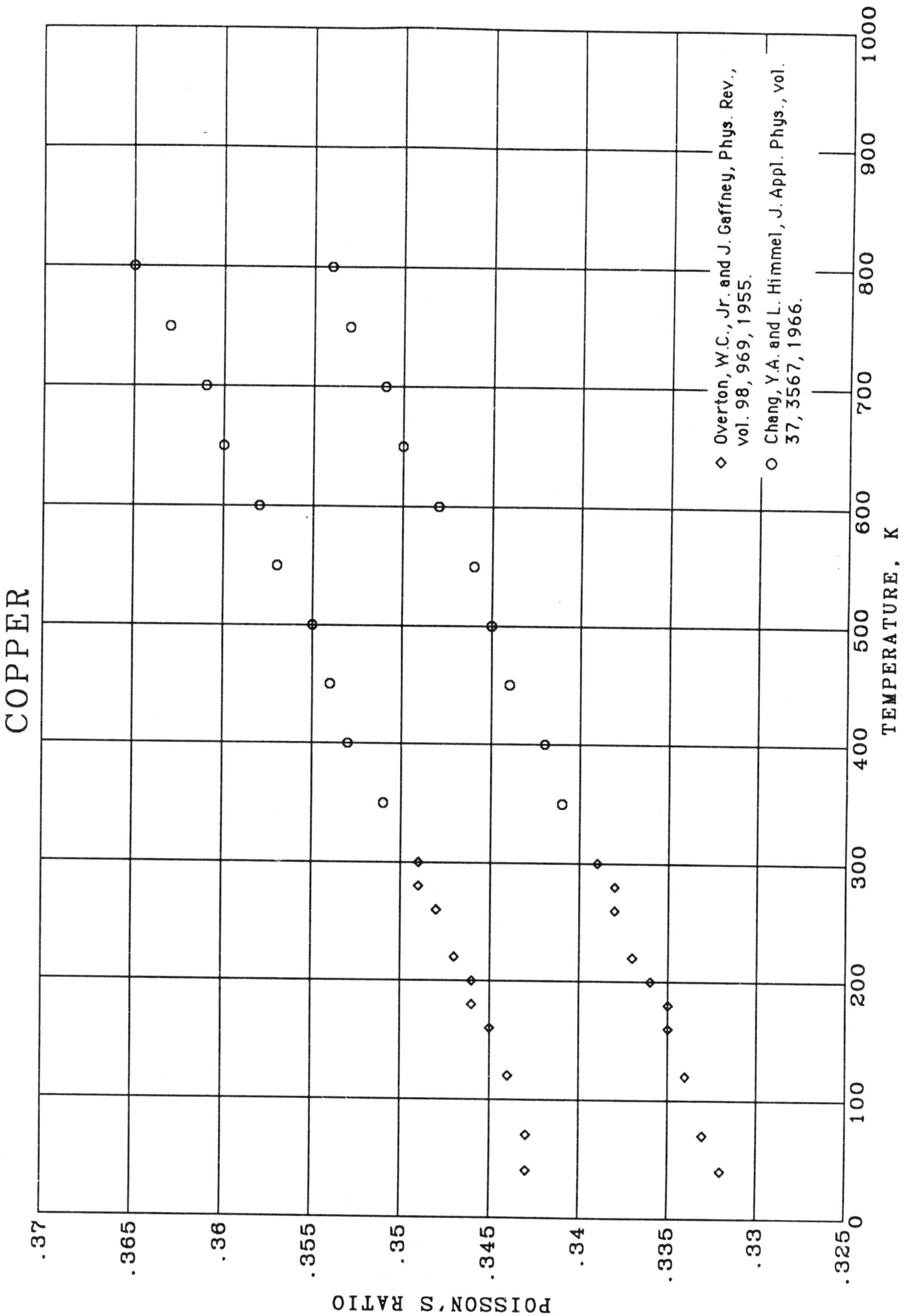

COPPER
POISSON'S RATIO
TEMPERATURE, K
.37
.365
.36
.355
.35
.345
.34
.335
.33
.325
0
100
200
300
400
500
600
700
800
900
1000
◇ Overton, W.C., Jr. and J. Gaffney, Phys. Rev., vol. 98, 969, 1955.
○ Chang, Y.A. and L. Himmel, J. Appl. Phys., vol. 37, 3567, 1966.

COPPER

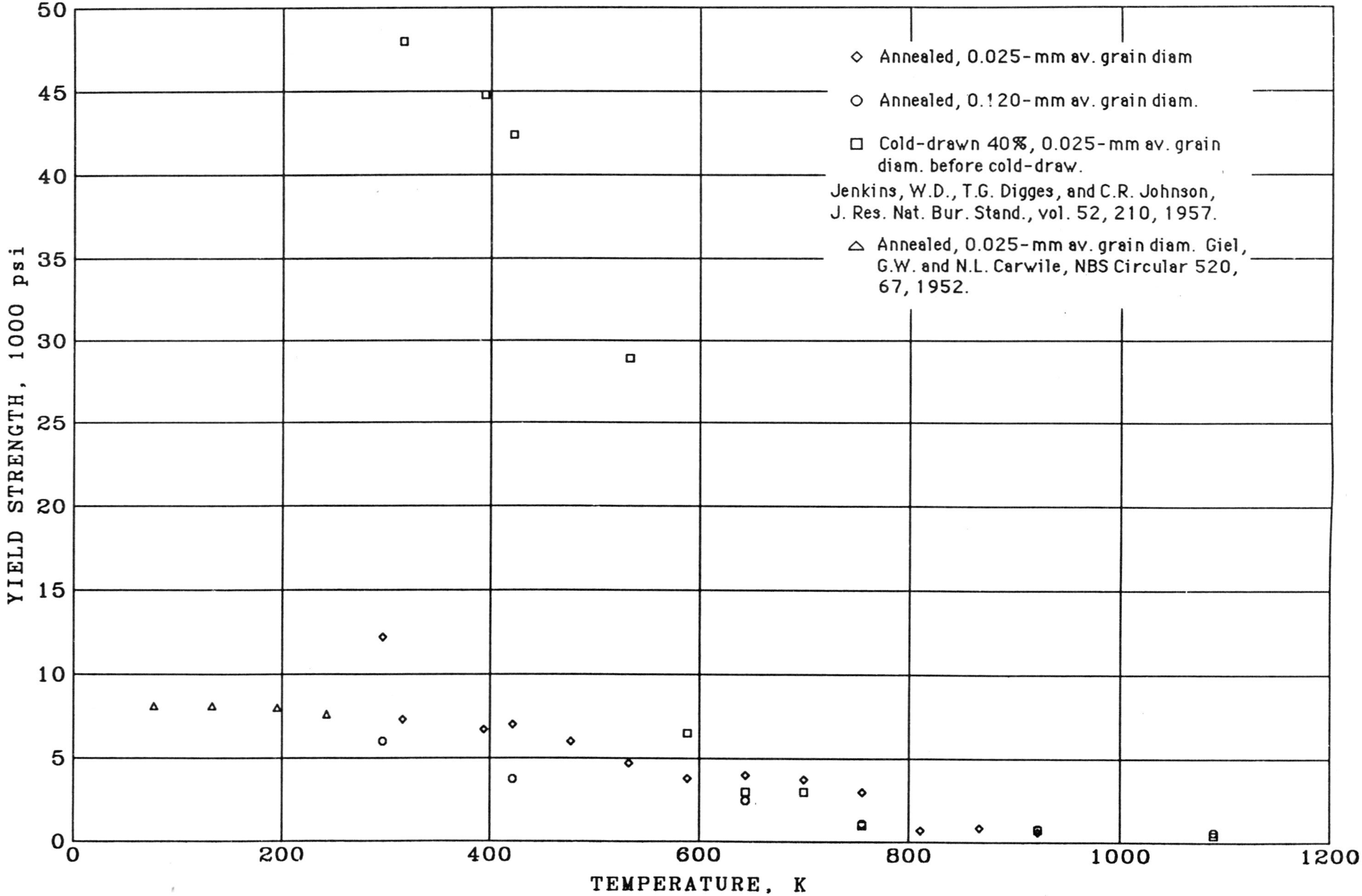

◇ Annealed, 0.025-mm av. grain diam
○ Annealed, 0.120-mm av. grain diam.
□ Cold-drawn 40%, 0.025-mm av. grain diam. before cold-draw.
Jenkins, W.D., T.G. Digges, and C.R. Johnson, J. Res. Nat. Bur. Stand., vol. 52, 210, 1957.
△ Annealed, 0.025-mm av. grain diam. Giel, G.W. and N.L. Carwile, NBS Circular 520, 67, 1952.
YIELD STRENGTH, 1000 psi
TEMPERATURE, K
0
5
10
15
20
25
30
35
40
45
50
0
200
400
600
800
1000
1200

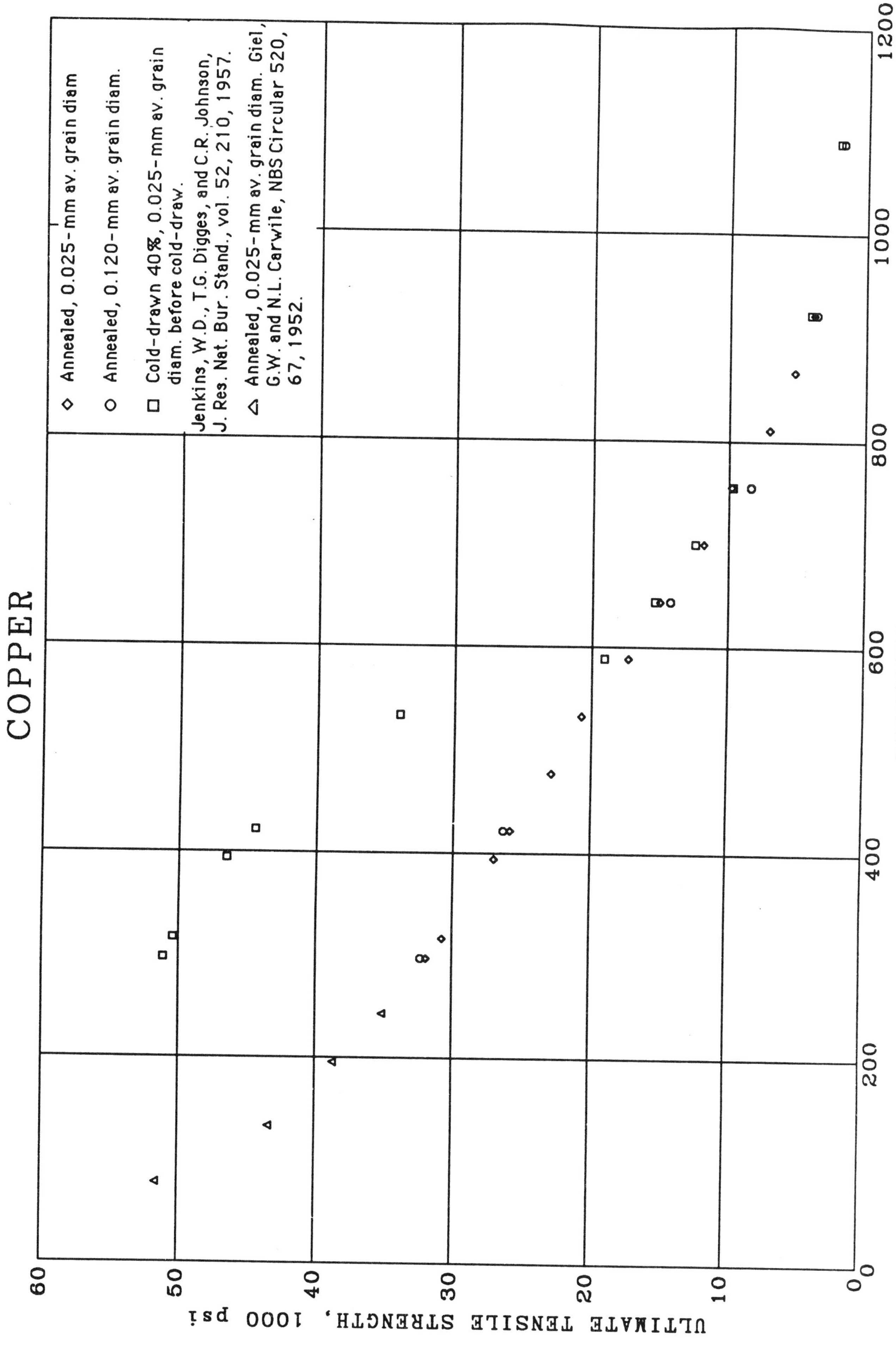
COPPER
◇ Annealed, 0.025-mm av. grain diam
○ Annealed, 0.120-mm av. grain diam.
□ Cold-drawn 40%, 0.025-mm av. grain diam. before cold-draw.
Jenkins, W.D., T.G. Digges, and C.R. Johnson, J. Res. Nat. Bur. Stand., vol. 52, 210, 1957.
△ Annealed, 0.025-mm av. grain diam. Giel, G.W. and N.L. Carwile, NBS Circular 520, 67, 1952.
ULTIMATE TENSILE STRENGTH, 1000 psi
TEMPERATURE, K
0
10
20
30
40
50
60
0
200
400
600
800
1000
1200

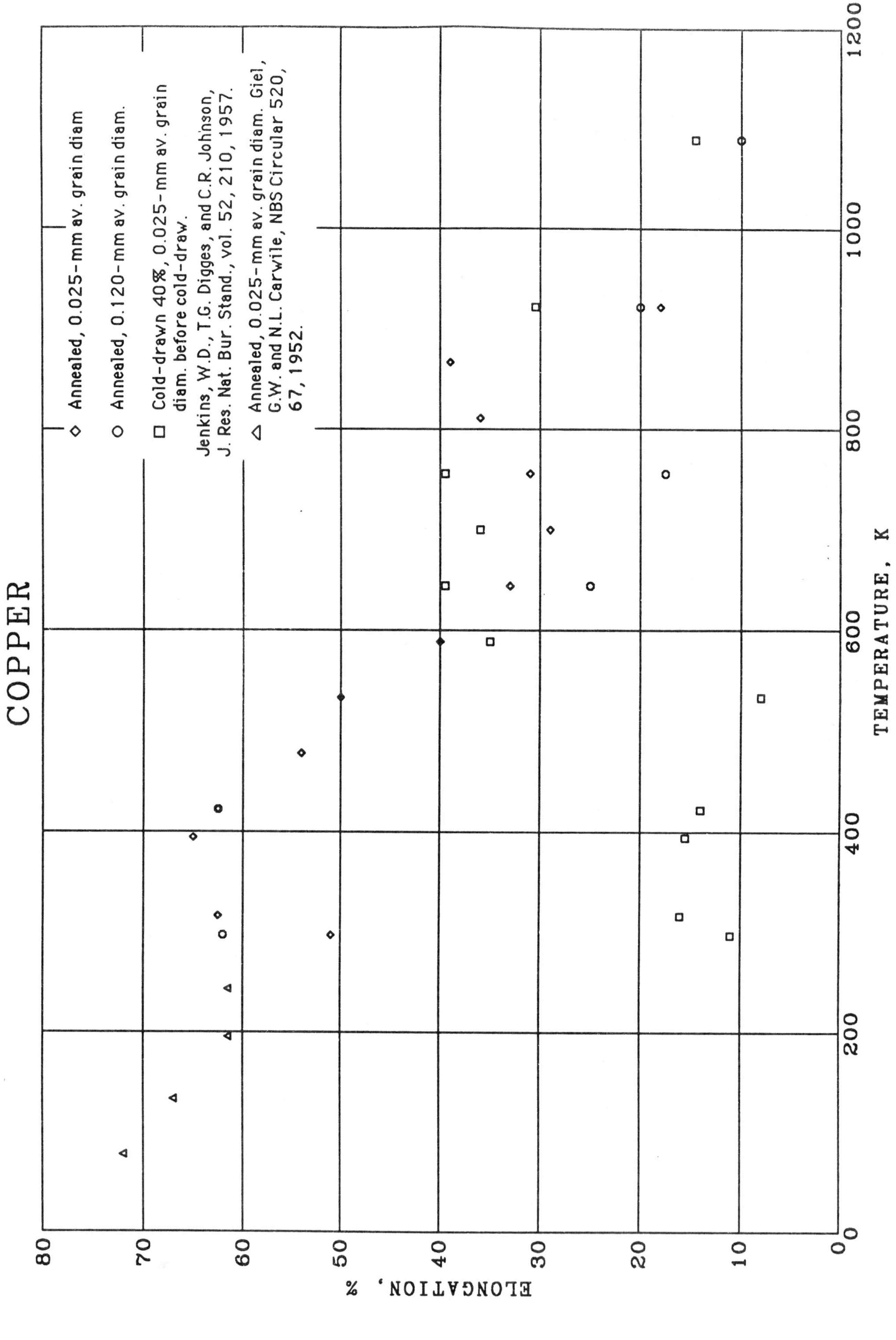

COPPER
◇ Annealed, 0.025-mm av. grain diam
○ Annealed, 0.120-mm av. grain diam.
□ Cold-drawn 40%, 0.025-mm av. grain diam. before cold-draw.
Jenkins, W.D., T.G. Digges, and C.R. Johnson, J. Res. Nat. Bur. Stand., vol. 52, 210, 1957.
△ Annealed, 0.025-mm av. grain diam. Giel, G.W. and N.L. Carwile, NBS Circular 520, 67, 1952.
ELONGATION, %
TEMPERATURE, K
0
10
20
30
40
50
60
70
80
0
200
400
600
800
1000
1200

COPPER

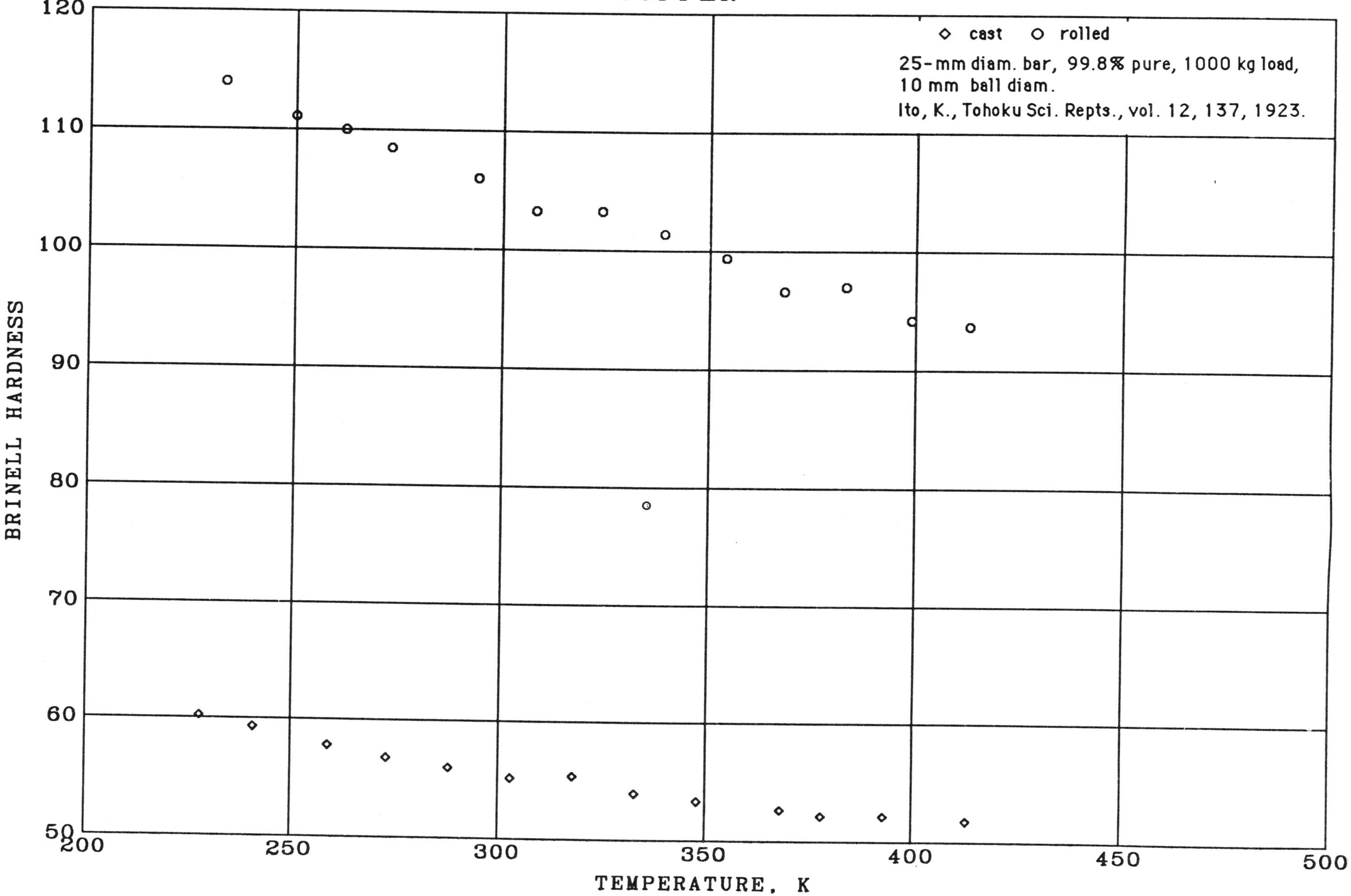

◇ cast ○ rolled
25-mm diam. bar, 99.8% pure, 1000 kg load,
10 mm ball diam.
Ito, K., Tohoku Sci. Repts., vol. 12, 137, 1923.
BRINELL HARDNESS
TEMPERATURE, K
120
110
100
90
80
70
60
50
200
250
300
350
400
450
500

GERMANIUM

PROPERTY	@ 300K	UNITS	REFERENCES
Density	5.32	g/cm^3	Ge1 Ge2 Ge3
Specific Heat	0.0761	cal/g-K	Ge4 Ge5 Ge6
Thermal Conductivity	0.143	cal/cm-s-K	Ge7 Ge8 Ge9
Thermal Exp Coef	5.7x10^{-6}	1/K	Ge10 Ge11 Ge12
Melting Point	1210.7	K	Ge13
Resistivity [1]	0.018	Ohm-cm	Ge14 Ge15 Ge16
TCR [1]	-1440	ppm/K	Ge14 Ge15 Ge16
Young's Modulus	1.913x10^7	psi	Ge17 Ge18 Ge19
Poisson's Ratio	0.207	-----	Ge17 Ge18 Ge19
Yield Strength [2]	60,000	psi	Ge21
Ult Tensile Strength	57,000	psi	Ge20 Ge21
Elongation at Break [2]	- 0 -	%	
Hardness [3]	1020	Knoop	Ge22 Ge23 Ge24

[1] The resistivity is a very strong function of the doping level. For the potential variations in the resistivity and TCR see the graph of resistivity -v- temperature. The values listed are intrinsic values.

[2] Below about 500C, Ge is a brittle solid. Tensile fracture is preceeded by no yielding or plastic deformation. The maximum breaking strength is determined by the sizes and distribution of flaws present in the material. The flaw distribution is a strong function of the volume. Smaller samples will have a narrow distribution of smaller flaws and therefore will be inherently stronger. When loaded in compression between 700 and 1000K, a definite yield point is displayed. The yield strength for a single crystal loaded in the <100> at 700K is listed.

[3] Hardness for the (111) face. For variation of hardness with orientation, see references.

GERMANIUM

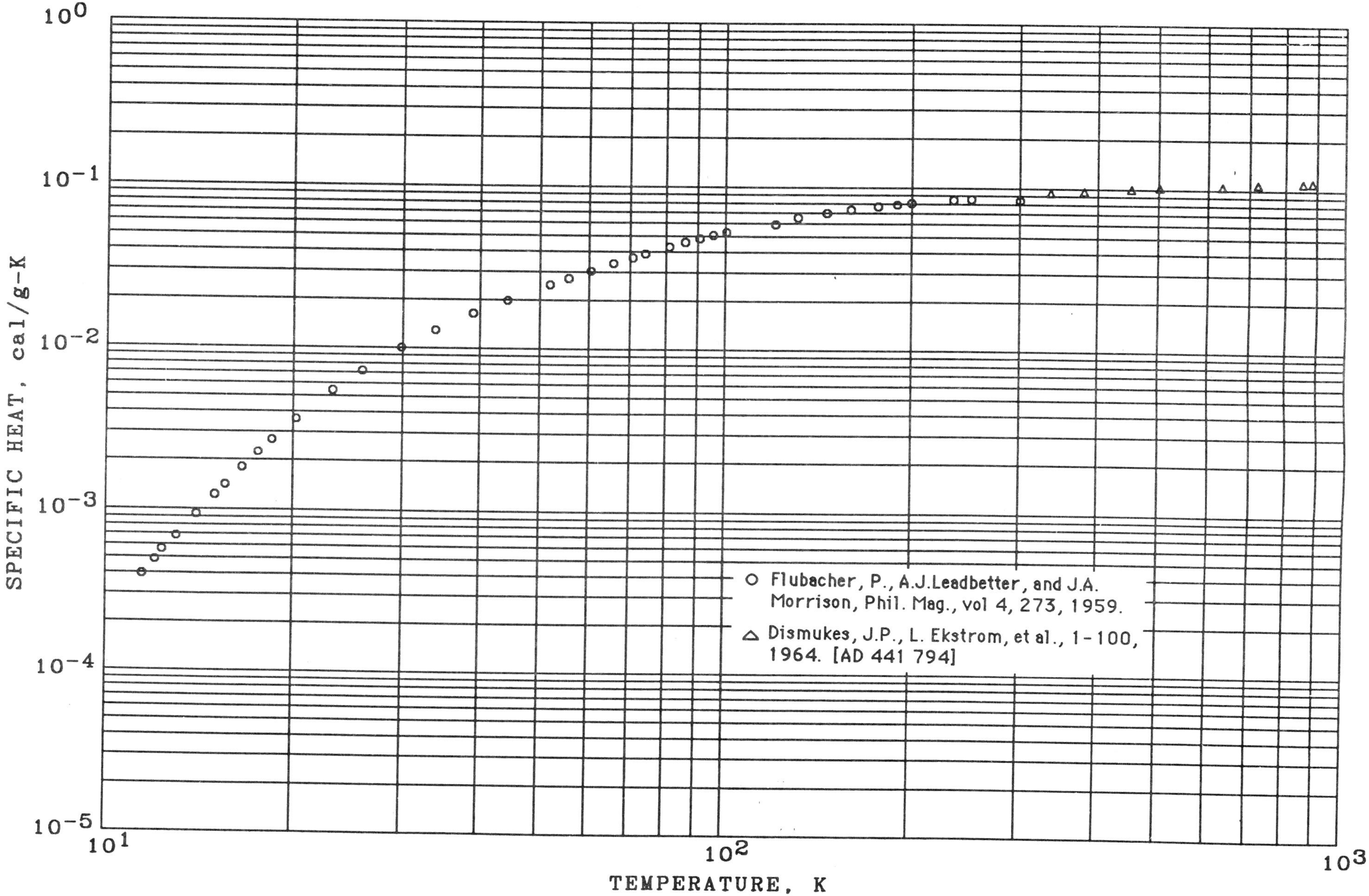

SPECIFIC HEAT, cal/g-K
10^0
10^-1
10^-2
10^-3
10^-4
10^-5
10^1
10^2
10^3
TEMPERATURE, K
○ Flubacher, P., A.J.Leadbetter, and J.A. Morrison, Phil. Mag., vol 4, 273, 1959.
△ Dismukes, J.P., L. Ekstrom, et al., 1-100, 1964. [AD 441 794]

GERMANIUM

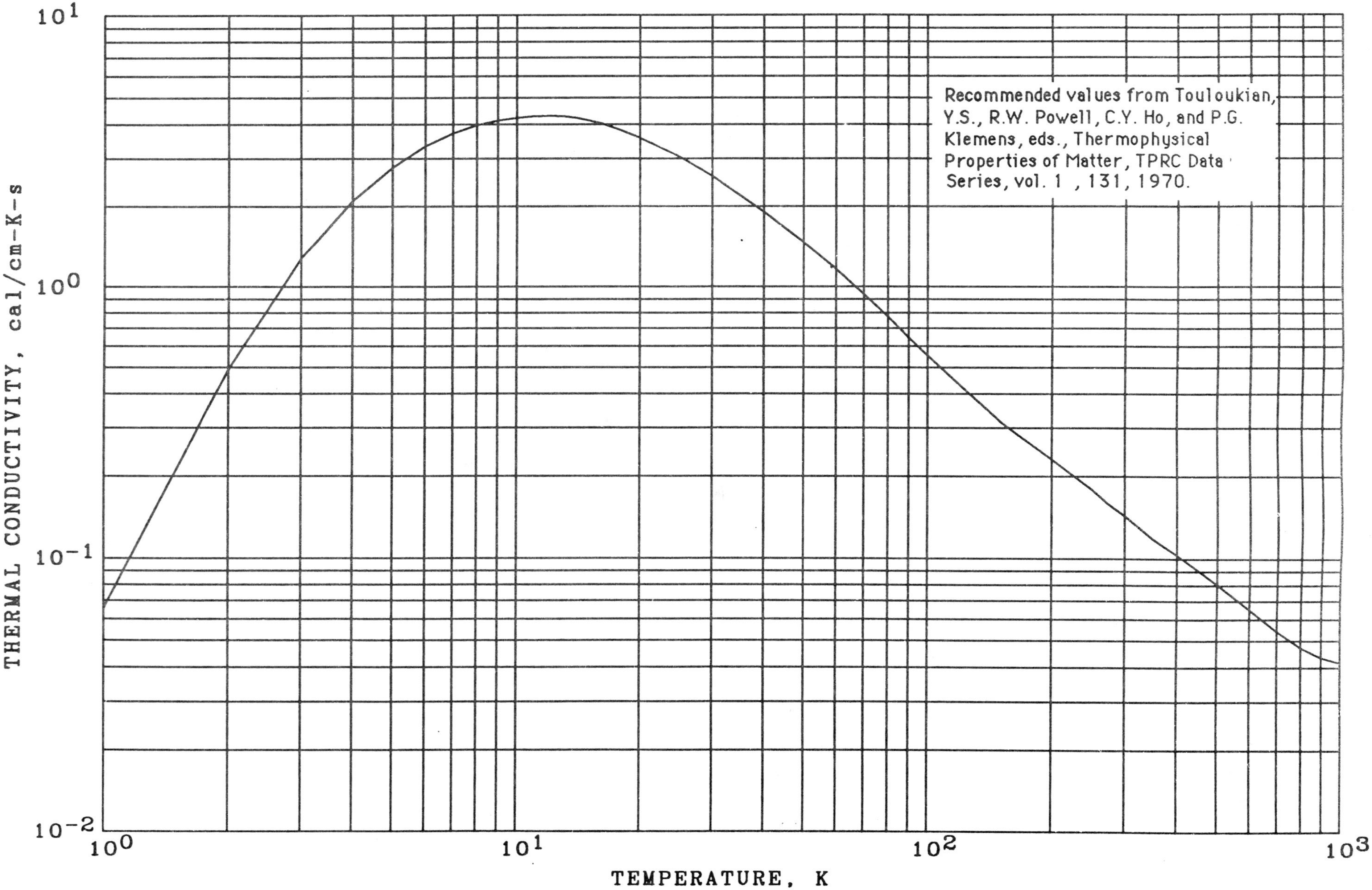

Recommended values from Touloukian, Y.S., R.W. Powell, C.Y. Ho, and P.G. Klemens, eds., Thermophysical Properties of Matter, TPRC Data Series, vol. 1 , 131, 1970.
THERMAL CONDUCTIVITY, cal/cm-K-s
TEMPERATURE, K
10^1
10^0
10^-1
10^-2
10^0
10^1
10^2
10^3

GERMANIUM

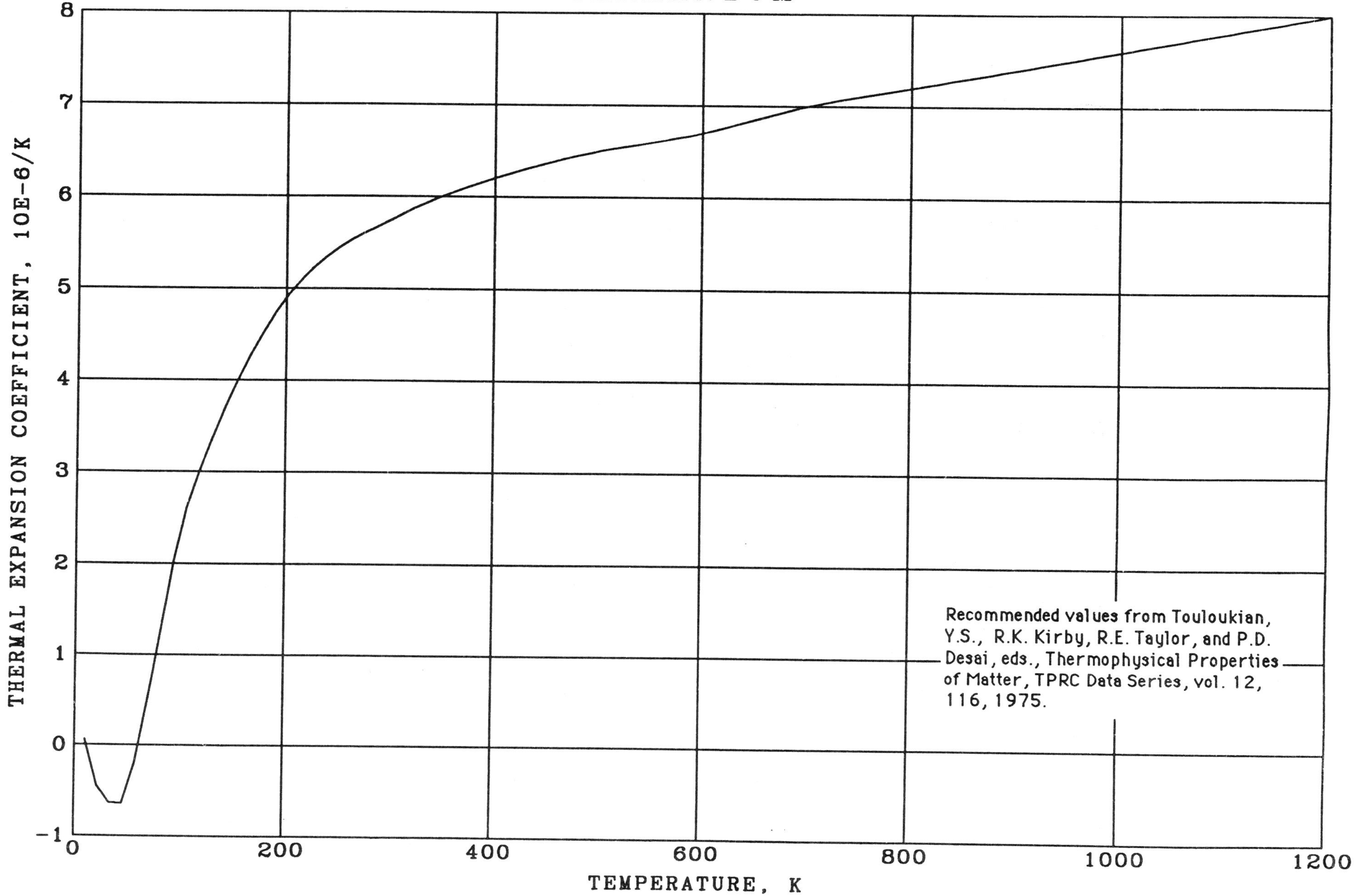

8
7
6
5
4
3
2
1
0
−1
0
200
400
600
800
1000
1200
THERMAL EXPANSION COEFFICIENT, 10E-6/K
TEMPERATURE, K
Recommended values from Touloukian, Y.S., R.K. Kirby, R.E. Taylor, and P.D. Desai, eds., Thermophysical Properties of Matter, TPRC Data Series, vol. 12, 116, 1975.

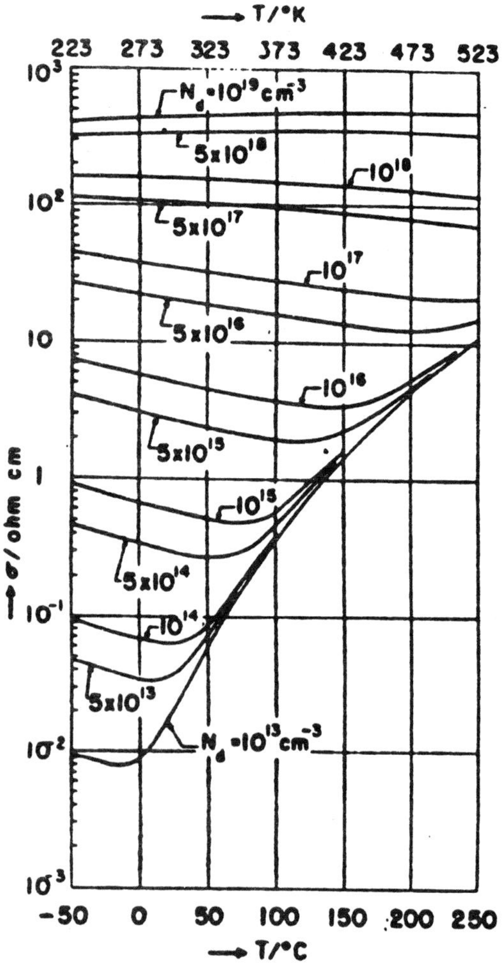

CONDUCTIVITY OF N-TYPE GERMANIUM

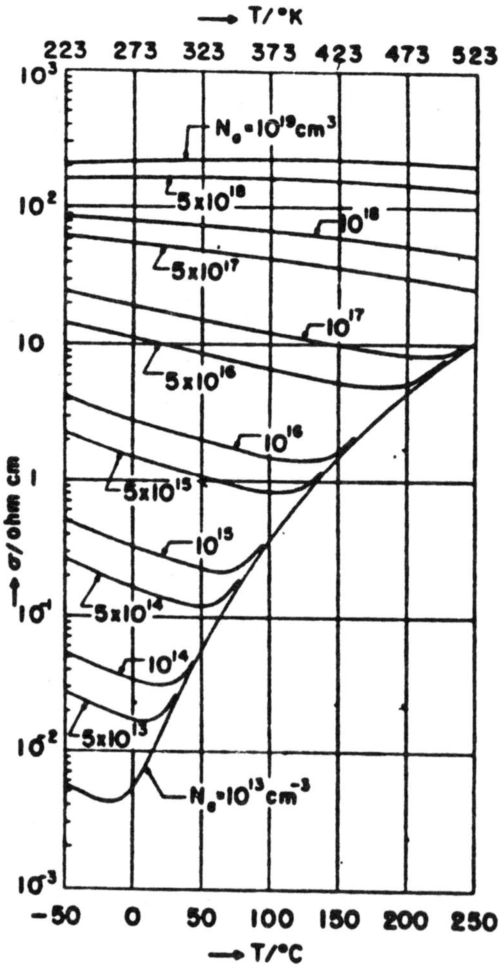

CONDUCTIVITY OF P-TYPE GERMANIUM

Reprinted from Gartner, W.W., Semiconductor Products, vol. 3, 29, 1960.

GERMANIUM

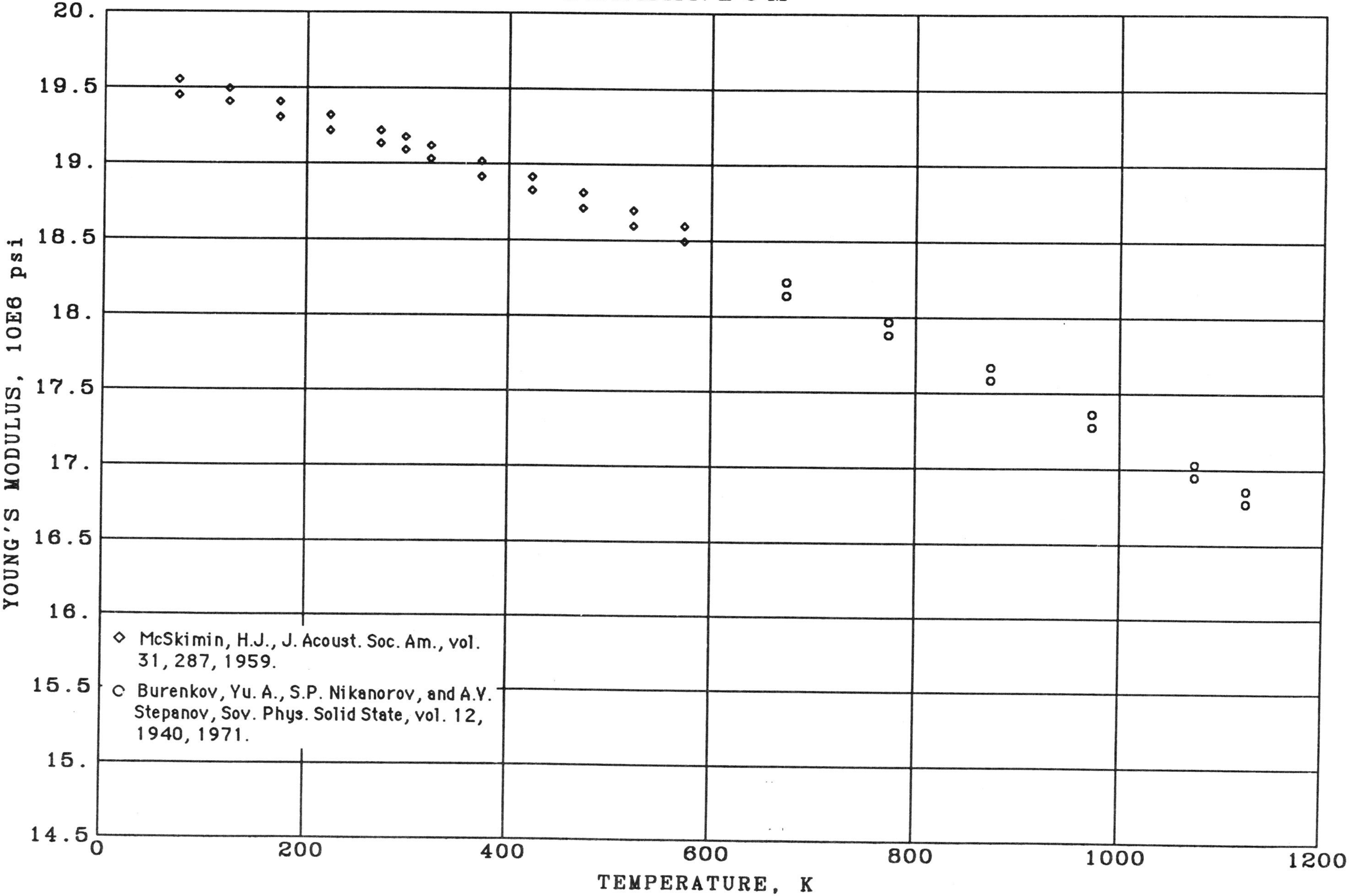

20.
19.5
19.
18.5
18.
17.5
17.
16.5
16.
15.5
15.
14.5
YOUNG'S MODULUS, 10E6 psi
0
200
400
600
800
1000
1200
TEMPERATURE, K
◇ McSkimin, H.J., J. Acoust. Soc. Am., vol. 31, 287, 1959.
○ Burenkov, Yu. A., S.P. Nikanorov, and A.V. Stepanov, Sov. Phys. Solid State, vol. 12, 1940, 1971.

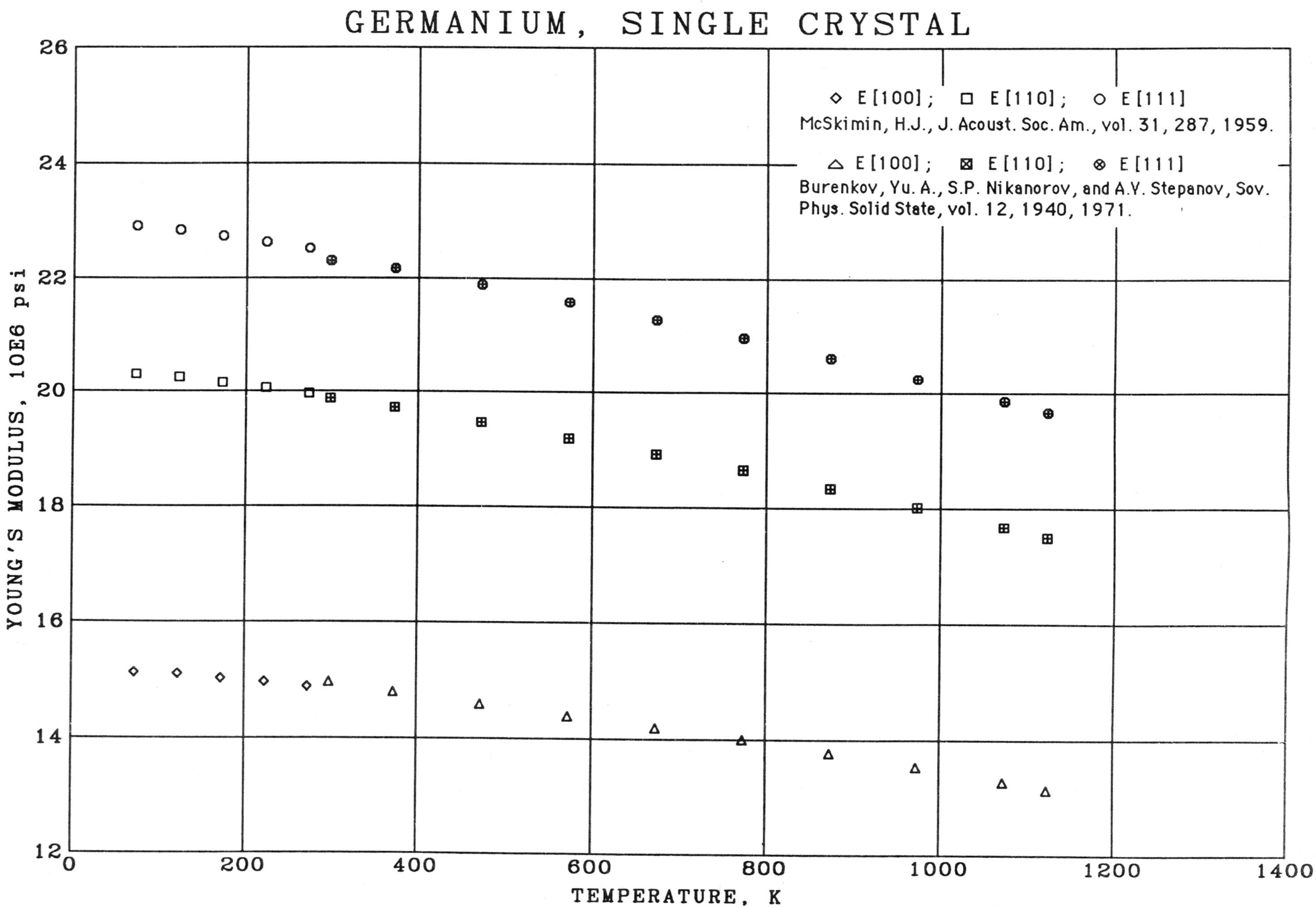
GERMANIUM, SINGLE CRYSTAL
◇ E[100]; □ E[110]; ○ E[111]
McSkimin, H.J., J. Acoust. Soc. Am., vol. 31, 287, 1959.
△ E[100]; ⊠ E[110]; ⊗ E[111]
Burenkov, Yu. A., S.P. Nikanorov, and A.Y. Stepanov, Sov. Phys. Solid State, vol. 12, 1940, 1971.
YOUNG'S MODULUS, 10E6 psi
TEMPERATURE, K
26
24
22
20
18
16
14
12
0
200
400
600
800
1000
1200
1400

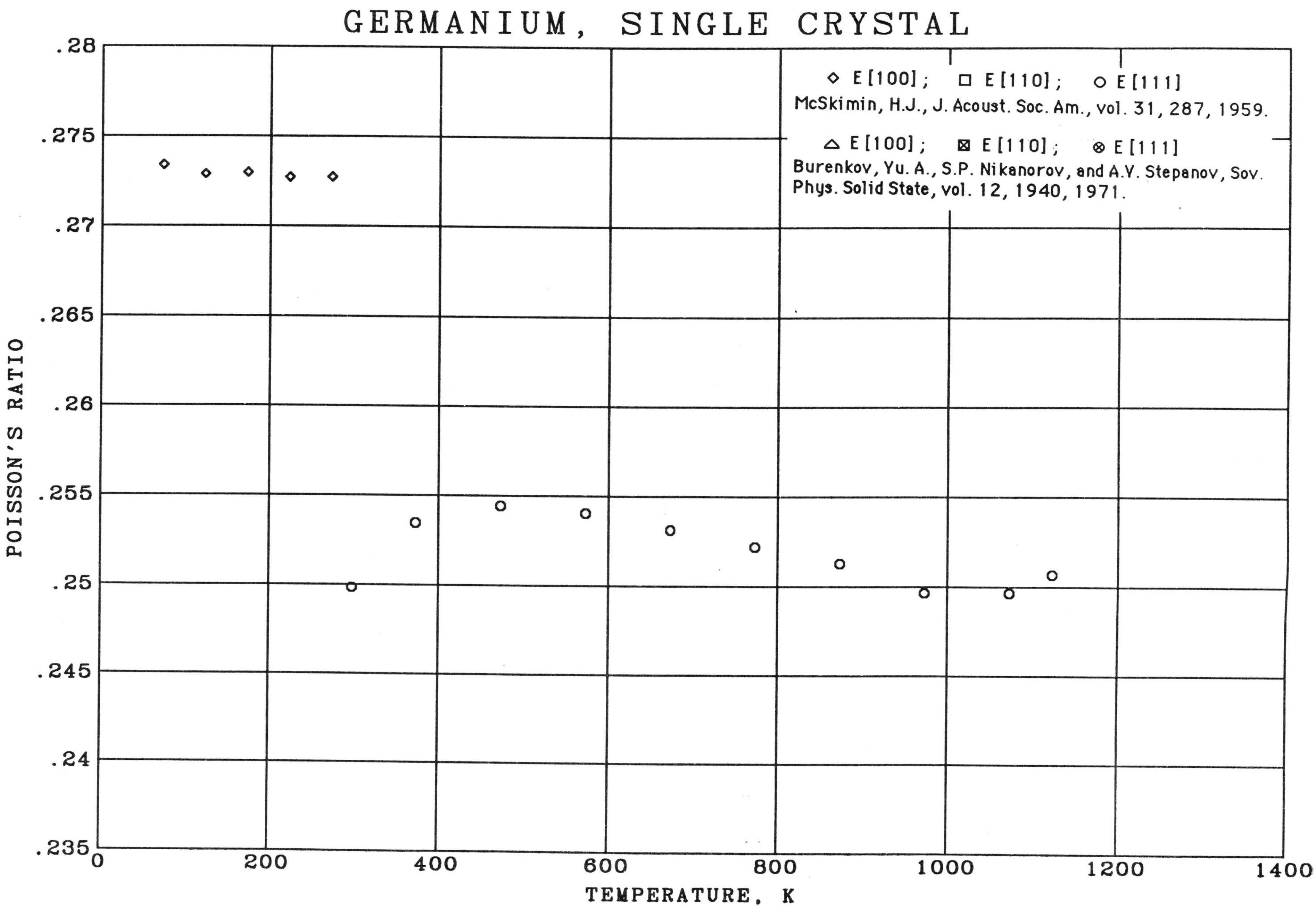
GERMANIUM, SINGLE CRYSTAL
◇ E[100]; □ E[110]; ○ E[111]
McSkimin, H.J., J. Acoust. Soc. Am., vol. 31, 287, 1959.
△ E[100]; ⊠ E[110]; ⊗ E[111]
Burenkov, Yu. A., S.P. Nikanorov, and A.Y. Stepanov, Sov. Phys. Solid State, vol. 12, 1940, 1971.
POISSON'S RATIO
.28
.275
.27
.265
.26
.255
.25
.245
.24
.235
TEMPERATURE, K
0
200
400
600
800
1000
1200
1400

GERMANIUM

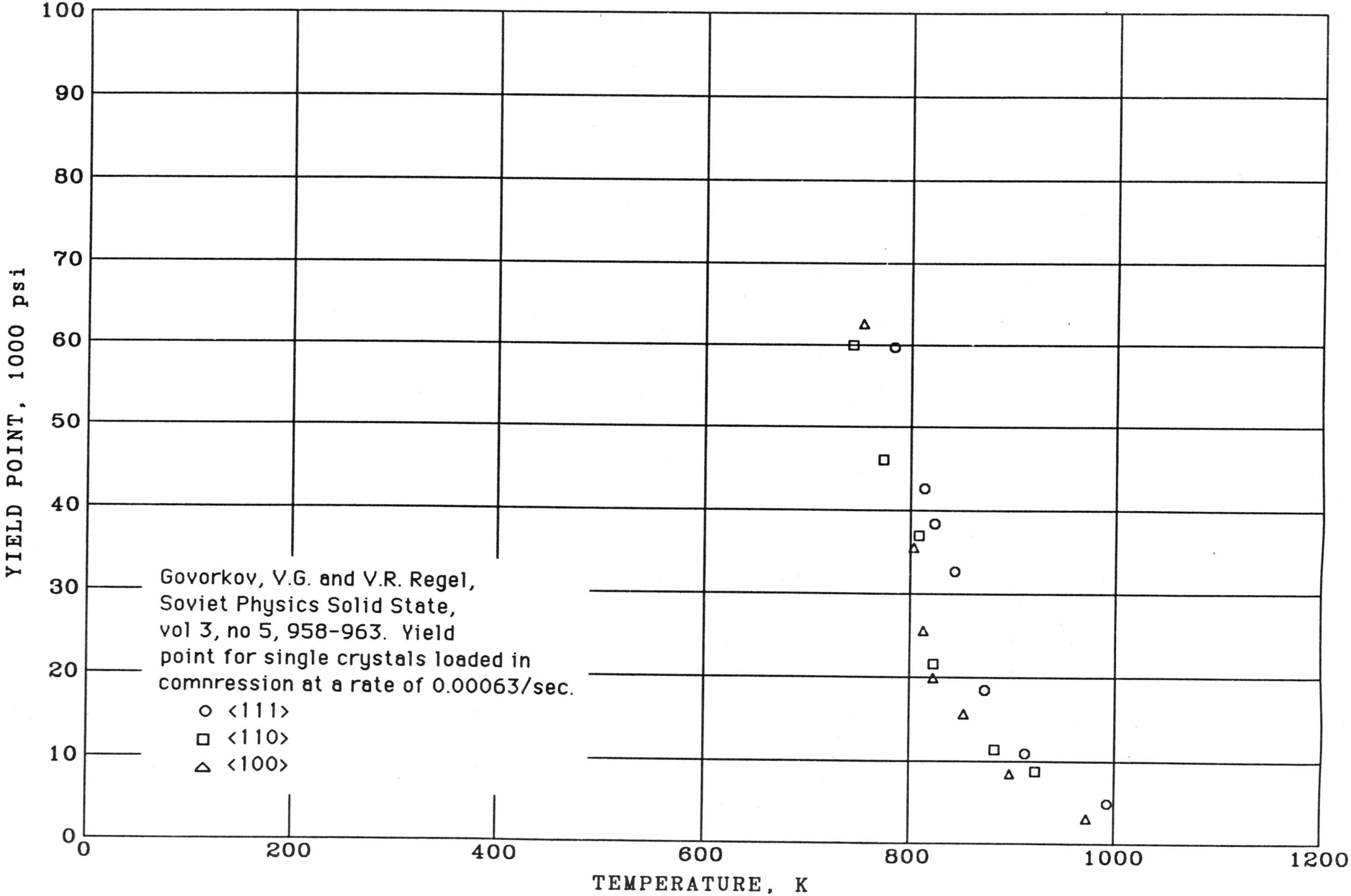

YIELD POINT, 1000 psi
100
90
80
70
60
50
40
30
20
10
0
0
200
400
600
800
1000
1200
TEMPERATURE, K
Govorkov, V.G. and V.R. Regel,
Soviet Physics Solid State,
vol 3, no 5, 958-963. Yield
point for single crystals loaded in
comnression at a rate of 0.00063/sec.
○ <111>
□ <110>
△ <100>

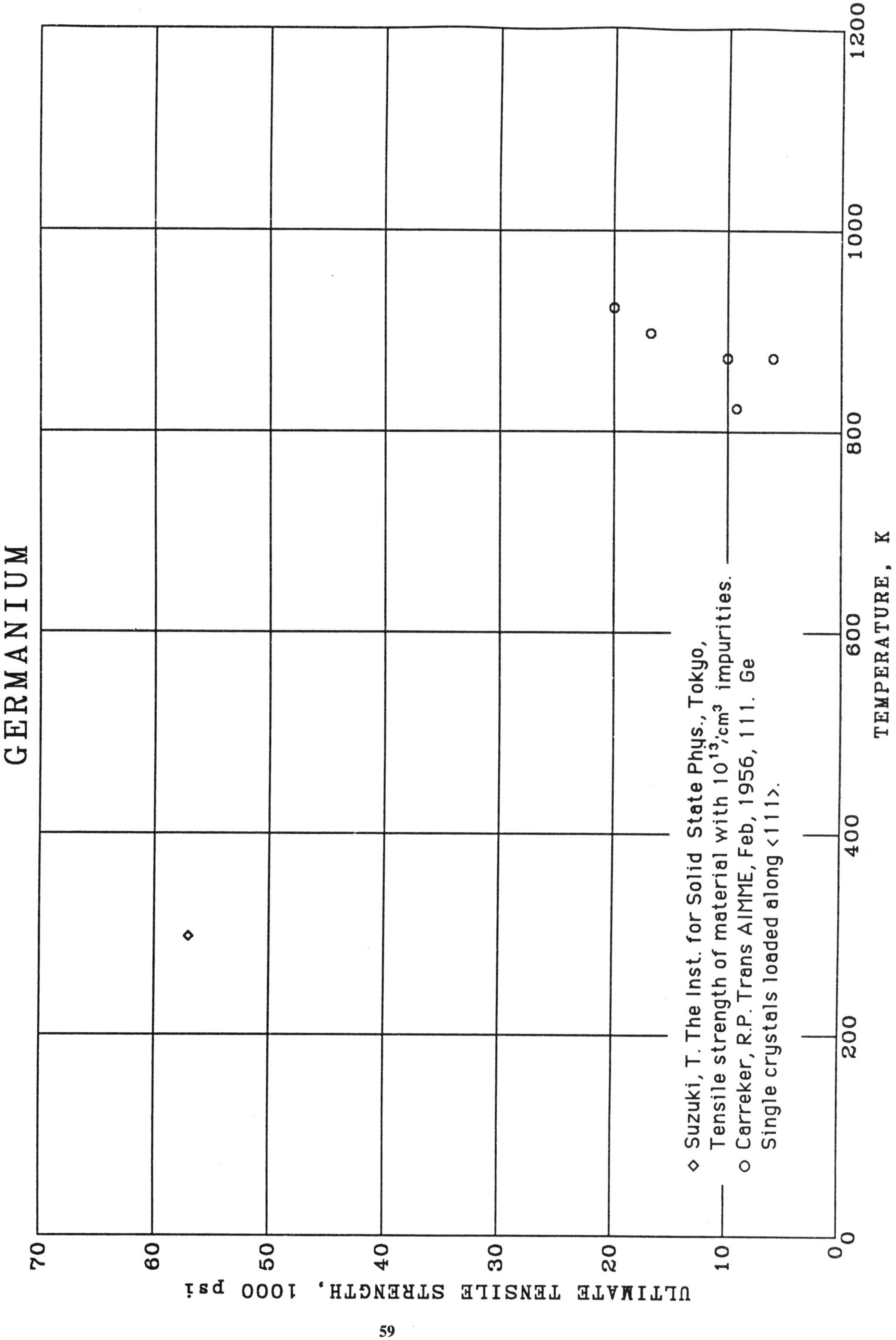
GERMANIUM
◇ Suzuki, T. The Inst. for Solid State Phys., Tokyo,
Tensile strength of material with 10^13/cm^3 impurities.
○ Carreker, R.P. Trans AIMME, Feb, 1956, 111. Ge
Single crystals loaded along <111>.
ULTIMATE TENSILE STRENGTH, 1000 psi
TEMPERATURE, K
0
10
20
30
40
50
60
70
200
400
600
800
1000
1200

GERMANIUM

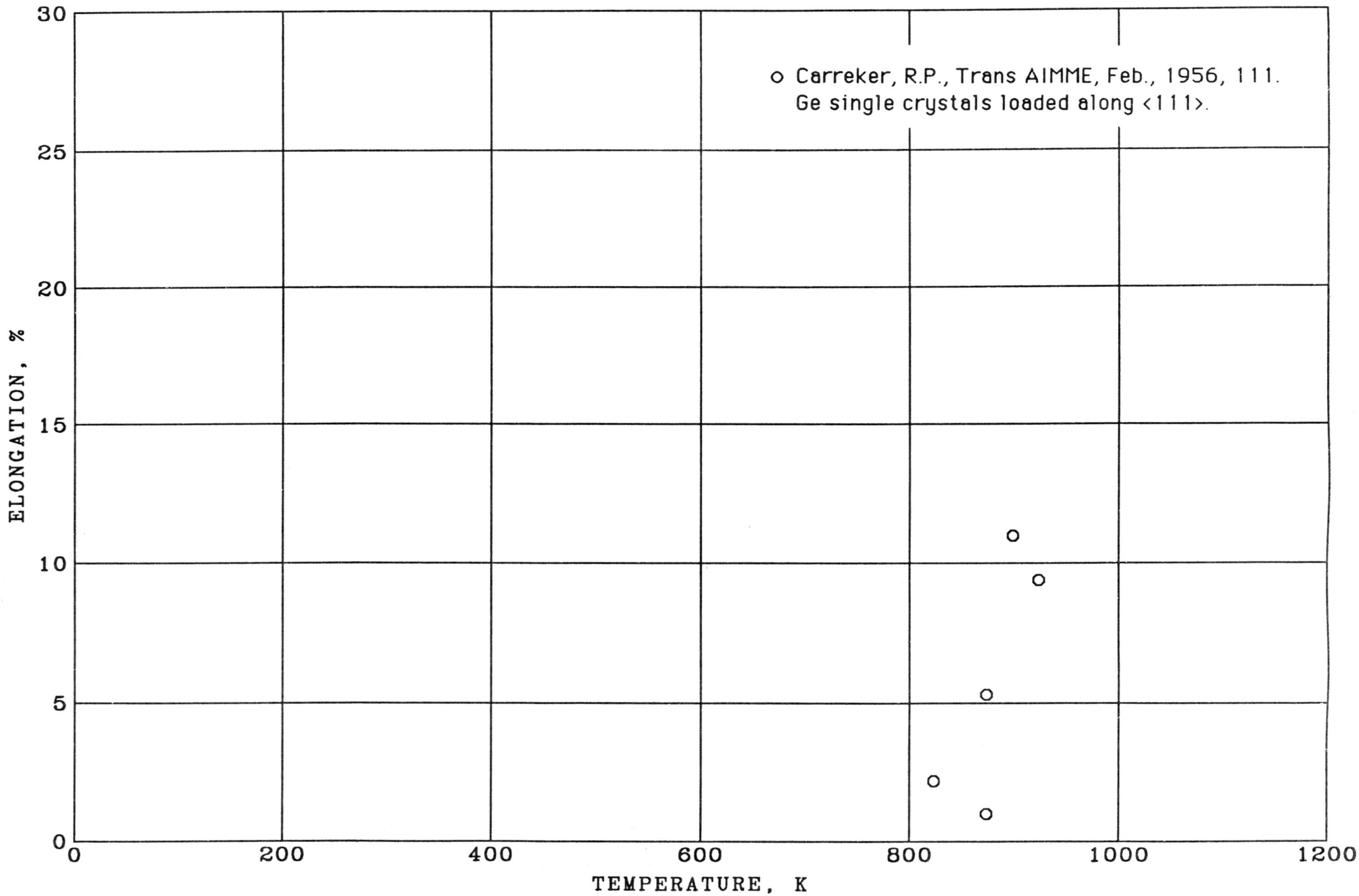

o Carreker, R.P., Trans AIMME, Feb., 1956, 111.
Ge single crystals loaded along <111>.
ELONGATION, %
TEMPERATURE, K
30
25
20
15
10
5
0
0
200
400
600
800
1000
1200

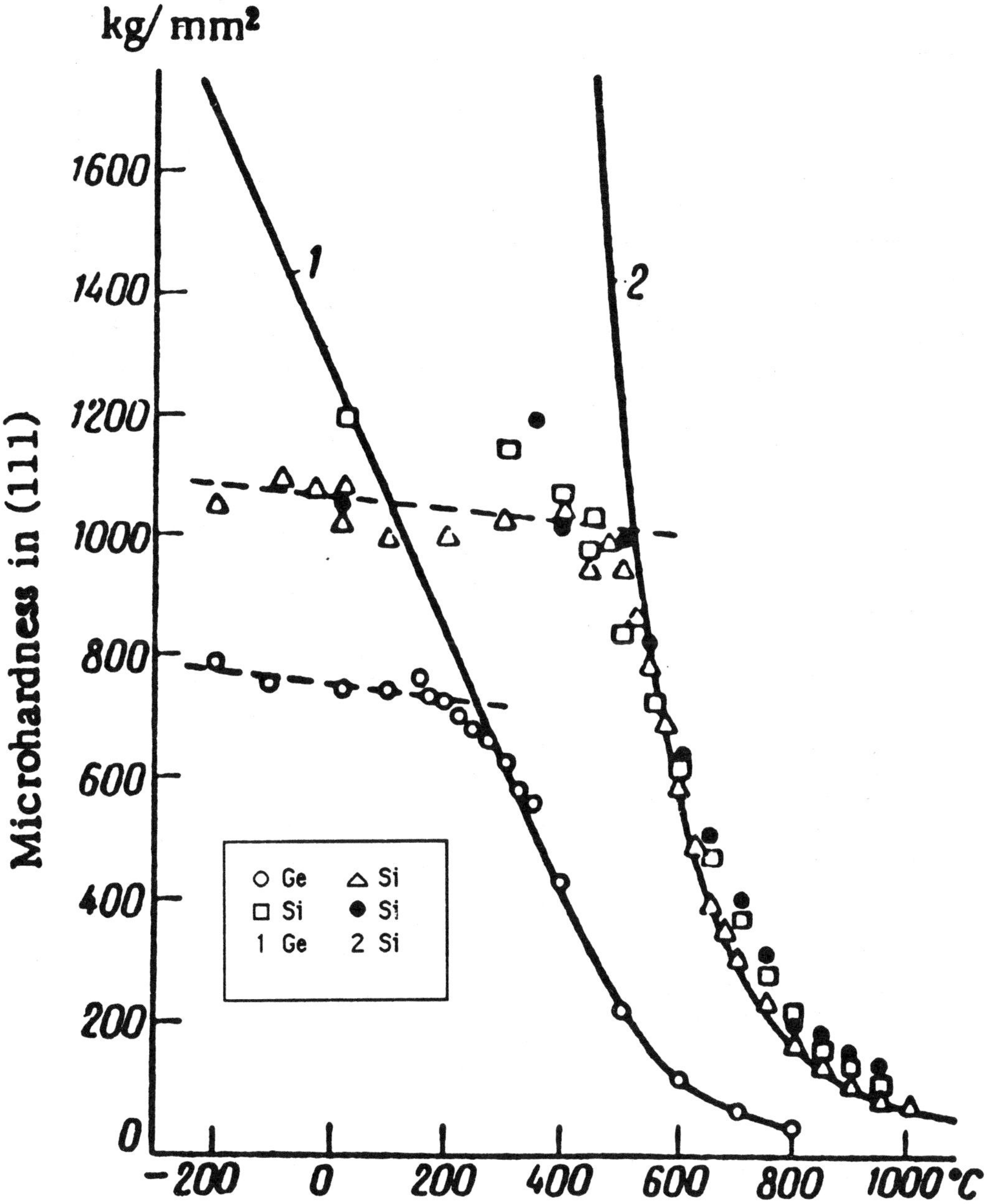

Microhardness of Si and Ge

Reprinted from Trefilov, V.I., and Yu V. Mil'man, Sov. Phys. Doklady, vol. 8, 1240, 1964.

GOLD

PROPERTY	@ 300K	UNITS	REFERENCES
Density	19.3	g/cm^3	Au1 Au2 Au3
Specific Heat	0.03088	cal/g-K	Au4 Au5 Au6
Thermal Conductivity	0.753	cal/cm-s-K	Au7 Au8 Au9
Thermal Exp Coef	$14.2x10^{-6}$	1/K	Au10 Au11 Au12
Melting Point	1337.58	K	Au13
Resistivity [1]	$2.249x10^{-6}$	Ohm-cm	Au14 Au15 Au16
TCR (218K to 398K)	3670	ppm/K	Au14 Au15 Au16
Young's Modulus	$1.12x10^7$	psi	Au17 Au18 Au19
Poisson's Ratio	0.425	-----	Au17 Au18 Au20
Yield Strength	250	psi	Au21 Au24 Au27
Ult Tensile Strength	17,000	psi	Au21 Au22 Au23
Elongation at Break	50	%	Au21 Au22 Au23
Hardness	18.5	Brinell	Au23 Au25 Au26

[1] The resistivity of sputtered gold films onto bismuth oxide shows resistivities in the range of 7 to 16 micro ohm-cm. Resistivity is dependent on thickness and substrate temperature. Annealed films with thicknesses > 200 angstroms deposited on substrates held at -183C provides resistivities nearest the bulk values.[Au28]

Mechanical properties listed are for fully annealed, bulk material. 60% cold work producing 0.05 inch dia wire will increase strength to 30,000 psi and decrease elongation to 4%.[Au30] Polycrystalline films demonstrate strengths on the order of 70,000 psi with 2 to 3% elongation.[Au29] Heating cold worked material above 150C causes recovery and recrystalization erasing the effects of cold work.

The mechanical properties fo very thin wires differ considerably from the values listed above. Recent work indicates room temperature properties for 0.001inch(25 micron) wire are: yield strength = 25,000 psi; tensile strength = 30,000 psi; elongation = 4%.[Au30]

GOLD

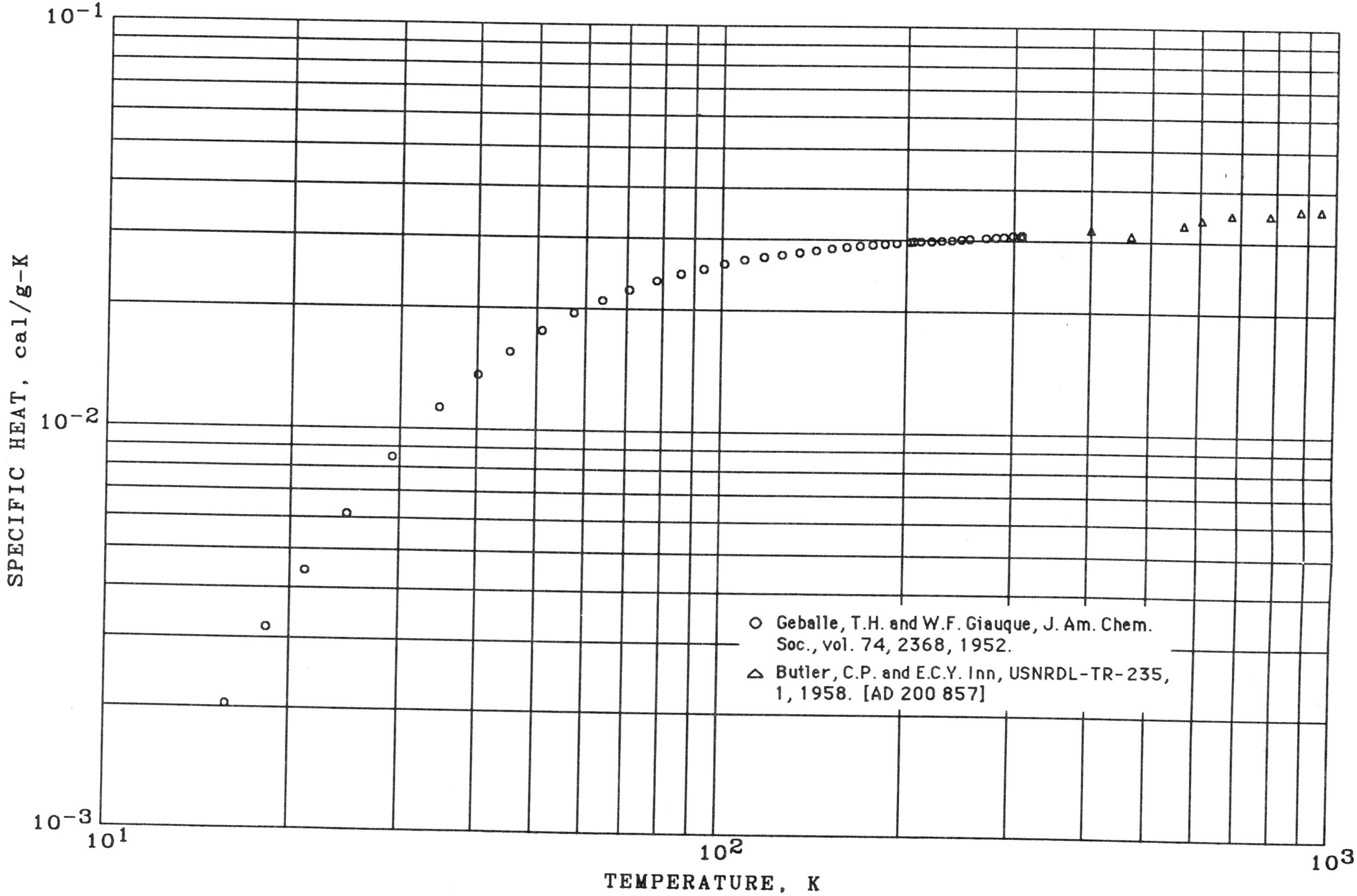

SPECIFIC HEAT, cal/g-K
10⁻¹
10⁻²
10⁻³
TEMPERATURE, K
10¹
10²
10³
○ Geballe, T.H. and W.F. Giauque, J. Am. Chem. Soc., vol. 74, 2368, 1952.
△ Butler, C.P. and E.C.Y. Inn, USNRDL-TR-235, 1, 1958. [AD 200 857]

GOLD

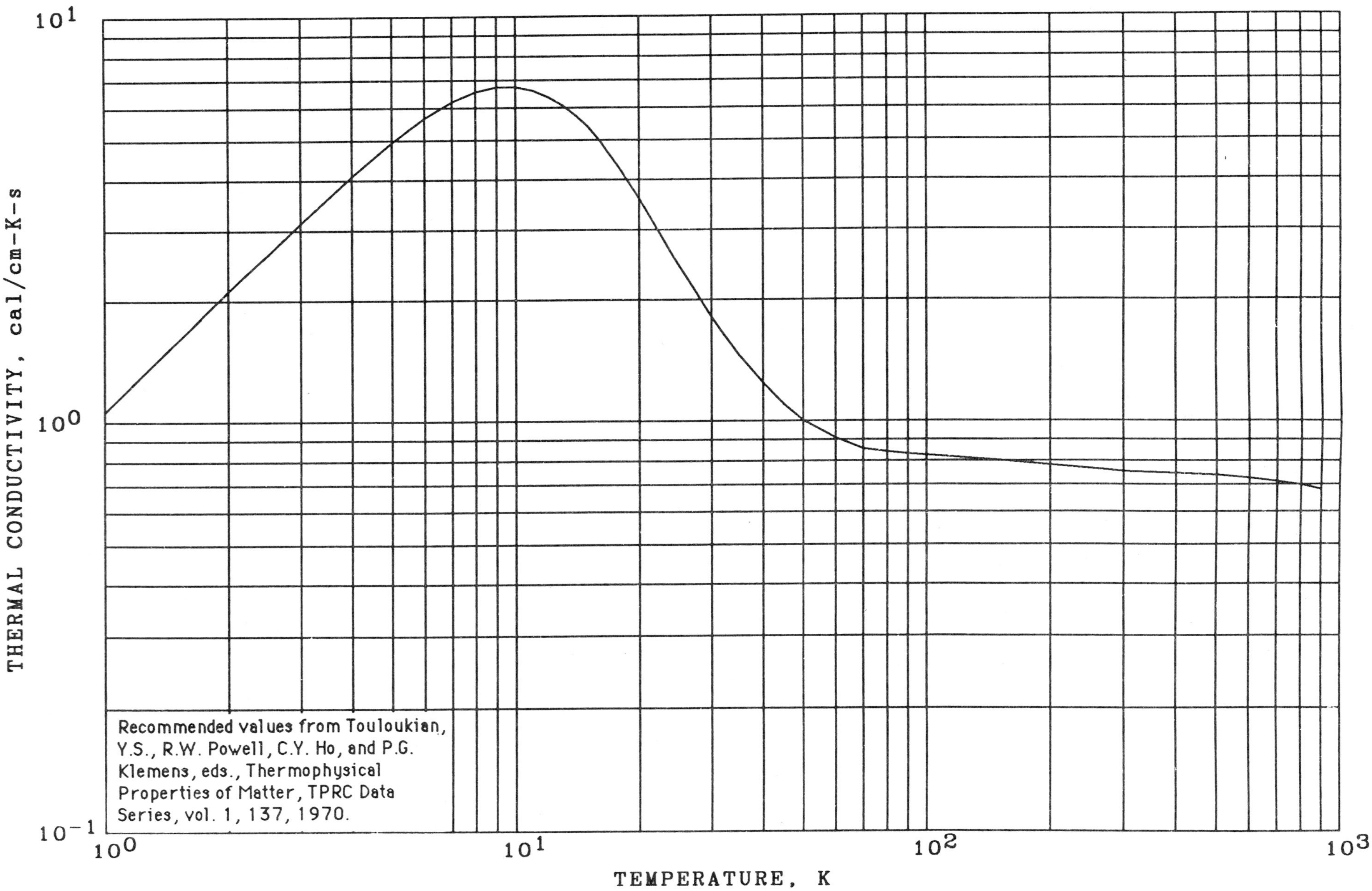

THERMAL CONDUCTIVITY, cal/cm-K-s
TEMPERATURE, K
10^1
10^0
10^-1
10^0
10^1
10^2
10^3
Recommended values from Touloukian, Y.S., R.W. Powell, C.Y. Ho, and P.G. Klemens, eds., Thermophysical Properties of Matter, TPRC Data Series, vol. 1, 137, 1970.

GOLD

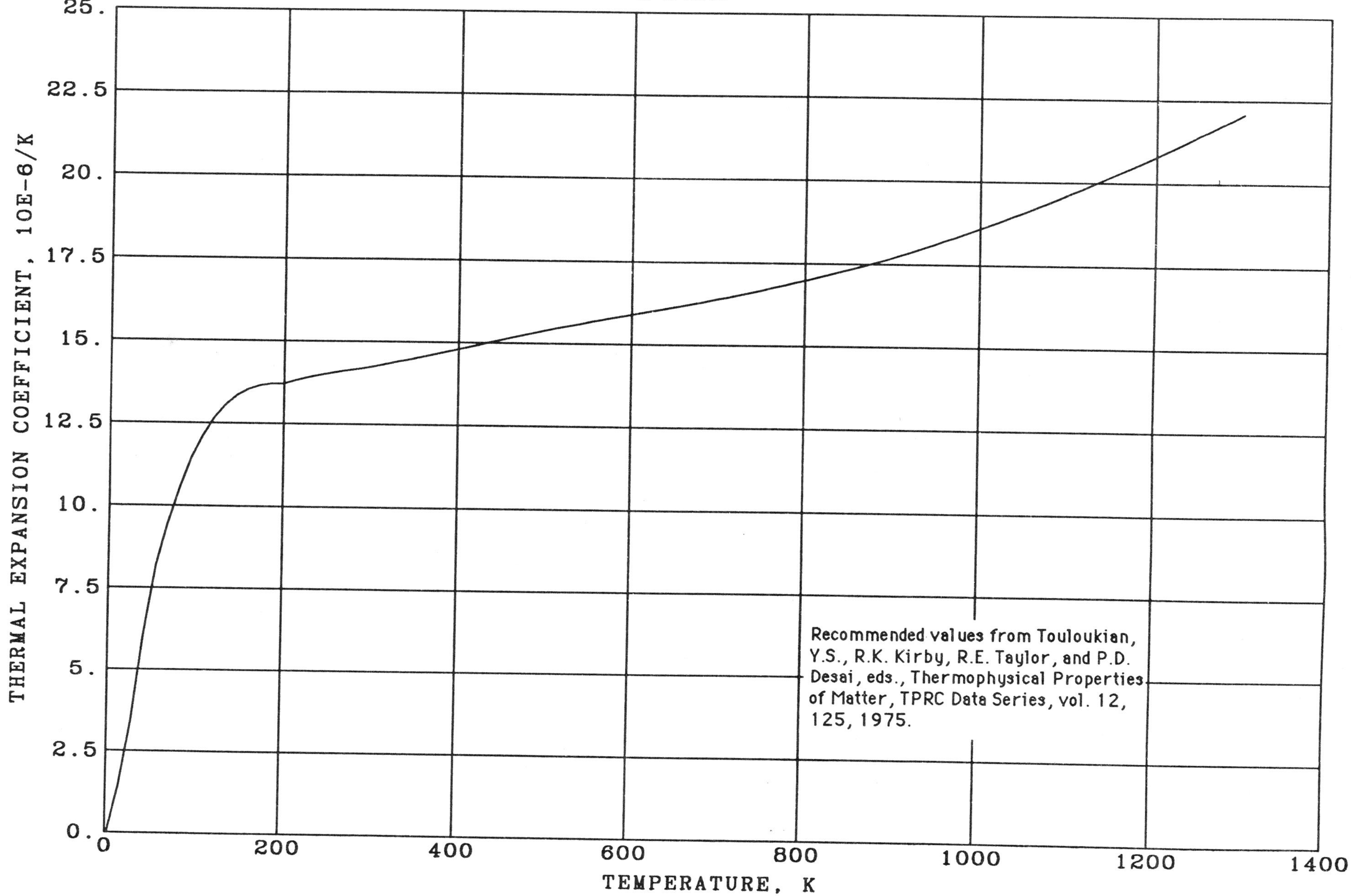

THERMAL EXPANSION COEFFICIENT, 10E-6/K
25.
22.5
20.
17.5
15.
12.5
10.
7.5
5.
2.5
0.
0
200
400
600
800
1000
1200
1400
TEMPERATURE, K
Recommended values from Touloukian, Y.S., R.K. Kirby, R.E. Taylor, and P.D. Desai, eds., Thermophysical Properties of Matter, TPRC Data Series, vol. 12, 125, 1975.

GOLD

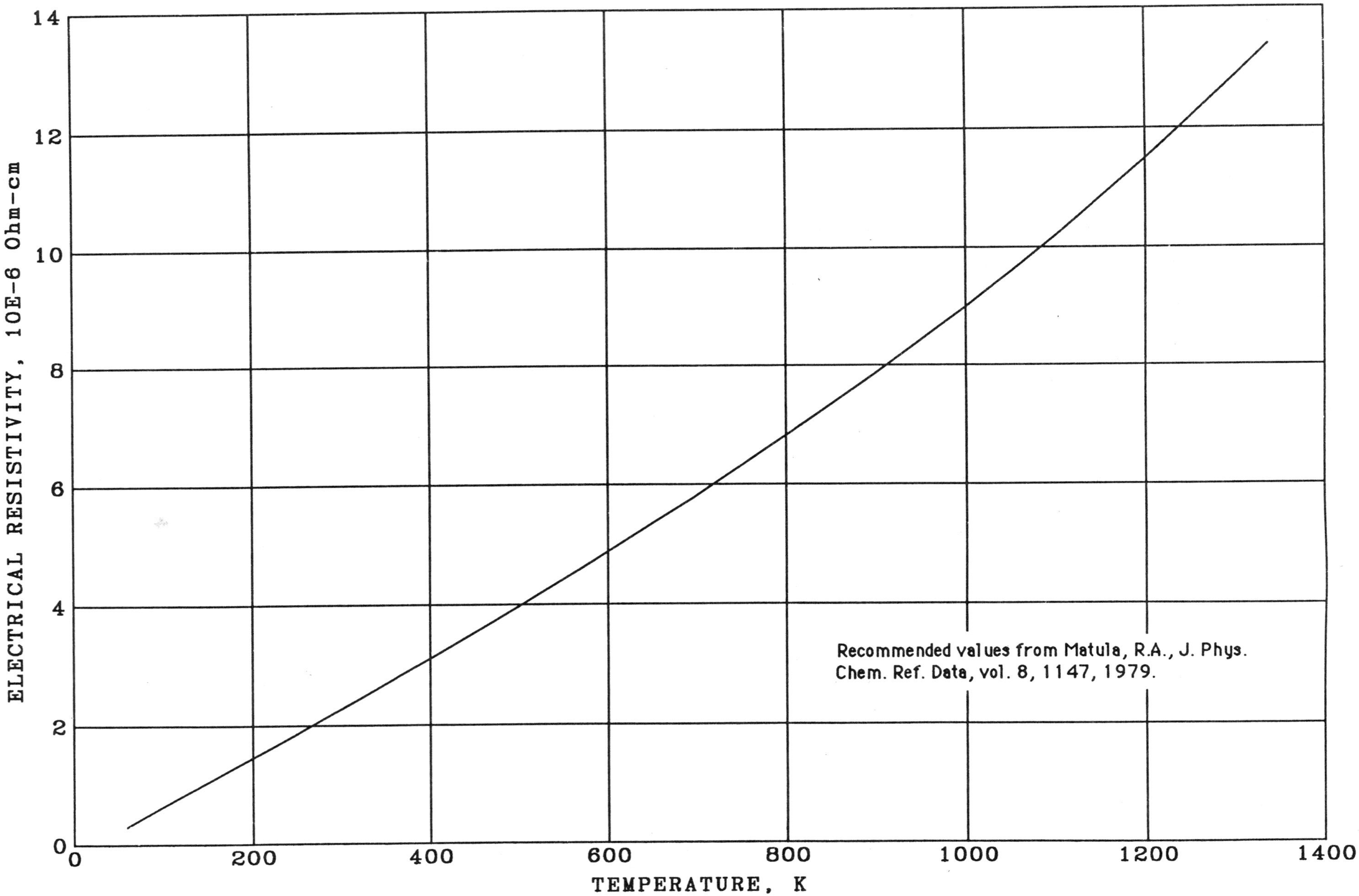

ELECTRICAL RESISTIVITY, 10E-6 Ohm-cm
0
2
4
6
8
10
12
14
TEMPERATURE, K
0
200
400
600
800
1000
1200
1400
Recommended values from Matula, R.A., J. Phys. Chem. Ref. Data, vol. 8, 1147, 1979.

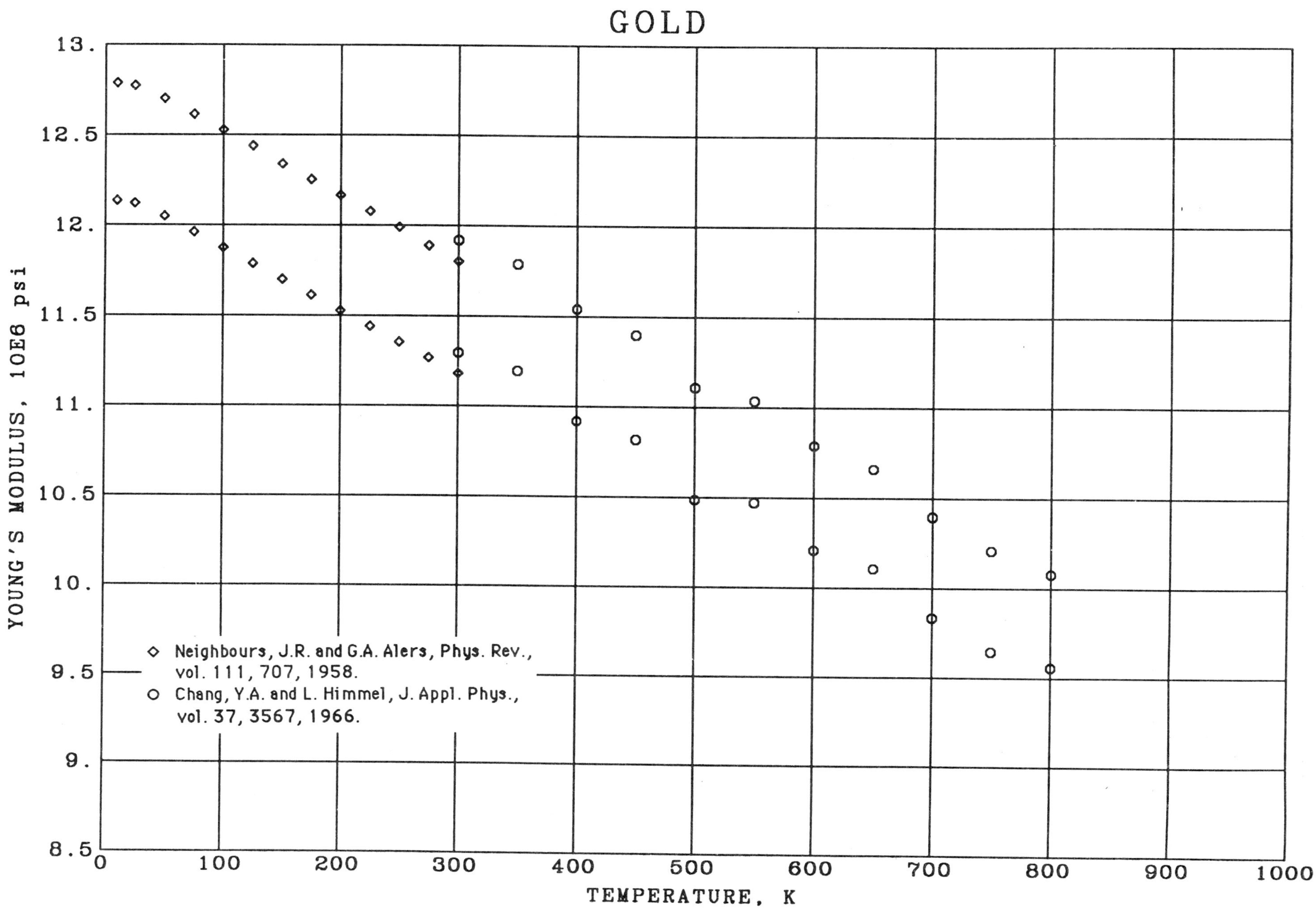
GOLD
13.
12.5
12.
11.5
11.
10.5
10.
9.5
9.
8.5
0
100
200
300
400
500
600
700
800
900
1000
YOUNG'S MODULUS, 10E6 psi
TEMPERATURE, K
◇ Neighbours, J.R. and G.A. Alers, Phys. Rev., vol. 111, 707, 1958.
○ Chang, Y.A. and L. Himmel, J. Appl. Phys., vol. 37, 3567, 1966.

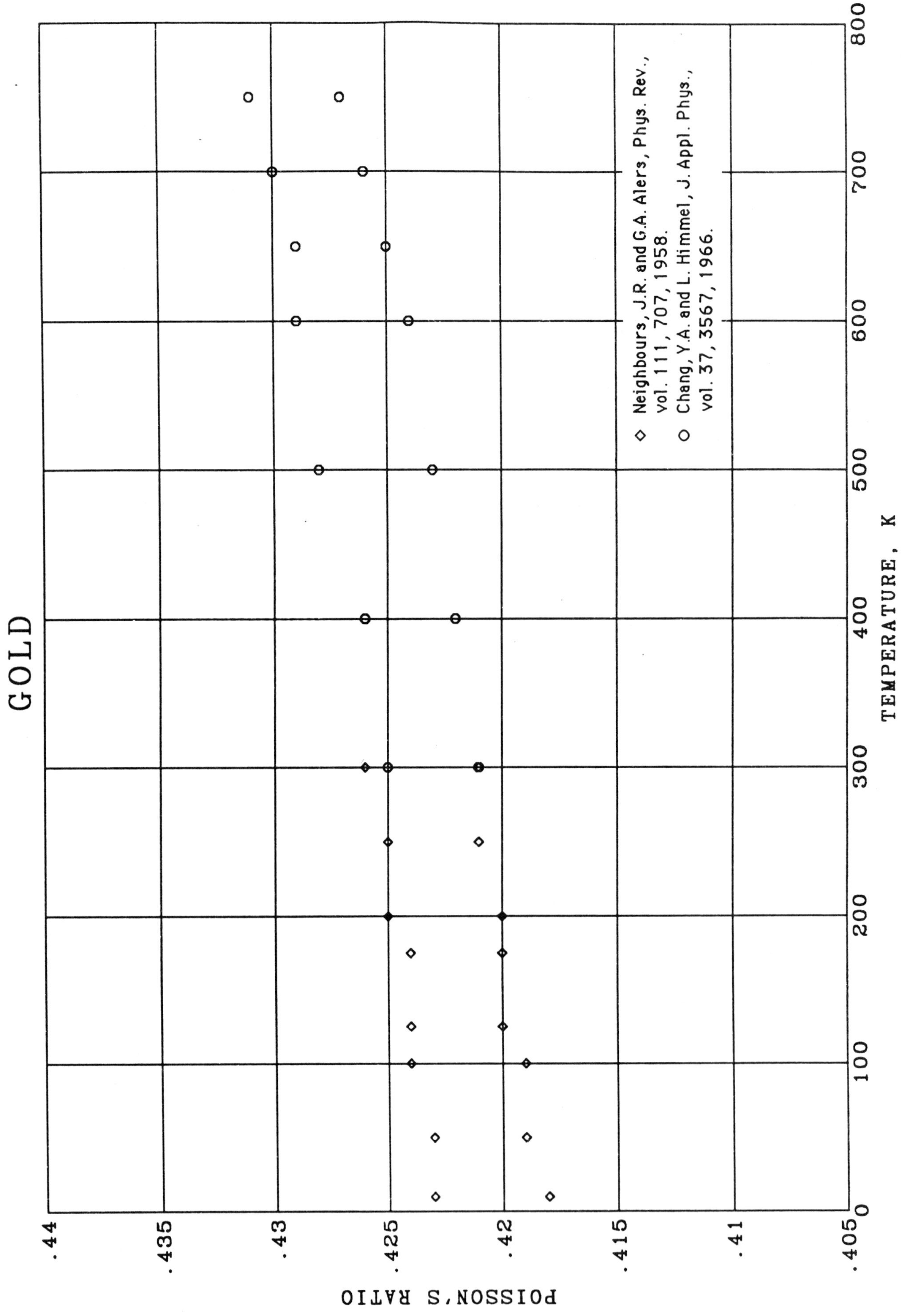
GOLD
POISSON'S RATIO
.405
.41
.415
.42
.425
.43
.435
.44
TEMPERATURE, K
0
100
200
300
400
500
600
700
800
◇ Neighbours, J.R. and G.A. Alers, Phys. Rev., vol. 111, 707, 1958.
○ Chang, Y.A. and L. Himmel, J. Appl. Phys., vol. 37, 3567, 1966.

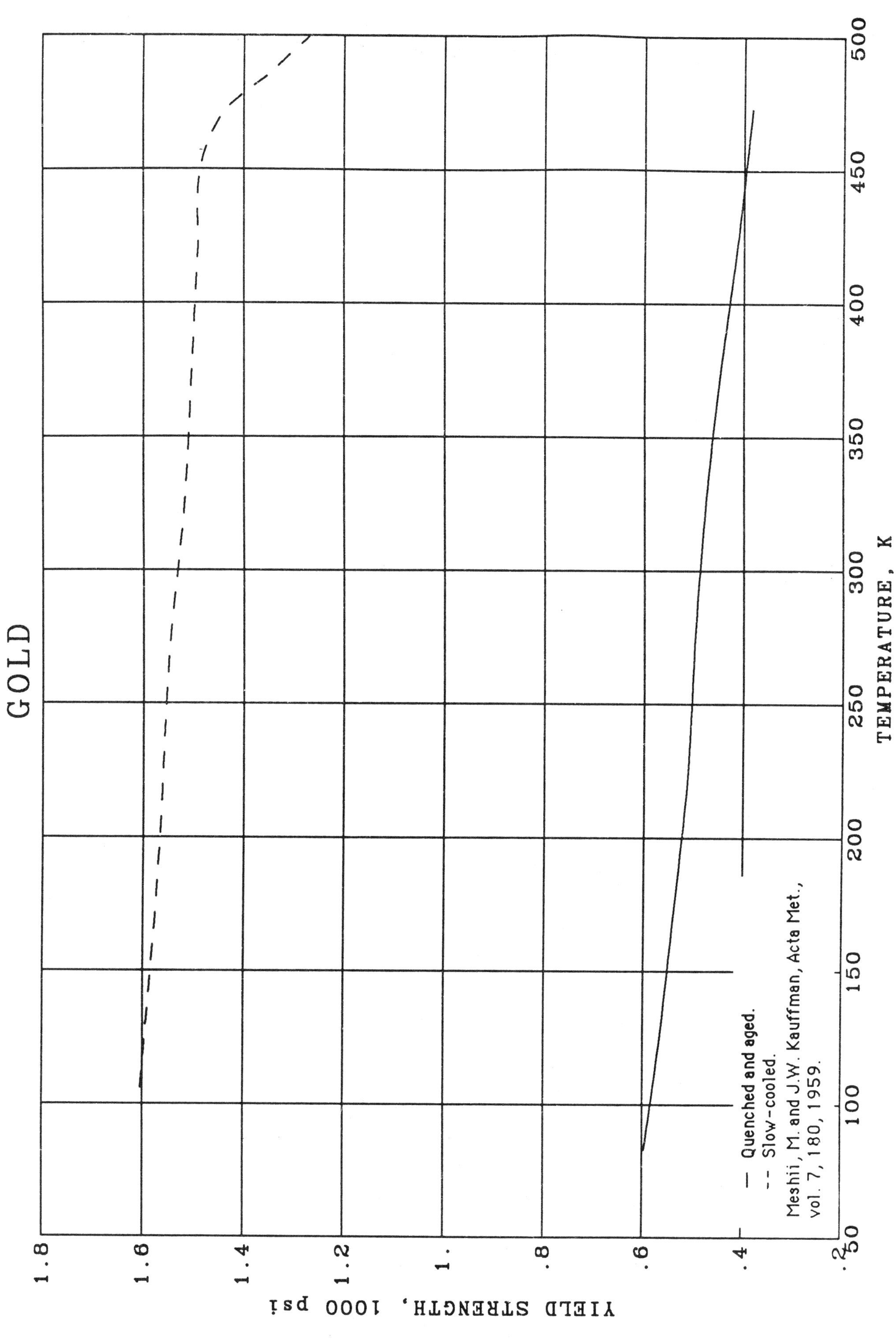
GOLD
YIELD STRENGTH, 1000 psi
1.8
1.6
1.4
1.2
1.
.8
.6
.4
.2
TEMPERATURE, K
50
100
150
200
250
300
350
400
450
500
— Quenched and aged.
-- Slow-cooled.
Meshii, M. and J.W. Kauffman, Acta Met.,
vol. 7, 180, 1959.

GOLD

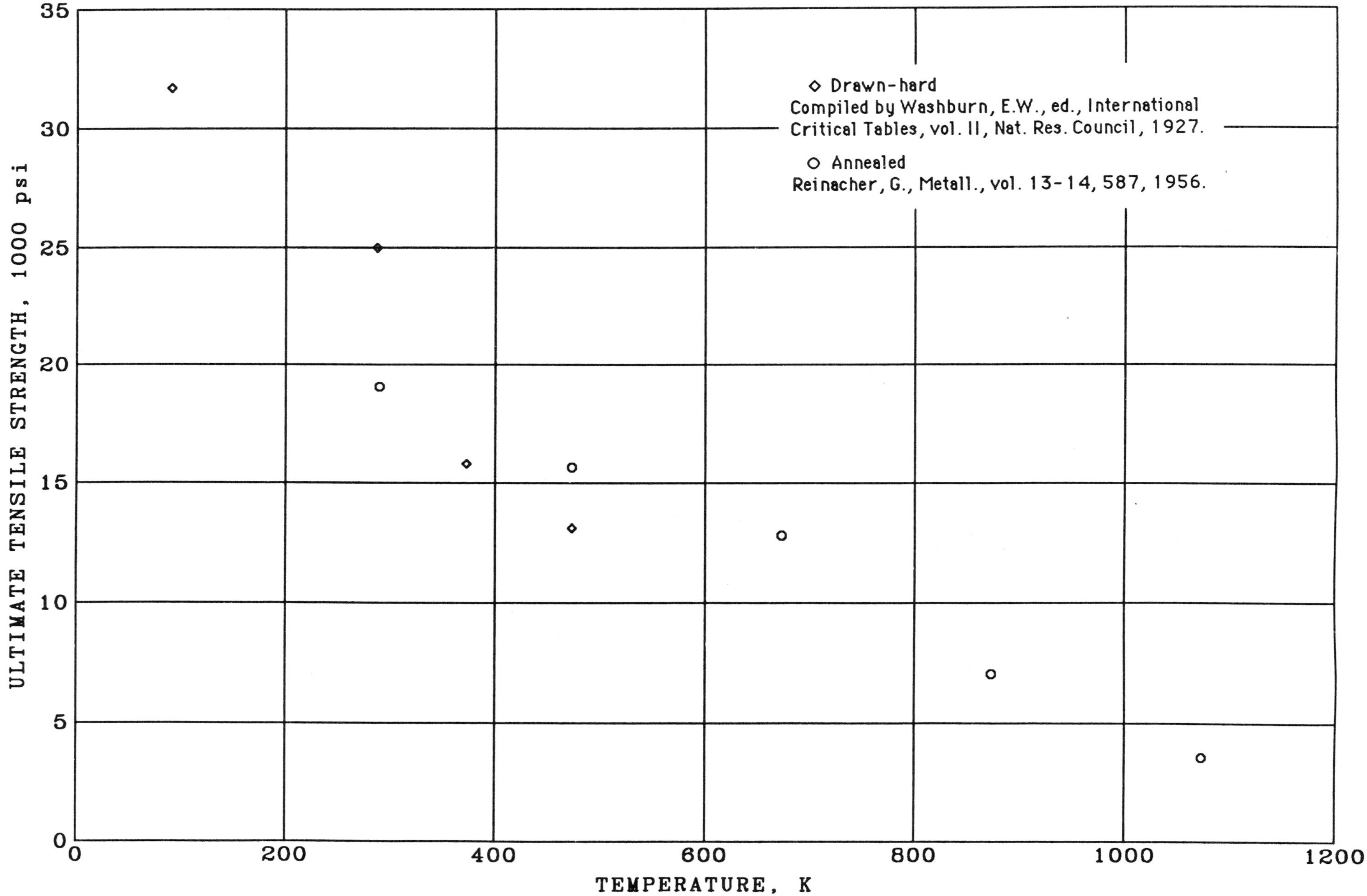

◇ Drawn-hard
Compiled by Washburn, E.W., ed., International Critical Tables, vol. II, Nat. Res. Council, 1927.
○ Annealed
Reinacher, G., Metall., vol. 13-14, 587, 1956.
ULTIMATE TENSILE STRENGTH, 1000 psi
0
5
10
15
20
25
30
35
TEMPERATURE, K
0
200
400
600
800
1000
1200

GOLD

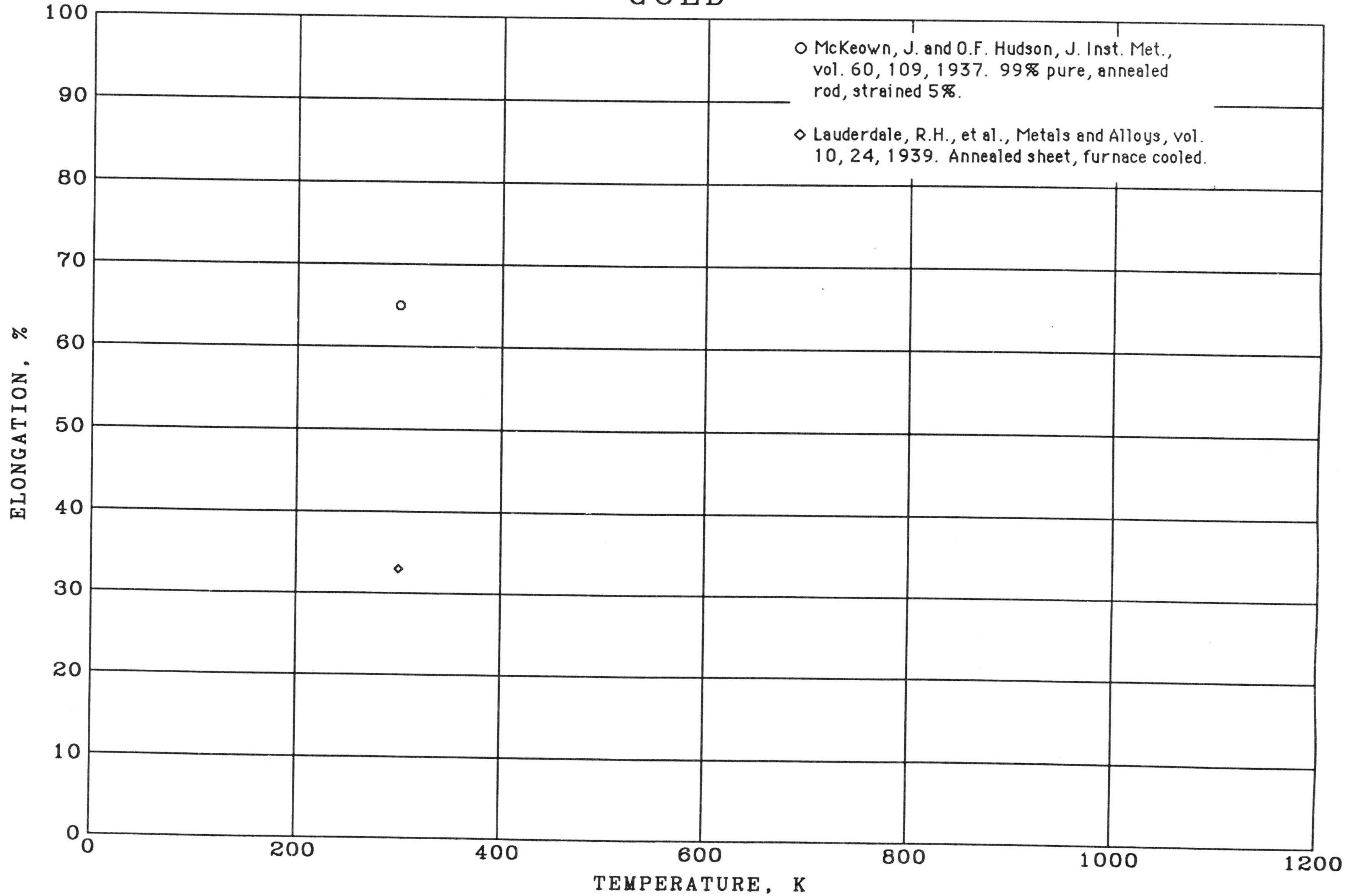

100
90
80
70
60
50
40
30
20
10
0
ELONGATION, %
0
200
400
600
800
1000
1200
TEMPERATURE, K
○ McKeown, J. and O.F. Hudson, J. Inst. Met., vol. 60, 109, 1937. 99% pure, annealed rod, strained 5%.
◇ Lauderdale, R.H., et al., Metals and Alloys, vol. 10, 24, 1939. Annealed sheet, furnace cooled.

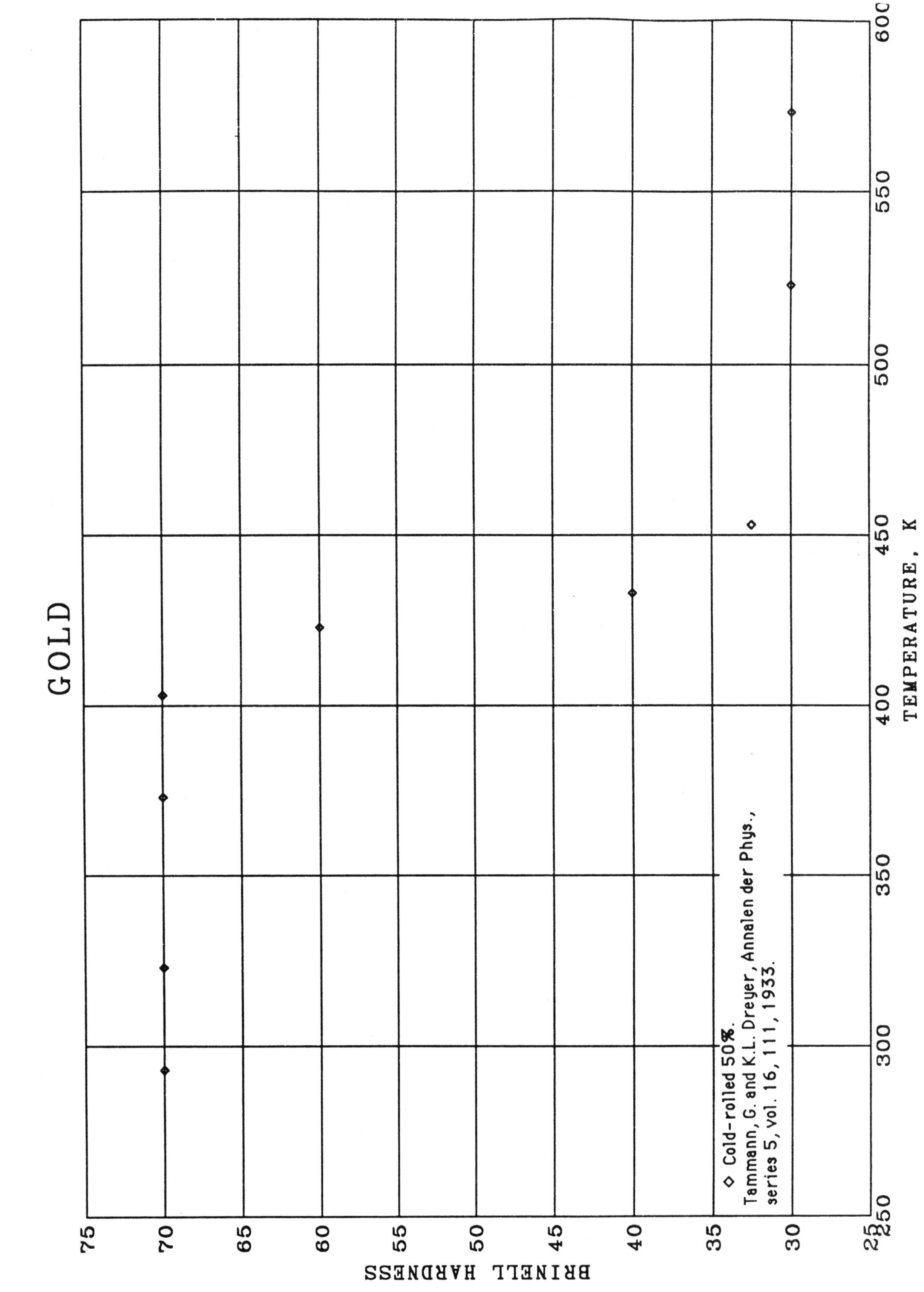
GOLD
BRINELL HARDNESS
75
70
65
60
55
50
45
40
35
30
25
TEMPERATURE, K
250
300
350
400
450
500
550
600
◇ Cold-rolled 50%.
Tammann, G. and K.L. Dreyer, Annalen der Phys., series 5, vol. 16, 111, 1933.

IRON

PROPERTY	@ 300K	UNITS	REFERENCES
Density	7.874	g/cm^3	Fe1 Fe2 Fe3
Specific Heat	0.1050	cal/g-K	Fe4 Fe5 Fe6
Thermal Conductivity	0.192	cal/cm-s-K	Fe7 Fe8 Fe9
Thermal Exp Coef	11.9×10^{-6}	1/K	Fe10 Fe11 Fe12
Melting Point	1808	K	Fe13
Resistivity	10.0×10^{-6}	Ohm-cm	Fe14 Fe15 Fe16
TCR (218K to 398K)	5670	ppm/K	Fe14 Fe15 Fe16
Young's Modulus	3.056×10^{7}	psi	Fe17 Fe18 Fe19
Poisson's Ratio	0.291	-----	Fe17 Fe18 Fe19
Yield Strength	6000	psi	Fe20 Fe22 Fe24
Ult Tensile Strength	20,000	psi	Fe20 Fe21 Fe22
Elongation at Break	20-50	%	Fe20 Fe21 Fe22
Hardness	49	Brinell	Fe20 Fe23 Fe25

Mechanical properties listed are for fully annealed, bulk specimens. The grain size and strain rate have a very strong influence on the mechanical properties. 80% cold work will increase the tensile strength of ingot iron (0.08% C, 0.024% P, 0.019% S, 0.007% Si) to 95 ksi, increase hardness to 210 and decrease elongation to 7% [Fe26]. Heating cold worked material to temperatures in the range of 300 to 500C will cause recovery and recrystallization erasing the effects of cold work.

IRON

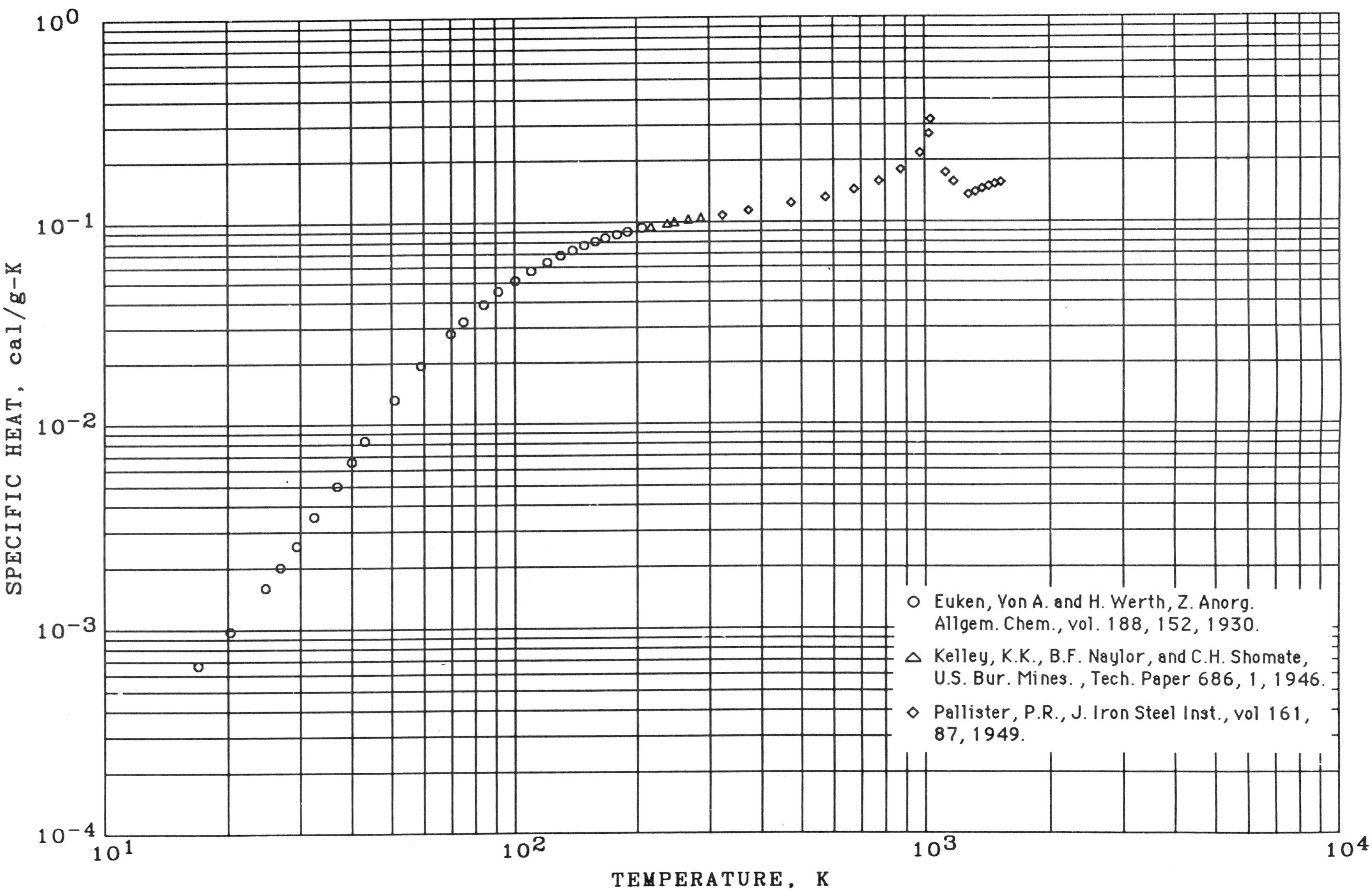

SPECIFIC HEAT, cal/g-K
TEMPERATURE, K
10^0
10^{-1}
10^{-2}
10^{-3}
10^{-4}
10^1
10^2
10^3
10^4
○ Euken, Von A. and H. Werth, Z. Anorg. Allgem. Chem., vol. 188, 152, 1930.
△ Kelley, K.K., B.F. Naylor, and C.H. Shomate, U.S. Bur. Mines., Tech. Paper 686, 1, 1946.
◇ Pallister, P.R., J. Iron Steel Inst., vol 161, 87, 1949.

IRON

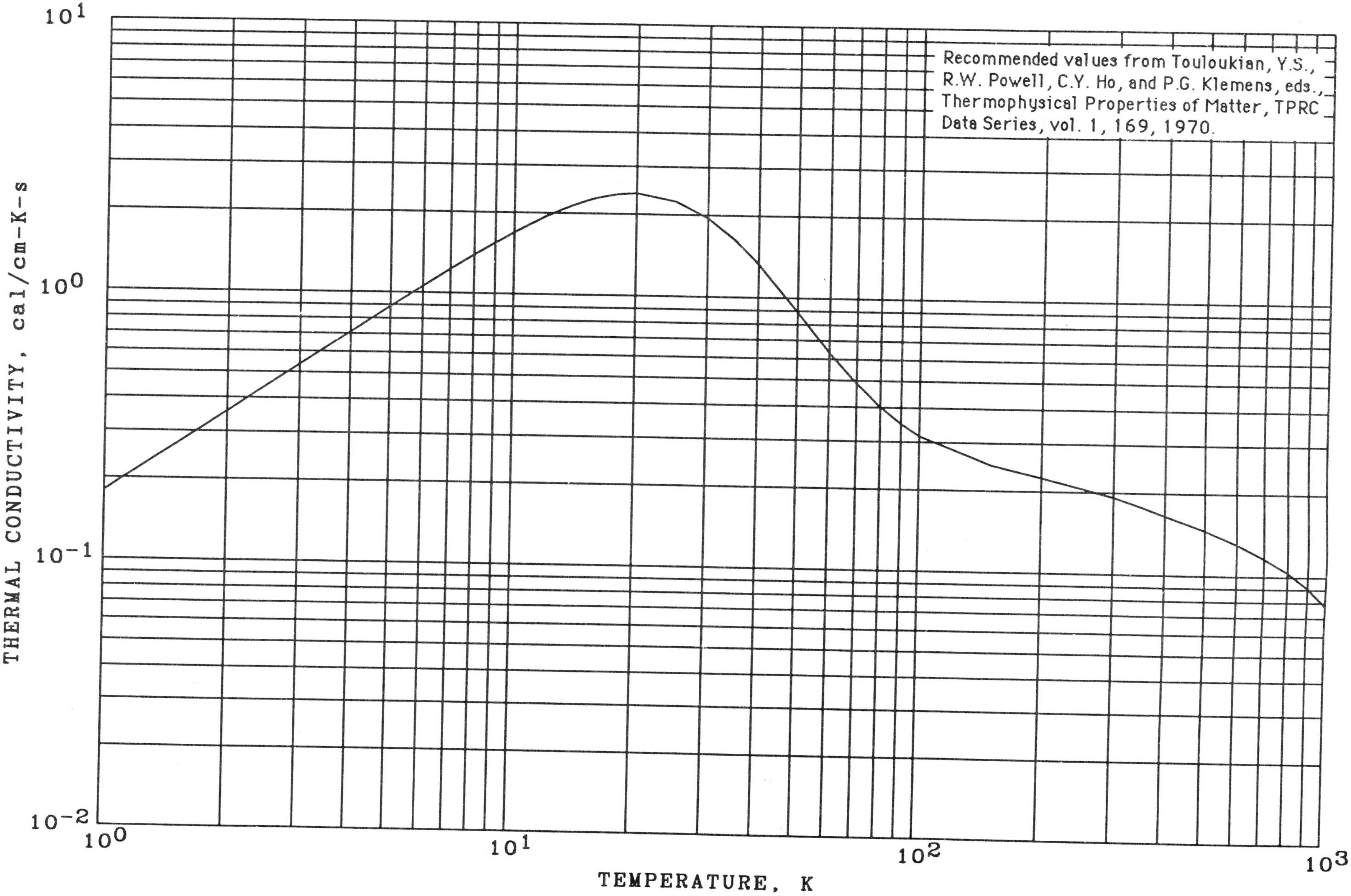

Recommended values from Touloukian, Y.S., R.W. Powell, C.Y. Ho, and P.G. Klemens, eds., Thermophysical Properties of Matter, TPRC Data Series, vol. 1, 169, 1970.
THERMAL CONDUCTIVITY, cal/cm-K-s
TEMPERATURE, K
10^1
10^0
10^-1
10^-2
10^0
10^1
10^2
10^3

IRON

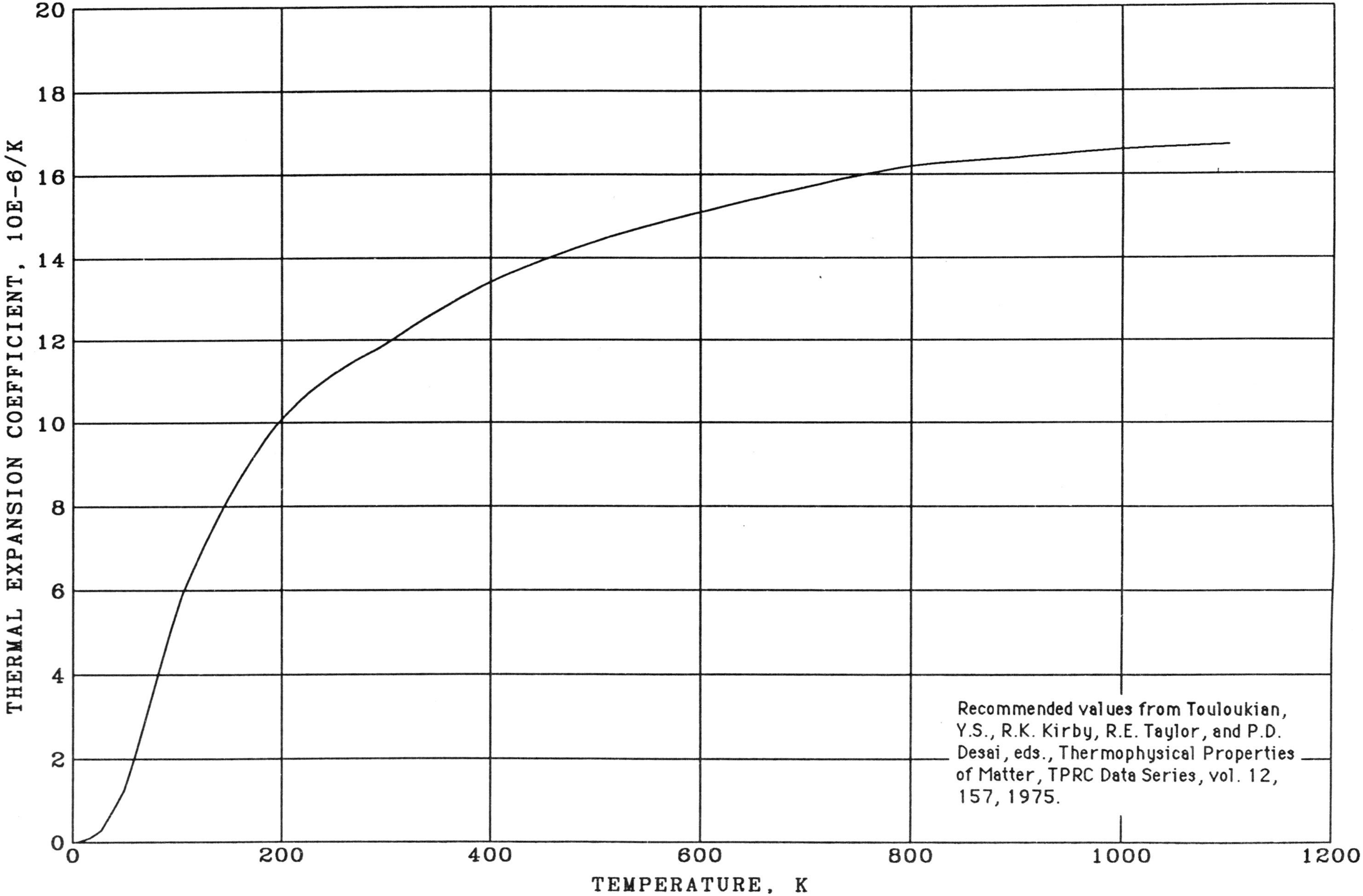
THERMAL EXPANSION COEFFICIENT, 10E-6/K
0
2
4
6
8
10
12
14
16
18
20
0
200
400
600
800
1000
1200
TEMPERATURE, K
Recommended values from Touloukian, Y.S., R.K. Kirby, R.E. Taylor, and P.D. Desai, eds., Thermophysical Properties of Matter, TPRC Data Series, vol. 12, 157, 1975.

IRON

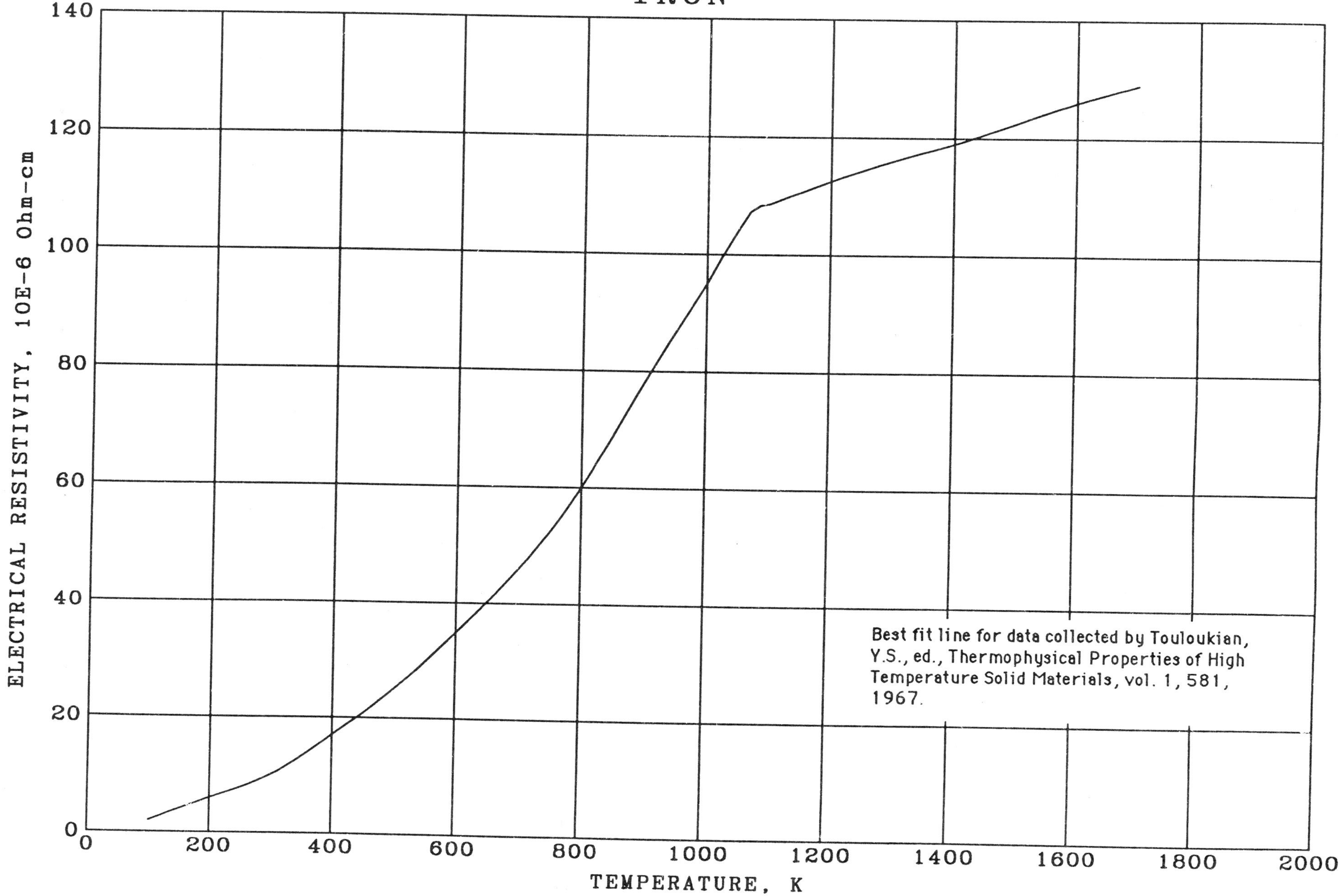
ELECTRICAL RESISTIVITY, 10E-6 Ohm-cm
0
20
40
60
80
100
120
140
TEMPERATURE, K
0
200
400
600
800
1000
1200
1400
1600
1800
2000
Best fit line for data collected by Touloukian, Y.S., ed., Thermophysical Properties of High Temperature Solid Materials, vol. 1, 581, 1967.

IRON

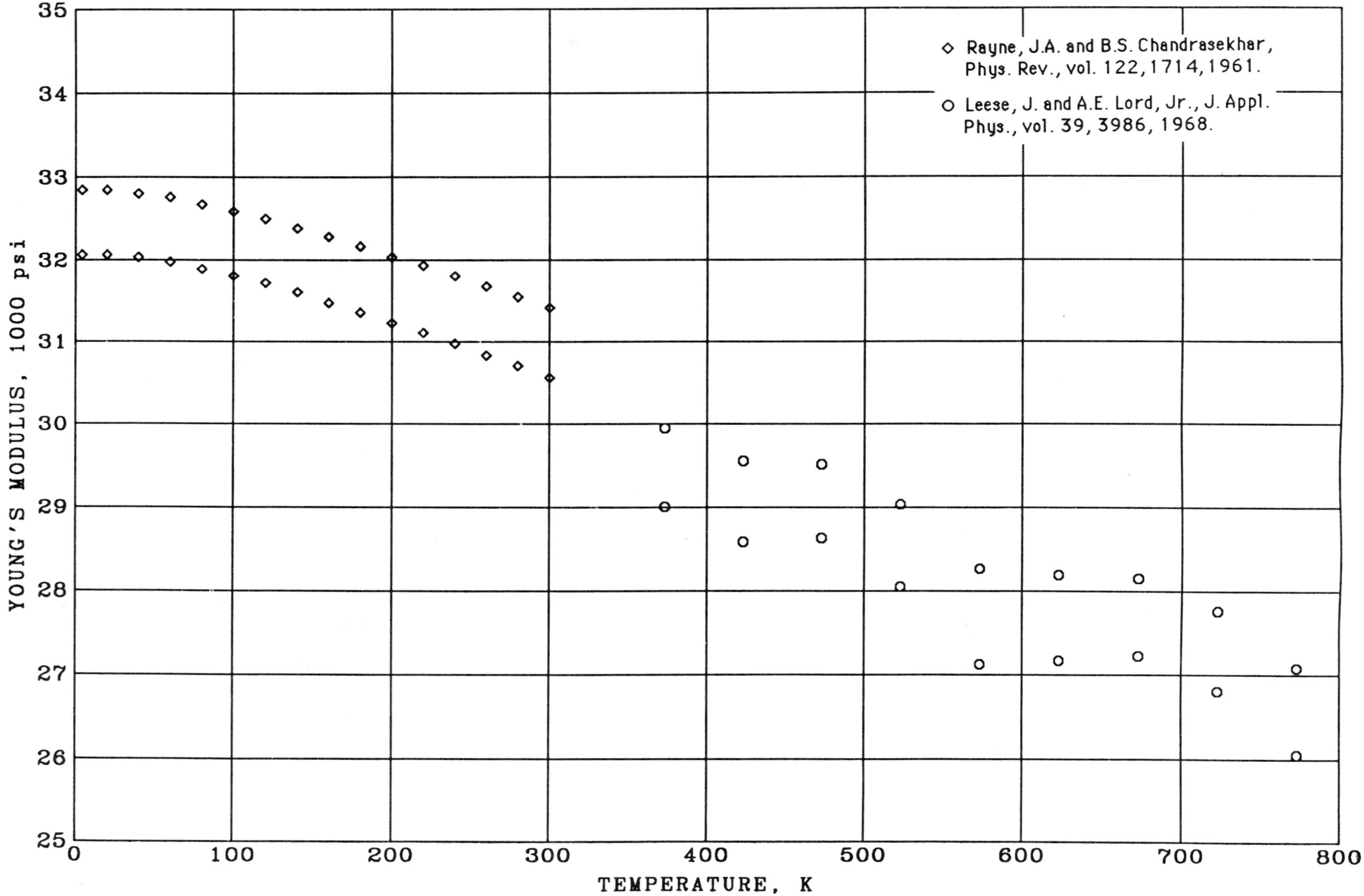

35
34
33
32
31
30
29
28
27
26
25
0
100
200
300
400
500
600
700
800
YOUNG'S MODULUS, 1000 psi
TEMPERATURE, K
◇ Rayne, J.A. and B.S. Chandrasekhar, Phys. Rev., vol. 122, 1714, 1961.
○ Leese, J. and A.E. Lord, Jr., J. Appl. Phys., vol. 39, 3986, 1968.

Poisson's Ratio for iron is virtually constant from -50C to above 600C.

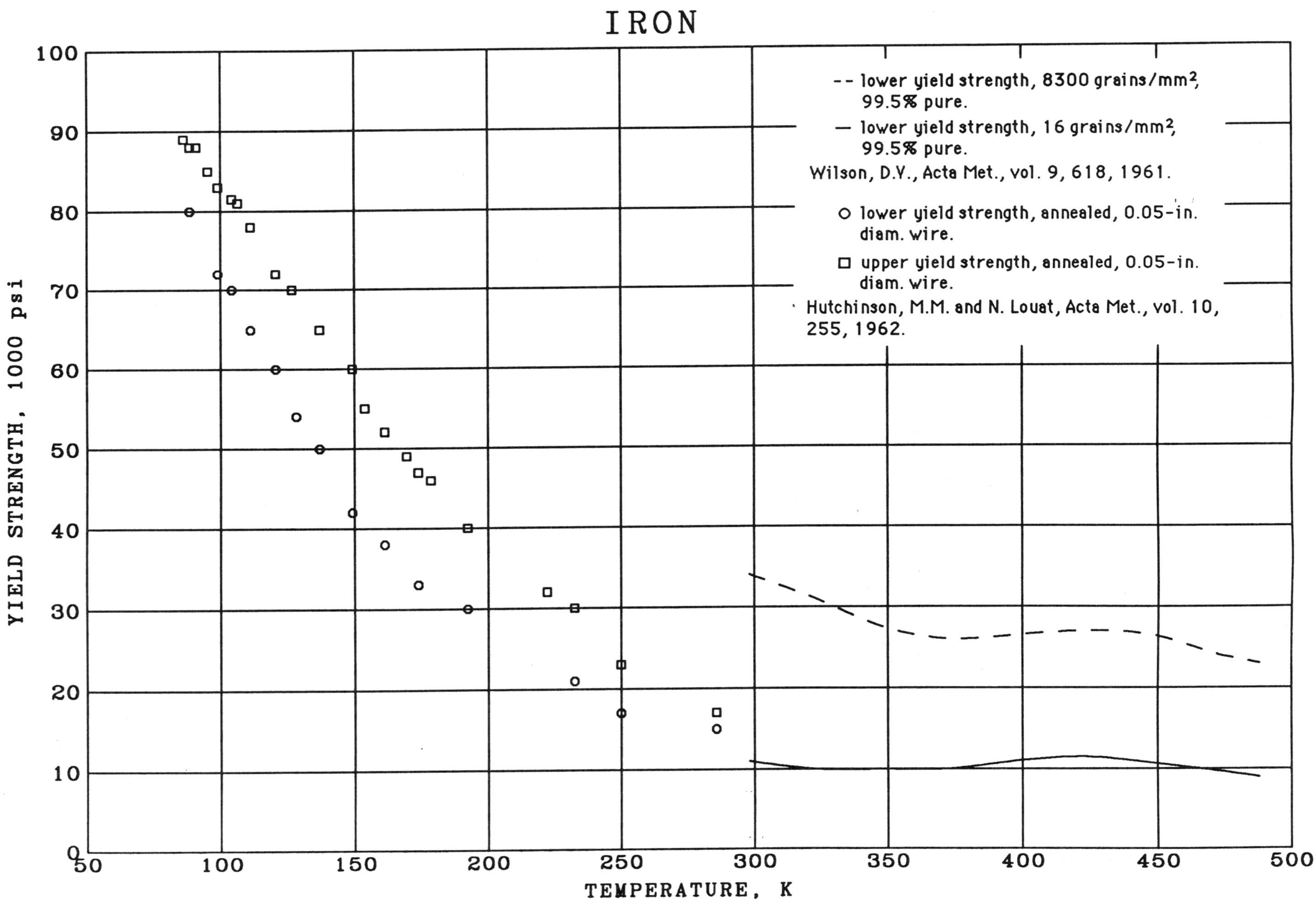
IRON
-- lower yield strength, 8300 grains/mm², 99.5% pure.
— lower yield strength, 16 grains/mm², 99.5% pure.
Wilson, D.V., Acta Met., vol. 9, 618, 1961.
○ lower yield strength, annealed, 0.05-in. diam. wire.
□ upper yield strength, annealed, 0.05-in. diam. wire.
Hutchinson, M.M. and N. Louat, Acta Met., vol. 10, 255, 1962.
YIELD STRENGTH, 1000 psi
TEMPERATURE, K
0
10
20
30
40
50
60
70
80
90
100
50
100
150
200
250
300
350
400
450
500

IRON

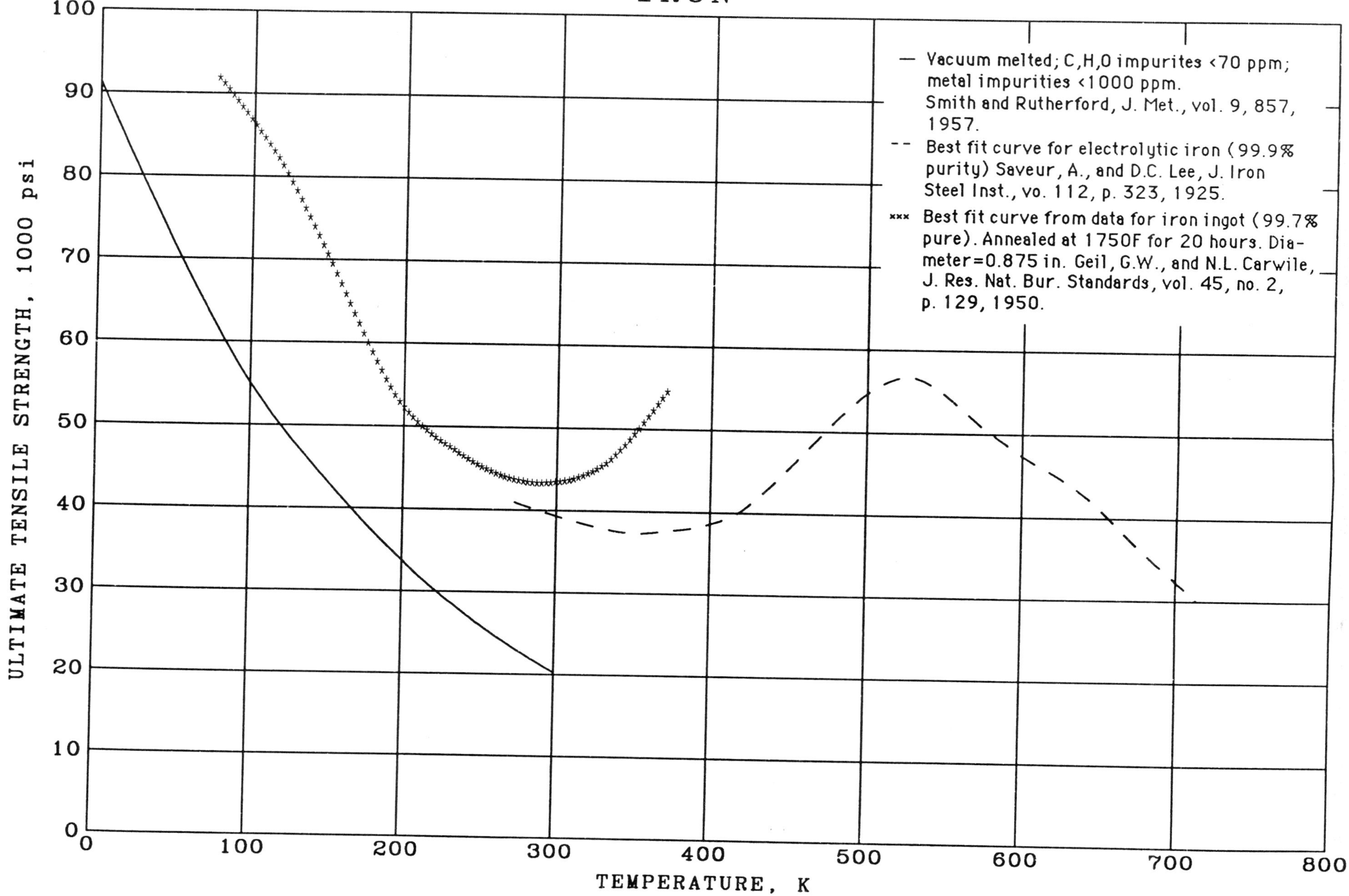

100
90
80
70
60
50
40
30
20
10
0
ULTIMATE TENSILE STRENGTH, 1000 psi
0
100
200
300
400
500
600
700
800
TEMPERATURE, K
— Vacuum melted; C,H,O impurites <70 ppm; metal impurities <1000 ppm. Smith and Rutherford, J. Met., vol. 9, 857, 1957.
-- Best fit curve for electrolytic iron (99.9% purity) Saveur, A., and D.C. Lee, J. Iron Steel Inst., vo. 112, p. 323, 1925.
xxx Best fit curve from data for iron ingot (99.7% pure). Annealed at 1750F for 20 hours. Diameter=0.875 in. Geil, G.W., and N.L. Carwile, J. Res. Nat. Bur. Standards, vol. 45, no. 2, p. 129, 1950.

IRON

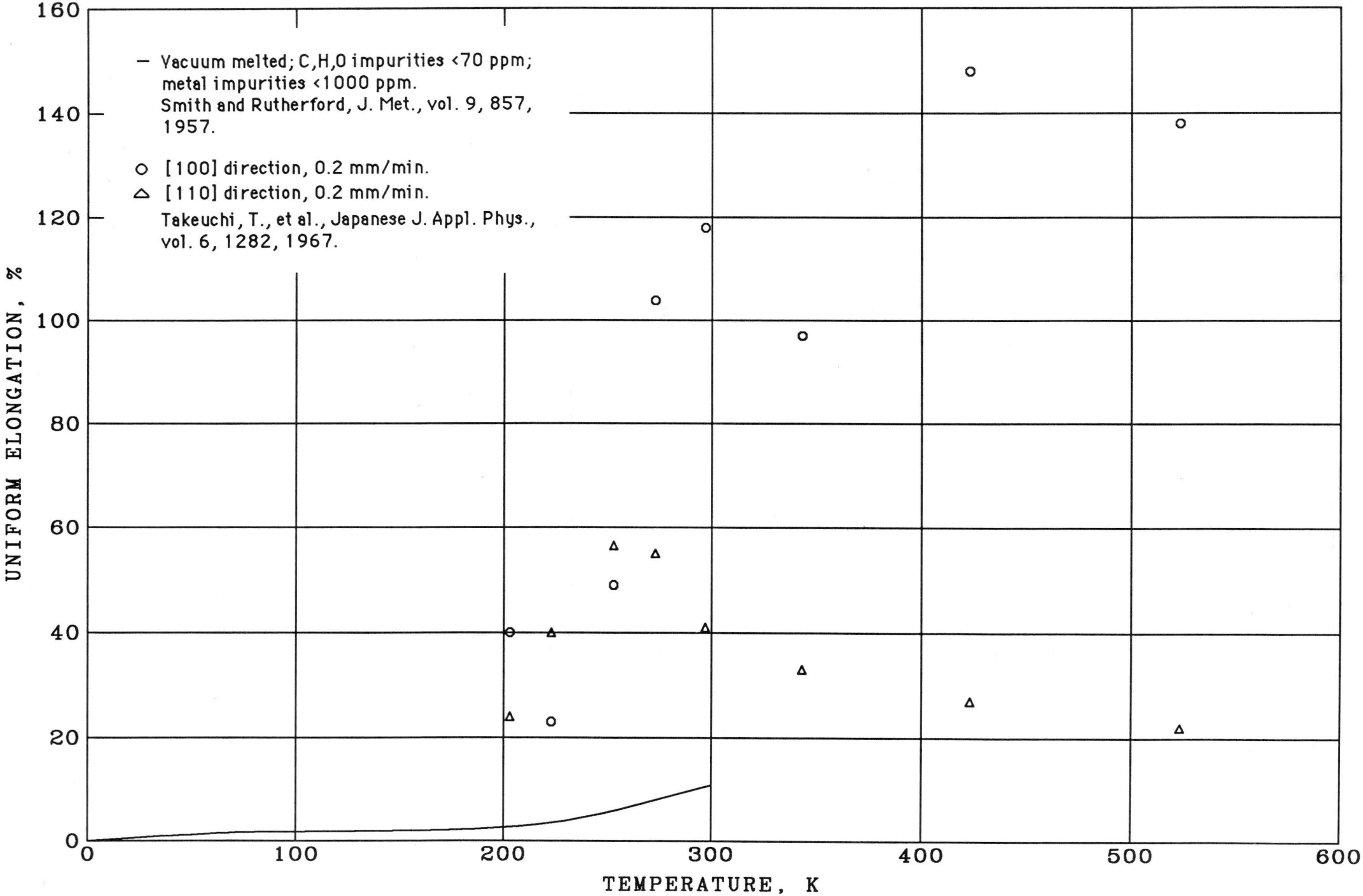

160
140
120
100
80
60
40
20
0
UNIFORM ELONGATION, %
0
100
200
300
400
500
600
TEMPERATURE, K
— Vacuum melted; C,H,O impurities <70 ppm; metal impurities <1000 ppm. Smith and Rutherford, J. Met., vol. 9, 857, 1957.
○ [100] direction, 0.2 mm/min.
△ [110] direction, 0.2 mm/min.
Takeuchi, T., et al., Japanese J. Appl. Phys., vol. 6, 1282, 1967.

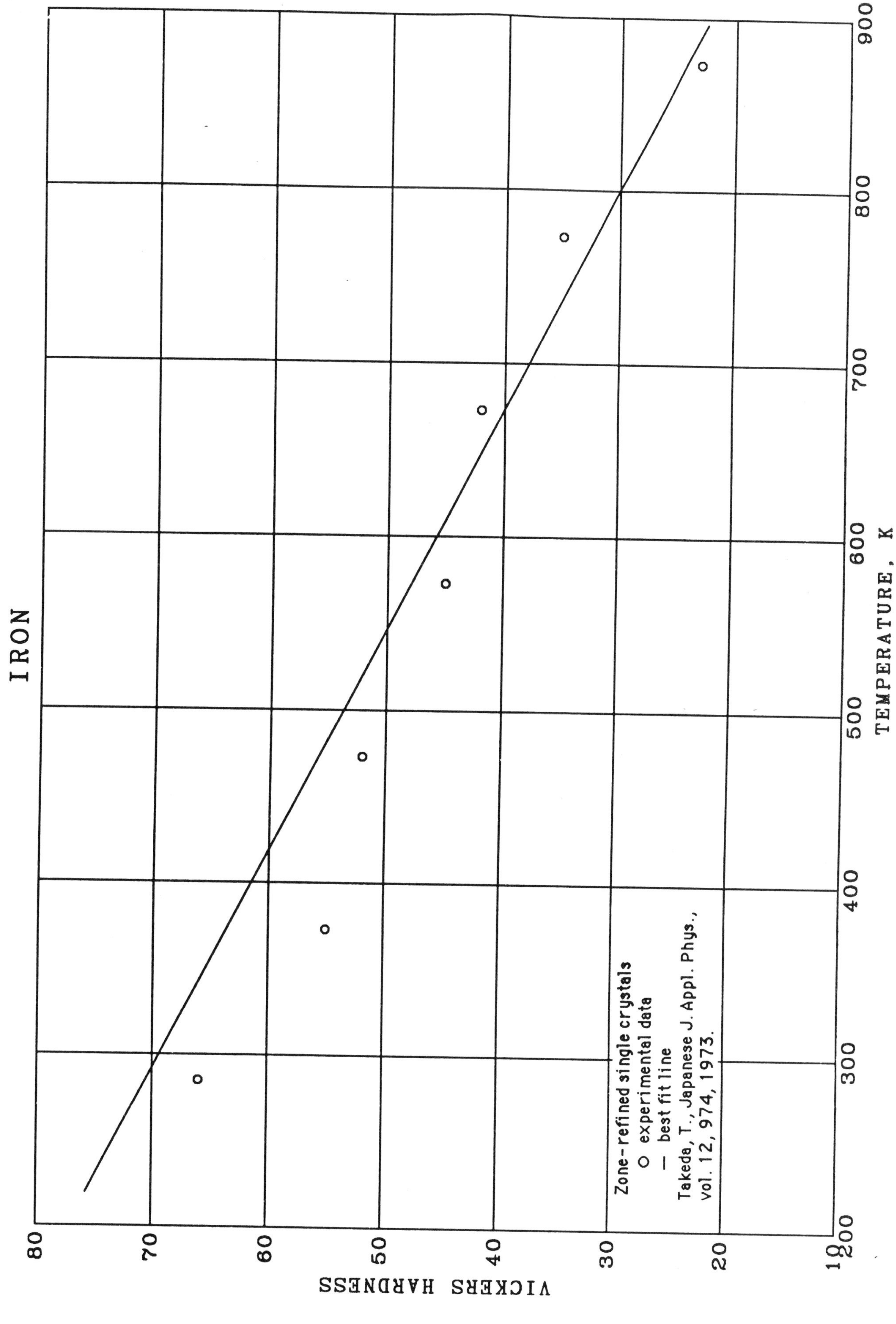
IRON
VICKERS HARDNESS
TEMPERATURE, K
10
20
30
40
50
60
70
80
200
300
400
500
600
700
800
900
Zone-refined single crystals
o experimental data
— best fit line
Takeda, T., Japanese J. Appl. Phys.,
vol. 12, 974, 1973.

MOLYBDENUM

PROPERTY	@ 300K	UNITS	REFERENCES
Density	10.24	g/cm^3	Mo1 Mo2 Mo3
Specific Heat	0.0602	cal/g-K	Mo4 Mo5 Mo6
Thermal Conductivity	0.334	cal/cm-s-K	Mo7 Mo8 Mo9
Thermal Exp Coef	4.8	1/K	Mo10 Mo11 Mo12
Melting Point	2890	K	Mo13
Resistivity 1	5.52x10^{-6}	Ohm-cm	Mo14 Mo15 Mo16
TCR (218K to 398K)	4450	ppm/K	Mo14 Mo15 Mo16
Young's Modulus	4.63x10^7	psi	Mo17 Mo18 Mo19
Poisson's Ratio	0.301	-----	Mo17 Mo18 Mo19
Yield Strength 2	80 to 95	ksi	Mo20 Mo21 Mo22
Ult Tensile Strength2	95 to 115	ksi	Mo20 Mo21 Mo22
Elongation at Break 2	1 to 18	%	Mo20 Mo21 Mo22
Hardness	160	Vickers	Mo23 Mo24 Mo25

1 Molybdenum evaporated onto substrates held at temperatures between 450 and 675C in vacuums <2x10^{-7} torr produces films with resistivities in the range of 6-11 micro-ohm centimeters. Films with thicknesses in the range of 6000 to 11000 angstroms have resistivities on the order of 0.1 ohms/square.[Mo26]

2 ASTM Std. B 386-82. Properties are provided for matl 365(99.97% pure) which has been stress relieved. Actual values depend on source, heat treatment and thickness. Cold work can increase strength to values greater than 200,000 psi. Cold work can also improve ductility. Recrystallization ocurrs between 900-1300C depending on the prior deformation. For a thorough review of the mechanical properties see "The Engineering Properties of Molybdenum and Molybdenum Alloys", September 20, 1963, Battelle Metals Information Center, Columbus Ohio.

MOLYBDENUM

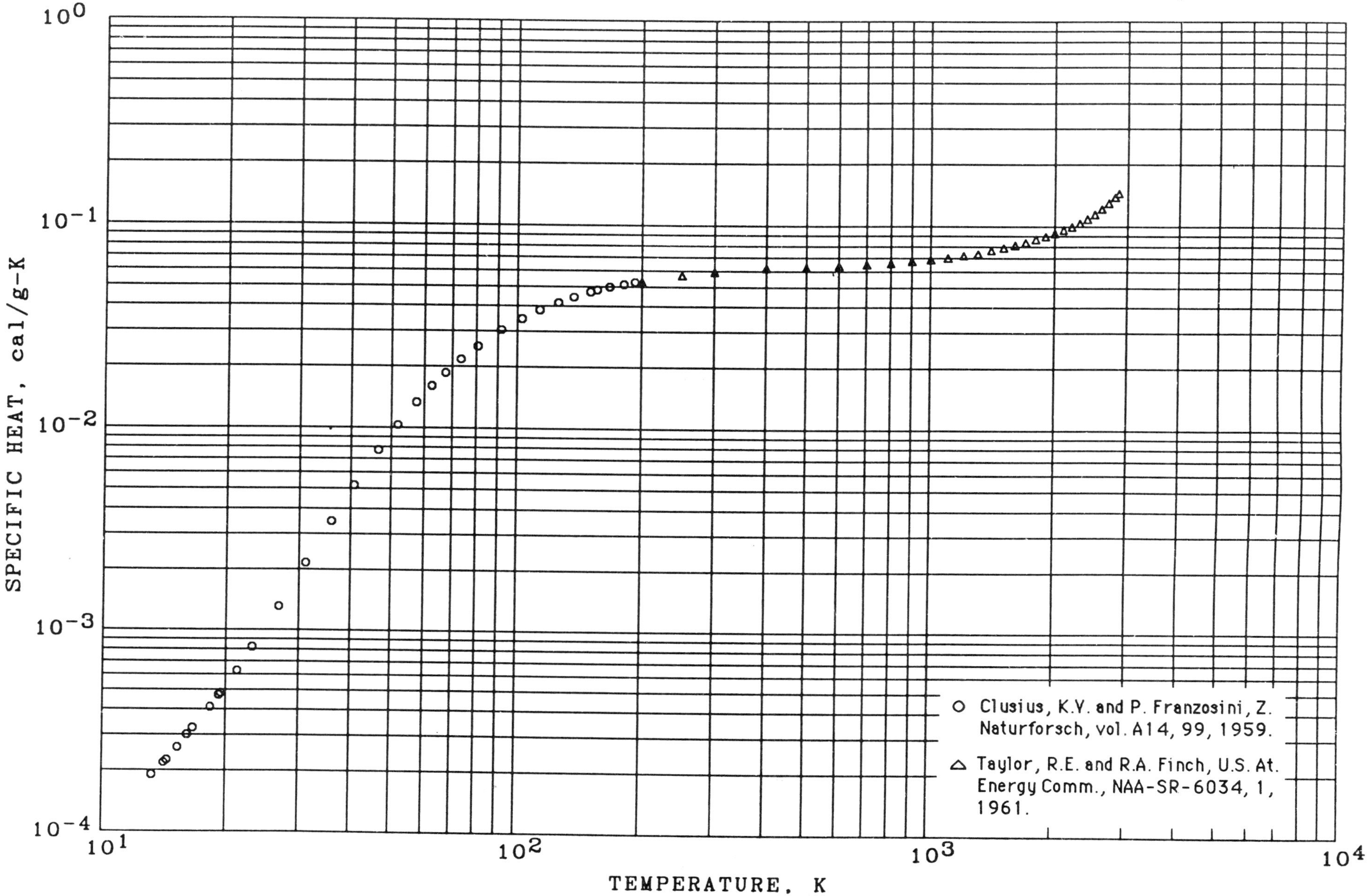

SPECIFIC HEAT, cal/g-K
TEMPERATURE, K
10^0
10^{-1}
10^{-2}
10^{-3}
10^{-4}
10^1
10^2
10^3
10^4
○ Clusius, K.Y. and P. Franzosini, Z. Naturforsch, vol. A14, 99, 1959.
△ Taylor, R.E. and R.A. Finch, U.S. At. Energy Comm., NAA-SR-6034, 1, 1961.

MOLYBDENUM

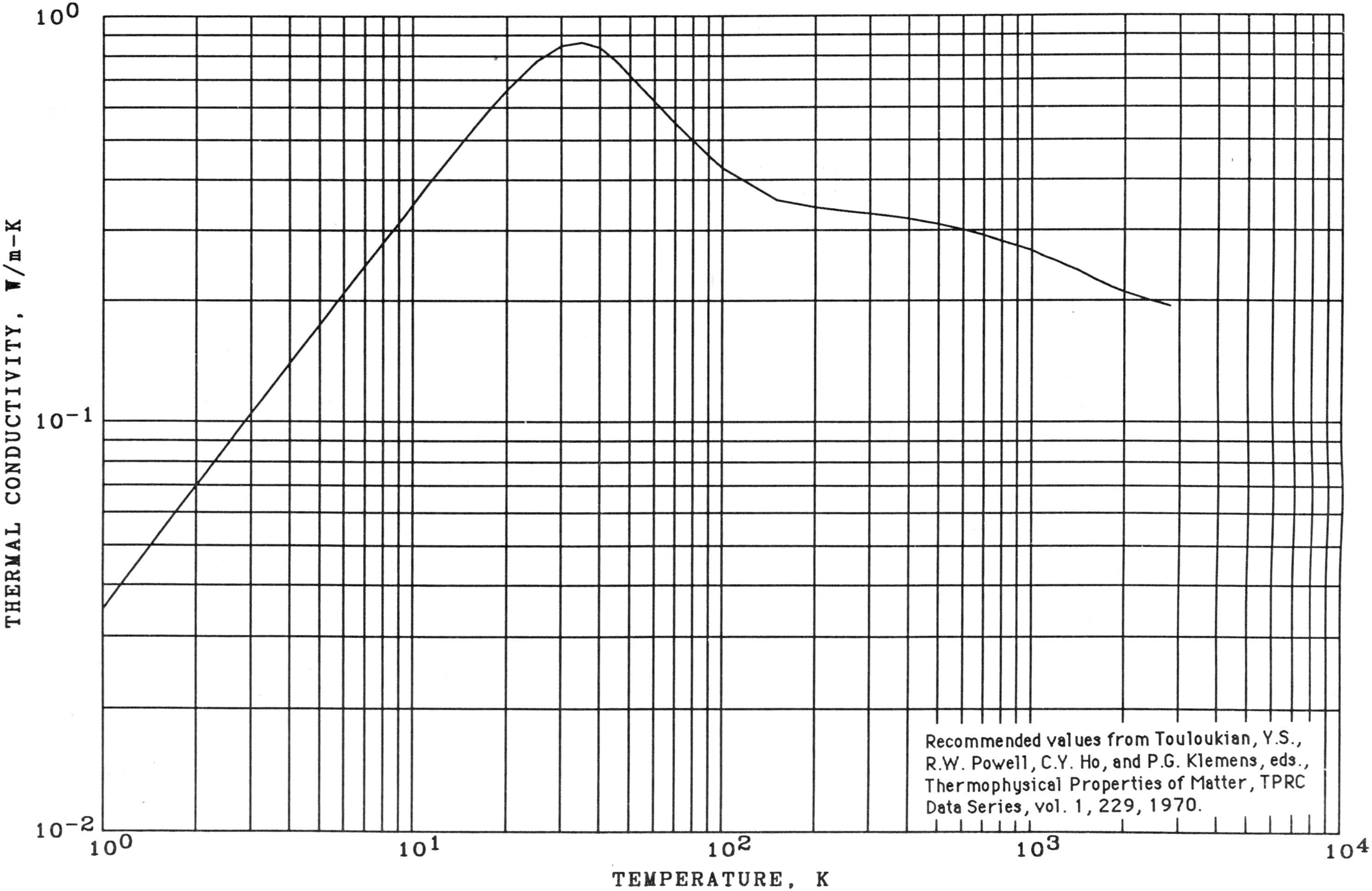

10^0
10^{-1}
10^{-2}
THERMAL CONDUCTIVITY, W/m-K
10^0
10^1
10^2
10^3
10^4
TEMPERATURE, K
Recommended values from Touloukian, Y.S., R.W. Powell, C.Y. Ho, and P.G. Klemens, eds., Thermophysical Properties of Matter, TPRC Data Series, vol. 1, 229, 1970.

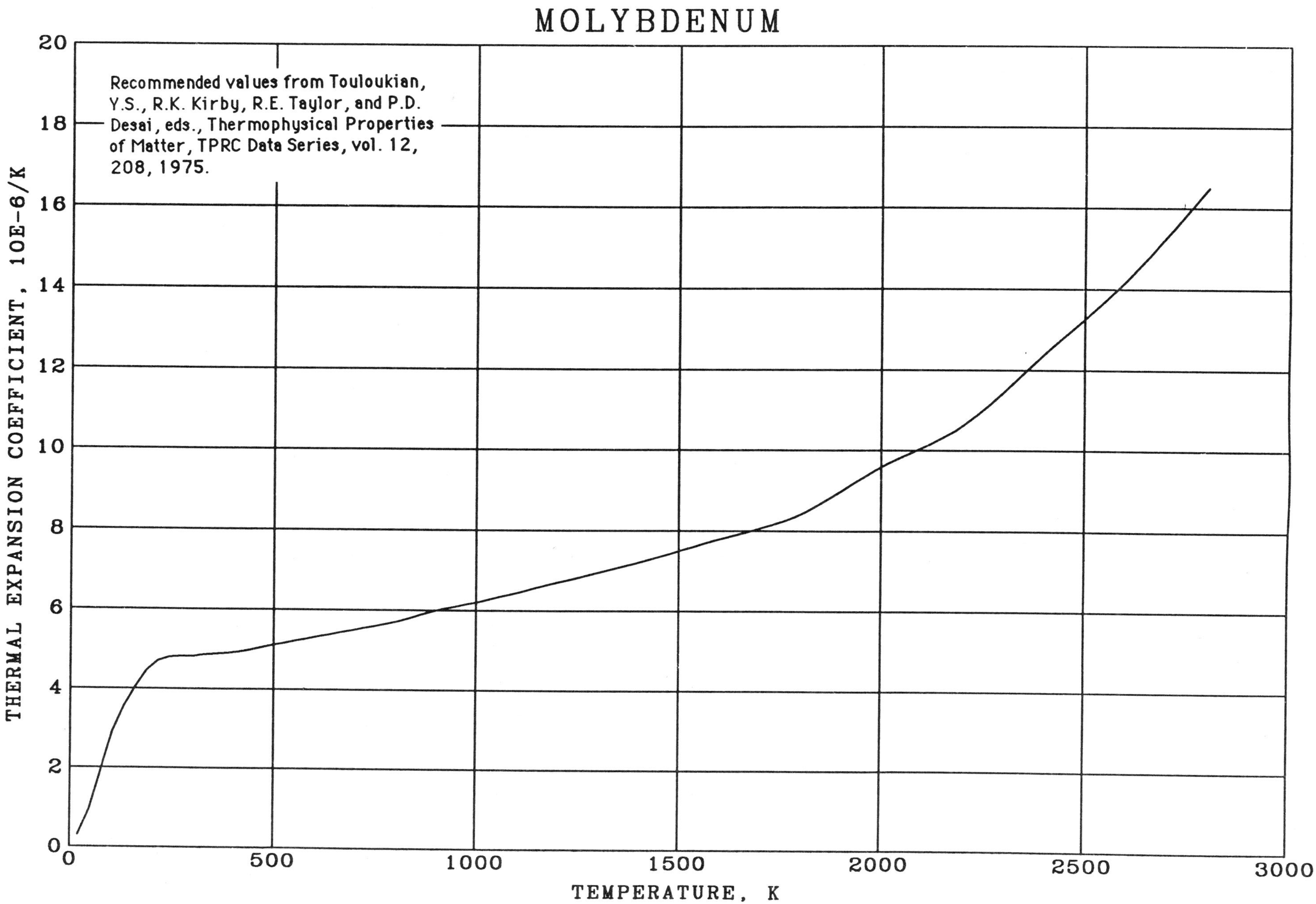
MOLYBDENUM
Recommended values from Touloukian, Y.S., R.K. Kirby, R.E. Taylor, and P.D. Desai, eds., Thermophysical Properties of Matter, TPRC Data Series, vol. 12, 208, 1975.
THERMAL EXPANSION COEFFICIENT, 10E-6/K
TEMPERATURE, K
0
2
4
6
8
10
12
14
16
18
20
0
500
1000
1500
2000
2500
3000

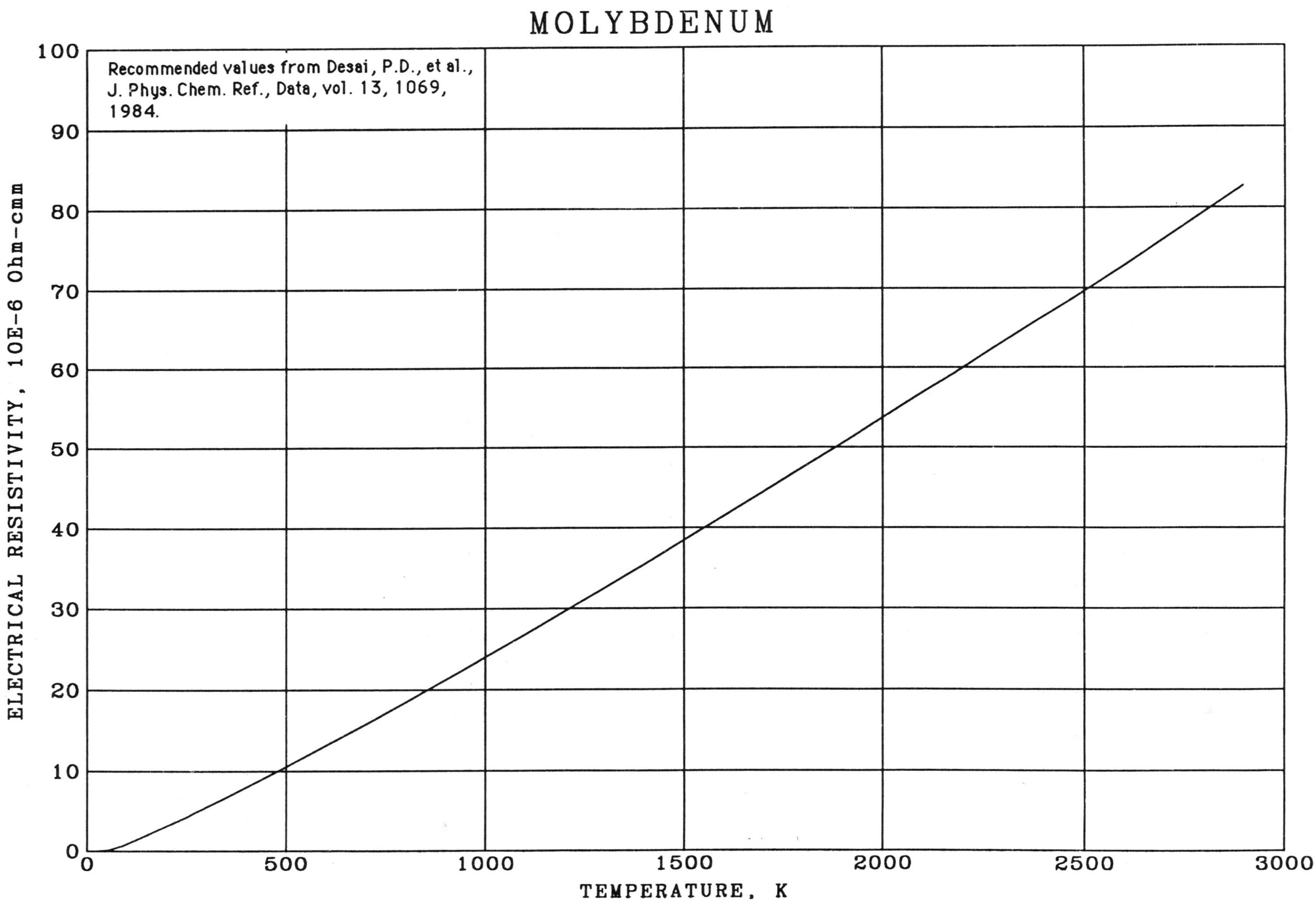

MOLYBDENUM
Recommended values from Desai, P.D., et al., J. Phys. Chem. Ref., Data, vol. 13, 1069, 1984.
ELECTRICAL RESISTIVITY, 10E-6 Ohm-cmm
100
90
80
70
60
50
40
30
20
10
0
0
500
1000
1500
2000
2500
3000
TEMPERATURE, K

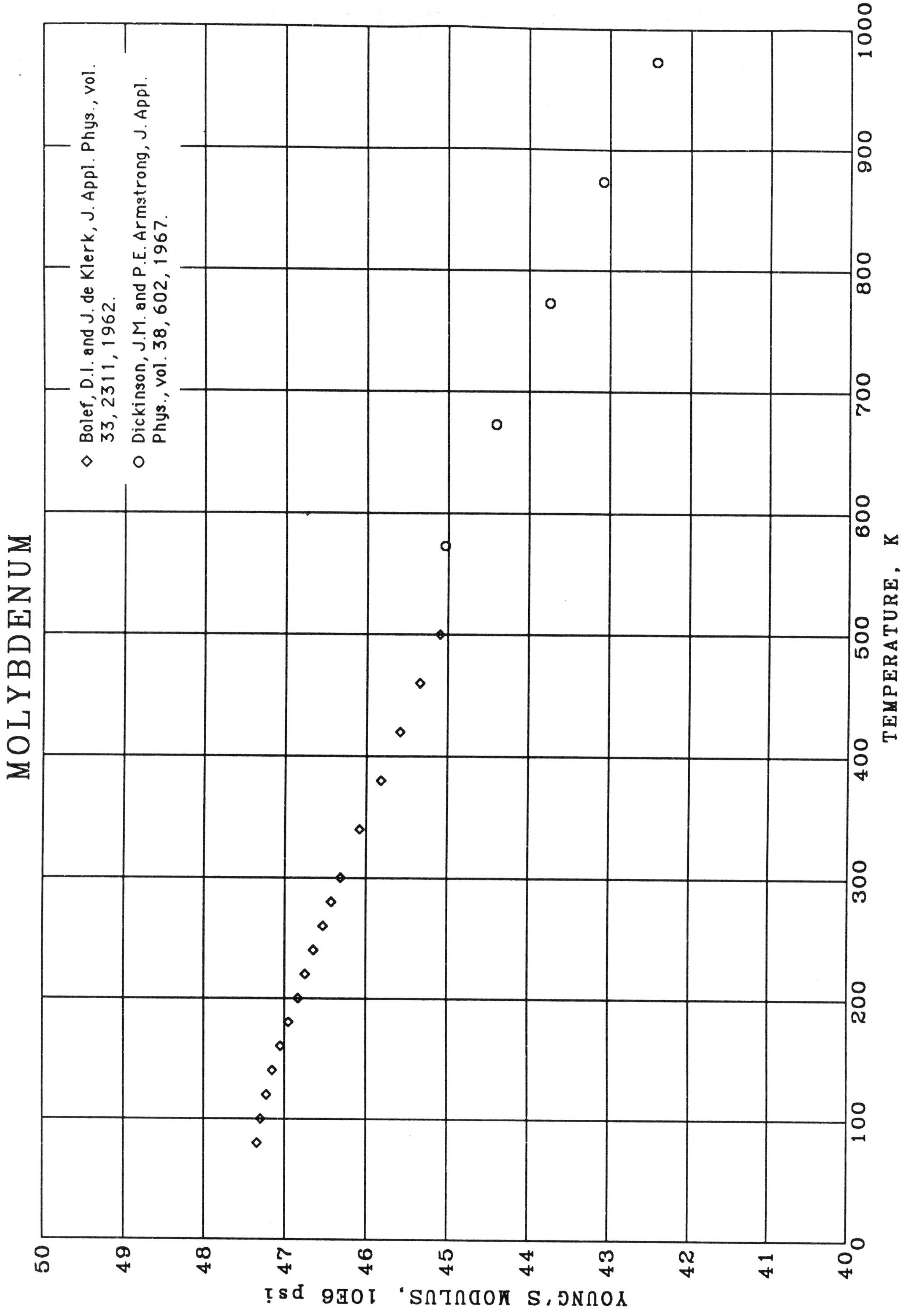
MOLYBDENUM
◇ Bolef, D.I. and J. de Klerk, J. Appl. Phys., vol. 33, 2311, 1962.
○ Dickinson, J.M. and P.E. Armstrong, J. Appl. Phys., vol. 38, 602, 1967.
YOUNG'S MODULUS, 10E6 psi
TEMPERATURE, K
0
100
200
300
400
500
600
700
800
900
1000
40
41
42
43
44
45
46
47
48
49
50

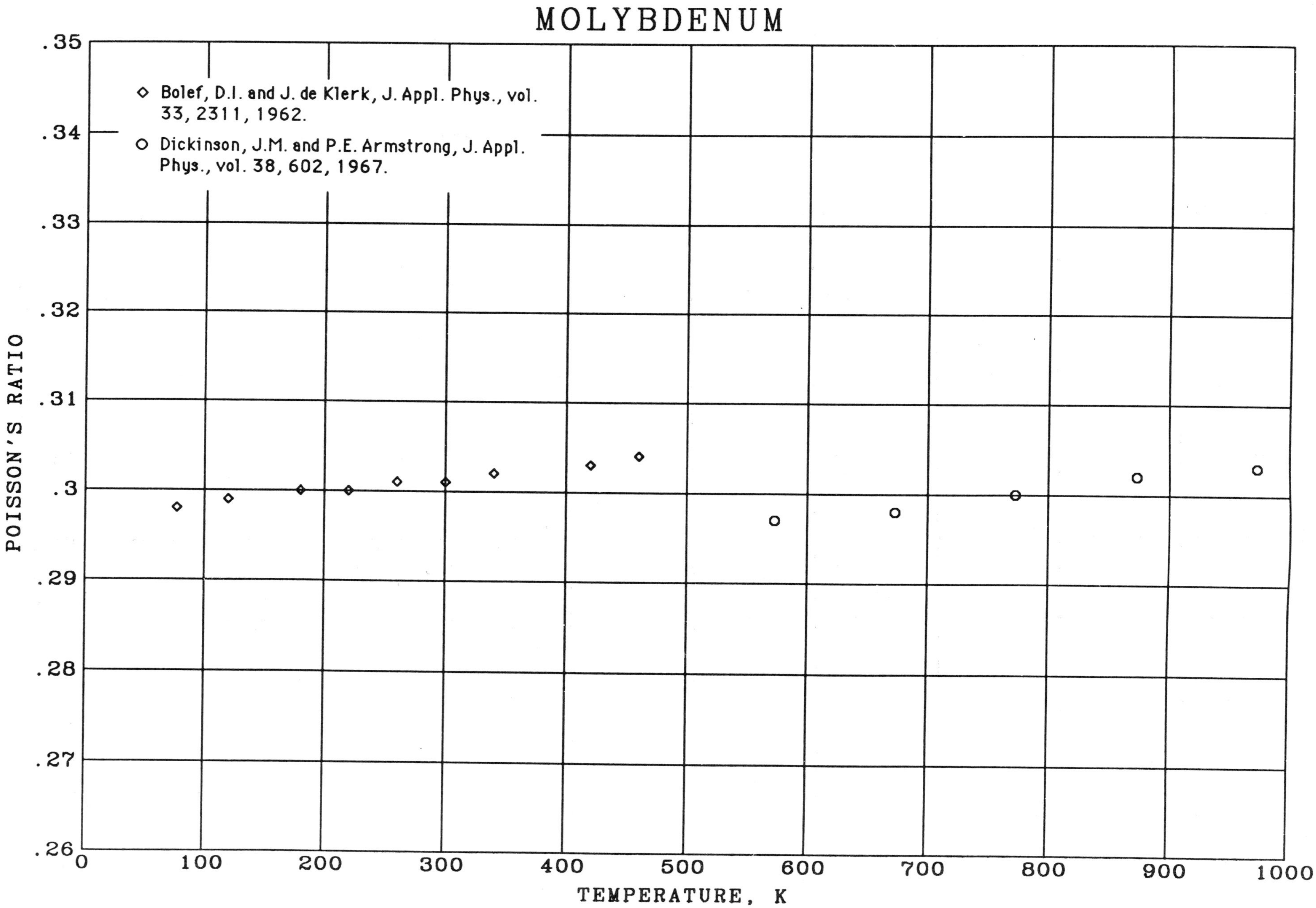
MOLYBDENUM
◇ Bolef, D.I. and J. de Klerk, J. Appl. Phys., vol. 33, 2311, 1962.
○ Dickinson, J.M. and P.E. Armstrong, J. Appl. Phys., vol. 38, 602, 1967.
POISSON'S RATIO
.35
.34
.33
.32
.31
.3
.29
.28
.27
.26
0
100
200
300
400
500
600
700
800
900
1000
TEMPERATURE, K

MOLYBDENUM

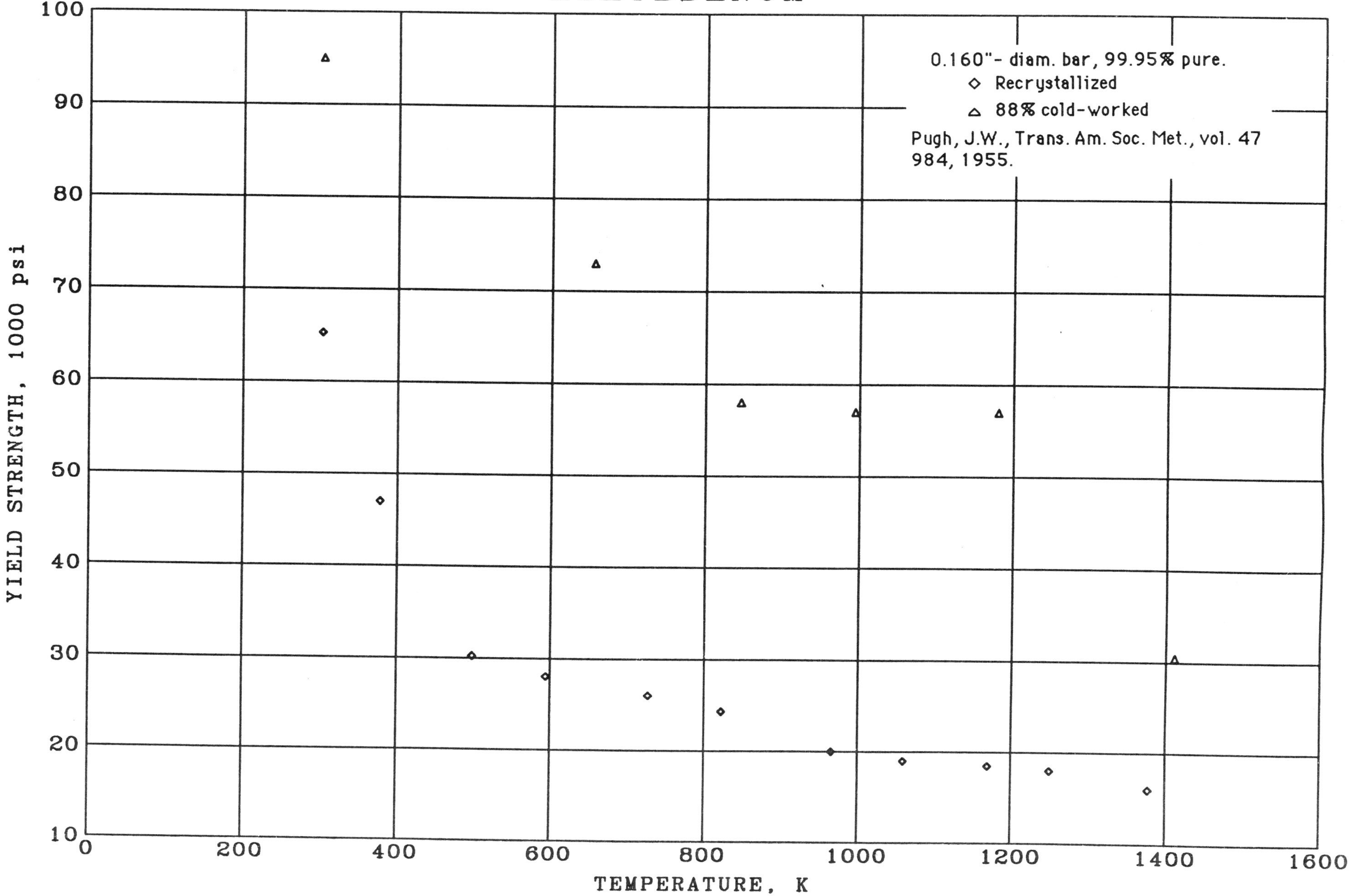
0.160"- diam. bar, 99.95% pure.
◇ Recrystallized
△ 88% cold-worked
Pugh, J.W., Trans. Am. Soc. Met., vol. 47
984, 1955.
YIELD STRENGTH, 1000 psi
TEMPERATURE, K
100
90
80
70
60
50
40
30
20
10
0
200
400
600
800
1000
1200
1400
1600

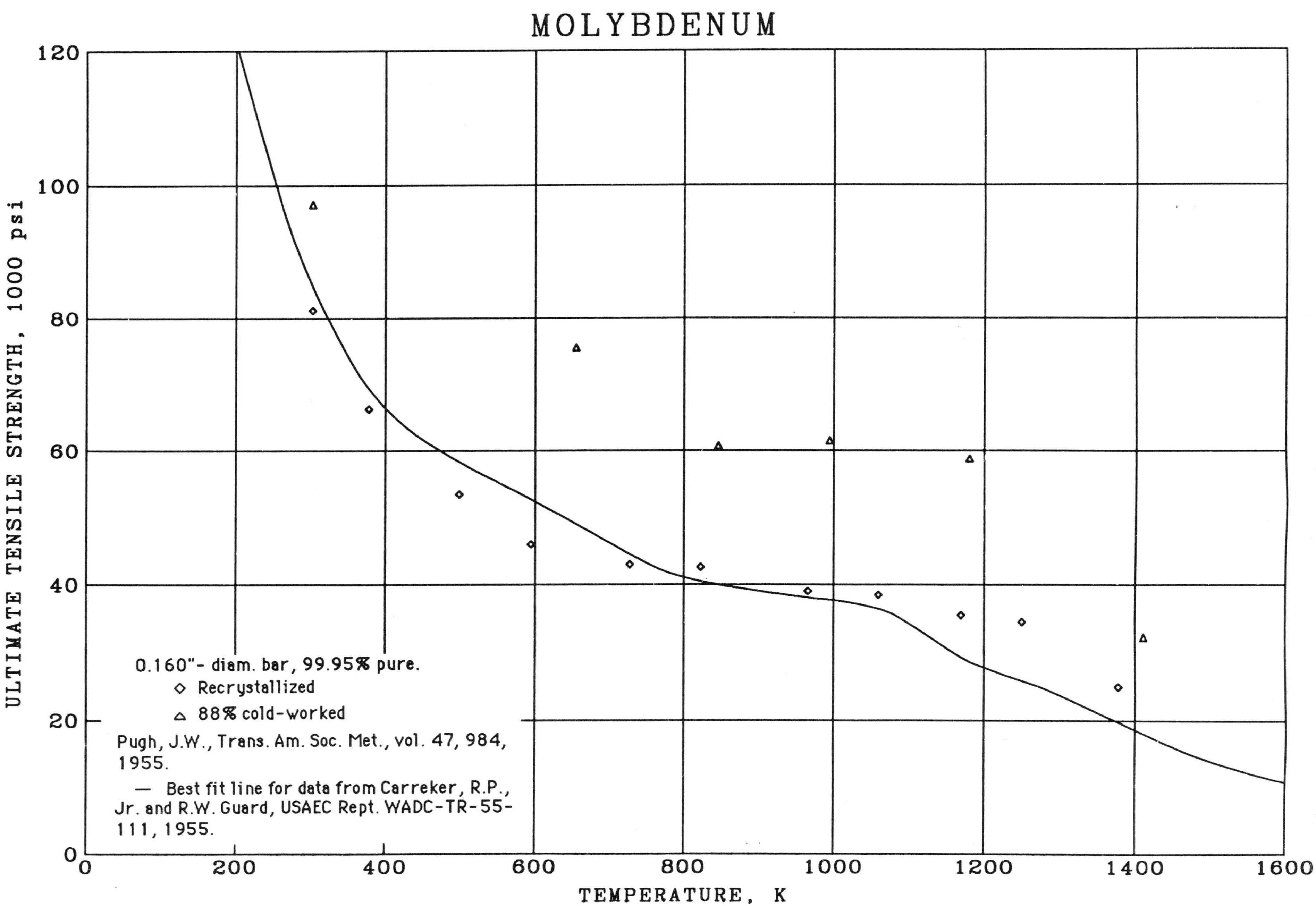
MOLYBDENUM
ULTIMATE TENSILE STRENGTH, 1000 psi
TEMPERATURE, K
0
20
40
60
80
100
120
0
200
400
600
800
1000
1200
1400
1600
0.160"- diam. bar, 99.95% pure.
◇ Recrystallized
△ 88% cold-worked
Pugh, J.W., Trans. Am. Soc. Met., vol. 47, 984, 1955.
— Best fit line for data from Carreker, R.P., Jr. and R.W. Guard, USAEC Rept. WADC-TR-55-111, 1955.

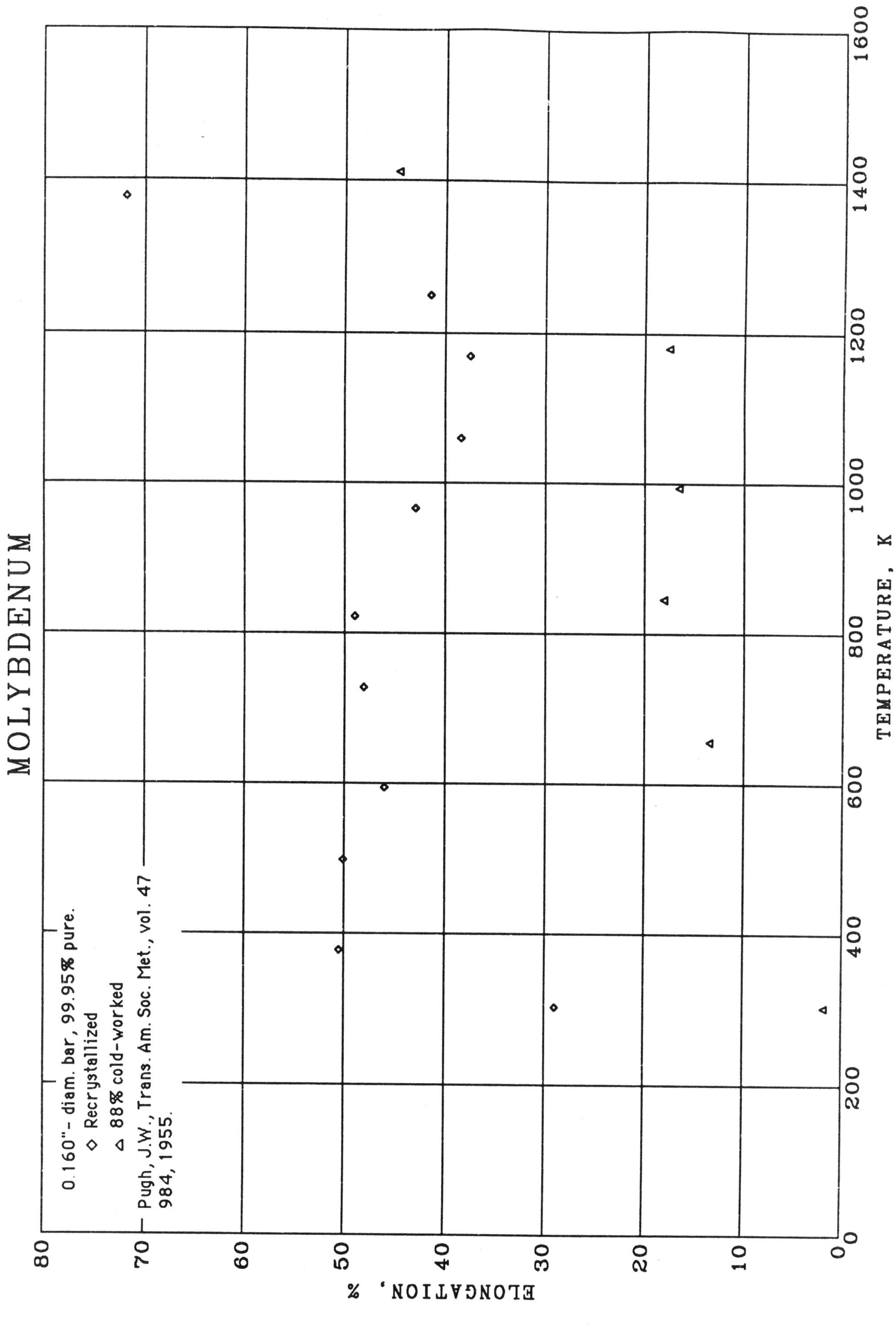
MOLYBDENUM
0.160"- diam. bar, 99.95% pure.
◇ Recrystallized
△ 88% cold-worked
Pugh, J.W., Trans. Am. Soc. Met., vol. 47
984, 1955.
ELONGATION, %
TEMPERATURE, K
0
10
20
30
40
50
60
70
80
0
200
400
600
800
1000
1200
1400
1600

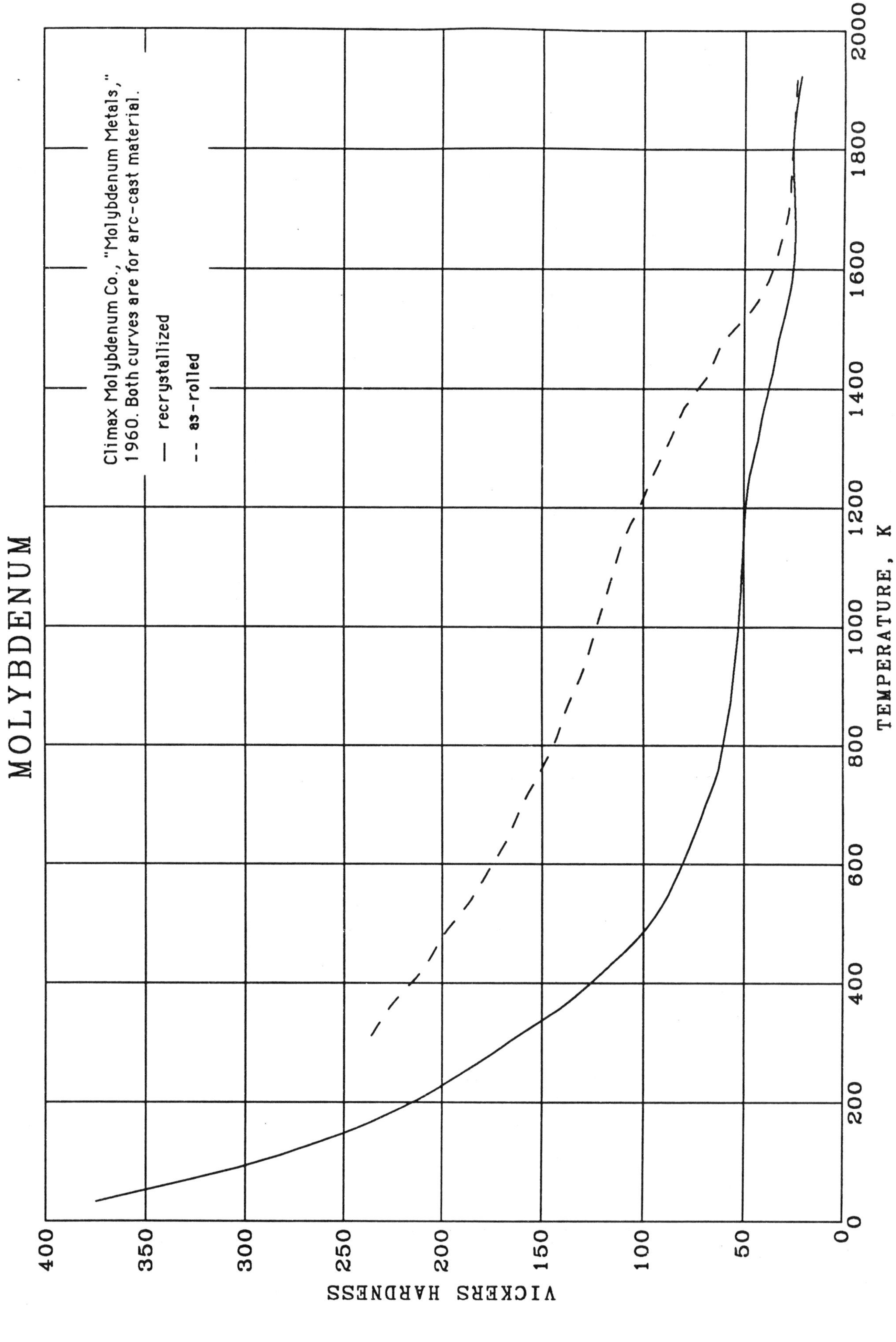
MOLYBDENUM
Climax Molybdenum Co., "Molybdenum Metals," 1960. Both curves are for arc-cast material.
— recrystallized
- - as-rolled
VICKERS HARDNESS
0
50
100
150
200
250
300
350
400
TEMPERATURE, K
0
200
400
600
800
1000
1200
1400
1600
1800
2000

NICKEL

PROPERTY	@ 300K	UNITS	REFERENCES
Density	8.90	g/cm^3	Ni1 Ni2 Ni3
Specific Heat	0.107	cal/g-K	Ni4 Ni5 Ni6
Thermal Conductivity	0.216	cal/cm-s-K	Ni7 Ni8 Ni9
Thermal Exp Coef	13.5	1/K	Ni10 Ni11 Ni12
Melting Point	1726	K	Ni13
Resistivity	$7.16x10^{-6}$	Ohm-cm	Ni14 Ni15 Ni16
TCR (218K to 398K)	5550	ppm/K	Ni14 Ni15 Ni16
Young's Modulus 1	$3.20x10^{7}$	psi	Ni17 Ni18 Ni19
Poisson's Ratio 1	0.299	-----	Ni17 Ni18 Ni20
Yield Strength	8500	psi	Ni21 Ni22 Ni23
Ult Tensile Strength	46,000	psi	Ni21 Ni22 Ni23
Elongation at Break	50	%	Ni21 Ni23 Ni24
Hardness	40 B	Rockwell	Ni24 Ni25 Ni26

1 The elastic constants of Ni depend on magnetization as well as temperature. Young's modulus of demagnetized Ni is 19.5% smaller than the modulus of saturated material. At room temperature the difference is about 7%.[Ni27]

Mechanical properties listed are for fully annealed, bulk material. Cold work will dramatically increase the strength and hardness and reduce the elongation. By suitable choice of hot rolling and cold drawing tensile strengths ranging from 65,000 to 115,000 psi can be achieved for commercially pure Ni(99.85%) in rods and bars, as high as 130 psi in strip and 160,000 psi in wire. [Ni23] Tensile strengths greater than 300,000 psi with 1.8% elongation have been reported for polycrystalline Ni films.[Ni28] Heating cold worked material into the temperature range 300-500C will result in recovery and recrystallization erasing the effects of cold work. For more information on the effects of cold work on mechanical properties, see graphs of mechanical properties-v-temperature.

NICKEL

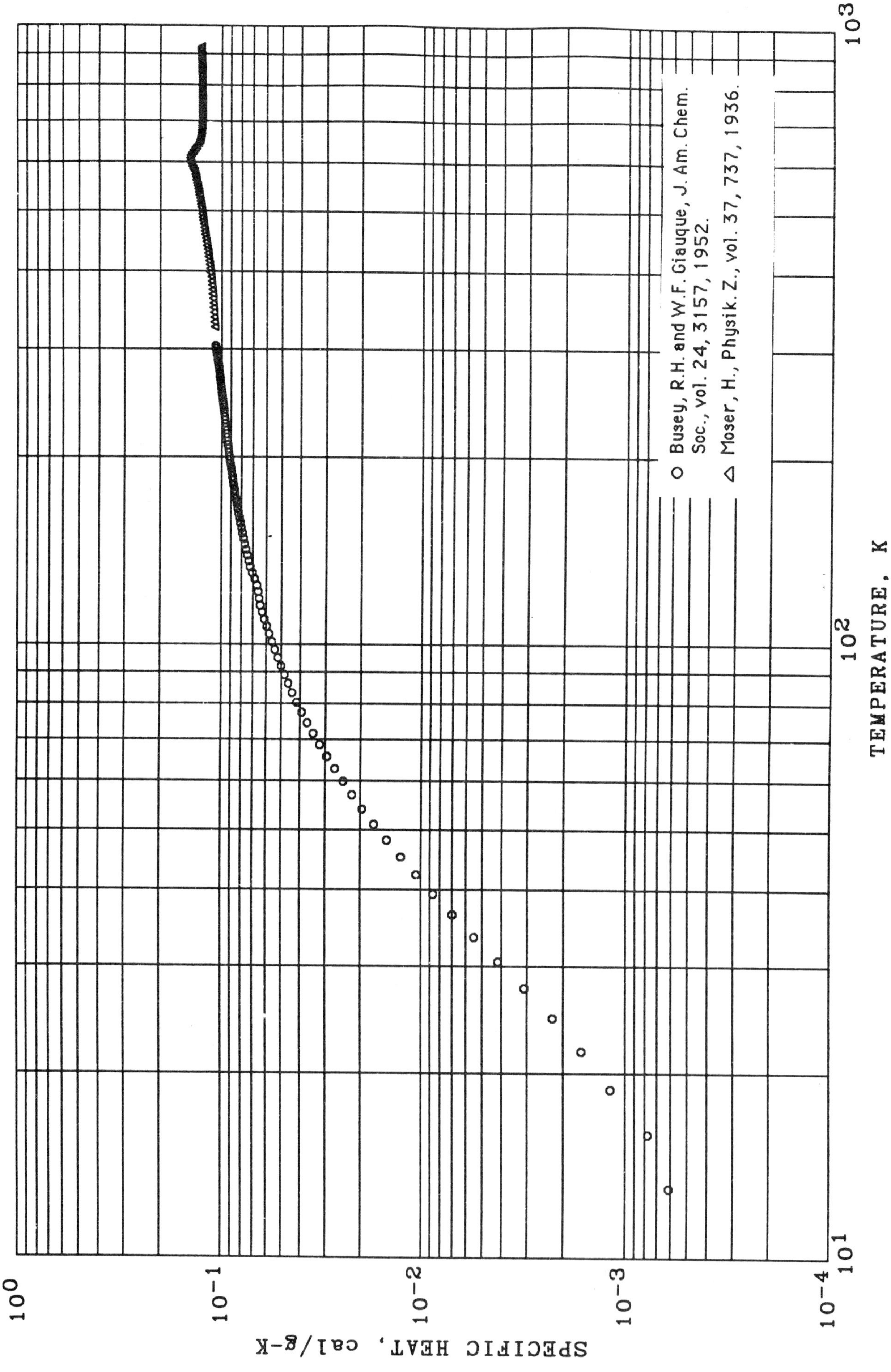

○ Busey, R.H. and W.F. Giauque, J. Am. Chem. Soc., vol. 24, 3157, 1952.
△ Moser, H., Physik. Z., vol. 37, 737, 1936.
TEMPERATURE, K
SPECIFIC HEAT, cal/g-K
10^0
10^{-1}
10^{-2}
10^{-3}
10^{-4}
10^1
10^2
10^3

NICKEL

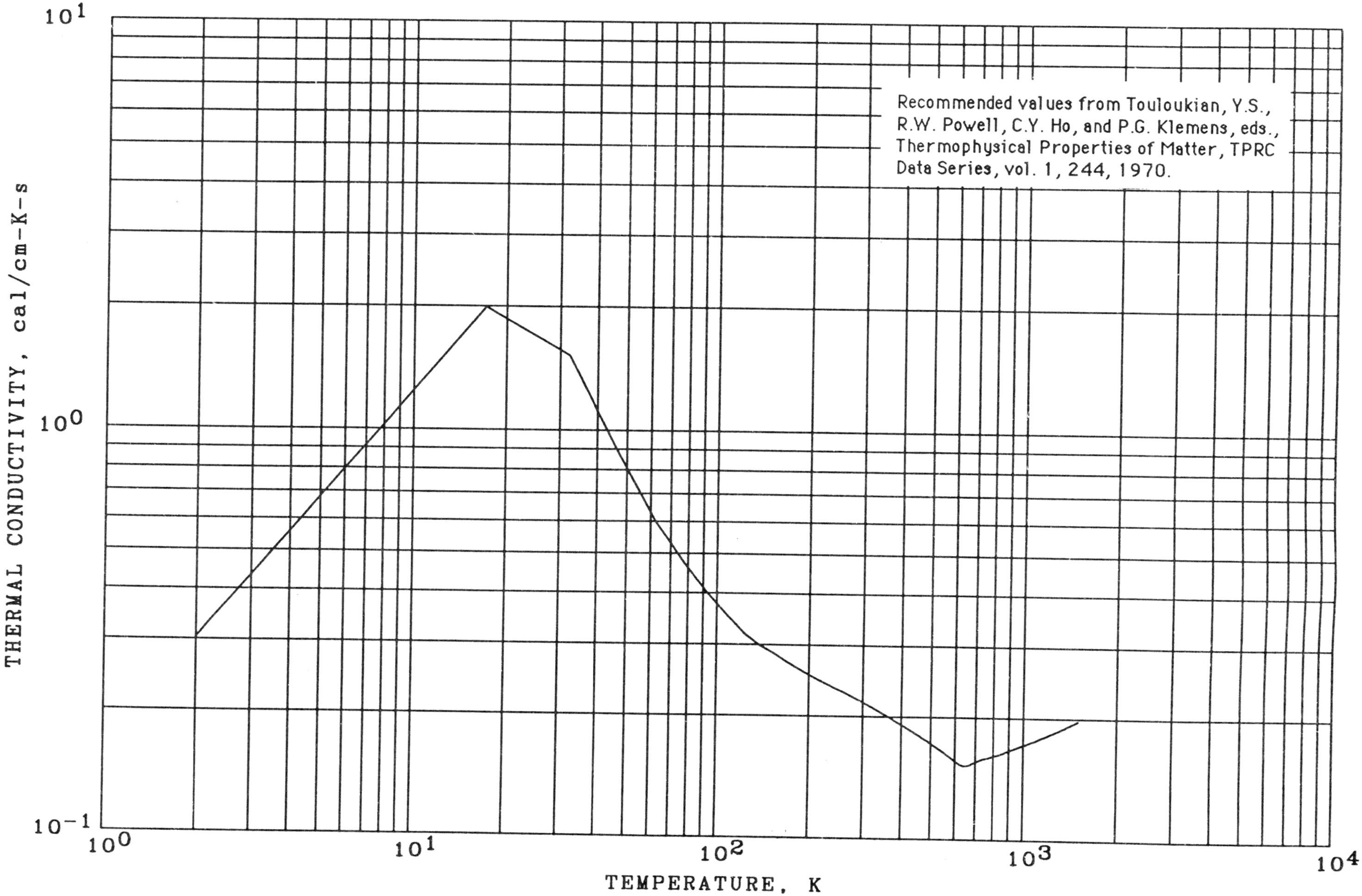
Recommended values from Touloukian, Y.S., R.W. Powell, C.Y. Ho, and P.G. Klemens, eds., Thermophysical Properties of Matter, TPRC Data Series, vol. 1, 244, 1970.
THERMAL CONDUCTIVITY, cal/cm-K-s
10^1
10^0
10^{-1}
10^0
10^1
10^2
10^3
10^4
TEMPERATURE, K

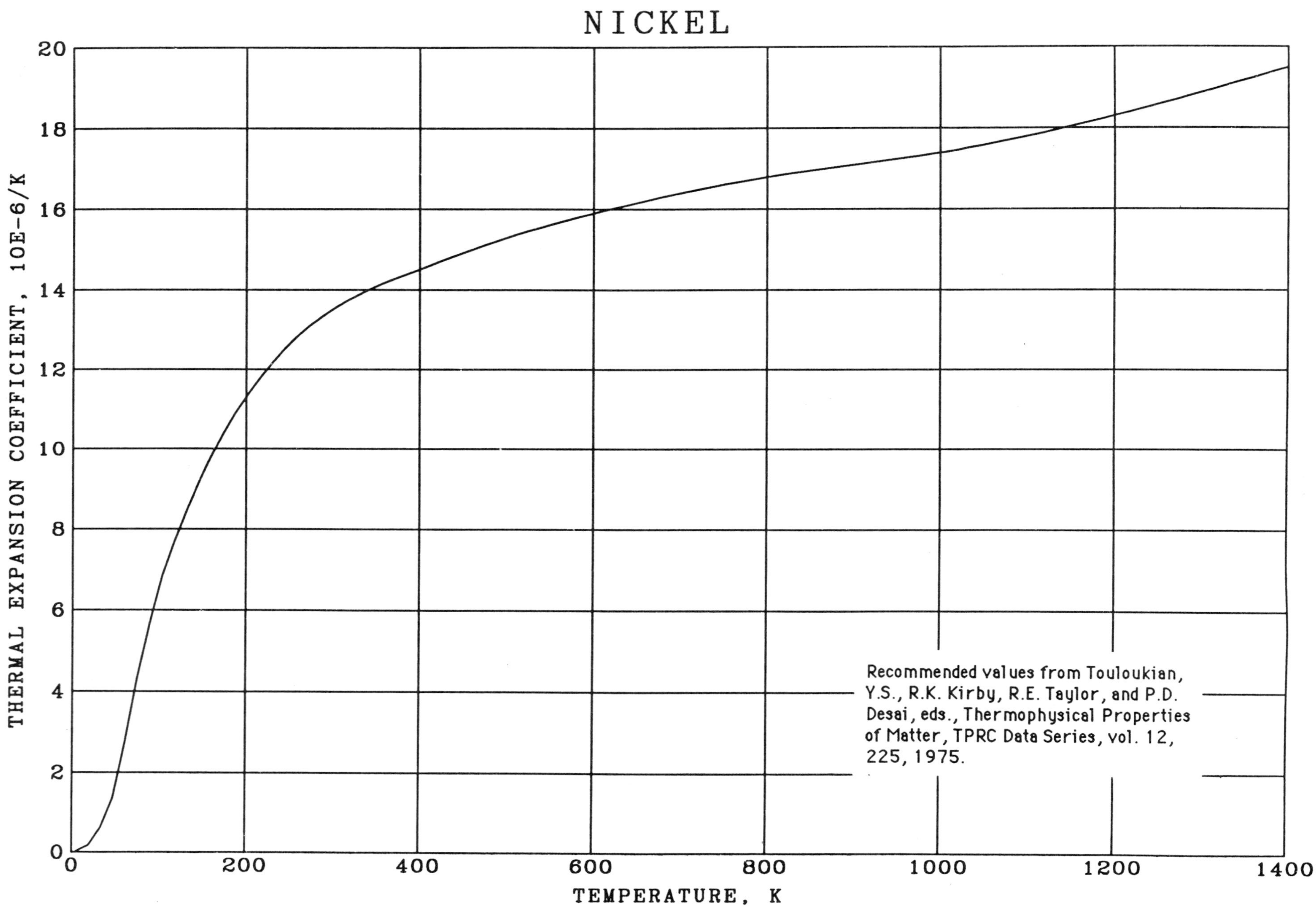
NICKEL
20
18
16
14
12
10
8
6
4
2
0
THERMAL EXPANSION COEFFICIENT, 10E-6/K
0
200
400
600
800
1000
1200
1400
TEMPERATURE, K
Recommended values from Touloukian, Y.S., R.K. Kirby, R.E. Taylor, and P.D. Desai, eds., Thermophysical Properties of Matter, TPRC Data Series, vol. 12, 225, 1975.

NICKEL

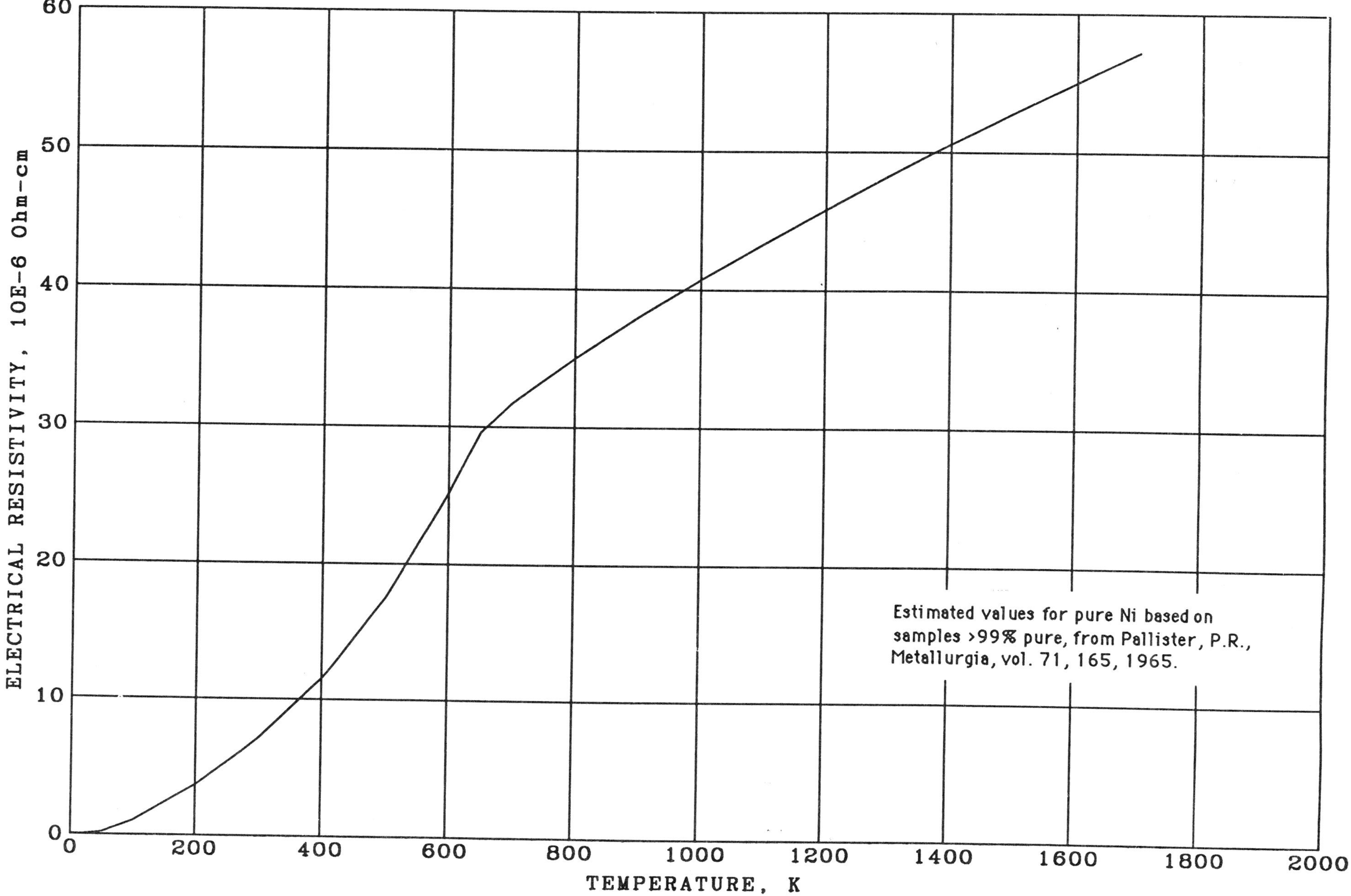
ELECTRICAL RESISTIVITY, 10E-6 Ohm-cm
0
10
20
30
40
50
60
TEMPERATURE, K
0
200
400
600
800
1000
1200
1400
1600
1800
2000
Estimated values for pure Ni based on samples >99% pure, from Pallister, P.R., Metallurgia, vol. 71, 165, 1965.

NICKEL

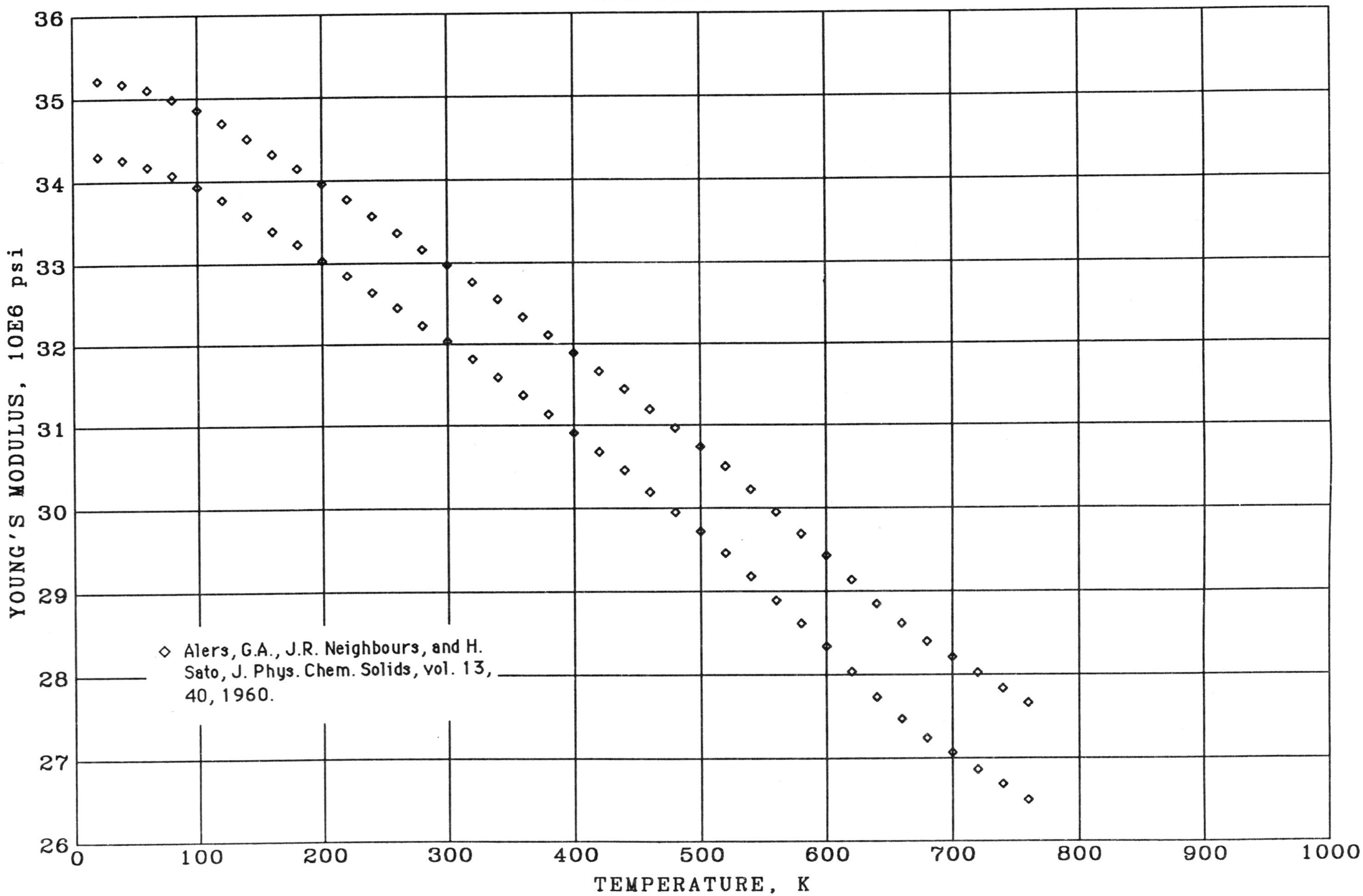

36
35
34
33
32
31
30
29
28
27
26
YOUNG'S MODULUS, 10E6 psi
0
100
200
300
400
500
600
700
800
900
1000
TEMPERATURE, K
◇ Alers, G.A., J.R. Neighbours, and H. Sato, J. Phys. Chem. Solids, vol. 13, 40, 1960.

NICKEL

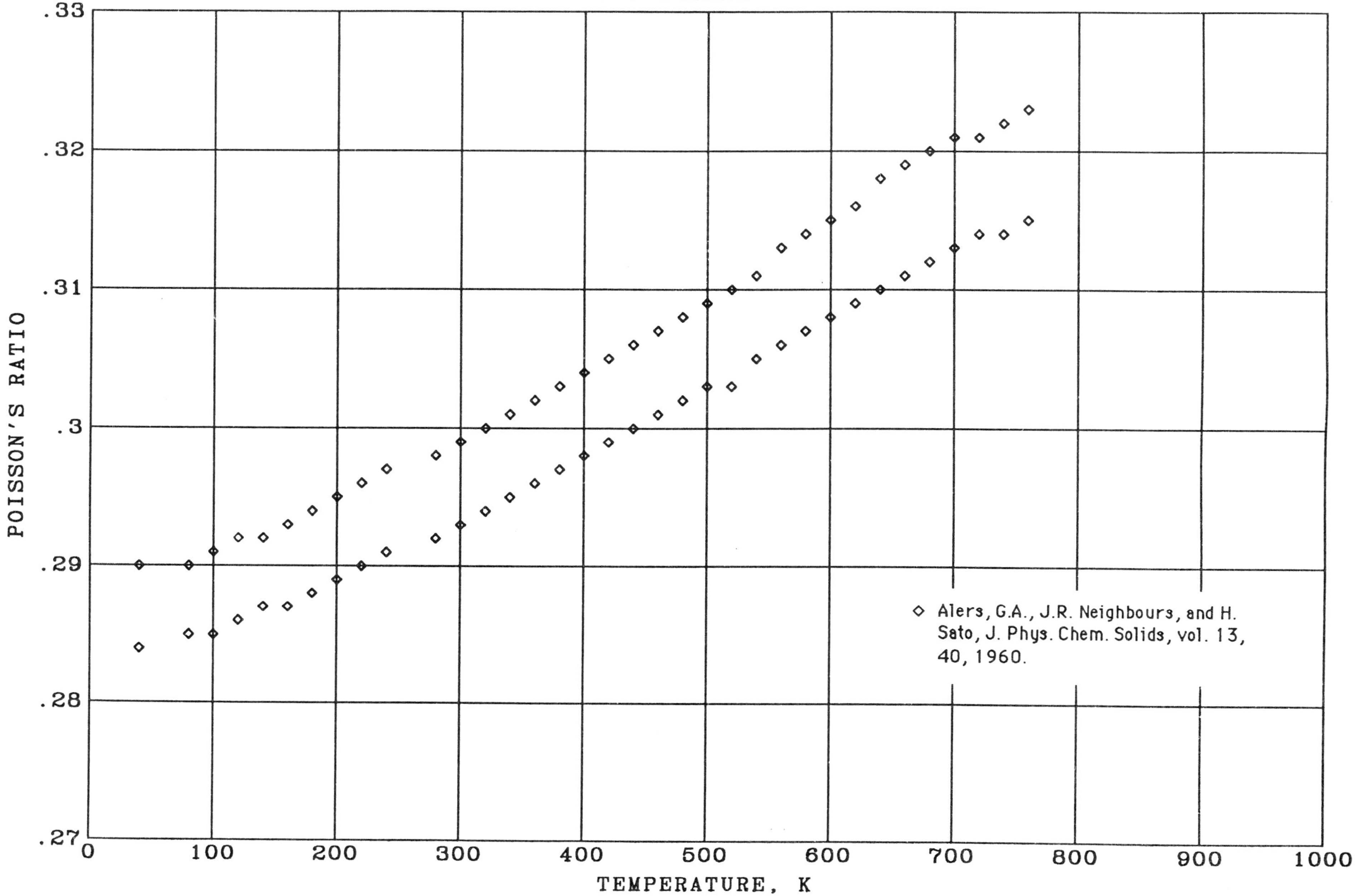

.33
.32
.31
.3
.29
.28
.27
0
100
200
300
400
500
600
700
800
900
1000
POISSON'S RATIO
TEMPERATURE, K
◇ Alers, G.A., J.R. Neighbours, and H. Sato, J. Phys. Chem. Solids, vol. 13, 40, 1960.

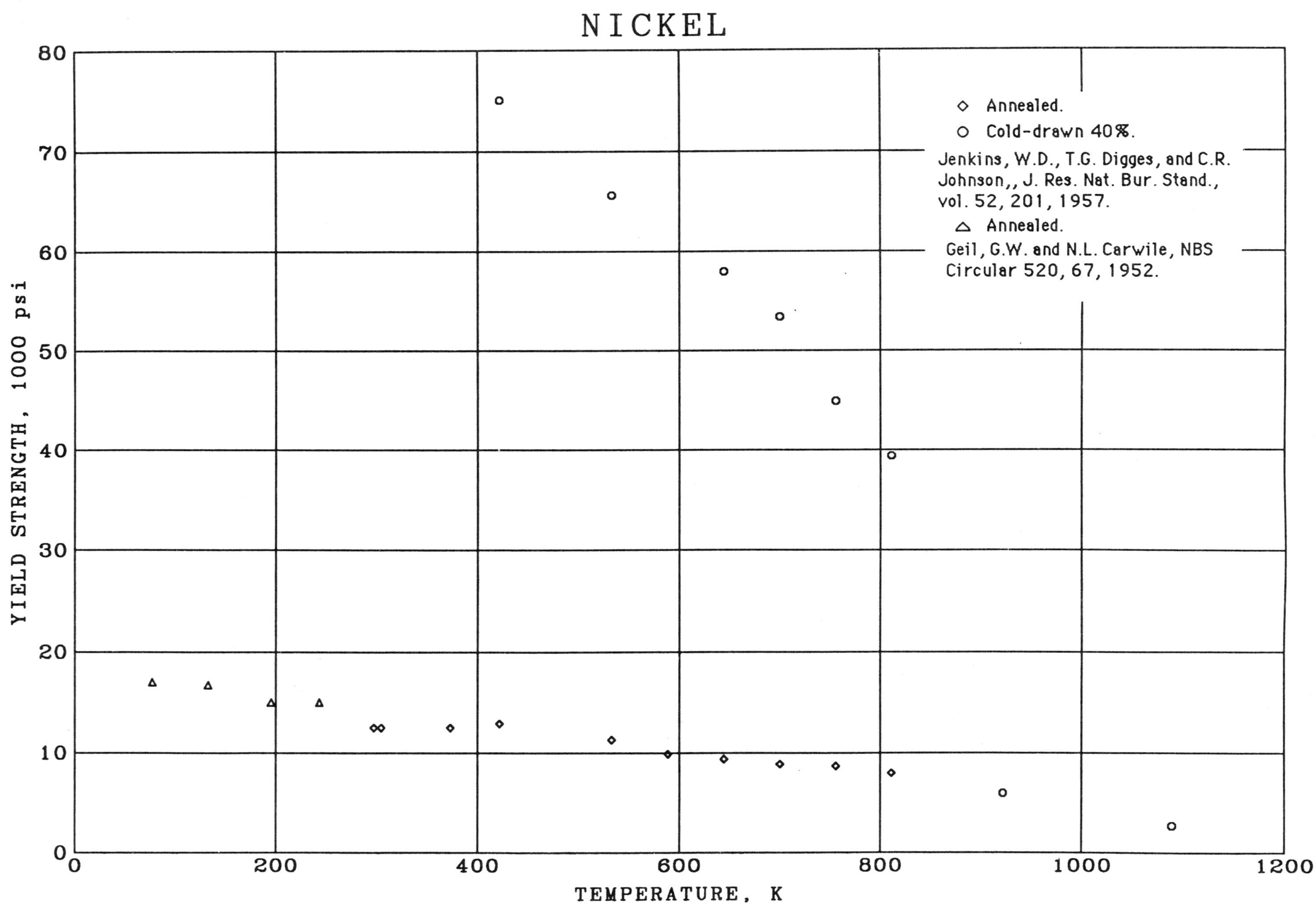

NICKEL
◇ Annealed.
○ Cold-drawn 40%.
Jenkins, W.D., T.G. Digges, and C.R. Johnson,, J. Res. Nat. Bur. Stand., vol. 52, 201, 1957.
△ Annealed.
Geil, G.W. and N.L. Carwile, NBS Circular 520, 67, 1952.
YIELD STRENGTH, 1000 psi
0
10
20
30
40
50
60
70
80
TEMPERATURE, K
0
200
400
600
800
1000
1200

NICKEL

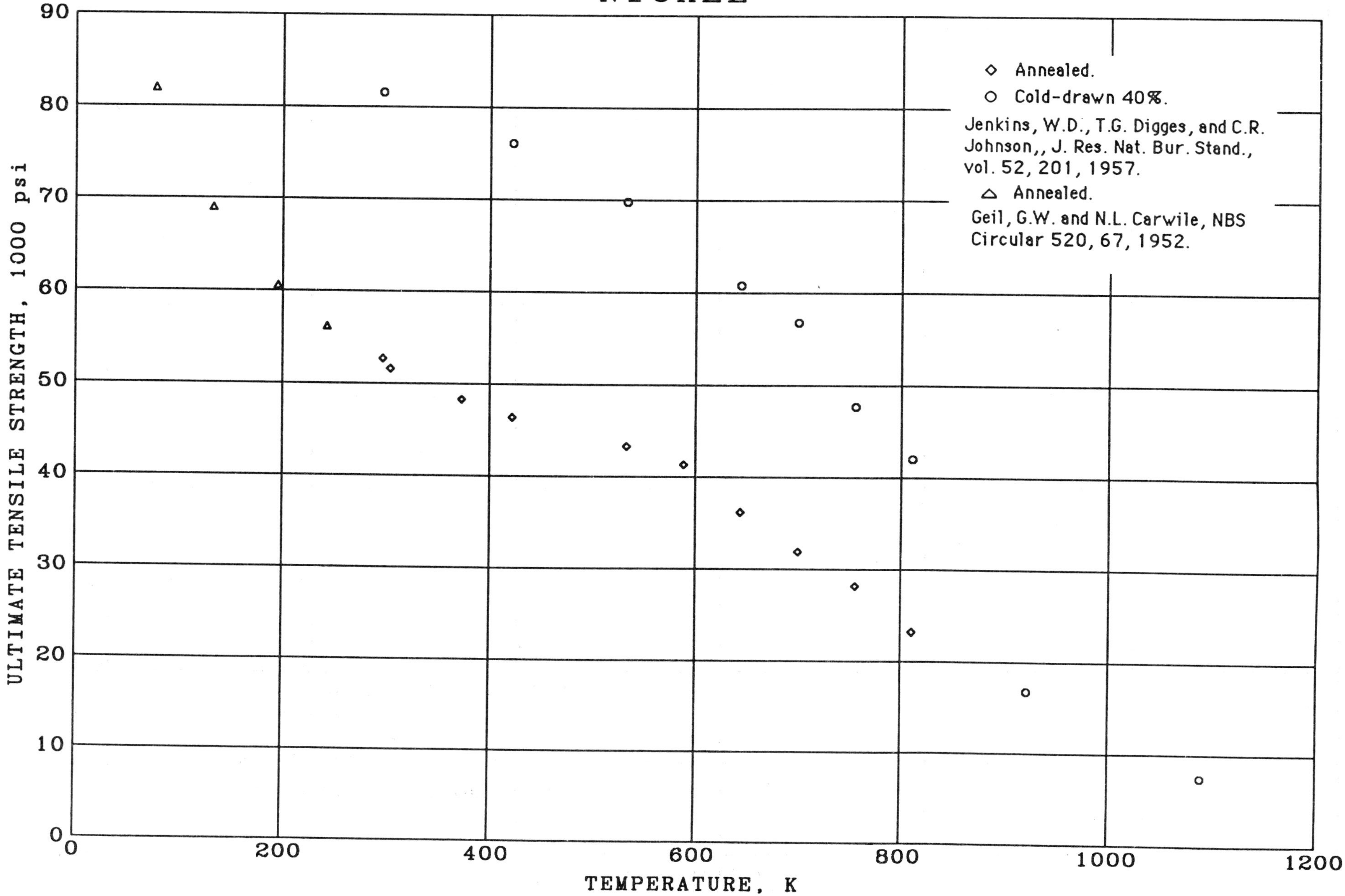

90
80
70
60
50
40
30
20
10
0
ULTIMATE TENSILE STRENGTH, 1000 psi
0
200
400
600
800
1000
1200
TEMPERATURE, K
◇ Annealed.
○ Cold-drawn 40%.
Jenkins, W.D., T.G. Digges, and C.R. Johnson,, J. Res. Nat. Bur. Stand., vol. 52, 201, 1957.
△ Annealed.
Geil, G.W. and N.L. Carwile, NBS Circular 520, 67, 1952.

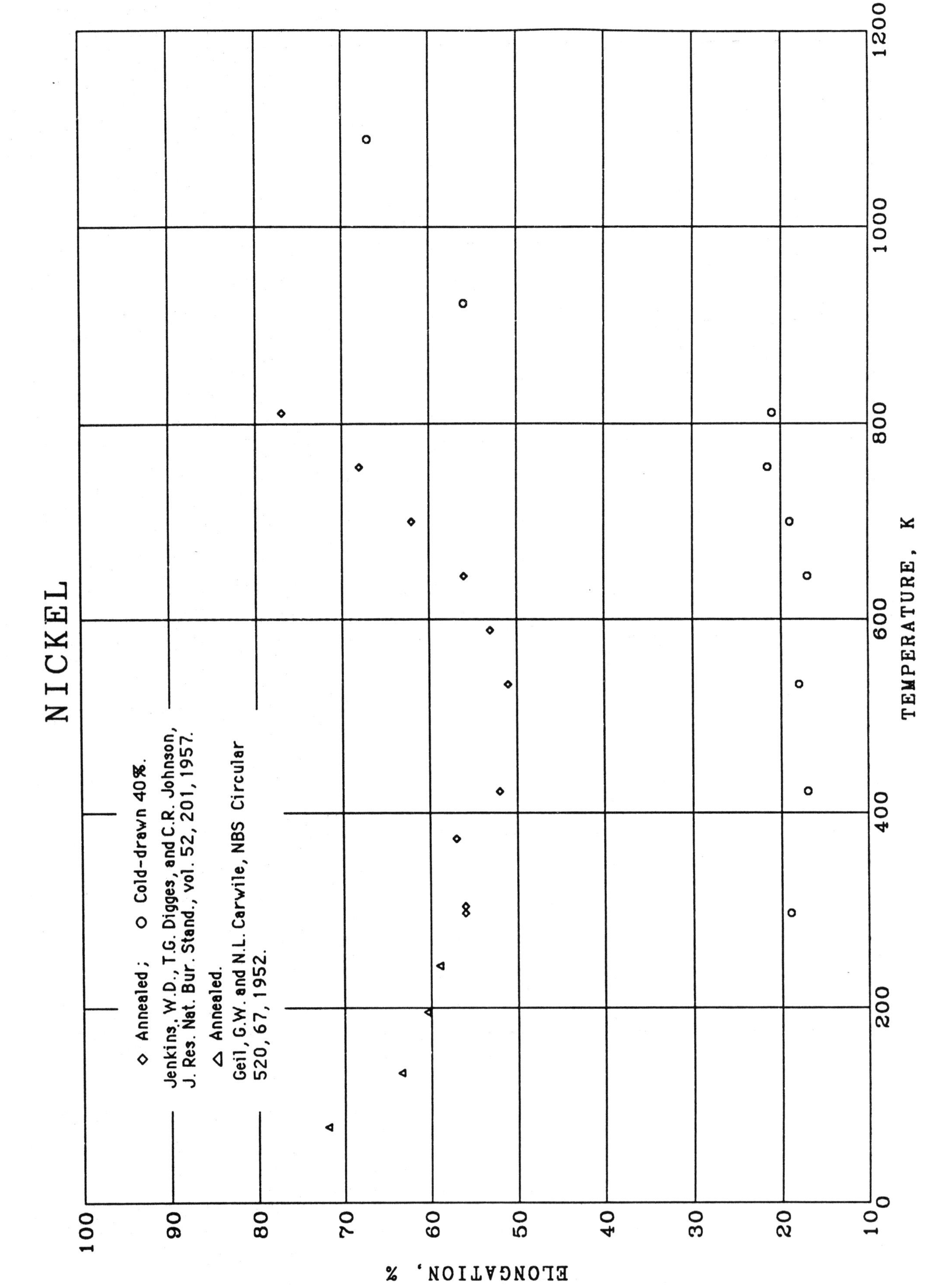
NICKEL
◇ Annealed ; ○ Cold-drawn 40%.
Jenkins, W.D., T.G. Digges, and C.R. Johnson, J. Res. Nat. Bur. Stand., vol. 52, 201, 1957.
△ Annealed.
Geil, G.W. and N.L. Carwile, NBS Circular 520, 67, 1952.
ELONGATION, %
TEMPERATURE, K
100
90
80
70
60
50
40
30
20
10
0
200
400
600
800
1000
1200

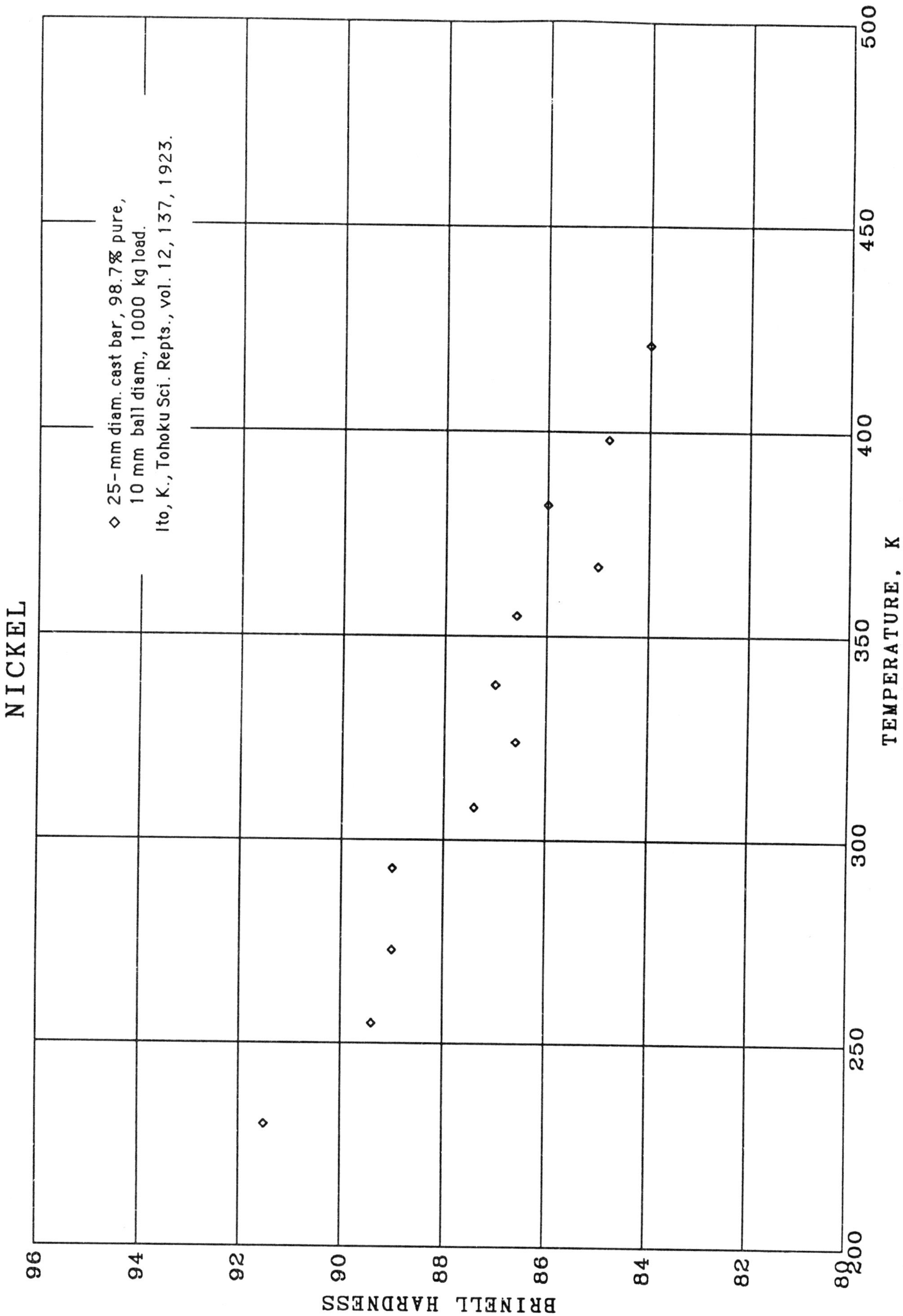

NICKEL
◇ 25-mm diam. cast bar, 98.7% pure,
10 mm ball diam., 1000 kg load.
Ito, K., Tohoku Sci. Repts., vol. 12, 137, 1923.
BRINELL HARDNESS
TEMPERATURE, K
80
82
84
86
88
90
92
94
96
200
250
300
350
400
450
500

SILICON

PROPERTY		@ 300K	UNITS	REFERENCES
Density		2.33	g/cm^3	Si1 Si2 Si3
Specific Heat		0.169	cal/g-K	Si4 Si5 Si6
Thermal Conductivity		0.354	cal/cm-s-K	Si7 Si8 Si9
Thermal Exp Coef		2.618	1/K	Si10 Si11 Si12
Melting Point		1683	K	Si13
Resistivity	1	316000	Ohm-cm	Si14 Si15 Si16
TCR		-15.068×10^6	ppm/K	Si14 Si15 Si16
Young's Modulus	2	2.356×10^7	psi	Si17 Si18 Si19
Poisson's Ratio	2	0.222	-----	Si17 Si18 Si19
Yield Strength	3	-----	psi	
Ult Tensile Strength	4	26,800	psi	Si20 Si21 Si22
Elongation at Break	3	-----	%	
Hardness		1050	Knoop	Si24 Si25 Si26

1 The resistivity is a very strong function of the doping level. For the potential variations in the resistivity and TCR see the graphs of resistivity -v- temperature. The values listed are intrinsic values and are calculated from the equation provided by reference Si27.

2 Values listed are for polycrystalline material. The values for single crystal material can be found in the graphs of modulus versus temperature.

3 Below about 600C, Si is a brittle solid. Fracture is preceeded by no yielding or plastic deformation. The maximum breaking strength is determined by the sizes and distribution of flaws present in the material. The flaw distribution is a strong function of the volume. Smaller samples will have a narrow distribution of smaller flaws and therefore will be inherently stronger.

4 Strength varies from 9046 psi (flexural strength of polycrystalline bar) to >500,000 psi (tensile strength of small diameter wiskers).

SILICON

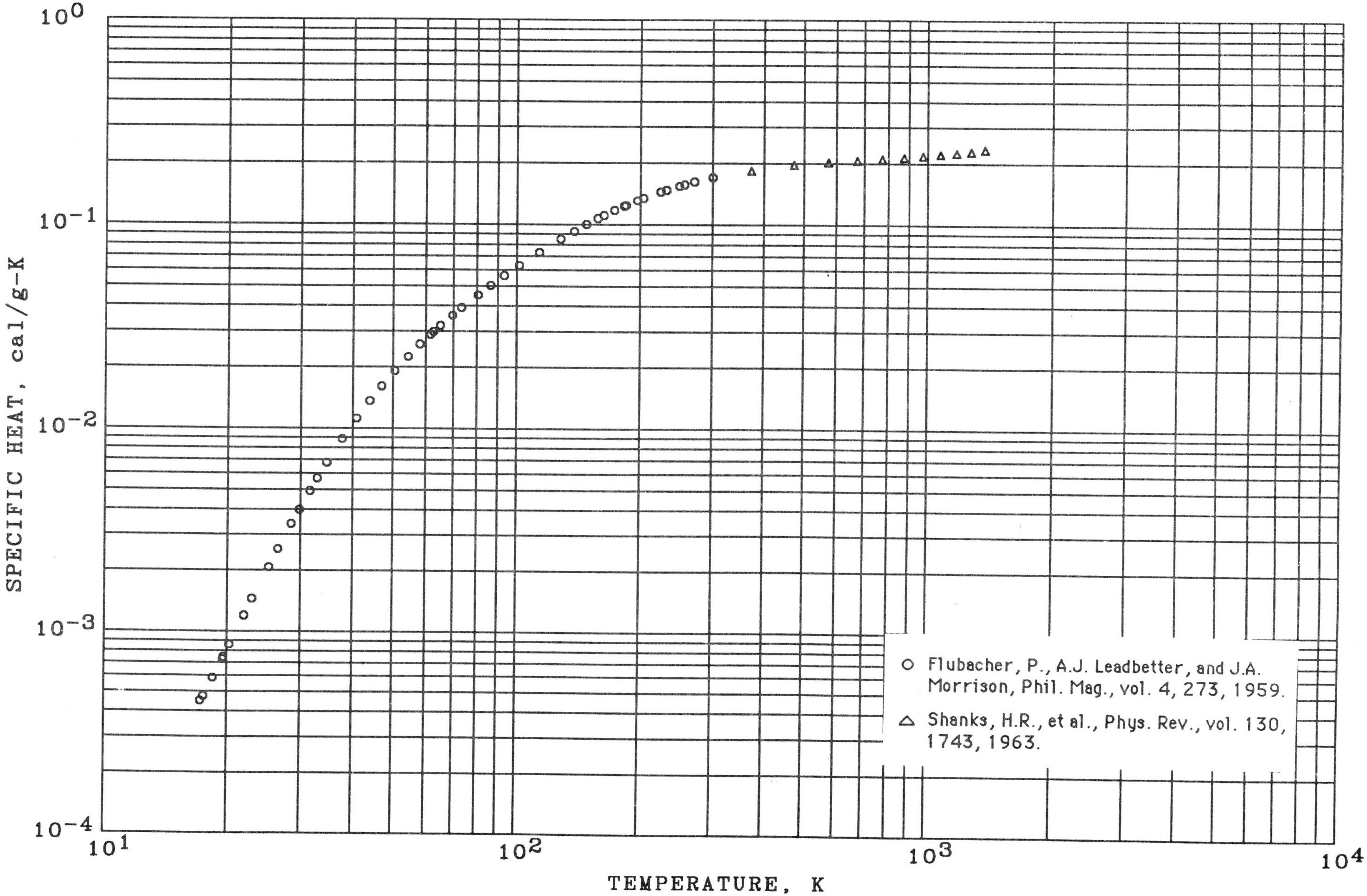

10^0
10^{-1}
10^{-2}
10^{-3}
10^{-4}
SPECIFIC HEAT, cal/g-K
10^1
10^2
10^3
10^4
TEMPERATURE, K
○ Flubacher, P., A.J. Leadbetter, and J.A. Morrison, Phil. Mag., vol. 4, 273, 1959.
△ Shanks, H.R., et al., Phys. Rev., vol. 130, 1743, 1963.

SILICON

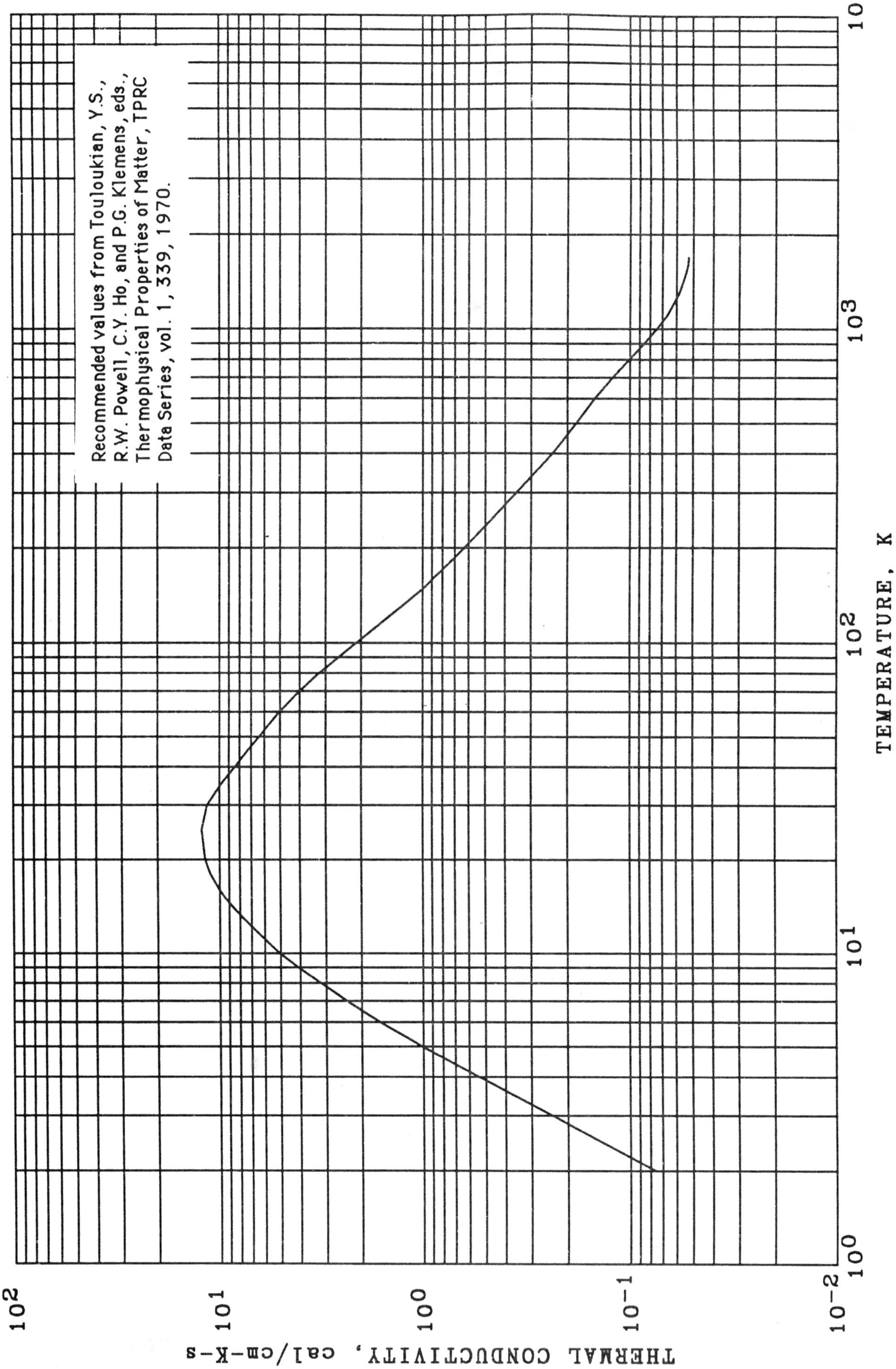

Recommended values from Touloukian, Y.S., R.W. Powell, C.Y. Ho, and P.G. Klemens, eds., Thermophysical Properties of Matter, TPRC Data Series, vol. 1, 339, 1970.
THERMAL CONDUCTIVITY, cal/cm-K-s
10^2
10^1
10^0
10^-1
10^-2
TEMPERATURE, K
10^0
10^1
10^2
10^3
10^4

SILICON

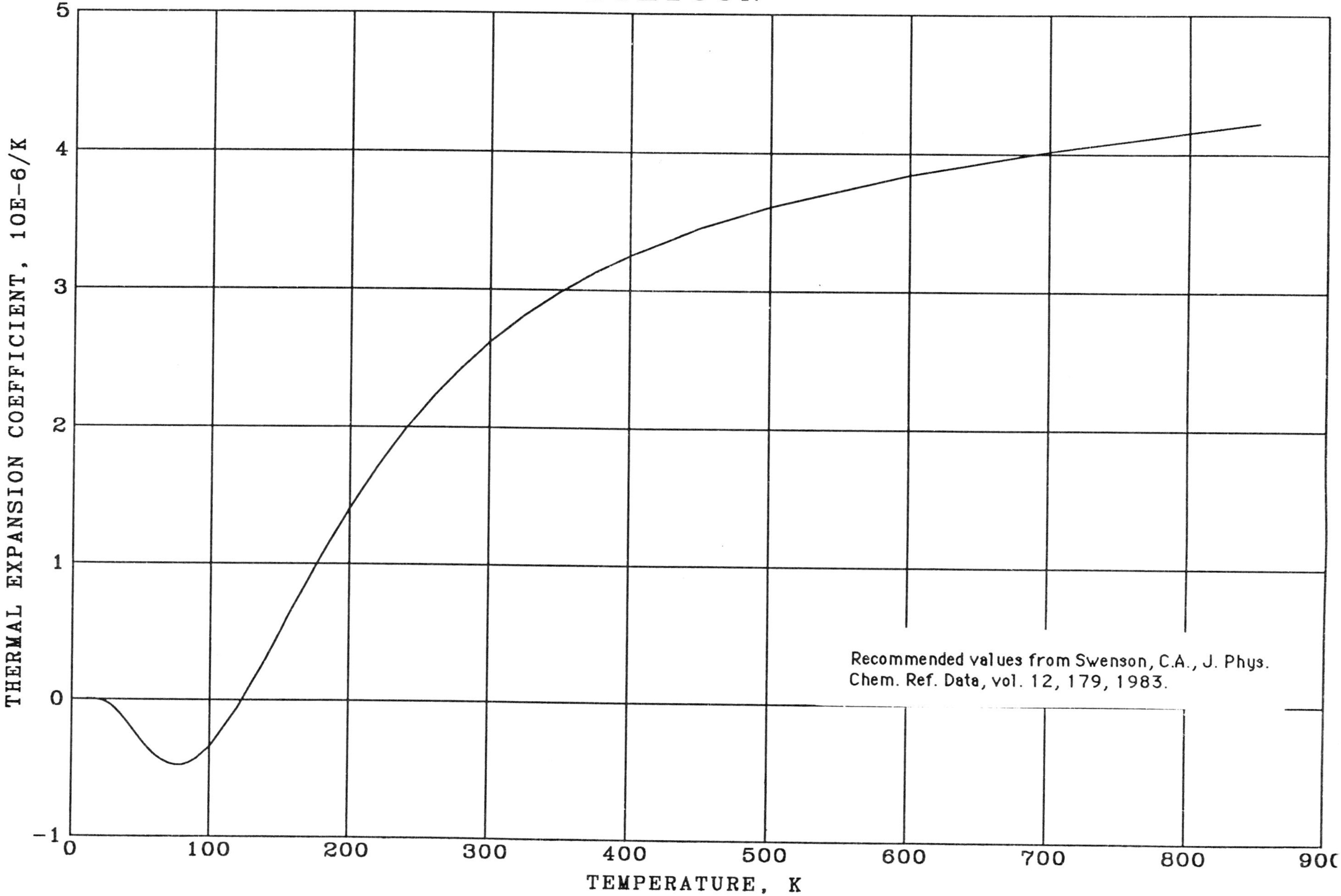

THERMAL EXPANSION COEFFICIENT, 10E-6/K
5
4
3
2
1
0
−1
0
100
200
300
400
500
600
700
800
900
TEMPERATURE, K
Recommended values from Swenson, C.A., J. Phys. Chem. Ref. Data, vol. 12, 179, 1983.

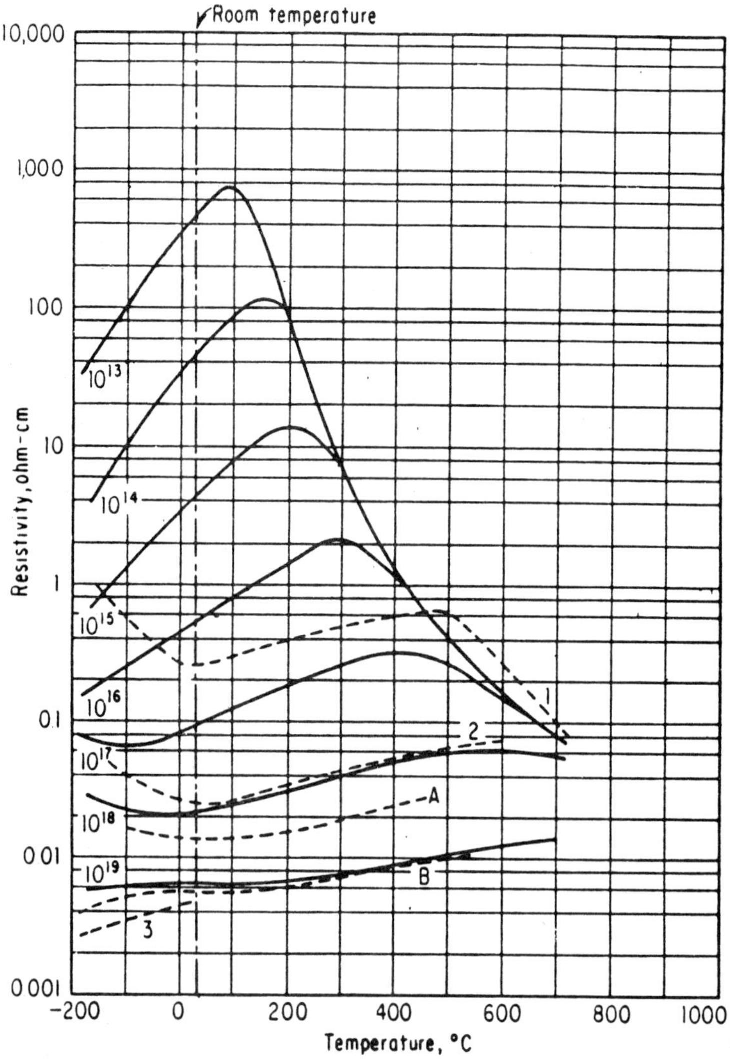

Reprinted from Runyan, W.R., Silicon Semiconductor Technology, 1965.

— = Sb-doped, theoretical values

"A" = As-doped, 1.7×10^{18}

"B" = P-doped, 1.1×10^{19}

Chapman, P.W., et al., J. Appl. Phys., vol. 34, 3291, 1963.

"1" = P-doped, 1.0×10^{17}

"2" = P-doped, 1.3×10^{18}

"3" = P-doped, 1.7×10^{19}

Pearson, G.L. and J.Bardeen, Phys. Rev., vol. 75, 865, 1949.

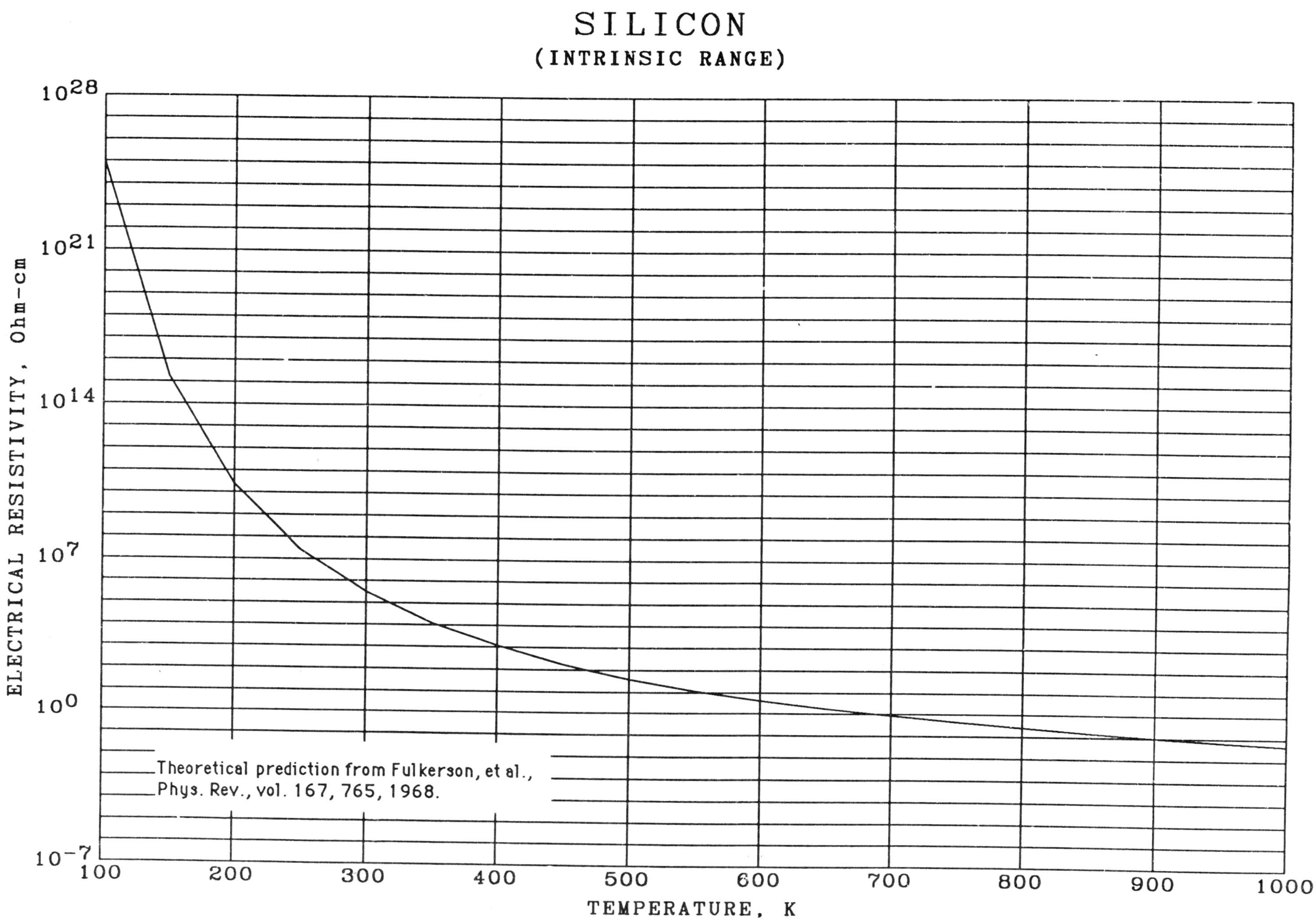
SILICON
(INTRINSIC RANGE)
ELECTRICAL RESISTIVITY, Ohm-cm
10^{28}
10^{21}
10^{14}
10^{7}
10^{0}
10^{-7}
TEMPERATURE, K
100
200
300
400
500
600
700
800
900
1000
Theoretical prediction from Fulkerson, et al.,
Phys. Rev., vol. 167, 765, 1968.

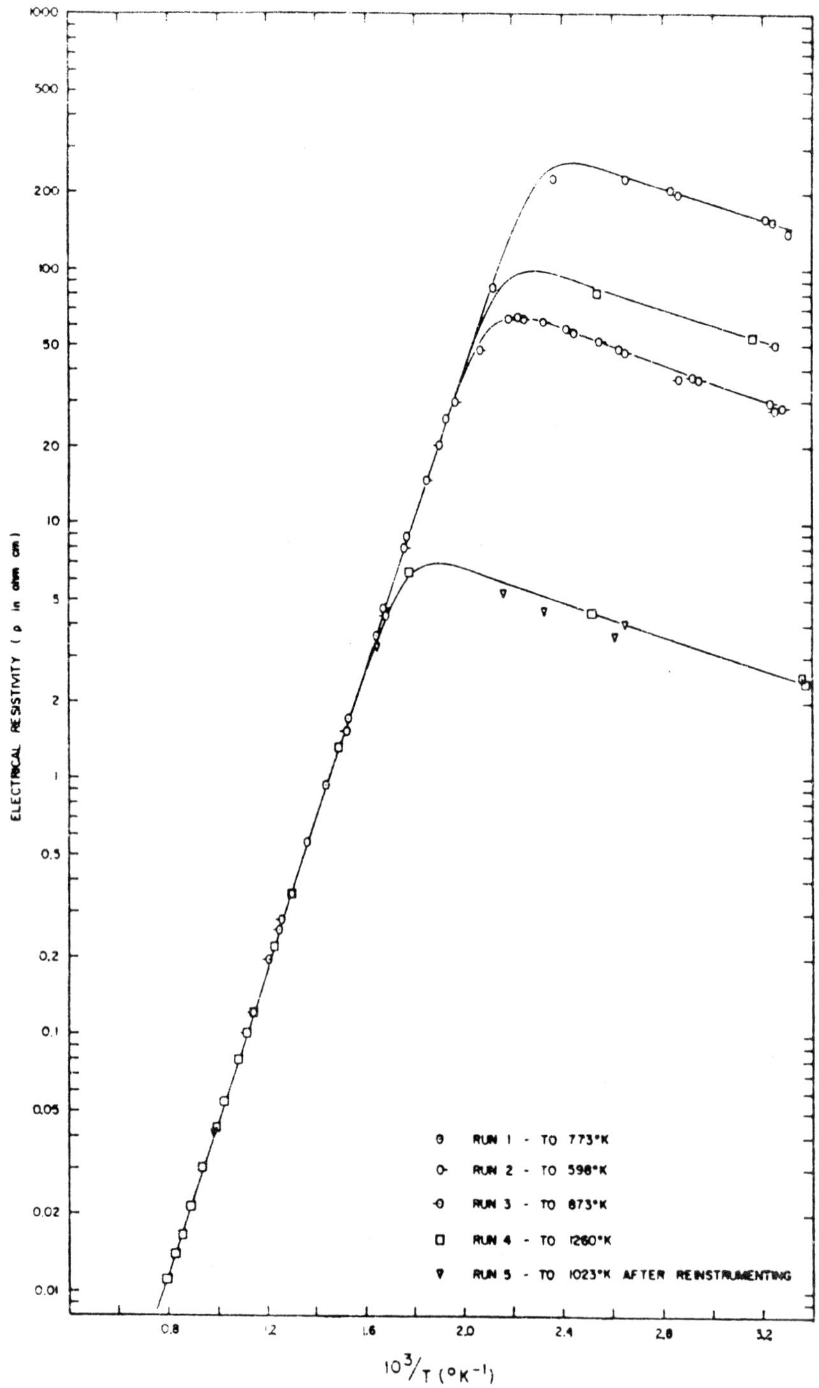

Reprinted from Fulkerson, W., et al., Phys. Rev., vol. 167, 765, 1968.

The Effect of Thermal Treatment on the Electrical Resistivity of n-type Silicon Single Crystals

SILICON

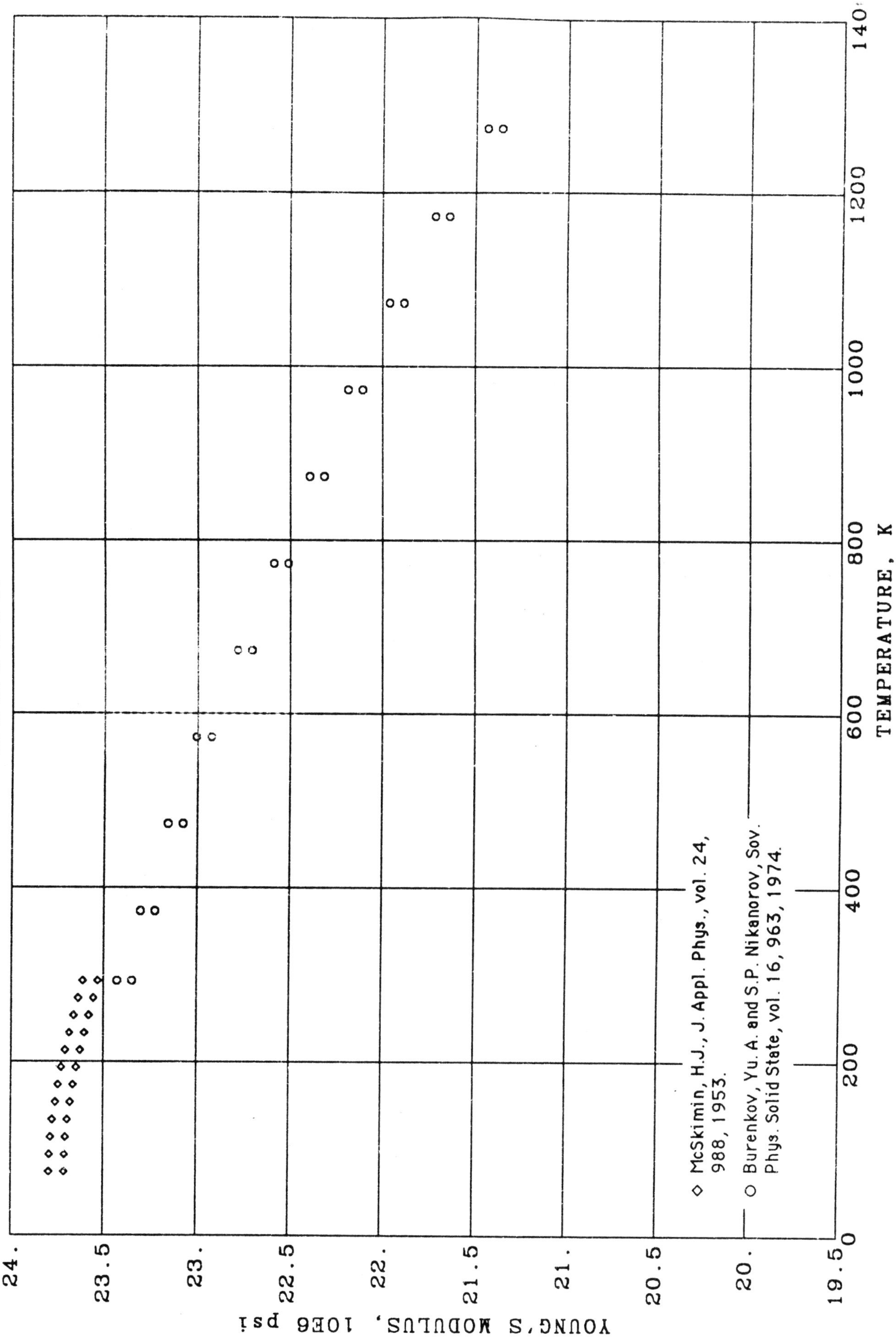

YOUNG'S MODULUS, 10E6 psi
24.
23.5
23.
22.5
22.
21.5
21.
20.5
20.
19.5
TEMPERATURE, K
0
200
400
600
800
1000
1200
1400
◇ McSkimin, H.J., J. Appl. Phys., vol. 24, 988, 1953.
○ Burenkov, Yu. A. and S.P. Nikanorov, Sov. Phys. Solid State, vol. 16, 963, 1974.

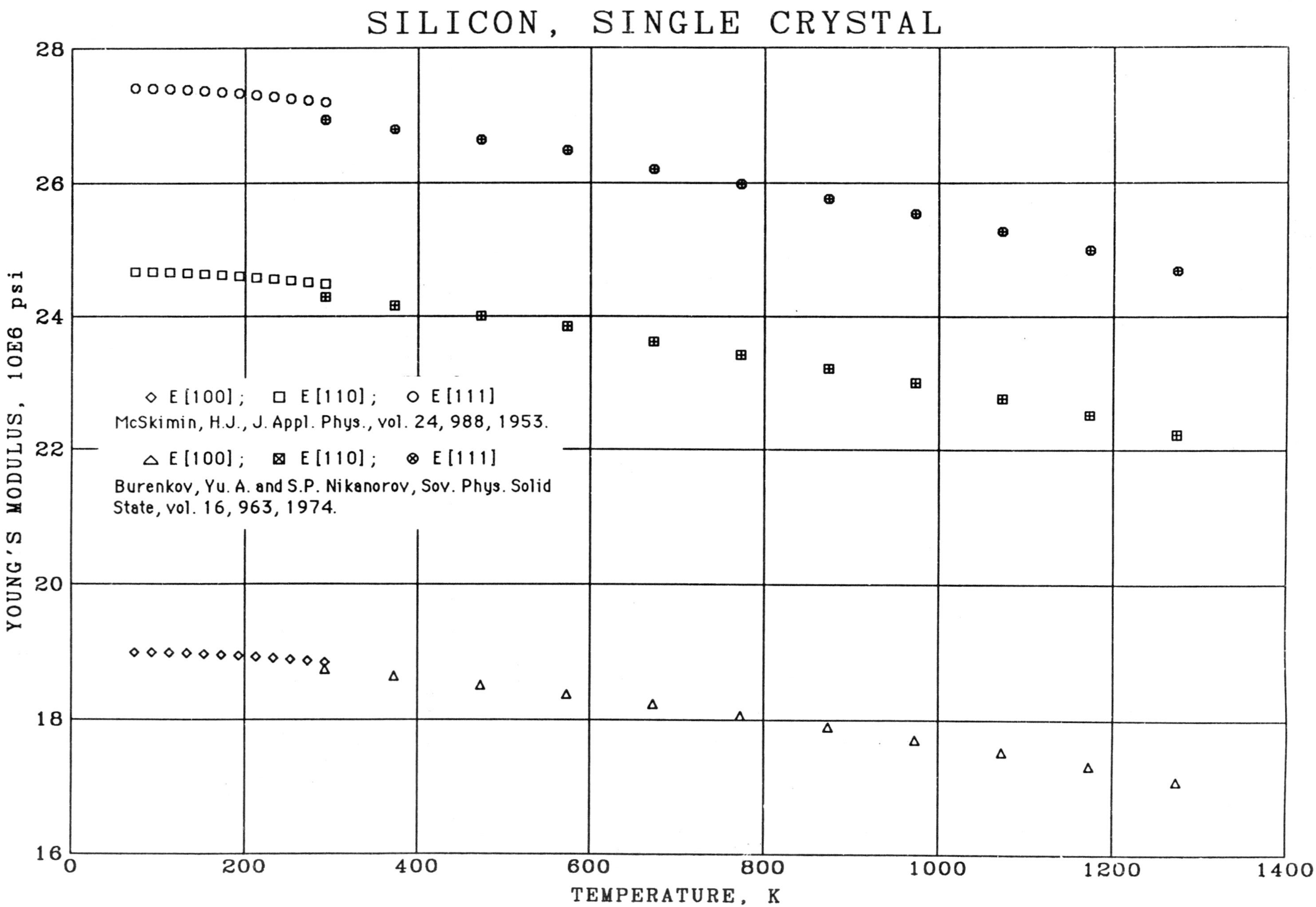

SILICON, SINGLE CRYSTAL
YOUNG'S MODULUS, 10E6 psi
TEMPERATURE, K
28
26
24
22
20
18
16
0
200
400
600
800
1000
1200
1400
◇ E[100]; □ E[110]; ○ E[111]
McSkimin, H.J., J. Appl. Phys., vol. 24, 988, 1953.
△ E[100]; ⊠ E[110]; ⊗ E[111]
Burenkov, Yu. A. and S.P. Nikanorov, Sov. Phys. Solid State, vol. 16, 963, 1974.

SILICON

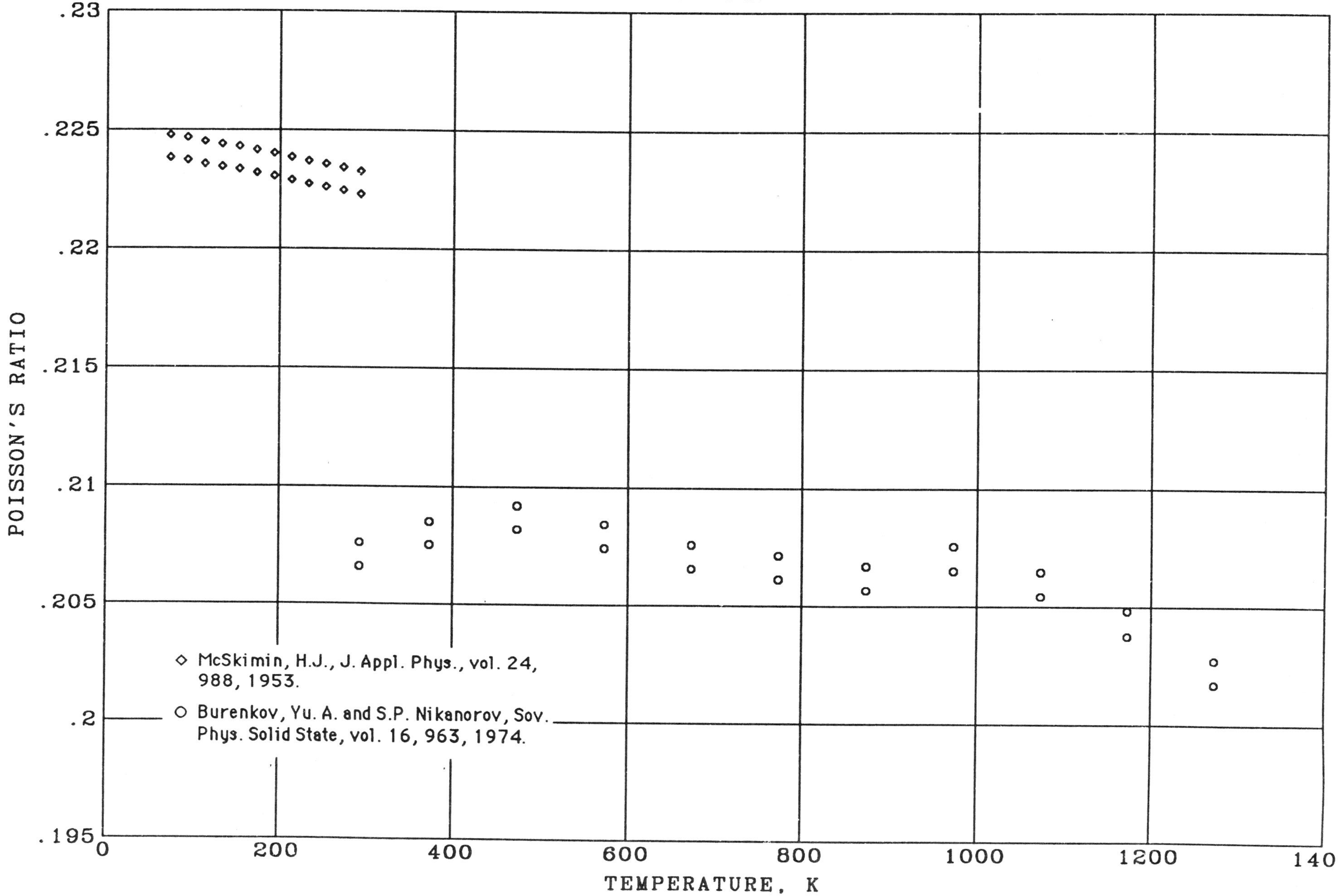

.23
.225
.22
.215
.21
.205
.2
.195
POISSON'S RATIO
0
200
400
600
800
1000
1200
1400
TEMPERATURE, K
◇ McSkimin, H.J., J. Appl. Phys., vol. 24, 988, 1953.
○ Burenkov, Yu. A. and S.P. Nikanorov, Sov. Phys. Solid State, vol. 16, 963, 1974.

SILICON

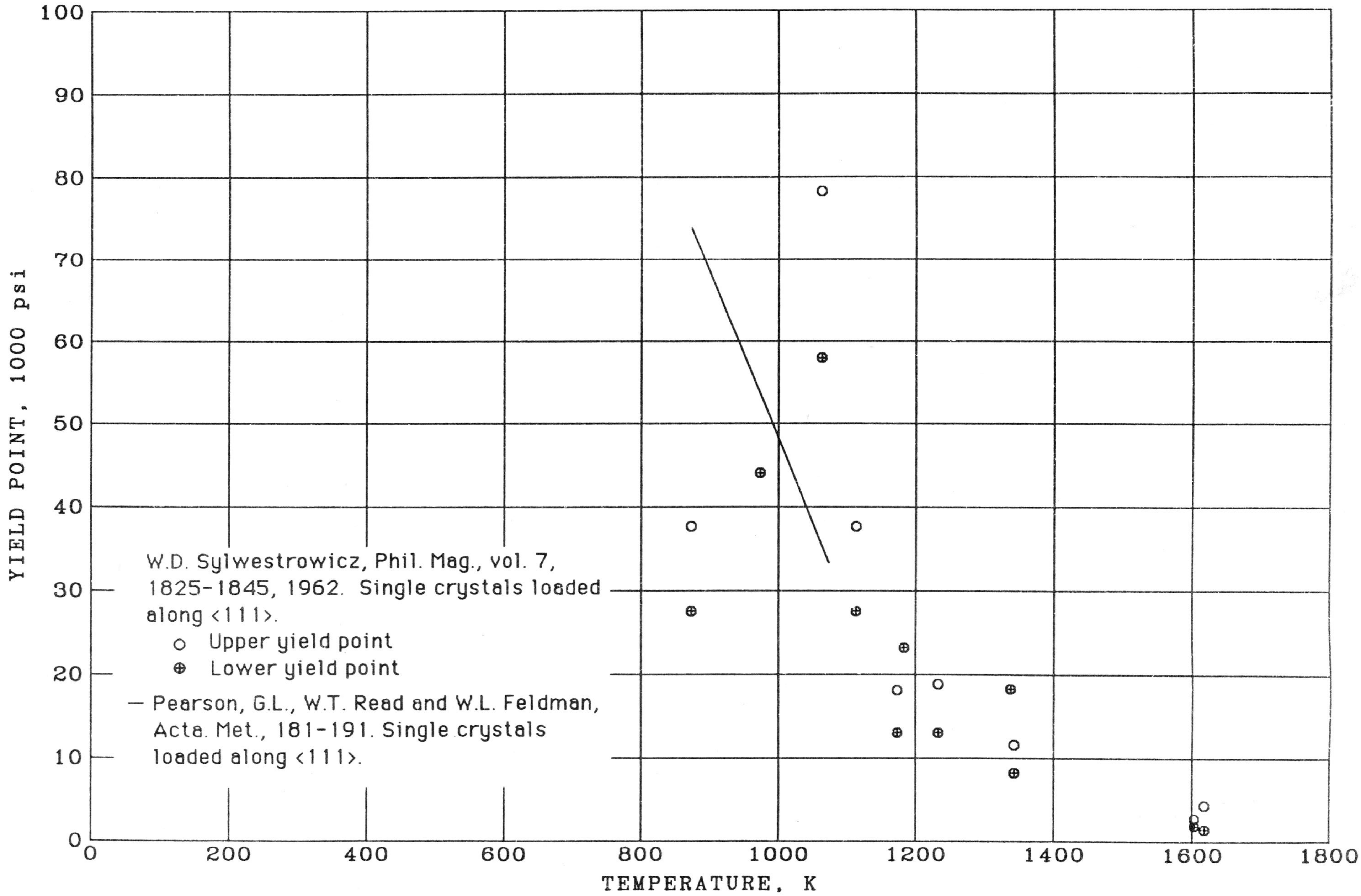

YIELD POINT, 1000 psi
100
90
80
70
60
50
40
30
20
10
0
0
200
400
600
800
1000
1200
1400
1600
1800
TEMPERATURE, K
W.D. Sylwestrowicz, Phil. Mag., vol. 7, 1825-1845, 1962. Single crystals loaded along <111>.
○ Upper yield point
⊕ Lower yield point
— Pearson, G.L., W.T. Read and W.L. Feldman, Acta. Met., 181-191. Single crystals loaded along <111>.

SILICON, SINGLE CRYSTAL

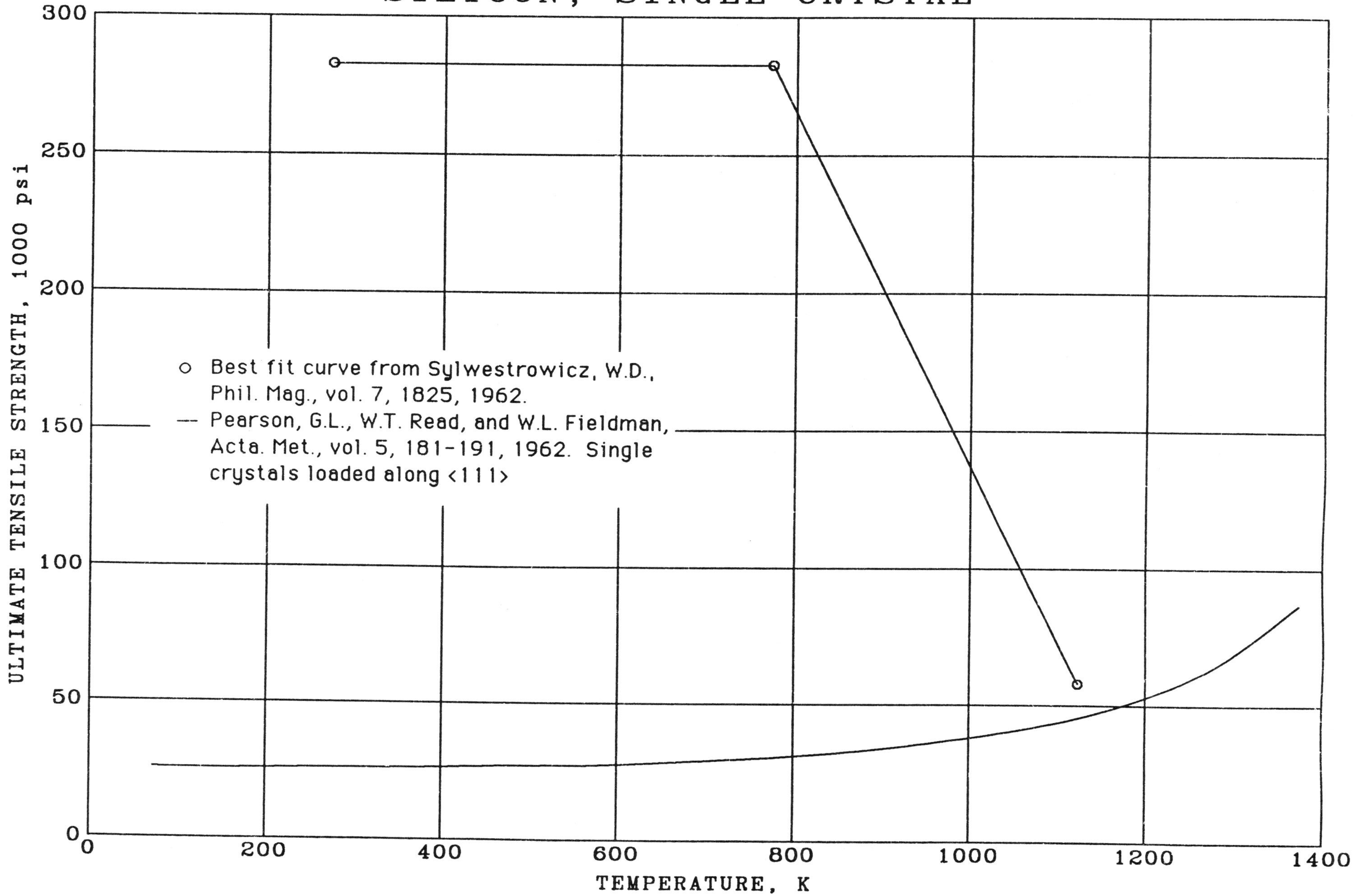

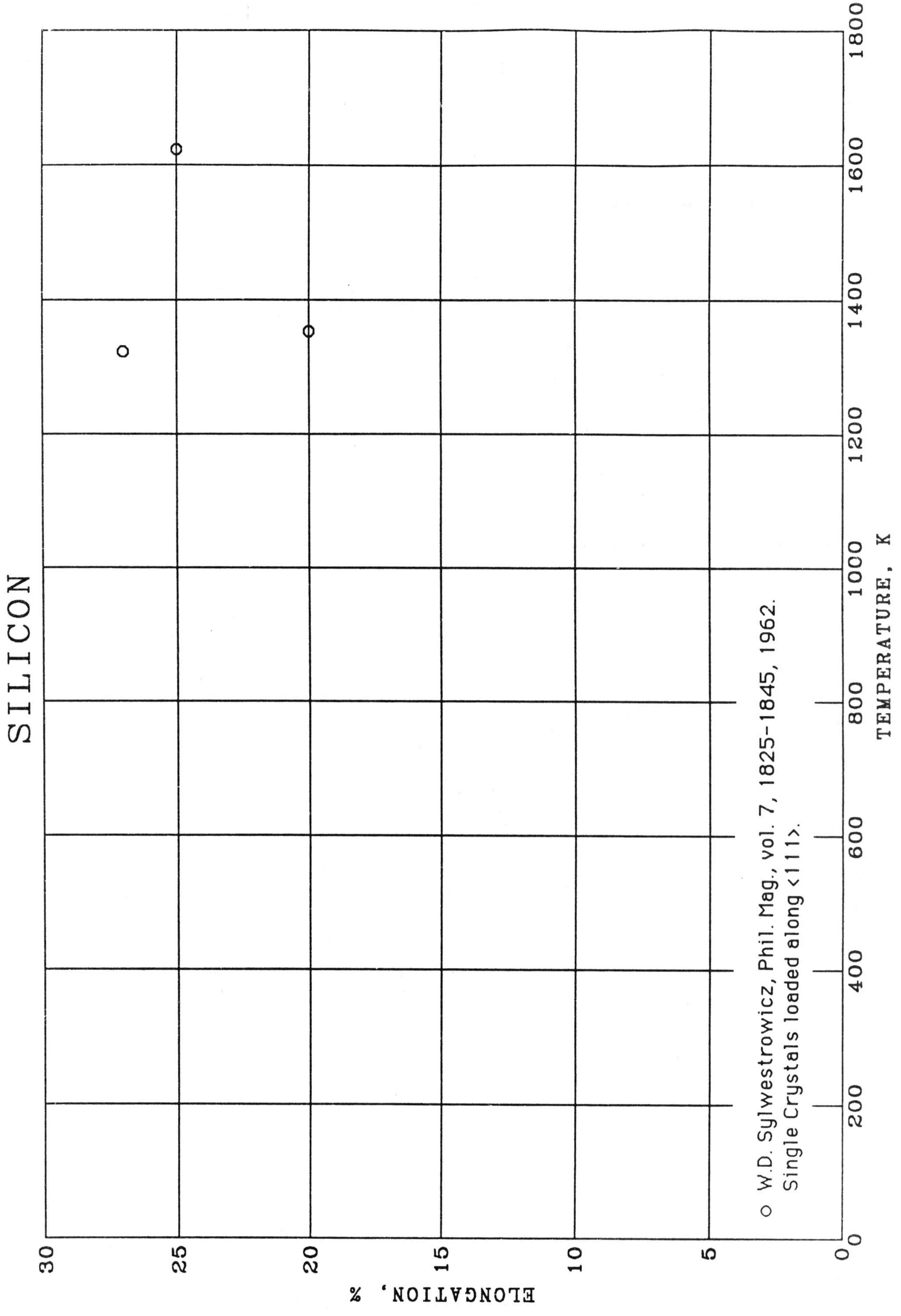
SILICON
○ W.D. Sylwestrowicz, Phil. Mag., vol. 7, 1825-1845, 1962.
Single Crystals loaded along <111>.
TEMPERATURE, K
ELONGATION, %
0
200
400
600
800
1000
1200
1400
1600
1800
0
5
10
15
20
25
30

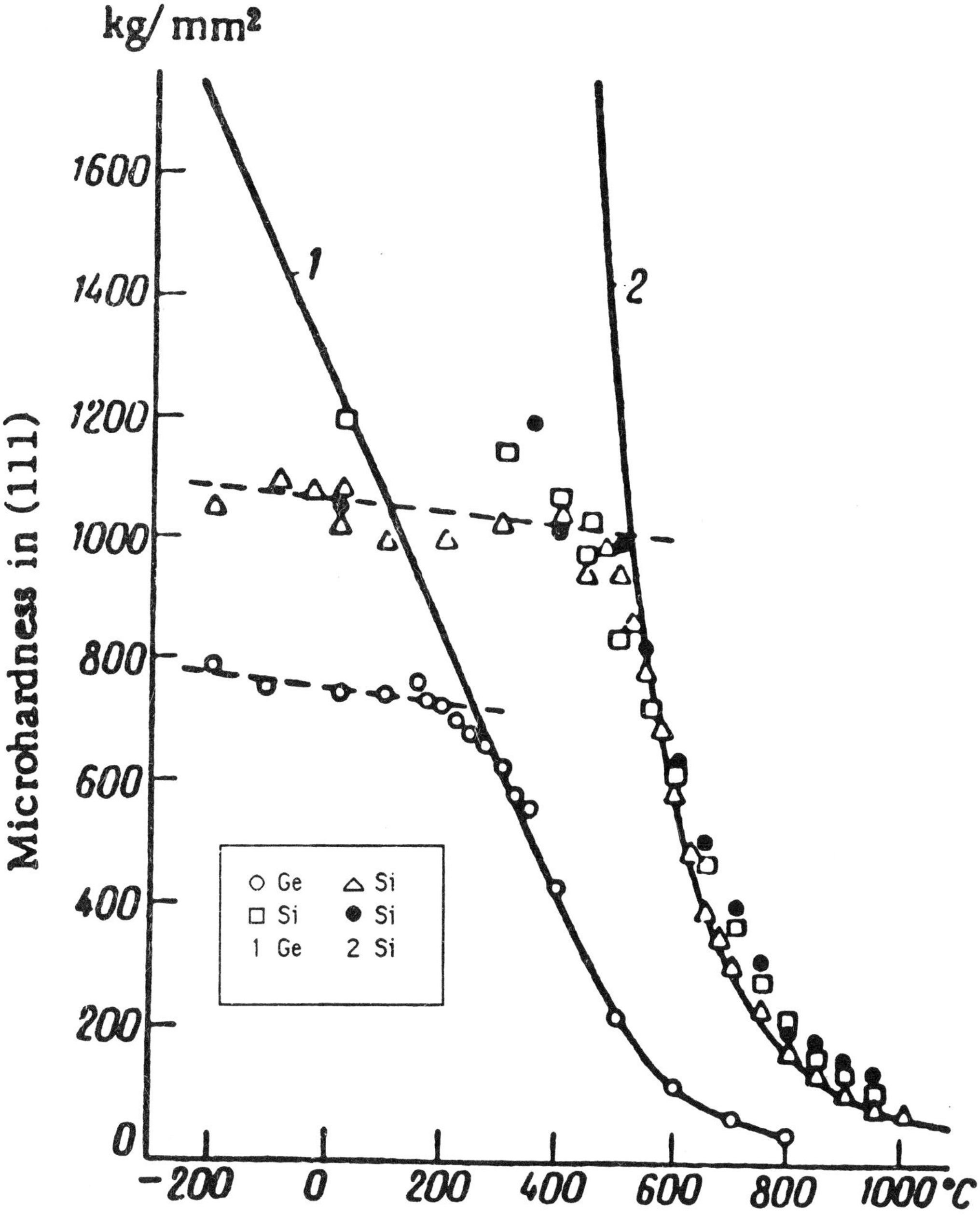

Microhardness of Si and Ge

Reprinted from Trefilov, V.I., and Yu V. Mil'man, Sov. Phys. Doklady, vol. 8, 1240, 1964.

SILVER

PROPERTY	@ 300K	UNITS	REFERENCES
Density	10.5	g/cm^3	Ag1 Ag2 Ag3
Specific Heat	0.0560	cal/g-K	Ag4 Ag5 Ag6
Thermal Conductivity	1.02	cal/cm-s-K	Ag7 Ag8 Ag9
Thermal Exp Coef	19.0×10^{-6}	1/K	Ag10 Ag11 Ag12
Melting Point	1235.08	K	Ag13
Resistivity 1	1.628×10^{-6}	Ohm-cm	Ag14 Ag15 Ag16
TCR (218K to 398K) 1	3730	ppm/K	Ag14 Ag15 Ag16
Young's Modulus	1.18×10^{7}	psi	Ag17 Ag18 Ag19
Poisson's Ratio	0.369	-----	Ag17 Ag18 Ag19
Yield Strength	1000	psi	Ag20 Ag21 Ag22
Ult Tensile Strength	21000	psi	Ag20 Ag21 Ag22
Elongation at Break	50	%	Ag20 Ag21 Ag22
Hardness	25	Brinell	Ag23 Ag24 Ag25

1 The resistivity and TCR of epitaxial grown silver films on rock salt >1700 angstroms thick are nearly equal to values for the bulk material above 200K.[Ag27]

Mechanical properties listed are for fully annealed, bulk material. 50% reduction producing 0.032 inch sheet will increase yield strength to 44,300 psi, increase tensile strength to 54,300 psi and reduce the elongation (2 inch gauge length) to 2.4% [Ag22]. Polycrystalline silver films have demonstrated tensile strengths greater than 90 ksi with elongations of 0.7%.[Ag26] Heating cold worked material into the temperature range of 400 to 600K will cause recovery and recrystallization erasing the effects of cold work.

SILVER

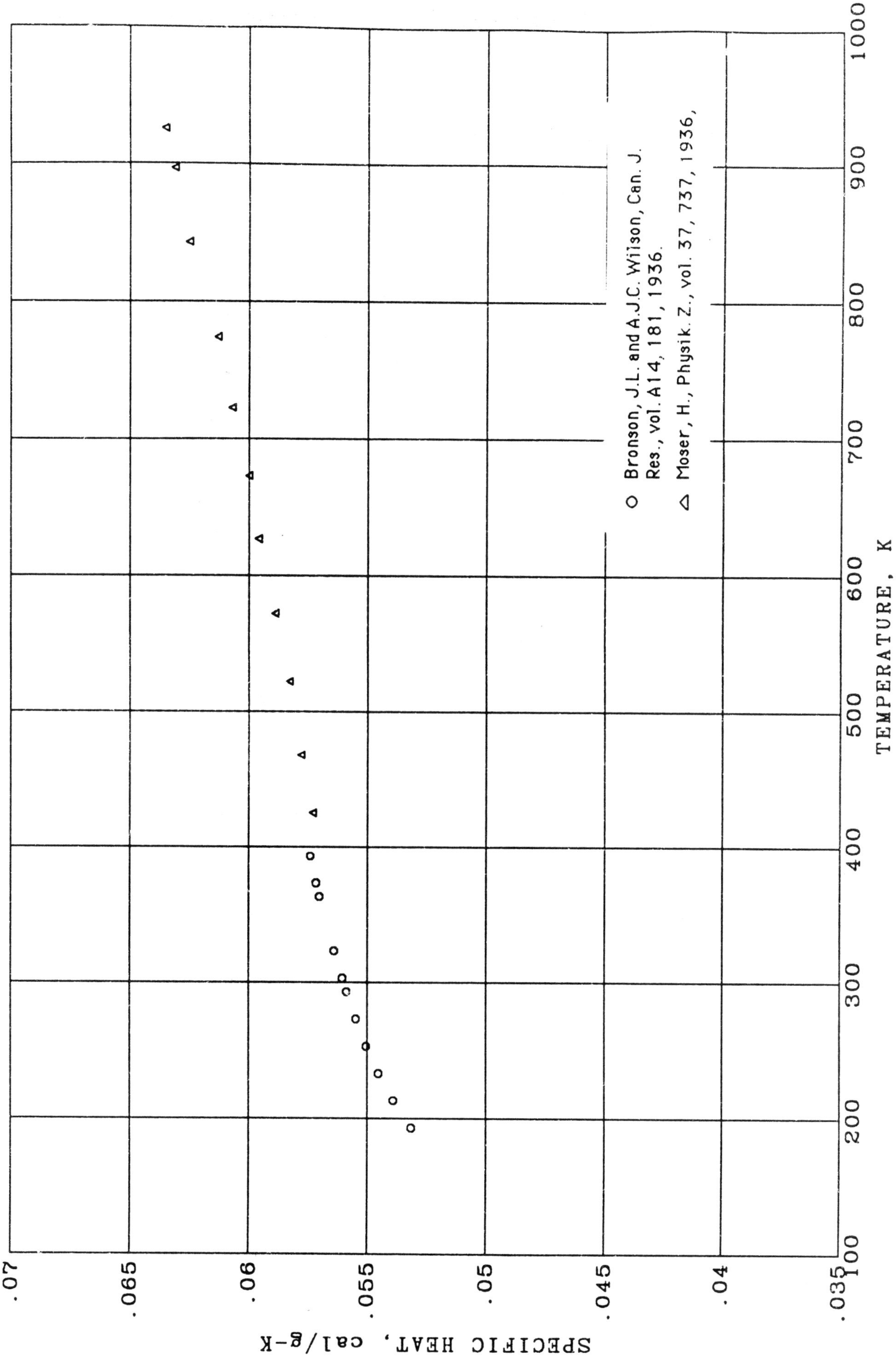

SPECIFIC HEAT, cal/g-K
.035
.04
.045
.05
.055
.06
.065
.07
TEMPERATURE, K
100
200
300
400
500
600
700
800
900
1000
○ Bronson, J.L. and A.J.C. Wilson, Can. J. Res., vol. A14, 181, 1936.
△ Moser, H., Physik. Z., vol. 37, 737, 1936,

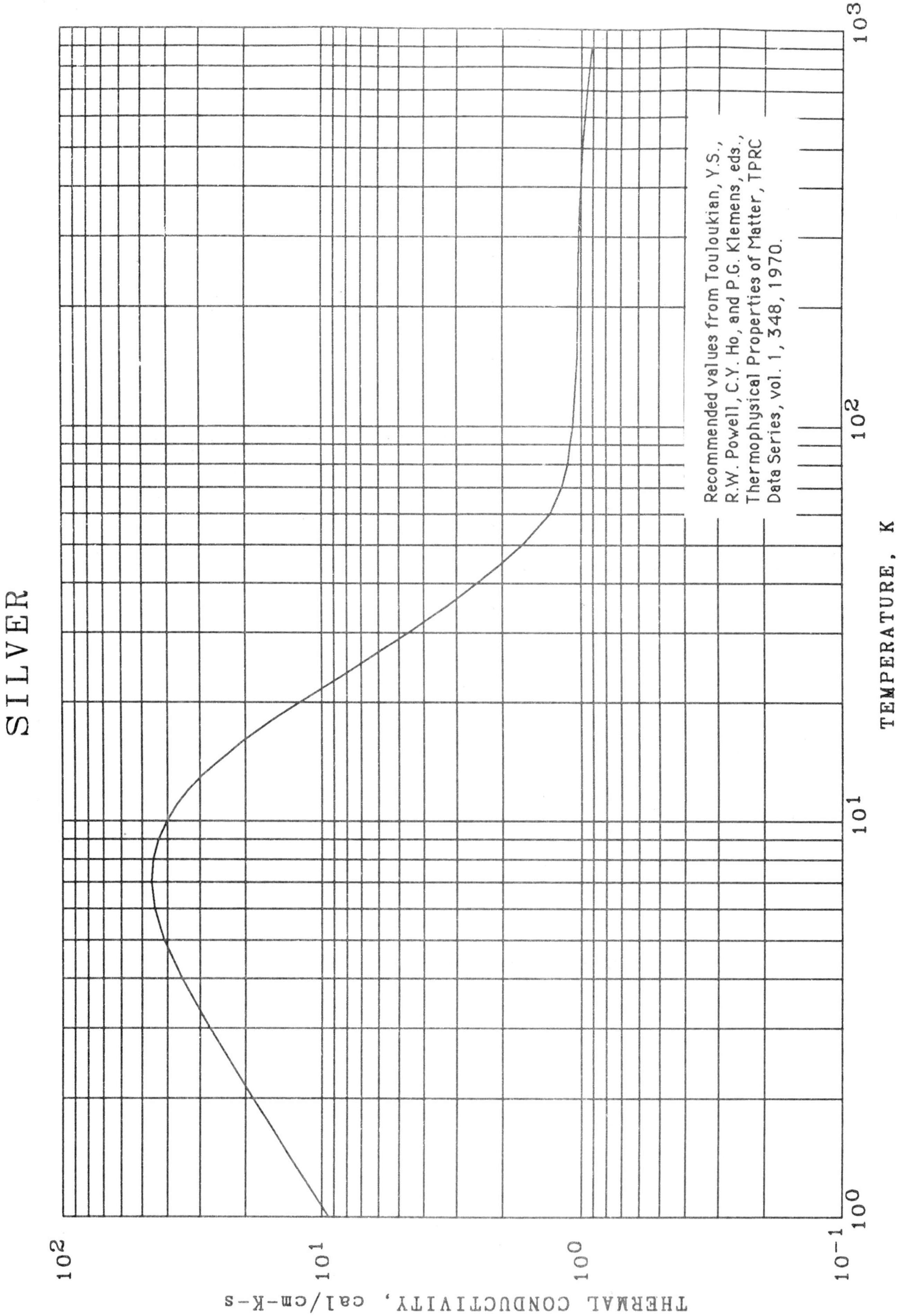
SILVER
THERMAL CONDUCTIVITY, cal/cm-K-s
TEMPERATURE, K
10^{-1}
10^{0}
10^{1}
10^{2}
10^{0}
10^{1}
10^{2}
10^{3}
Recommended values from Touloukian, Y.S., R.W. Powell, C.Y. Ho, and P.G. Klemens, eds., Thermophysical Properties of Matter, TPRC Data Series, vol. 1, 348, 1970.

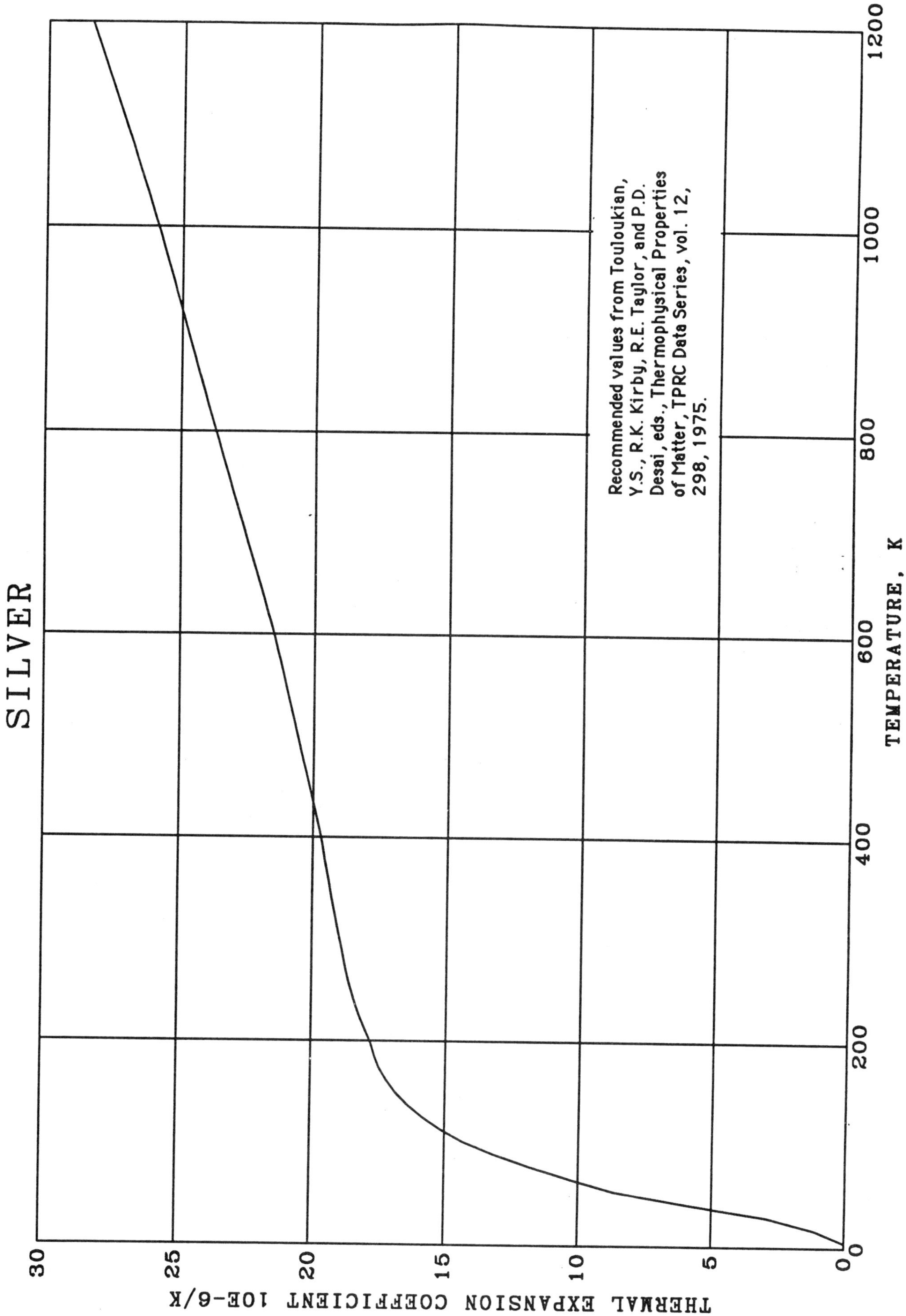
SILVER
THERMAL EXPANSION COEFFICIENT 10E-6/K
TEMPERATURE, K
0
5
10
15
20
25
30
0
200
400
600
800
1000
1200
Recommended values from Touloukian, Y.S., R.K. Kirby, R.E. Taylor, and P.D. Desai, eds., Thermophysical Properties of Matter, TPRC Data Series, vol. 12, 298, 1975.

SILVER

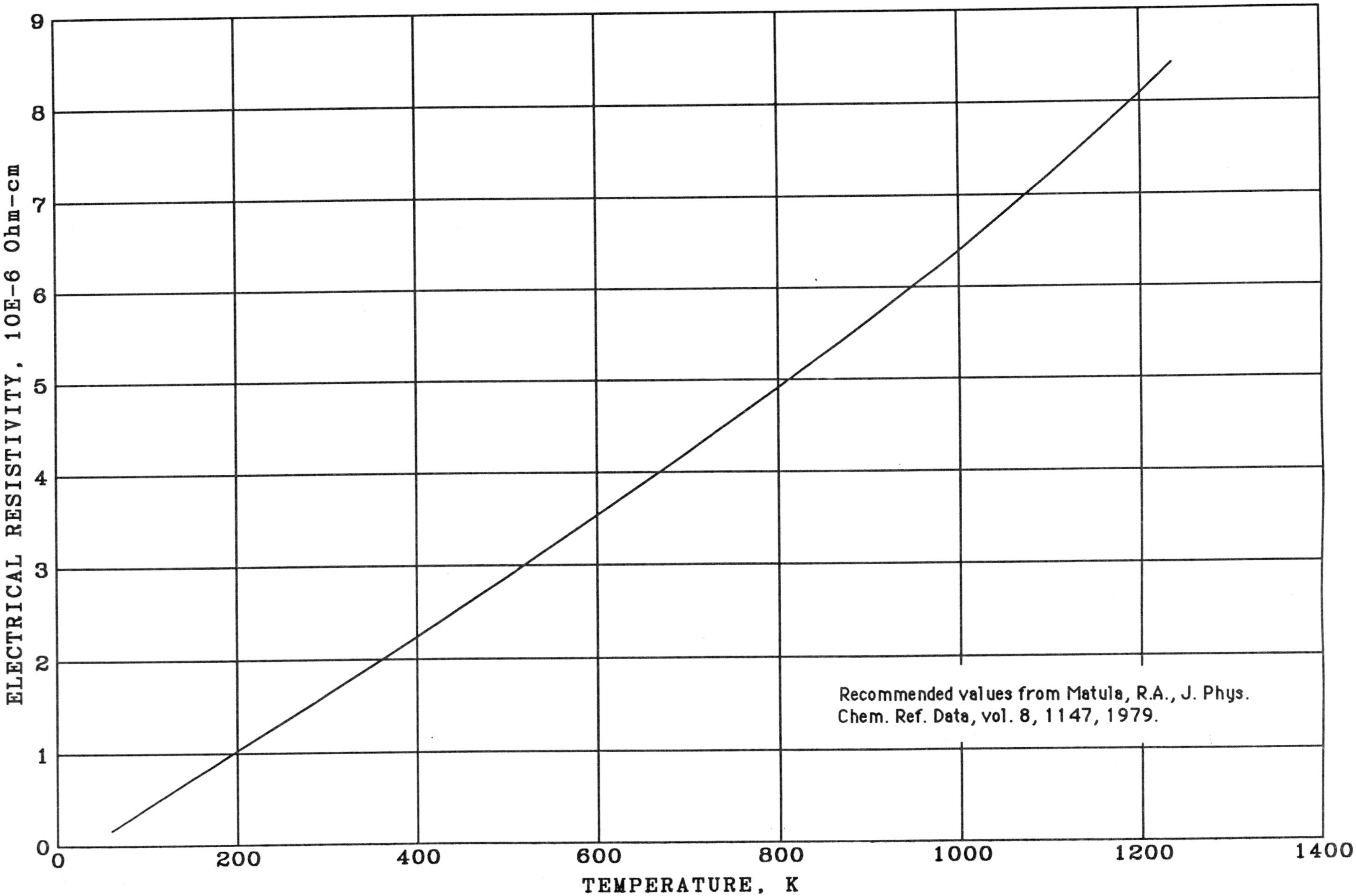

ELECTRICAL RESISTIVITY, 10E-6 Ohm-cm
0
1
2
3
4
5
6
7
8
9
0
200
400
600
800
1000
1200
1400
TEMPERATURE, K
Recommended values from Matula, R.A., J. Phys. Chem. Ref. Data, vol. 8, 1147, 1979.

SILVER

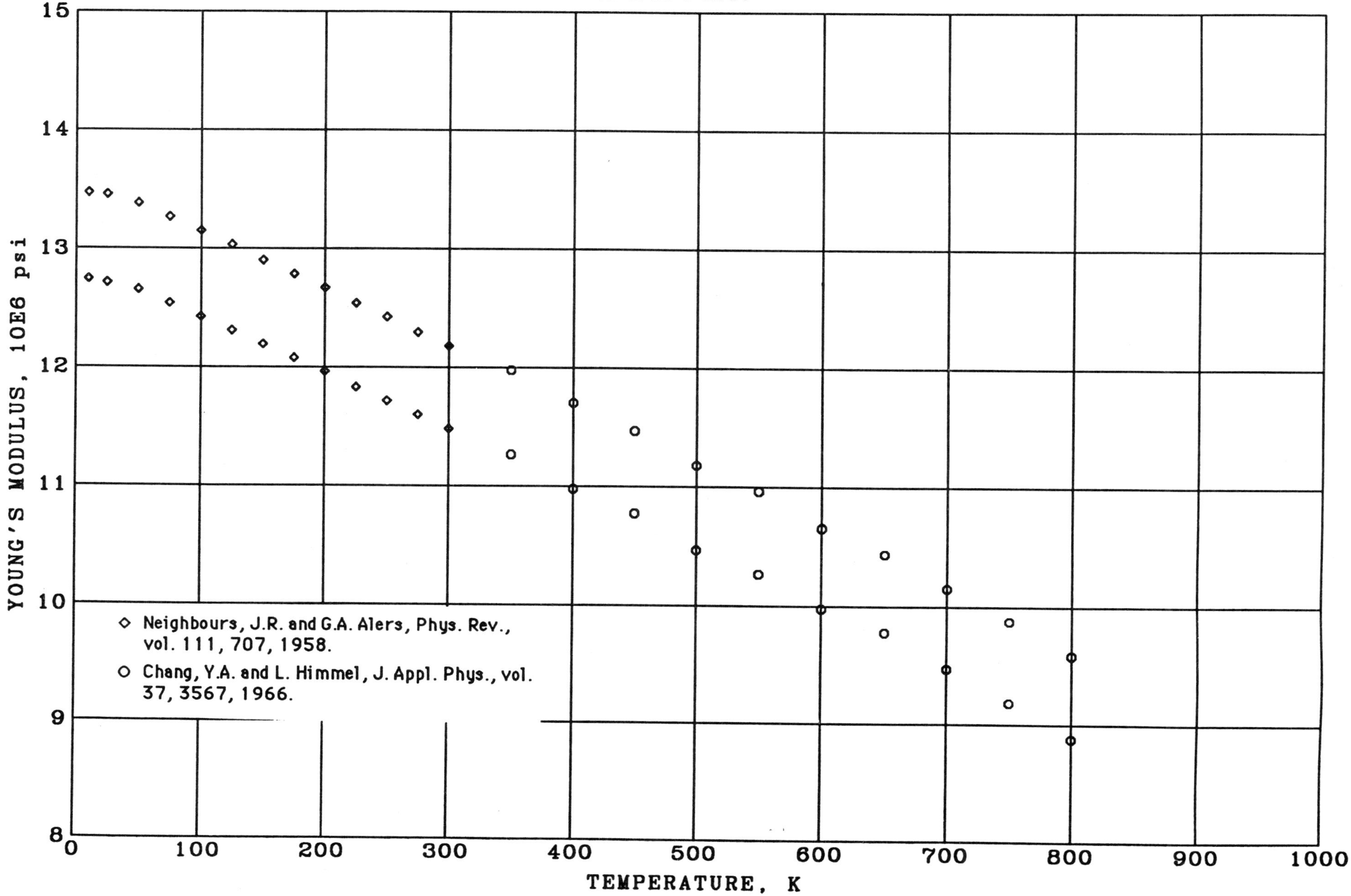

15
14
13
12
11
10
9
8
0
100
200
300
400
500
600
700
800
900
1000
YOUNG'S MODULUS, 10E6 psi
TEMPERATURE, K
◇ Neighbours, J.R. and G.A. Alers, Phys. Rev., vol. 111, 707, 1958.
○ Chang, Y.A. and L. Himmel, J. Appl. Phys., vol. 37, 3567, 1966.

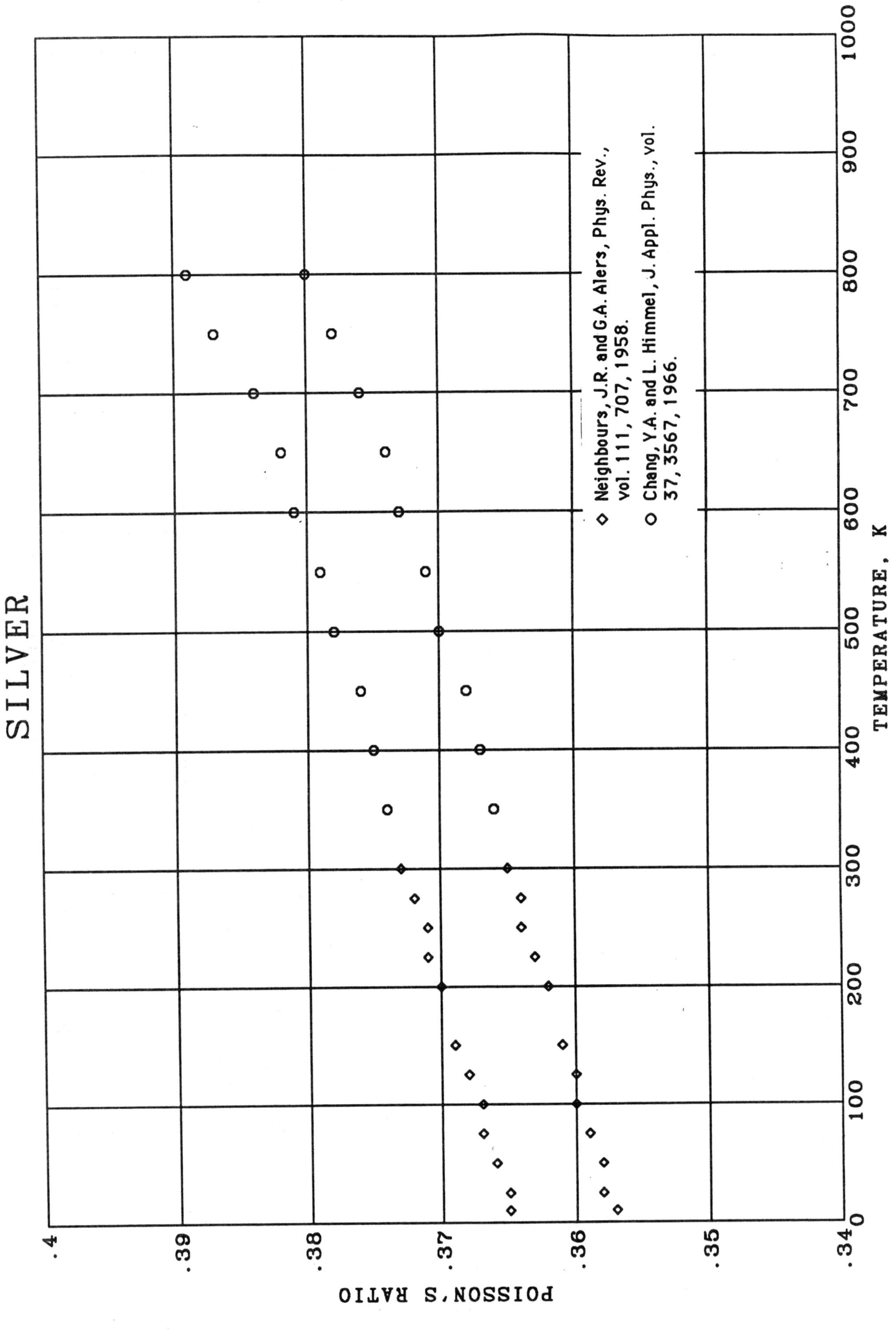

SILVER
POISSON'S RATIO
TEMPERATURE, K
◇ Neighbours, J.R. and G.A. Alers, Phys. Rev., vol. 111, 707, 1958.
○ Chang, Y.A. and L. Himmel, J. Appl. Phys., vol. 37, 3567, 1966.
.34
.35
.36
.37
.38
.39
.4
0
100
200
300
400
500
600
700
800
900
1000

SILVER

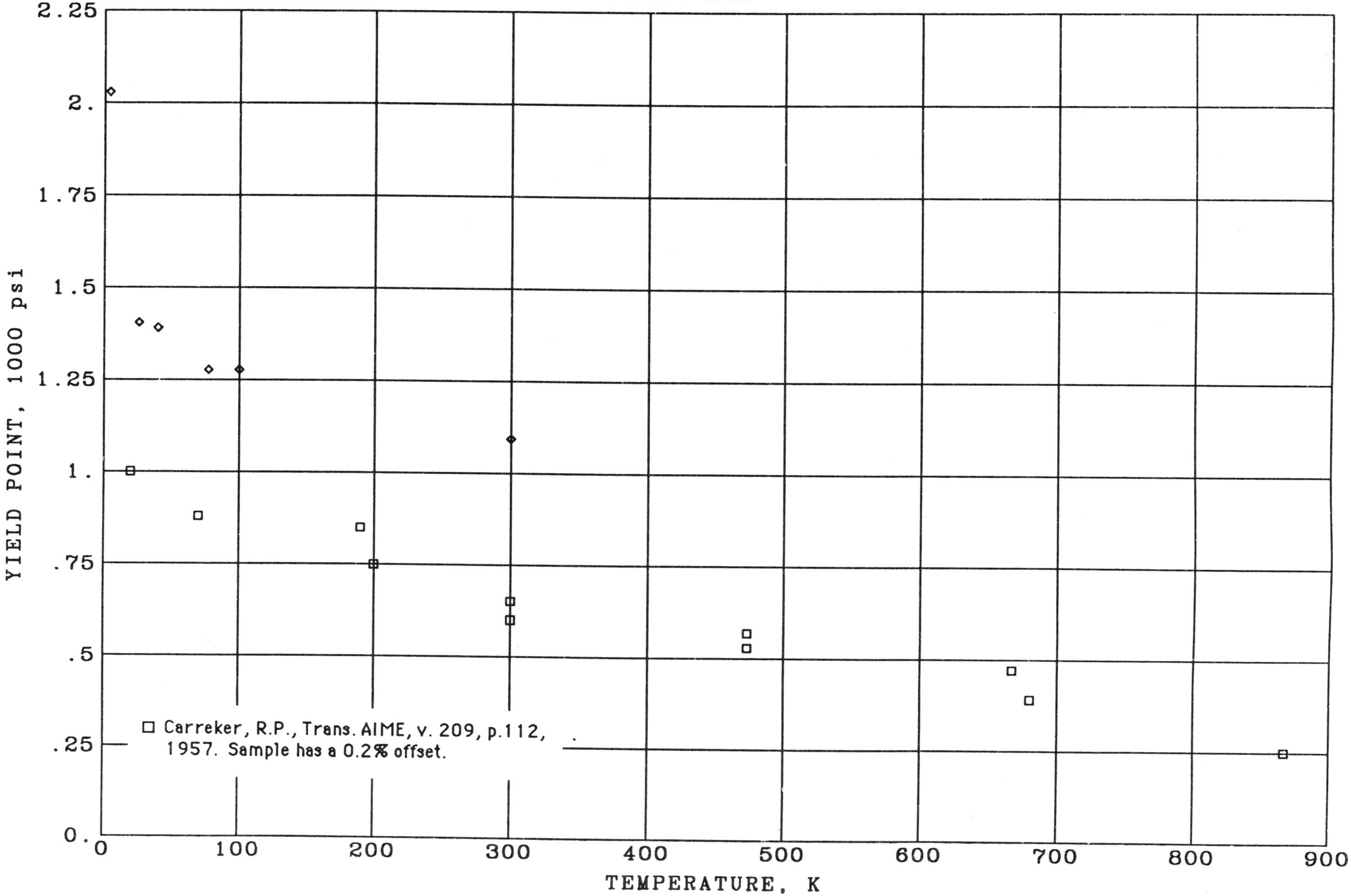
2.25
2.
1.75
1.5
1.25
1.
.75
.5
.25
0.
YIELD POINT, 1000 psi
0
100
200
300
400
500
600
700
800
900
TEMPERATURE, K
Carreker, R.P., Trans. AIME, v. 209, p.112, 1957. Sample has a 0.2% offset.

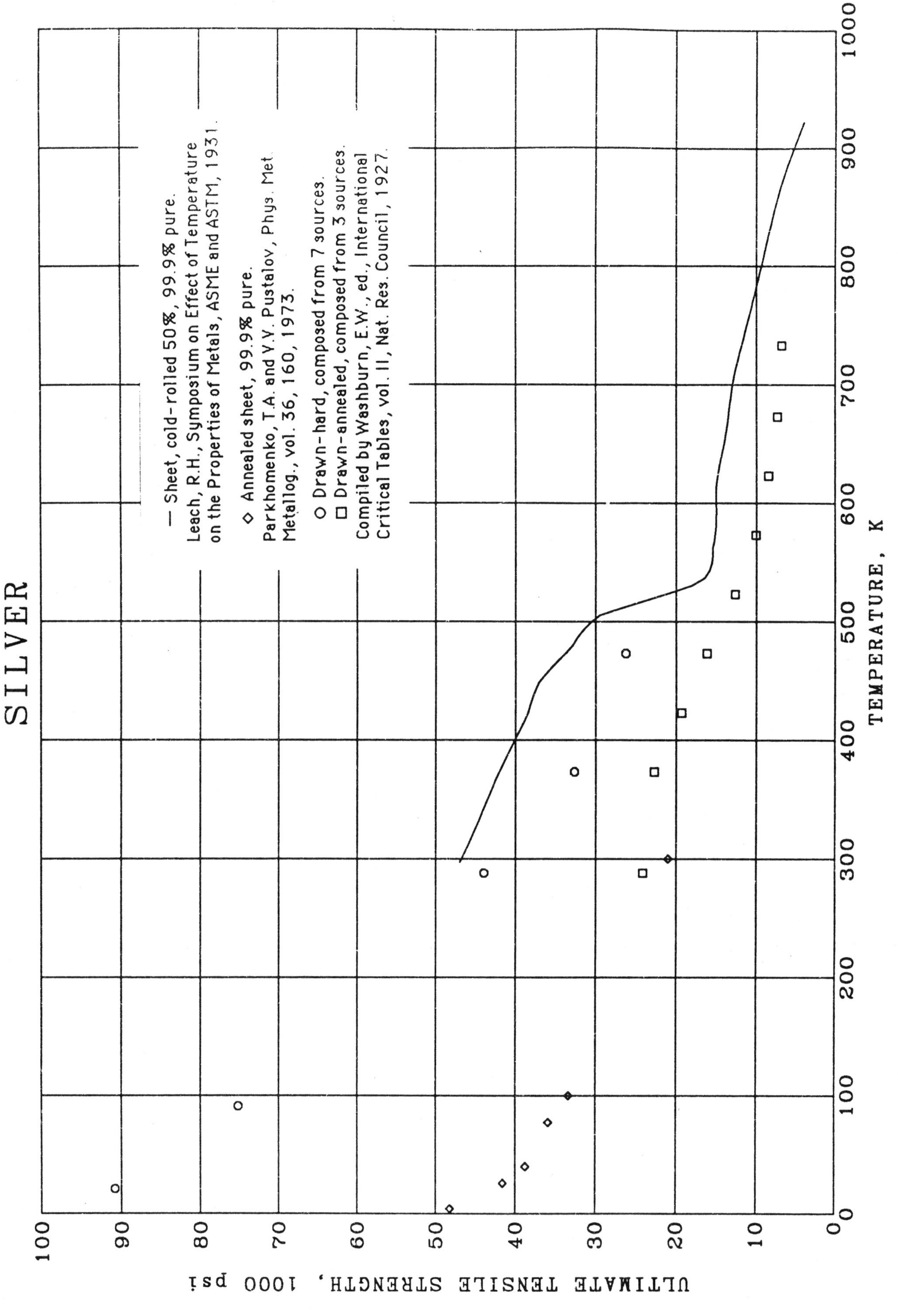
SILVER
— Sheet, cold-rolled 50%, 99.9% pure.
Leach, R.H., Symposium on Effect of Temperature on the Properties of Metals, ASME and ASTM, 1931.
◇ Annealed sheet, 99.9% pure.
Parkhomenko, T.A. and V.Y. Pustalov, Phys. Met. Metallog., vol. 36, 160, 1973.
○ Drawn-hard, composed from 7 sources.
□ Drawn-annealed, composed from 3 sources.
Compiled by Washburn, E.W., ed., International Critical Tables, vol. II, Nat. Res. Council, 1927.
ULTIMATE TENSILE STRENGTH, 1000 psi
TEMPERATURE, K
0
10
20
30
40
50
60
70
80
90
100
0
100
200
300
400
500
600
700
800
900
1000

SILVER

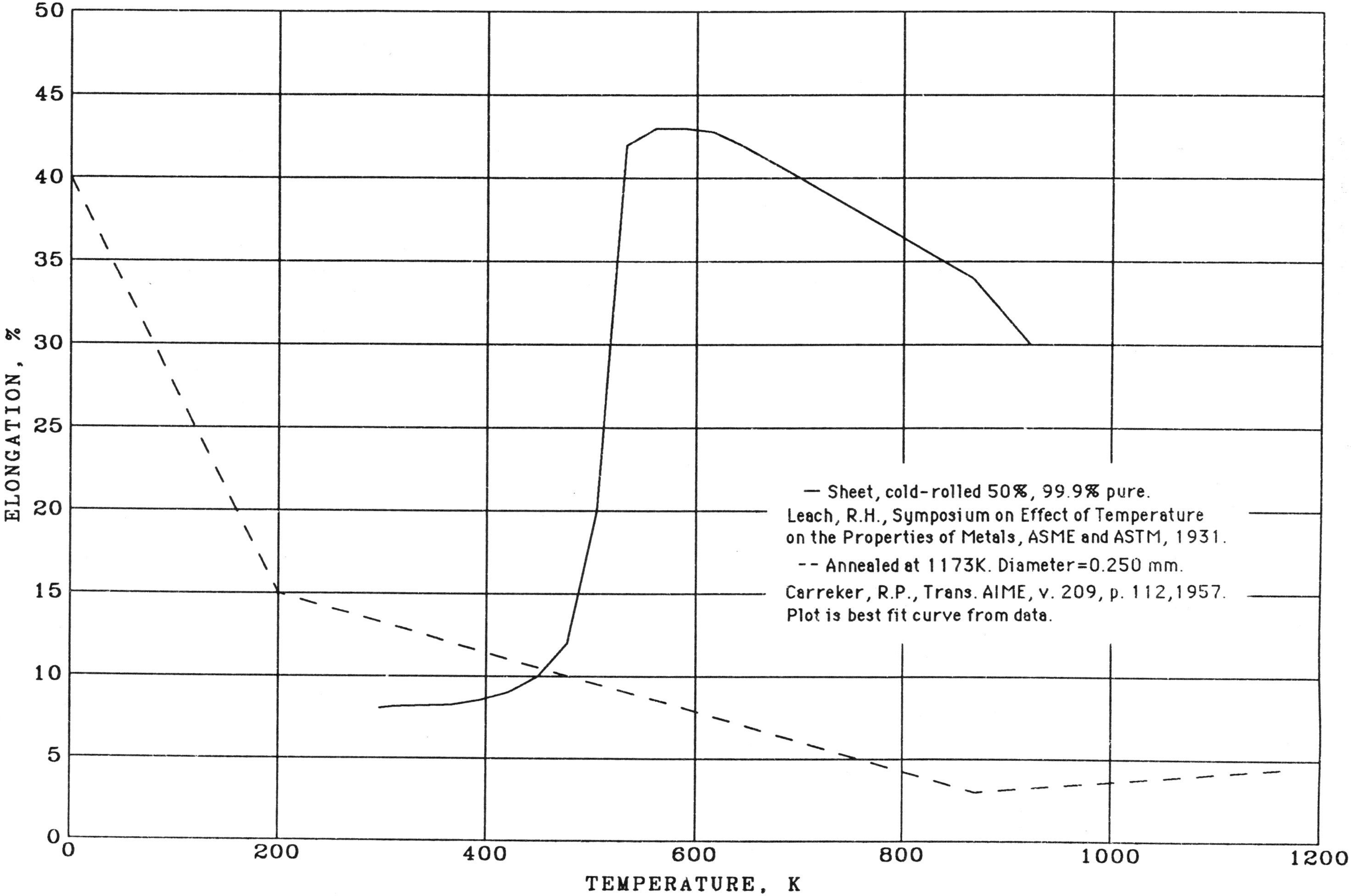
ELONGATION, %
TEMPERATURE, K
— Sheet, cold-rolled 50%, 99.9% pure.
Leach, R.H., Symposium on Effect of Temperature on the Properties of Metals, ASME and ASTM, 1931.
-- Annealed at 1173K. Diameter=0.250 mm.
Carreker, R.P., Trans. AIME, v. 209, p. 112,1957.
Plot is best fit curve from data.

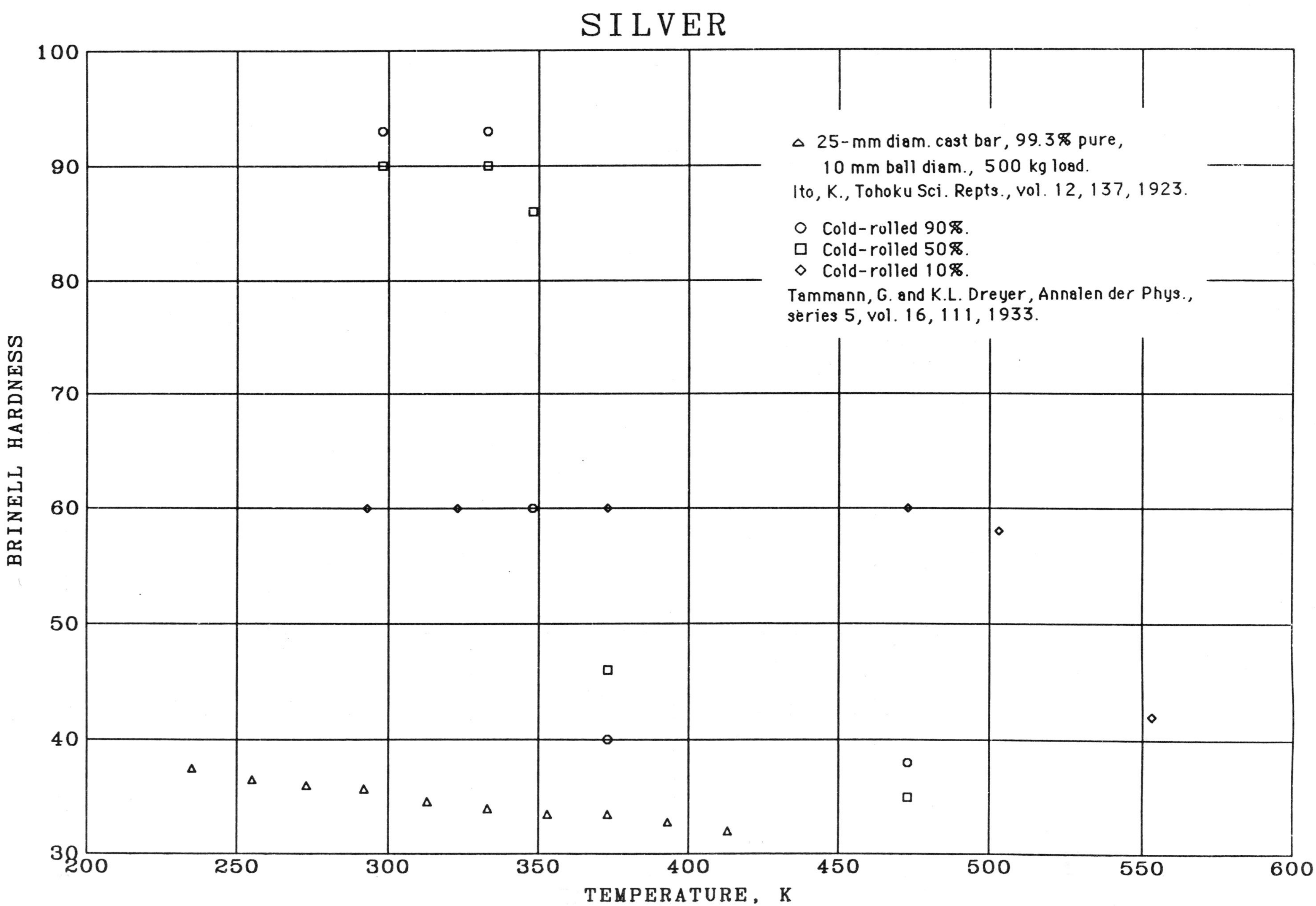
SILVER
100
90
80
70
60
50
40
30
200
250
300
350
400
450
500
550
600
BRINELL HARDNESS
TEMPERATURE, K
△ 25-mm diam. cast bar, 99.3% pure,
10 mm ball diam., 500 kg load.
Ito, K., Tohoku Sci. Repts., vol. 12, 137, 1923.
○ Cold-rolled 90%.
□ Cold-rolled 50%.
◇ Cold-rolled 10%.
Tammann, G. and K.L. Dreyer, Annalen der Phys.,
series 5, vol. 16, 111, 1933.

TANTALUM

PROPERTY	@ 300K	UNITS	REFERENCES
Density	16.6	g/cm^3	Ta1 Ta2 Ta3
Specific Heat	0.03306	cal/g-K	Ta4 Ta5 Ta6
Thermal Conductivity	0.137	cal/cm-s-K	Ta7 Ta8 Ta9
Thermal Exp Coef	$6.3x10^{-6}$	1/K	Ta10 Ta11 Ta12
Melting Point	3269	K	Ta13
Resistivity	$13.48x10^{-6}$	Ohm-cm	Ta14 Ta15 Ta16
TCR (218K to 398K)	3540	ppm/K	Ta14 Ta15 Ta16
Young's Modulus	$2.656x10^{7}$	psi	Ta17 Ta18 Ta19
Poisson's Ratio	0.341	-----	Ta17 Ta18 Ta19
Yield Strength	30 - 60	ksi	Ta20 Ta21 Ta22
Ult Tensile Strength	30 - 70	ksi	Ta20 Ta21 Ta22
Elongation at Break	30 - 50	%	Ta20 Ta21 Ta22
Hardness	80	Vickers	Ta23 Ta24 Ta25

Mechanical properties listed are for fully annealed, bulk material. Cold work can increase the yield strength to 100 ksi, the tensile strength to 180 ksi and reduce the elongation to 1%. Heating cold worked material into the temperature range of 1000 to 1400C causes recovery and recrystallization erasing the effects of cold work. For a complete review of mechanical properties, see Schmidt and Ogden, DMIC Reports 133 & 189, 1963, Battelle Metals Information Center, Columbus, Ohio.

TANTALUM

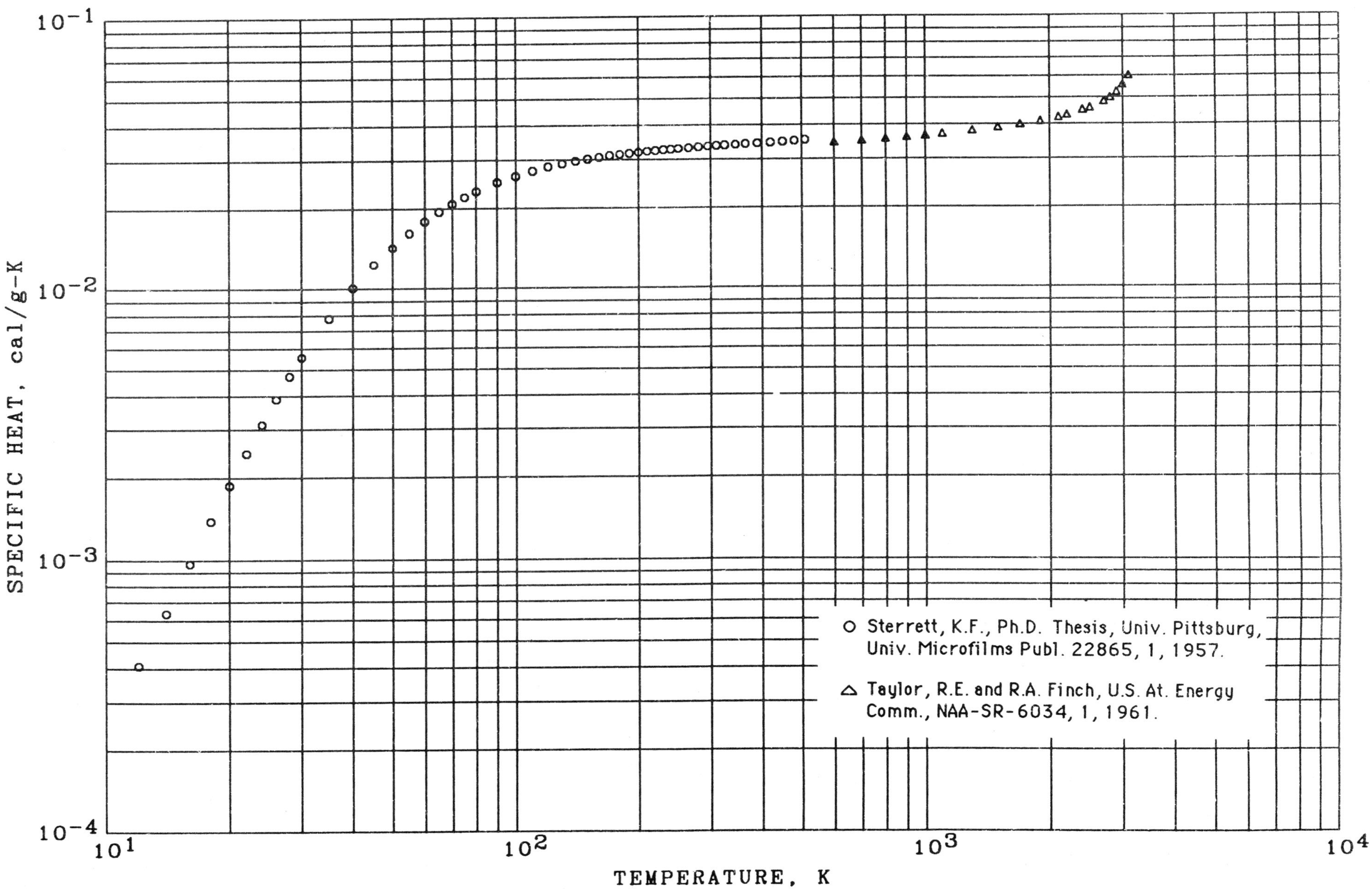
SPECIFIC HEAT, cal/g-K
10^-1
10^-2
10^-3
10^-4
10^1
10^2
10^3
10^4
TEMPERATURE, K
○ Sterrett, K.F., Ph.D. Thesis, Univ. Pittsburg, Univ. Microfilms Publ. 22865, 1, 1957.
△ Taylor, R.E. and R.A. Finch, U.S. At. Energy Comm., NAA-SR-6034, 1, 1961.

TANTALUM

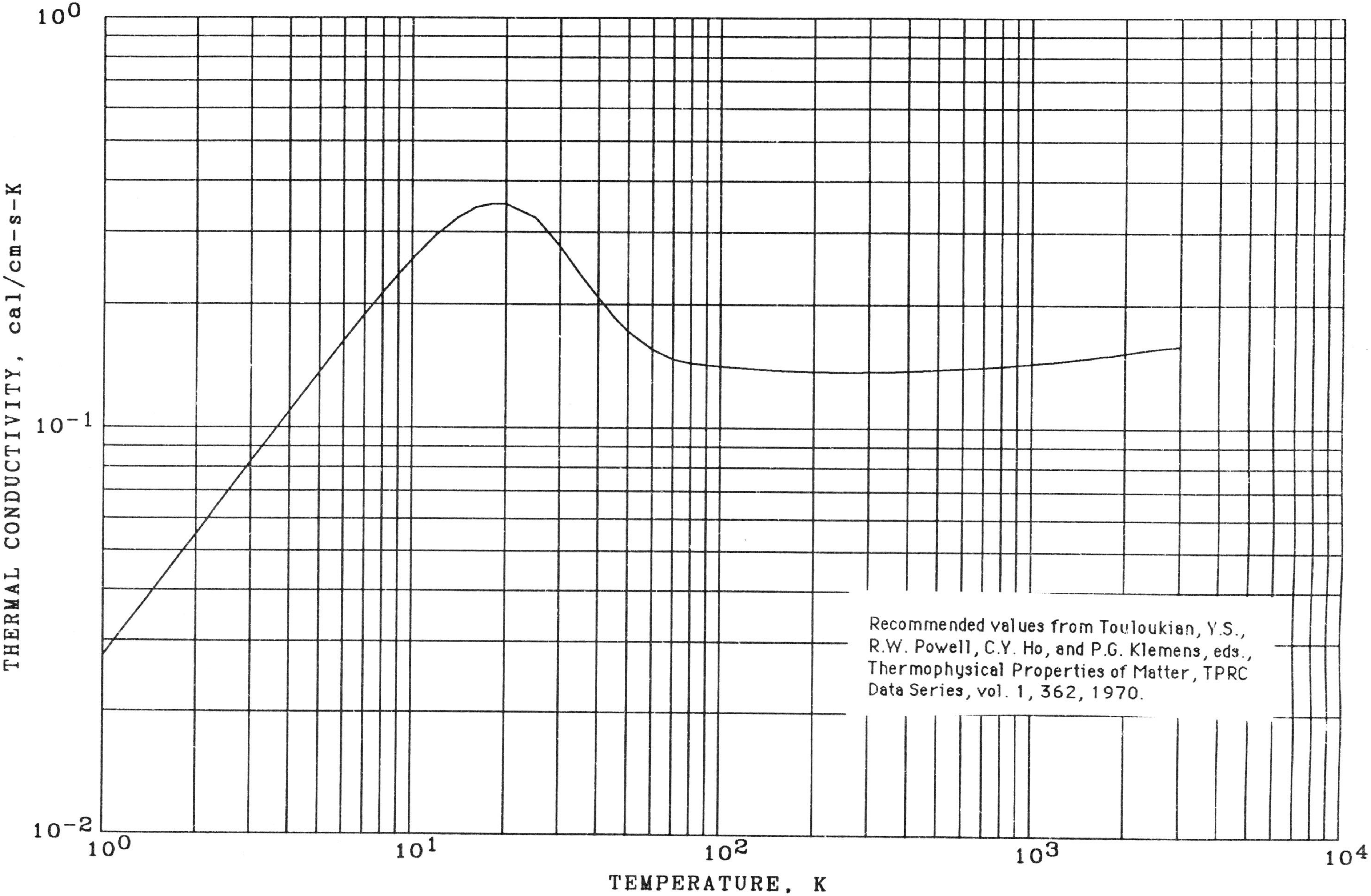
1 0^0
1 0^-1
1 0^-2
THERMAL CONDUCTIVITY, cal/cm-s-K
1 0^0
1 0^1
1 0^2
1 0^3
1 0^4
TEMPERATURE, K
Recommended values from Touloukian, Y.S., R.W. Powell, C.Y. Ho, and P.G. Klemens, eds., Thermophysical Properties of Matter, TPRC Data Series, vol. 1, 362, 1970.

TANTALUM

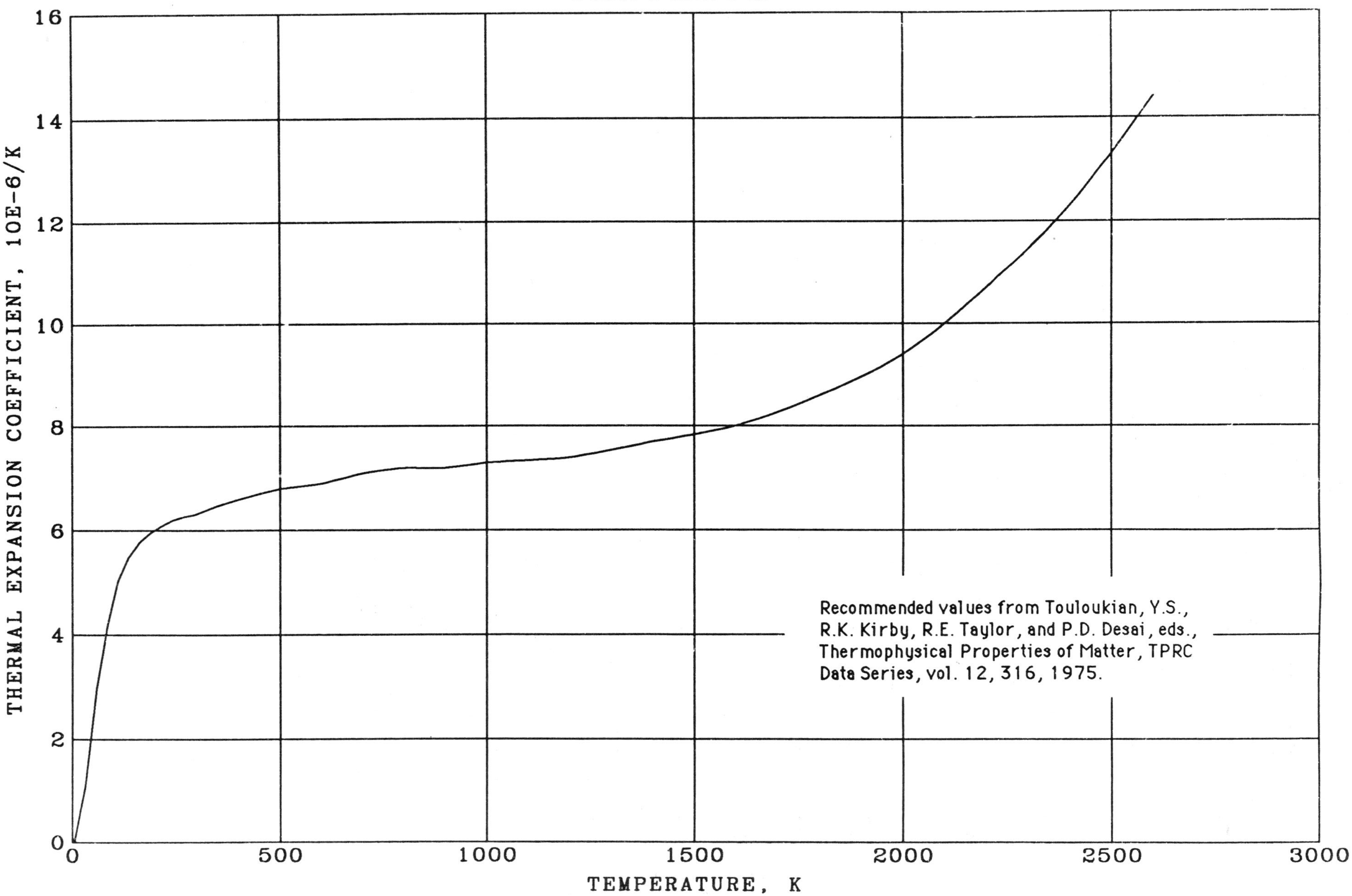
THERMAL EXPANSION COEFFICIENT, 10E-6/K
0
2
4
6
8
10
12
14
16
0
500
1000
1500
2000
2500
3000
TEMPERATURE, K
Recommended values from Touloukian, Y.S., R.K. Kirby, R.E. Taylor, and P.D. Desai, eds., Thermophysical Properties of Matter, TPRC Data Series, vol. 12, 316, 1975.

TANTALUM

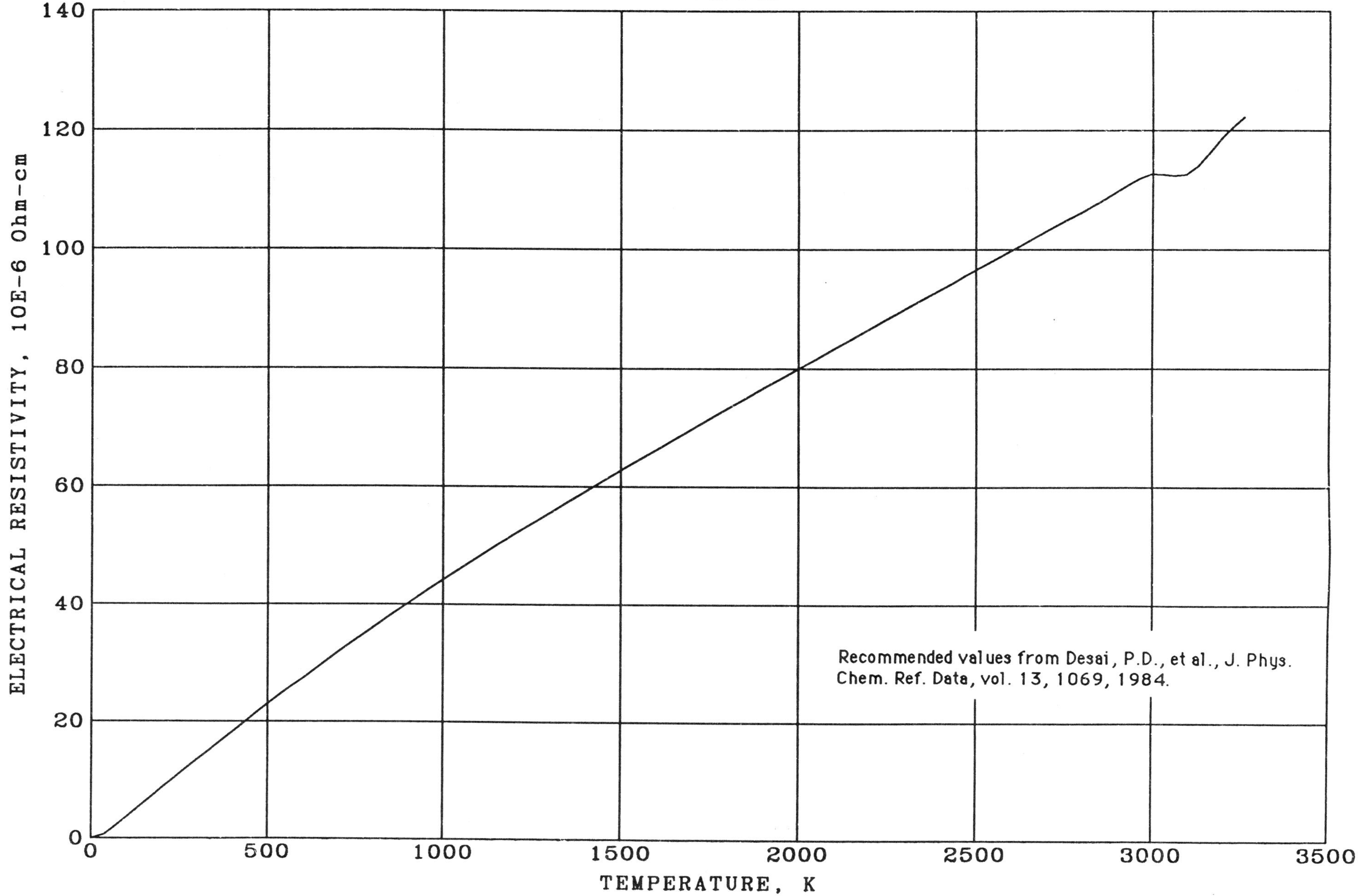

140
120
100
80
60
40
20
0
ELECTRICAL RESISTIVITY, 10E-6 Ohm-cm
0
500
1000
1500
2000
2500
3000
3500
TEMPERATURE, K
Recommended values from Desai, P.D., et al., J. Phys. Chem. Ref. Data, vol. 13, 1069, 1984.

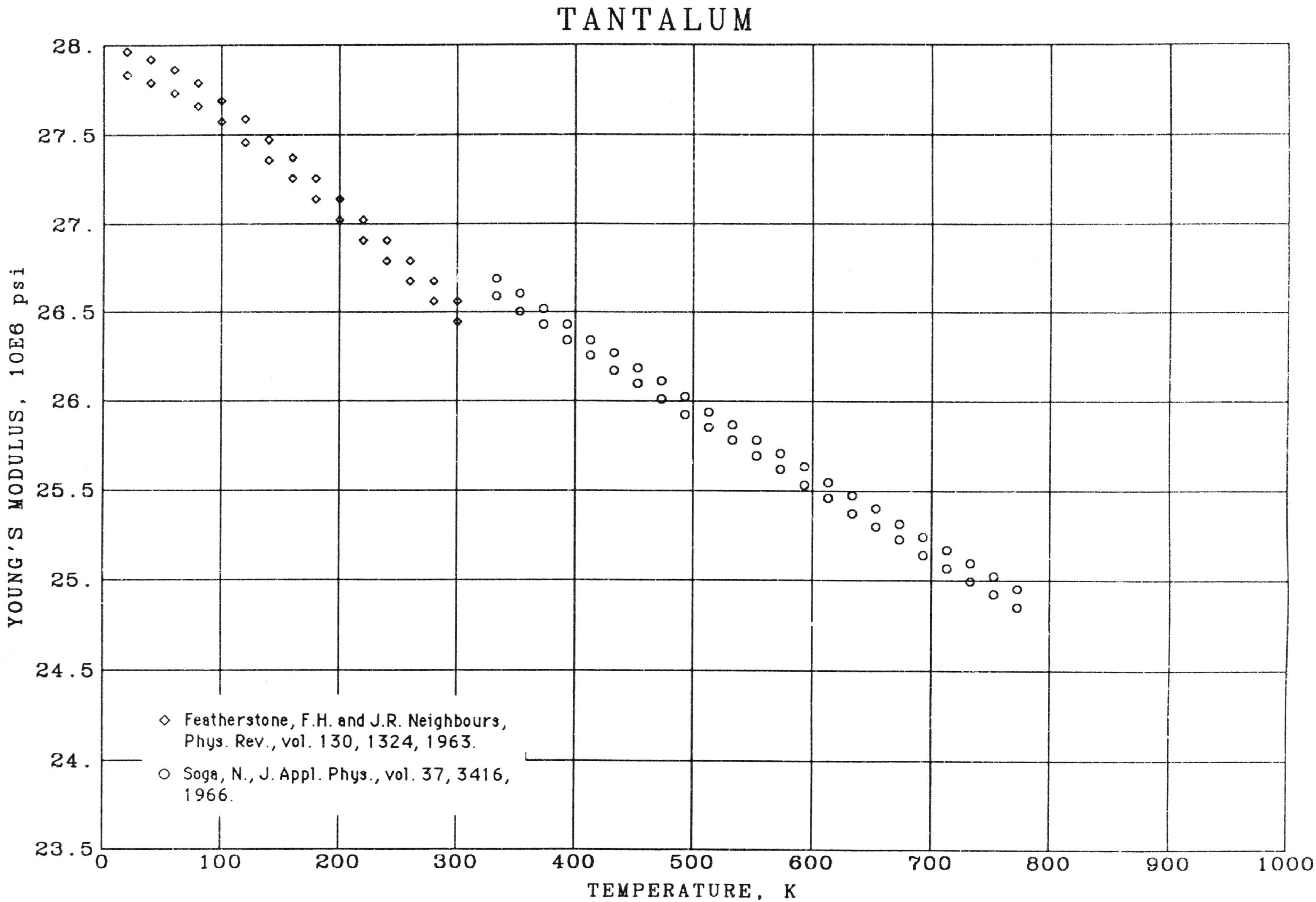
TANTALUM
YOUNG'S MODULUS, 10E6 psi
28.
27.5
27.
26.5
26.
25.5
25.
24.5
24.
23.5
0
100
200
300
400
500
600
700
800
900
1000
TEMPERATURE, K
◇ Featherstone, F.H. and J.R. Neighbours, Phys. Rev., vol. 130, 1324, 1963.
○ Soga, N., J. Appl. Phys., vol. 37, 3416, 1966.

TANTALUM

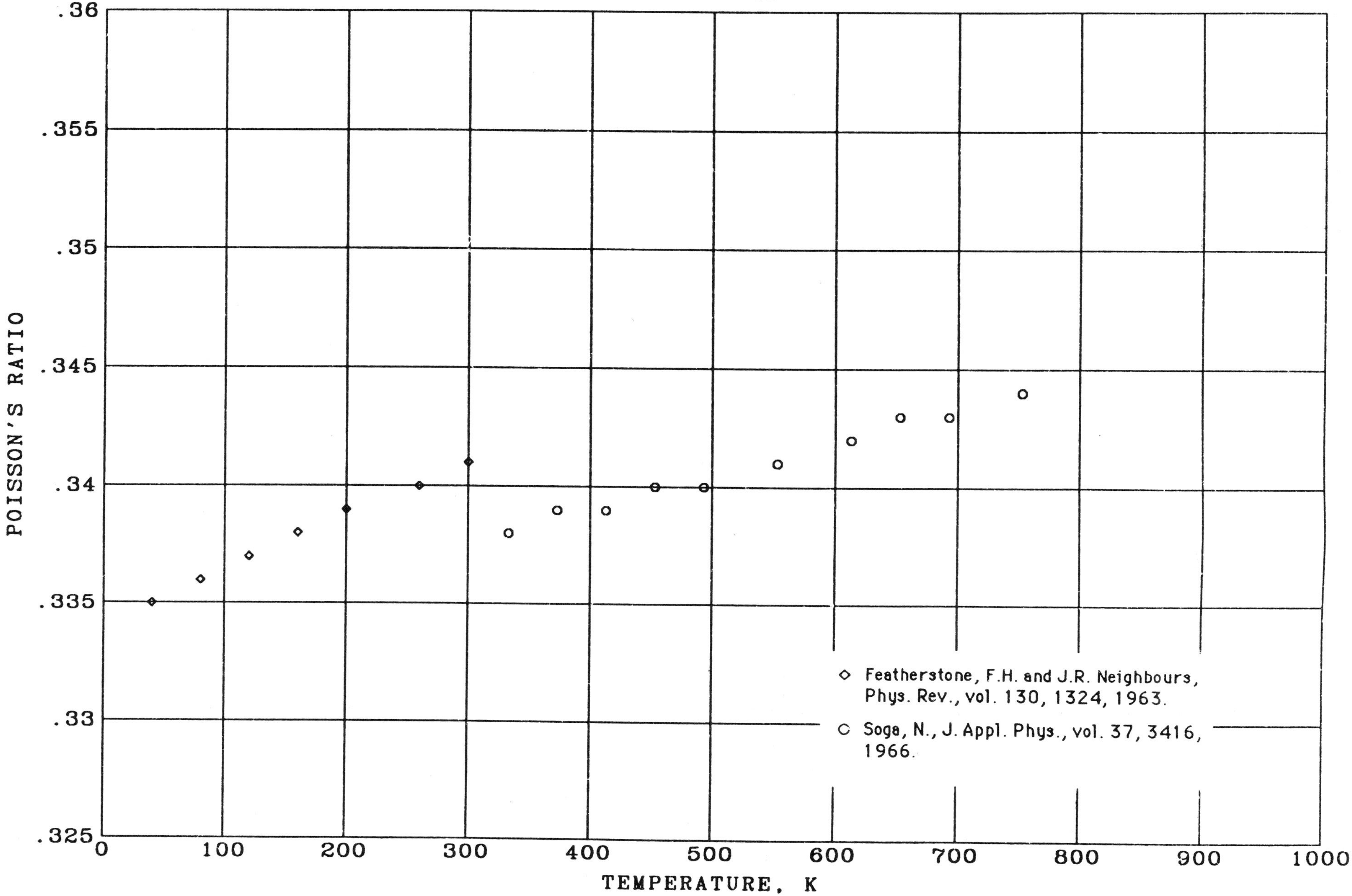

.36
.355
.35
.345
.34
.335
.33
.325
POISSON'S RATIO
0
100
200
300
400
500
600
700
800
900
1000
TEMPERATURE, K
◇ Featherstone, F.H. and J.R. Neighbours, Phys. Rev., vol. 130, 1324, 1963.
○ Soga, N., J. Appl. Phys., vol. 37, 3416, 1966.

TANTALUM

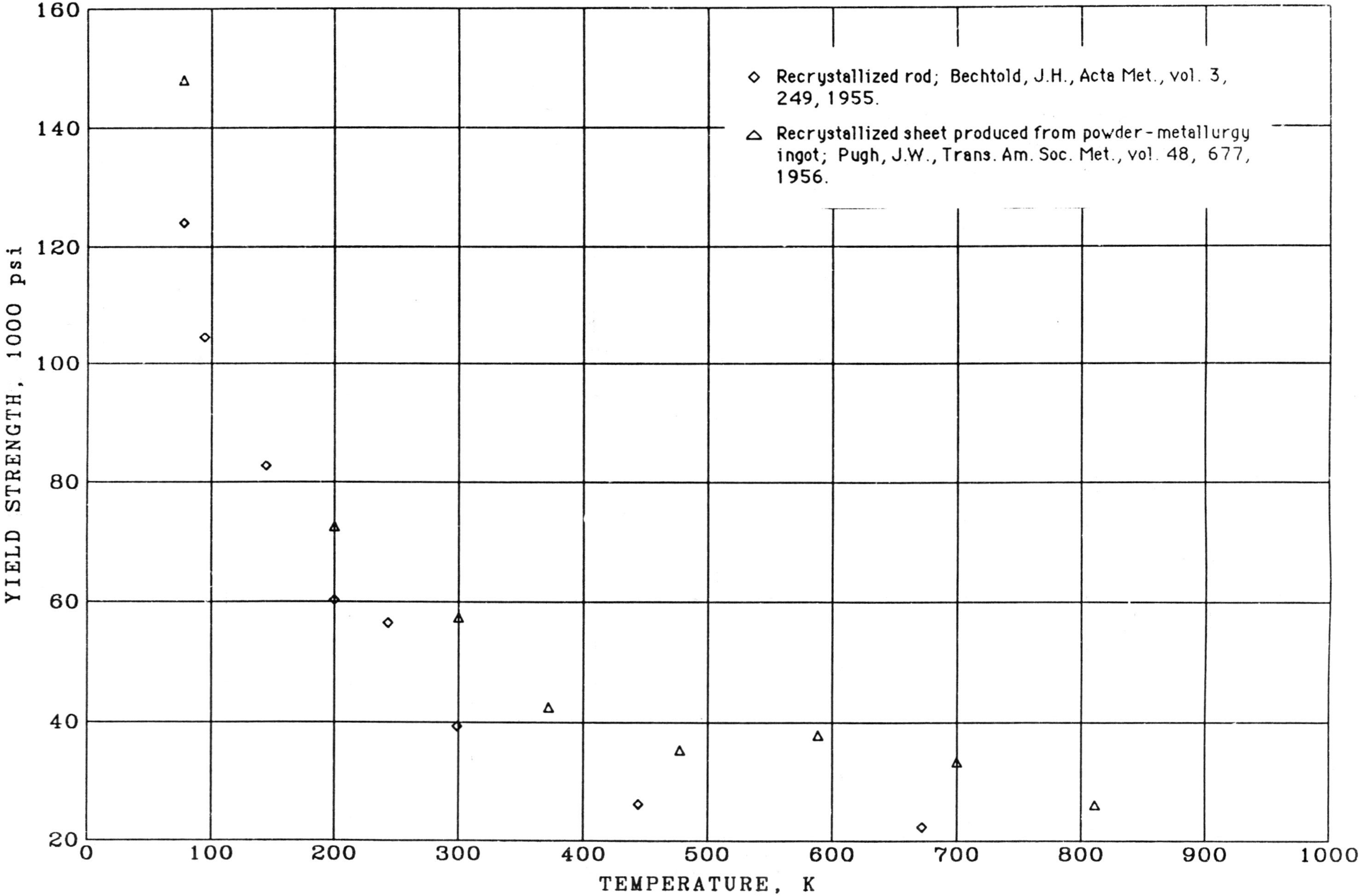

YIELD STRENGTH, 1000 psi
160
140
120
100
80
60
40
20
0
100
200
300
400
500
600
700
800
900
1000
TEMPERATURE, K
◇ Recrystallized rod; Bechtold, J.H., Acta Met., vol. 3, 249, 1955.
△ Recrystallized sheet produced from powder-metallurgy ingot; Pugh, J.W., Trans. Am. Soc. Met., vol. 48, 677, 1956.

TANTALUM

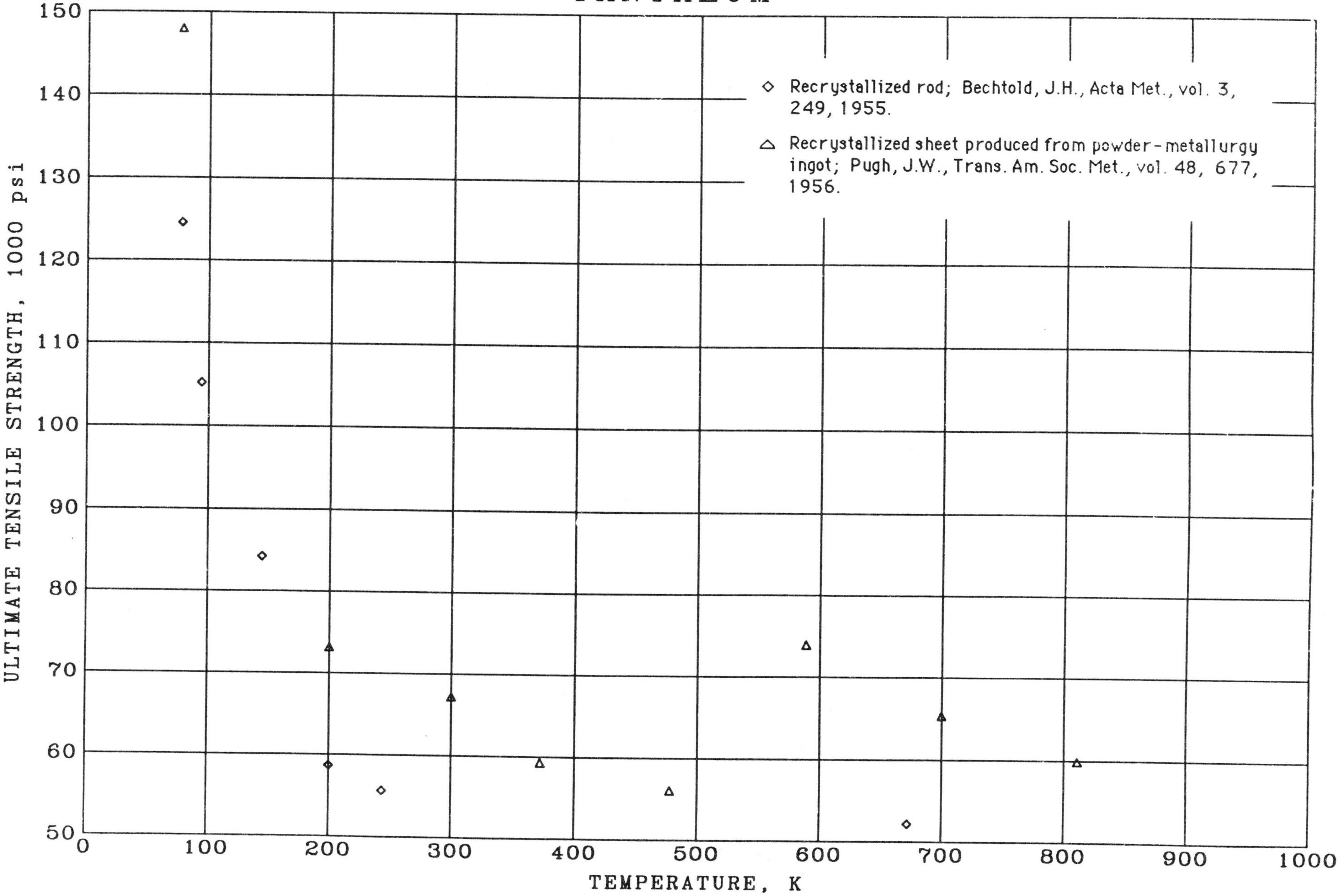

ULTIMATE TENSILE STRENGTH, 1000 psi
150
140
130
120
110
100
90
80
70
60
50
0
100
200
300
400
500
600
700
800
900
1000
TEMPERATURE, K
◇ Recrystallized rod; Bechtold, J.H., Acta Met., vol. 3, 249, 1955.
△ Recrystallized sheet produced from powder-metallurgy ingot; Pugh, J.W., Trans. Am. Soc. Met., vol. 48, 677, 1956.

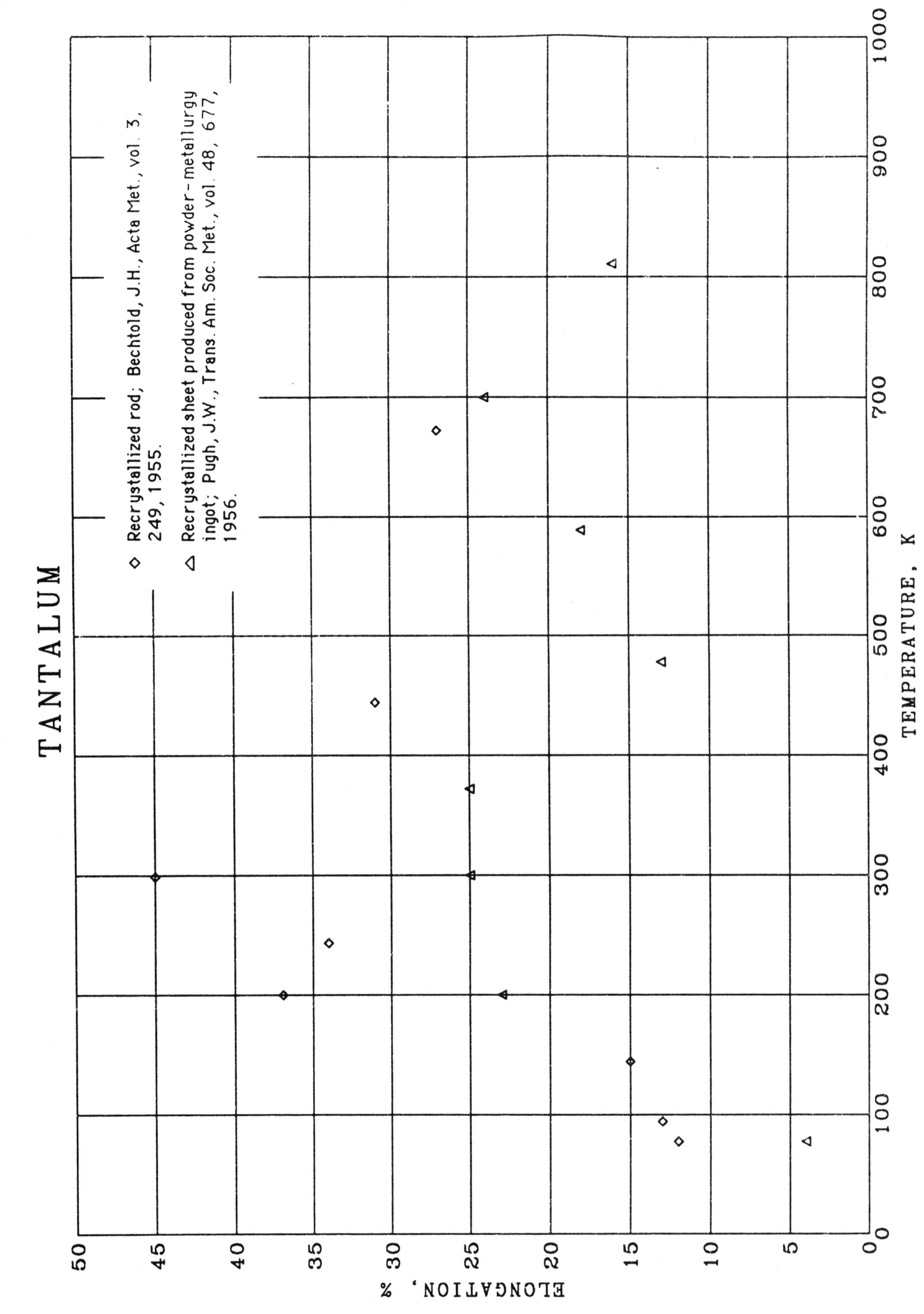
TANTALUM
◇ Recrystallized rod; Bechtold, J.H., Acta Met., vol. 3, 249, 1955.
△ Recrystallized sheet produced from powder-metallurgy ingot; Pugh, J.W., Trans. Am. Soc. Met., vol. 48, 677, 1956.
ELONGATION, %
TEMPERATURE, K
0
5
10
15
20
25
30
35
40
45
50
0
100
200
300
400
500
600
700
800
900
1000

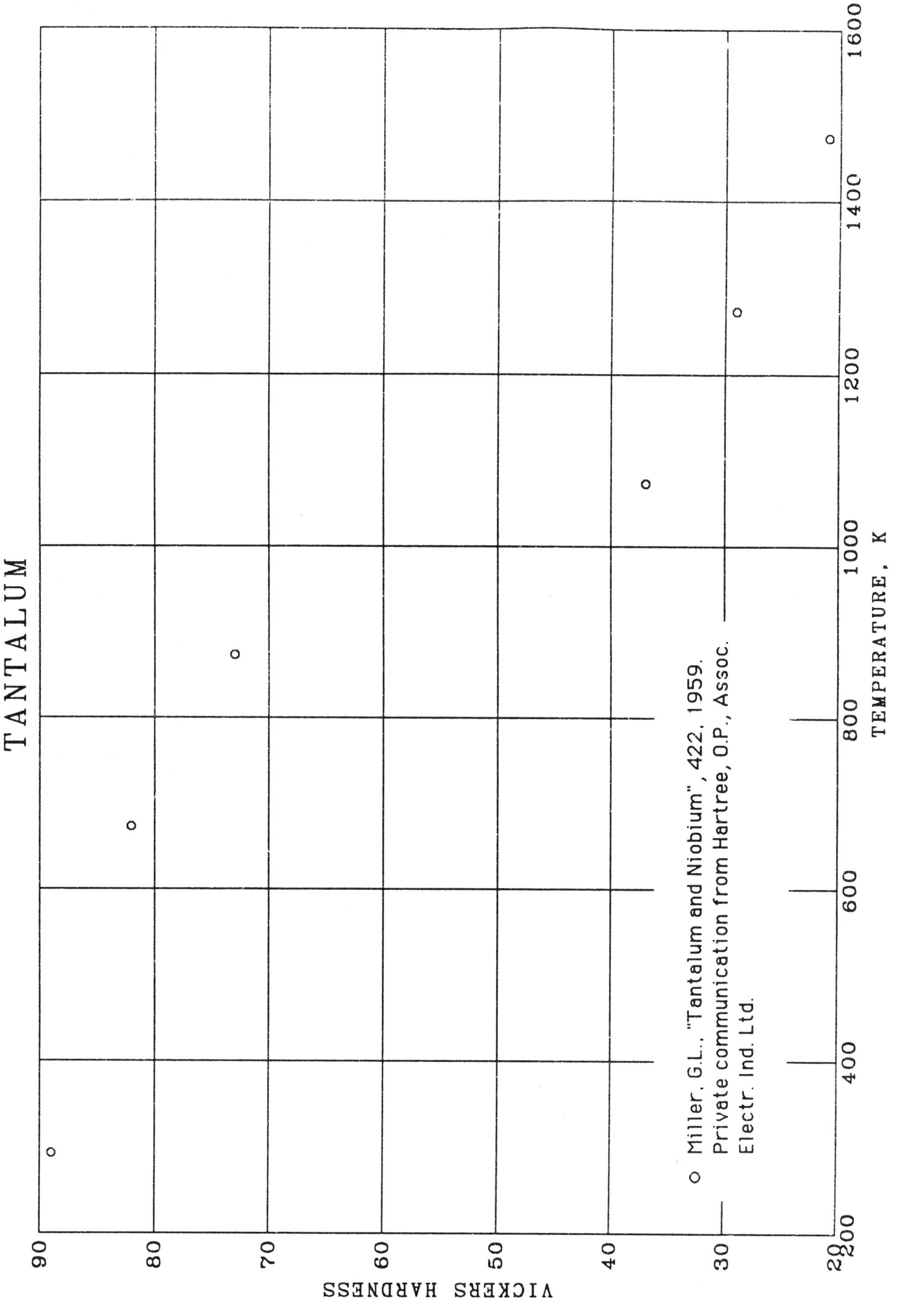
TANTALUM
VICKERS HARDNESS
TEMPERATURE, K
90
80
70
60
50
40
30
20
200
400
600
800
1000
1200
1400
1600
○ Miller, G.L., "Tantalum and Niobium", 422, 1959.
Private communication from Hartree, O.P., Assoc.
Electr. Ind. Ltd.

TIN

PROPERTY	@ 300K	UNITS	REFERENCES
Density	7.29	g/cm^3	Sn1 Sn2 Sn3
Specific Heat	0.055	cal/g-K	Sn4 Sn5 Sn6
Thermal Conductivity	0.159	cal/cm-s-K	Sn7 Sn8 Sn9
Thermal Exp Coef	22.2×10^{-6}	1/K	Sn10 Sn11 Sn12
Melting Point	505.12	K	Sn13
Resistivity	13.194×10^{-6}	Ohm-cm	Sn14 Sn15 Sn16
TCR (218K to 398K)	2550	ppm/K	Sn14 Sn15 Sn16
Young's Modulus 1	6.89×10^{6}	psi	Sn17 Sn18 Sn19
Poisson's Ratio 1	0.368	-----	Sn17 Sn18 Sn20
Yield Strength	1300	psi	Sn21
Ult Tensile Strength	2000	psi	Sn22 Sn23 Sn24
Elongation at Break	50-80	%	Sn22 Sn23
Hardness	3.7	Brinell	Sn24 Sn25 Sn26

1 Average of Voigt and Reuss methods taken from the compilation by Simmons and Wang, "Single Crystal Constants and Calculated Aggregrate Properties", The MIT Press, 1971.

Mechanical properties are for fully annealed, bulk material. 30% cold work producing 0.040 inch sheet increases yield strength to 2000 psi, increases tensile strength to 2800 and reduces elongation to 35%.[Sn27] Heating cold worked material to near 100C will cause recovery and recrystallization erasing the effects of cold work.

WHITE TIN

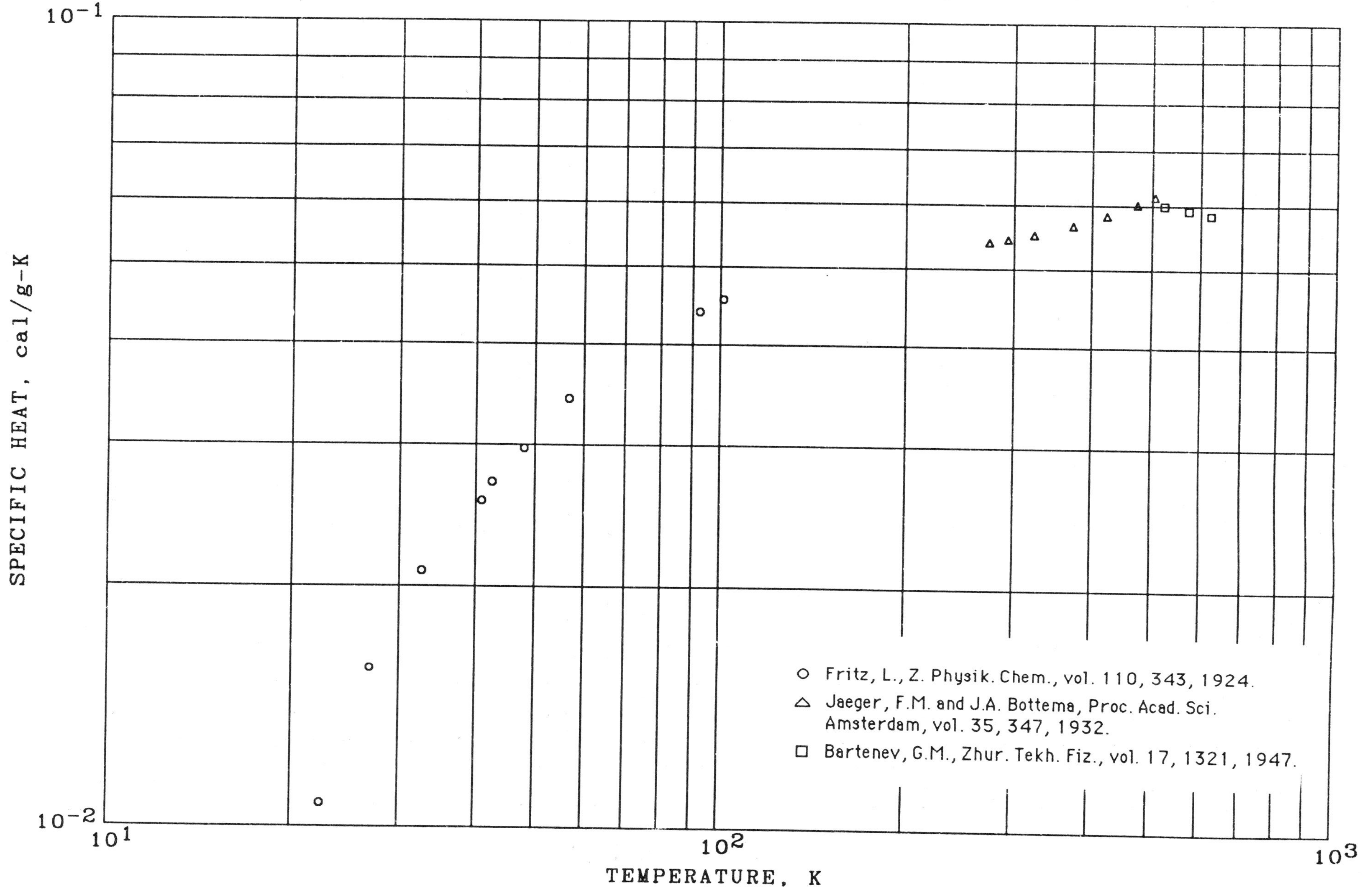

10^{-1}
10^{-2}
SPECIFIC HEAT, cal/g-K
10^{1}
10^{2}
10^{3}
TEMPERATURE, K
○ Fritz, L., Z. Physik. Chem., vol. 110, 343, 1924.
△ Jaeger, F.M. and J.A. Bottema, Proc. Acad. Sci. Amsterdam, vol. 35, 347, 1932.
□ Bartenev, G.M., Zhur. Tekh. Fiz., vol. 17, 1321, 1947.

GREY TIN

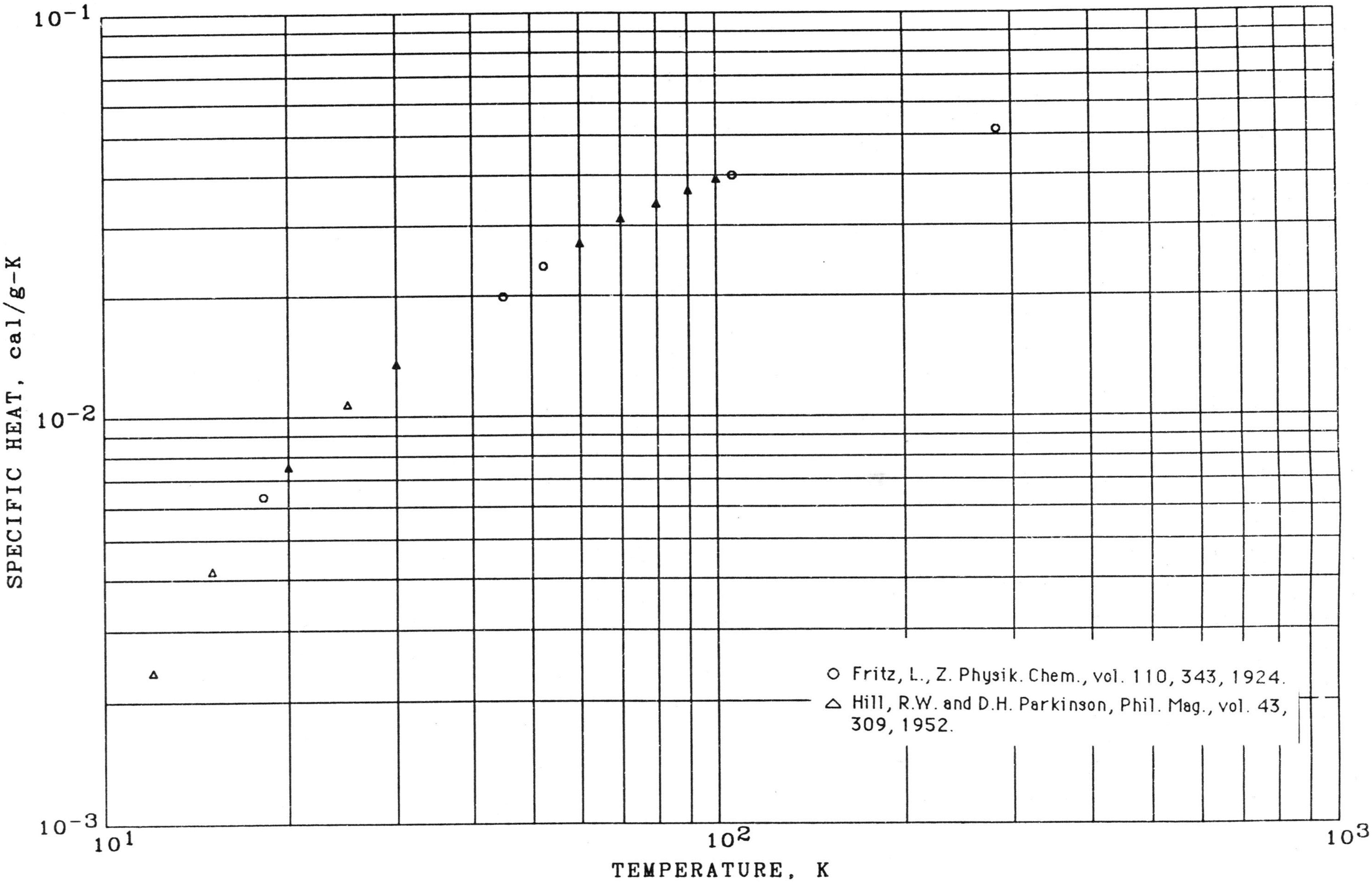

10^{-1}
10^{-2}
10^{-3}
SPECIFIC HEAT, cal/g-K
10^{1}
10^{2}
10^{3}
TEMPERATURE, K
○ Fritz, L., Z. Physik. Chem., vol. 110, 343, 1924.
△ Hill, R.W. and D.H. Parkinson, Phil. Mag., vol. 43, 309, 1952.

POLYCRYSTALLINE TIN

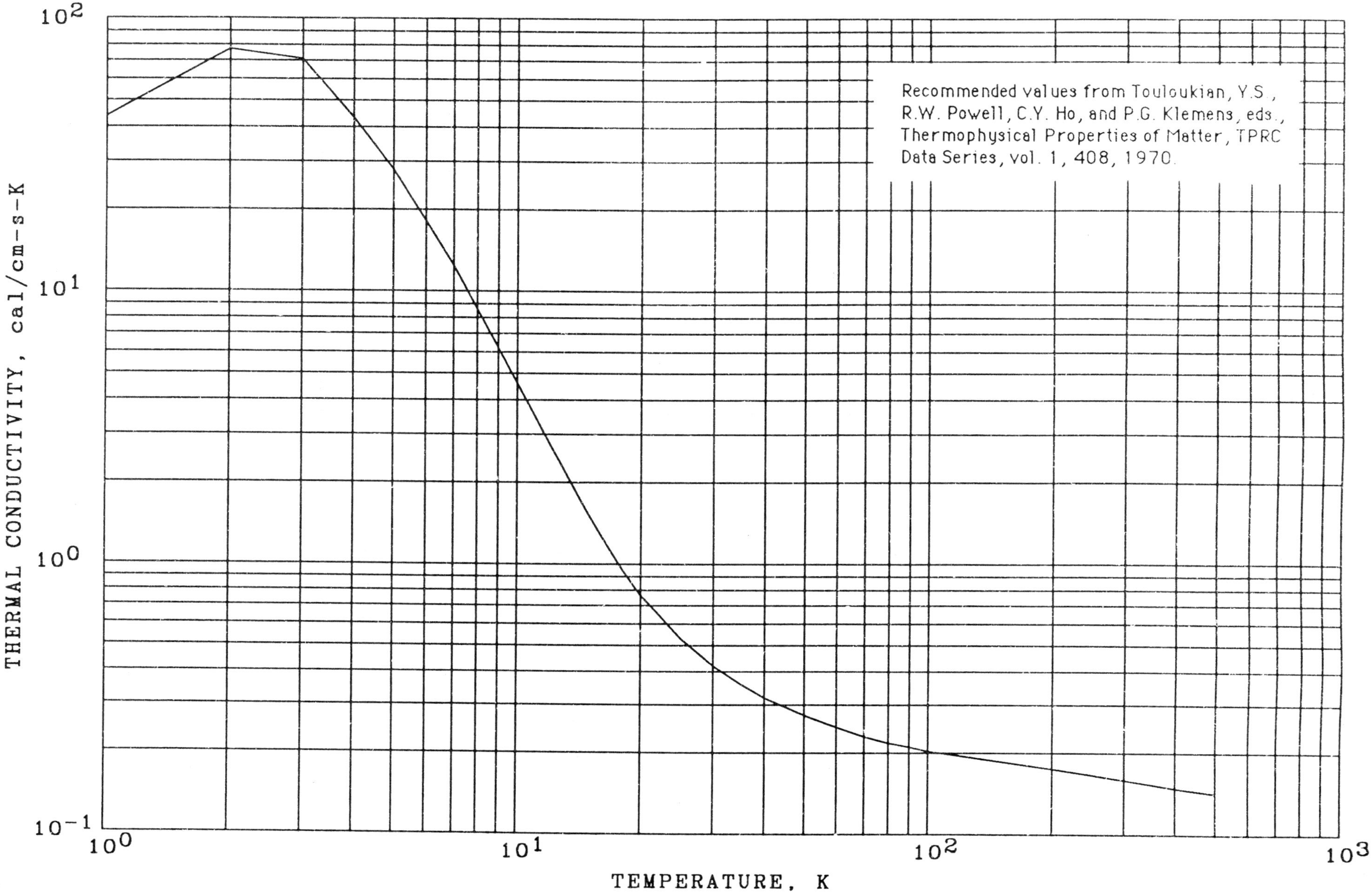

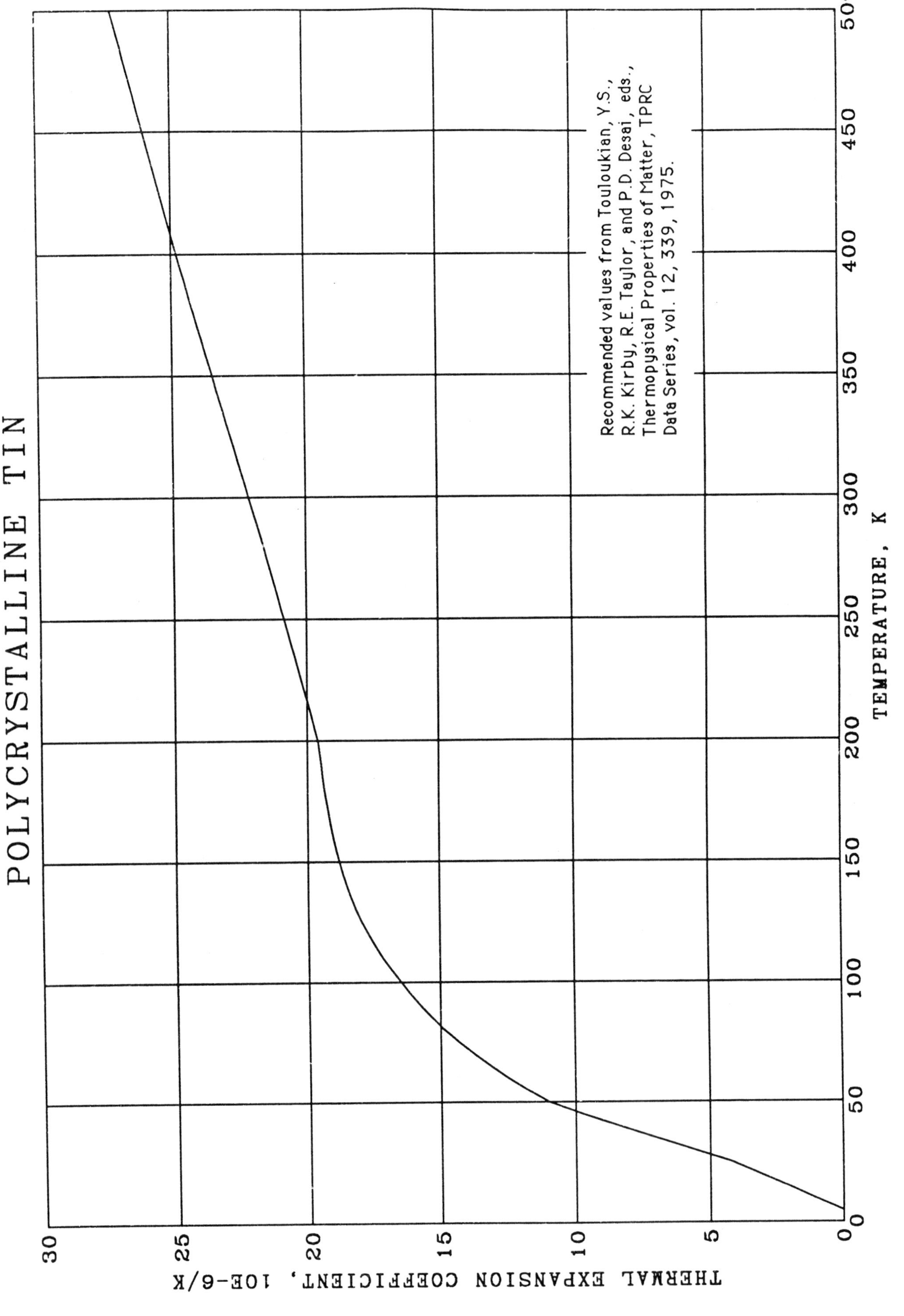
POLYCRYSTALLINE TIN
Recommended values from Touloukian, Y.S., R.K. Kirby, R.E. Taylor, and P.D. Desai, eds., Thermopysical Properties of Matter, TPRC Data Series, vol. 12, 339, 1975.
THERMAL EXPANSION COEFFICIENT, 10E-6/K
TEMPERATURE, K
0
5
10
15
20
25
30
0
50
100
150
200
250
300
350
400
450
500

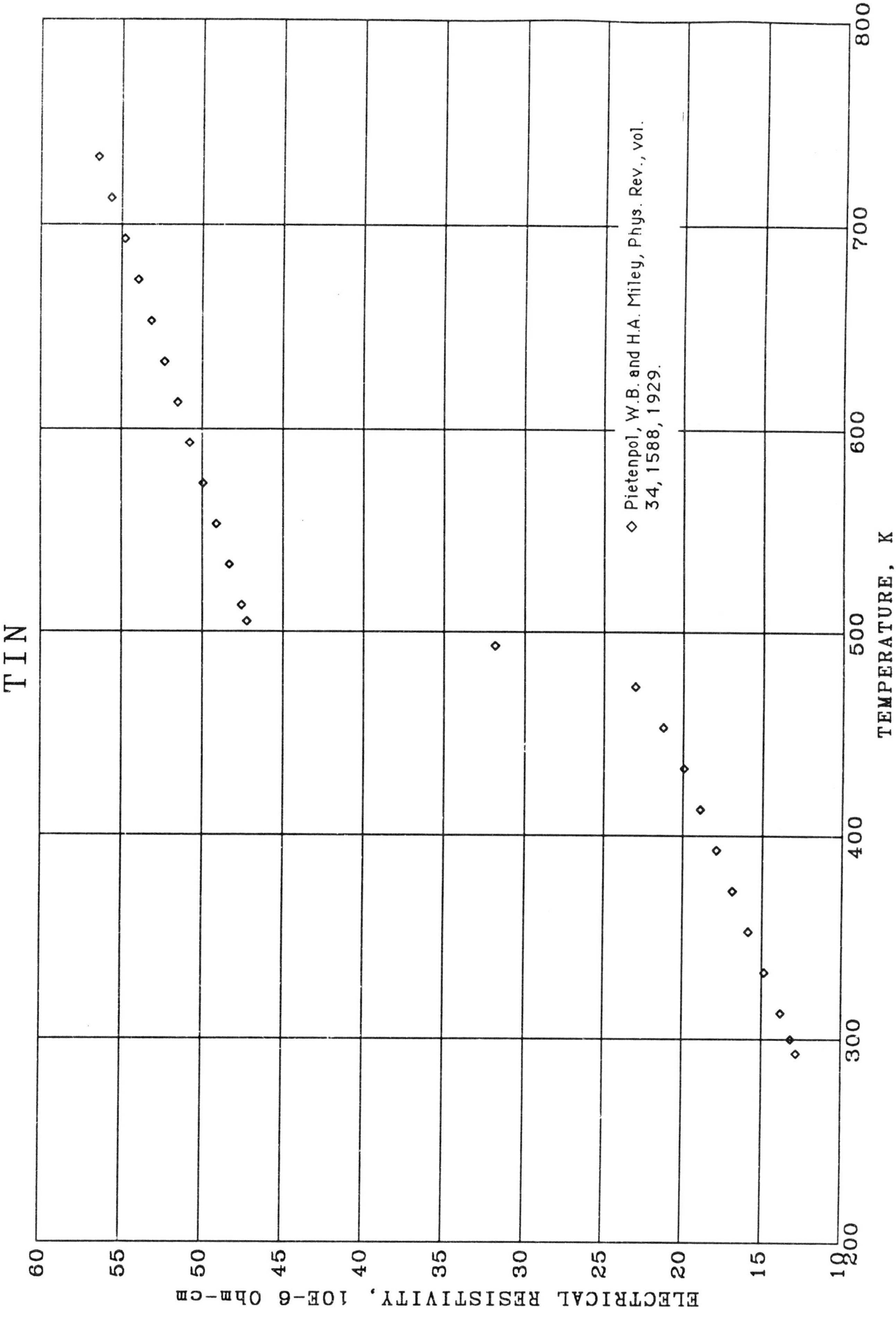
TIN
ELECTRICAL RESISTIVITY, 10E-6 Ohm-cm
TEMPERATURE, K
◇ Pietenpol, W.B. and H.A. Miley, Phys. Rev., vol. 34, 1588, 1929.
60
55
50
45
40
35
30
25
20
15
10
200
300
400
500
600
700
800

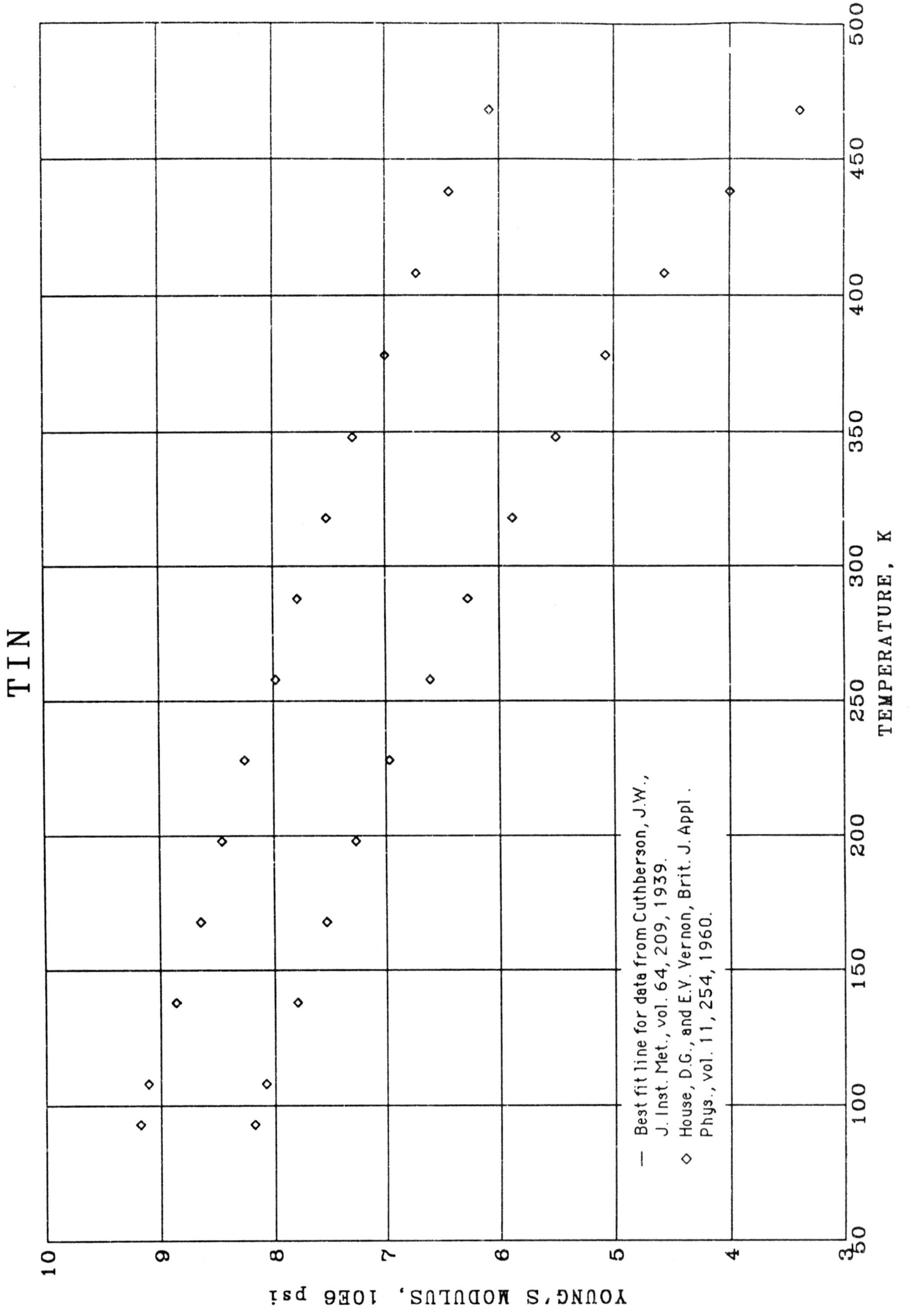
TIN
YOUNG'S MODULUS, 10E6 psi
TEMPERATURE, K
— Best fit line for data from Cuthberson, J.W., J. Inst. Met., vol. 64, 209, 1939.
◇ House, D.G., and E.Y. Vernon, Brit. J. Appl. Phys., vol. 11, 254, 1960.
10
9
8
7
6
5
4
3
50
100
150
200
250
300
350
400
450
500

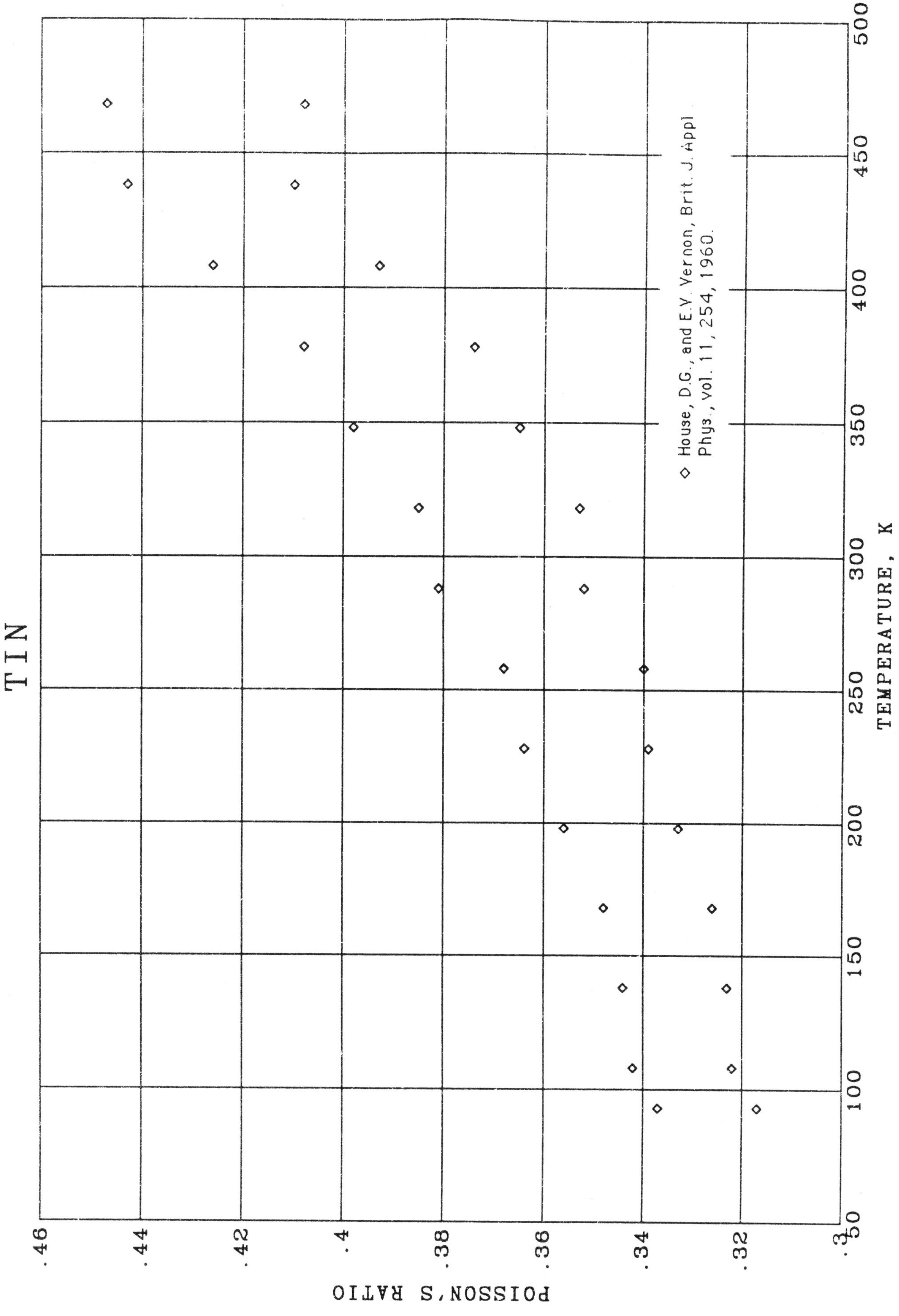
TIN
POISSON'S RATIO
.3
.32
.34
.36
.38
.4
.42
.44
.46
TEMPERATURE, K
50
100
150
200
250
300
350
400
450
500
◇ House, D.G., and E.V. Vernon, Brit. J. Appl. Phys., vol. 11, 254, 1960.

TIN

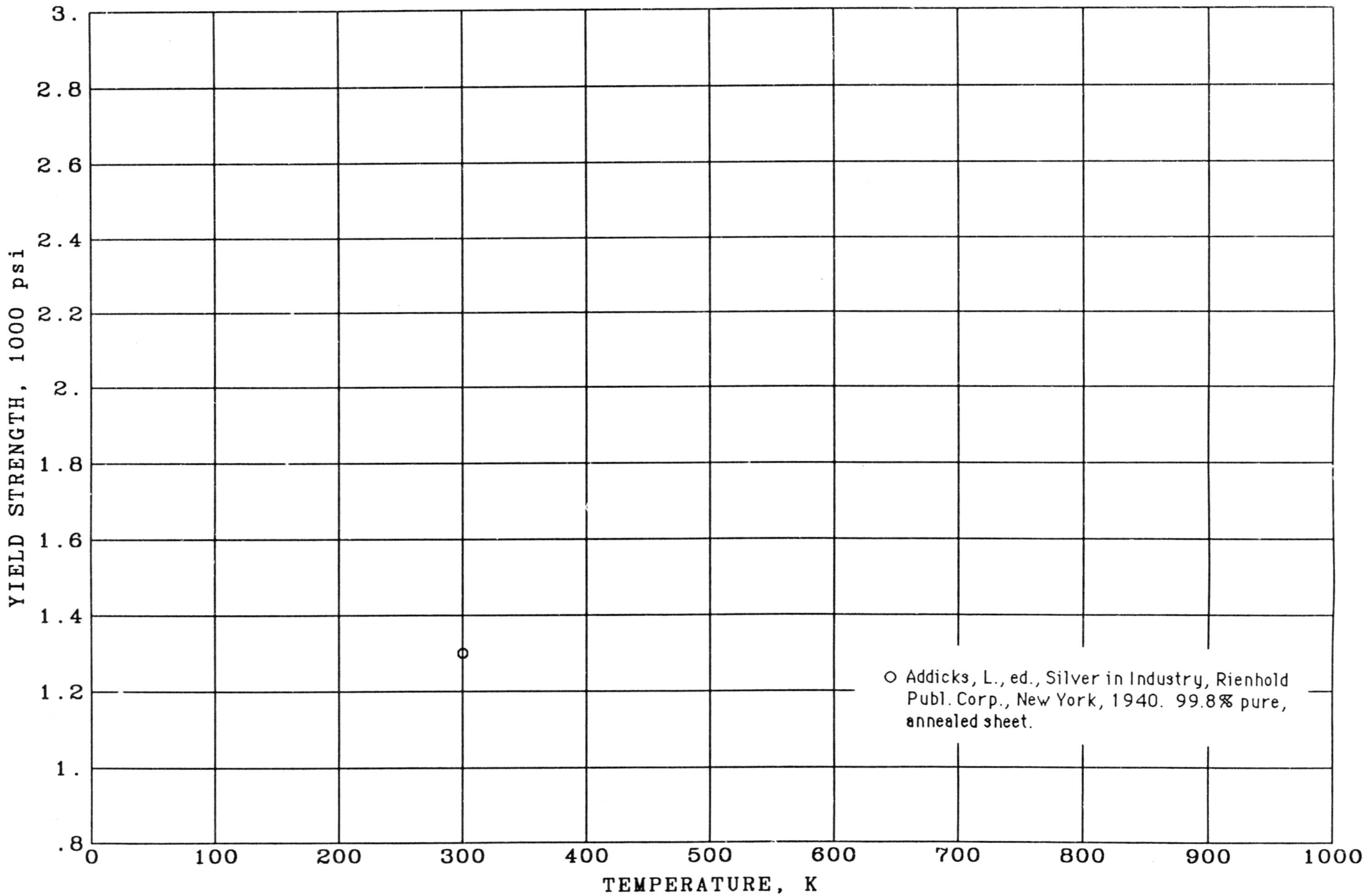

3.
2.8
2.6
2.4
2.2
2.
1.8
1.6
1.4
1.2
1.
.8
YIELD STRENGTH, 1000 psi
0
100
200
300
400
500
600
700
800
900
1000
TEMPERATURE, K
○ Addicks, L., ed., Silver in Industry, Rienhold Publ. Corp., New York, 1940. 99.8% pure, annealed sheet.

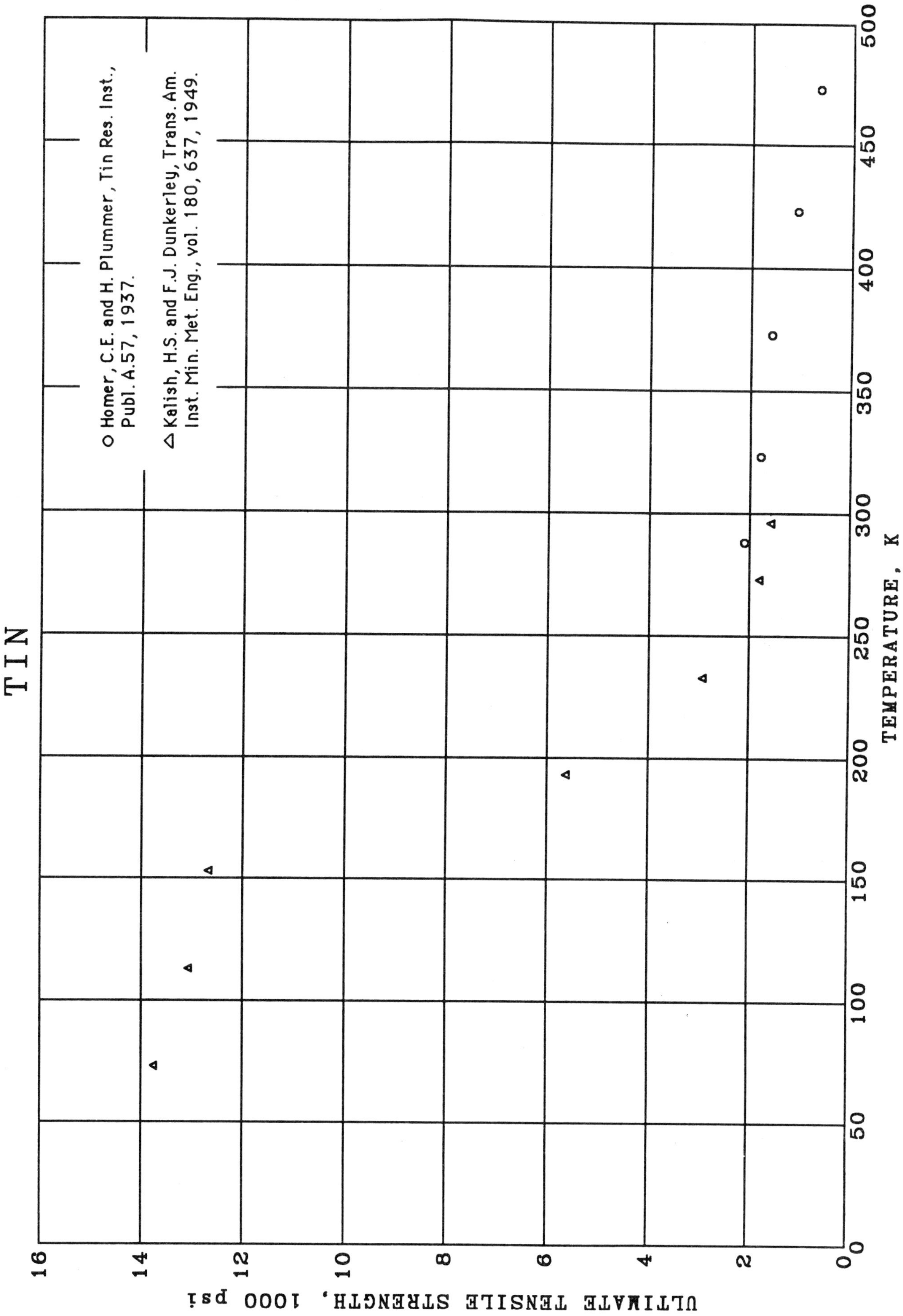
TIN
O Homer, C.E. and H. Plummer, Tin Res. Inst., Publ. A.57, 1937.
△ Kalish, H.S. and F.J. Dunkerley, Trans. Am. Inst. Min. Met. Eng., vol. 180, 637, 1949.
TEMPERATURE, K
ULTIMATE TENSILE STRENGTH, 1000 psi
0
50
100
150
200
250
300
350
400
450
500
2
4
6
8
10
12
14
16

TIN

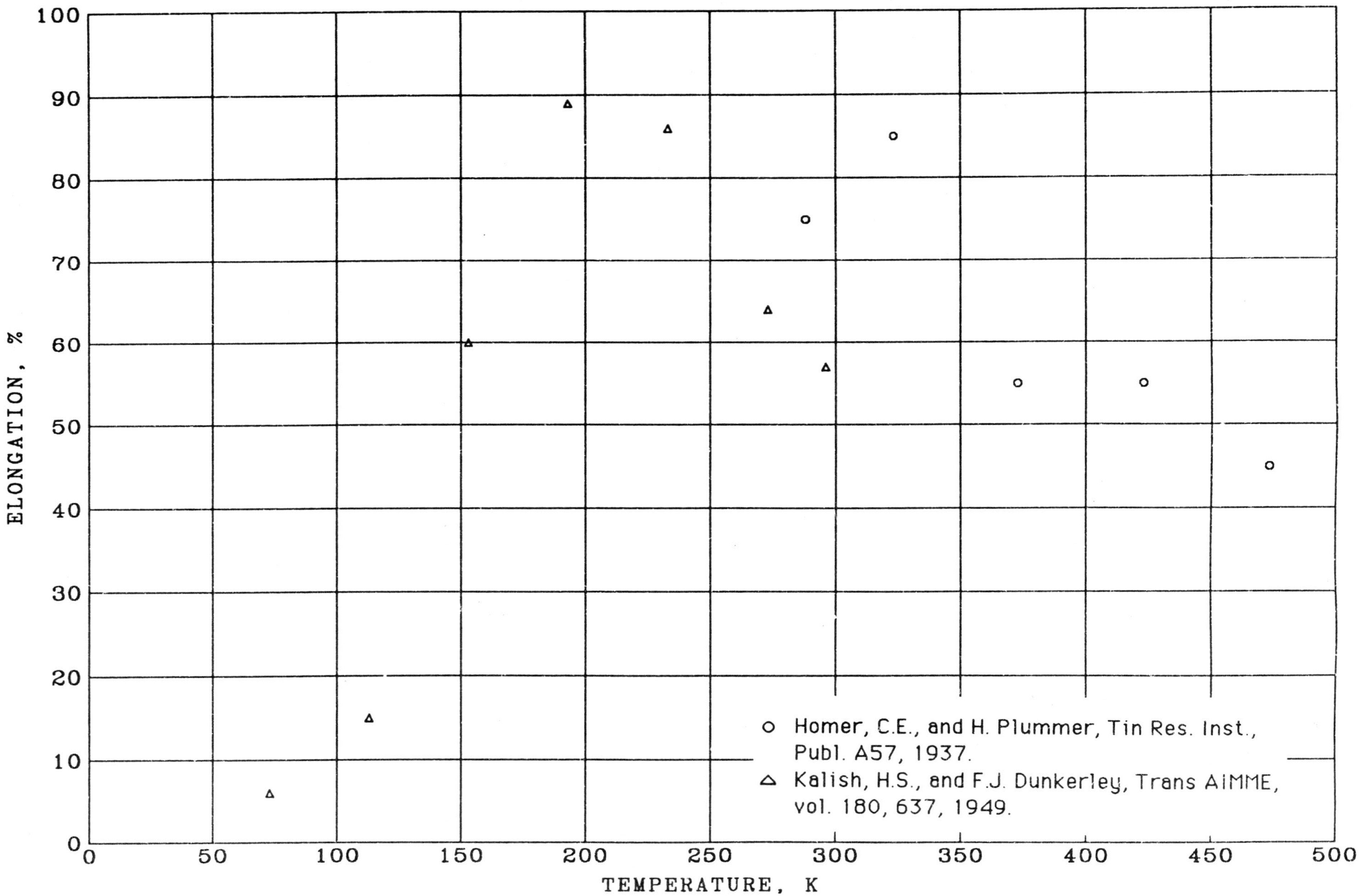
100
90
80
70
60
50
40
30
20
10
0
ELONGATION, %
0
50
100
150
200
250
300
350
400
450
500
TEMPERATURE, K
○ Homer, C.E., and H. Plummer, Tin Res. Inst., Publ. A57, 1937.
△ Kalish, H.S., and F.J. Dunkerley, Trans AIMME, vol. 180, 637, 1949.

TIN

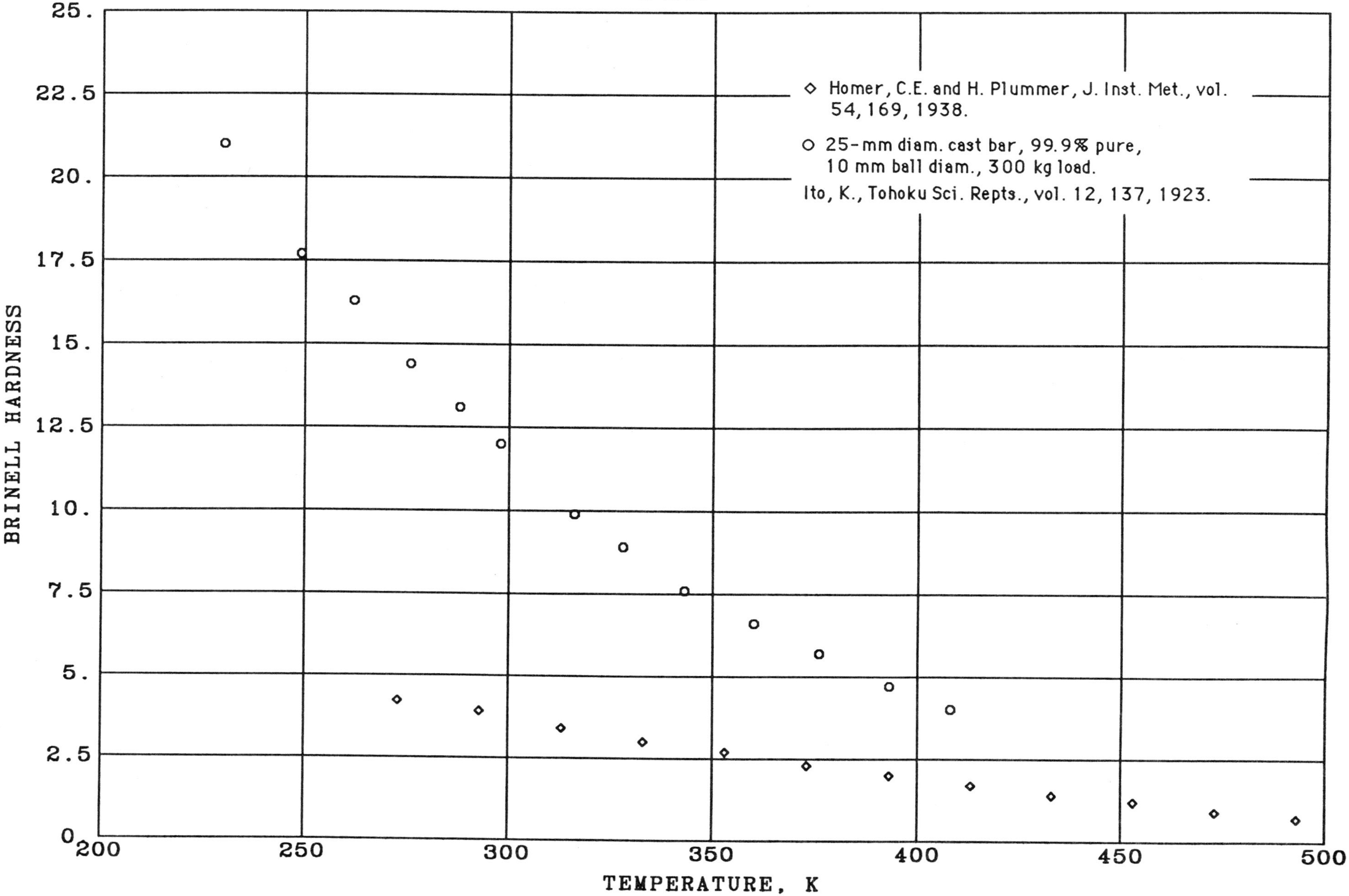

25.
22.5
20.
17.5
15.
12.5
10.
7.5
5.
2.5
0
200
250
300
350
400
450
500
BRINELL HARDNESS
TEMPERATURE, K
◇ Homer, C.E. and H. Plummer, J. Inst. Met., vol. 54, 169, 1938.
○ 25-mm diam. cast bar, 99.9% pure, 10 mm ball diam., 300 kg load.
Ito, K., Tohoku Sci. Repts., vol. 12, 137, 1923.

TUNGSTEN

PROPERTY	@ 300K	UNITS	REFERENCES
Density	19.3	g/cm^3	W1 W2 W3
Specific Heat	0.03169	cal/g-K	W4 W5 W6
Thermal Conductivity	0.425	cal/cm-s-K	W7 W8 W9
Thermal Exp Coef	4.5×10^{-6}	1/K	W10 W11 W12
Melting Point	3683	K	W13
Resistivity	5.44×10^{-6}	Ohm-cm	W14 W15 W16
TCR (218K to 398K)	4290	ppm/K	W14 W15 W16
Young's Modulus	5.939×10^{7}	psi	W17 W18 W19
Poisson's Ratio	0.280	-----	W17 W18 W19
Yield Strength [1]	------	ksi	W20 W22 W23
Ult Tensile Strength	80-100	ksi	W20 W21 W22
Elongation at Break [1]	0	%	W20 W21 W22
Hardness	351	Vickers	W24 W25 W26

Mechanical properties are for fully recrystallized, bulk material. Wrought material has significantly higher strengths than those listed. For additional information see the graphs of mechanical properties -v- temperature.

Like most bcc metals, the mechanical properties are strongly affected by the interstitial composition. Material consolidated from powder with 50 wppm oxygen has 1/2 the yield strength and 2/3 the tensile strength of material with 4 wppm oxygen.

[1] Fully recrystallized tungsten demonstrates very little permanent deformation at room temperature. There is therefore no true yield strength although the range of 80-100 ksi has been reported. At higher temperatures, tungsten undergoes sufficient plastic deformation to exhibit a yield strength.

TUNGSTEN

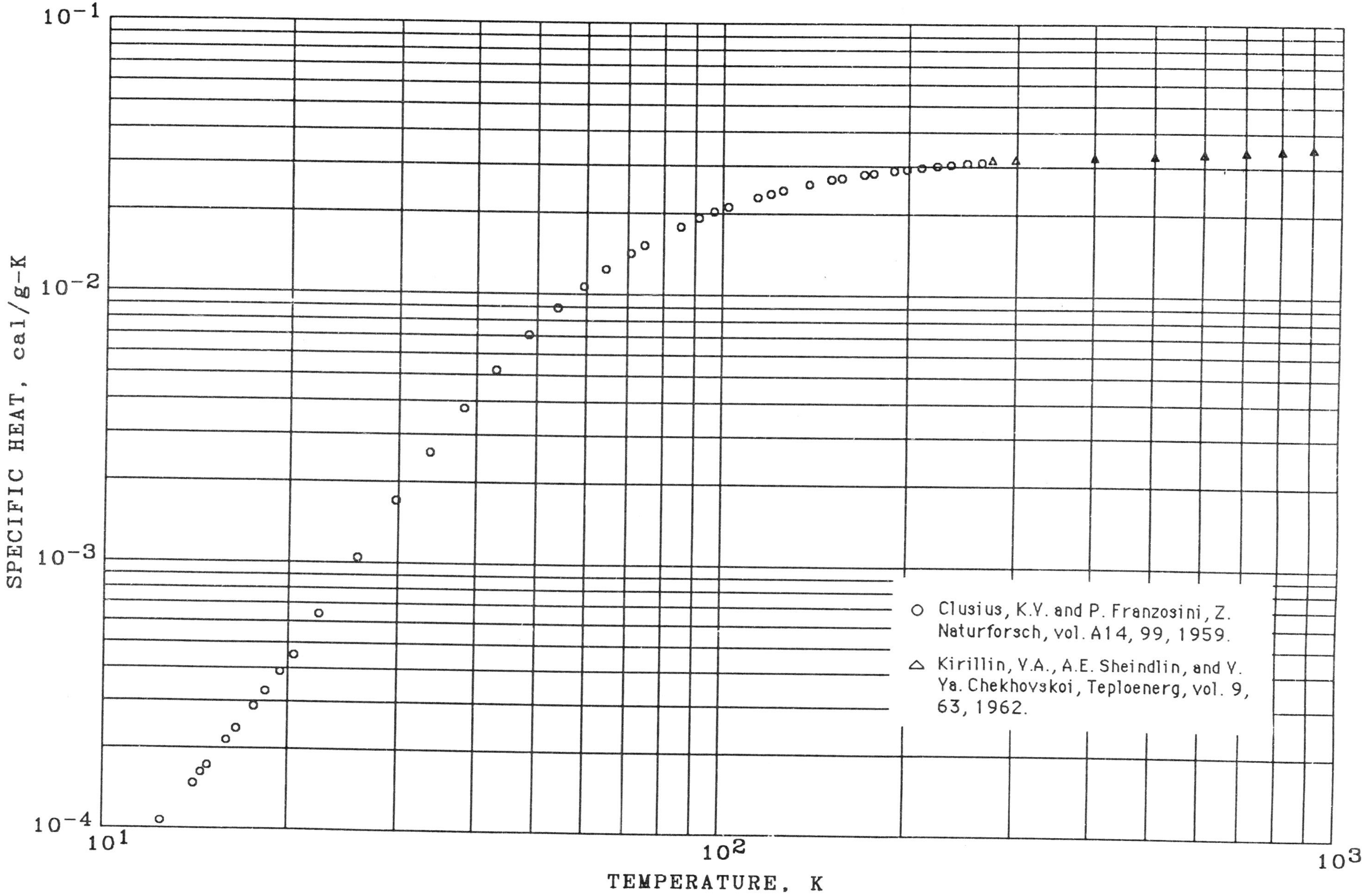

10−1
10−2
10−3
10−4
SPECIFIC HEAT, cal/g-K
10¹
10²
10³
TEMPERATURE, K
○ Clusius, K.Y. and P. Franzosini, Z. Naturforsch, vol. A14, 99, 1959.
△ Kirillin, V.A., A.E. Sheindlin, and V. Ya. Chekhovskoi, Teploenerg, vol. 9, 63, 1962.

TUNGSTEN

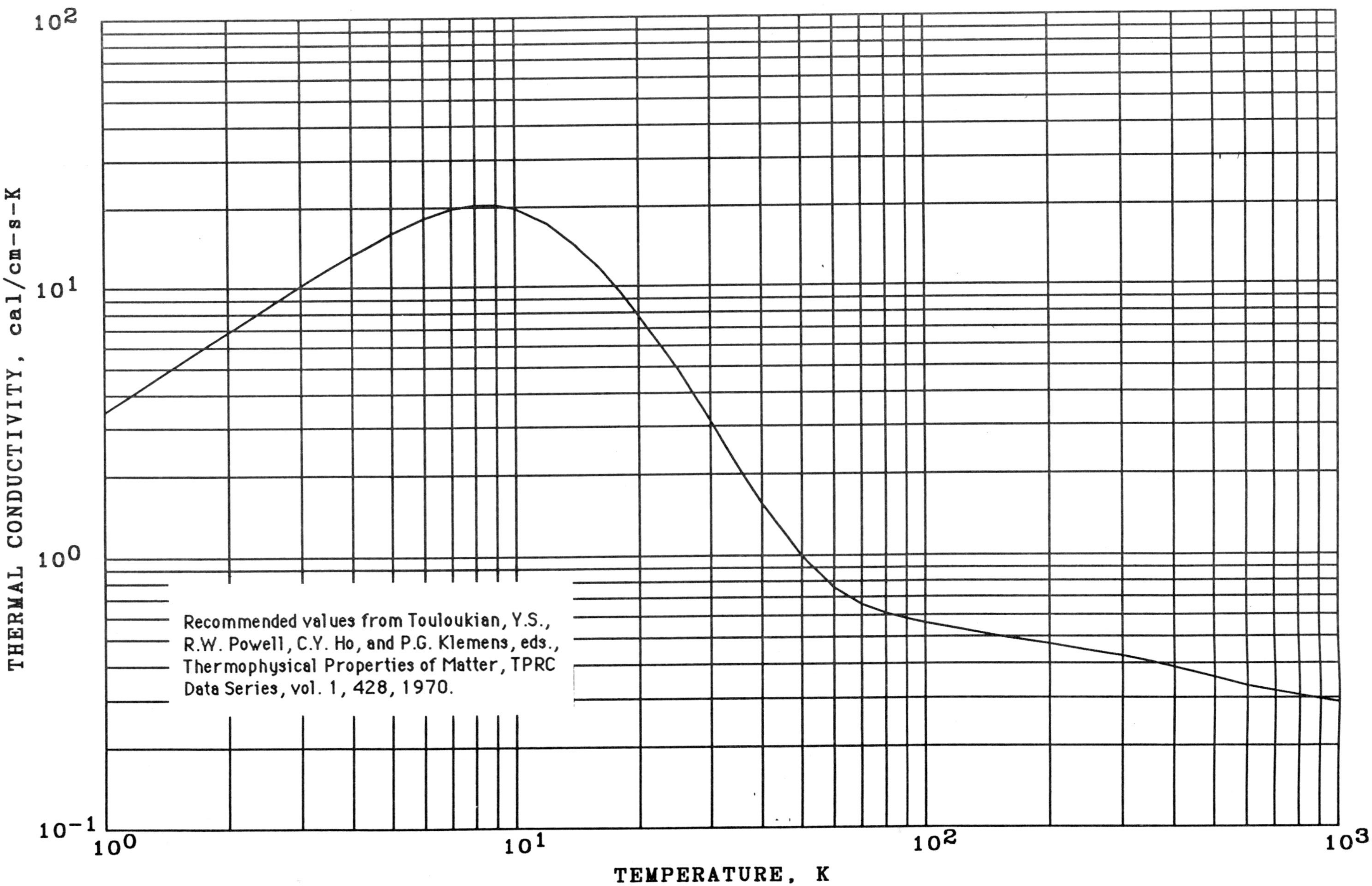

10^2
10^1
10^0
10^{-1}
THERMAL CONDUCTIVITY, cal/cm-s-K
10^0
10^1
10^2
10^3
TEMPERATURE, K
Recommended values from Touloukian, Y.S., R.W. Powell, C.Y. Ho, and P.G. Klemens, eds., Thermophysical Properties of Matter, TPRC Data Series, vol. 1, 428, 1970.

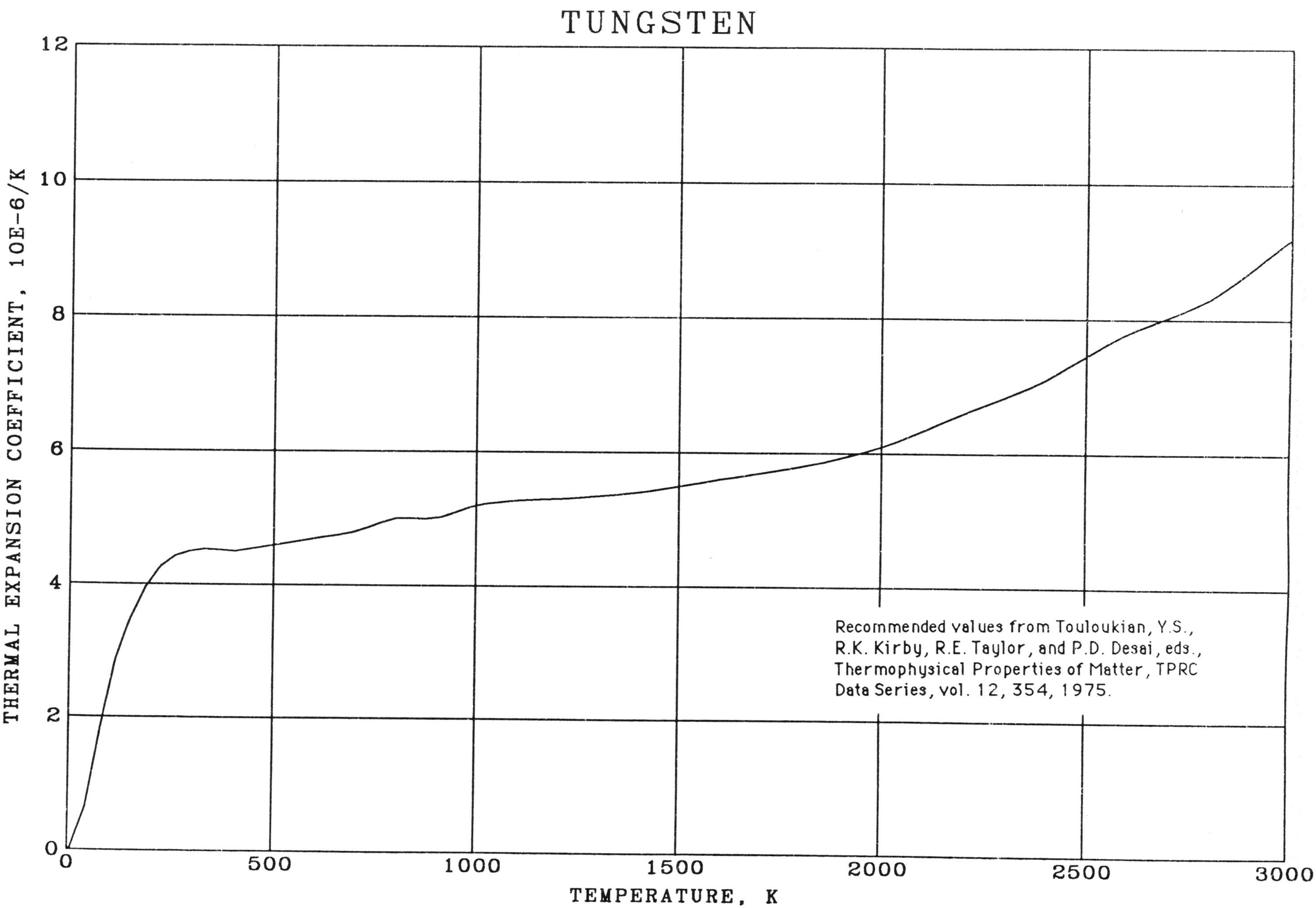
TUNGSTEN
THERMAL EXPANSION COEFFICIENT, 10E-6/K
TEMPERATURE, K
0
2
4
6
8
10
12
0
500
1000
1500
2000
2500
3000
Recommended values from Touloukian, Y.S., R.K. Kirby, R.E. Taylor, and P.D. Desai, eds., Thermophysical Properties of Matter, TPRC Data Series, vol. 12, 354, 1975.

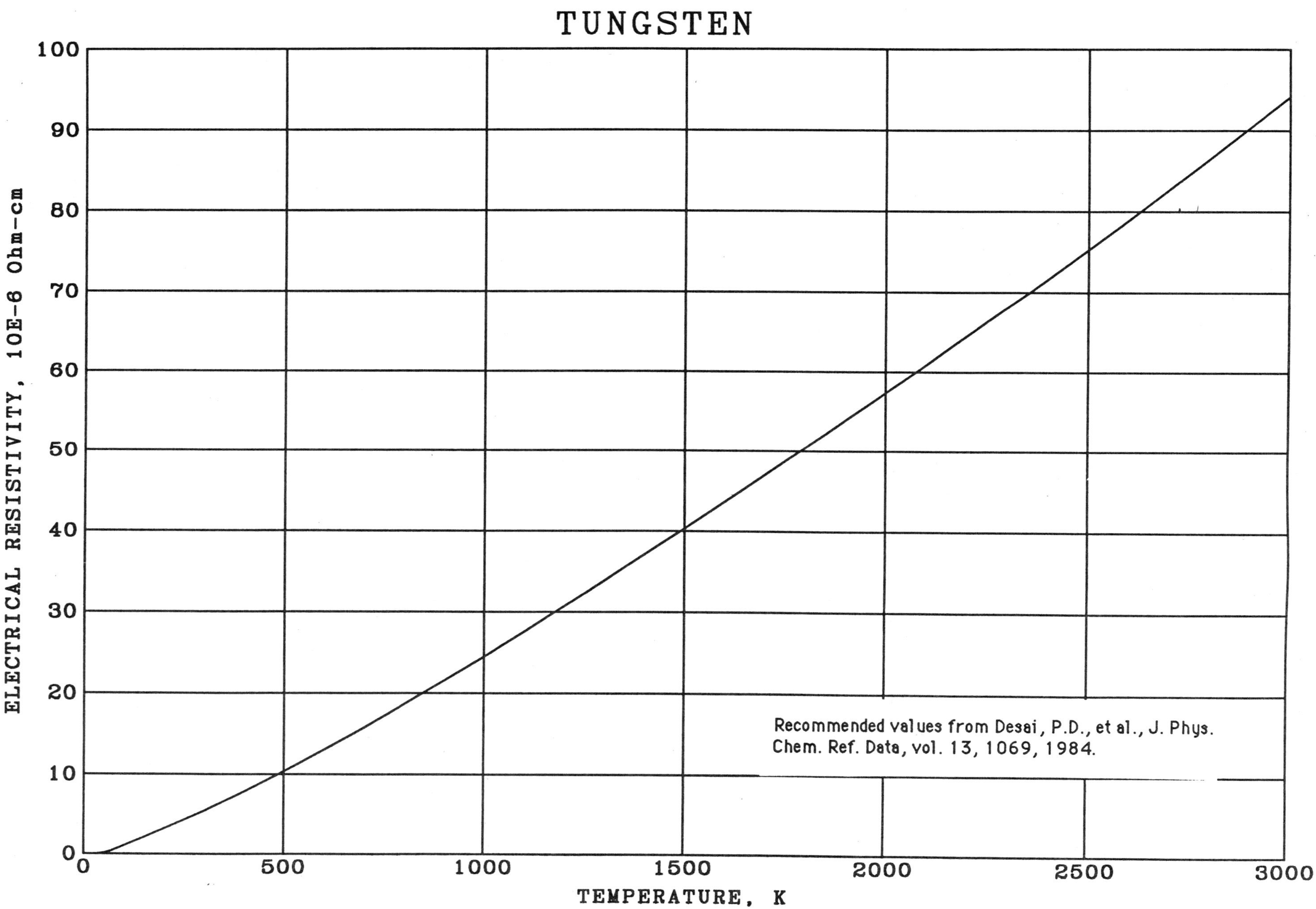
TUNGSTEN
ELECTRICAL RESISTIVITY, 10E-6 Ohm-cm
100
90
80
70
60
50
40
30
20
10
0
0
500
1000
1500
2000
2500
3000
TEMPERATURE, K
Recommended values from Desai, P.D., et al., J. Phys. Chem. Ref. Data, vol. 13, 1069, 1984.

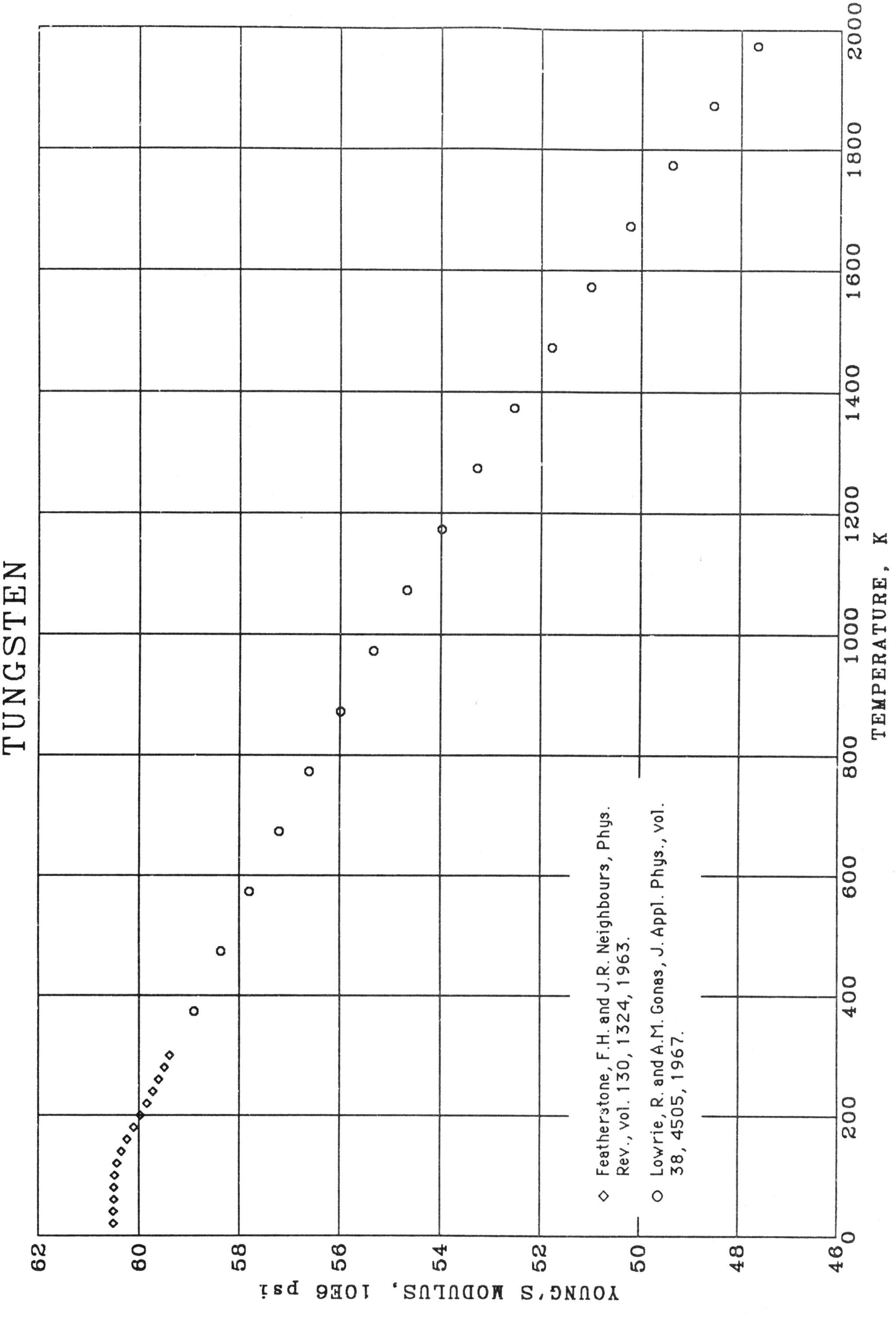
TUNGSTEN
YOUNG'S MODULUS, 10E6 psi
TEMPERATURE, K
62
60
58
56
54
52
50
48
46
0
200
400
600
800
1000
1200
1400
1600
1800
2000
◇ Featherstone, F.H. and J.R. Neighbours, Phys. Rev., vol. 130, 1324, 1963.
○ Lowrie, R. and A.M. Gonas, J. Appl. Phys., vol. 38, 4505, 1967.

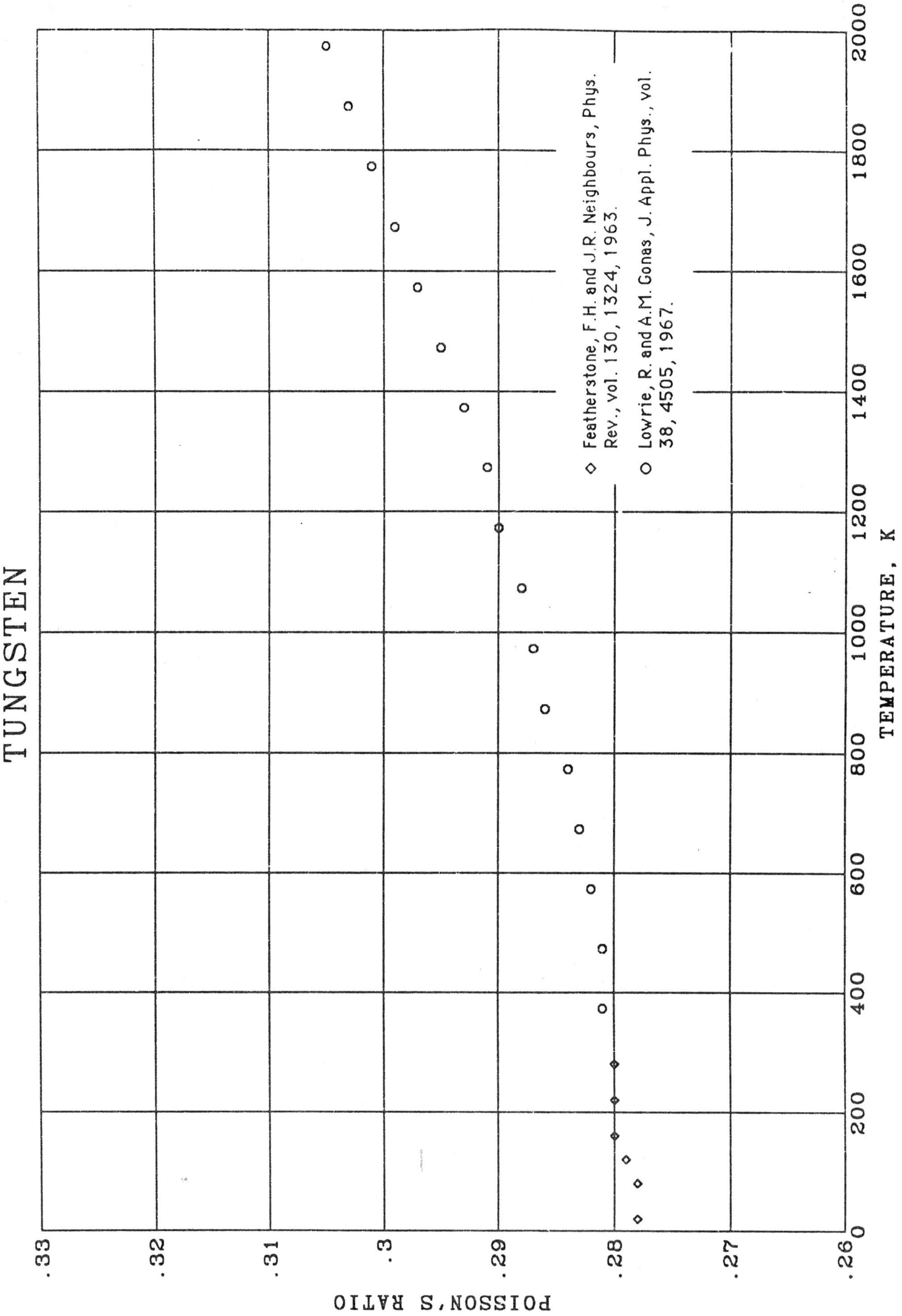
TUNGSTEN
POISSON'S RATIO
TEMPERATURE, K
.33
.32
.31
.3
.29
.28
.27
.26
0
200
400
600
800
1000
1200
1400
1600
1800
2000
◇ Featherstone, F.H. and J.R. Neighbours, Phys. Rev., vol. 130, 1324, 1963.
○ Lowrie, R. and A.M. Gonas, J. Appl. Phys., vol. 38, 4505, 1967.

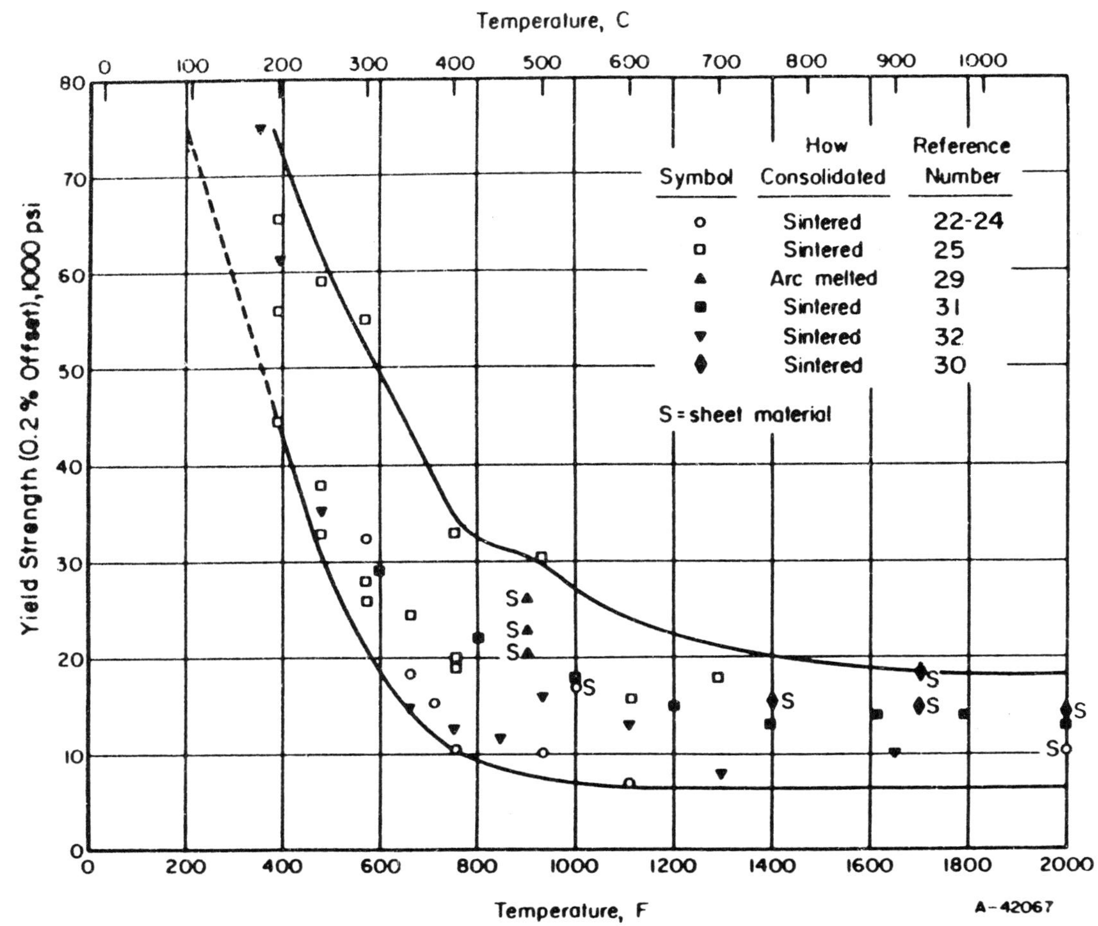

Reprinted from Schmidt, F F and H R Ogden, Battelle Memorial Inst , DMIC Rept 191, 1963

References

22 Imgram, A G , et al , Battelle Memorial Inst , WADD TR 60-278, 1960

23 Ratliff, J L , et al , Battelle Memorial Inst , Summary Rept on Contract NOw-61-0677-C, 1962

24 Ratliff, J L , Battelle Memorial Inst , unpublished data, 1962

25 Union Carbide Metals Co , WADD TR 60-144, 1960

29 Schoenfeld, W J , Universal Cyclops Steel Corp , USAF, interim repts 3 and 4, 1962

30 Japka J E and G S Hoppin, General Electric Co , final rept on Contract NOrd 88119

31 Pugh, J W , Proc ASTM, vol 57, 906, 1957

32 Bechtold, J H , Trans ASM, vol 46, 397, 1954

EFFECT OF TEMPERATURE ON THE TENSILE PROPERTIES OF RECRYSTALLIZED TUNGSTEN

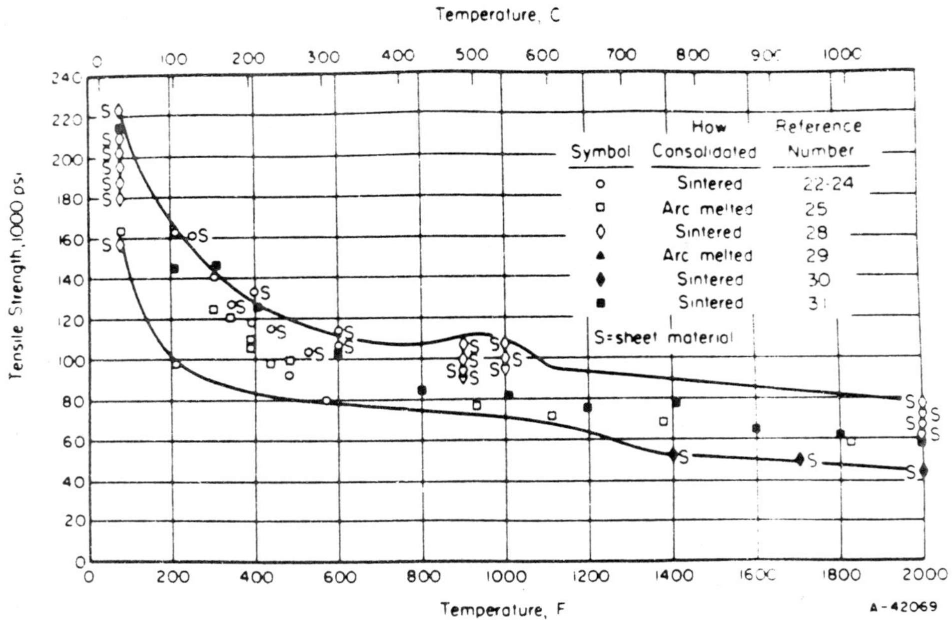

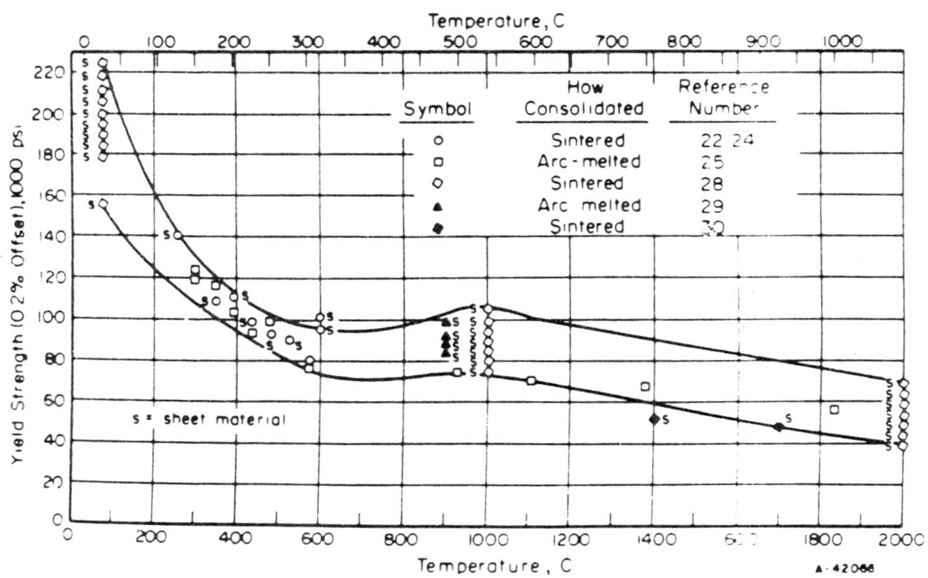

Reprinted from Schmidt, F F and H R Ogden, Battelle Memorial Inst, DMIC Rept 191, 1963

References

22 Imgram, A G, et al, Battelle Memorial Inst, WADD TR 60-278, 1960

23 Ratliff, J L, et al, Battelle Memorial Inst, Summary Rept on Contract NOw-61-0677-C, 1962

24 Ratliff, J L, Battelle Memorial Inst, unpublished data, 1962

25 Union Carbide Metals Co, WADD TR 60-144, 1960

28 Bodine, G C, Fansteel Metallurg Co, Contract NOw 60-0621-C, interim repts 6 and 9, 1961-62

29 Schoenfeld, W J, Universal Cyclops Steel Corp, USAF, interim repts 3 and 4, 1962

30 Japka, J E and G S Hoppin, General Electric Co, final rept on Contract NOrd 88119

31 Pugh, J W, Proc ASTM, vol 57, 906, 1957

EFFECT OF TEMPERATURE ON THE TENSILE PROPERTIES OF WROUGHT OR STRESS-RELIEVED TUNGSTEN FROM 0 TO 2000 F

Reprinted from Schmidt, F.F. and H.R. Ogden, Battelle Memorial Inst., DMIC Rept. 191, 1963.

References.

22 Imgram, A.G., et al., Battelle Memorial Inst., WADD TR 60-278, 1960.

23 Ratliff, J.L., et al., Battelle Memorial Inst., Summary Rept. on Contract NOw-61-0677-C, 1962.

24 Ratliff, J.L., Battelle Memorial Inst., unpublished data, 1962.

25 Union Carbide Metals Co., WADD TR 60-144, 1960

29 Schoenfeld, W.J., Universal Cyclops Steel Corp., USAF, interim repts. 3 and 4, 1962

30 Japka, J.E. and G.S. Hoppin, General Electric Co., final rept. on Contract NOrd 88119.

31 Pugh, J.W., Proc ASTM, vol 57, 906, 1957

32 Bechtold, J.H., Trans ASM, vol 46, 397, 1954

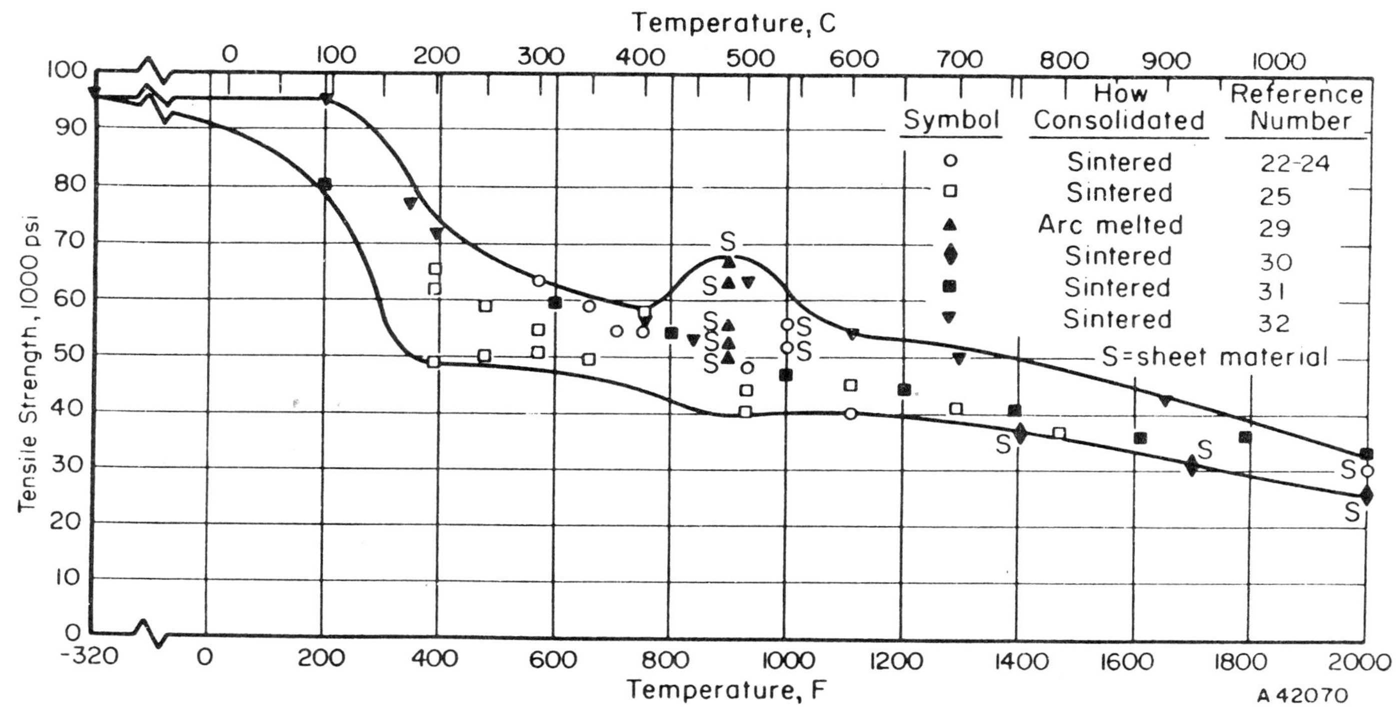

EFFECT OF TEMPERATURE ON THE TENSILE PROPERTIES OF RECRYSTALLIZED TUNGSTEN

Reprinted from Schmidt, F F and H R Ogden, Battelle Memorial Inst., DMIC Rept 191, 1963

References
22 Imgram, A G , et al , Battelle Memorial Inst , WADD TR 60-278, 1960
23 Ratliff, J L , et al , Battelle Memorial Inst , Summary Rept on Contract NOw-61-0677-C, 1962
25 Union Carbide Metals Co , WADD TR 60-144, 1960
29 Schoenfeld, W J , Universal Cyclops Steel Corp , USAF, interim repts 3 and 4, 1962
30 Japka, J E and G S Hoppin, General Electric Co , final rept on Contract NOrd 88119
31 Pugh, J W., Proc ASTM, vol 57, 906, 1957
32 Bechtold, J H , Trans ASM, vol 46, 397, 1954

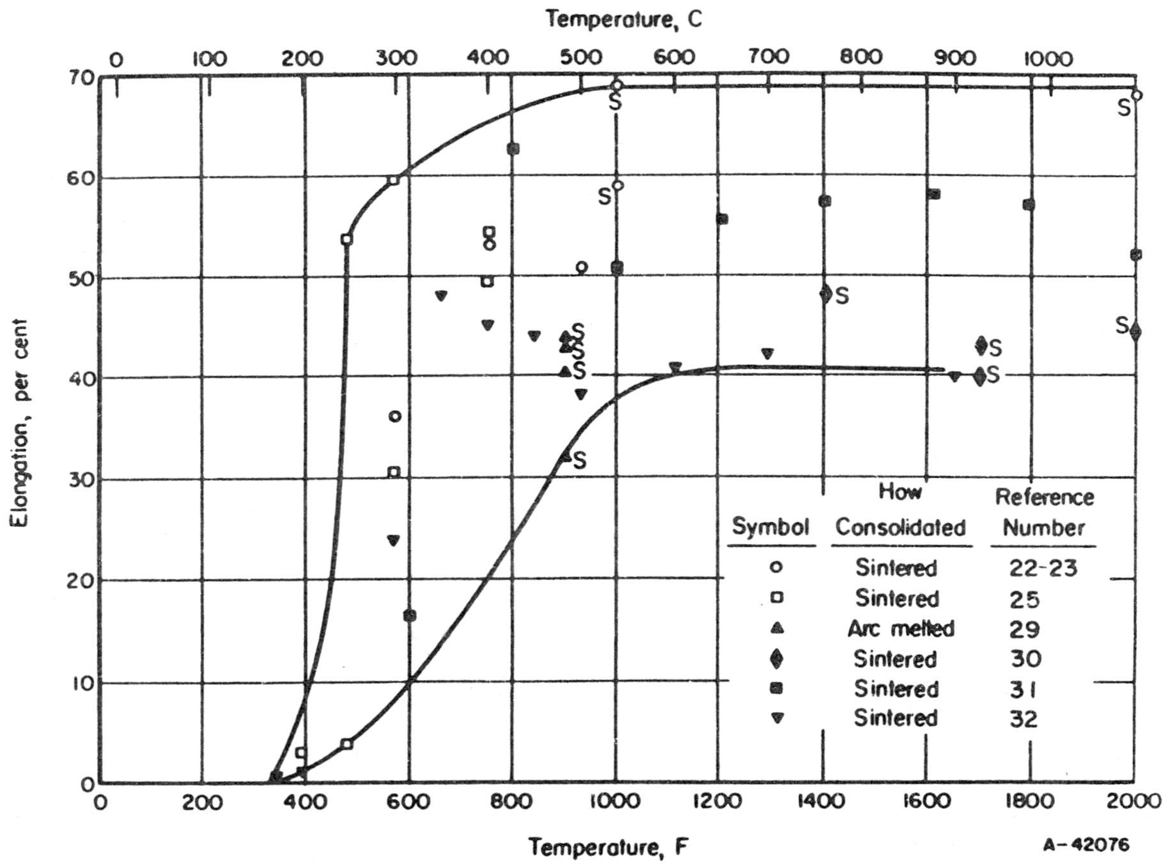

EFFECT OF TEMPERATURE ON THE TENSILE PROPERTIES OF RECRYSTALLIZED TUNGSTEN

Reprinted from Schmidt, F F and H R Ogden, Battelle Memorial Inst , DMIC Rept 191, 1963

References

22 Imgram, A G , et al , Battelle Memorial Inst , WADD TR 60-278, 1960

23 Ratliff, J L , et al , Battelle Memorial Inst , Summary Rept on Contract NOw-61-0677-C, 1962

24 Ratliff, J L , Battelle Memorial Inst , unpublished data, 1962

25 Union Carbide Metals Co , WADD TR 60-144, 1960

28 Bodine, G C , Fansteel Metallurg Co , Contract NOw 60-0621-C, interim repts 6 and 9, 1961-62

29 Schoenfeld, W J , Universal Cyclops Steel Corp , USAF, interim repts 3 and 4, 1962

30 Japka, J E and G S Hoppin, General Electric Co , final rept on Contract NOrd 88119

31 Pugh, J W , Proc ASTM, vol 57, 906, 1957

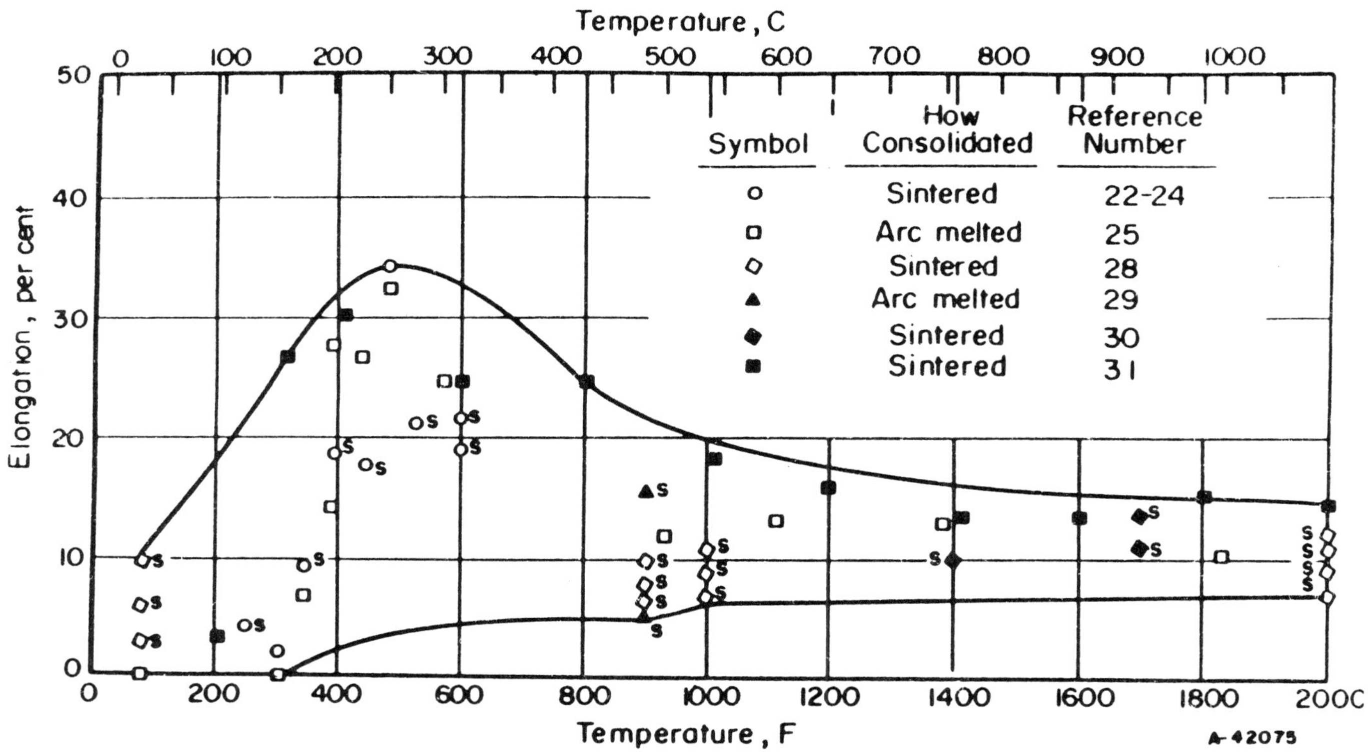

EFFECT OF TEMPERATURE ON THE TENSILE PROPERTIES OF WROUGHT OR STRESS-RELIEVED TUNGSTEN FROM 0 TO 2000 F

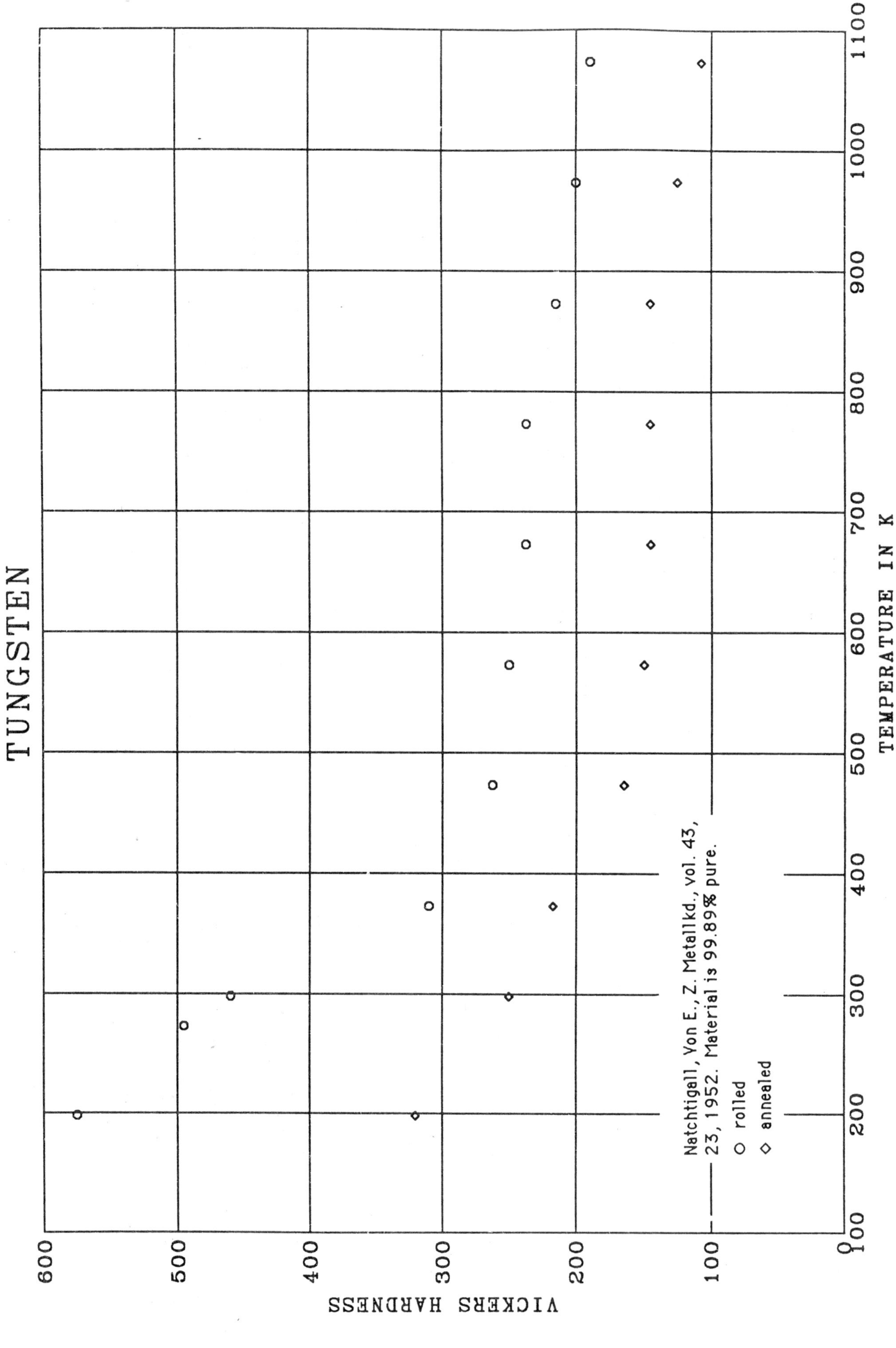

TUNGSTEN
Natchtigall, Von E., Z. Metallkd., vol. 43, 23, 1952. Material is 99.89% pure.
○ rolled
◇ annealed
VICKERS HARDNESS
TEMPERATURE IN K
0
100
200
300
400
500
600
700
800
900
1000
1100

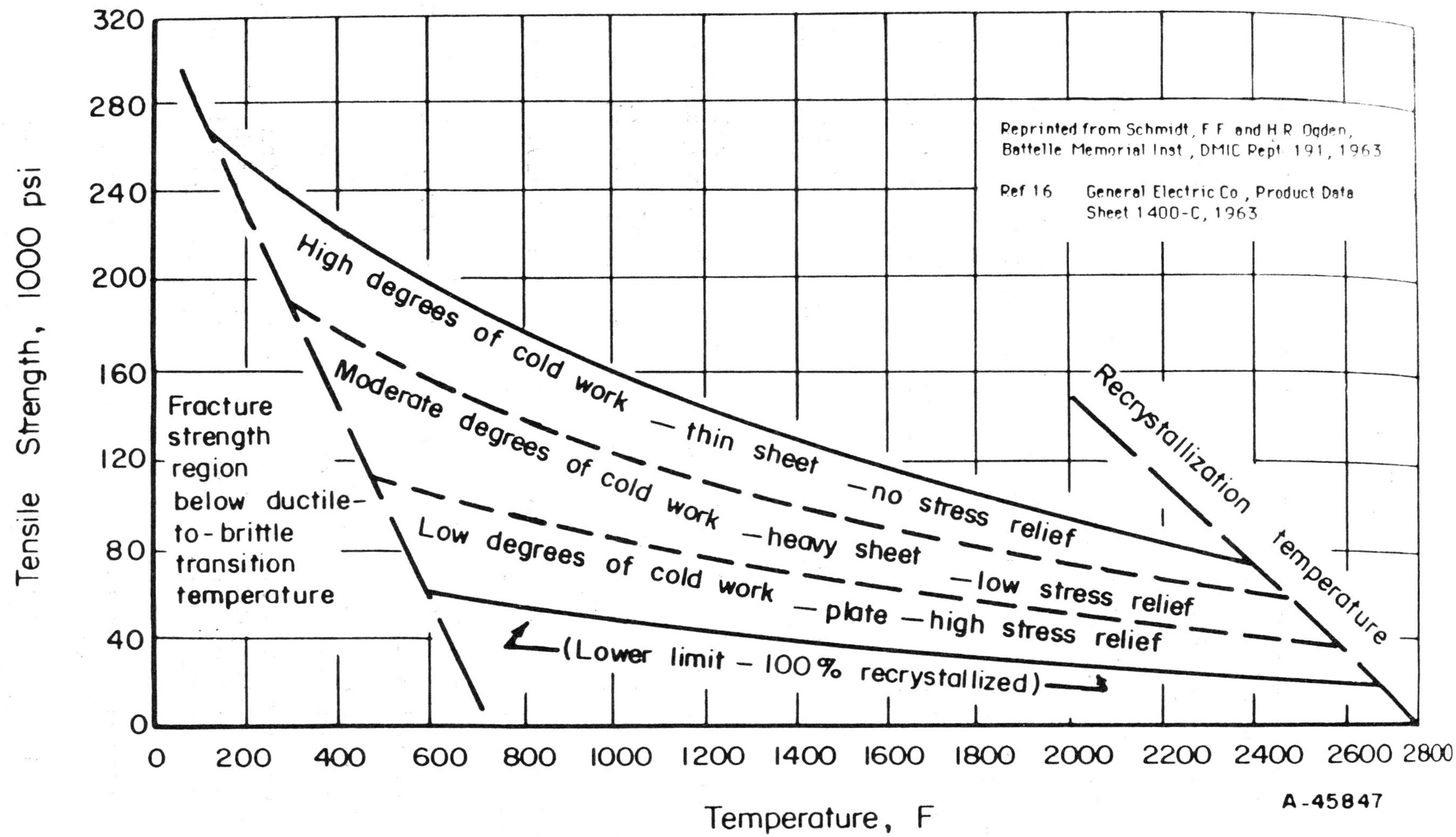

TYPICAL TENSILE-STRENGTH RANGES OF COMMERCIAL-PURITY, POWDER-METALLURGY, TUNGSTEN FLAT-ROLLED PRODUCTS(16)

REFERENCES

Aluminum

Al1. Powell, H. and Evans, E. J., Phil Mag., vol. 34(230), 145, 1943. Density at 295K = 2.71 g/cm^3.

Al2. Ehrlich, P., Z. Anorg. Chem., vol. 259, 1, 1949. Density at 298K = 2.704 g/cm^3.

Al3. Taylor, C.S., et al., Metals and Alloys, vol. 9, 189, 1938. Density at 298K = 2.6978 g/cm^3.

Al4. Giauque, W.F. and P.F. Meads, J. Am. Chem. Soc., vol. 63, 1897, 1941. Specific heat at 300K = 0.2164 cal/g-K, interpolated between 295.94K and 301.60K.

Al5. Pochapsky, T.E., Acta. Met., vol. 1, 747, 1953. Specific heat at 300K = 0.205 cal/g-K, interpolated between 273.15K and 373K.

Al6. Maier, C.G. and C.T. Anderson, J. Chem. Phys., vol. 2, 513, 1934. Specific heat at 300K = 0.213 cal/g-K, extrapolated from 296.3K.

Al7. Bidwell, C.C. and Hogan, C.L., J. Appl. Phys., vol. 18, 776, 1947. Thermal conductivity at 298.2K = 0.538 cal/cm-s-K.

Al8. Hogan, C.L., Lehigh Univ. Doctoral Dissertation, 1, 1950. Thermal conductivity at 300K = 0.569 cal/cm-s-K, interpolated between 273.2K and 373.2K.

Al9. Powell, R.W., J. Sci. Instr., vol. 34, 485, 1957. Thermal Conductivity at 298.2K = 0.571 cal/cm-s-K.

Al10. Fraser, D.B. and A.C. Hollis-Hallett, Proc. Int. Conf. Low Temp. Phys., Seventh, 689, 1961. Thermal expansion coefficient at 300K = 23.3x10^{-6} K^{-1}.

Al11. Pathak, P.D. and N.G. Vasavada, J. Phys. C, vol. 3, L44, 1970. Thermal expansion coefficient at 300K = 23.4x10^{-6} K^{-1}.

Al12. Strelkov, P.G. and S.I. Novikova, Prib. Tekh. Eksp. (USSR), vol. 5, 105, 1957; English Translation: Sandia Laboratories, Library Operations Division, Technical Translation Service, SC-T-69-1054, 1969. Thermal expansion coefficient at 300K = 24.1x10^{-6} K^{-1}.

Al13. CRC Handbook of Chemistry and Physics, 53rd ed., B-1. Melting point = 933.52K.

Al14. Cook, J.G., J.P. Moore, T. Matsumura, and M.P. Van der Meer, ORNL Report No. ORNL-5079, 1975. Electrical resistivity at 300K = 2.733 $\mu\Omega$-cm.

Al15. Katawa, S. and T. Kino, J. Phys. Soc. Japan, vol. 39, 684, 1975. Electrical resistivity 300K = 2.733 $\mu\Omega$-cm.

Al16. Boehm, R. and E. Wachtel, Z. Metallkd., vol. 60, 505, 1969. Electrical resistivity at 298.2K = 2.72 $\mu\Omega$-cm.

Al17. Gerlich, D. and E.S. Fisher, J. Phys. Chem. Solids, vol. 30, 1197, 1969. Young's modulus at 300K = 1.02×10^7 psi. Poisson's ratio at 300K = 0.346.

Al18. Kamm, G.N. and G.A. Alers, J. Appl. Phys., vol. 35, 327, 1964. Young's modulus at 300K = 1.02×10^7 psi. Poisson's ratio at 300K = 0.347.

Al19. Vallin, J. M. Mongy, K. Salama, and O. Beckman, J. Appl. Phys., vol. 35, 1825, 1964. Young's modulus at 300K = 1.02×10^7 psi. Poisson's ratio at 300K = 0.346.

Al20. Technical Report No. C-6, The Aluminum Assn., 1978. Annealed, 99.99% pure: Yield strength at 300K = 1500 psi. Ultimate tensile strength at 300K = 6500 psi. Elongation at 300K = 50%.

Al21. The Aluminum Data Book, Reynolds Aluminum Co., 1957. Annealed, 99.45% pure: Yield strength at 300K = 4000 psi. Ultimate tensile strength at 300K = 12000 psi.

Al22. Martin, T., J. Inst. Metals, vol. 31, 121, 1924. Fully annealed rod, 99.57% pure: Ultimate tensile strength at 300K = 9950 psi, interpolated between 289K and 373K. Elongation at 300K = 52.7%, interpolated between 289K and 373K. Cold-worked rod, 99.57% pure: Ultimate tensile strength at 300K = 14400 psi, interpolated between 289K and 373K. Elongation at 300K = 28.9%, interpolated between 289K and 373K.

Al23. Edwards, J.D., F.C. Frary, and Z. Jeffries, The Aluminum Industry, 1930. Annealed, 99.95% pure: Yield strength (0.2% offset) at room temp. = 3000 psi. Elongation (2 in. gauge) at room temp. = 60%.

Al24. Schwinning, Von W. and F.Fischer, Z. Metallkd., vol. 22, 1, 1930. 8-mm rod, 99.5% pure: Brinell hardness at 300K = 30.

Al25. Ludwik, P., Z. Physik. Chem., vol. 91, 232, 1916. Brinell hardness at 300K = 40.

Al26. Taylor, C.S., et.al., Metals and Alloys, vol. 9, 189, 1938. Brinell hardness using 12.61 Kg load and 1/16 inch ball at room temp. = 17.

Al27. Reinhard Glang, J. Vac. Sci. & Tech., vol. 3, 37, {?yr}

Al28. M. Janda, Thin Solid Films, vol. 112, 219, 1984.

A129. H. Slick, Aluminum, vol. 20, 75, 1938.

A130. J.E. Gordon, Growth an Perfection in Crystals, Doremus Roberts and Turnbull, eds., Wiley and Sons, New York, 219, 1958.

Chromium

Cr1. Lucks, C.F. and H.W. Deem, Battelle Memorial Inst., WADC TR 55-496, 1, 1956. [AD97185] Density at 297K = 7.16 g/cm^3.

Cr2. Hidnert, P.J., Res. Natl. Bur. Std., vol. 26, 81, Res. Paper No. 1361. Density at 298K = 7.10 g/cm^3.

Cr3. Neuburger, M.C., Z. Krist., vol. 86, 395, 1933. Density at 290K = 7.188 g/cm^3.

Cr4. Koenig, J.H., Rutgers Univ. N.J. Ceran. Research Sta. Progr. Rept. 6, 1, 1952, [ATI 163519]. Specific heat at 299.443K = 0.1085 cal/g-K.

Cr5. Plietz, G.L., A.T. Muccigrosso, and C.W. Krystyniak, USAEC Rept. KAPL-1918, 1, 1958. Specific heat at 300K = 0.1067 cal/g-K, interpolated between 273.15K and 373K.

Cr6. Nishima, T., T. Kawada, and A. Ishihata, J. Am. Ceram. Soc., vol. 48, 31, 1965. Specific heat at 291.1K = 0.1066 cal/g-K.

Cr7. Sochtig, H., Ann. Physik, vol. 38, 97, 1940. Thermal conductivity at 299.2K = 0.06355 cal/cm-s-K.

Cr8. Powell, R.W. and R.P. Tye, J. Inst. Metals, vol. 85, 185, 1957. Thermal conductivity at 323.2K = 0.2054 cal/cm-s-K.

Cr9. Coblentz, W.W. and R. Stair, Nat. Bur. Std. Research Paper no. 39, 1928. Thermal conductivity at 293K = 0.16 cal/cm-s-K.

Cr10. White, G.K., Aust. J. Phys., vol. 14, 359, 1961. Thermal expansion coefficient at 300K = 4.44x10^{-6} K^{-1}.

Cr11. Kondorskii, E.I., T.I. Kostina, and L.N. Ekonomova, J. Phys. (Paris), Colloq., Part 1, C1-417-C1-8, 1971. Thermal expansion coefficient at 300K = 9.6x10^{-6} K^{-1}.

Cr12. Fine, M.E., E.S. Greiner, and W.C. Ellis, J. Metals, vol. 191, 56, 1951. Thermal expansion coefficient at 293K = 4.1x10^{-6} K^{-1}.

Cr13. CRC Handbook of Chemistry and Physics, 53rd ed., B-1. Melting point = 2130K.

Cr14. Fine, M.E., E.S. Greiner, and W.C. Ellis, J.Metals, vol. 191, 56, 1951. Electrical resistivity at 298K = 13.5 $\mu\Omega$-cm.

Cr15. Harper, A.F.A., W.R.G. Kemp, P.G. Klemens, R.J. Tainsh, and G.K. White, Phil Mag., vol. 2, 577, 1957. Electrical resistivity at 300K = 12.9 $\mu\Omega$-cm.

Cr16. Moore, J.P., R.K. Williams and R.S. Graves, J. Appl. Phys., vol. 48, 610, 1977. Electrical resistivity at 300K = 12.7 $\mu\Omega$-cm.

Cr17. Bolef, D.I., J. Appl. Phys., vol. 33, 2311, 1962. Young's modulus at 298K = 4.049x10^7 psi. Poisson's ratio at 300K = 0.212.

Cr18. Palmer, S.B. and E.W. Lee, Phil. Mag., vol. 24, 311, 1971. Young's modulus at 300K = 4.034x10^7 psi. Poisson's ratio at 300K = 0.215.

Cr19. Sumer, A. and J.F. Smith, J. Appl. Phys., vol. 34, 2691, 1963. Young's modulus at 298K = 3.962x10^7 psi. Poisson's ratio at 298K = 0.201.

Cr20. Henderson, F., F.P. Bullen, and H.L. Wain, J. Inst. Metals, vol. 98, 65, 1970. 99.7% cold-worked: Yield strength at 300K = 227000 psi, interpolated between 273K and 323K. Ultimate tensile strength at 298K = 253000 psi. Elongation at 300K = 2.1%.

Cr21. Allen, B.C., D.J Maykuth, and R.I. Jaffee, Trans. AIME, vol. 227, 724, 1963. Wrought, furnace-cooled: Upper yield point at 298K = 65500 psi. Ultimate tensile strength at 298K = 64200 psi. Elongation at 298K = 57%.

Cr22. Marcinkowski, M.J. and Lipsit, H.A., Acta Met., vol. 10, 95, 1962. Fully recrystallized: Lower yield strength at 300K = 33000 psi.

Cr23. Weaver, C.W., J. Inst. Metals, vol. 89, 385, 1960-61. Arc-melted, fully recrystallized, 350 to 400 grains/mm^2, 0.002% N_2 contg.: Fracture stress at 300K = 38400 psi. Elongation at 300K = 0%.

Cr24. Gilbert, H.L., H.L. Johansen, and R.G. Nelson, U.S. Bur. Mines, Rept. Invest., No. 4905, 1952. Brinell hardness at 300K = 89, interpolated between 273K and 373K.

Cr25. Bens, F., Trans. ASM, vol. 38, 505, 1947. Vickers hardness at 300K = 135.

Cr26. Allen, B.C., et.al., Trans. AIME, vol. 227, 724, 1963. Hardness in the fully annealed condition at room temperature lies between 120 and 130 VHN.

Cr27. See Al27.

Cr28. Allen, B.C., Trans. AIME, vol. 227, 724, 1963.

Copper

Cu1. Lucks, C.F. and H.W. Deem, Battelle Memorial Inst., WADC TR 55-496, 1, 1956. [AD97185] Density at 297K = 8.92 g/cm^3.

Cu2. Pawlek, F. and K. Reichel, Z. Metallkd., vol. 47, 347, 1956. Density at 298K = 8.94 g/cm^3.

Cu3. Giauque, W.F. and P.F. Meads, J. Am. Chem. Soc., vol. 63, 1897, 1941. Density at 298K = 8.95 g/cm^3.

Cu4. Giauque, W.F. and P.F. Meads, J. Am. Chem. Soc., vol. 63, 1897, 1941. Specific heat at 300K = 0.09227 cal/g-K, interpolated between 294.76K and 300.15K.

Cu5. Martin, D.L., Can J. Phys., vol. 38, 17, 1960. Specific heat at 300K = 0.09192 cal/g-K.

Cu6. Martin, D.L., Can J. Phys., vol. 40, 1166, 1962. Specific heat at 300K = 0.08935 cal/g-K, interpolated between 299.72 and 300.70.

Cu7. Vernotte, P. and A. Jeufroy, Compt. Rend., vol. 192, 612, 1931. Thermal Conductivity at 300K = 0.9236 cal/cm-s-K.

Cu8. Lyubimov, A.P. and D.K. Belashchenko, Sbor. Moskovskogo Inst. Stali, Bol 33, 3, 1955. Thermal Conductivity at 303.2 K = 0.9412 cal/cm-s-K.

Cu9. Larson, K.B. and K Koyama, J. Appl. Phys., vol. 39, 4408, 1968. Thermal Conductivity at 298.2K = 0.9078 cal/cm-s-K.

Cu10. Fraser, D.B. and A.C. Hollis-Hallett, Proc. Int. Conf. Low Temp. Phys., Seventh, 689, 1961. Thermal expansion coefficient at 300K = 16.4×10^{-6} K^{-1}.

Cu11. Strelkov, P.G. and S.I. Novikova, Prib. Tekh. Eksp. (USSR), vol. 5, 105, 1957; English Translation: Sandia Laboratories, Library Operations Division, Technical Translation Service, SC-T-69-1054, 1969. Thermal expansion coefficient at 301K = 16.07×10^{-6} K^{-1}.

Cu12. Nasekovskii, A.P., Ukr. Fiz. Zh., vol. 12, 1352, 1967. Thermal expansion coefficient at 300K = 15.81×10^{-6} K^{-1}.

Cu13. CRC Handbook of Chemistry and Physics, 53rd ed., B-1. Melting point = 1356.6K.

Cu14. Radenac, A., M. Lacoste and C. Roux, Rev. Int. Hautes Temp. Refract., vol. 7, 389, 1970. Electrical resistivity at 300K = 1.73 $\mu\Omega$-cm.

Cu15. Moore, J.P., D.L. McElroy, and R.S. Graves, Can. J. Phys., vol. 45, 3849, 1967. Resistivity at 300K = 1.725 $\mu\Omega$-cm.

Cu16. Svensson, B., Ann. Phys. (Leipzig), vol. 5, 263, 1936. Electrical resistivity at 300K = 1.73 μΩ-cm, interpolated between 273K and 323K.

Cu17. Chang, Y.A., and L. Himmel, J. Appl. Phys., vol. 37, 3567, 1966. Young's modulus at 300K = 1.932×10^7 psi. Poisson's ratio at 300K = 0.339.

Cu18. Overton, W.C., Jr., and J. Gaffney, Phys. Rev., vol. 34, 2494, 1963. Young's modulus at 300K = 1.917×10^7 psi. Poisson's ratio at 300K = 0.339.

Cu19. Swartz, K.D. and A.V. Granato, J. Acoust. Soc. Am., vol. 38, 824, 1965. Young's modulus at 300K = 1.919×10^7 psi. Poisson's ratio at 300K = 0.339.

Cu20. Jenkins, W.D., T.G. Digges, and C.R. Johnson, J. Res. Nat. Bur. Std., vol. 52, 201, 1957. Annealed, 0.025-mm av. grain diam.: Yield strength at 297K = 12200 psi. Ultimate tensile strength at 297K = 31900 psi. Elongation at 297K = 51%. Annealed, 0.120-mm av. grain diam.: Yield strength at 297K = 6000 psi. Ultimate tensile strength at 297K = 32300 psi. Elongation at 297K = 62%.

Cu21. Standards Hdbk, pt.2, Copper Development Assn., Inc., 1985. 0.050-mm tempered rod: Yield strength at 300K = 10000 psi. Ultimate tensile strength at 300K = 32000 psi. Elongation at 300K = 55%. Rockwell hardness at 300K = 40F.

Cu22. Carreker, R.P., Jr. and W.R. Hibbard, Acta Met., vol. 1, 654, 1953. Annealed, 0.045-mm grain size, 99.999% pure: Yield strength at 300K = 600 psi. Annealed, 0.030-in. diam., 99.999% pure: Ultimate tensile strength at 300K = 32000 psi.

Cu23. Colbeck, E.W. and W.E. MacGillivray, Trans. Inst. Chem. Engrs., vol. 11, 107, 1933. Annealed, 0.504-in. diam., 99.98% pure: Elongation at 300K = 51%.

Cu24. Ito, Kiyosi, Tohoku Sci, Repts., vol. 12, 137, 1923. 25-mm diam. bar, 99.8% pure. Cast bar: Brinell hardness at 300K = 55.4, interpolated between 288K and 303 K. Rolled bar: Brinell hardness at 300K = 104.8, interpolated between 294K and 308K.

Cu25. Kurth, A., Vereines Deut. Ingen., vol. 53, 85, 209, 1909. Brinell hardness with 10 mm ball and 400 kg load = 37 at 300K.

Cu26. F. Pester, Z. Metallk., vol. 22, 261, 1930.

Cu27. F. Pester, Z. Metallk., vol. 24, 115, 1932.

Cu28. Vook, R.W. and F. Witt, J Vacuum Sci. Tech., vol. 2:49,243, 1965.

Cu29. Oding, A. and I.T. Aleksanyan, Soviet Phys. Doklady, vol. 8, 818, 1964.

Germanium

Ge1. Straumanis, M.E. and E.Z. Aka, J. Appl. Phys., vol. 23, 330, 1952. Density at 298K = 5.32 g/cm^3.

Ge2. CRC Handbook of Chemistry and Physics, 53rd ed., B-92. Density = 5.35 g/cm^3.

Ge3. Weiting, R., Z. Metallkd., vol. 43, 158, 1952. Density = 5.3234 g/cm^3.

Ge4. Flubacher, P., A.J. Leadbetter, and J.A. Morrison, Phil. Mag., vol. 4, 273, 1959. Specific heat at 300.01K = 0.07726 cal/g-K.

Ge5. Gul'tyaev, P.V. and A.V. Petrov, Soviet Phys. Solid State, vol. 1, 330, 1959. Specific heat at 300K = 0.0753 cal/g-K.

Ge6. Sommelet, P. and R.L. Orr, J. Chem. Eng. Data, vol. 11, 64, 1966. Specific heat at 298.15K = 0.0758 cal/g-K.

Ge7. McCarthy, K.A. and S.S. Ballard, Phys. Rev., vol. 99, 1104, 1955. Thermal conductivity at 300K = 0.138 cal/cm-s-K, interpolated between 294.4K and 308.7K.

Ge8. Devyatkova, E.D. and I.A. Smirnov, Soviet Phys. Tech. Phys., vol. 2, 1805, 1957. Thermal conductivity at 300K = 0.154 cal/cm-s-K, interpolated between 275K and 305K.

Ge9. Slack, G.A. and C. Glassbrenner, Phys. Rev., vol. 120, 782, 1960. Thermal conductivity at 300K = 0.143 cal/cm-s-K.

Ge10. Zhdanova, V.V. and T.A. Kontorova, Soviet Phys. Solid State, vol. 7, 2685, 1966. Thermal expansion coefficient at 299K = $5.83x10^{-6}$ K^{-1}.

Ge11. Novikova, S.I., Soviet Phys. Solid State, vol. 2, 37, 1960. Thermal expansion coefficient at 301K = $6.00x10^{-6}$ K^{-1}.

Ge12. Nasekovskii, A.P., Ukr. Fiz. Zh., vol. 12, 1352, 1967. Thermal expansion coefficient at 300K = $5.81x10^{-6}$ K^{-1}.

Ge13. CRC Handbook of Chemistry and Physics, 53rd ed., B-1. Melting point = 1210.7 K.

Ge14. Kurova, I. A., et. al., Sov. Phys. Solid State, vol. 1, 1353, 1959. Resistivity varies from 0.80 to 22 with carrier concentrations in the range $3x10^{13}/cm^3$ to $3x10^{15}$(n-type). TCR varies wildely as the carrier concentration varies.

Ge15. Morin, F. J., et. al., Phys. Rev., vol. 94, 1525, 1954. Resistivity of intrinsic Ge at 300K = 0.018 ohm-cm. TCR = -1440 ppm/K.

Ge16. Gartner, W. W., Semiconductor Products, vol. 3, 29, 1960. Resistivity varies from 0.015 to 450 ohm-cm as the carrier concentration varies from 10^{13} to $10^{19}/cm^3$ for n-type at 300K. P-type varies from 0.014 to 300 for the same carrier concentration ranges. TCR varies wildy with the doping level.

Ge17. Bogardus, E.H., J. Appl. Phys., vol. 36, 2504, 1965. Young's modulus at 298K = $1.902x10^7$ psi. Poisson's ratio at 298K = 0.208.

Ge18. McSkimin, H.J., J. Acoust. Soc Am., vol. 31, 287, 1959. Young's modulus at 298K = $1.913x10^7$ psi. Poisson's ratio at 298K = 0.207.

Ge19. Burenkov, Yu.A., S.P. Nikanorov, and A.V. Stepanov, Sov. Phys. Solid State, vol. 12, 1940, 1971. Young's modulus at 298K = $1.910x10^7$ psi. Poisson's ratio at 298K = 0.181.

Ge20. Suzuki, T., The Inst. for Solid State Physic, Tokyo University and Dayano, H. and J. Watanabe, The research Institute for Iron, Steel and Other Metals, Tokyo University, Sendai, Japan. Increasing the concentration of doping impurities from 10^{13} to 5 x $10^{19}/cm^3$ reduces the fracture stress from 57,000 to 11,400 psi.

Ge21. Govorkov, V.G. and V.R. Regel, Sov. Phys. Solid State, vol. 3, 958, 1961. Maximum compressive stress at 293K = 64,000 psi for <111> direction and strain rate of $6.3x10^{-4}$/sec.

Ge22. Wolff, G. A., et. al., Semiconductors and Phosphorus, Proc., Internat. Colloquium, 1956, Garmisch-Partenkirchen. Schon, M. and H. Welker, eds., N.Y. Interscience, 463, 1958. Knoop microhardness at 298K = 780 kg/mm^2(110) and 845 kg/mm^2(111).

Ge23. Ablova, M. S., Sov. Phys. Solid State, vol. 3, 1320, 1961. Microhardness with diamond pyramid indenter, 200 gram load applied for 15-20 seconds = 860 (110) and 930 (100) and 1020 (111) $kg\backslash mm^2$ at room temperature.

Ge24. Trefilov, V.I., et.al., Sov. Phys. Doklady, vol. 8, 1240, 1964. Microhardness on the (111) is 770 kg/mm^2 at room temperature.

Gold

Au1. Seybold, A.U., USAEC Publ. AECD-2434, 1. Density at 298K = 19.30 g/cm^3.

Au2. Neighborurs, J.R. and G. A. Alers, Phys Rev., vol. 111, 707, 1958. Density = 19.3 g/cm^3 at 300K.

Au3. Chang, Y.A., et.al., J. Appl Phys., vol. 37, 3567, 1966. Density at 300K = 19.300 g/cm^3.

Au4. Geballe, T.H. and W.F. Giauque, J.A. Chem. Soc., vol. 74, 2368, 1952. Specific heat at 298.46 K = 0.03088 cal/g-K.

Au5. Kelley, K.K., U.S. Bur. Mines, Bull. 371, 476, 1934. Specific heat at 298K = 0.03081 cal/g-K.

Au6. Eastman, E.D., A.M. Williams, and T.F. Young, J. Am. Soc., vol. 46, 1184, 1924. Specific heat at 300K = 0.0306 cal/g-K, interpolated between 273K and 323K.

Au7. Meissner, W., Ann. Physik., vol. 4, 1001, 1915. Thermal conductivity at 300K = 0.743 cal/cm-s-K.

Au8. Schulze, F.A., Physik. Z., vol. 12, 1028, 1911. Thermal conductivity at 298.2K = 0.760 cal/cm-s-K.

Au9. Masumoto, H., Sci. Rept. Tohoku Imp. Univ., vol. 16, 417, 1927. Thermal conductivity at 297.2K = 0.7126 cal/cm-s-K.

Au10. Fraser, D.B. and A.C. Hollis-Hallett, Proc. Int. Conf. Low Temp. Phys., Seventh, 689, 1961. Thermal expansion coefficient at 300K = 14.4×10^{-6} K^{-1}.

Au11. Leksina, I.E. and S.I. Novikova, Soviet Phys. Solid State, vol. 5, 798, 1963. Thermal expansion coefficient at 300K = 14.06×10^{-6} K^{-1}.

Au12. White, G.K. and J.G. Collins, J. Low Temp. Phys., vol. 7, 43, 1972. Thermal expansion coefficient at 283K = 14.08×10^{-6} K^{-1}.

Au13. CRC Handbook of Chemistry and Physics, 53rd ed., B-1. Melting point = 1337.58K.

Au14. Domenicali, C.A. and E.L. Christenson, J. Appl. Phys., vol. 32, 2450, 1961. Electrical resistivity at 299K = 2.33 $\mu\Omega$-cm.

Au15. Damon, D.H., M.P. Mathur and P.G. Klemens, Phys. Rev., vol. 176, 876, 1968. Electrical resistivity at 300K = 2.238 $\mu\Omega$-cm.

Au16. Laubitz, M.J., Can. J. Phys., vol. 47, 2633, 1969. Electrical resistivity at 300K = 2.271 $\mu\Omega$-cm, interpolated between 282K and 307K.

Au17. Chang, Y.A., and L. Himmel, J.Appl. Phys., vol. 37, 3567, 1966. Young's modulus at 300K = 1.13×10^7 psi. Poisson's ratio at 300K = 0.425.

Au18. Neighbours, J.R. and G.A. Alers, Phys. Rev., vol. 111, 707, 1958. Young's modulus at 300K = 1.12×10^7 psi. Poisson's ratio at 300K = 0.426.

Au19. Koster, W., Z. Metallkd., vol. 39, 1, 1948. Young's modulus at 293K = 1.12×10^7 psi.

Au20. Gruneisen, E., Ann. Phys., vol. [4] 25 [1908] 825/51,845. Poisson's ratio = 0.420 for 99.8% purity at room temperature.

Au21. McKeown, J. and O.F. Hudson, J. Inst. Met., vol. 60, 109, 1937. 99.9% pure, 0.8-in. diam. rod, annealed 10 min. at 500C and then strained 5%.: Ultimate tensile strength at room temp. = 17900 psi. Elongation (2 in. gauge) at room temp. = 65%. 0.01% proof stress at room temp. = 8100 psi.

Au22. Coleman, R.L., J. Res. Nat. Bur. Std., vol. 1, 867, 1928. 1/8-in. diam. cast bar: Ultimate tensile strength at room temp. = 16500 psi. Elongation (3 in. gauge) at room temp. = 34%.

Au23. Lauderdale, R.H., et.al., Metals and Alloys, vol. 10, 24, 1939. 0.035 in. sheet, annealed 30 min. at 260°C and furnace cooled. Ultimate tensile strength at room temp. = 17700 psi. Elongation (1 in. gauge) at room temp. = 33%. Brinell hardness at room temp. = 28.

Au24. Meshii, M., Acta Met, vol. 7, 180, 1959. The yield strength of 99.999% gold wire annealed and slow cooled = 500 psi at room temperature. For quenched material, the yeild strength = 1500 psi.

Au25. Sachs, G., L.B.V. Erg.-Bd., 1927, S.49. Brinell hardness = 18.5 at room temperature.

Au26. Shell, J.H., Metals and Alloys, 15[1942] 236/240, 237. Brinell hardness = 16 to 20 at room temperature.

Au27. Sachs, G. and J. Weert, Z. Phys., vol. 62, 473, 1930. The yield strength was found to be between 253 to 335 psi at room temperature.

Au28. Gillham, E.J., Phil. Mag., vol. 46, 1051, 1955.

Au29. Beans, J.W., Structure and Properties of Thin Films, John Wiley and Sons, 183, 1959.

Au30. Simo-Pekka Hannula, John Wangel and Che-Yu-Li, IEEE Trans Components and Hybrids, vol CHMT-6, no 4, 494-502, 1983.

Iron

Fe1. Cleaves, H.E. and J.M. Hiegel, J. Res. Natl. Bur. Std. (Res. paper No. 1472), vol. 28, 643, 1942. Density at 293K = 7.874 g/cm^3.

Fe2. Pattee, H.E. and R.M. Evans, DMIC, DMIC memo. No. 153, 1, 1962. [AD278193] Density at 293K = 7.87 g/cm^3.

Fe3. Pallister, P.R., J. Oron Steel Inst., vol. 161, 87, 1949. Density at 298K = 7.875 g/cm^3.

Fe4. Pallister, P.R., J. Oron Steel Inst., vol. 161, 87, 1949. Specific heat at 300K = 0.106 cal/g-K, interpolated between 293K and 323K.

Fe5. Wallace, D.C., P.H. Sidles, and G.C. Danielson, J. Appl. Phys., vol. 31, 168, 1960. Specific heat at 298K = 0.104 cal/g-K.

Fe6. Anderson, P.D. and R. Hultgren, Trans. Met. Soc. AIME, vol. 224, 842, 1962. Specific heat at 298.15K = 0.1074 cal/g-K.

Fe7. Kohlhaas, R. and W. Kierspe, Arch. Eisenhuttenw., vol. 36, 301, 1965. Thermal conductivity at 300K = 0.19 cal/cm-s-K.

Fe8. Richter, F. and R. Kohlhaas, Z. Naturforsch., vol. 19A, 1632, 1964. Thermal conductivity at 300K = 0.192 cal/cm-s-K, interpolated between 293.2K and 323.2K.

Fe9. Lucks, C.F., H.B. Thompson, A.R. Smith, F.P. Curry, H.W. Deem, and G.F. Bing, USAF TR 6145, I, 1, 1951. [ATI 117715] Thermal conductivity at 300K = 0.181 cal/cm-s-K, interpolated between 265.3K and 344.8K.

Fe10. Fraser, D.B. and A.C. Hollis-Hallett, Proc. Int. Conf. Low Temp. Phys., Seventh, 689, 1961. Thermal expansion coefficient at 300K = 11.7×10^{-6} K^{-1}.

Fe11. Austin, J.B. and R.H.H. Pierce, Jr., Physics, vol. 4, 409, 1933. Thermal expansion coefficient at 293K = 11.58×10^{-6} K^{-1}.

Fe12. Arbuzov, M.P. and I.A. Zelenkov, Phys. Metals Metallogr., vol. 18, 149, 1964. Thermal expansion coefficient at 293K = 11.65×10^{-6} K^{-1}.

Fe13. CRC Handbook of Chemistry and Physics, 53rd ed., B-1. Melting point = 1808K.

Fe14. Domenicali, C.A. and F.A. Offer, J. Appl. Phys., vol. 26, 377, 1955. Electrical resistivity at 300K = 10 $\mu\Omega$-cm.

Fe15. White, G.K. and S.B. Woods, Phil. Trans. Roy. Soc. London, vol. A251, 273, 1959. Electrical resistivity at 295K = 9.8 $\mu\Omega$-

cm.

Fe16. Weiner, L, P. Chiotti, and H.A. Wilhelm, USAEC Publ. ISC-305, 1. [AD 18042] Electrical resistivity at 300K = 9.5 μΩ-cm.

Fe17. Rayne, J.A. and B.S. Chandresekhar, Phys. Rev., vol. 122, 1714, 1961. Young's modulus at 300K = 3.098×10^7 psi. Poisson's ratio at 300K = 0.288.

Fe18. Leese, J. and A.E. Lord, Jr., J. Appl. Phys., vol. 39, 3986, 1968. Young's modulus at 298K = 2.946×10^7 psi. Poisson's ratio at 298K = 0.299.

Fe19. Lord, A.E., Jr., and D.N. Beshers, J. Appl. Phys., vol. 36, 1620, 1965. Young's modulus at 300K = 3.052×10^7 psi. Poisson's ratio at 300K = 0.286.

Fe20. Cleaves, H.E. and J.M. Hiegel, J. Res. Nat. Bur. Std., vol. 28, 643, 1942. Swaged rod, 1/4-in. diam., annealed 4 hr at 1600°F, 99.99% pure: Yield strength at room temp. = 8100-8600 psi. Ultimate tensile strength at room temp. = 28500-29700 psi. Elongation at room temp. = 36-46%. Brinell hardness at room temp. = 60.

Fe21. Adcock, F. and C.A. Bristow, Proc.Roy. Soc. London, vol. A153, 172, 1935-36. Rod, 1/4-in. diam., 99.989% pure: Ultimate tensile strength at room temp. = 20400-26700 psi. Elongation at room temp. = 36-49%.

Fe22. Besnard, S., New Physical and Chemical Properties of Metals of Very High Purity, Gordon and Breach, 1965, Science Publishers, Library of Congress 65-16833. The yield point of zone refined iron varies from 7800 to 10,700 psi. The ultimate tensile strength varies from 12,800 to 19,900 psi. The elongation is 22%. All properties at room temperature.

Fe23. Takeda, T., Jap. J. Appl. Phys., vol. 12, 974, 1973. Zone-purified single crystals: Vickers hardness at 300K = 65.

Fe24. Matsui, H., et.al., Trans. Japan Inst Met, vol. 19, 163, 1978. 0.1% yield strength (RRR = 3600 at 300K, strain rate 8.33×10^{-5}) = 5800 psi.

Molybdenum

Mo1. Lucks, C.F. and H.W. Deem, Battelle Memorial Inst., WADC TR 55-496, 1, 1956. [AD97185] Density at 297K = 10.24 g/cm^3.

Mo2. Fieldhouse, I.B., J.C. Hedge, and T.E. Waterman, Armour Res. Foundation, WADC TR 55-495, I, 1, 1956. [AD110404] Density at 298K = 10.24 g/cm^3.

Mo3. Pattee, H.E. and R.M. Evans, DMIC, DMIC memo. No. 153, 1, 1962. [AD278193] Density at 293K = 10.22 g/cm^3.

Mo4. Cape, J.A., Atomics International, AI-6127, 1, 1961. [AD 251317] Specific heat at 300K = 0.0614 cal/g-K.

Mo5. Taylor, R.E. and Finch, R.A., U.S. At. Energy Comm., NAA-SR-6034, 1, 1961. Specific heat at 300K = 0.0592 cal/g-K.

Mo6. Cooper, C. and G.O. Langstroth, Phys. Rev., vol. 33, 243, 1929. Specific heat at 298K = 0.0597 cal/g-K.

Mo7. Kannuluik, W.G., C.E. Eddy, and T.H. Oddie, Proc. Roy. Soc. (London), vol. A141, 159, 1933. Thermal conductivity at 300K = 0.3290 cal/cm-s-K.

Mo8. Backlund, N.G., J. Phys. Chem. Solids, vol. 28, 2219, 1967. Thermal conductivity at 300K = 0.3278 cal/cm-s-K, interpolated between 280.7K and 333.4K.

Mo9. Cutler, M., H.R. Snodgrass, G.T. Cheney, J. Appel, C.E. Mallon, and C.H. Meyer, Jr., USAEC Rept., GA-1939, 1, 1961. Thermal conductivity at 290K = 0.3392 cal/cm-s-K.

Mo10. Lisovskii, Yu.A., Soviet Phys. Solid State, vol. 14, 2015, 1973. Thermal expansion coefficient at 300K = 4.82×10^{-6} K^{-1}.

Mo11. Arbuzov, M.P. and I.A. Zelekov, Phys. Met. Metallog., vol. 18, 149, 1964. Thermal expansion coefficient at 293K = 3.51×10^{-6} K^{-1}.

Mo12. Shodhan, R.P., Univ. of Missouri at Rolla, M.S. Thesis, 1965. Thermal expansion coefficient at 300K = 4.71×10^{-6} K^{-1}.

Mo13. CRC Handbook of Chemistry and Physics, 53rd ed., B-1. Melting point = 2890K.

Mo14. Moore, J.P., R.K Williams, and R.S. Graves, Rev. Sci. Inst., vol. 45, 87, 1974. Electrical resistivity at 300K = 5.5058 $\mu\Omega$-cm.

Mo15. White, G.K. and S.B. Woods, Phil. Trans. Roy. Soc. London, vol. A251, 273, 1959. Electrical resistivity at 295K = 5.3 $\mu\Omega$-cm.

Mo16. Worthing, A.G., Phys. Rev., vol. 28, 190, 1926. Electrical resistivity at 300K = 5.78 $\mu\Omega$-cm.

Mo17. Bolef, D.I., J. Appl. Phys., vol. 33, 2311, 1962. Young's modulus at 300K = 4.628x10^7 psi. Poisson's ratio at 300K = 0.302.

Mo18. Featherstone, F.H. and J.R. Neighbours, Phys. Rev., vol. 130, 1324, 1963. Young's modulus at 300K = 4.734x10^7 psi. Poisson's ratio at 300K = 0.292.

Mo19. Dickinson, J.M. and P.E. Armstrong, J. Appl. Phys., vol. 38, 602, 1967. Young's modulus at 300K = 4.667x10^7 psi, interpolated between 273K and 373K. Poisson's ratio at 300K = 0.293, interpolated between 273K and 373K.

Mo20. Imgram, A.G., et al., Battelle Memorial Inst., WADD TR 60-278, 1960. Recrystallized bar, 7/16-in. diam., 0.002% C contg.: Ultimate tensile strength at 300K = 67200 psi. Yield strength (0.2% offset) at 300K = 65900 psi. Elongation at 300K = 23%.

Mo21. Molybdenum Metal, Climax Molybdenum Co., 1960. Stress-relieved rod, 5/8-in. diam., 0.015% C contg.: Ultimate tensile strength at 300K = 97200 psi. Yield strength (0.1% offset) at 300K = 82900 psi. Elongation at 300K = 42%.

Mo22. Pugh, J.W., Trans. ASM, vol. 47, 984, 1955. Cold-swaged 88%, 99.95% pure: Ultimate tensile strength at 300K = 97100 psi. Yield strength (0.2% offset) at 300K = 95000 psi. Elongation at 300K = 18%.

Mo23. Bens, F., Trans. ASM, vol. 38, 505, 1947. Vickers hardness at 300K = 206.

Mo24. Climax Molybdenum Co., Molybdenum Metal, 1960, p. 37. Vickers hardnes number = 170 for annealed and 235 for cold worked material at about 300K.

Mo25. Northcot, L., Molybdenum, Butterworths Scientific Publishers London (Metallurgy of the Rarer Metals No. 5), 69, 1956. VPN for annealed metal at room temperature = 160.

Nickel

Ni1. Fieldhouse, I.B., J.C. Hedge, et al., Armour Res. Foundation, WADC TR 55-495. I, 1, 1956. [AD110404] Density at 298K = 8.944 g/cm^3.

Ni2. Yamamoto, M. Sci. Rept. Res. Inst. Tohoku Univ. Ser. A2, 871, 1950. Density at 298K = 8.96 g/cm^3.

Ni3. Weeks, J.L. and R.A. Seifert, Rev. Sci. Instr., vol. 24, 1054, 1953. Density at 298K = 8.79 g/cm^3.

Ni4. Busey, R.J. and W.F. Giauque, J. Am. Chem. Soc., vol. 74, 3157, 1952. Specific heat at 300K = 0.1067 cal/g-K, interpolated between 293.99K and 302.98K.

Ni5. Lapp, Ch., Compt. Rend., vol. 186, 1104, 1928. Specific heat at 300K = 0.1079 cal/g-K, interpolated between 294K and 323K.

Ni6. Grew, K.E., Proc. Roy. Soc., vol. A145, 509, 1934. Specific heat at 0.1073 cal/g-K, interpolated between 296.9K and 385.0K.

Ni7. Fritz, W. and K.H. Bode, Z. angew. Phys., vol. 12, 121, 1960. Thermal conductivity at 303.2K = 0.198 cal/cm-s-K.

Ni8. Powell, R.W., R.P. Tye, and M.J. Hickman, Int. J. Heat Mass Transfer, vol. 8, 679, 1965. Thermal conductivity at 323K = 0.196 cal/cm-s-K.

Ni9. Euken, A. and K. Dittrich, Z. physik. Chem., vol. 125, 211, 1927. Thermal conductivity at 273K = 0.200 cal/cm-s-K.

Ni10. Fraser, D.B. and A.C. Hollis-Hallett, Proc. Int. Conf. Low Temp. Phys., Seventh, 689, 1961. Thermal expansion coefficient at 300K = 12.9×10^{-6} K^{-1}.

Ni11. Nix. F.C. and D. MacNair, Phys. Rev., vol. 60, 597, 1941. Thermal expansion coefficient at 300K = 12.59×10^{-6} K^{-1}, interpolated between 287K and 312K.

Ni12. Altman, H.W., T.Rubin, and H.L. Johnston, Ohio State Univ., Cryogenic Laboratory Rept. OSU-TR-264-27, 1, 1954. Thermal expansion coefficient at 300K = 12.9×10^{-6} K^{-1}.

Ni13. CRC Handbook of Chemistry and Physics, 53rd ed., B-1. Melting point = 1726K.

Ni14. Schindler, A.I, et al., J.Phys. Chem, Solids, vol. 1, 39, 1956. Electrical resistivity at 300K = 7.0 $\mu\Omega$-cm.

Ni15. Pallister, P.R., J. Iron and Steel Inst., No.II, 1946. Electrical resistivity at 300K = 7.31 $\mu\Omega$-cm.

Ni16. White, G.K. and S.B. Woods, Phil. Trans. Roy. Soc. London, vol. 251A, 273, 1959. Electrical resistivity at 300K = 7.20 $\mu\Omega$-cm.

Ni17. Alers, G.A., J.R. Neighbours, and H. Sato, J.Phys. Chem. Solids, vol. 13, 40, 1960. Young's modulus at 300K = 3.20x10^7 psi. Poisson's ratio at 300K = 0.299.

Ni18. Epstein, S.G., and O.N. Carlson, Acta. Met., vol. 13, 487, 1965. Young's modulus at 298K = 3.12x10^7 psi. Poisson's ratio at 298K = 0.307.

Ni19. Koster, W., Z. Metallkd., vol. 39, 1, 1948. Young's modulus at 293K = 2.80x10^7 psi.

Ni20. Bradfield, G., Notes on Applied Science No.30, H.M.S.O., London, 1964. Poisson's ratio at 293K = 0.306.

Ni21. Jenkins, W.D., T.G. Digges, and C.R. Johnson, J. Res. Nat. Bur. Std., vol. 52, 201, 1957. Annealed: Yield strength at 297K = 12500 psi. Ultimate tensile strength at 297K = 52700 psi. Elongation at 297K = 56%. 40% cold-worked: Yield strength at 297K = 80100 psi. Ultimate tensile strength at 297K = 81600 psi. Elongation at 297K = 19%.

Ni22. Geil, G.W. and N.L. Carwile, NBS Circular 520, 67, 1952. Annealed: Yield strength at 297K = 11100 psi. Ultimate tensile strength at 297K = 51800 psi.

Ni23. Wise, E.M. and R.H. Schaefer, Metals and Alloys, vol. 16, 424, 1942. Annealed, 99.99% pure, Unfused electrolytic wire, 0.063 in. diam., annealed 28 hours in hydrogen at >1300K, cold drawn 50%, annealed 1/2 hour at 1033K: Yield strength at room temp. = 8500 psi. Ultimate tensile strength at room temp. = 46000 psi. Elongation at room temp. = 30%. Hardness = 40 R_B. Test performed at Research Lab, International Nickel Company.

Ni24. Catlin, C.F. and W.A. Mudge, Proc. Am. Soc. Test. Mat., vol. 38, 269, 1938. 1-in. diam. rod, hot-rolled, 99.5% pure: Elongation (2 in. gauge) at room temp. = 44%. Brinell hardness at room temp. = 107.

Ni25. Ito, Kiyosi, Tohoku Sci, Repts., vol. 12, 137, 1923. 25-mm diam. cast bar, 98.7% pure: Brinell hardness at 300K = 88.3, interpolated between 293K and 308K.

Ni26. International Nickel Co, Nickel and Nickel Alloys, New York, 1941. Hardness reported for material with 20 ksi yield and 65 ksi ultimate strength = 45 Rockwell B.

Si1. Straumanis, M.E. and E.Z. Aka, J. Appl. Phys., vol. 23, 330, 1952. Density at 293K = 2.3283 g/cm^3.

Silicon

Si2. Horn, F.H., Phys. Rev., vol. 97, 1521, 1955. Density at 298K = 2.3305 g/cm^3.

Si3. Oak Ridge Rational Lab., USAEC Publ. ANL-5717, 1067, 1957. Density at 295K = 2.330 g/cm^3.

Si4. Flubacher, P., A.J. Leadbetter, and J.A. Morrison, Phil. Mag., vol. 4, 273, 1959. Specific heat at 298.41K = 0.1703 cal/g-K.

Si5. Gul'tyaev, P.V. and A.V. Petrov, Soviet Phys. Solid State, vol. 1, 330, 1959. Specific heat at 300K = 0.166 cal/g-K.

Si6. Anderson, C.T., J. Am. Chem. Soc., vol. 52, 2301, 1930. Specific heat at 296.3K = 0.1656 cal/g-K.

Si7. Caruthers, J. A. , T.H. Geballe, H.M. Rosenberg, adn J.M. Ziman, Proc. Roy. Soc. London, vol. A238, 502, 1957. Thermal conductivity at 300K = 0.342 cal/cm-s-K.

Si8. Holland, M.G., Proc. Intern. Conf. Low Temp. Phys., 7th, Toronto, Canada, 1960, 280, 1961. Thermal conductivity at 299.5K = 0.358 cal/cm-s-K.

Si9. Glassbrenner, C.J. and G.A. Slack, Phys. Rev., vol. 134, A 1058, 1964. Thermal conductivity at 299.8K = 0.373 cal/cm-s-K.

Si10. Lyon, K.G., G.L. Salinger and G.K. White, Philos. Mag., vol. 12, 157, 1965. Thermal expansion coefficient at 300K = 2.618×10^{-6} K^{-1}.

Si11. Iback, H., Phys. Status Solidi, vol. 31, 625, 1969. Thermal expansion coefficient at 300K = 2.56×10^{-6} K^{-1}.

Si12. Batchelder, D.N. and R.O. Simmons, J.Chem. Phys., vol. 41, 2324, 1964. Thermal expansion coefficient at 300K = 2.64×10^{-6} K^{-1}.

Si13. CRC Handbook of Chemistry and Physics, 53rd ed., B-1. Melting point = 1683K.

Si14. Morin, J. and J.P. Maita, Phys. Rev, vol. 94, 1525, 1954. Intrinsic electrical resistivity at 300K = 245 Ω-cm.

Si15. Chapman, P.W., O.N. Tufte, J.D. Zook, and D.Long. J. Appl. Phys., vol. 34, 3291, 1963. Arsenic, 1.7×10^{18}: Electrical resistivity at 300K = 0.013 Ω-cm. Phosphorus, 1.1×10^{19}: Electrical resistivity at 300K = 0.0055 Ω-cm.

Si16. Pearson, G.L. and J. Bardeen, Phys. Rev., vol. 75, 865,

1949. Phosphorus, $1x10^{17}$: Electrical resistivity at 300K = 0.25 Ω-cm. Phosphorus, $1.3x10^{18}$: Electrical resistivity at 300K = 0.025 Ω-cm. Phosphorus, $1.7x10^{19}$: Electrical resistivity at 300K = 0.005 Ω-cm.

Si17. McSkimin, H.J., J. Appl. Phys., vol. 24, 988, 1953. Polycrystalline: Young's modulus at 293K = $2.357x10^7$ psi. Poisson's ratio at 293K = 0.222. Single crystal: E[100] at 293K = $1.885x10^7$ psi. E[110] at 293K = $2.449x10^7$ psi. E[111] at 293K = $2.720x10^7$ psi. Poisson's ratio at 293K = 0.278.

Si18. Burenkov, Yu.A. and S.P. Nikanorov, Sov. Phys. Solid State, vol. 16, 963, 1974. Polycrystalline: Young's modulus at 293K = $2.339x10^7$ psi. Poisson's ratio at 293K = 0.208. Single crystal: E[100] at 293K = $1.874x10^7$ psi. E[110] at 293K = $2.429x10^7$ psi. E[111] at 293K = $2.694x10^7$ psi. Poisson's ratio at 293K = 0.265.

Si19. Wortman, J.J. and R.A. Evans, J. Appl. Phys., vol. 36, 215, 1965. Young's modulus for [111] = 24.5 x 10^6 psi at 298K. The calculated Poisson's ratios are 0.358 [111] and 0.279 [100] at 298K.

Si20. Sylwestrowicz, W.D., Phil. Mag. vol. 7, 1825, 1962. Fracture stress = 26,800 psi at room temperature. Samples were high purity single crystals 7 cm long by 1.5 mm wide with the tensile axis aligned with <111> and {110} and {112} lateral faces.

Si21. Eisner, R.L., Acta Met., vol. 3, 414, 1955. Room temperature tensile strength of silicon wisker (1 micron diam.) = 550,000 psi.

Si22. Pearson, G.L., et al., Acta Met, vol. 5, 181, 1957. Bending strength of crystals with longitudinal axis aligned with the <111> direction varied from 319,000 psi for samples with 10^{-6} cm^2 cross-section to 43,500 psi for samples with 10^{-2} cm_2 cross-section at room temperature.

Si23. Kinzel, A.B., et al., Trans. Am. Inst. Min. Met. Eng. 137[140] 425/9. Room temperature bend strength of 99.84% pure cast bar .254 in. square and 3 inches long = 9046 psi.

Si24. Wolf, G.A., et al., Semiconductors and Phosphorus Products, Proc., Intern. Colloquium, 1956. Garmisch-Partenkirchen. Schon, M. and H. Welker., eds., N.Y. Interscience, 463, 1958. Knoop microhardness at 298K = 1150 kg/mm^2.

Si25. Trefilov, V.I., et al., Sov. Phys. Doklady, vol. 8, 1240, 1964. Microhardness on the (111) is 1050 kg/mm^2 at room temperature.

Si26. Templin, R.L., Metals and Alloys, 3[1932], 136/7, 152. Hardness at room temp. = 105.6 Rockwell C.

Silver

Ag1. Stockdale, D., J.Inst. Metals, vol. 66, 287, 1940. Density at 293K = 10.4994 g/cm^3.

Ag2. Powell, H. and E.J. Evans, Phil. Mag., vol. 34, 145, 1943. Density at 298K = 10.49 g/cm^3.

Ag3. Neighbors, J. R. and G.A. Alers, Phys. Rev., vol. 111, 707, 1958. Density at 300K = 10.5000 g/cm^3.

Ag4. Bronson, H.L. and A.J.C. Wilson, Can. J. Res., vol. 14A, 181, 1936. Specific heat at 300K = 0.05599 cal/g-K, interpolated between 293.15K and 303.15K.

Ag5. Meads, P.F., W.R. Forsythe, and W.F. Giauque, J. Am. Chem. Soc., vol. 63, 1902, 1941. Specific heat at 297.81K = 0.05652 cal/g-K.

Ag6. Jaeger. F.M., J.E. Rosenvohm, and W.A. Veenstra, Prok. K. Akad. Wet. Amsterdam, vol. 36, 291, 1933. Specific heat at 300K = 0.05578 cal/g-K, interpolated between 273.15K and 373.15K.

Ag7. Lucks, C.F. and H.W. Deem, Battelle, Memorial Inst., WADC TR 55-496, 1, 1956. [AD97185] Thermal conductivity at 300K = 0.9658 cal/cm-s-K, interpolated between 291.2K and 306.2K.

Ag8. Kannuluik, W.G., C.E. Eddy, and T.H. Oddie, Proc. Roy Soc. (London), vol. A141, 159, 1933. Thermal conductivity at 300K = 0.9983 cal/cm-s-K, interpolated between 273.2K and 373.2K.

Ag9. Parker, W.J., R.J. Jenkins, C.P. Butler, and G.L. Abbott, J.Appl. Phys., vol. 32, 1679, 1961. Thermal conductivity at 295.2K = 0.9496 cal/cm-s-K.

Ag10. Fraser, D.B. and A.C. Hollis-Hallett, Proc. Int. Conf. Low Temp. Phys., Seventh, 689, 1961. Thermal expansion coefficient at 300K = 19.0×10^{-6} K^{-1}.

Ag11. Pathak, P.D. and N.G. Vasavada, J.Phys. C, vol. 3, L44, 1970. Thermal expansion coefficient at 300K = 18.8×10^{-6} K^{-1}.

Ag12. Leksina, I.E. and S.I. Novikova, Soviet Phys. Solid State, vol. 5, 798, 1963. Thermal expansion coefficient at 300K = 18.73×10^{-6} K^{-1}.

Ag13. CRC Handbook of Chemistry and Physics, 53rd ed., B-1. Melting point = 1235.08K.

Ag14. Weinberg, I., Phys. Rev., vol. 157, 564, 1967. Electrical resistivity at 295K = 1.83 $\mu\Omega$-cm.

Ag15. Schultze, F.A., Phys. Z., vol. 12, 1028, 1911. Electrical resistivity at 298K = 1.744 $\mu\Omega$-cm.

Ag16. Pravoverlov, N.L. and I. A. Tribunskaya, Russ. Metall., vol. 1, 75, 1967. Electrical resistivity at 298K = 1.63 μΩcm.

Ag17. Neighbours, J.R. and G.A. alers, Phys. Rev, vol. 111, 707, 1958. Young's modulus at 300K = 1.18×10^7 psi. Poisson's ratio at 300K = 0.369.

Ag18. Chang, Y.A., and L. Himmel, J.Appl. Phys., vol. 37, 3567, 1966. Young's modulus at 300K = 1.18×10^7 psi. Poisson's ratio at 300K = 0.369.

Ag19. Purwins, H.G., R.Labusch, and P. Haasen, Z. Metallkd., vol. 57, 867, 1966. Young's modulus at 300K = 1.18×10^7 psi. Poisson's ratio at 300K = 0.366.

Ag20. Parkhomenko, T.A. and V.V. Pustovalov, Phys. Met. Metallog., vol. 36(4), 160, 1973. Annealed, 1-mm av. grain size, 99.99% pure: Yield point at 300K = 1090 psi. Ultimate tensile strength at 300K = 21000 psi. Elongation at 300K = 39%.

Ag21. McKeown, J. and O.F. Hudson, J. Inst. Met., vol. 60, 109, 1937. Rolled rod, 0.8-in. diam., annealed 10 min at 930 °F: Yield strength (0.1% offset) at room temp. = 1500 psi. Ultimate tensile strength at room temp. = 21500 psi. Elongation (2 in. gauge) at room temp. = 65%.

Ag22. Rogers, B.A., I.C. Schoonover, and L. Jordan, NBS Cir. C412, 1936. 50% rolled sheet, 0.032 in., 99.9% pure: Yield point at room temp. = 44300 psi. Ultimate tensile strength at room temp. = 54300 psi. Elongation (2 in. gauge) at room temp. = 2.4%.

Ag23. Ito, K., Tohoku Sci. Repts., vol. 12, 137, 1923. 25-mm diam. cast bar, 99.3% pure: Brinell hardness at 300K = 35.3, interpolated between 292K and 313K.

Ag24. Saeftel, F. and G. Sachs, Z. Metallkd. 17[1925] 155/61, 294/8. H^B = 24 to 30 when heat treated from 500 to 700C.

Ag25. Tabor, D., The Hardness of Metals, Oxford 1951. Brinell hardness = 25 at room temperature.

Tantalum

Ta1. Pattee, H.E. and R.M. Evans, DMIC, DMIC memo. No. 153, 1, 1962. [AD278193] Density at 293K = 16.6 g/cm^3.

Ta2. Keesom, B.K. and M. Desirant, Physica, vol. 8, 273, 1941. Density at 291K = 16.57 g/cm^3.

Ta3. Fieldhouse, I.B., J.C. Hedge, and T.E. Waterman, Armour Res. Foundaton, WADC TR 55-495, I, 1, 1956. [AD110404] Density at 298K = 16.47 g/cm^3.

Ta4. Taylor, R.E. and Finch, R.A., U.S. At. Energy Comm., NAA-SR-6034, 1, 1961. Specific heat at 300K = 0.0328 cal/g-K.

Ta5. Lehman, G.W., WADD TR 60-581, 1019, 1960. [AD 247411] Specific heat at 300K = 0.0330 cal/g-K.

Ta6. Sterrett, K.F., Ph.D. Thesis, Univ. of Pittsburgh, Univ. Microfilms Publ. 22865, 1, 1957. Specific heat at 300K = 0.03343 cal/g-K.

Ta7. Cutler, M. H.R. Snodgrass, G.T. Cheney, J. Appel, C.E. Mallon, and C.H. Meyer, Jr., USAEC Rept., GA-1939, 1, 1961. Thermal conductivity at 299K = 0.129 cal/cm-s-K.

Ta8. Barratt, T. Proc. Phys. Soc. (London), vol. 26, 347, 1913. Thermal conductivity at 300K = 0.130 cal/cm-s-K, interpolated between 290.2K and 373.2K.

Ta9. Cutler, M., Advanced Energy Conversion, vol. 2, 29, 1962. Thermal conductivity at 300K = 0.118 cal/cm-s-K.

Ta10. Nic, F.C. and D. MacNair, Phys. Rev., vol. 61, 74, 1942. Thermal expansion coefficient at 293K = 6.55x10^{-6}K.

Ta11. Lisovskii, Yu.A., Soviet, Phys. (Solid state), vol. 14, 2015, 1973. Thermal expansion coefficient at 300K = 6.08x10^{-6} K^{-1}.

Ta12. Schmidt, F.F., Battelle Memorial Inst., DMIC Rept. 133, 1960. Thermal expansion coefficient at 300K = 6.61x10^{-6} K^{-1}.

Ta13. CRC Handbook of Chemistry and Physics, 53rd ed., B-1. Melting point = 3269K.

Ta14. White, G.K. and S.B. Woods, Phil. Trans. Roy. Soc. London, vol. A251, 273, 1959. Electrical resistivity at 295K = 13.1 $\mu\Omega$-cm.

Ta15. Williams, R.K., R.S. Graves, T.L. Hebble, D.L. McElroy, and J.P. Moore, Phys. Rev. B, vol. 26, 2932, 1982. Electrical resistivity at 300K = 13.35 $\mu\Omega$-cm.

Ta16. Khanna, S.N. and A. Jain, J. Phys. F, vol. 4, 1982, 1974. Electrical resistivity at 300K = 13.9 $\mu\Omega$-cm.

Ta17. Featherstone, F.H. and J.R. Neighbours, Phys. Rev., vol. 130, 1324, 1963. Young's modulus at 300K = 2.65×10^7 psi. Poisson's ratio at 300K = 0.341.

Ta18. Soga, N., J. Appl. Phys., vol. 37, 3416, 1966. Young's modulus at 298K = 2.68×10^7 psi. Poisson's ratio at 298K = 0.338.

Ta19. Bolef, D.I., J. Appl. Phys., vol. 33, 2311, 1962. Young's modulus at 300K = 2.688×10^7 psi. Poisson's ratio at 300K = 0.342.

Ta20. Bechtold, J.H., Acta Met., vol. 3, 249, 1955. Recrystallized rod: Elastic limit at 299K = 39300 psi. Ultimate tensile strength at 299K = 49800 psi. Elongation at 299K = 28%.

Ta21. Pugh, J.W., Trans. ASM, vol. 48, 677, 1956. Recrystallized sheet produced from powder-metallurgy ingot: Yield strength at 300K = 57400 psi. Ultimate tensile strength at 300K = 67100 psi. Elongation at 300K = 25%.

Ta22. Schmidt, F.F., Batelle Memorial Inst., WADD TR 61-106, 1961. Recrystallized sheet, 0.040-in. thick: Ultimate tensile strength at 300K = 29400 psi. Yield strength at 300K = 26300 psi. Elongation at 300K = 36%.

Tin

Sn1. Helfrich, W.J. and R.A. Dodd, Acta Met., vol. 12, 667, 1964. Density at 298K = 7.289 g/cm^3.

Sn2. Rayne, J.A. and B.S. Chandresekhar, Phys. Rev., vol. 120, 1658, 1960. Density at 300K = 7.279 g/cm^3.

Sn3. M. C. Neuburger, Z. Kirst., 93, 1936, p. 1/36,22. Density at room temperature = 7.287 g/cm^3.

Sn4. Jaeger, F.M. and J.A. Bottema, Proc. Acad. Sci. Amsterdam, vol. 35, 347, 1932. Specific heat at 300K = 0.054571 cal/g-K, interpolated between 293K and 323K.

Sn5. Jenkins, R.J. and W.J. Parker, WADD TR 61-95, 1, 1961. [AD 268752] Specific heat at 295K = 0.057 cal/g-K.

Sn6. Bartenev, G.M., Zhur. Tekh. Fiz., vol. 17, 1321, 1947. Specific heat at 300K = 0.054 cal/g-K, extrapolated from 323.15K.

Sn7. Lees, C.H., Phil. Trans. Roy. Soc. London, vol. A208, 381, 1908. Thermal conductivity at 300K = 0.156 cal/cm-s-K, interpolated between 291.2K and 303.2K.

Sn8. Brown, W.B., Phys. Rev., vol. 22, 171, 1923. Thermal conductivity at 300K = 0.144 cal/cm-s-K, extrapolated from 323.2K.

Sn9. Ruh, E., J. Am. Ceram. Soc., vol. 37, 224, 1954. Thermal conductivity at 306.4K = 0.153 cal/cm-s-K.

Sn10. Deshpande, V.T. and D.B. Sirdeshmukh, Acta Cryst., vol. 15, 294, 1962. Thermal expansion coefficient at 303K = 16.5x10^{-6} K^{-1}.

Sn11. Vernon, E.V and S. Weintroub, Proc. Phys. Soc. London, vol. 66B, 887, 1953. Thermal expansion coefficient at 301K = 18.81x10^{-6} K^{-1}.

Sn12. Childs, B.G. and S. Weintroub, Proc. Phys. Soc. London, vol. 63B, 267, 1950. Thermal expansion coefficient at 304K = 24.10x10^{-6} K^{-1}.

Sn13. CRC Handbook of Chemistry and Physics, 53rd ed., B-1. Melting point = 505.12K.

Sn14. Pietenpol, W.B. and H.A. Miley, Phys. Rev., vol. 34, 1588, 1929. Electrical resistivity at 300K = 13.194 $\mu\Omega$-cm, interpolated between 293K and 313K.

Sn15. Onnes, H.K. and W. Tuyn, Proc. Acad. Amsterdam, vol. 25, 443, 1923. Electrical resistivity at 295K = 11.0 $\mu\Omega$-cm.

Sn16. Aleksandrov, B.N. and I.G. D'Yakov, Sov. Phys. JETP, vol. 16, 603, 1963. Single crystal electrical resistivity at 293K: parallel to c-axis = 9.85 μΩ-cm; perpendicular to c-axis = 14.3 μΩ-cm.

Sn17. House, D.G. and E.V. Vernon, Brit. J. Appl. Phys., vol. 11, 254, 1960. Young's modulus at 300K = 6.89x10^6 psi, interpolated between 288K and 318K. Poisson's ratio at 300K = 0.368, interpolated between 288K and 318K.

Sn18. Rayne, J.A. and B.S. Chandrasekhar, Phys. Rev., vol. 120, 1658, 1960. Young's modulus at 300K = 7.02x10^6 psi. Poisson's ratio at 300K = 0.353.

Sn19. Cuthberson, J.W., J. Inst. Met., vol. 64, 209, 1939. Young's modulus at 300K = 6.4x10^6 psi.

Sn20. Reddy, P.J. and S.V. Subrahmanyam, Nature, vol. 185, 29, 1960. Poisson's ratio at 298K = 0.455.

Sn21. Addicks, L., ed., Silver in Industry, Reinhold Publ. Corp., New York, 1940. 99.8%, 0.040 in tin sheet annealed 1 hour at 188C. Yield Strength (0.2% offset) = 1300 psi. UTS = 2200 psi. Elongation in 2 inches = 45%. The same material after 30% reduction has yield strength = 2000 psi, UTS = 2.8 psi and elongation = 35%.

Sn22. Homer, C.E. and H. Plummer, Tin Res. Inst., Publ. A.57, 1937. Ultimate tensile strength at 300K = 1990 psi, interpolated between 288K and 323K. Elongation (22 mm. gauge) at 300K = 78%, interpolated between 288K and 323K.

Sn23. Kalish, H.S. and F.J. Dunkerley, Trans. Am. Inst. Min. Met. Eng., vol. 180, 637, 1949. Ultimate tensile strength at 296K = 1590 psi. Elongation (25 mm. gauge) at 296K = 57%.

Sn24. Coe, J. Inst. Met., vol. 30, 309, 1923. Pressed and annealed at 100° C: Ultimate tensile strength at room temp. = 2620 psi. Brinell hardness at room temp. = 5.0.

Sn25. Homer, C.E. and H. Plummer, J. Inst. Met., vol. 64, 169, 1938. Brinell hardness at 293K = 3.9.

Sn26. Ito, K., Tohoku Sci. Repts., vol. 12, 137, 1923. 25-mm diam. cast bar, 99.9% pure: Brinell hardness at 298K = 12.0.

Tungsten

W1. Pattee, H.E. and R.M. Evans, DMIC, DMIC memo. No. 153, 1, 1962. [AD278193] Density at 293K = 19.3 g/cm^3.

W2. Neel, D.S., C.D. Pears, and S. Oglesby, Jr., Southern Res. Inst., WADD TR 60-924, 58, 1962. [AD 275536;Nb2-12987] Density at 293K = 18.88 g/cm^3.

W3. Pelzel, E., Metall., vol. 11, 954, 1957. Density at 293K = 18.98 g/cm^3.

W4. Kirillin, V.A., A.E. Sheindlin, and V.Ya. Chekhovskoi, Teploenerg., vol. 9, 63, 1962. Specific heat at 300K = 0.03185 cal/g-K.

W5. Jaeger, F.M. and E. Rosenbohm, Rec. Trav. Chim., vol. 51, 1, 1932. Specific heat at 300K = 0.03212 cal/g-K, interpolated between 273.15K and 373.15K.

W6. Clusius, K.V. and P. Franzosini, Z. Naturforsch., vol. A14, 99, 1959. Specific heat at 273.84K = 0.03111 cal/g-K.

W7. Tye, R.P., J. Less-Common Metals, vol. 3, 13, 1961. Thermal conductivity at 300K = 0.436 cal/cm-s-K, extrapolated from 323.2K.

W8. Cox, M., Phys. Rev., vol. 64, 241, 1943. Thermal conductivity at 300K = 0.408 cal/cm-s-K, interpolated between 273.2K and 372.8K.

W9. Langmuir, I. and J.B. Taylor, Phys. Rev., vol. 2, 68, 1936. Thermal conductivity at 300K = 0.386 cal/cm-s-K.

W10. Nic, F.C. and D. MacNair, Phys. Rev, vol. 61, 74, 1942. Thermal expansion coefficient at 293K = 4.64x10^{-6} K^{-1}.

W11. Lisovskii, Yu.A., Soviet Phys. Solid State, vol. 14, 2015, 1973. Thermal expansion coefficient at 300K = 4.37x10^{-6} K^{-1}.

W12. Worthing, A.G., Phys. Rev., vol. 10, 638, 1917. Coefficient of thermal expansion at 300K = 4.44x10^{-6} K^{-1}.

W13. CRC Handbook of Chemistry and Physics, 53rd ed., B-1. Melting point = 3683K.

W14. Hust, J.G., High Temp.-High Press., vol. 8, 377, 1976. ELectrical resistivity at 300K = 5.48 $\mu\Omega$-cm.

W15. White, G.K. and S.B. Woods, Phil. Trans. Roy. Soc. London, vol. A251, 273, 1959. Electrical resistivity at 295K = 5.3 $\mu\Omega$-cm.

W16. Williams, R.K., J. Appl. Phys., vol. 46, 475, 1975.

Electrical resistivity at 298.15K = 5.388 μΩ-cm.

W17. Bolef, D.I and J. de Klerk, J. Appl. Phys., vol. 33, 100, 1962. Young's modulus at 300K = 5.932x10^7 psi. Poisson's ratio at 300K = 0.279.

W18. Featherstone, F.H. and J.R. Neighbours, Phys. Rev., vol. 130, 1324, 1963. Young's modulus at 300K = 5.939x10^7 psi. Poisson's ratio at 300K = 0.280.

W19. Lowrie, R. and A.M. Gonas, J. Appl. Phys., vol. 38, 4505, 1967. Young's modulus at 297K = 5.935x10^7 psi. Poisson's ratio at 297K = 0.280.

W20. J.W. Pugh, ASTM Preprint No. 71 (1957). UTS for commercial pure 0.160 rod annealed 1 hour at 1600C = 80,000psi at room temperature. Sample exhibited no elongation or yield strength.

W21. Tungsten, Sylvania Electric Products, Inc., 1962. UTS for recrystallized wire = 80,000 psi. Elongation = 0%.

W22. Fansteel Data Sheet for Cb, Ta and W, 1987. UTS at room temperature = 100,000 psi.

W24. Bens, F., Trans. ASM, vol. 38, 505, 1947. Vickers hardness at 300K = 351.

W25. General Electric Co. Product Data Sheet 1400-C (March 4, 1963). Diamond pyramid hardness varies from 360-400 for 100% recrystallized flat rolled products.

W26. Natchtigall, Von E., Z. Metallkd, vol. 43, 23, 1952. Vickers microhardness for 99.89% pure annealed material = 250 Kg/mm^2. Hardness for rolled material = 425 kg/mm^2.

ALLOYS

96.5Sn 3.5Ag

PROPERTY	@ 300K	UNITS	REFERENCES
Density	7.4	g/cm^3	2.1.A-C
Specific Heat	0.073	cal/g-K	2.2.A-B
Thermal Conductivity	0.187	cal/cm-s-K	2.3.A-C
Thermal Expansion	$30.2x10^{-6}$	1/K	2.4.A-B
Melting Point	494	K	2.5.A-D
Resistivity	$12.7x10^{-6}$	Ohm-cm	2.6.A-C
TCR	4300	ppm/K	2.7.A
Young's Modulus	$7.8x10^6$	psi	2.8.A
Poisson's Ratio	0.36	------	2.9.A
Yield Strength	3600	psi	2.10.A-B
Ult Tensile Strength	5000-7000	psi	2.11.A-C
Elongation at Break	37.0	%	2.12.A
Hardness	14.8HV, 40	Brinell	2.13.A-C

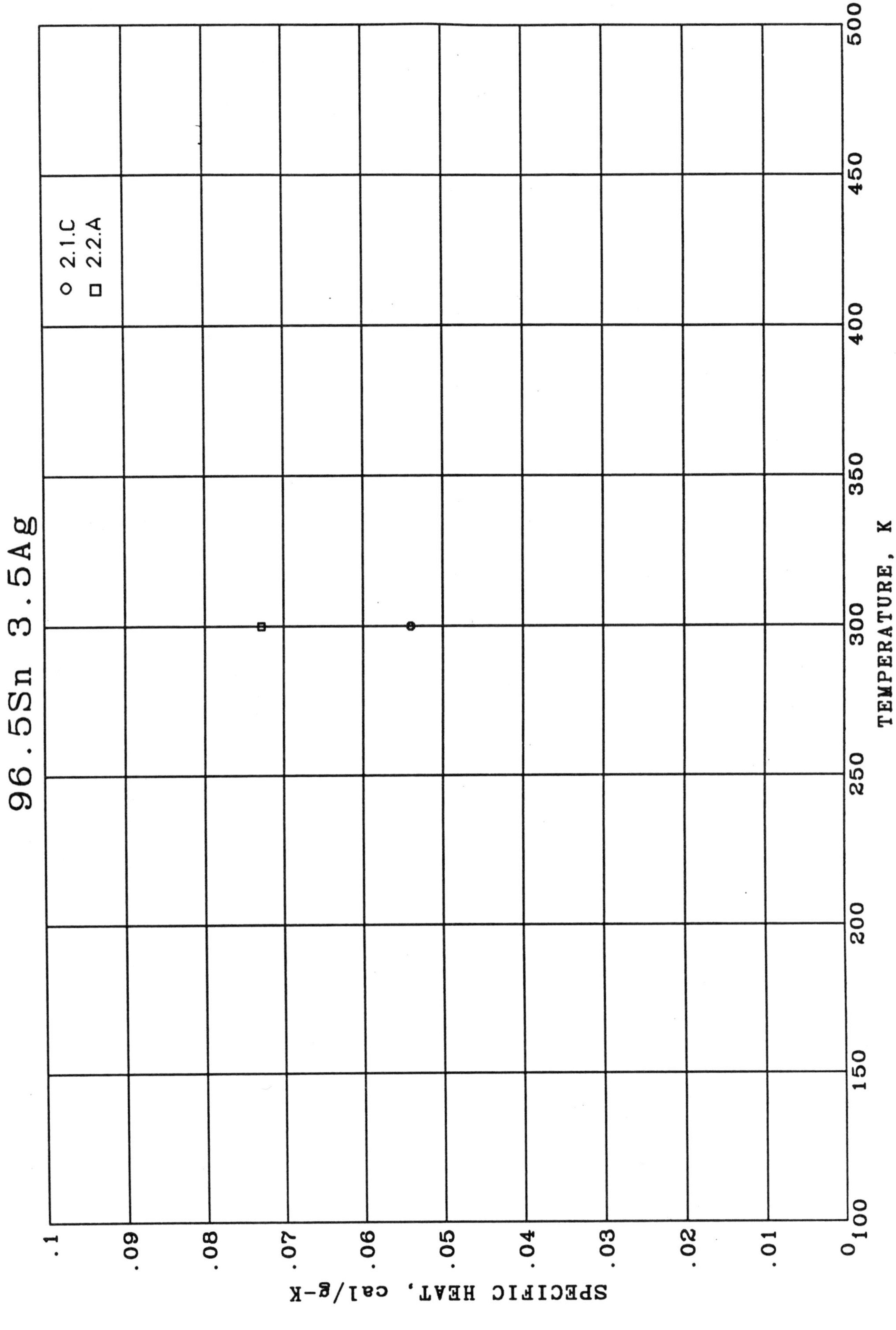
96.5Sn 3.5Ag
○ 2.1.C
□ 2.2.A
SPECIFIC HEAT, cal/g-K
TEMPERATURE, K
0
.01
.02
.03
.04
.05
.06
.07
.08
.09
.1
100
150
200
250
300
350
400
450
500

96.5Sn 3.5Ag

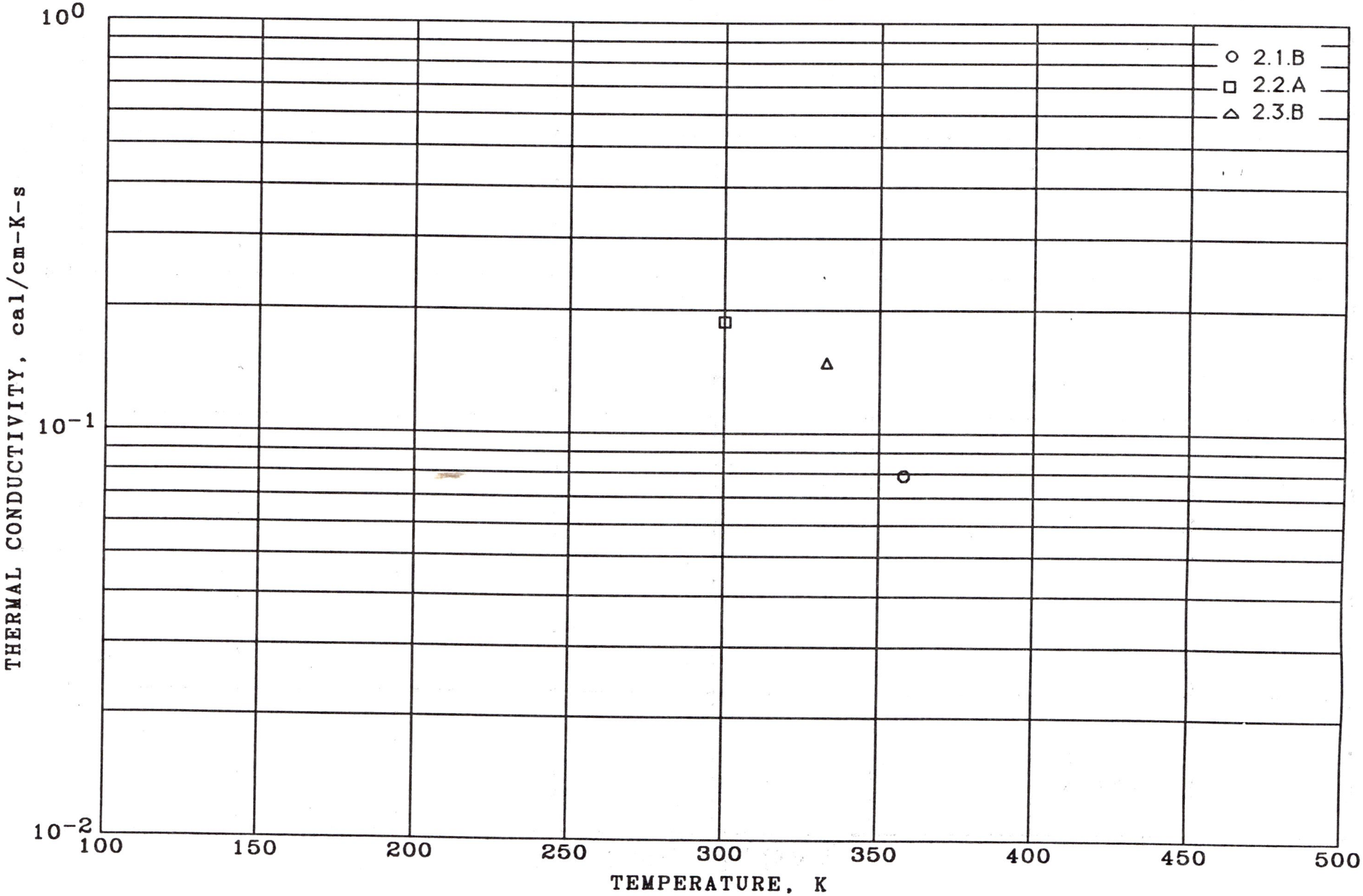

2.1.B
2.2.A
2.3.B
10^0
10^{-1}
10^{-2}
THERMAL CONDUCTIVITY, cal/cm-K-s
100
150
200
250
300
350
400
450
500
TEMPERATURE, K

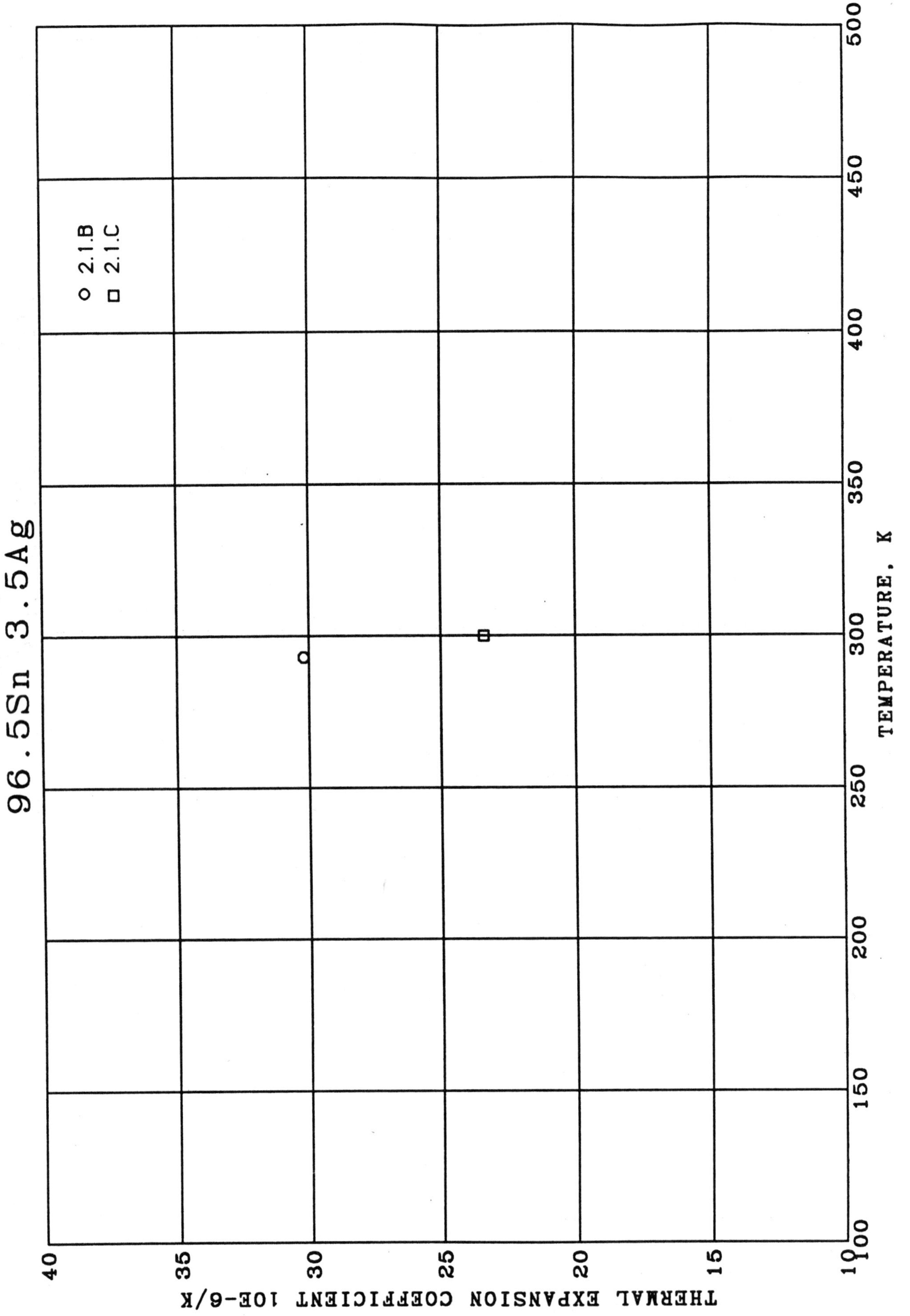
96.5Sn 3.5Ag
○ 2.1.B
□ 2.1.C
THERMAL EXPANSION COEFFICIENT 1OE-6/K
TEMPERATURE, K
40
35
30
25
20
15
10
100
150
200
250
300
350
400
450
500

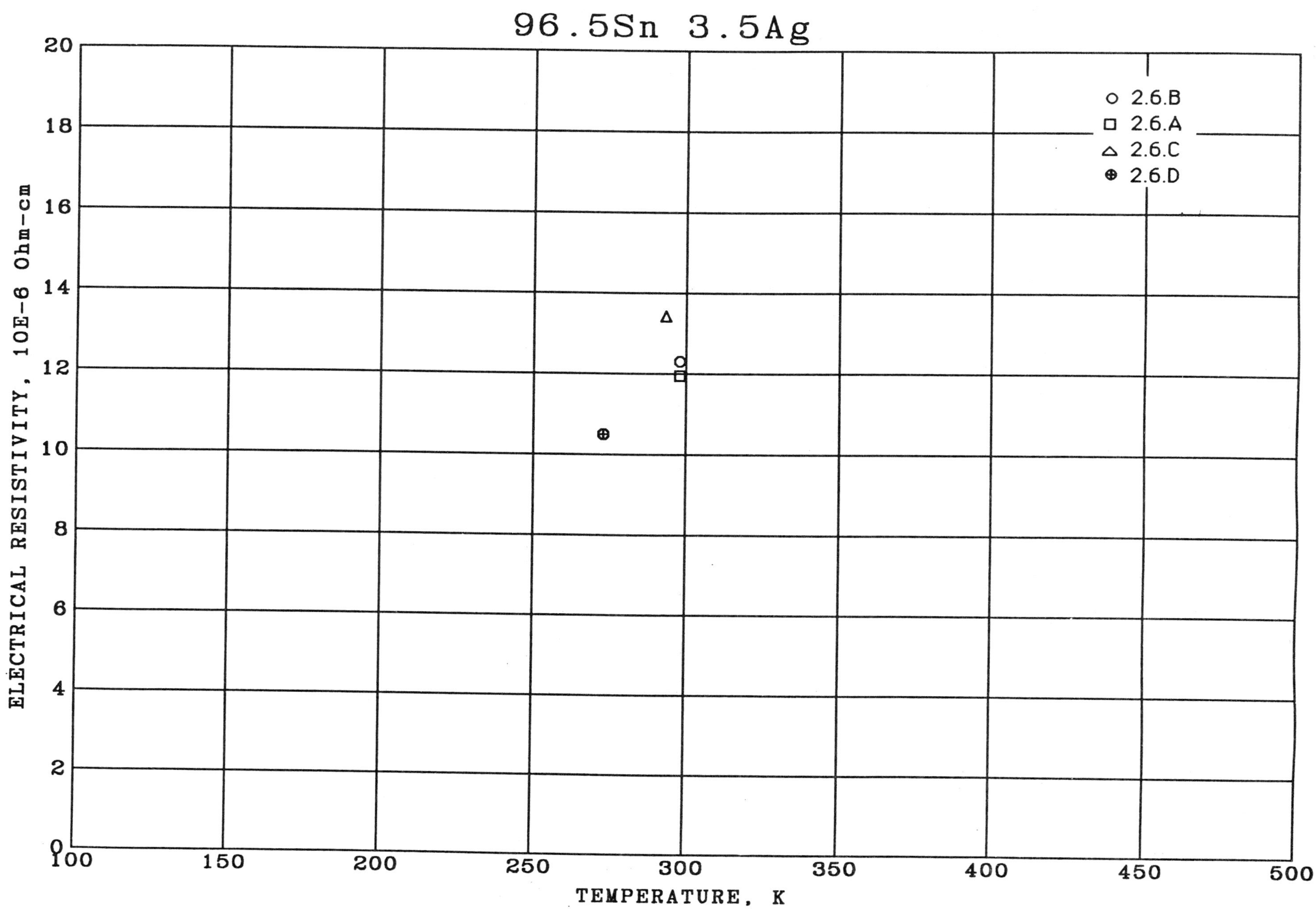
96.5Sn 3.5Ag
○ 2.6.B
□ 2.6.A
△ 2.6.C
⊕ 2.6.D
ELECTRICAL RESISTIVITY, 10E-6 Ohm-cm
0
2
4
6
8
10
12
14
16
18
20
100
150
200
250
300
350
400
450
500
TEMPERATURE, K

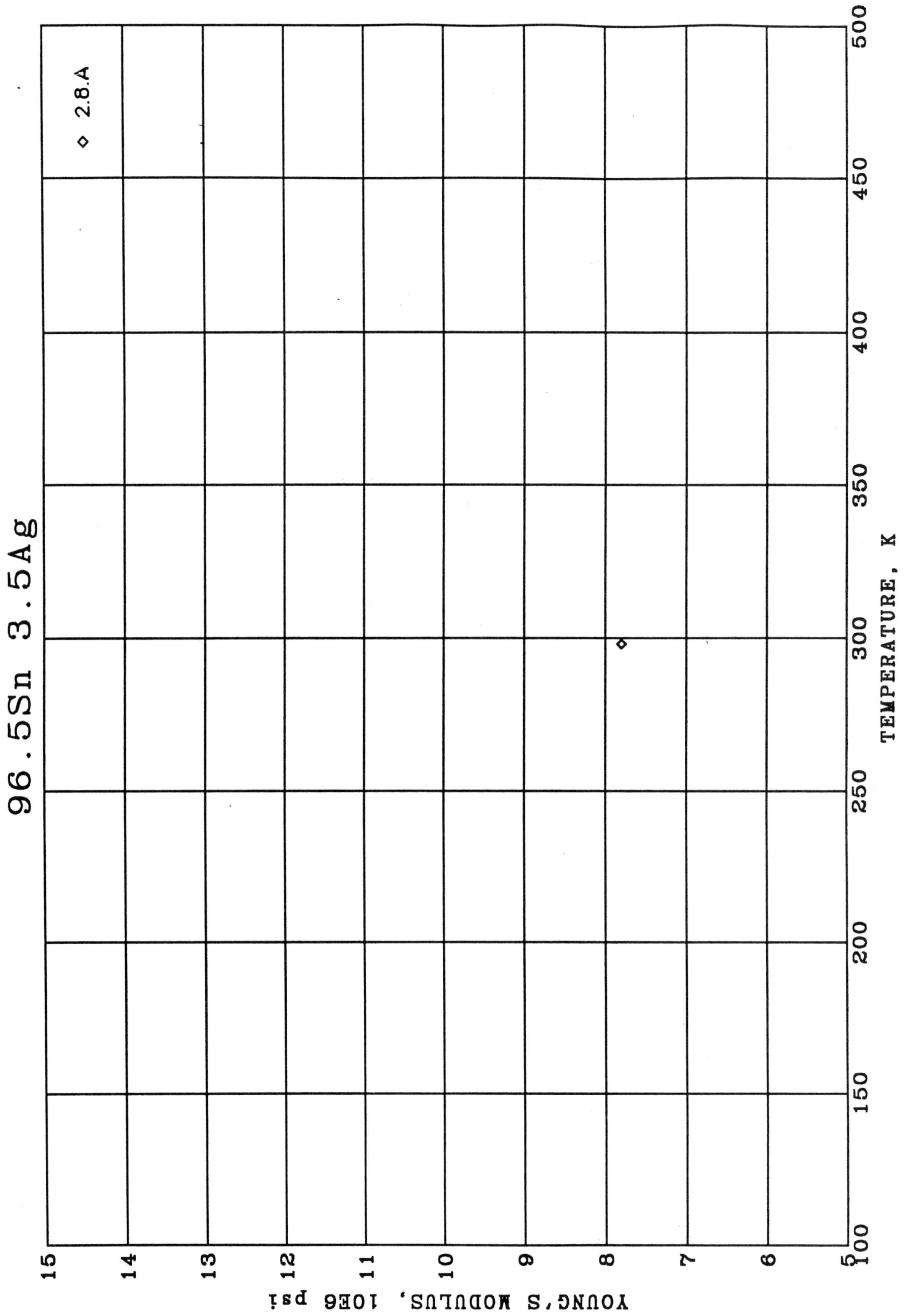
96.5Sn 3.5Ag
2.8.A
YOUNG'S MODULUS, 10E6 psi
TEMPERATURE, K
15
14
13
12
11
10
9
8
7
6
5
100
150
200
250
300
350
400
450
500

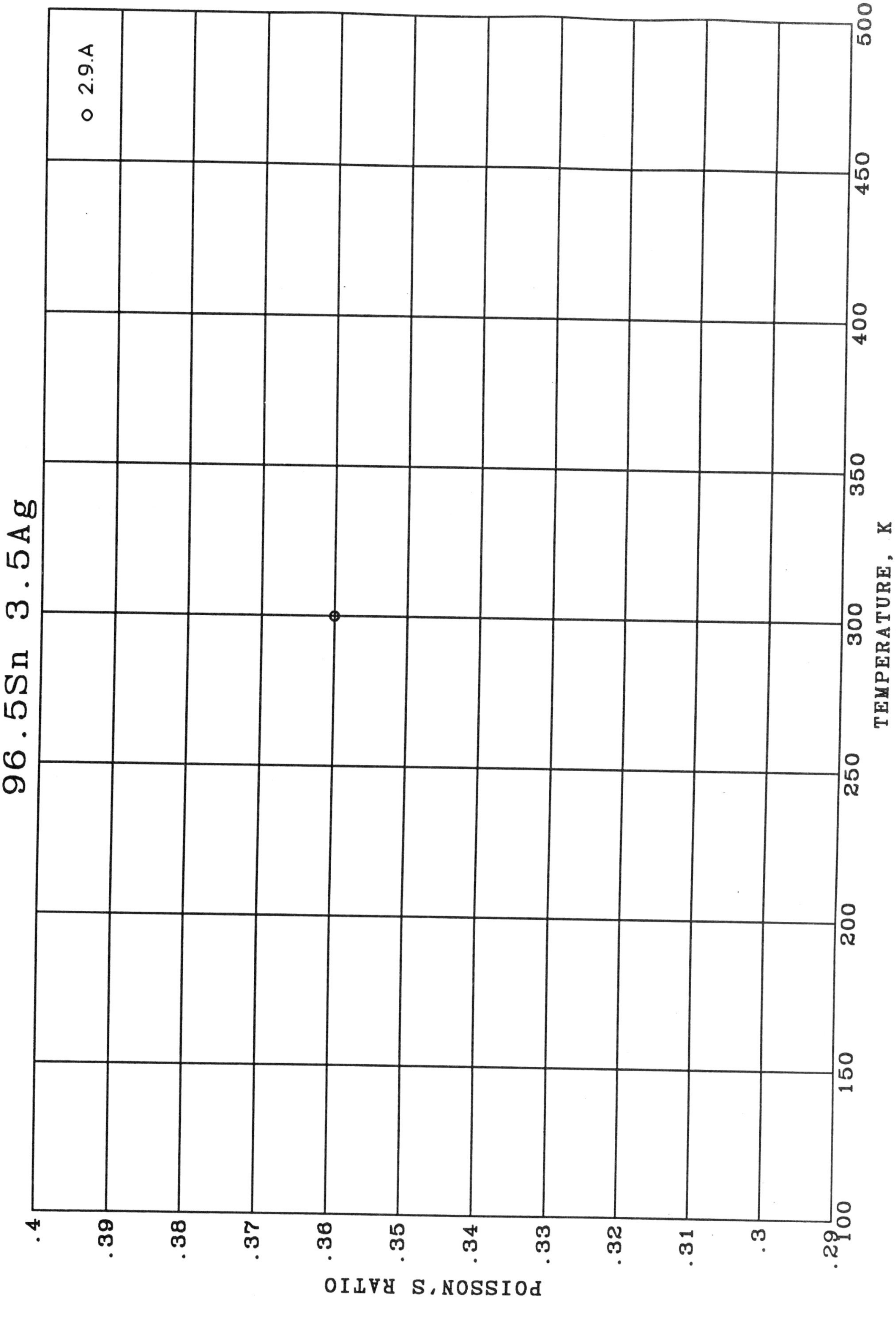
96.5Sn 3.5Ag
○ 2.9.A
POISSON'S RATIO
.4
.39
.38
.37
.36
.35
.34
.33
.32
.31
.3
.29
TEMPERATURE, K
100
150
200
250
300
350
400
450
500

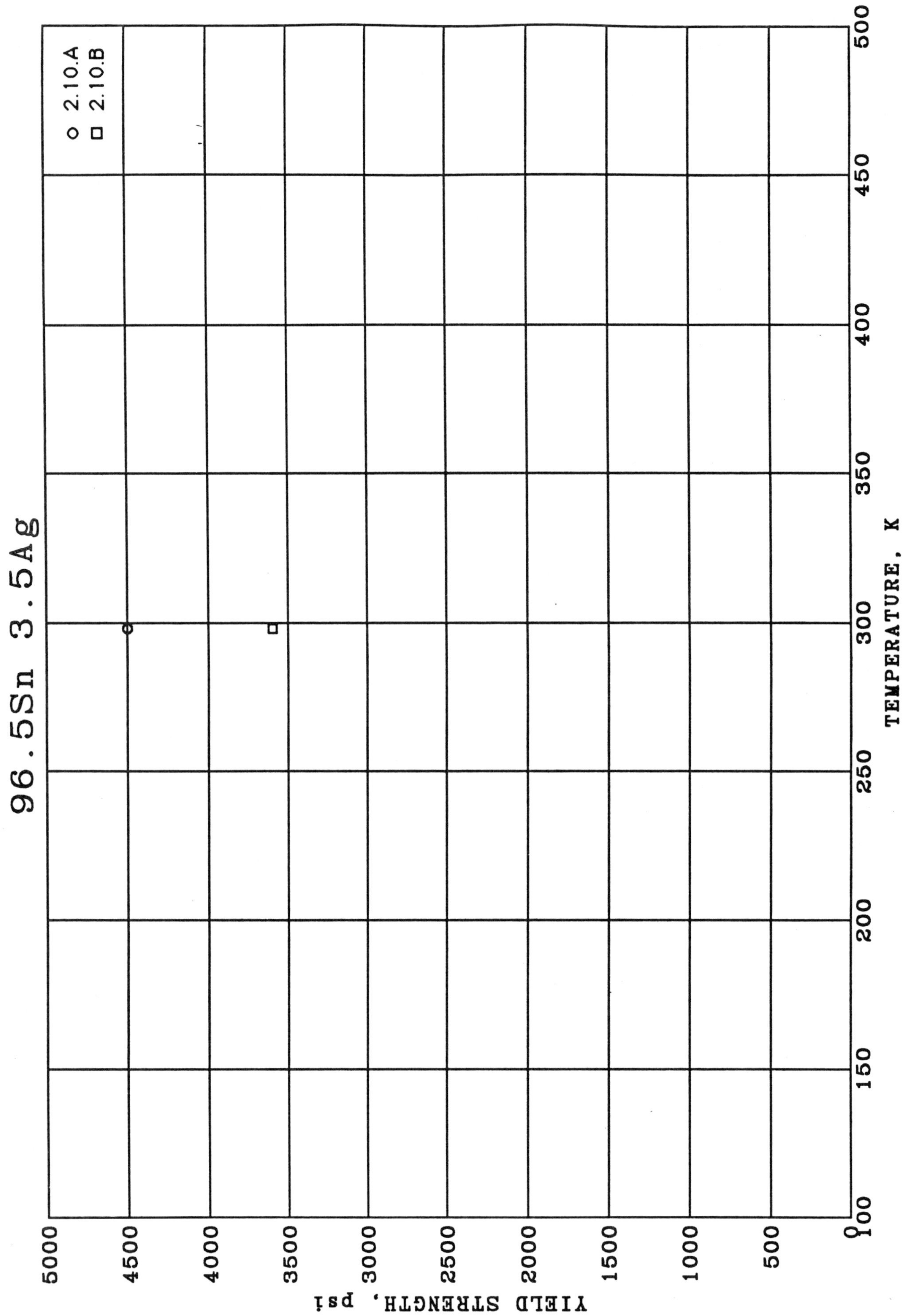
96.5Sn 3.5Ag
○ 2.10.A
□ 2.10.B
YIELD STRENGTH, psi
TEMPERATURE, K
0
500
1000
1500
2000
2500
3000
3500
4000
4500
5000
100
150
200
250
300
350
400
450
500

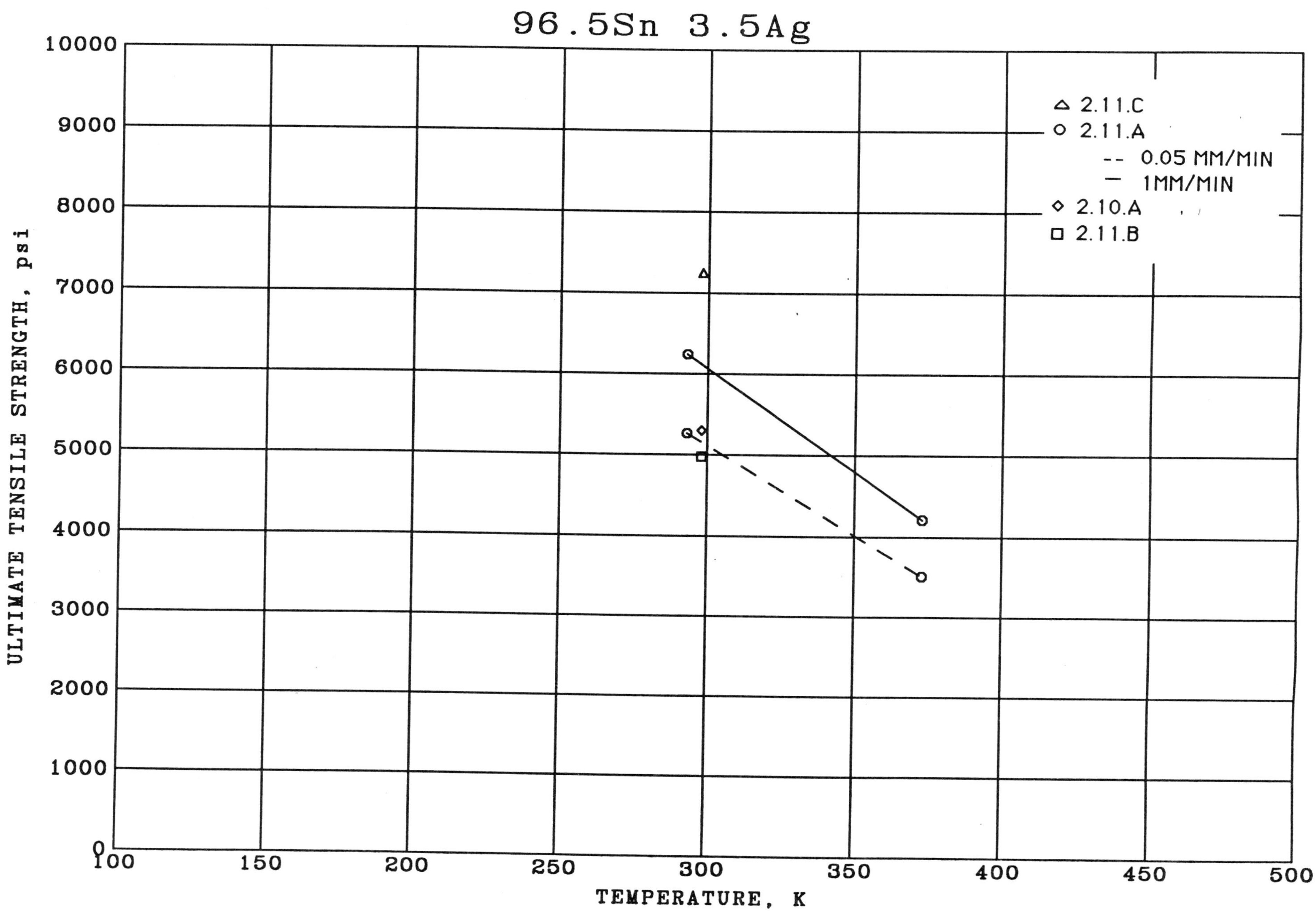
96.5Sn 3.5Ag
ULTIMATE TENSILE STRENGTH, psi
TEMPERATURE, K
2.11.C
2.11.A
-- 0.05 MM/MIN
— 1MM/MIN
2.10.A
2.11.B
10000
9000
8000
7000
6000
5000
4000
3000
2000
1000
0
100
150
200
250
300
350
400
450
500

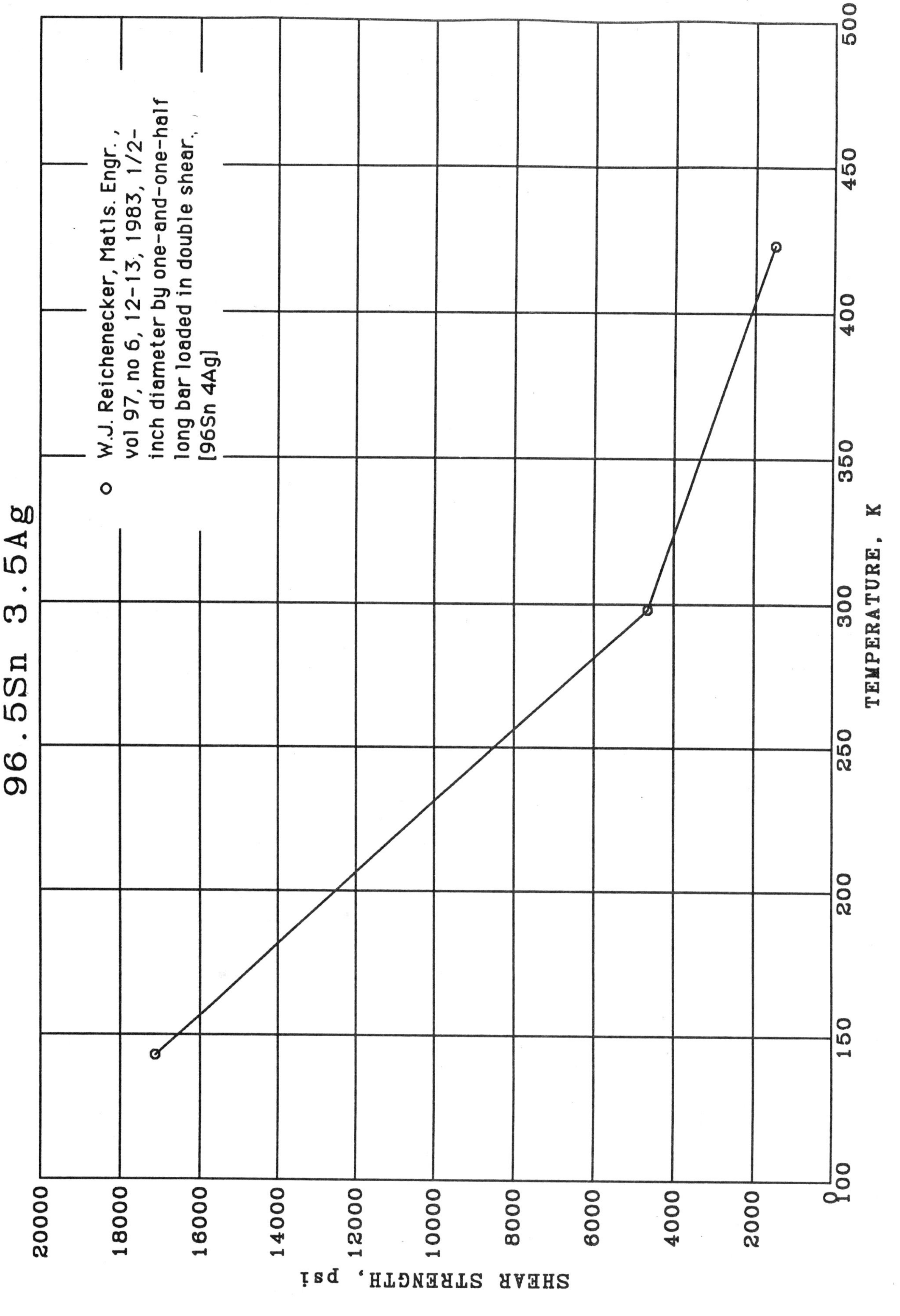
96.5Sn 3.5Ag
○ W.J. Reichenecker, Matls. Engr., vol 97, no 6, 12-13, 1983, 1/2-inch diameter by one-and-one-half long bar loaded in double shear. [96Sn 4Ag]
SHEAR STRENGTH, psi
TEMPERATURE, K
0
2000
4000
6000
8000
10000
12000
14000
16000
18000
20000
100
150
200
250
300
350
400
450
500

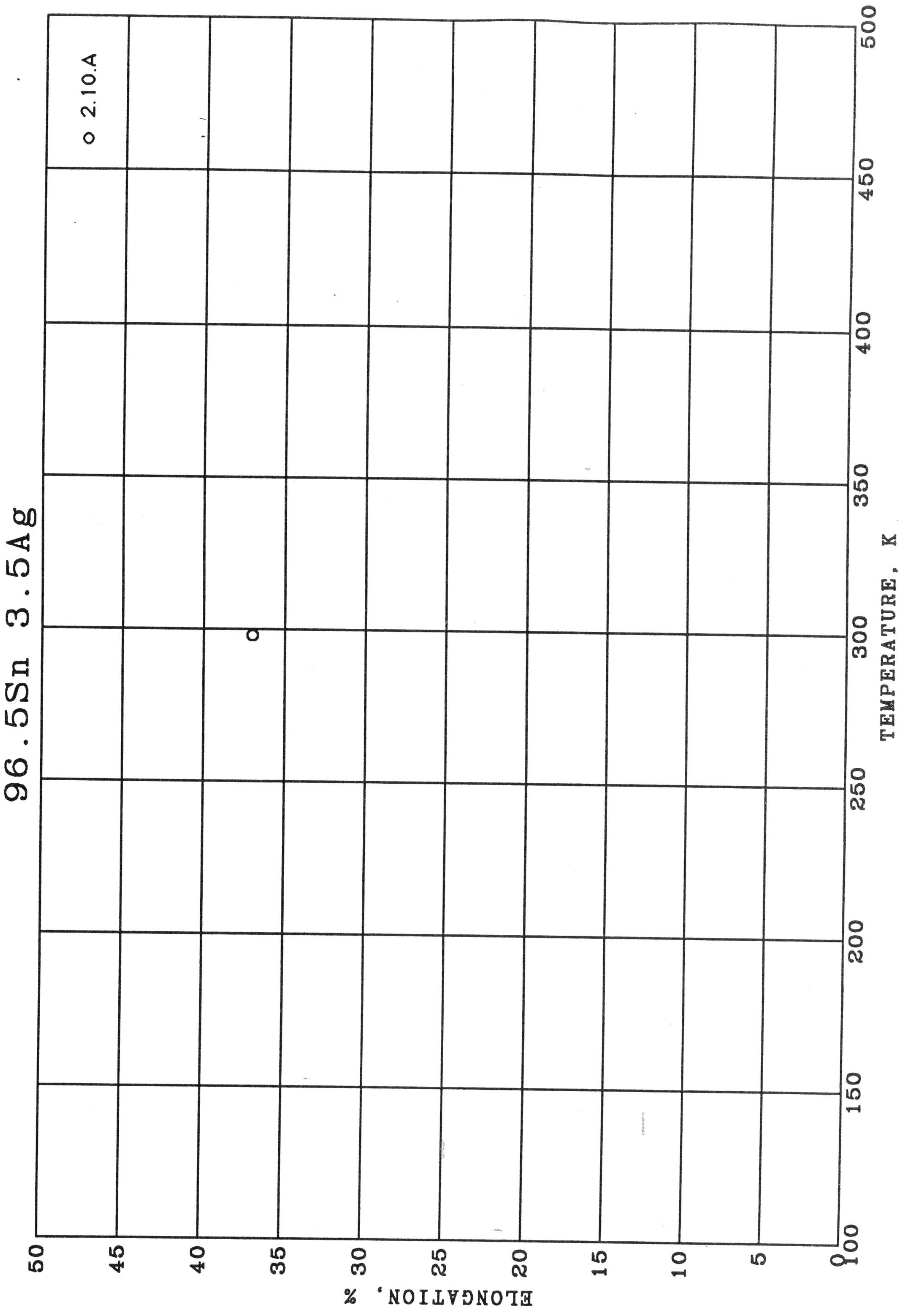
96.5Sn 3.5Ag
o 2.10.A
ELONGATION, %
0
5
10
15
20
25
30
35
40
45
50
TEMPERATURE, K
100
150
200
250
300
350
400
450
500

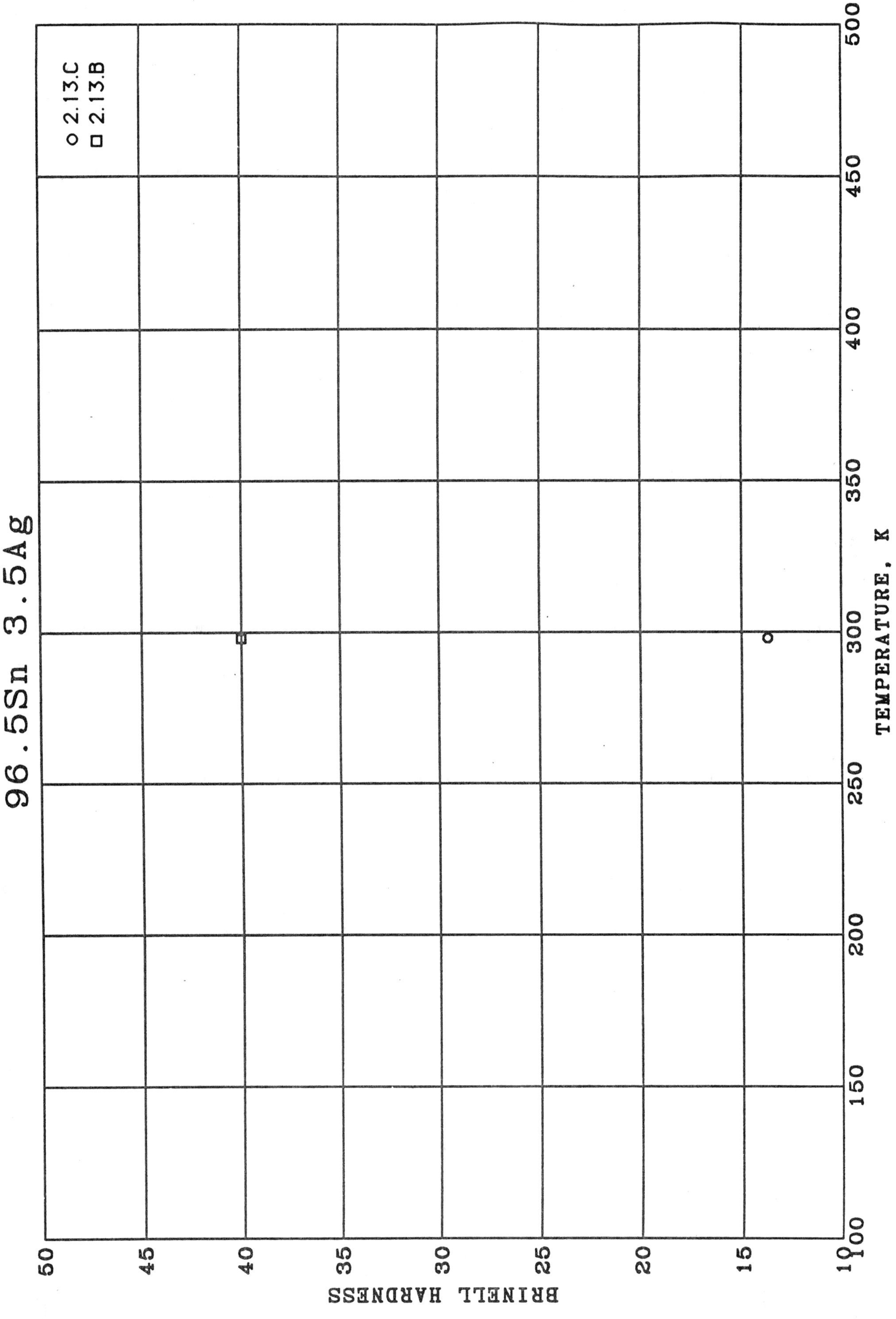
96.5Sn 3.5Ag
○ 2.13.C
□ 2.13.B
BRINELL HARDNESS
50
45
40
35
30
25
20
15
10
TEMPERATURE, K
100
150
200
250
300
350
400
450
500

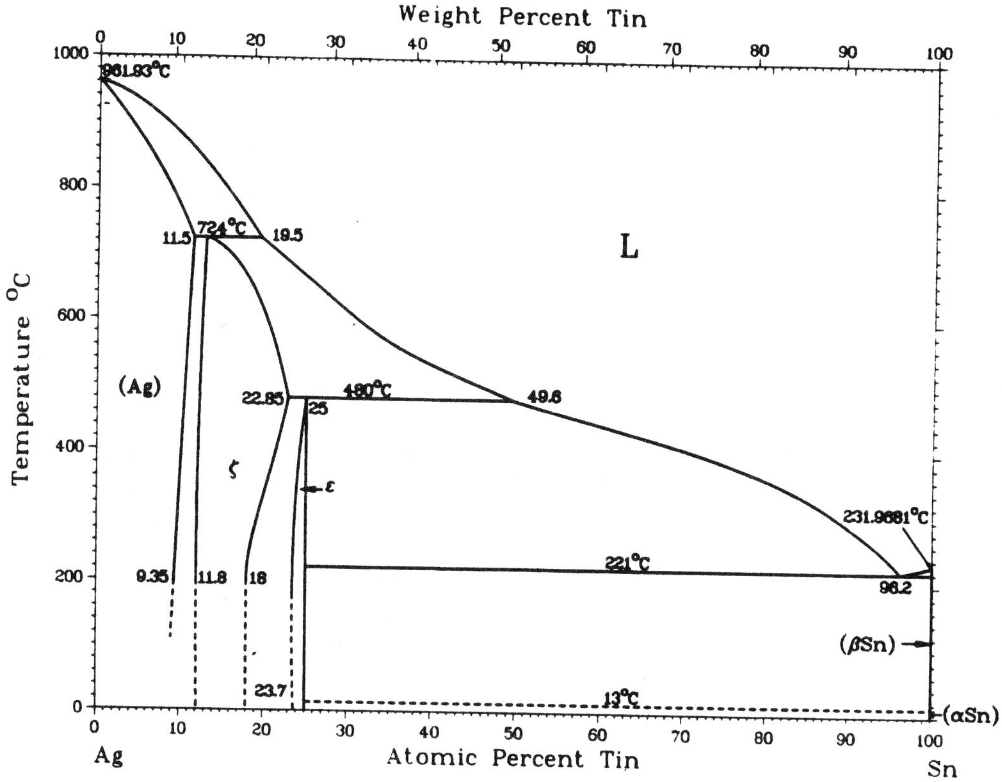

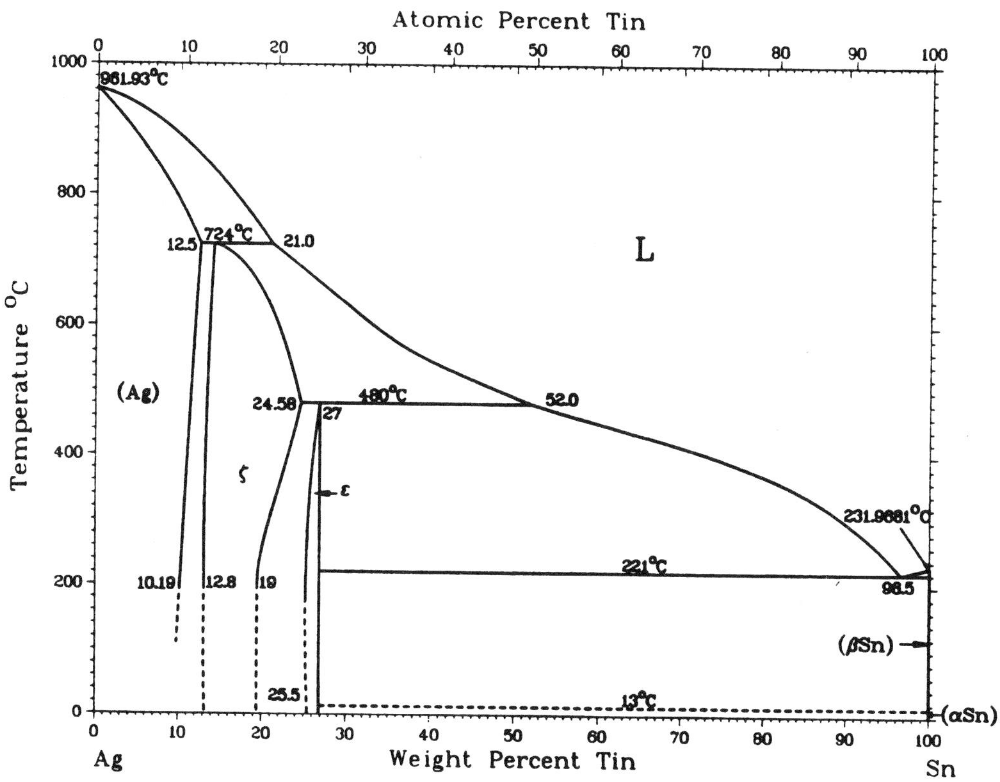

M. Hansen & K. Anderko, Constitution Binary Alloys, McGraw Hill, New York, 1958.

95Sn 5Ag

PROPERTY	@ 300K	UNITS	REFERENCES
Density	7.39	g/cm^3	1.1.A-C
Specific Heat	0.055	cal/g-K	1.2.A
Thermal Conductivity	0.079	cal/cm-s-K	1.3.A-B
Thermal Expansion	23.2	1/K	1.4.A
Solidus/Liquidus	494/518	K	1.5.A-C
Resistivity	12.9×10^{-6}	Ohm-cm	1.6.A-D
TCR	4250	ppm/K	1.7.A
Young's Modulus	8.2×10^{6}	psi	1.8.A
Poisson's Ratio	0.37	------	1.9.A
Yield Strength	3600	psi	1.10.A-B
Ult Tensile Strength	4500	psi	1.11.A-D
Elongation at Break	49	%	1.12.A-C
Hardness	18	Brinell	1.13.A-B

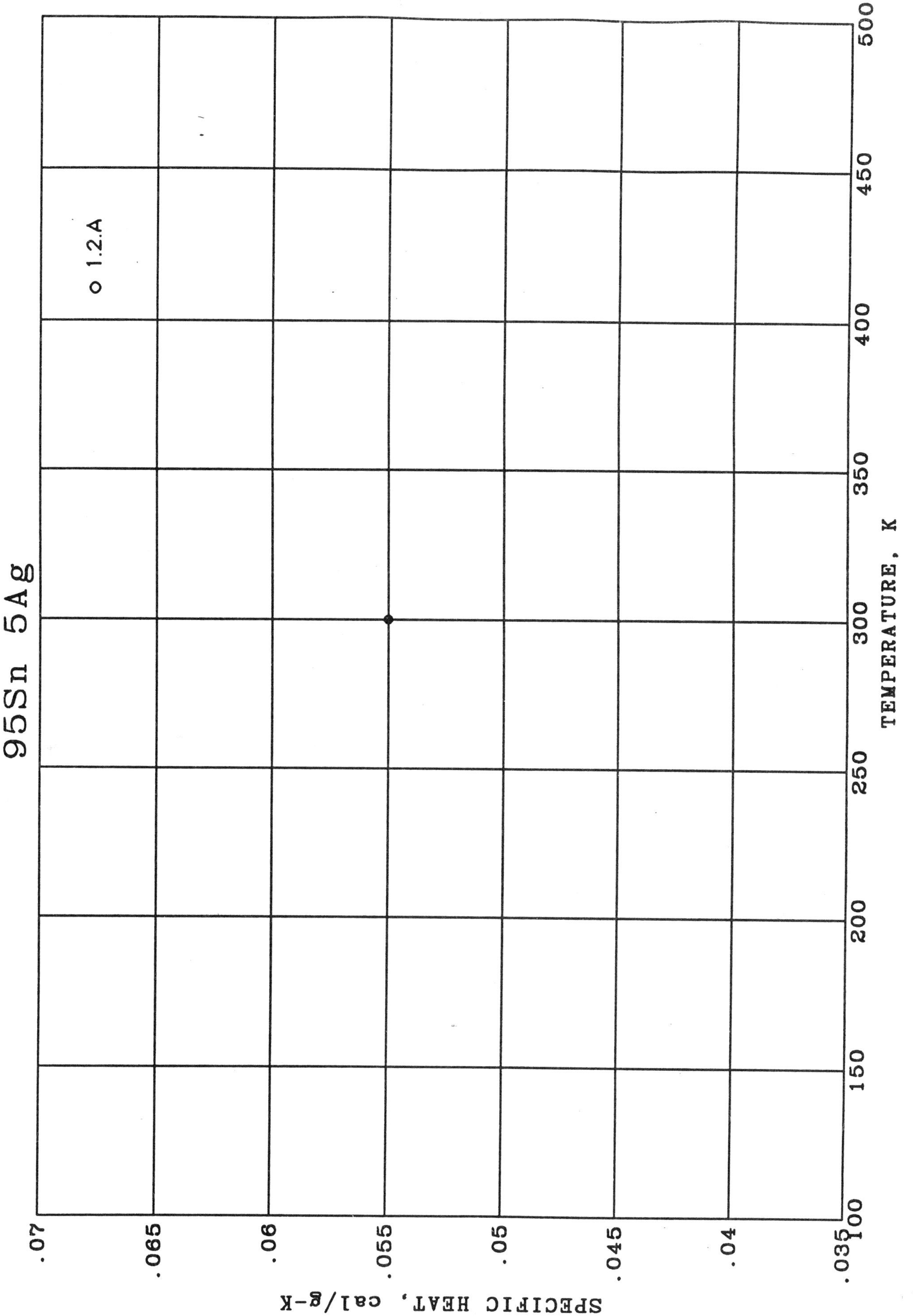
95Sn 5Ag
o 1.2.A
TEMPERATURE, K
100
150
200
250
300
350
400
450
500
SPECIFIC HEAT, cal/g-K
.035
.04
.045
.05
.055
.06
.065
.07

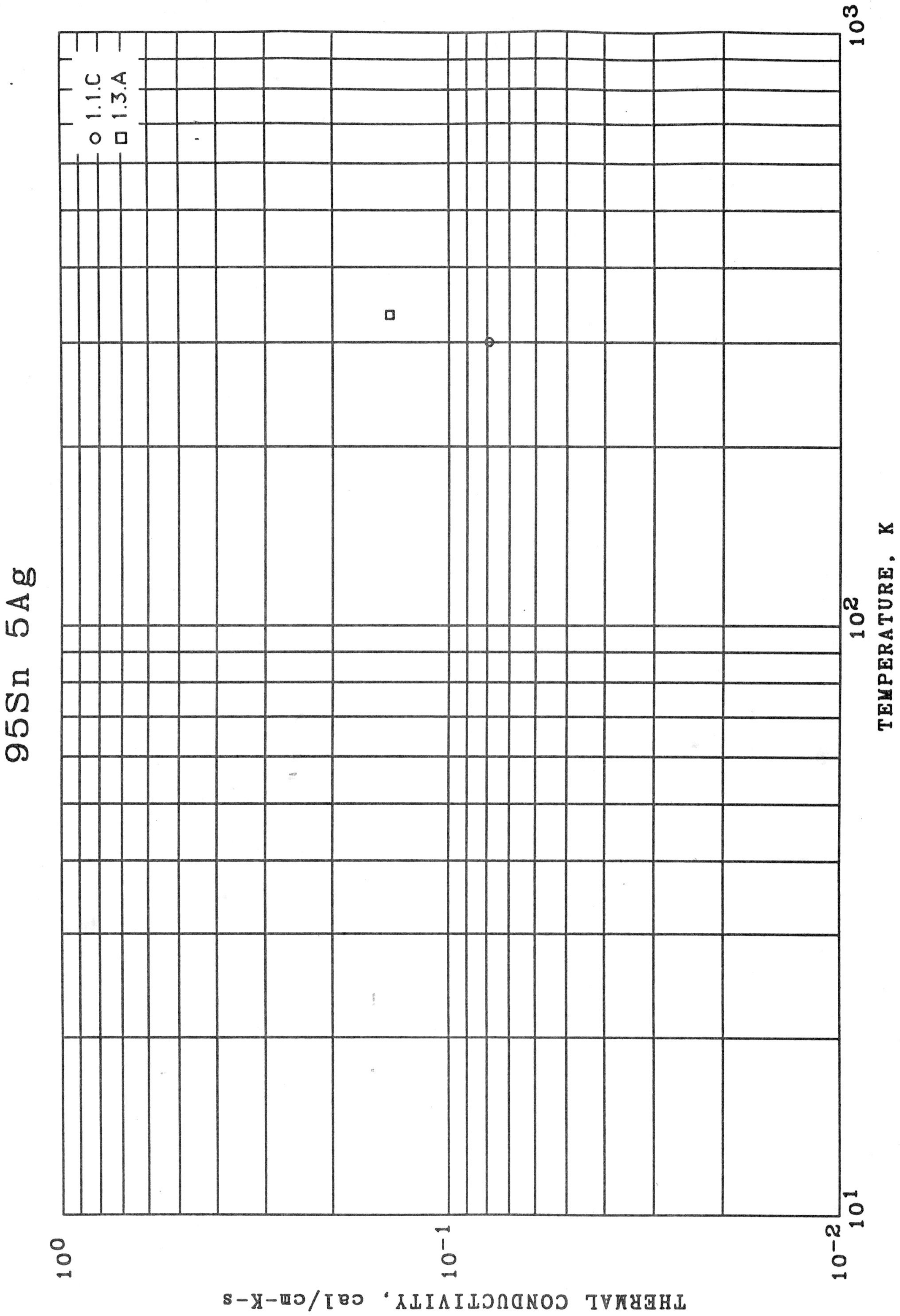
95Sn 5Ag
○ 1.1.C
□ 1.3.A
10^0
10^{-1}
10^{-2}
THERMAL CONDUCTIVITY, cal/cm-K-s
10^1
10^2
10^3
TEMPERATURE, K

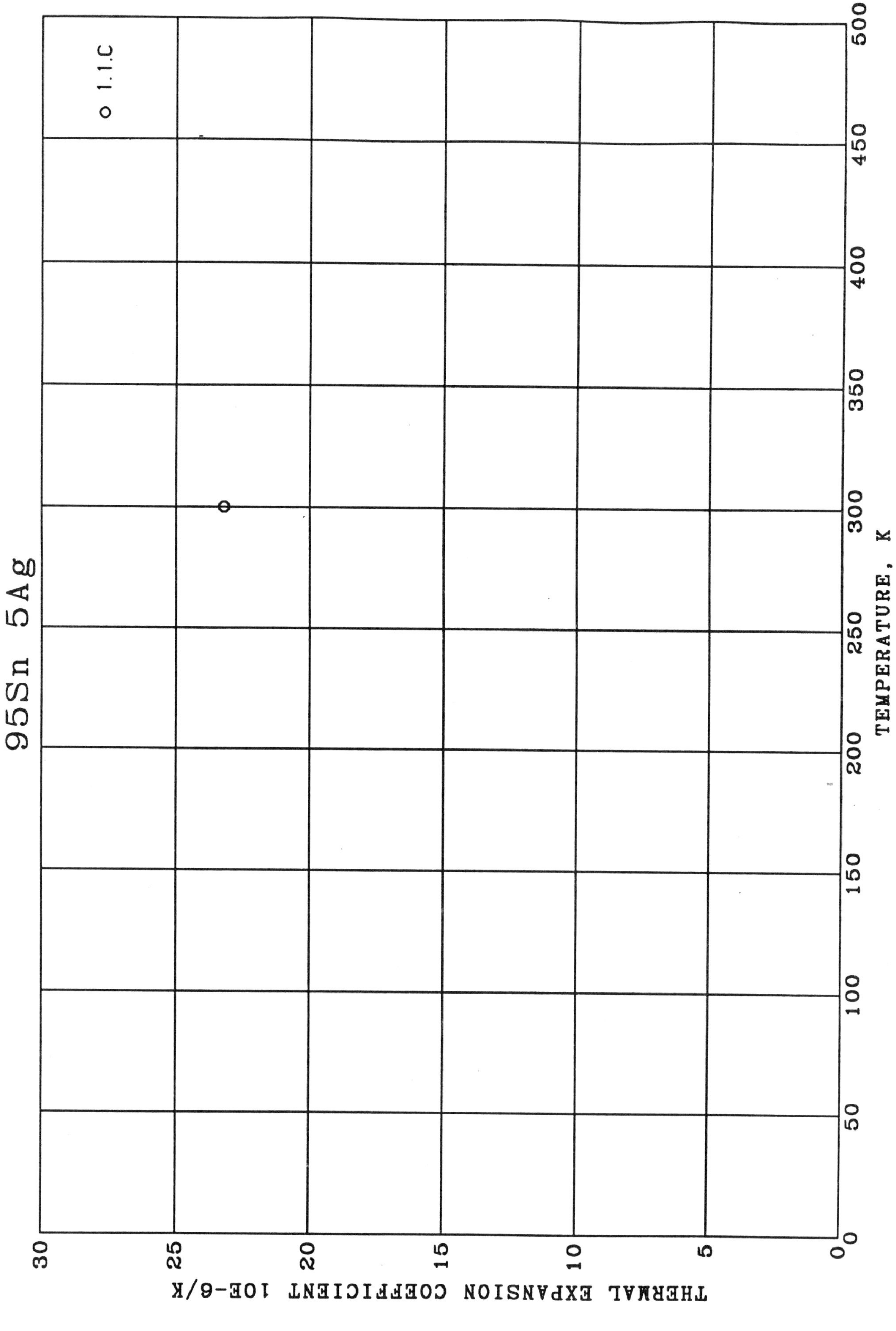

95Sn 5Ag
○ 1.1.C
THERMAL EXPANSION COEFFICIENT 10E-6/K
0
5
10
15
20
25
30
TEMPERATURE, K
0
50
100
150
200
250
300
350
400
450
500

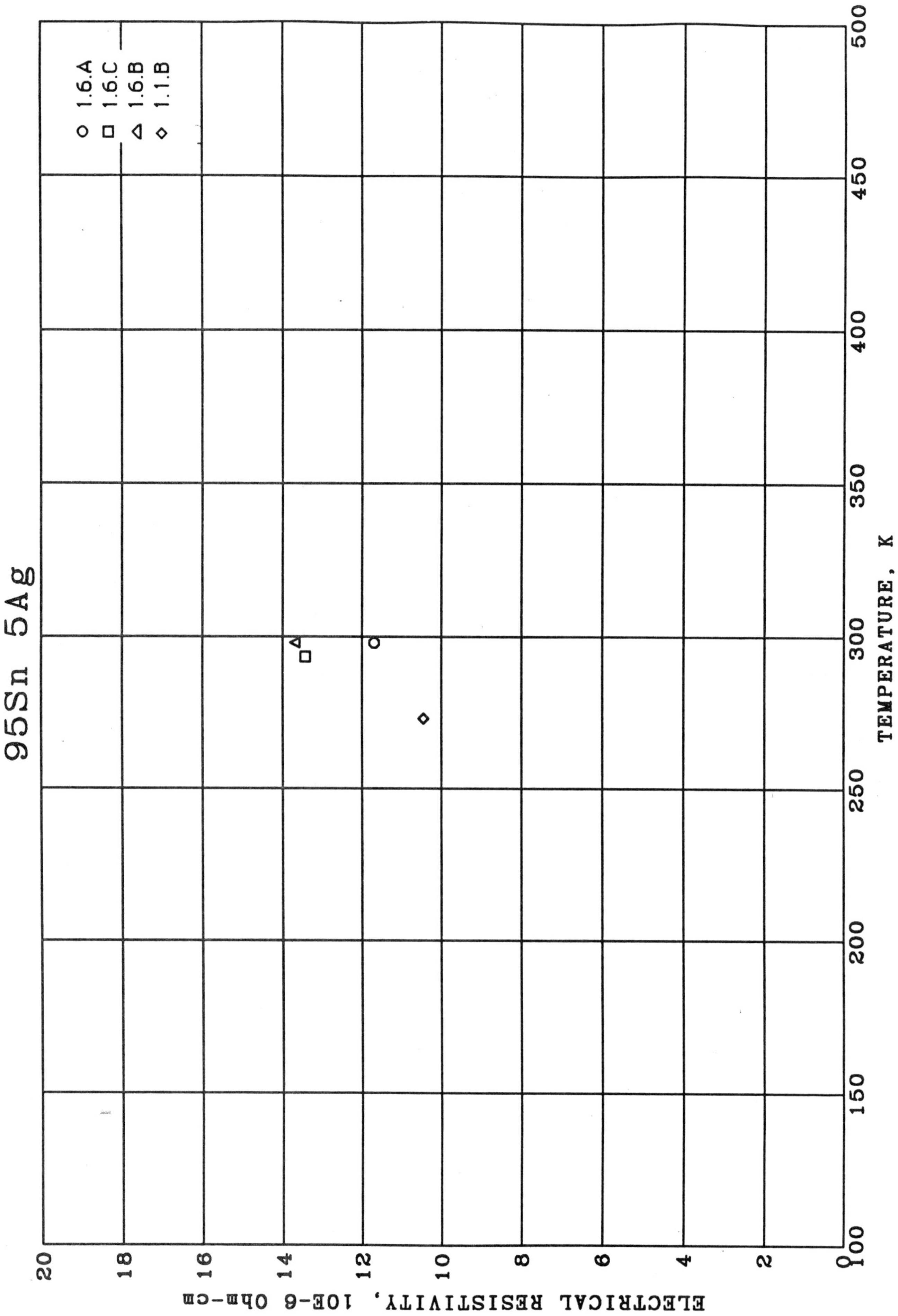
95Sn 5Ag
1.6.A
1.6.C
1.6.B
1.1.B
TEMPERATURE, K
100
150
200
250
300
350
400
450
500
ELECTRICAL RESISTIVITY, 10E-6 Ohm-cm
0
2
4
6
8
10
12
14
16
18
20

95Sn 5Ag

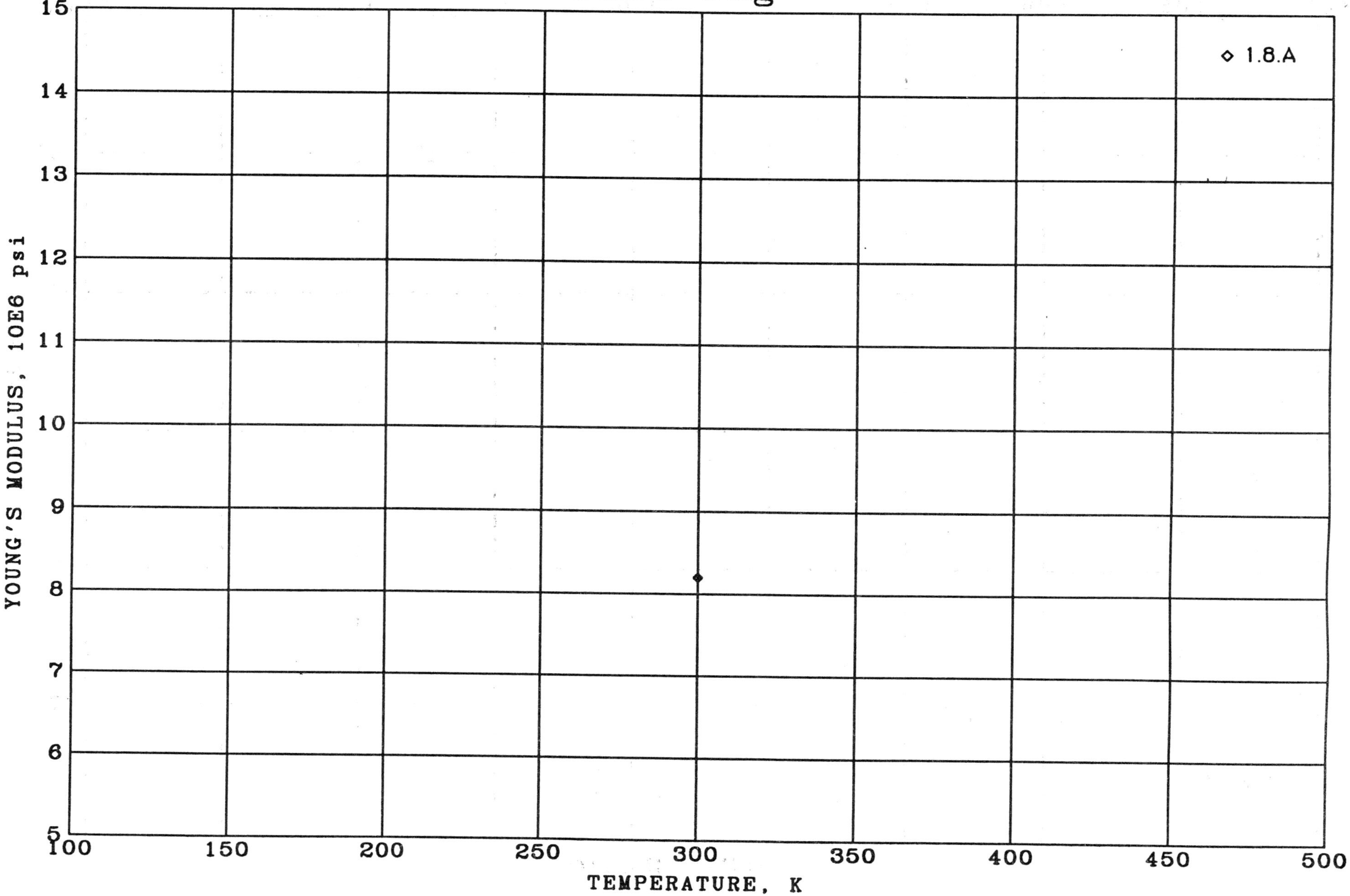
15
14
13
12
11
10
9
8
7
6
5
YOUNG'S MODULUS, 10E6 psi
◇ 1.8.A
100
150
200
250
300
350
400
450
500
TEMPERATURE, K

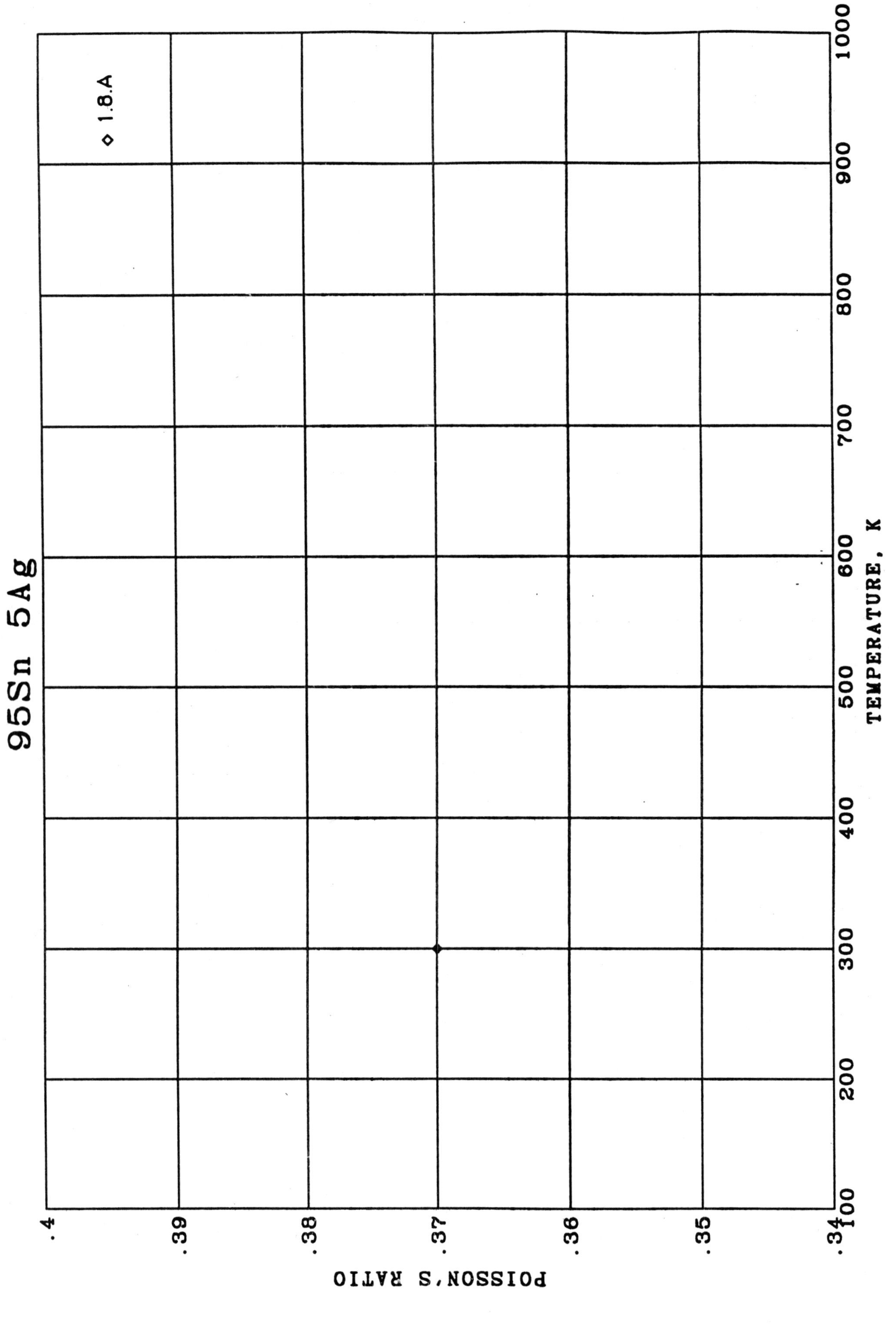
95Sn 5Ag
◇ 1.8.A
POISSON'S RATIO
.4
.39
.38
.37
.36
.35
.34
TEMPERATURE, K
100
200
300
400
500
600
700
800
900
1000

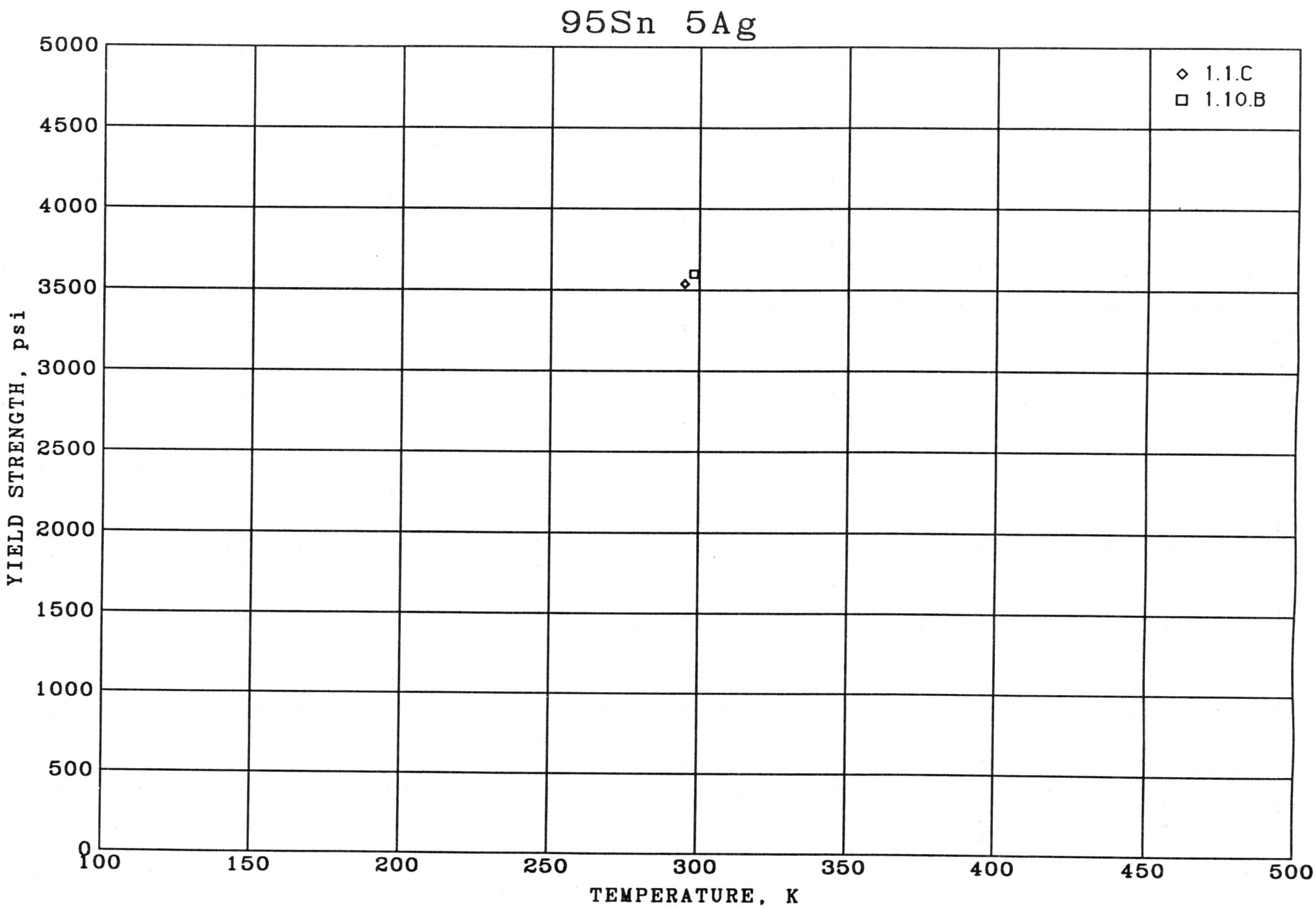
95Sn 5Ag
◇ 1.1.C
□ 1.10.B
YIELD STRENGTH, psi
5000
4500
4000
3500
3000
2500
2000
1500
1000
500
0
100
150
200
250
300
350
400
450
500
TEMPERATURE, K

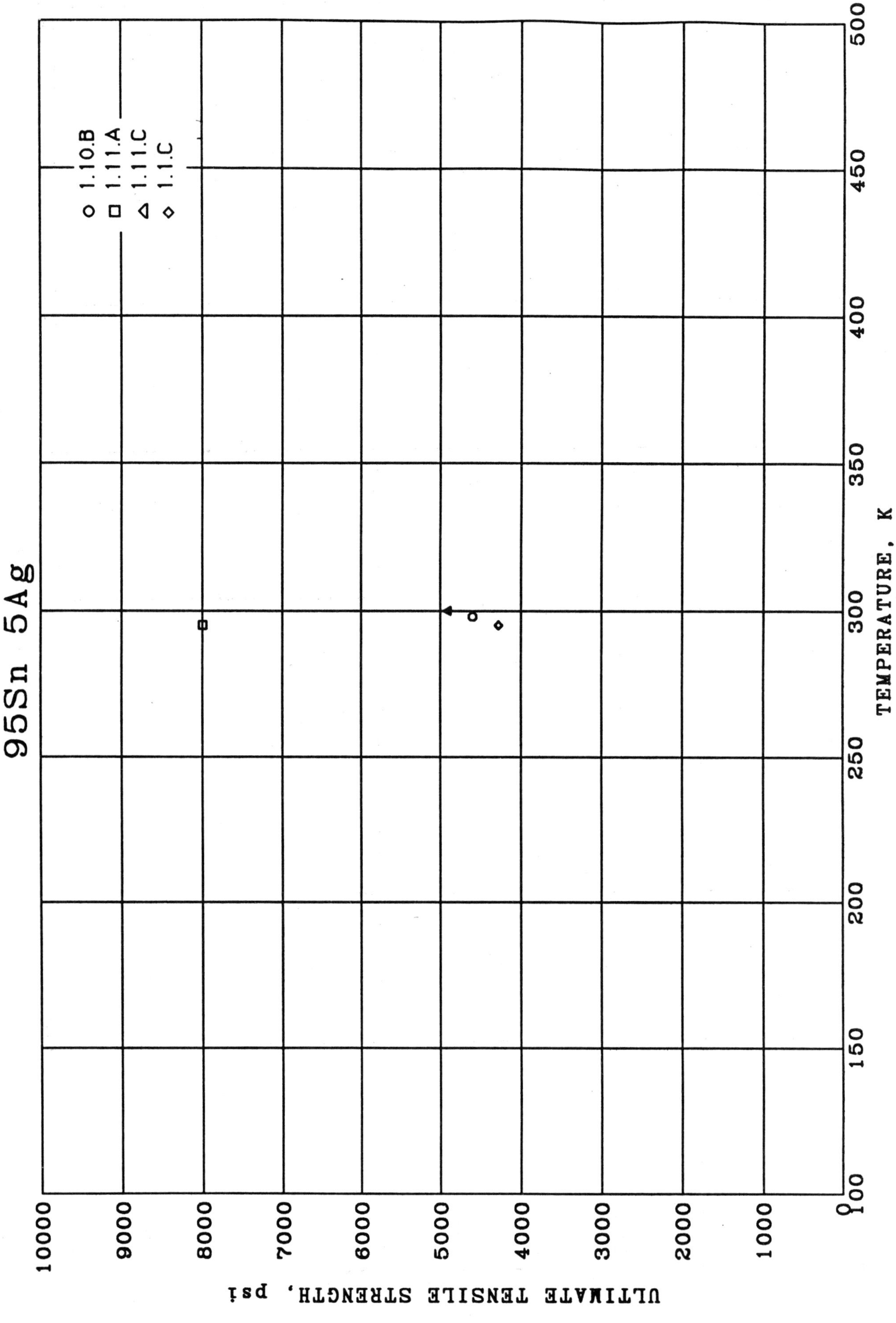

95Sn 5Ag
○ 1.10.B
□ 1.11.A
△ 1.11.C
◇ 1.1.C
ULTIMATE TENSILE STRENGTH, psi
TEMPERATURE, K
0
1000
2000
3000
4000
5000
6000
7000
8000
9000
10000
100
150
200
250
300
350
400
450
500

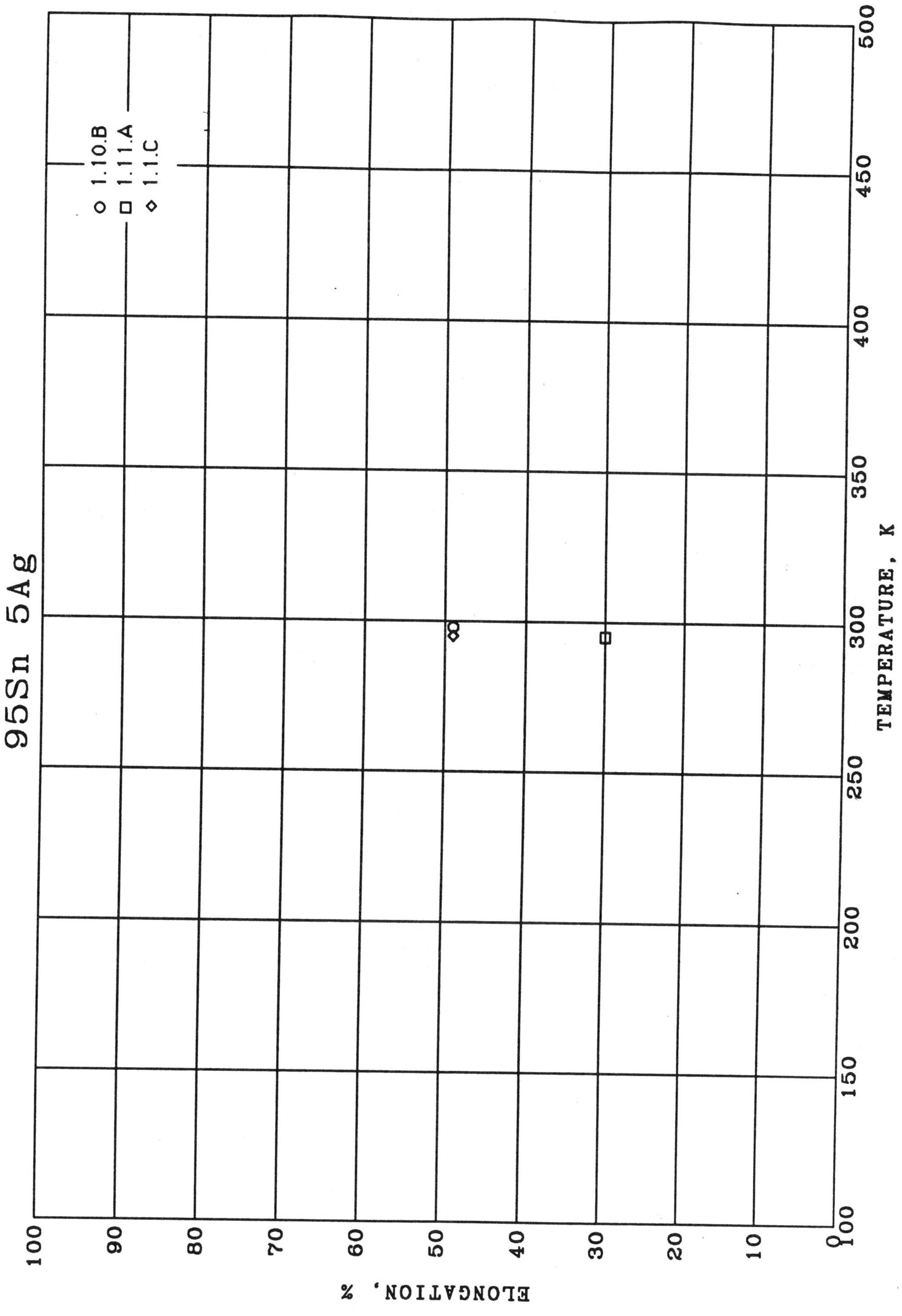
95Sn 5Ag
○ 1.10.B
□ 1.11.A
◇ 1.1.C
ELONGATION, %
0
10
20
30
40
50
60
70
80
90
100
TEMPERATURE, K
100
150
200
250
300
350
400
450
500

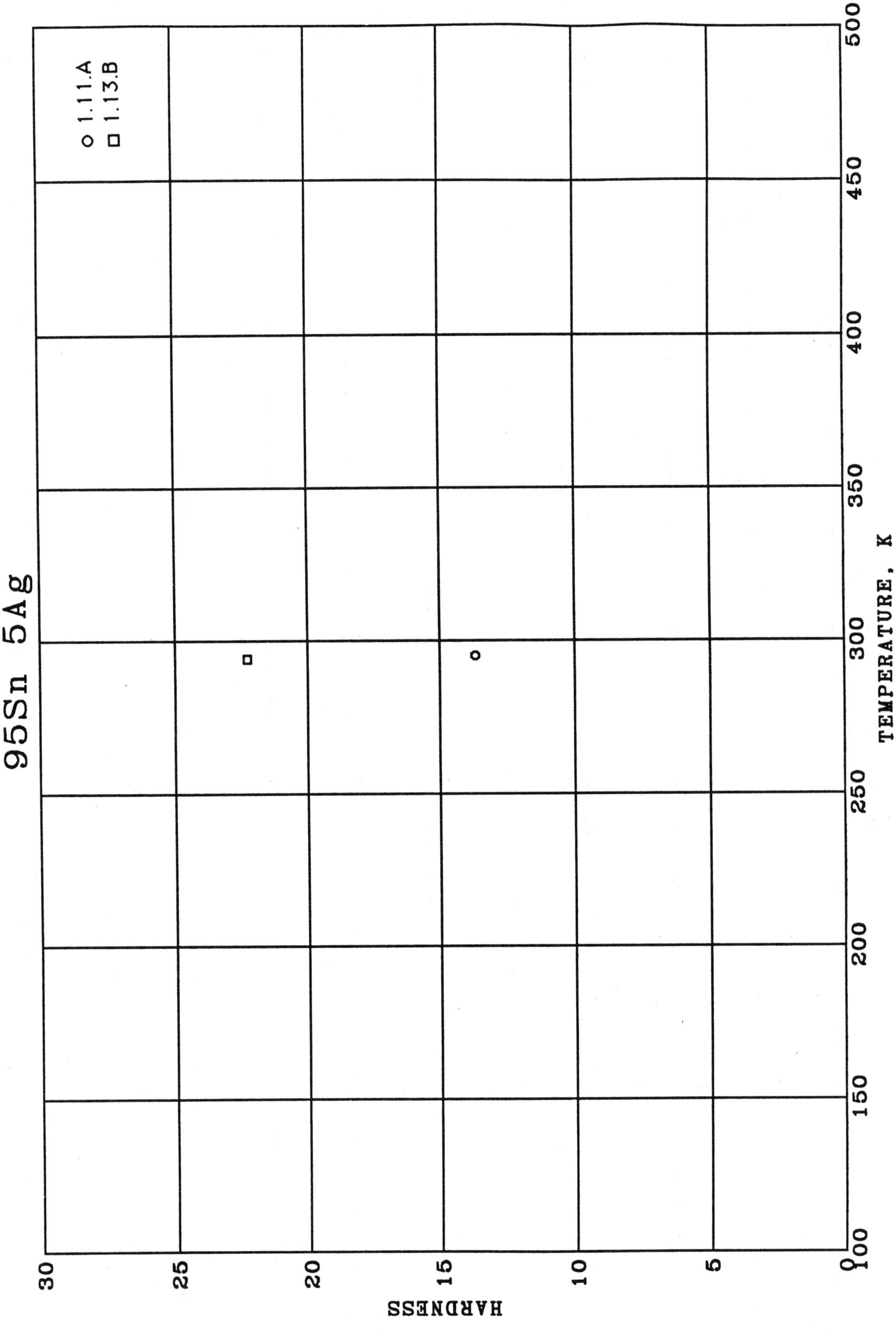
95Sn 5Ag
1.11.A
1.13.B
HARDNESS
0
5
10
15
20
25
30
TEMPERATURE, K
100
150
200
250
300
350
400
450
500

88Au 12Ge

PROPERTY	@ 300K	UNITS	REFERENCES
Density	14.7	g/cm^3	3.1.A-C
Specific Heat	0.036	cal/g-K	3.2.A
Thermal Conductivity	0.105	cal/cm-s-K	3.3.A-C
Thermal Expansion	14×10^{-6}	1/K	3.4.A-D
Melting Point	629	K	3.5.A-C
Resistivity	30×10^{-6}	Ohm-cm	3.6.A
TCR		ppm/K	
Young's Modulus	10550	ksi	3.8.A
Poisson's Ratio		------	
Yield Strength	26825	psi	3.10.A
Ult Tensile Strength	26825	psi	3.11.A-C
Elongation at Break	<2	%	3.12.A
Hardness	39	kg/mm^2	3.13.A-B

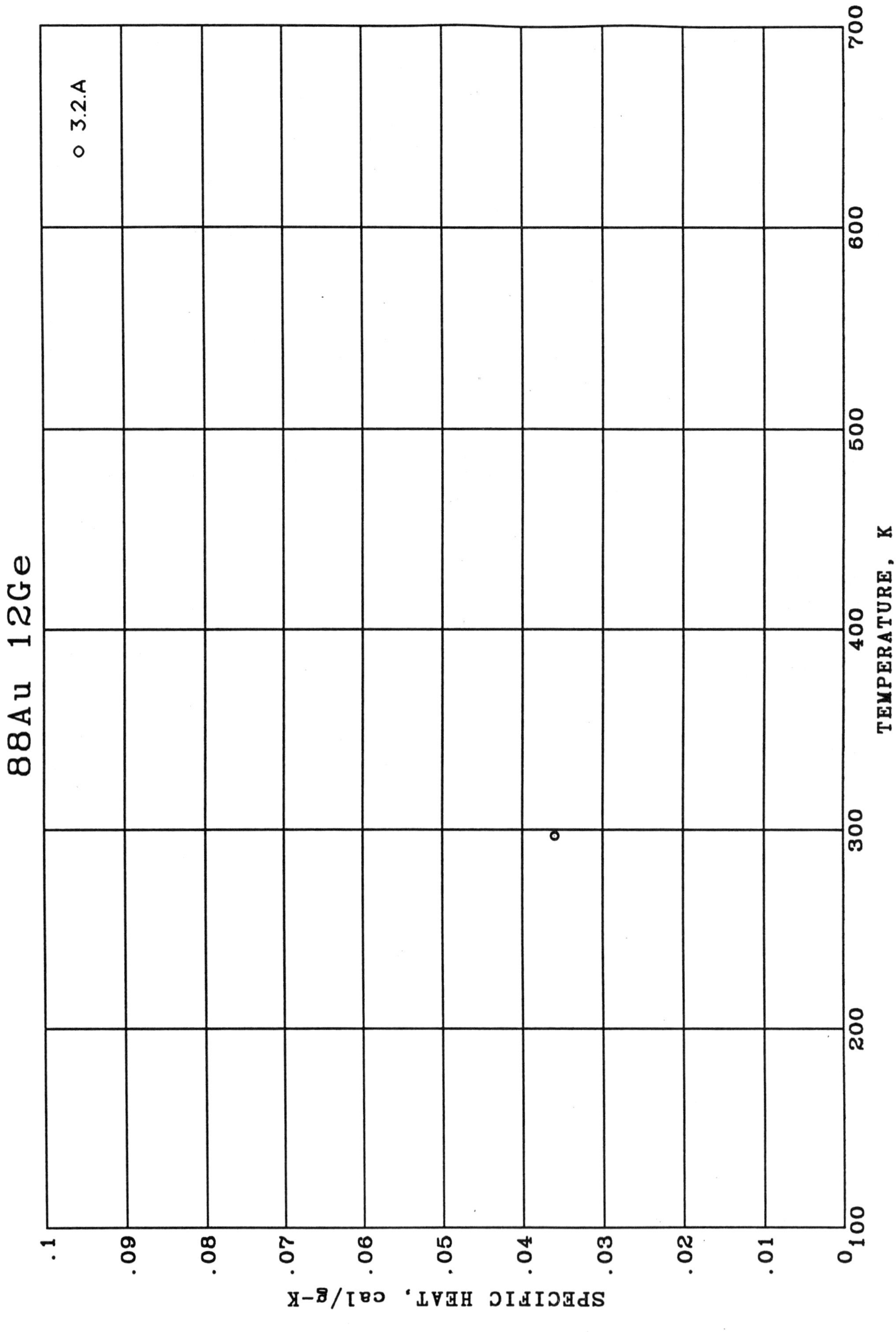

88Au 12Ge
○ 3.2.A
SPECIFIC HEAT, cal/g-K
TEMPERATURE, K
.1
.09
.08
.07
.06
.05
.04
.03
.02
.01
0
100
200
300
400
500
600
700

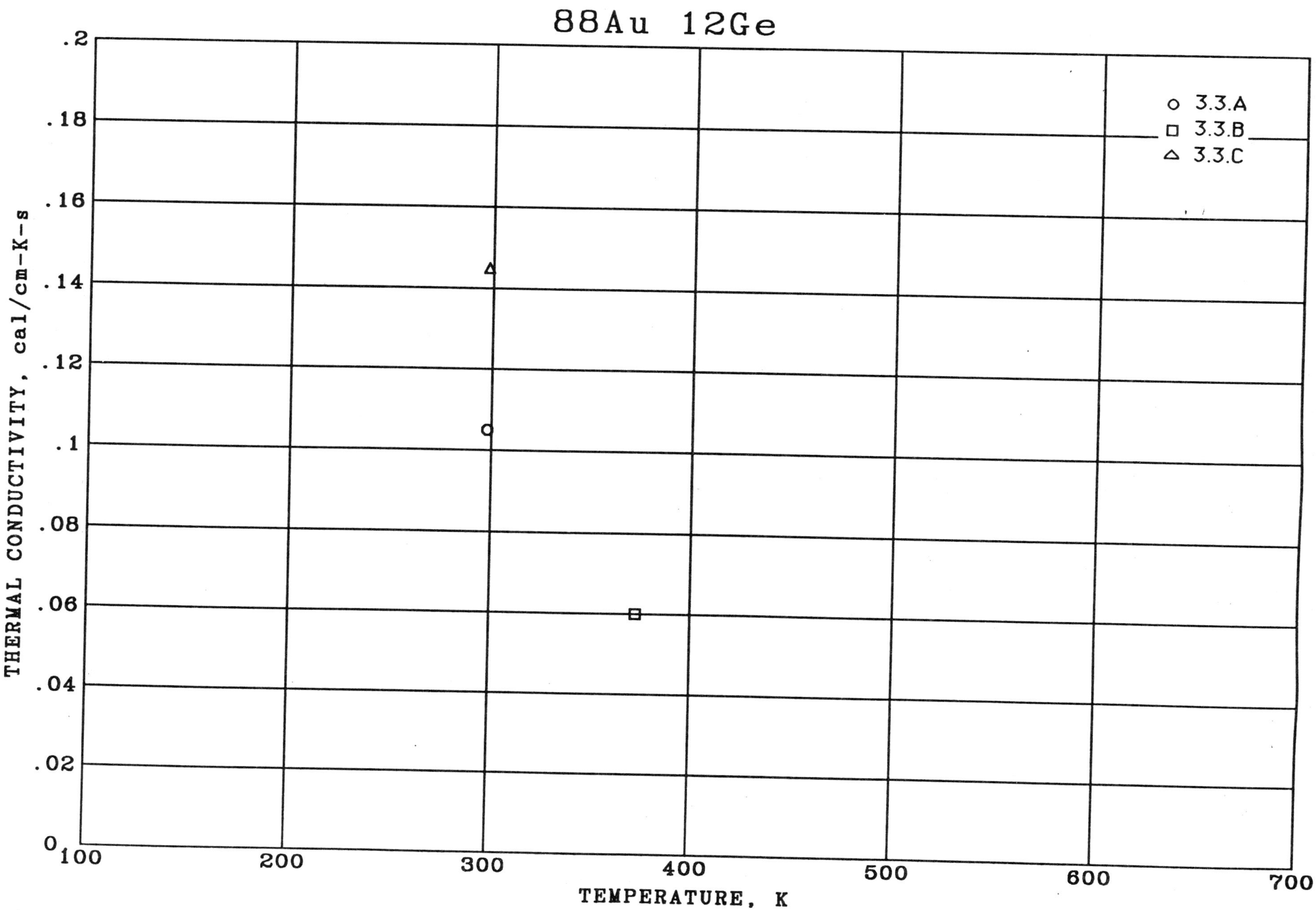

88Au 12Ge
○ 3.3.A
□ 3.3.B
△ 3.3.C
THERMAL CONDUCTIVITY, cal/cm-K-s
.2
.18
.16
.14
.12
.1
.08
.06
.04
.02
0
100
200
300
400
500
600
700
TEMPERATURE, K

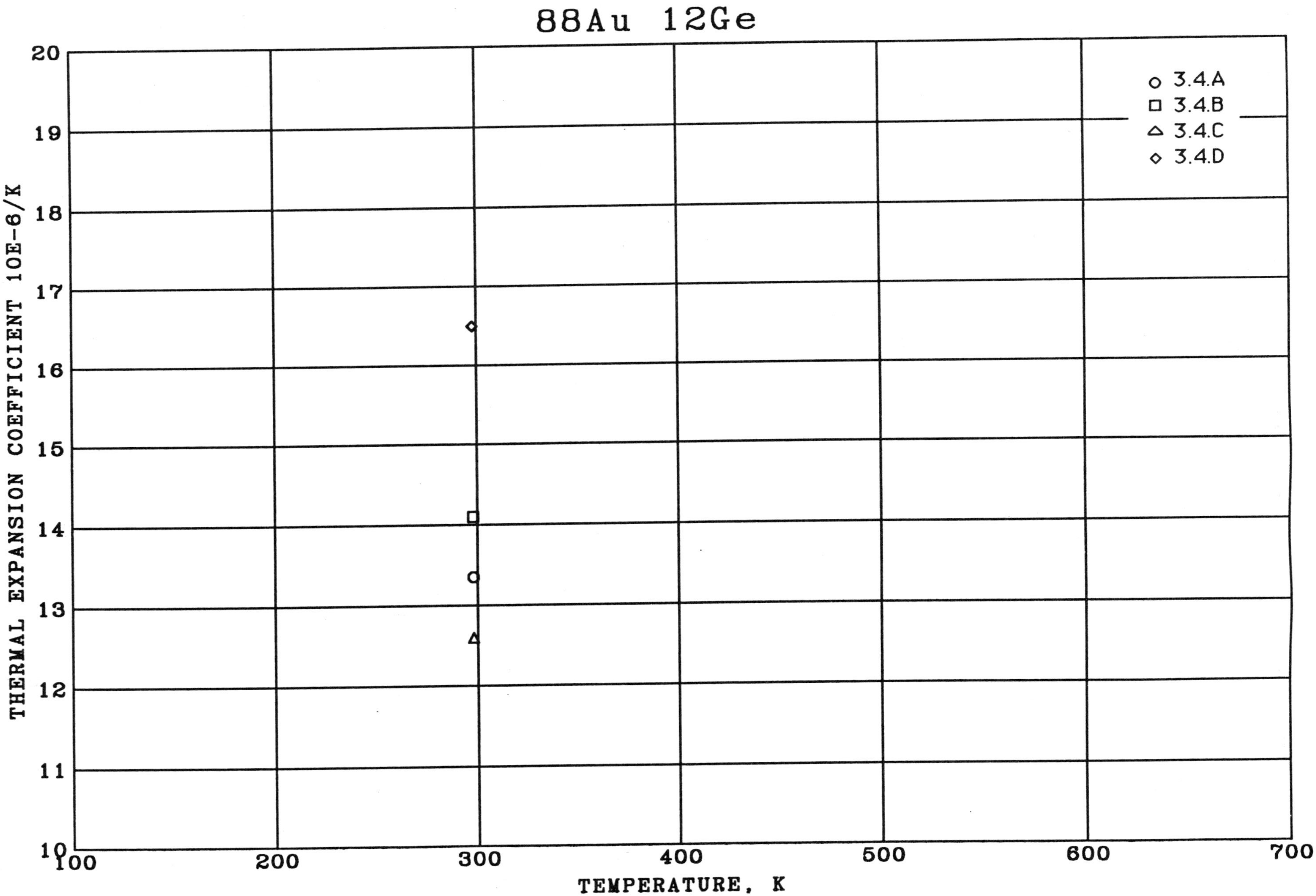
88Au 12Ge
THERMAL EXPANSION COEFFICIENT 10E-6/K
TEMPERATURE, K
10
11
12
13
14
15
16
17
18
19
20
100
200
300
400
500
600
700
3.4.A
3.4.B
3.4.C
3.4.D

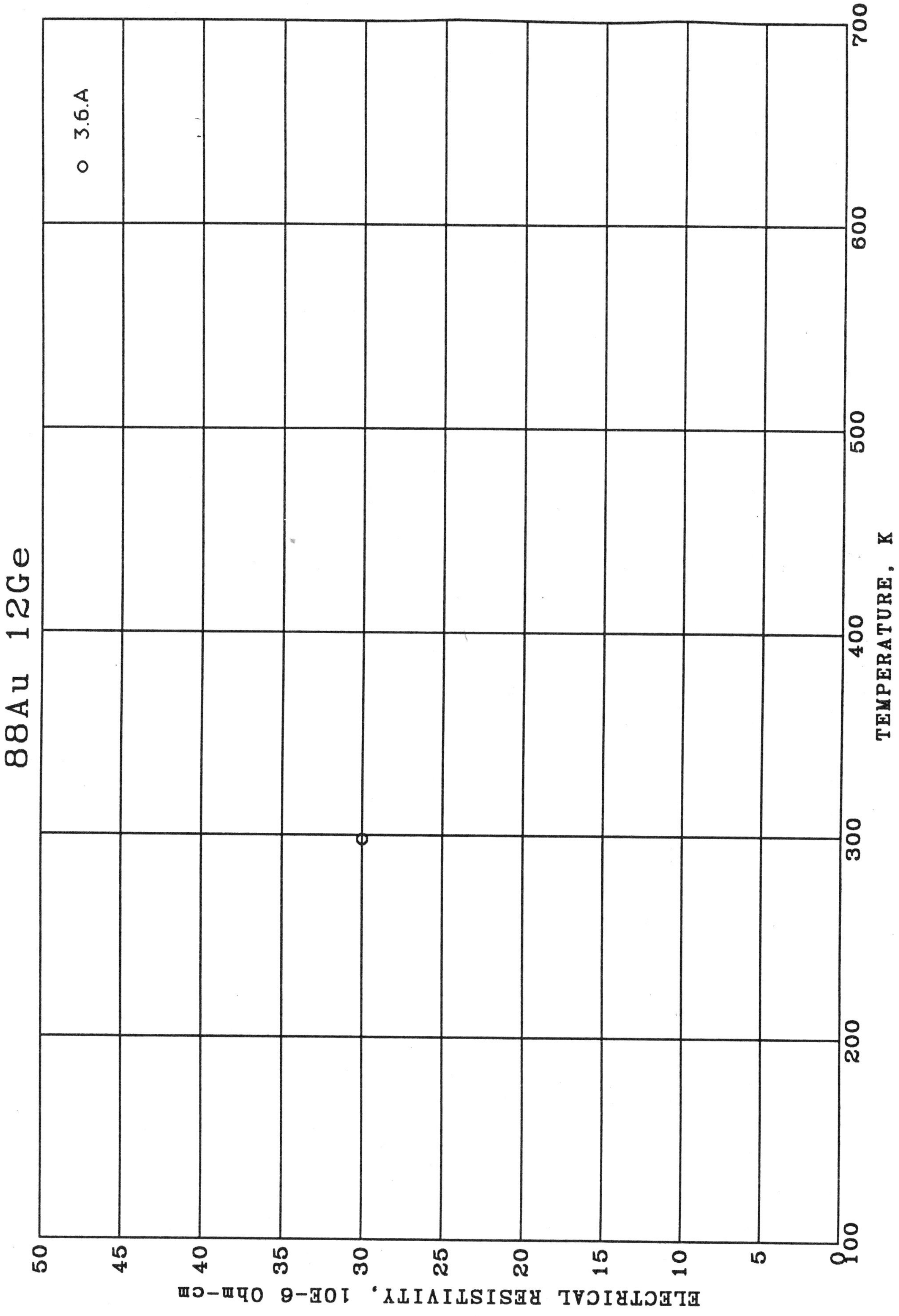

88Au 12Ge
○ 3.6.A
ELECTRICAL RESISTIVITY, 10E-6 Ohm-cm
0
5
10
15
20
25
30
35
40
45
50
TEMPERATURE, K
100
200
300
400
500
600
700

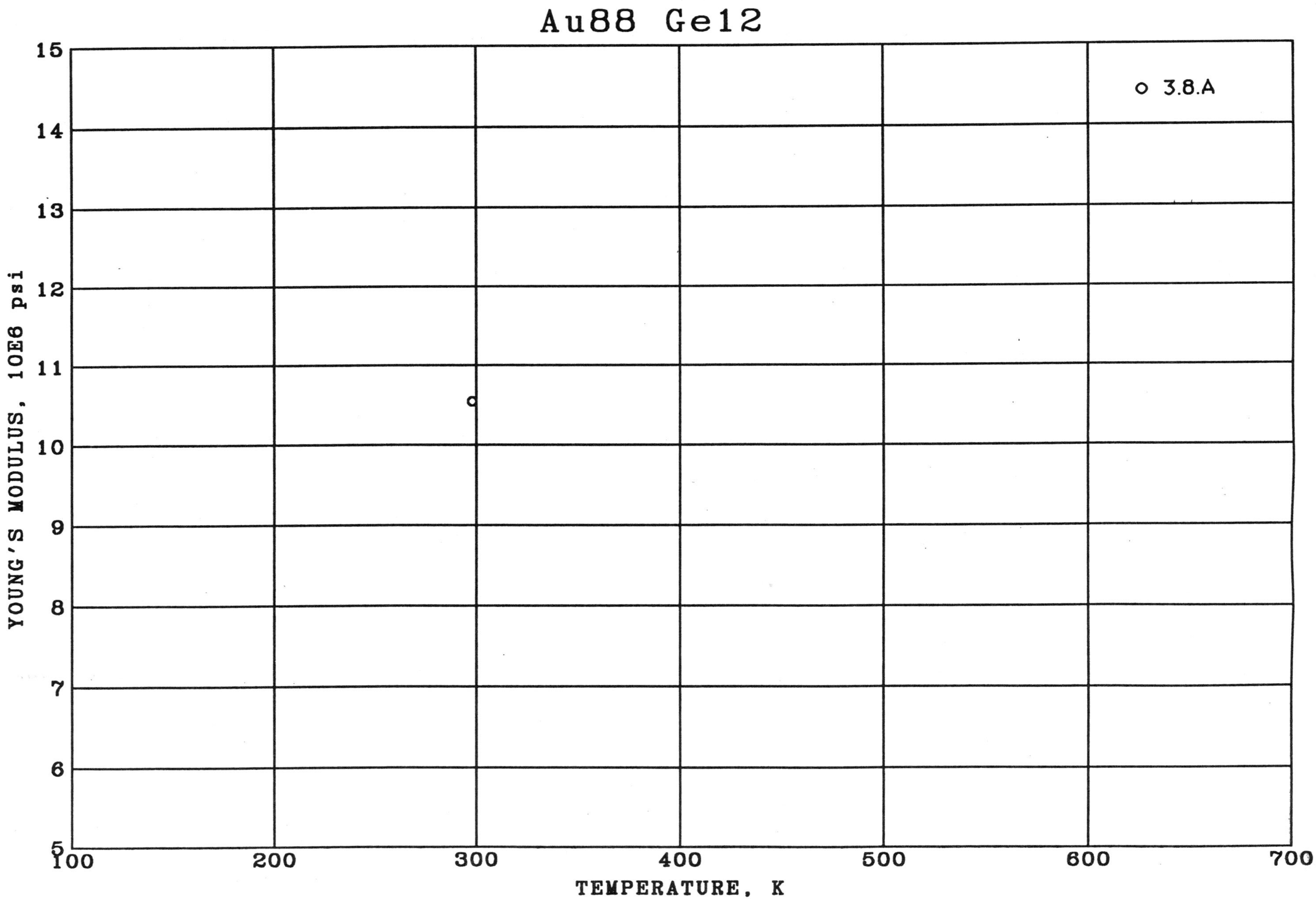
Au88 Ge12
o 3.8.A
YOUNG'S MODULUS, 10E6 psi
15
14
13
12
11
10
9
8
7
6
5
100
200
300
400
500
600
700
TEMPERATURE, K

88Au 12Ge

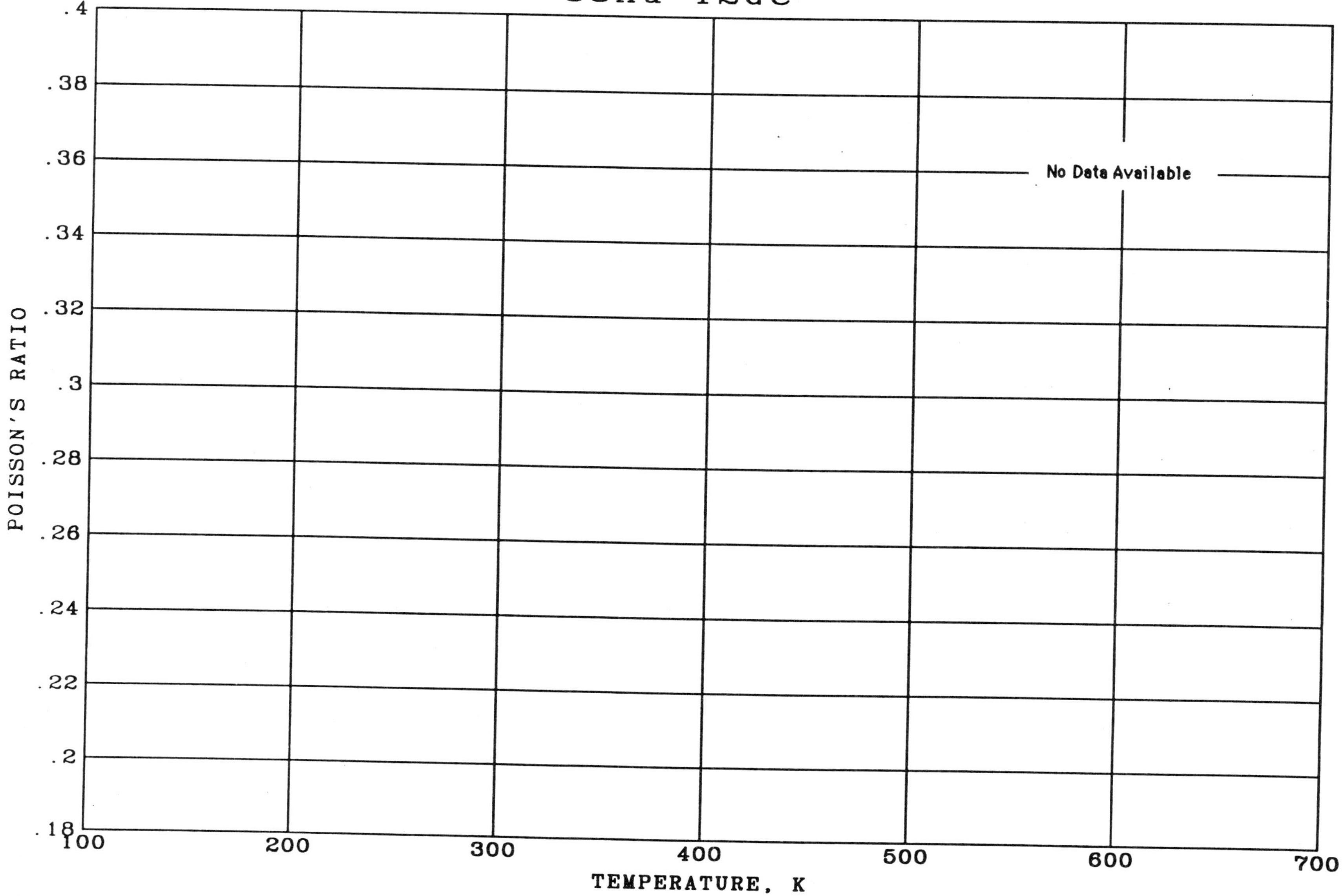

POISSON'S RATIO
.4
.38
.36
.34
.32
.3
.28
.26
.24
.22
.2
.18
100
200
300
400
500
600
700
TEMPERATURE, K
No Data Available

88Au 12Ge

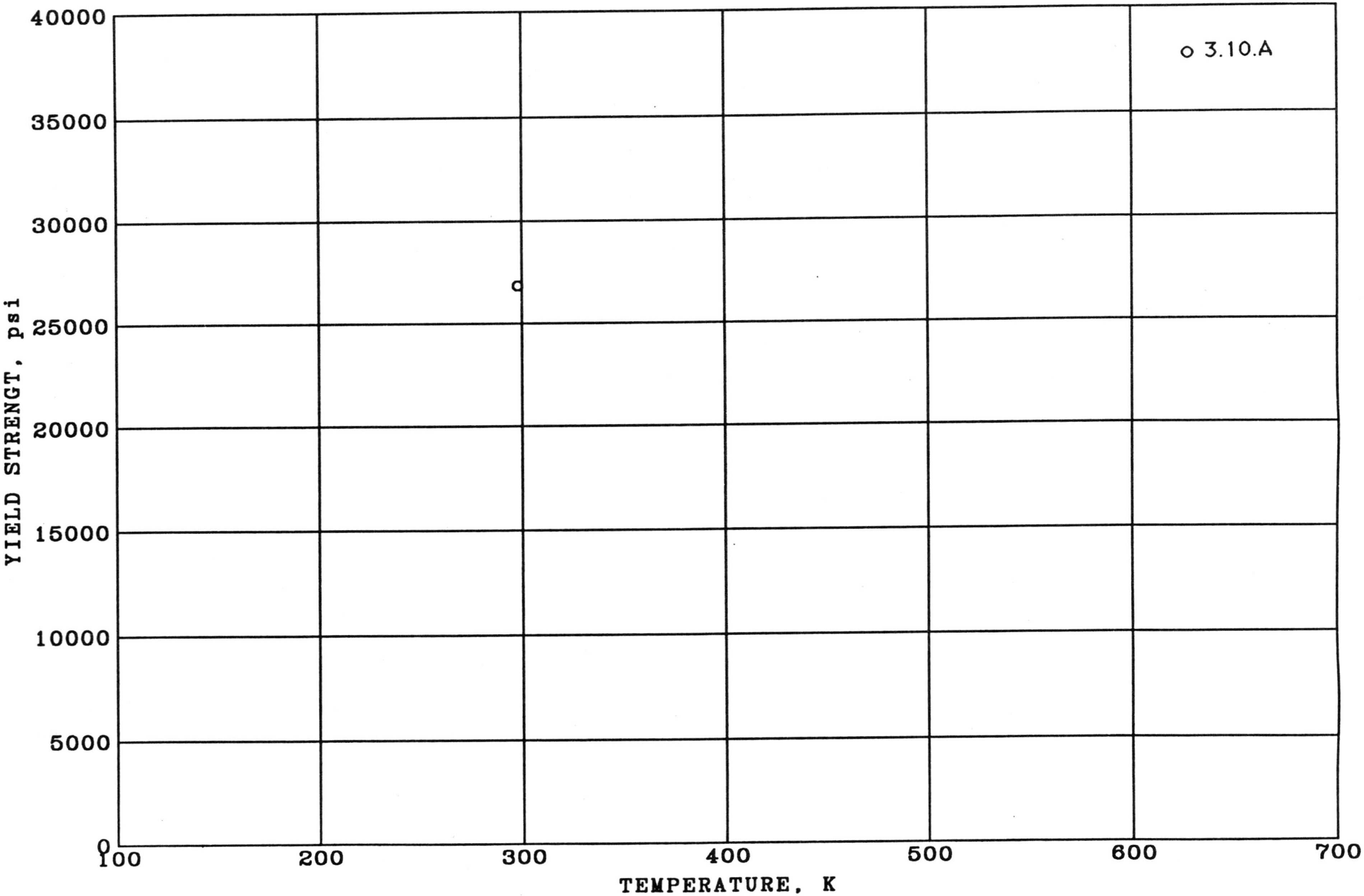

40000
35000
30000
25000
20000
15000
10000
5000
0
YIELD STRENGT, psi
100
200
300
400
500
600
700
TEMPERATURE, K
o 3.10.A

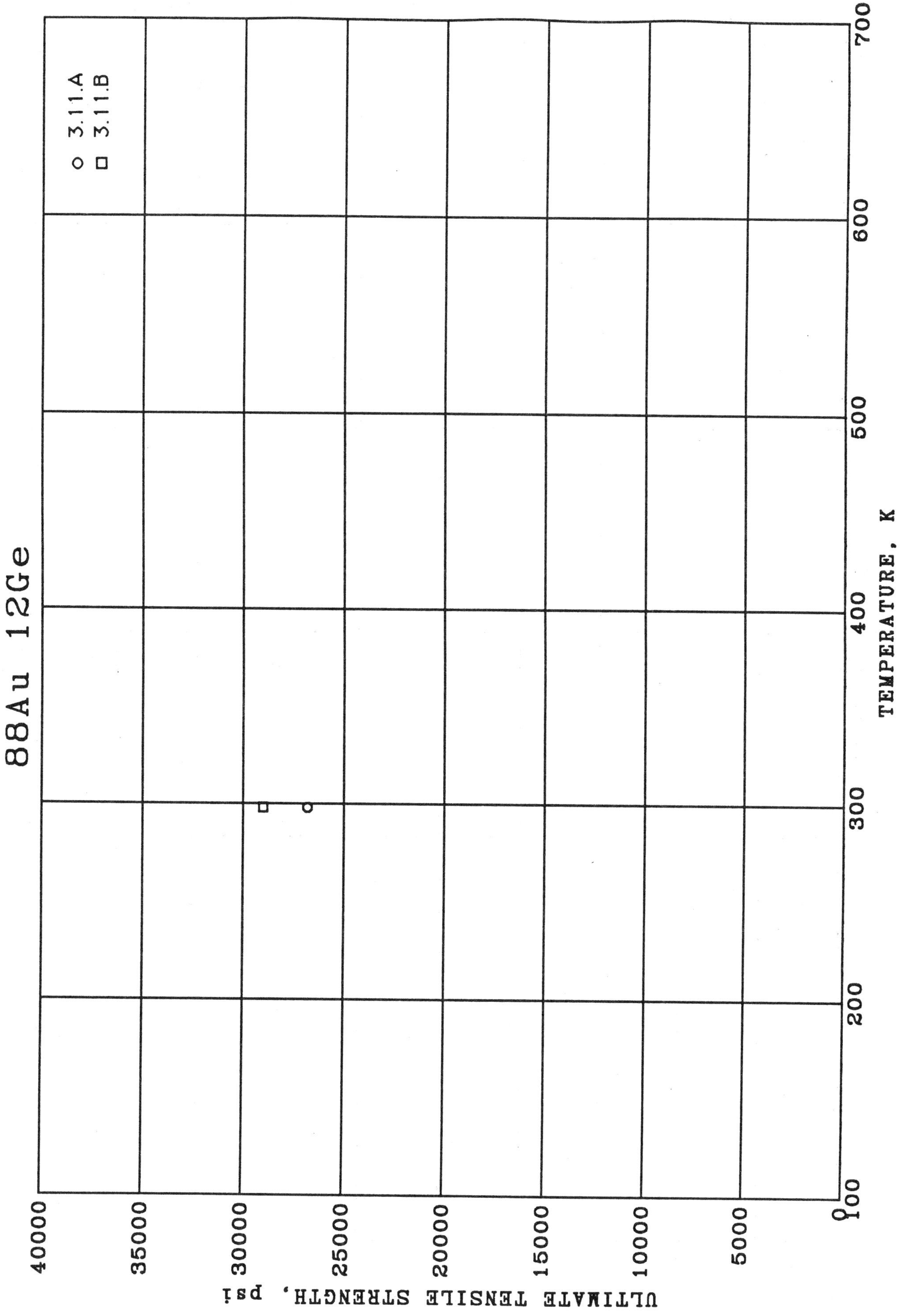

88Au 12Ge
○ 3.11.A
□ 3.11.B
ULTIMATE TENSILE STRENGTH, psi
0
5000
10000
15000
20000
25000
30000
35000
40000
TEMPERATURE, K
100
200
300
400
500
600
700

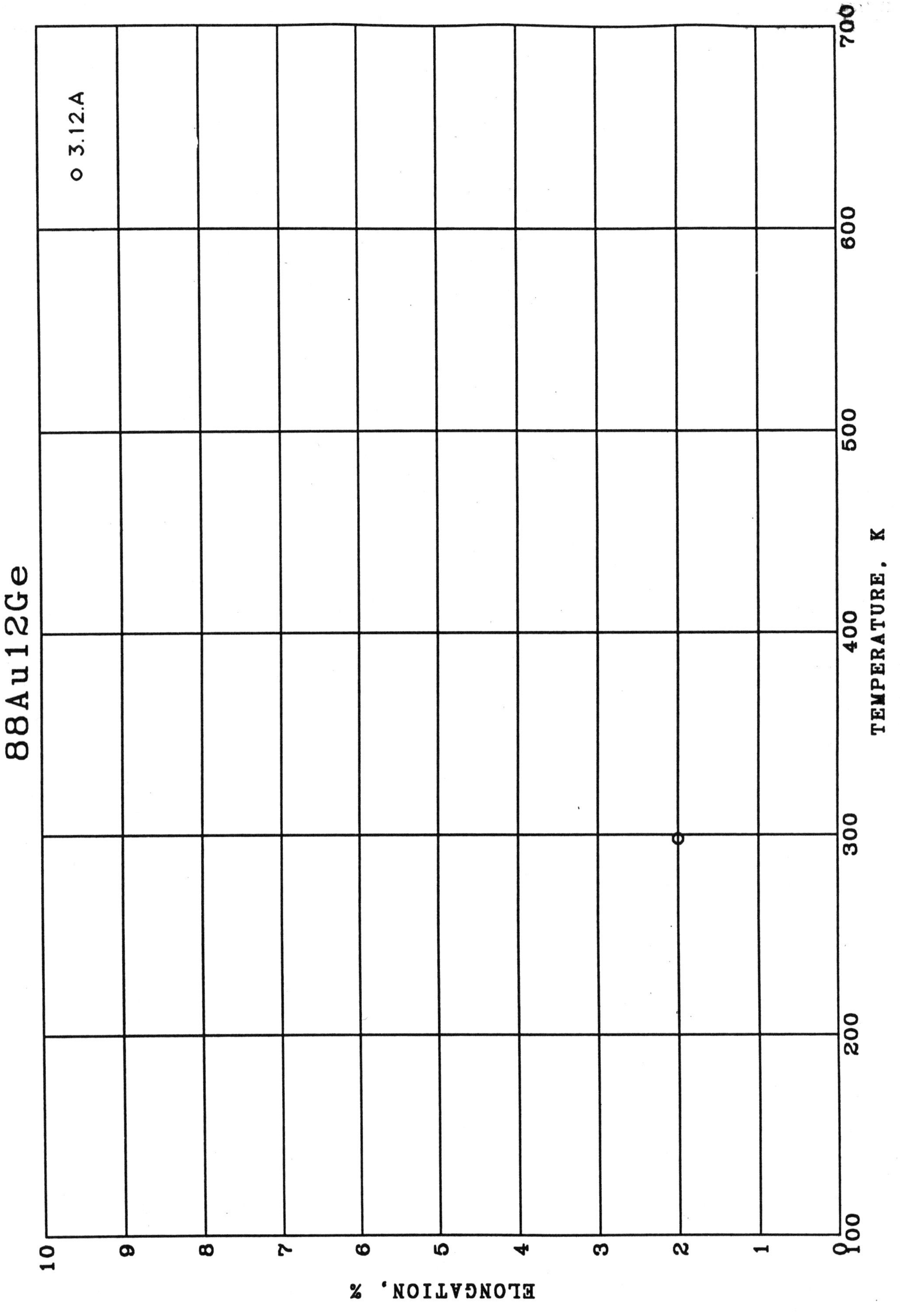
88Au12Ge
o 3.12.A
TEMPERATURE. K
ELONGATION, %
100
200
300
400
500
600
700
0
1
2
3
4
5
6
7
8
9
10

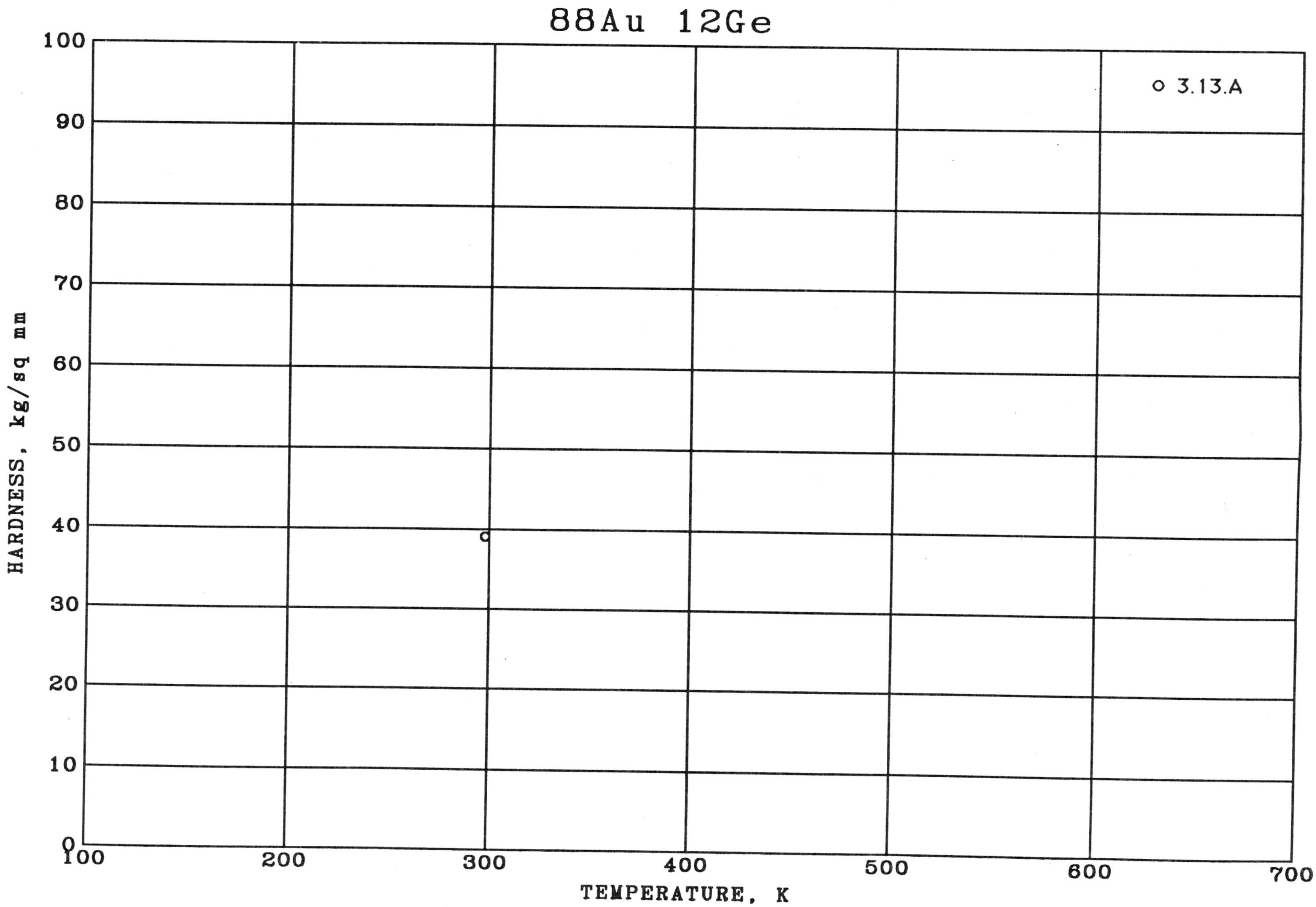
88Au 12Ge
○ 3.13.A
HARDNESS, kg/sq mm
0
10
20
30
40
50
60
70
80
90
100
TEMPERATURE, K
100
200
300
400
500
600
700

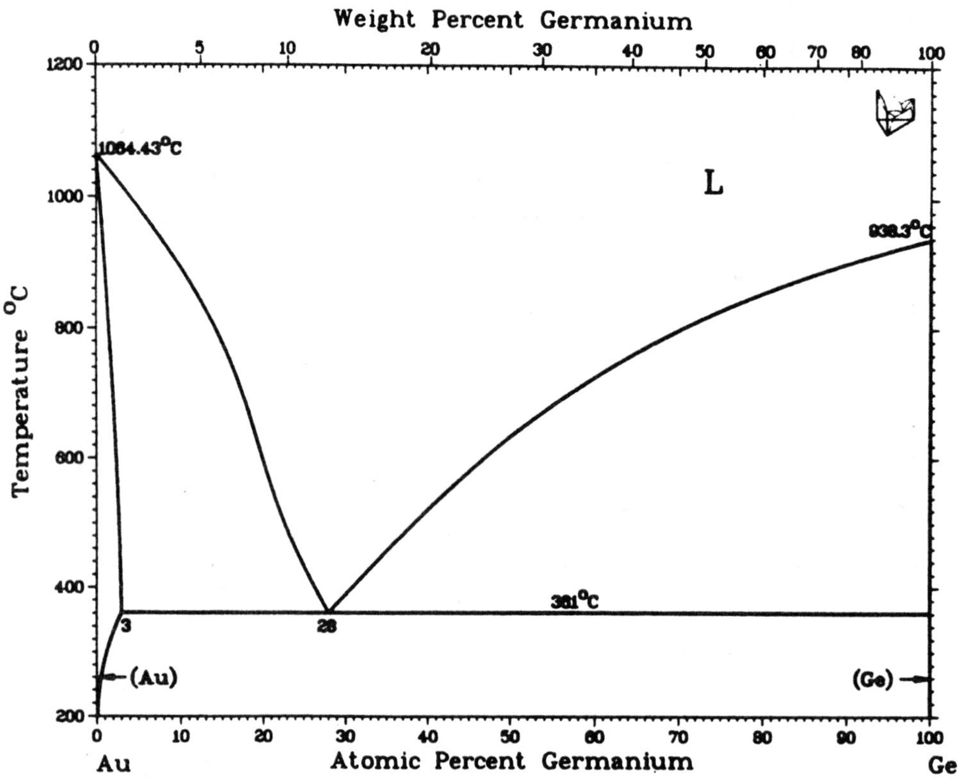
Weight Percent Germanium
Temperature °C
1084.43°C
L
938.3°C
361°C
3
28
(Au)
(Ge)
Atomic Percent Germanium
Au
Ge

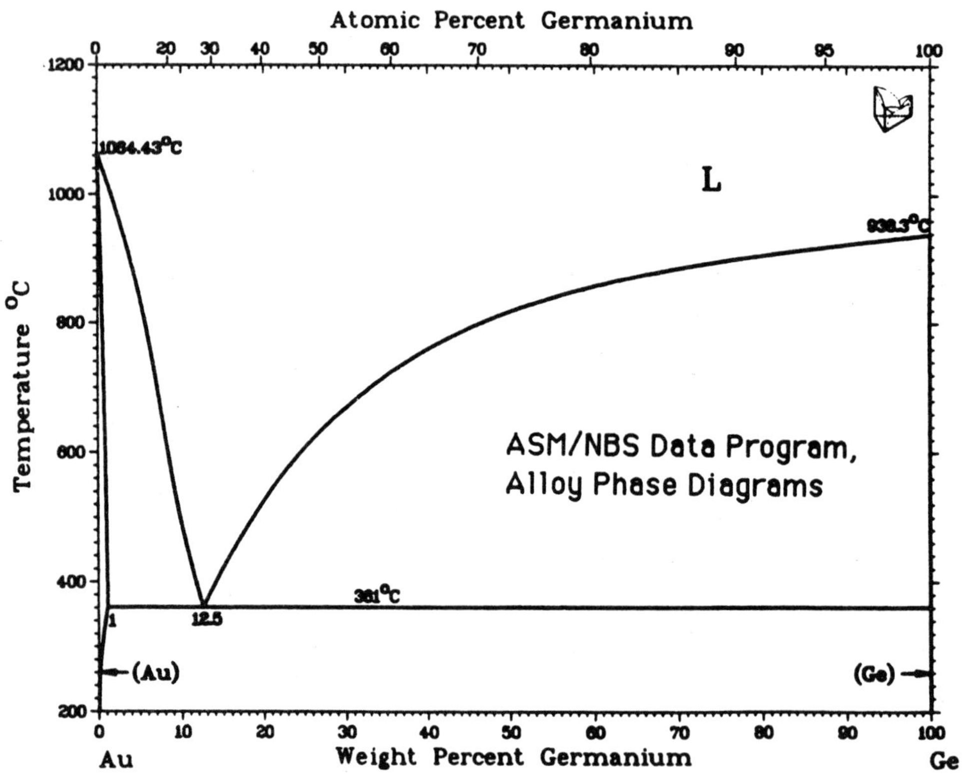
Atomic Percent Germanium
Temperature °C
1084.43°C
L
938.3°C
ASM/NBS Data Program,
Alloy Phase Diagrams
361°C
1
12.5
(Au)
(Ge)
Weight Percent Germanium
Au
Ge

97Au 3Si

PROPERTY	@ 300K	UNITS	REFERENCES
Density	15.4	g/cm^3	4.1.A-B
Specific Heat	0.0350	cal/g-K	4.2.A
Thermal Conductivity	0.065	cal/cm-s-K	4.3.A-C
Thermal Expansion	13×10^{-6}	1/K	4.4.A-D
Melting Point	636	K	4.5.A-B
Resistivity (77K) 1	117.2×10^{-6}	Ohm-cm	4.6.A
TCR (4.2 to 77K) 1	-150	ppm/K	4.7.A
Young's Modulus	11.7×10^{6}	psi	4.8.A-B
Poisson's Ratio		------	
Yield Strength	31,900	psi	4.10.A
Ult Tensile Strength	36,975	psi	4.11.A
Elongation at Break		%	
Hardness	56	Brinell	4.13.A

1 Data listed is for amorphous films of the eutectic composition. The material tested was getter sputtered in an Argon atmosphere onto saphire substrates held at 77K. The TCR listed is for material annealled at room temperature.

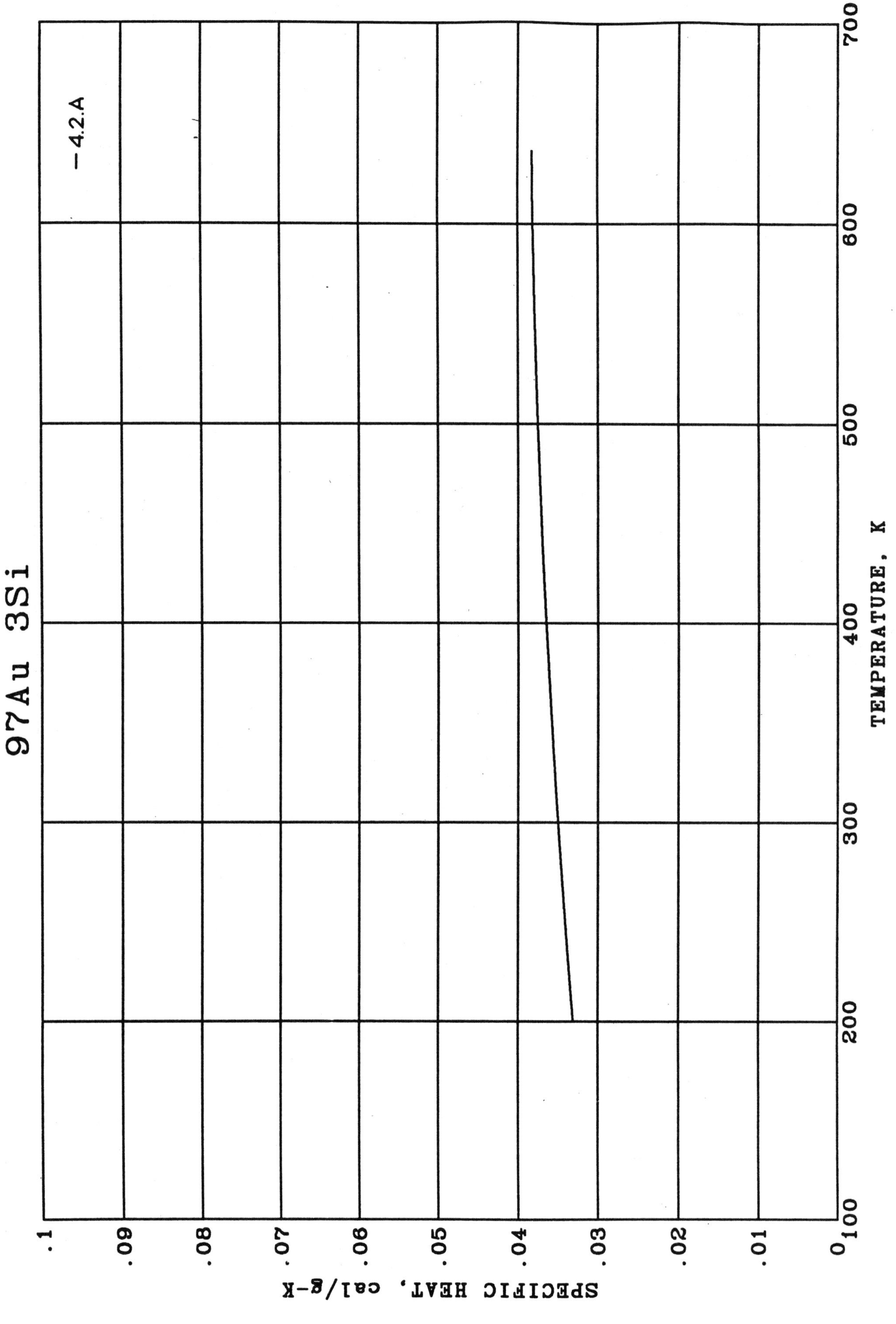

97Au 3Si
-4.2.A
SPECIFIC HEAT, cal/g-K
TEMPERATURE, K
.1
.09
.08
.07
.06
.05
.04
.03
.02
.01
0
100
200
300
400
500
600
700

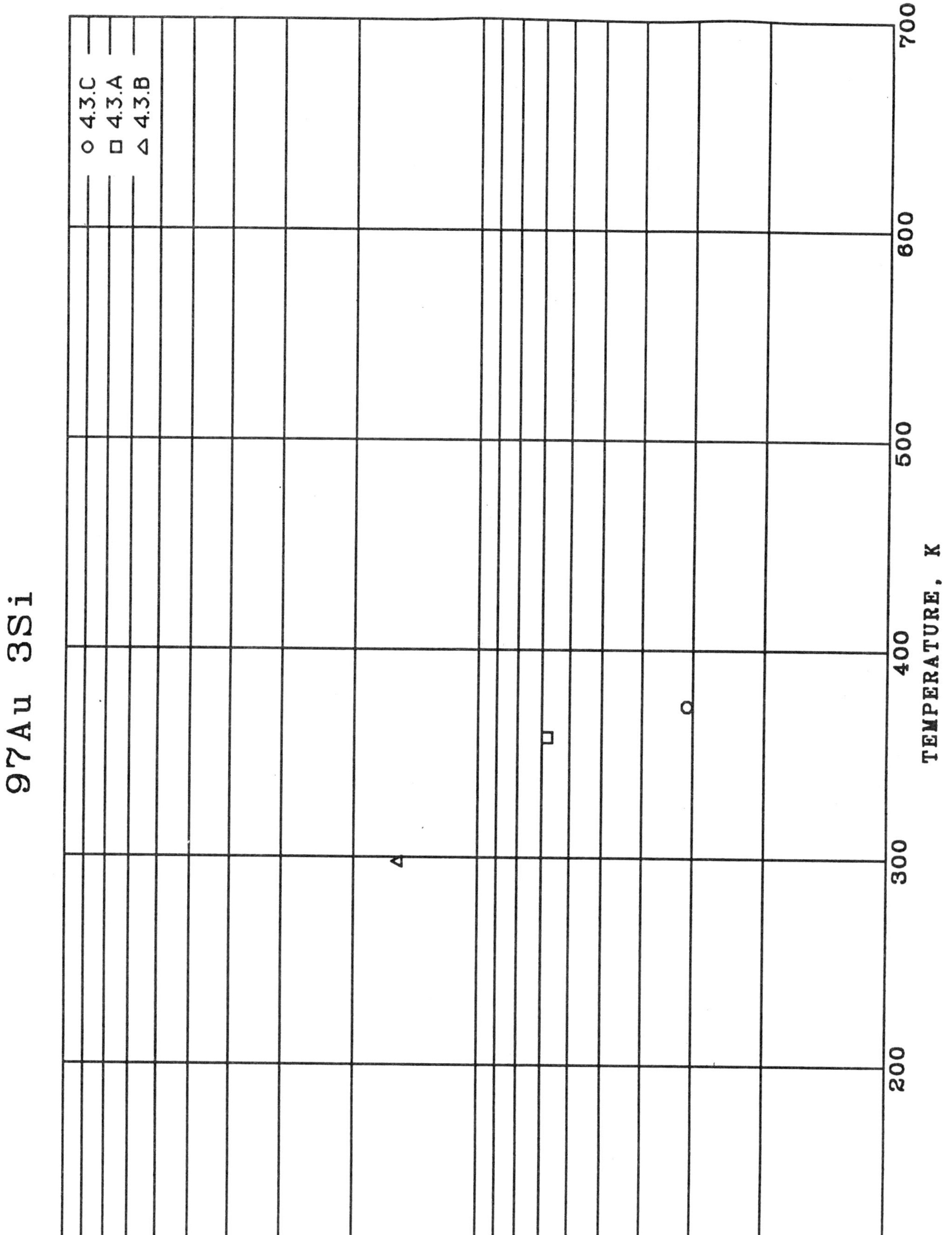

97Au 3Si
○ 4.3.C
□ 4.3.A
△ 4.3.B
THERMAL CONDUCTIVITY, cal/cm-K-s
10^0
10^{-1}
10^{-2}
TEMPERATURE, K
100
200
300
400
500
600
700

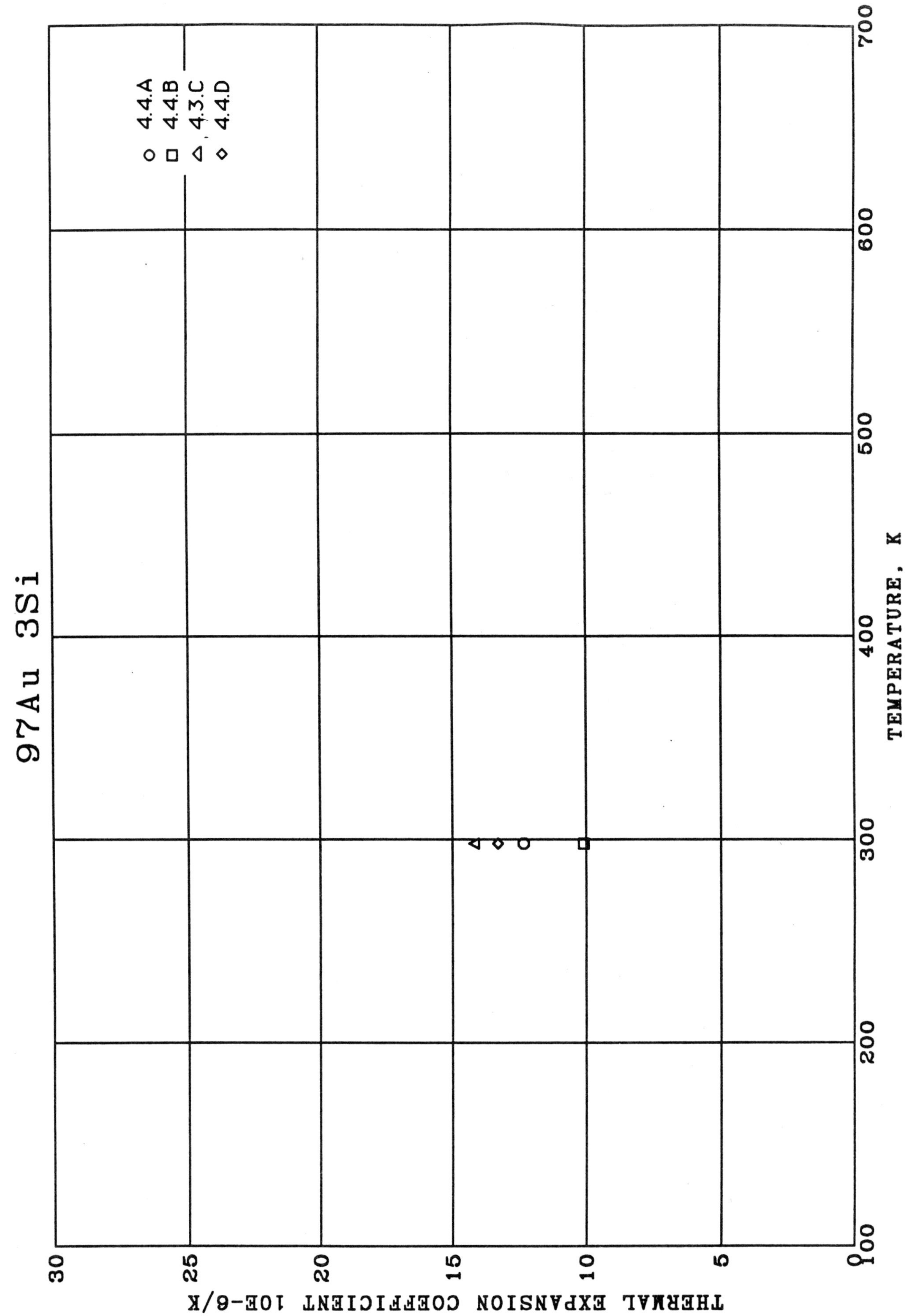
97Au 3Si
4.4.A
4.4.B
4.3.C
4.4.D
TEMPERATURE, K
THERMAL EXPANSION COEFFICIENT 10E-6/K
100
200
300
400
500
600
700
0
5
10
15
20
25
30

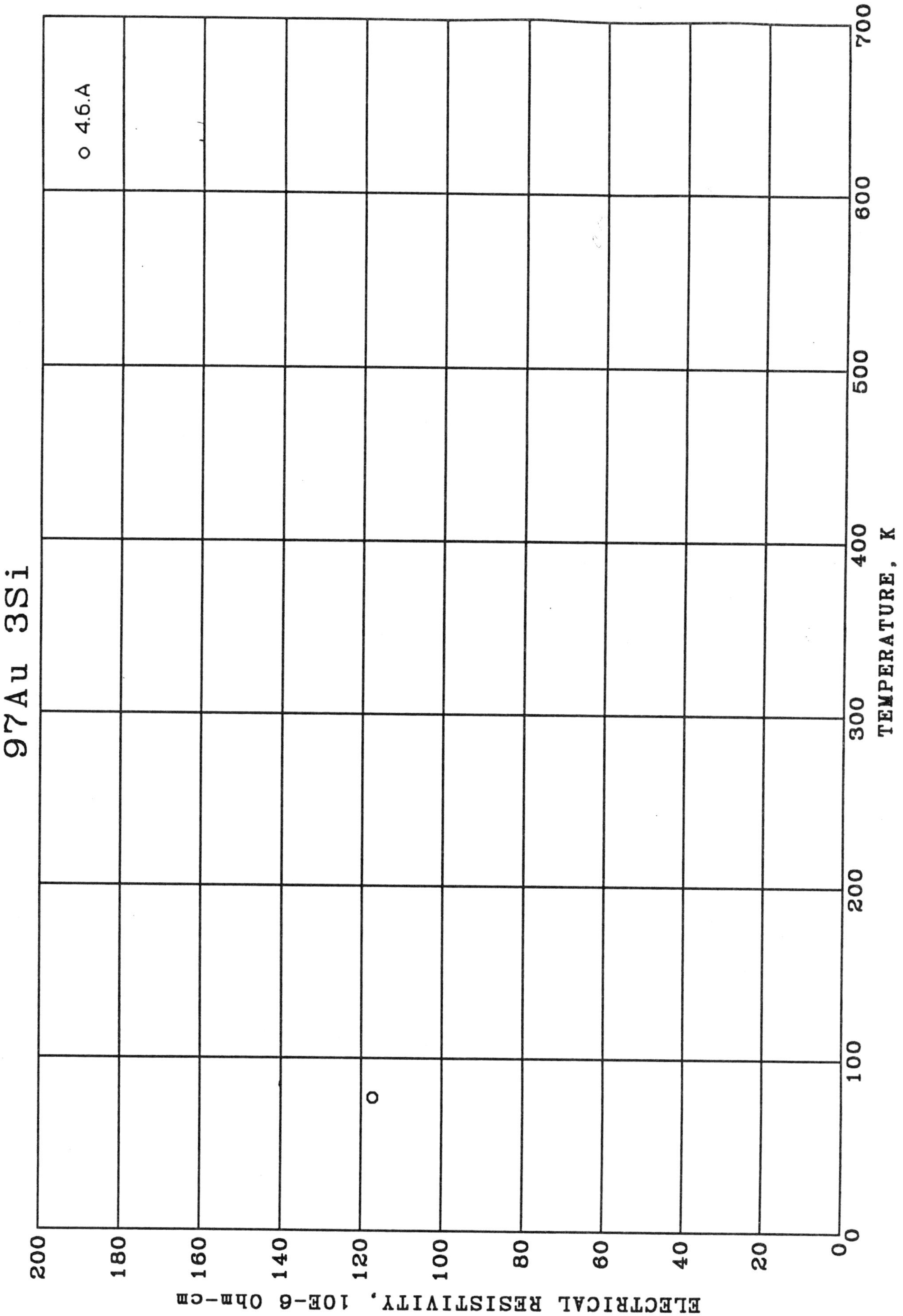
97Au 3Si
o 4.6.A
ELECTRICAL RESISTIVITY, 10E-6 Ohm-cm
0
20
40
60
80
100
120
140
160
180
200
TEMPERATURE, K
0
100
200
300
400
500
600
700

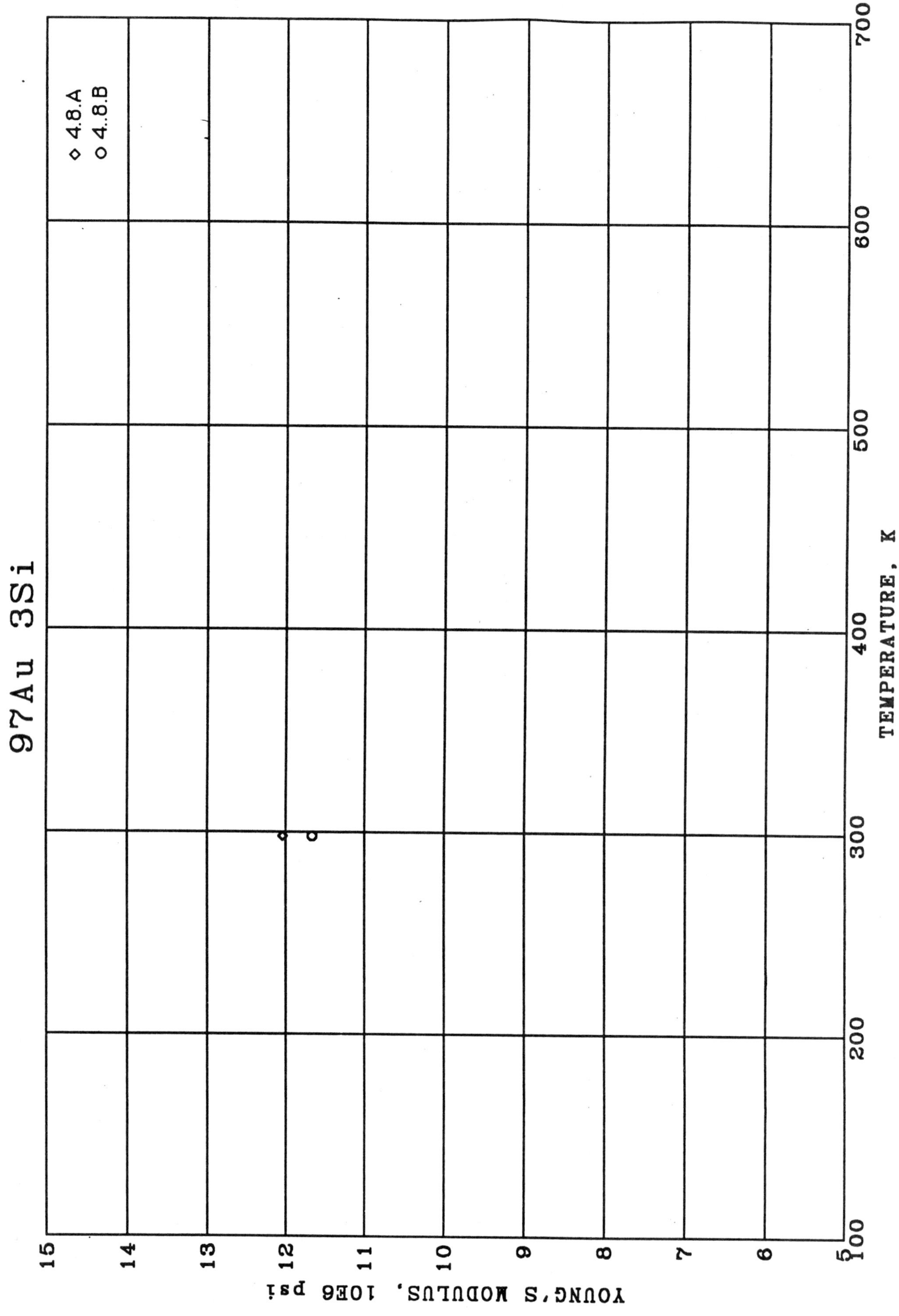

97Au 3Si
◇ 4.8.A
○ 4.8.B
YOUNG'S MODULUS, 10E6 psi
TEMPERATURE, K
5
6
7
8
9
10
11
12
13
14
15
100
200
300
400
500
600
700

97Au 3Si

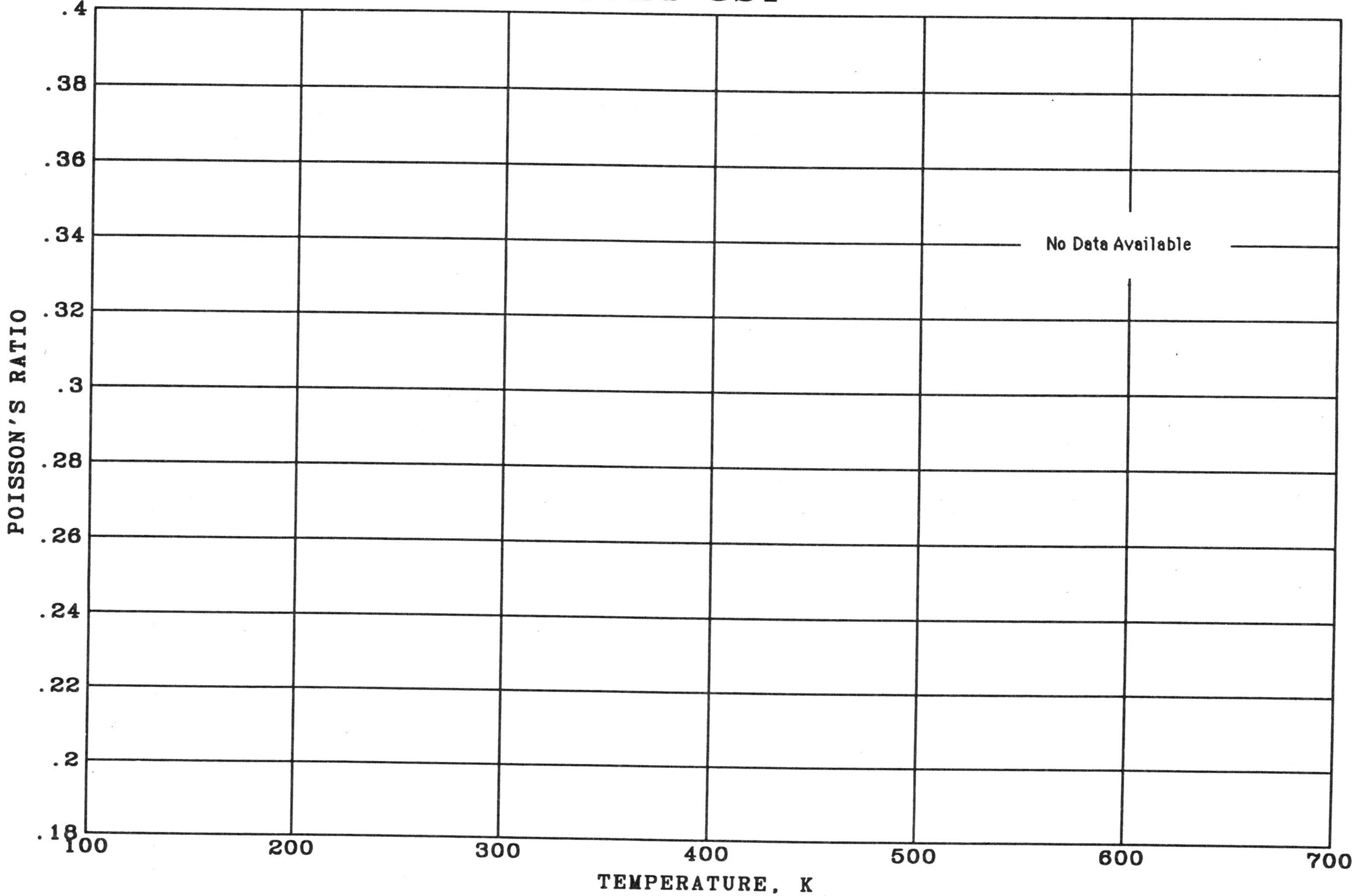

No Data Available
POISSON'S RATIO
.4
.38
.36
.34
.32
.3
.28
.26
.24
.22
.2
.18
TEMPERATURE, K
100
200
300
400
500
600
700

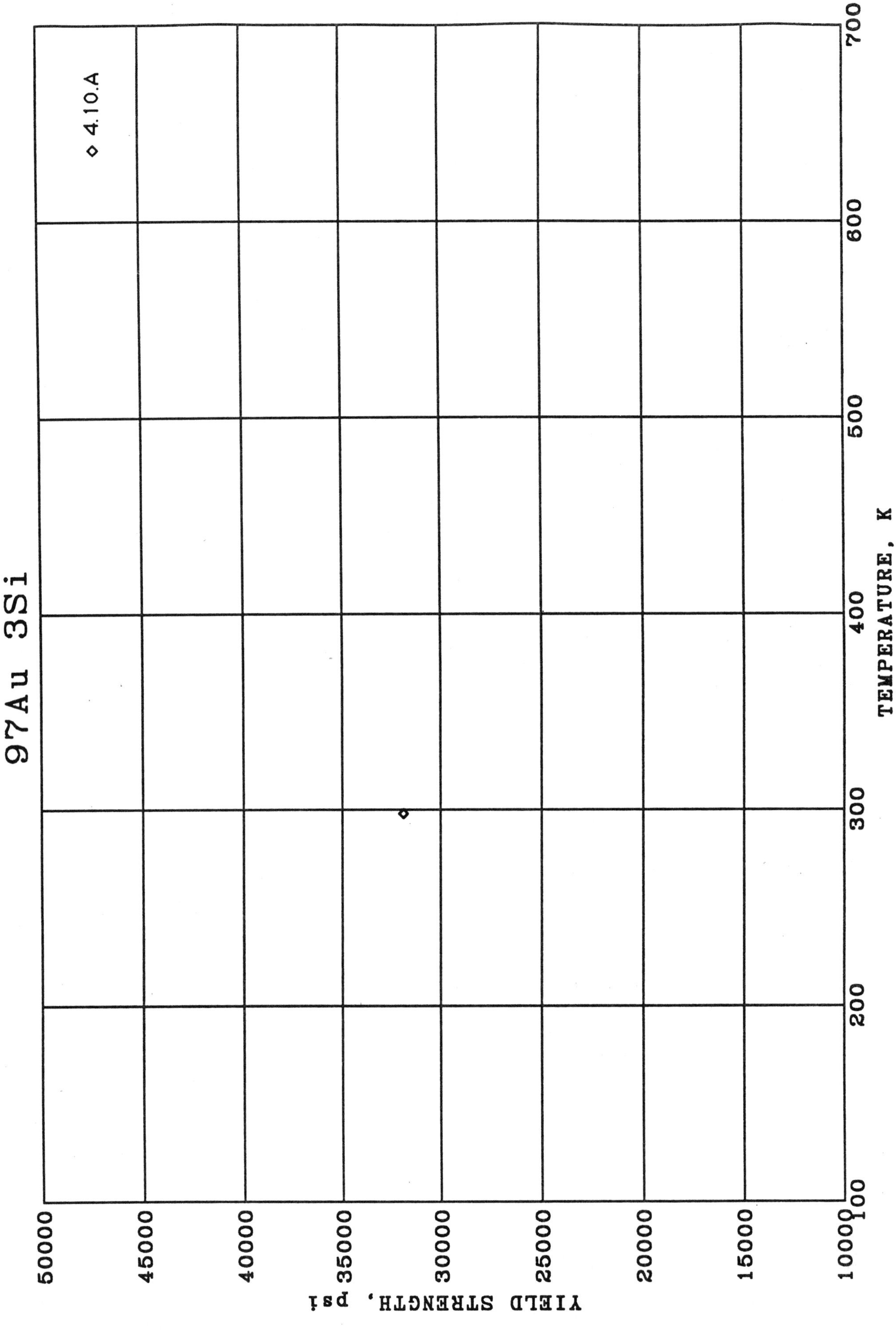

97Au 3Si
◇ 4.10.A
YIELD STRENGTH, psi
50000
45000
40000
35000
30000
25000
20000
15000
10000
TEMPERATURE, K
100
200
300
400
500
600
700

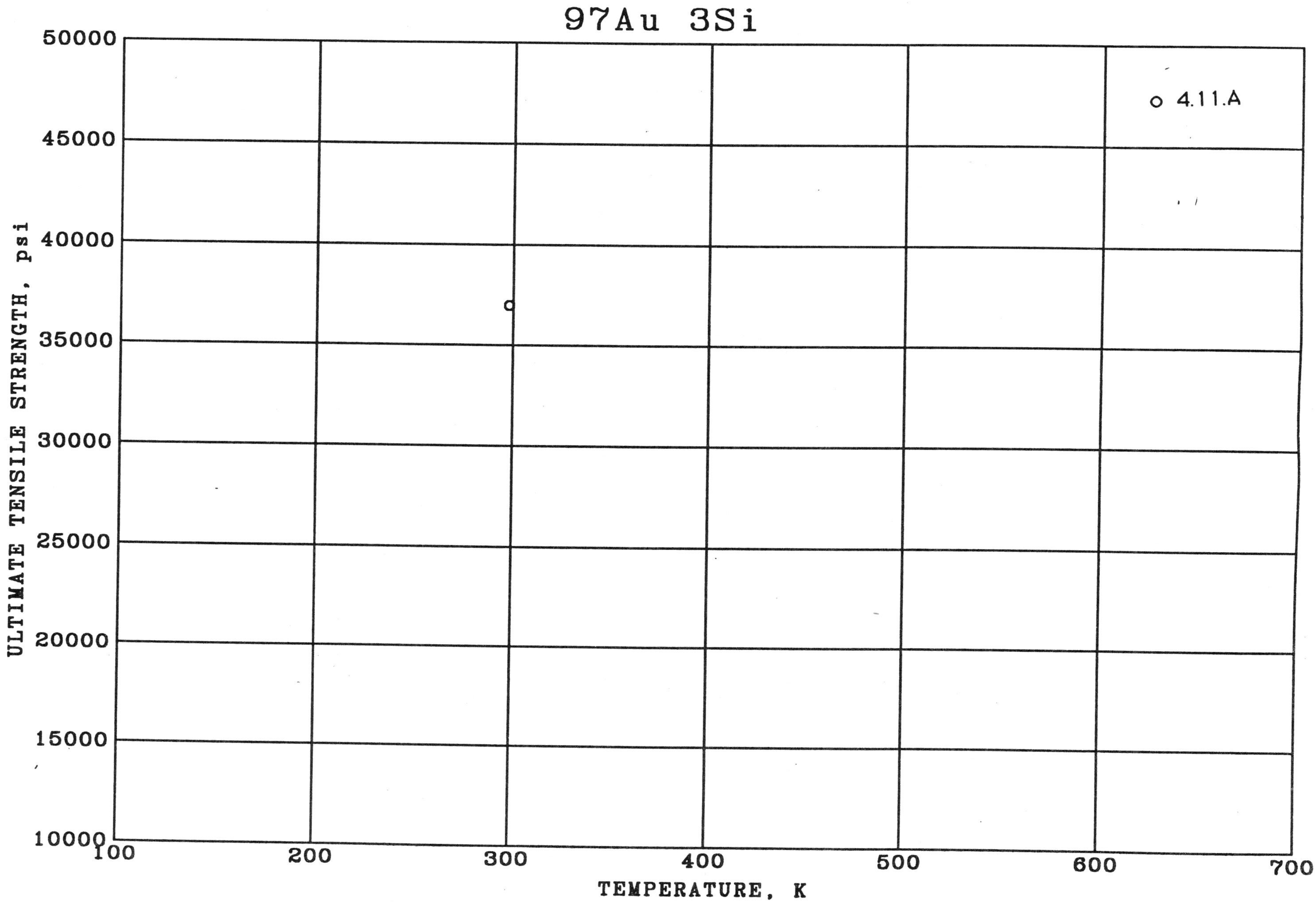
97Au 3Si
4.11.A
ULTIMATE TENSILE STRENGTH, psi
50000
45000
40000
35000
30000
25000
20000
15000
10000
100
200
300
400
500
600
700
TEMPERATURE, K

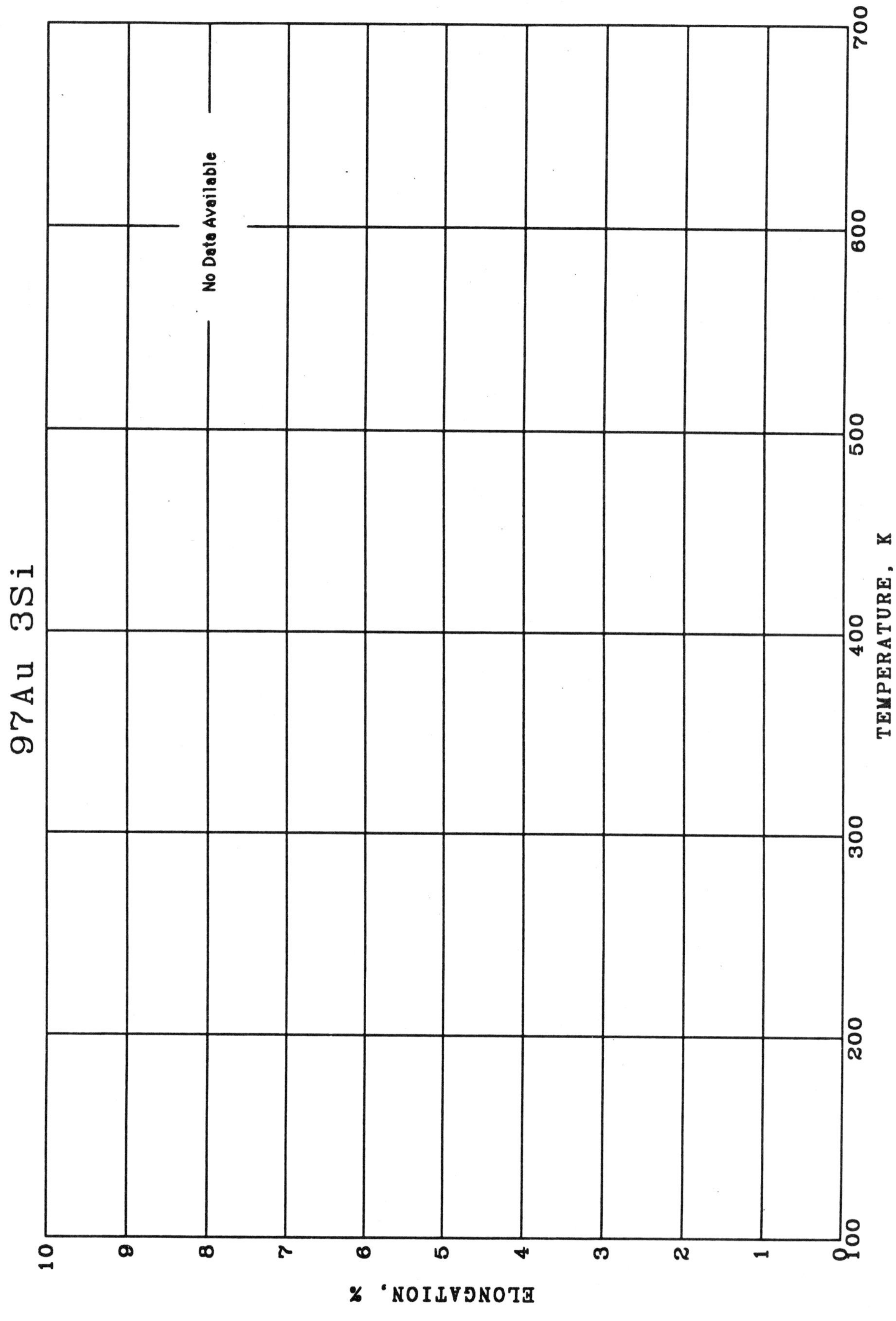
97Au 3Si
No Data Available
ELONGATION, %
0
1
2
3
4
5
6
7
8
9
10
TEMPERATURE, K
100
200
300
400
500
600
700

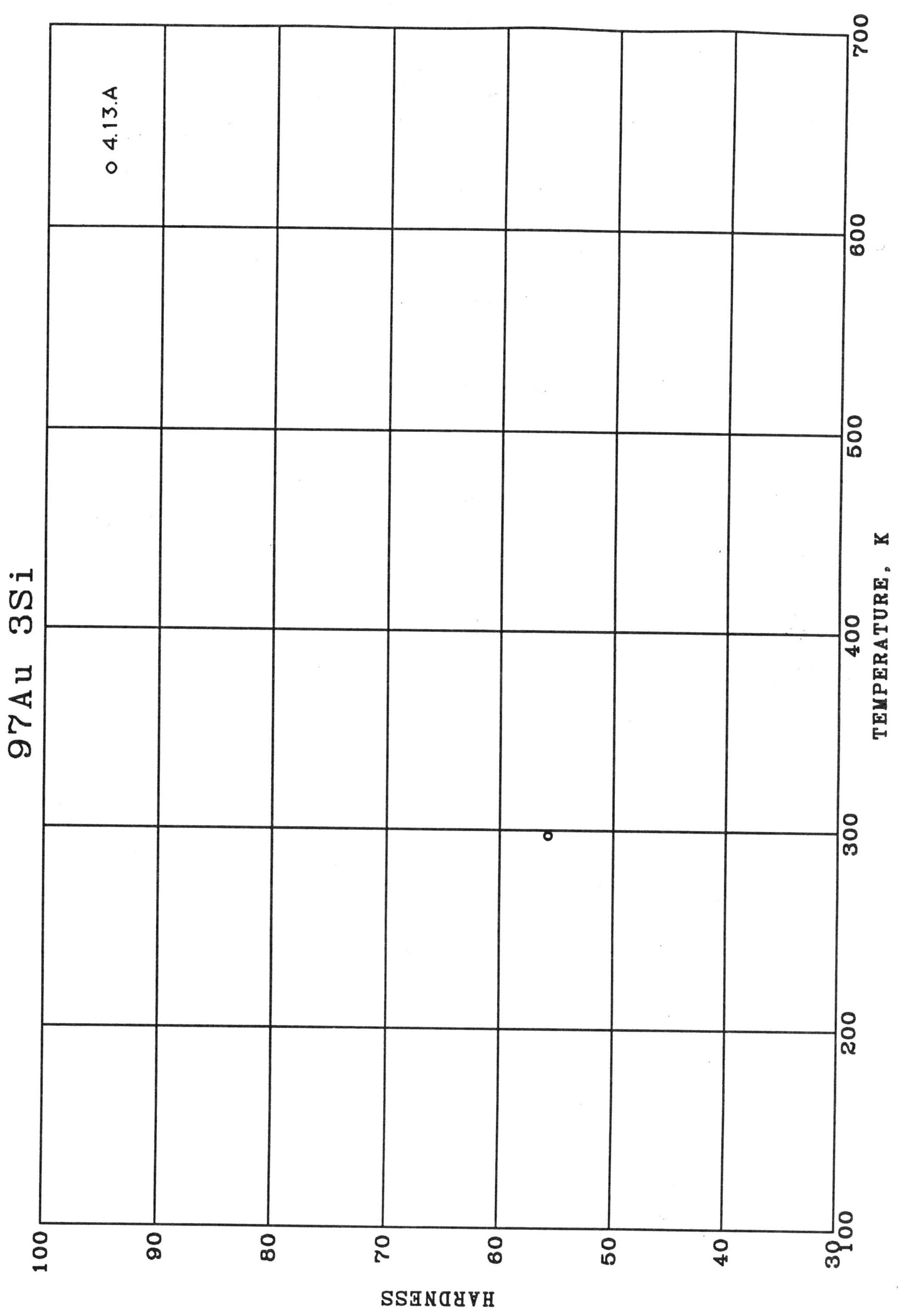

97Au 3Si
o 4.13.A
TEMPERATURE, K
100
200
300
400
500
600
700
HARDNESS
30
40
50
60
70
80
90
100

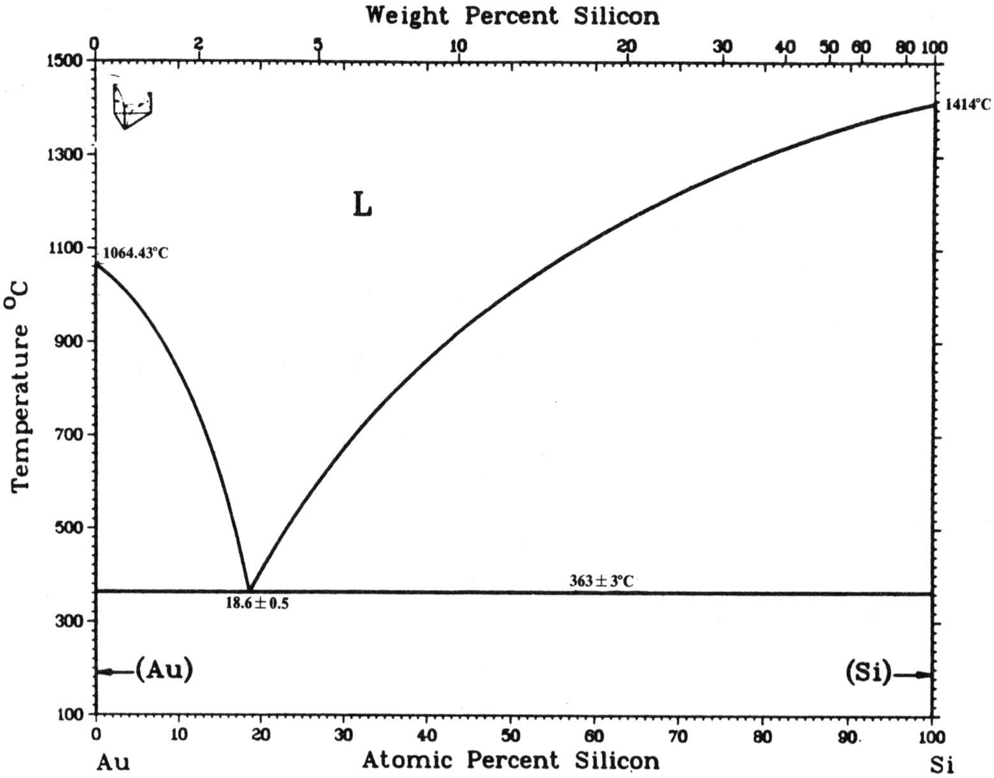
Weight Percent Silicon
L
1064.43°C
1414°C
363 ± 3°C
18.6 ± 0.5
(Au)
(Si)
Temperature °C
Au
Atomic Percent Silicon
Si

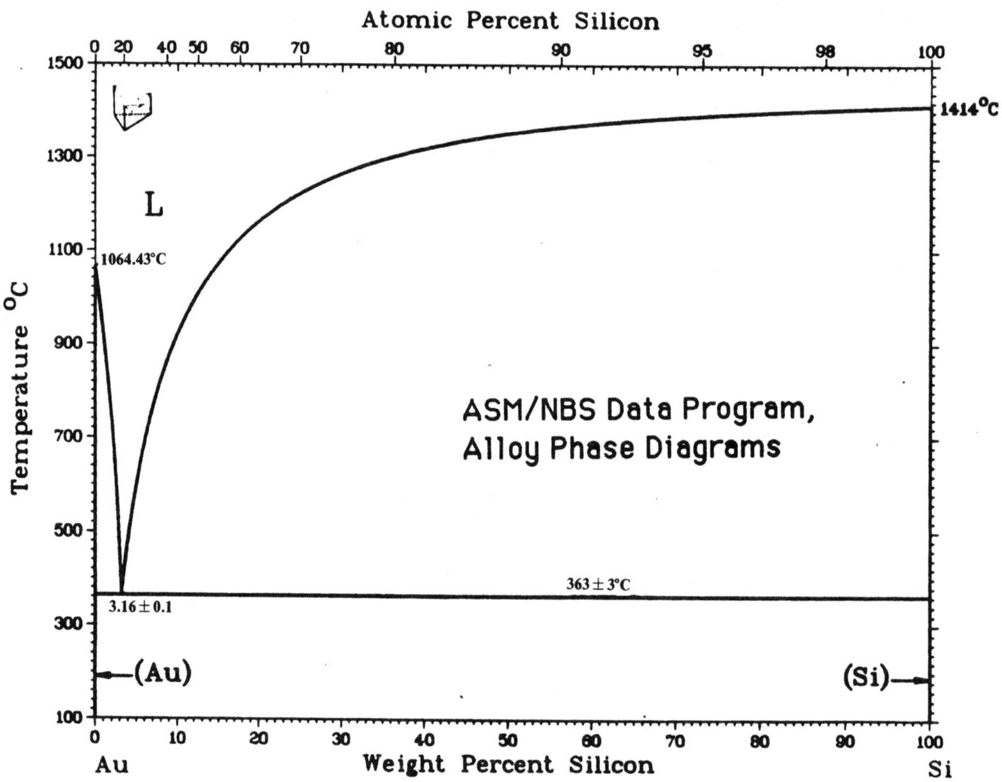
Atomic Percent Silicon
L
1064.43°C
1414°C
ASM/NBS Data Program,
Alloy Phase Diagrams
363 ± 3°C
3.16 ± 0.1
(Au)
(Si)
Temperature °C
Au
Weight Percent Silicon
Si

80Au 20Sn

PROPERTY	@ 300K	UNITS	REFERENCES
Density	14.52	g/cm^3	5.1.A-C
Specific Heat	0.036	cal/g-K	5.2.A-B
Thermal Conductivity	0.14	cal/cm-s-K	5.3.A-D
Thermal Expansion	16.5×10^{-6}	1/K	5.4.A-D
Melting Point	553	K	5.5.A-C
Resistivity	17.35×10^{-6}	Ohm-cm	5.6.A
TCR		ppm/K	
Young's Modulus	8.57×10^6	psi	5.8.A-B
Poisson's Ratio		------	
Yield Strength	40000	psi	5.10.A-B
Ult Tensile Strength	40000	psi	5.11.A-B
Elongation at Break	<2	%	5.12.A-B
Hardness	115	Brinell	5.13.A

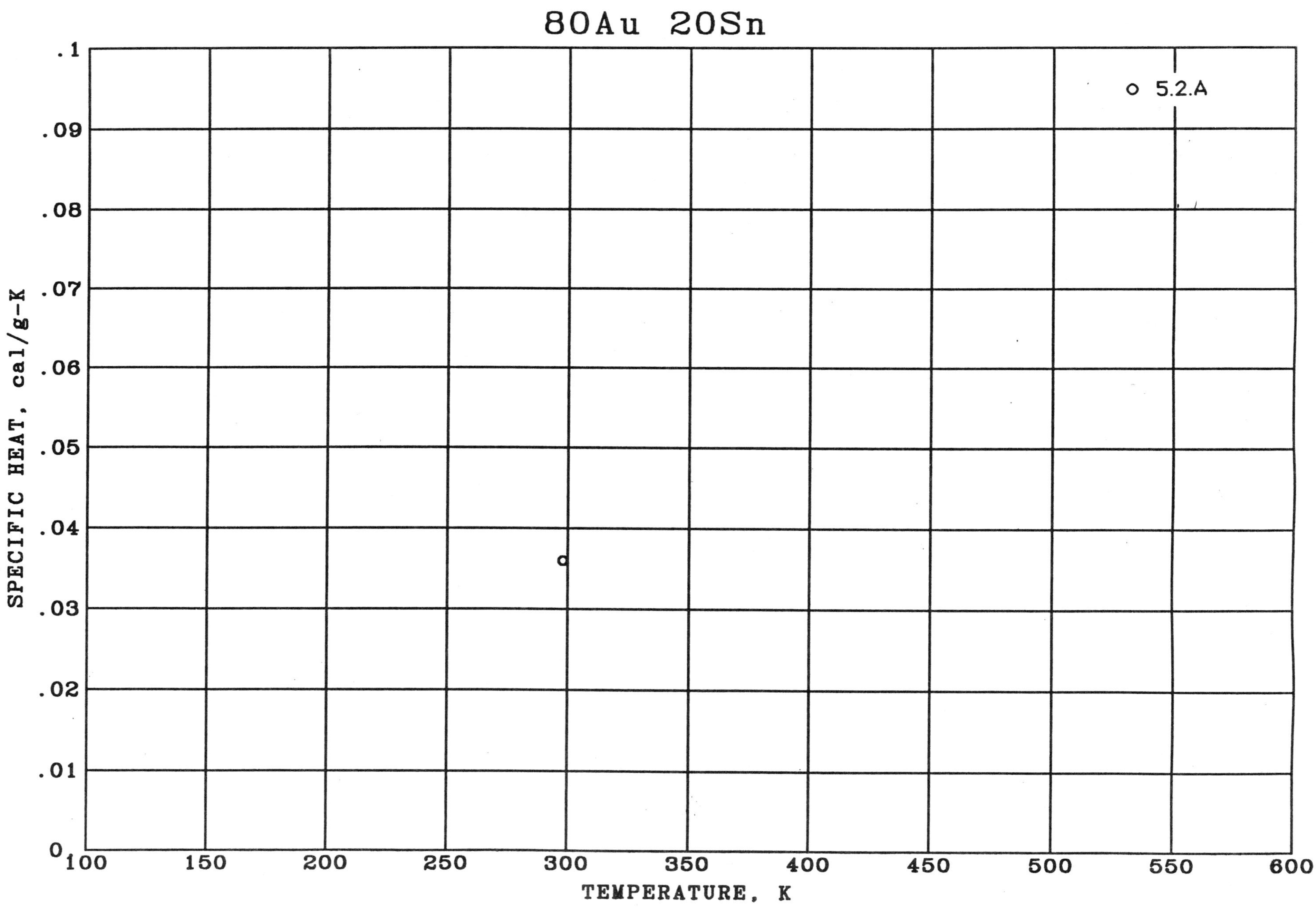
80Au 20Sn
5.2.A
SPECIFIC HEAT, cal/g-K
.1
.09
.08
.07
.06
.05
.04
.03
.02
.01
0
100
150
200
250
300
350
400
450
500
550
600
TEMPERATURE, K

80Au 20Sn

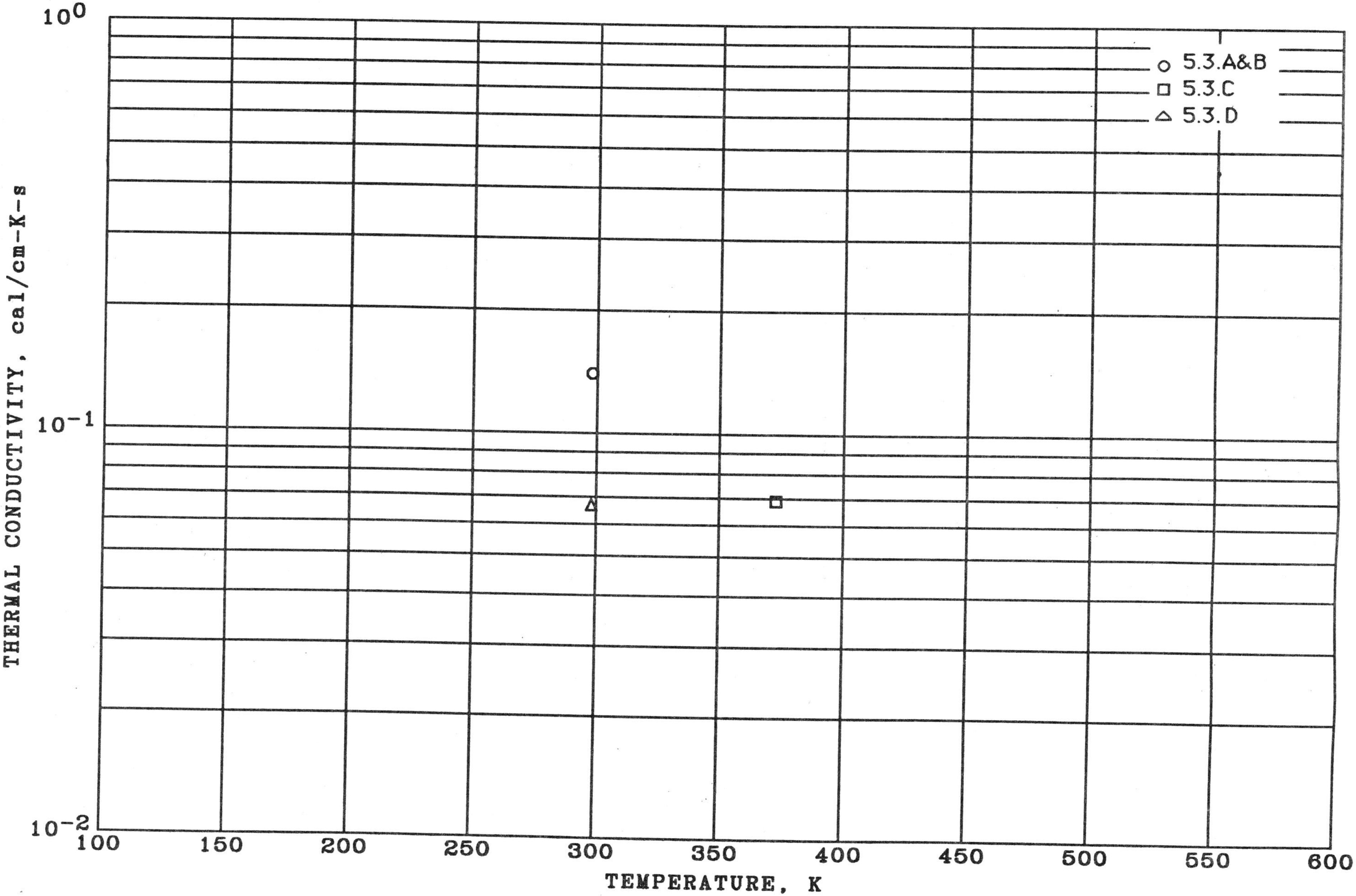

10^0
10^-1
10^-2
THERMAL CONDUCTIVITY, cal/cm-K-s
100
150
200
250
300
350
400
450
500
550
600
TEMPERATURE, K
○ 5.3.A&B
□ 5.3.C
△ 5.3.D

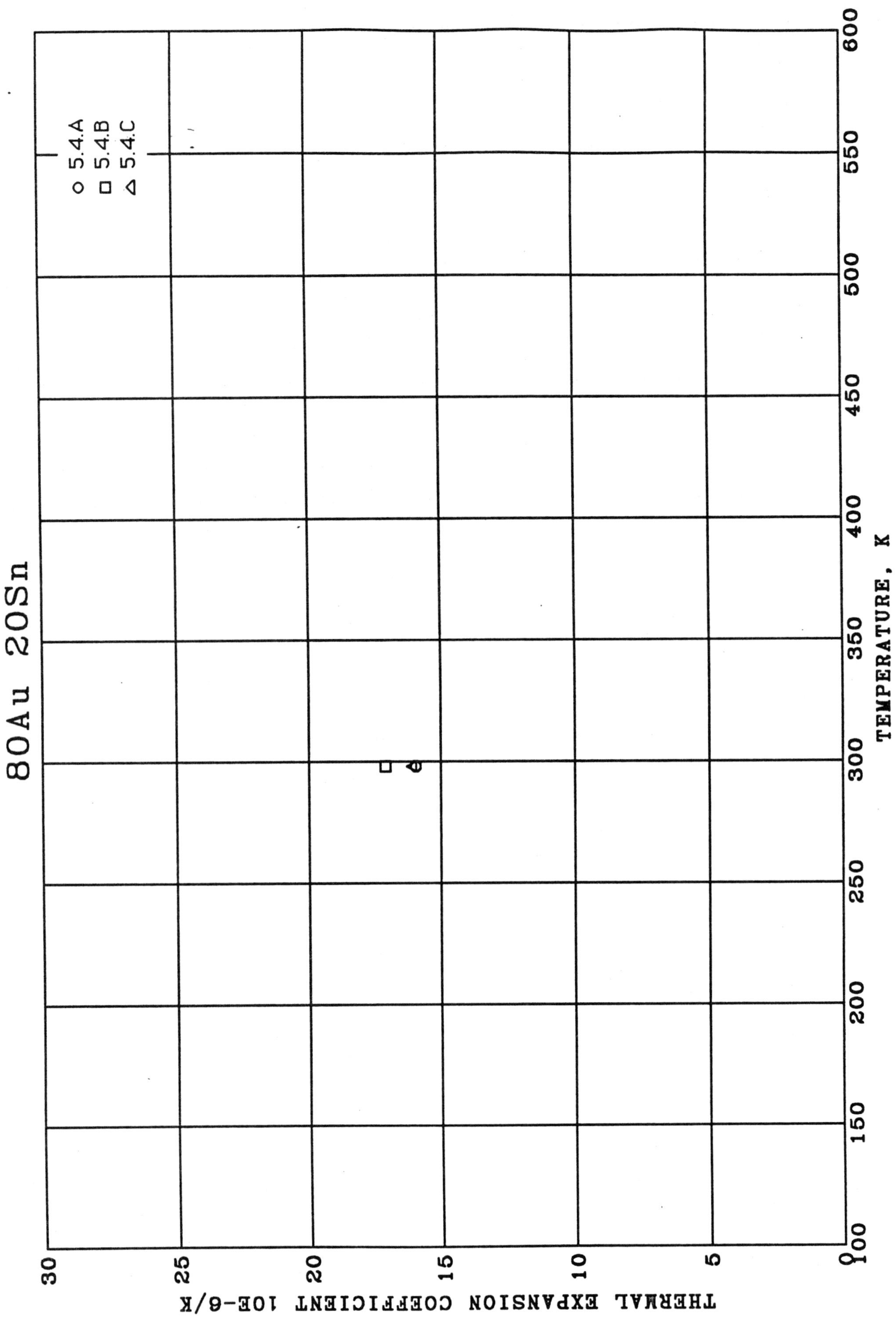
80Au 20Sn
○ 5.4.A
□ 5.4.B
△ 5.4.C
THERMAL EXPANSION COEFFICIENT 10E-6/K
TEMPERATURE, K
0
5
10
15
20
25
30
100
150
200
250
300
350
400
450
500
550
600

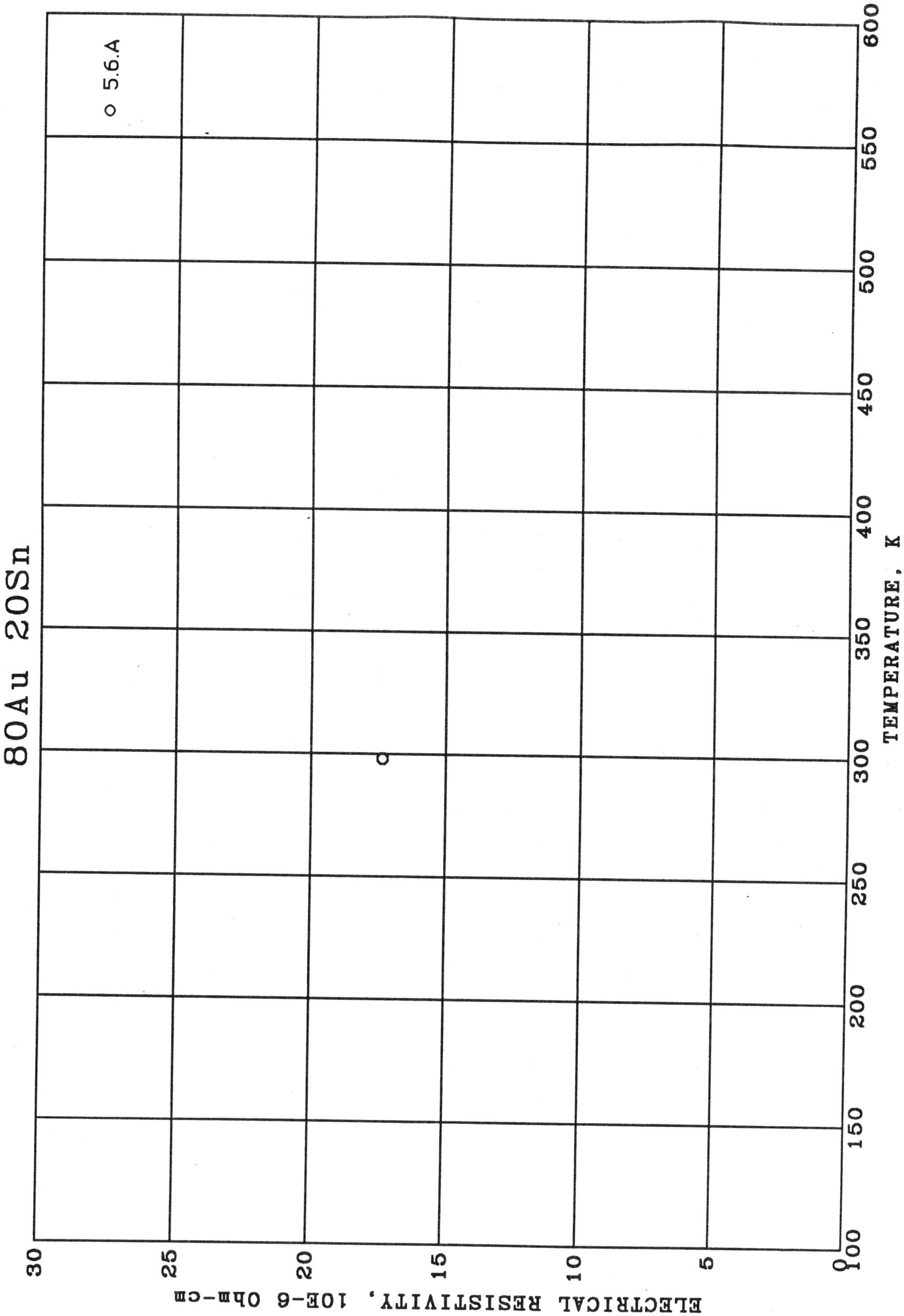
80Au 20Sn
5.6.A
ELECTRICAL RESISTIVITY, 10E-6 Ohm-cm
TEMPERATURE, K
0
5
10
15
20
25
30
100
150
200
250
300
350
400
450
500
550
600

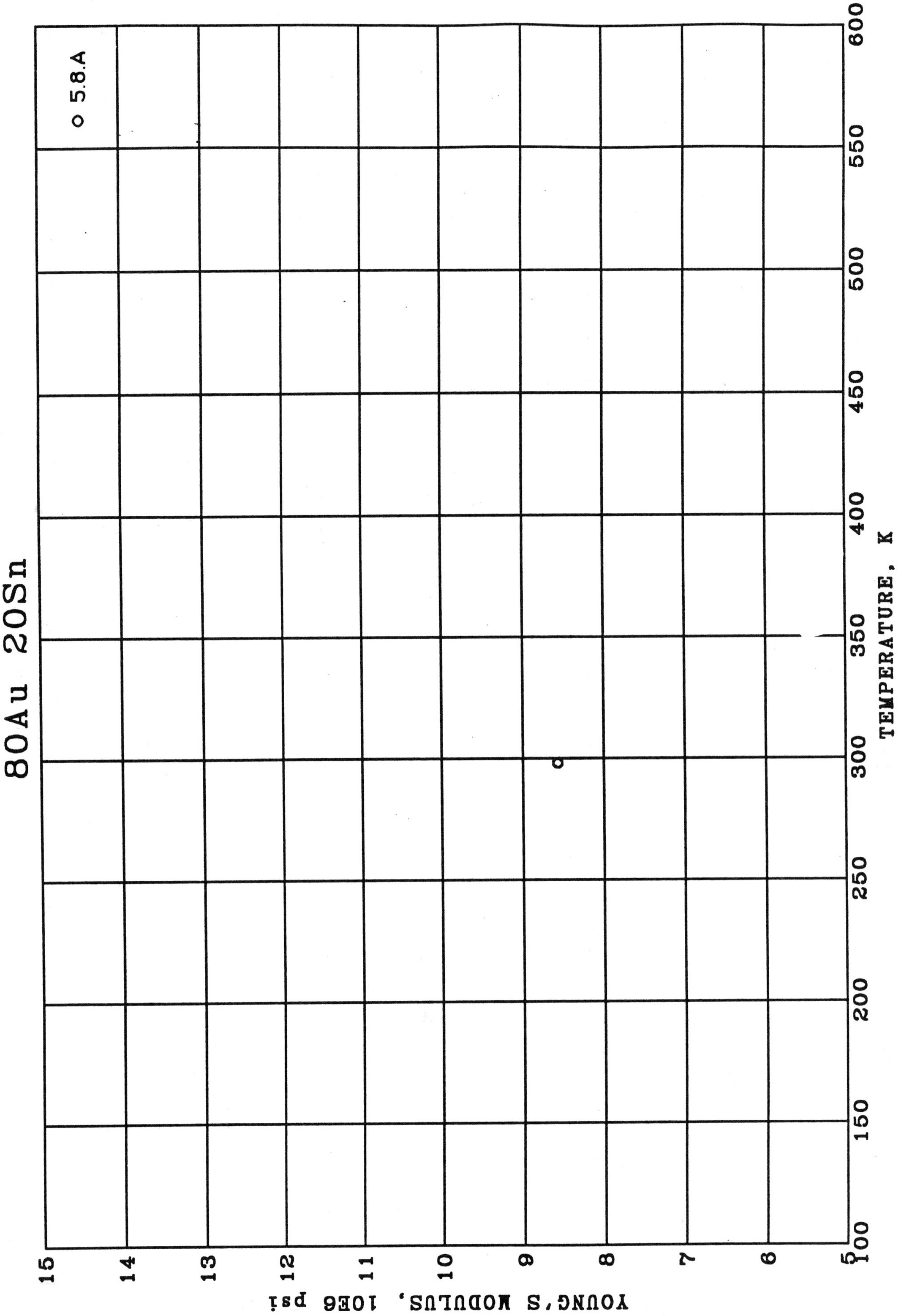
80Au 20Sn
o 5.8.A
YOUNG'S MODULUS, 10E6 psi
5
6
7
8
9
10
11
12
13
14
15
TEMPERATURE, K
100
150
200
250
300
350
400
450
500
550
600

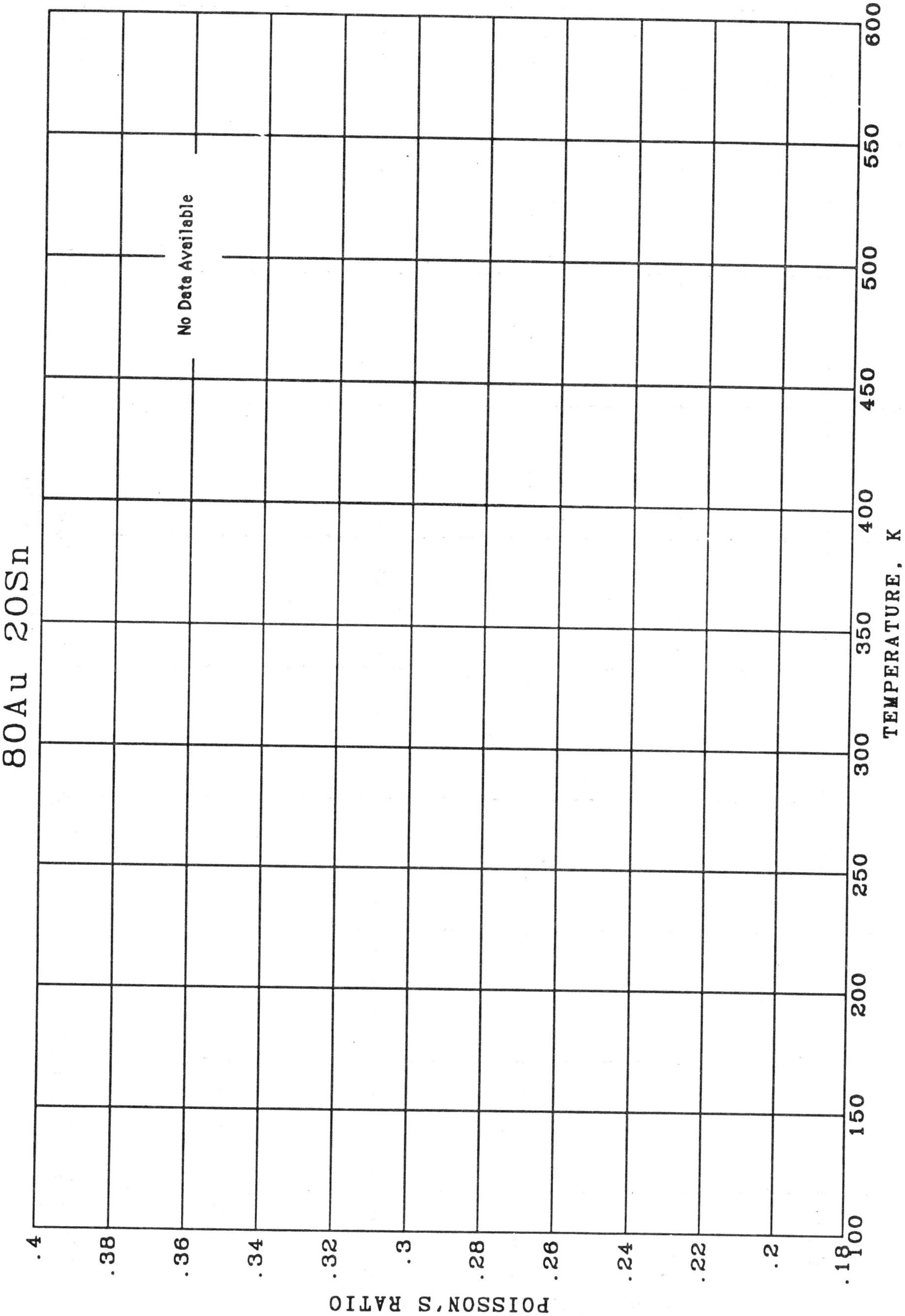
80Au 20Sn
No Data Available
TEMPERATURE, K
100
150
200
250
300
350
400
450
500
550
600
POISSON'S RATIO
.18
.2
.22
.24
.26
.28
.3
.32
.34
.36
.38
.4

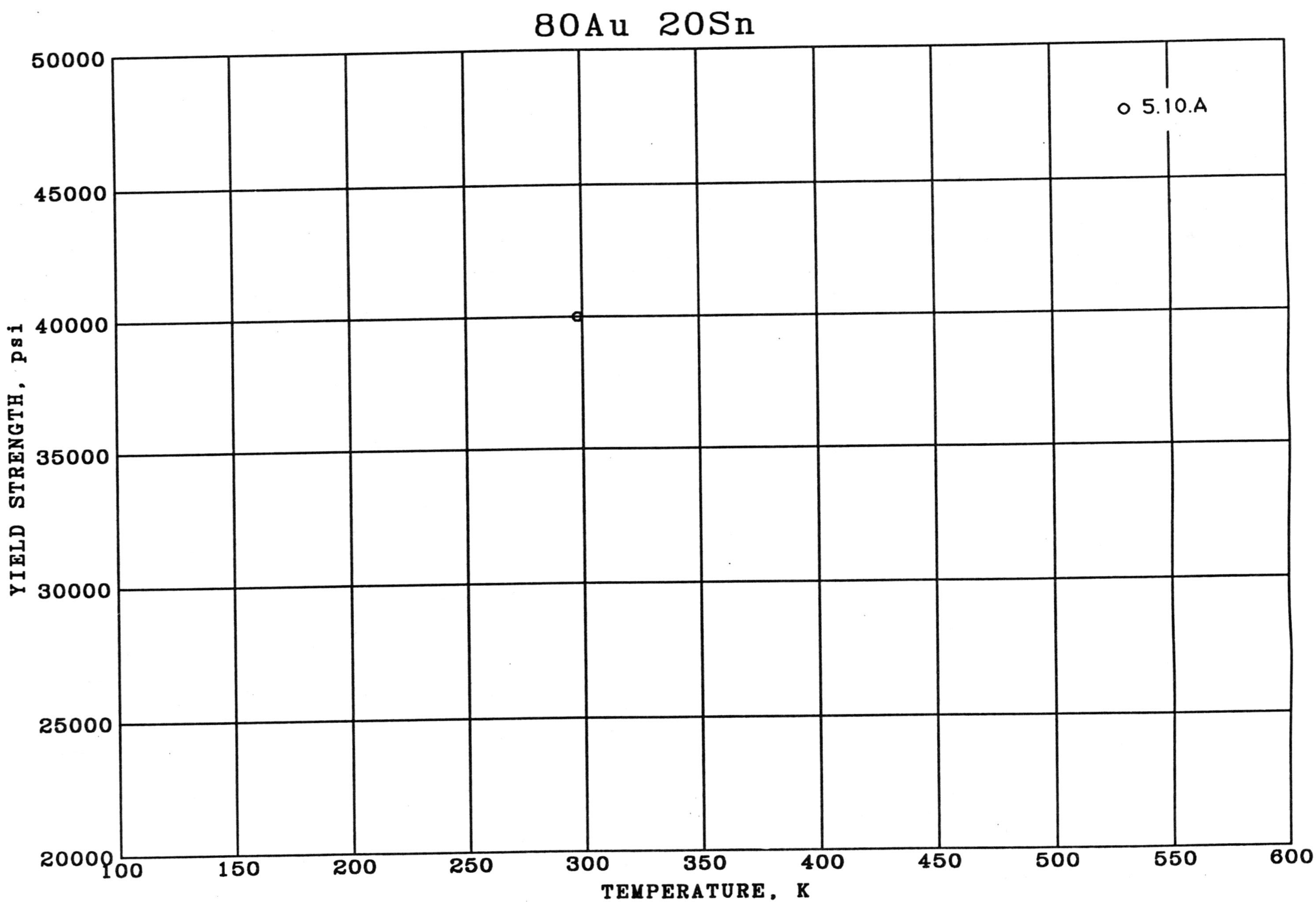
80Au 20Sn
o 5.10.A
50000
45000
40000
35000
30000
25000
20000
YIELD STRENGTH, psi
100
150
200
250
300
350
400
450
500
550
600
TEMPERATURE, K

80Au 20Sn

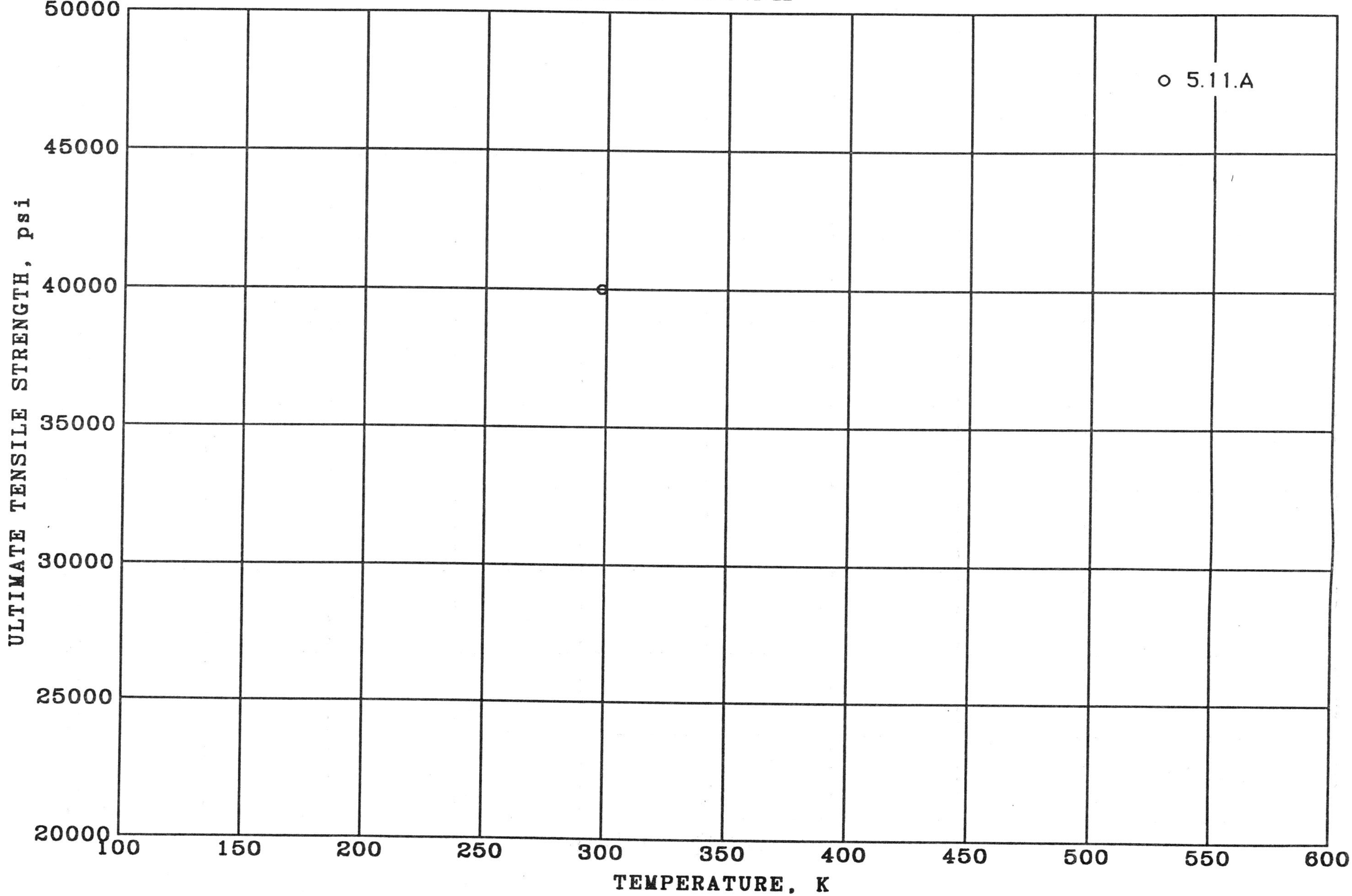

50000
45000
40000
35000
30000
25000
20000
100
150
200
250
300
350
400
450
500
550
600
ULTIMATE TENSILE STRENGTH, psi
TEMPERATURE, K
5.11.A

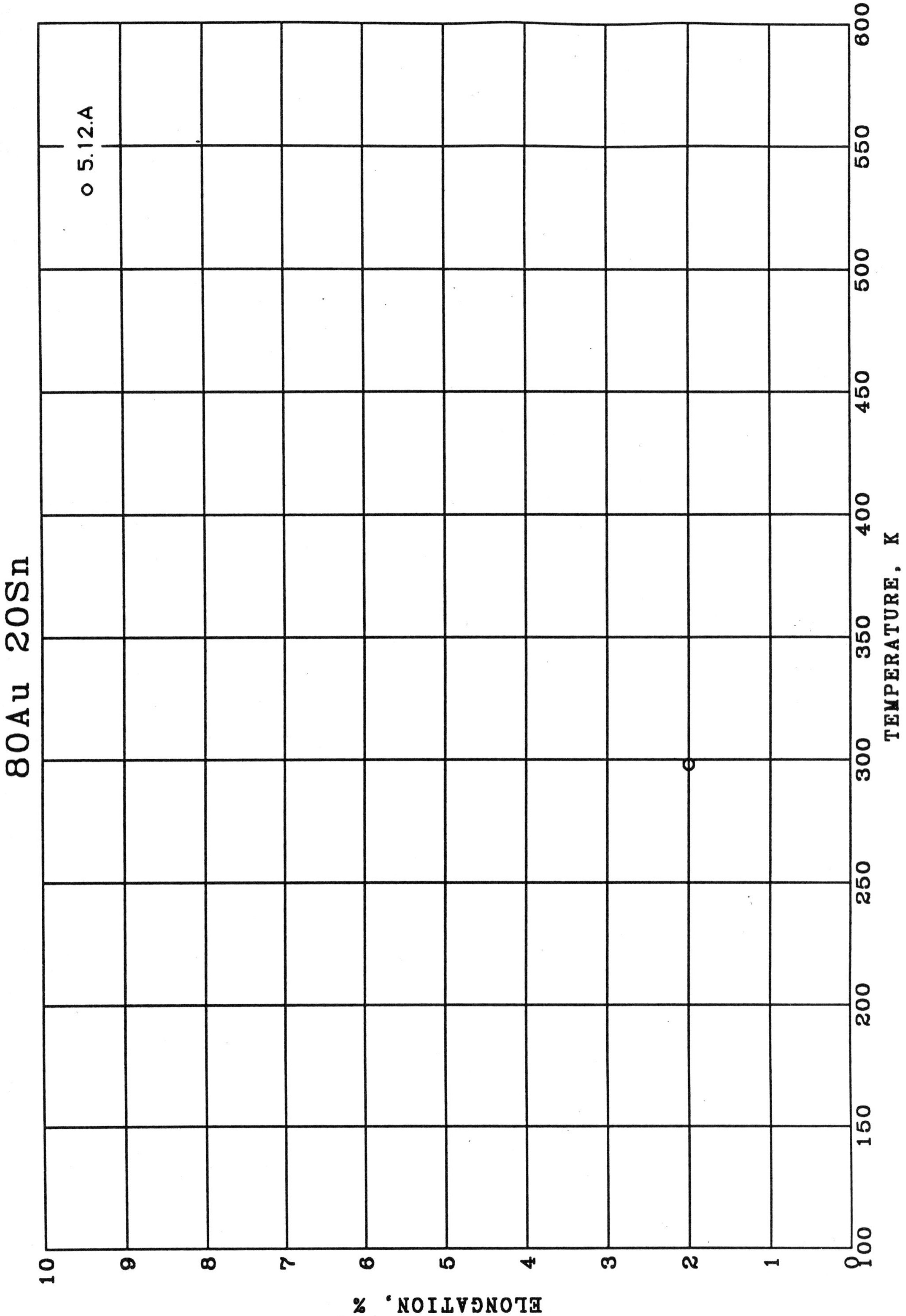
80Au 20Sn
o 5.12.A
ELONGATION, %
TEMPERATURE, K
0
1
2
3
4
5
6
7
8
9
10
100
150
200
250
300
350
400
450
500
550
600

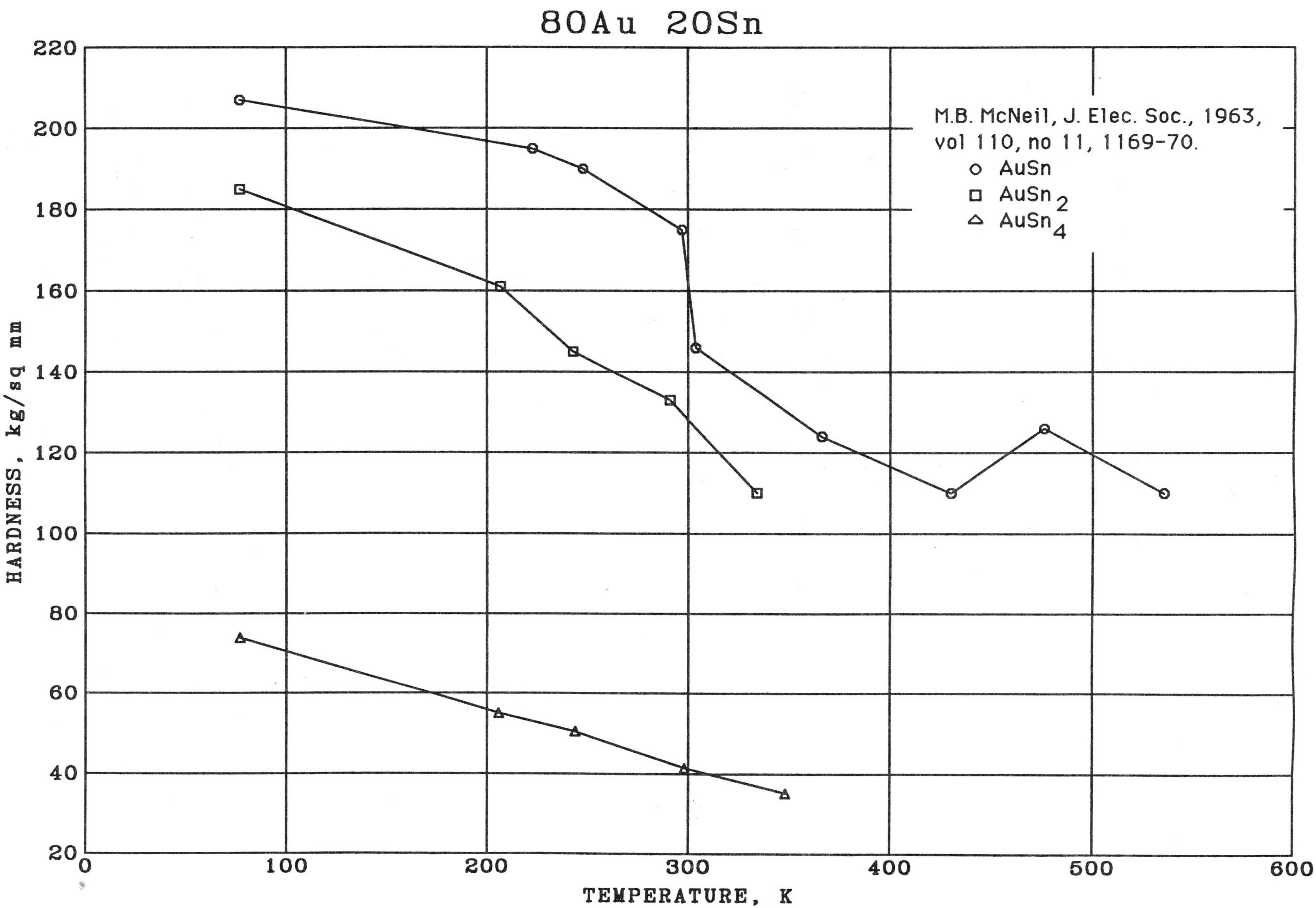
80Au 20Sn
HARDNESS, kg/sq mm
TEMPERATURE, K
M.B. McNeil, J. Elec. Soc., 1963,
vol 110, no 11, 1169-70.
AuSn
AuSn 2
AuSn 4
220
200
180
160
140
120
100
80
60
40
20
0
100
200
300
400
500
600

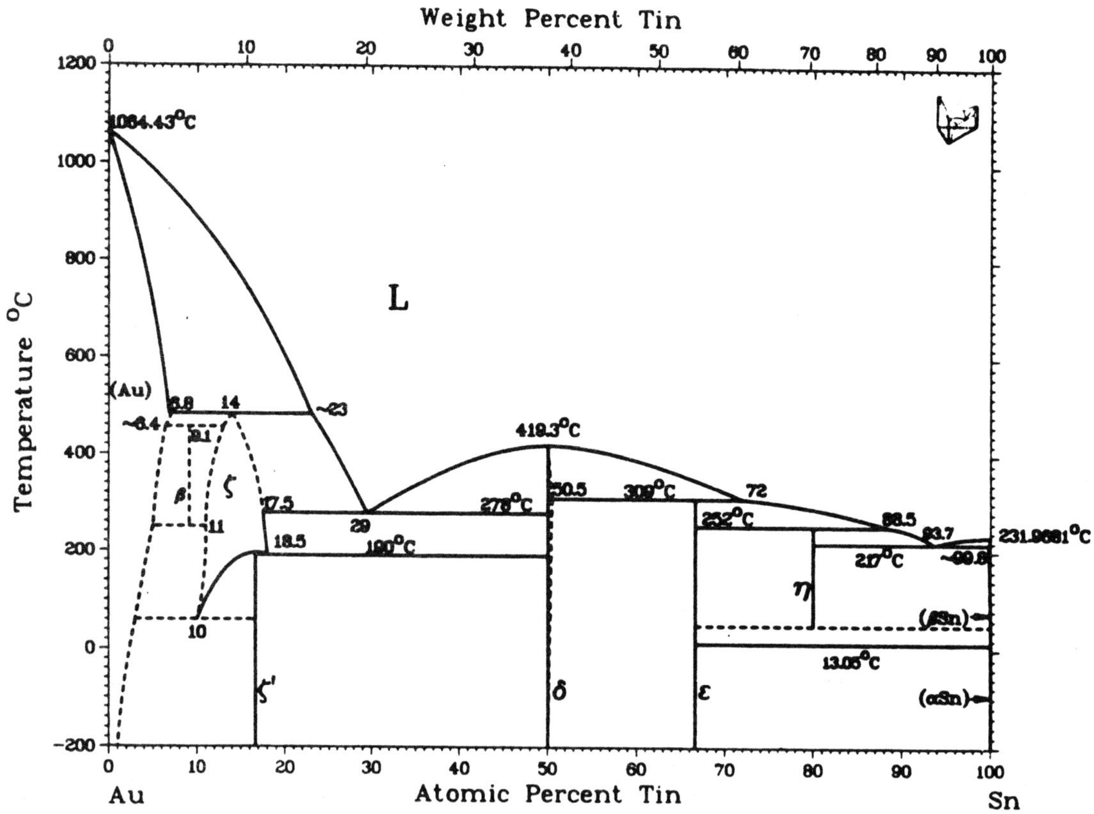
Weight Percent Tin
Temperature °C
1084.43°C
L
(Au)
6.8
14
~23
~6.4
9.1
β
ζ
11
10
419.3°C
17.5
29
278°C
18.5
190°C
50.5
309°C
72
252°C
88.5
93.7
231.9681°C
217°C
~99.8
η
(βSn)
13.05°C
ζ'
δ
ε
(αSn)
Atomic Percent Tin
Au
Sn

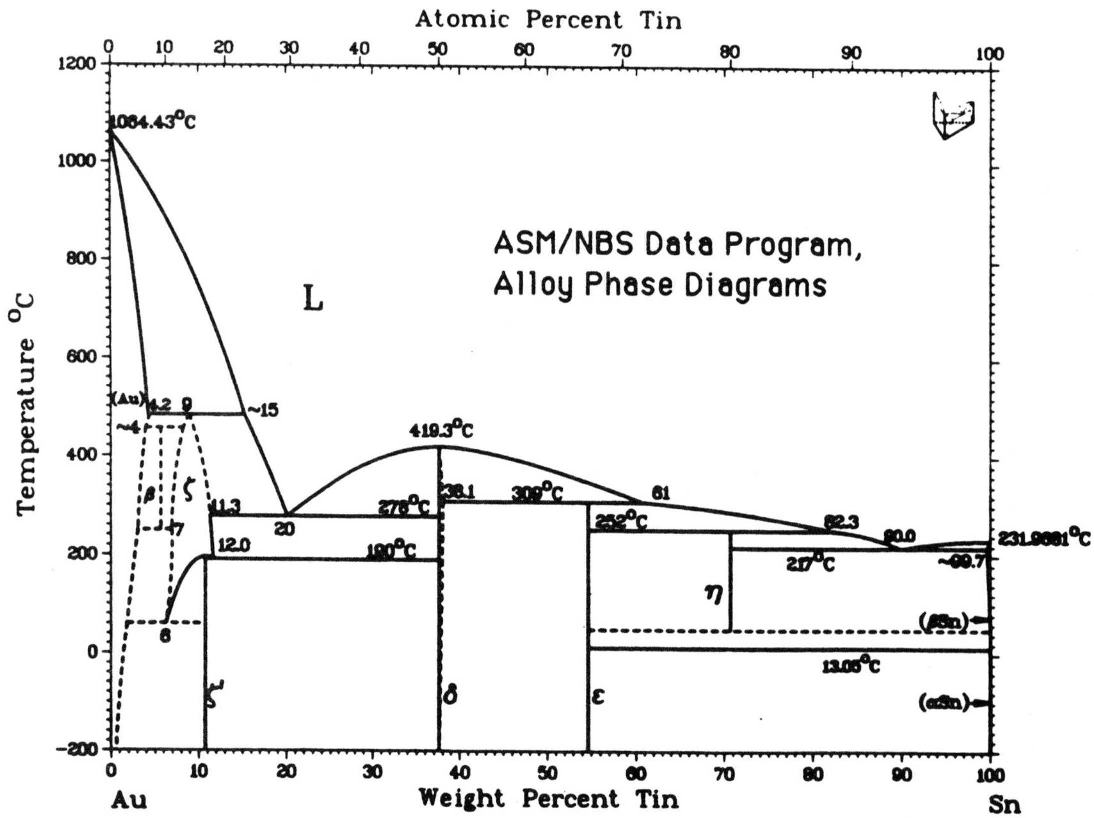
Atomic Percent Tin
Temperature °C
1084.43°C
ASM/NBS Data Program,
Alloy Phase Diagrams
L
(Au)
4.2
9
~15
~4
β
ζ
7
6
419.3°C
11.3
20
278°C
12.0
190°C
38.1
309°C
61
252°C
82.3
90.0
231.9681°C
217°C
~99.7
η
(βSn)
13.05°C
ζ'
δ
ε
(αSn)
Weight Percent Tin
Au
Sn

99.9Fe 0.1C

PROPERTY	@ 300K	UNITS	REFERENCES
Density	7.86	g/cm^3	6.1.A-B
Specific Heat	0.110	cal/g-K	6.2.A-B,D-E
Thermal Conductivity	0.625	cal/cm-s-K	6.3.A-D
Thermal Expansion	$10.9x10^{-6}$	1/K	6.4.A-C
Melting Point	1800	K	6.5.A
Resistivity	$13x10^{-6}$	Ohm-cm	6.6.A-C
TCR (298-405K)	7000	ppm/K	6.7.A-C
Young's Modulus	29650	ksi	6.8.A-D
Poisson's Ratio	0.292	------	6.9.A-B
Yield Strength	35,000	psi	6.10.A-D
Ult Tensile Strength	60,000	psi	6.11.A-D
Elongation at Break	40	%	6.12.A-C
Hardness	111	Brinell	6.13.A-C

LOW CARBON STEEL (0.1C-FE)

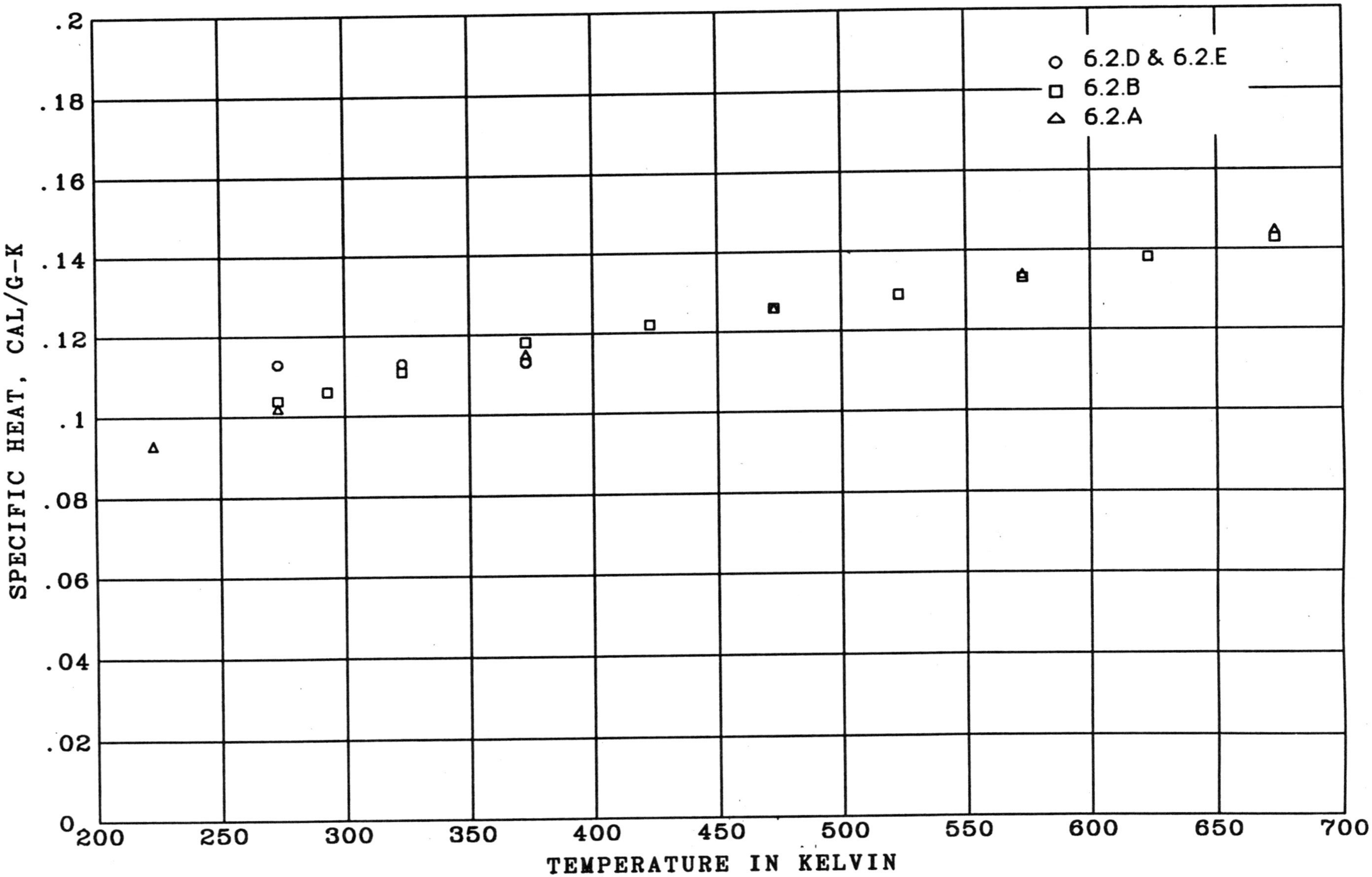

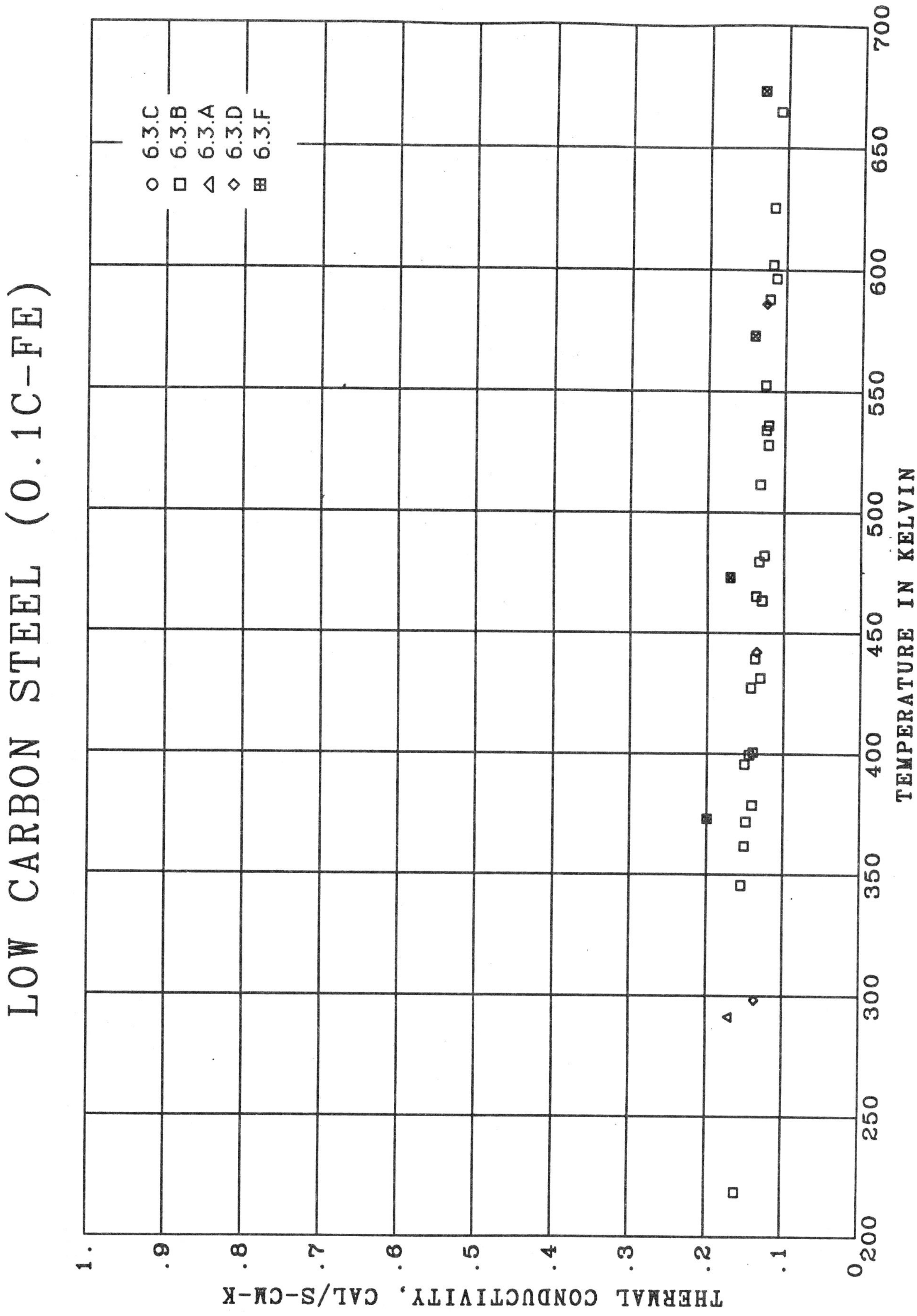
LOW CARBON STEEL (0.1C-FE)
○ 6.3.C
□ 6.3.B
△ 6.3.A
◇ 6.3.D
⊞ 6.3.F
THERMAL CONDUCTIVITY, CAL/S-CM-K
0
.1
.2
.3
.4
.5
.6
.7
.8
.9
1.
200
250
300
350
400
450
500
550
600
650
700
TEMPERATURE IN KELVIN

LOW CARBON STEEL (0.1C-FE)

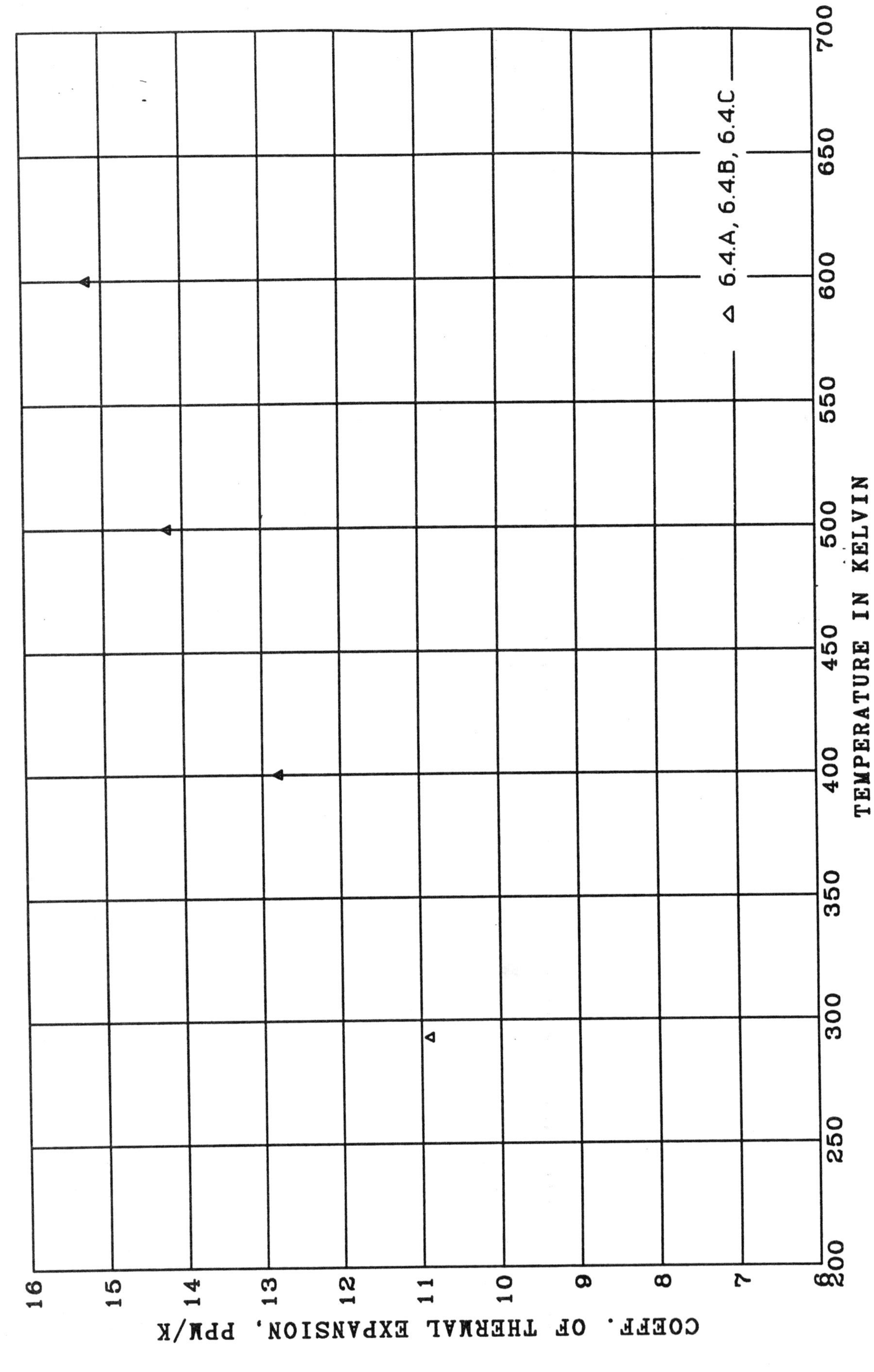

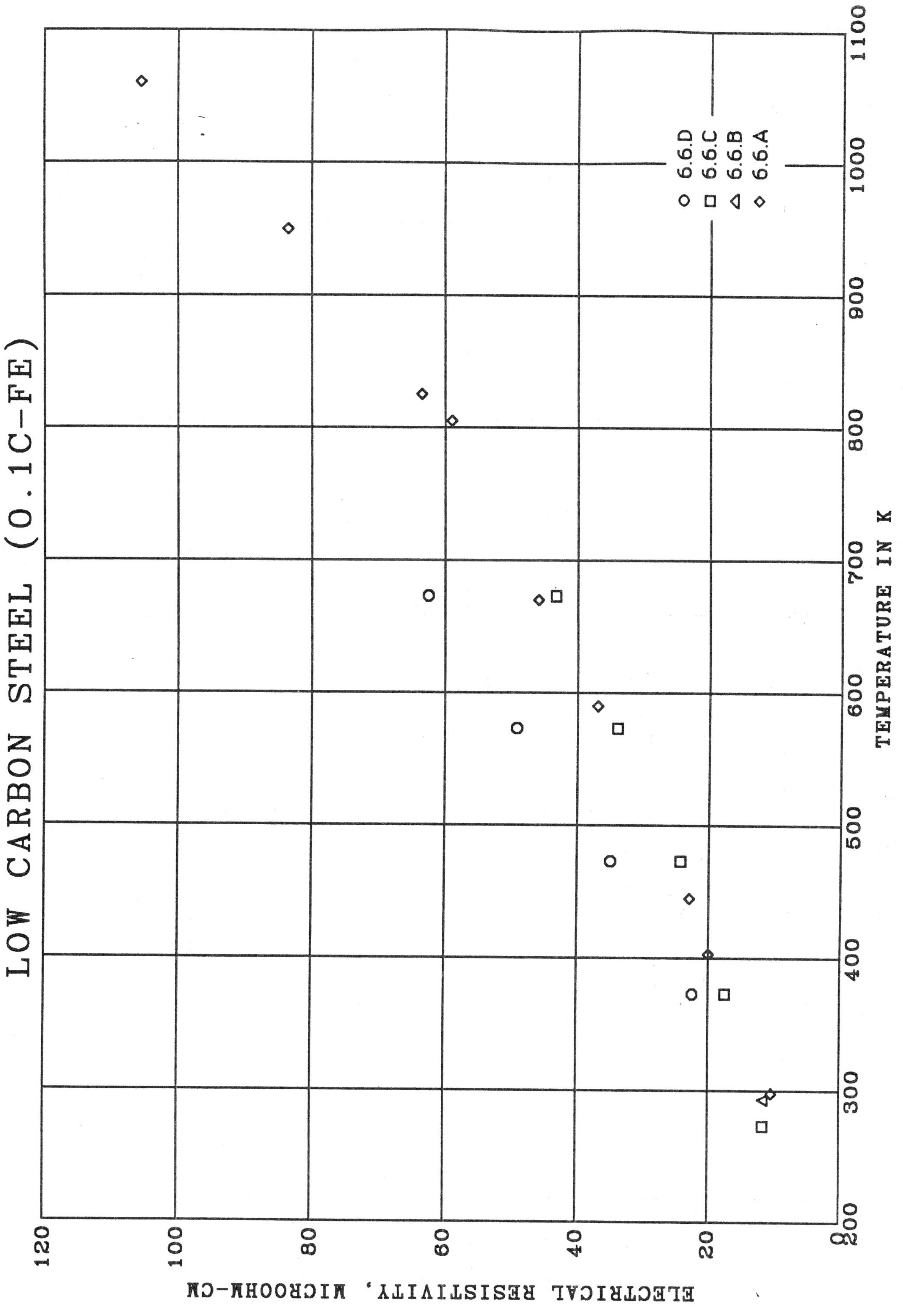
LOW CARBON STEEL (0.1C-FE)
ELECTRICAL RESISTIVITY, MICROOHM-CM
TEMPERATURE IN K
0
20
40
60
80
100
120
200
300
400
500
600
700
800
900
1000
1100
○ 6.6.D
□ 6.6.C
△ 6.6.B
◇ 6.6.A

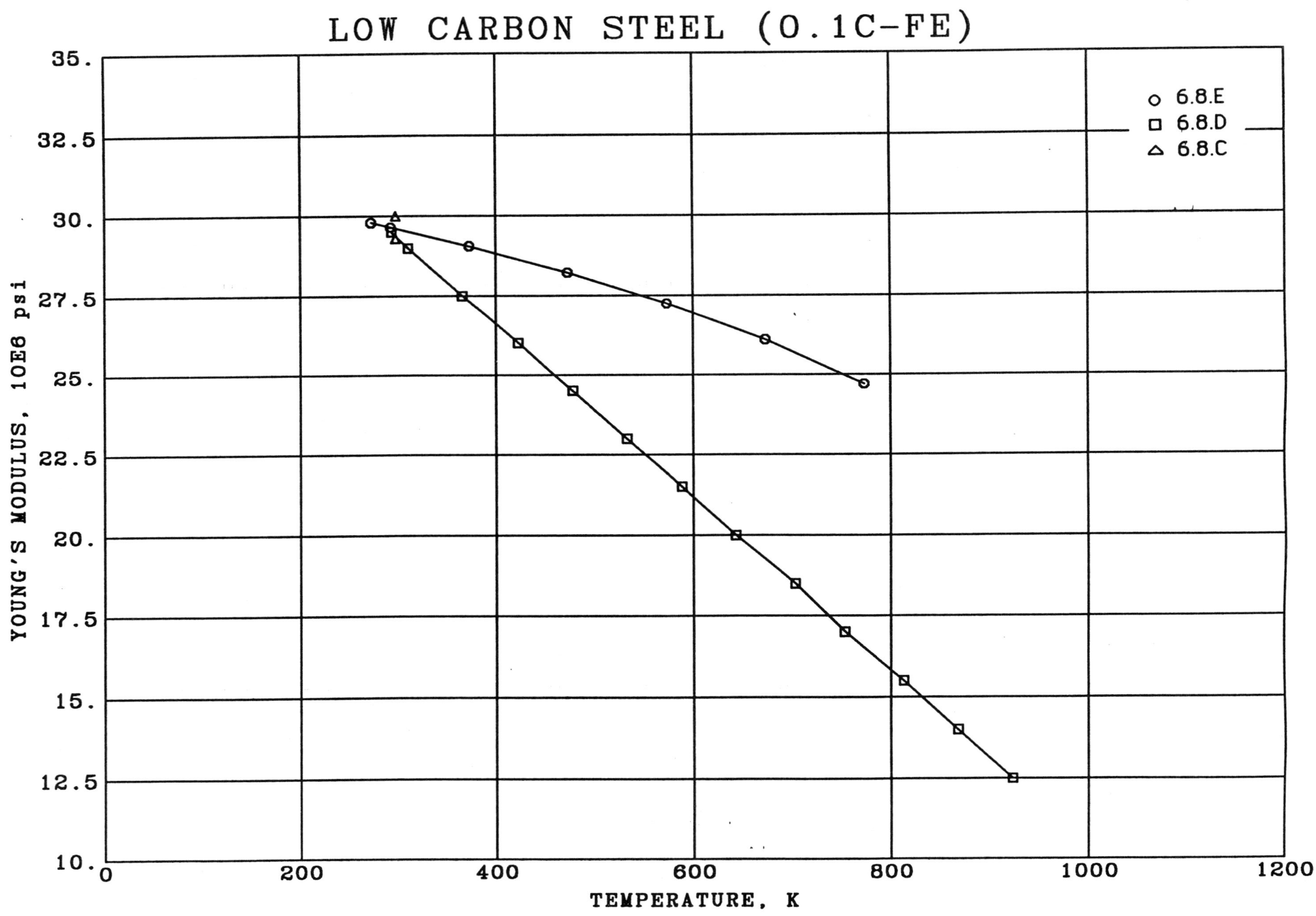
LOW CARBON STEEL (0.1C-FE)
○ 6.8.E
□ 6.8.D
△ 6.8.C
YOUNG'S MODULUS, 10E6 psi
TEMPERATURE, K
35.
32.5
30.
27.5
25.
22.5
20.
17.5
15.
12.5
10.
0
200
400
600
800
1000
1200

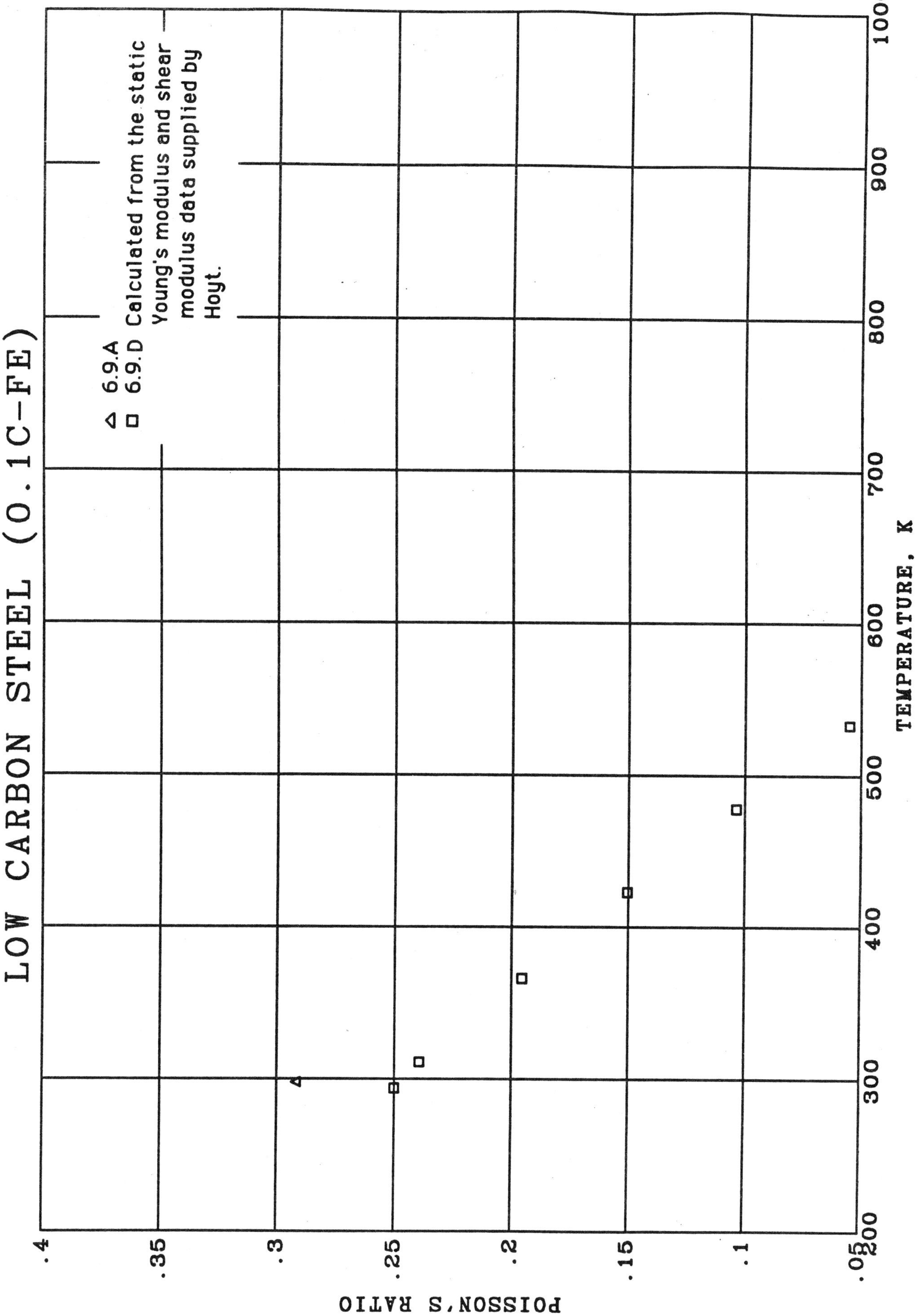
LOW CARBON STEEL (0.1C-FE)
△ 6.9.A
□ 6.9.D Calculated from the static Young's modulus and shear modulus data supplied by Hoyt.
POISSON'S RATIO
.05
.1
.15
.2
.25
.3
.35
.4
TEMPERATURE, K
200
300
400
500
600
700
800
900
1000

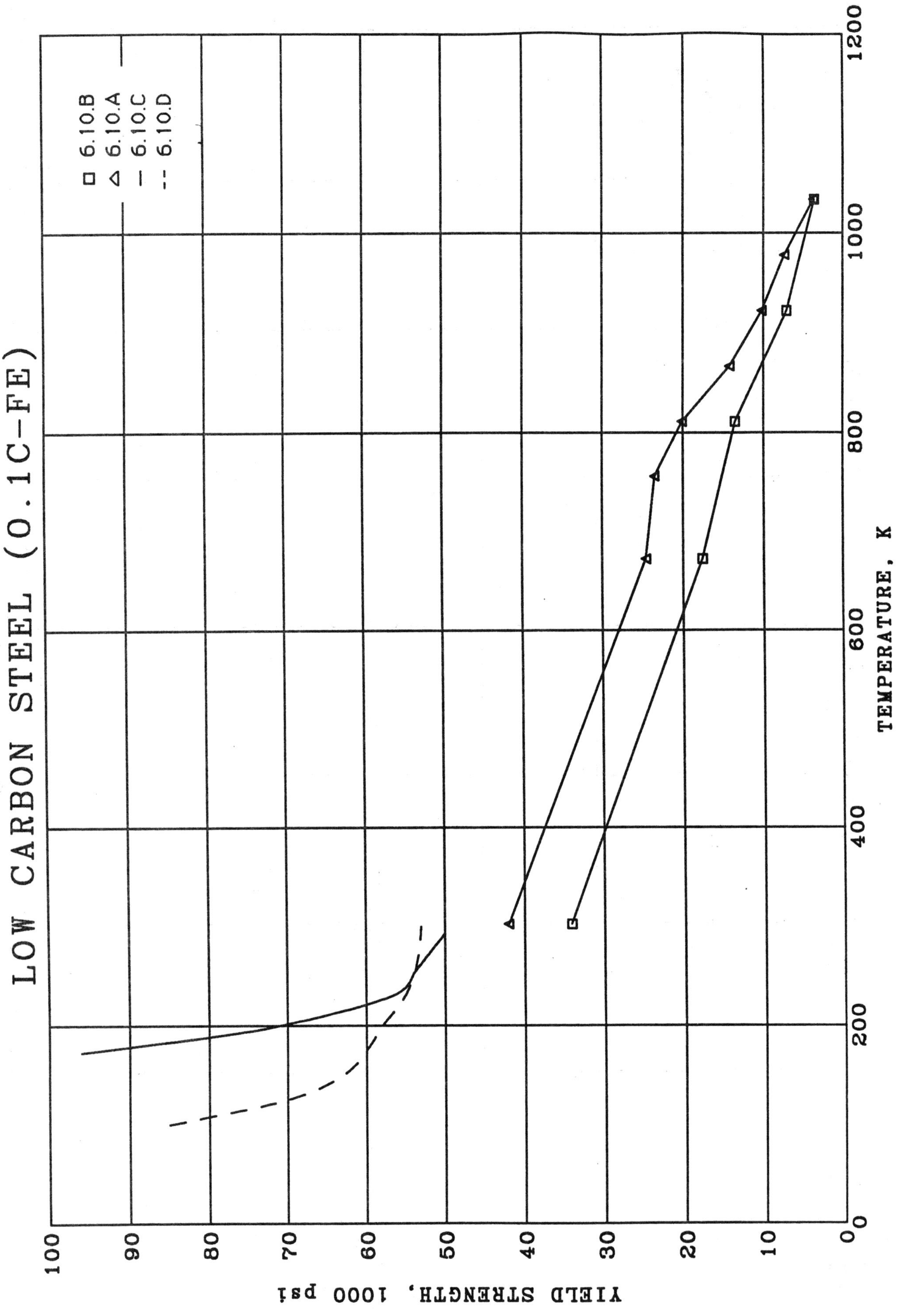
LOW CARBON STEEL (O.1C-FE)
6.10.B
6.10.A
6.10.C
6.10.D
TEMPERATURE, K
YIELD STRENGTH, 1000 psi
0
200
400
600
800
1000
1200
0
10
20
30
40
50
60
70
80
90
100

LOW CARBON STEEL (0.1C-FE)

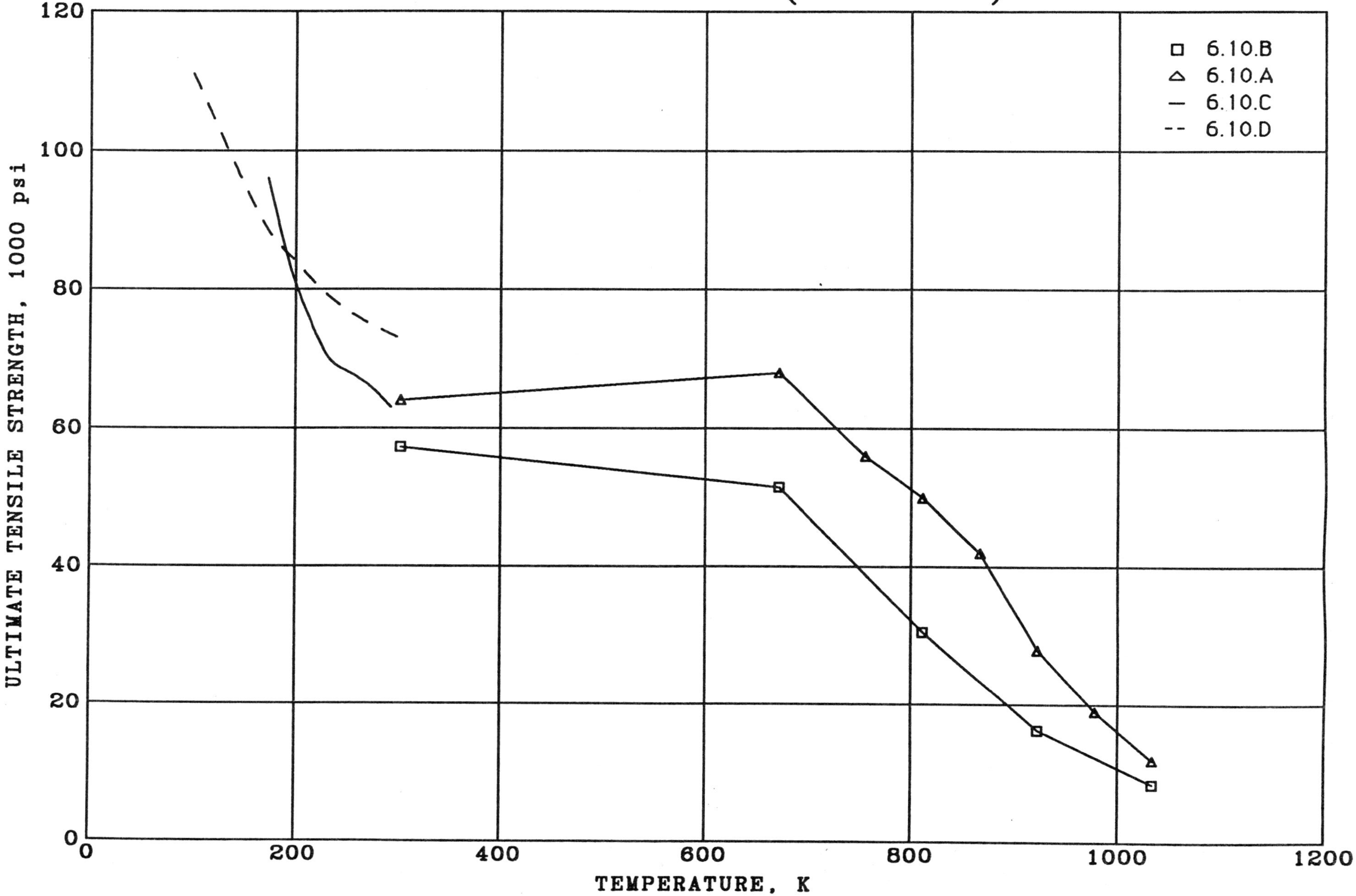

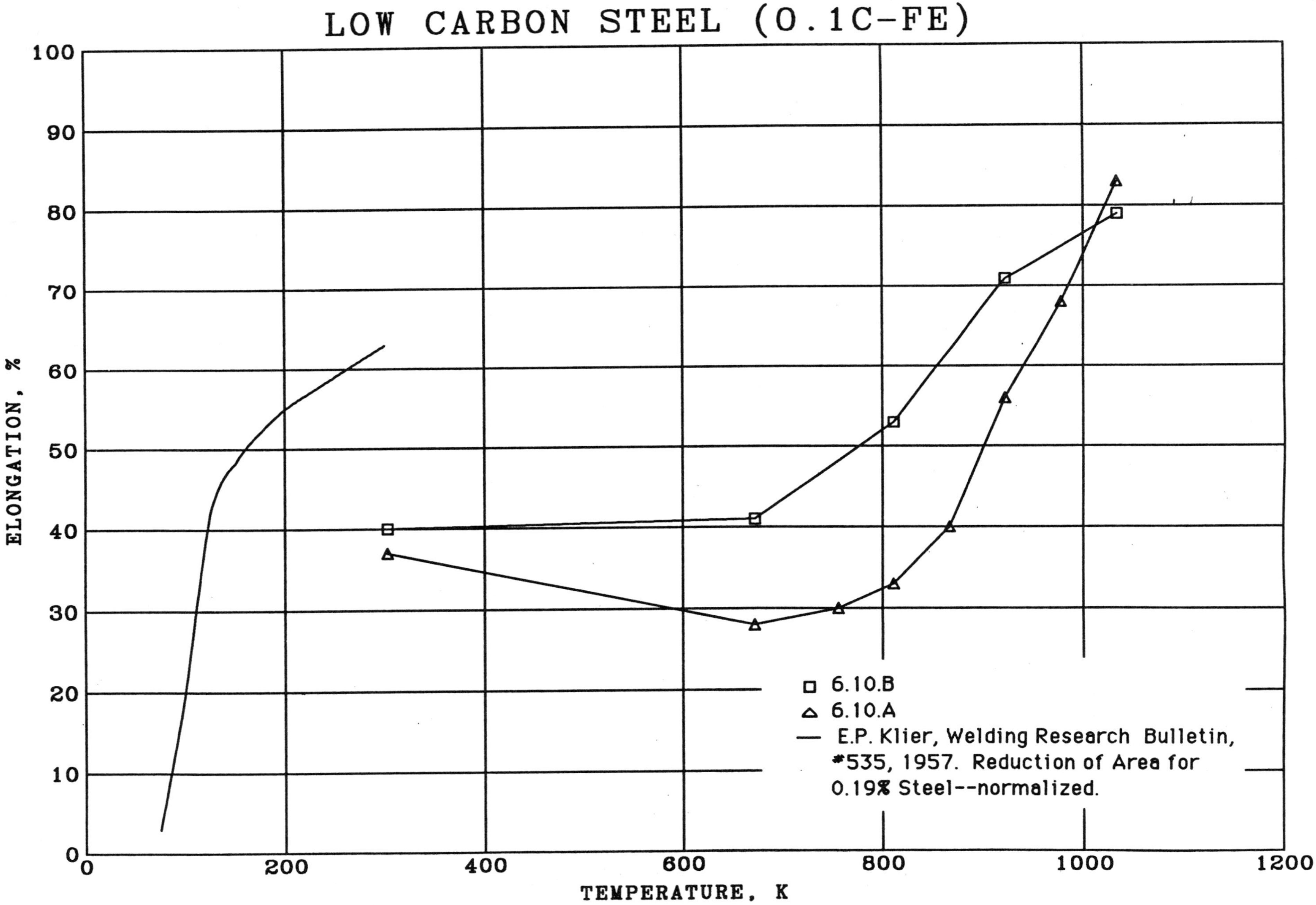
LOW CARBON STEEL (0.1C-FE)
ELONGATION, %
TEMPERATURE, K
0
10
20
30
40
50
60
70
80
90
100
0
200
400
600
800
1000
1200
6.10.B
6.10.A
E.P. Klier, Welding Research Bulletin, #535, 1957. Reduction of Area for 0.19% Steel--normalized.

LOW CARBON STEEL (0.1C–FE)

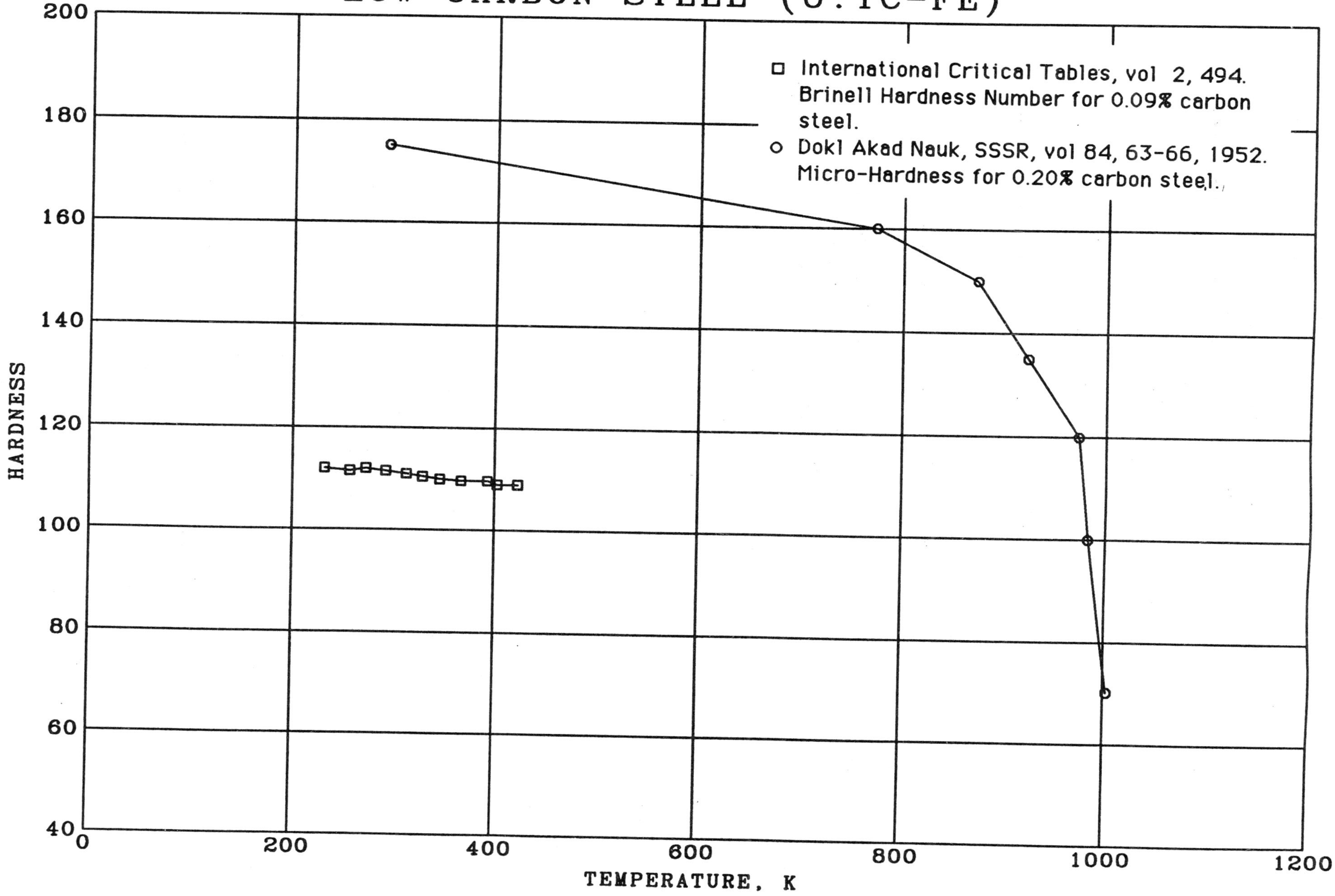

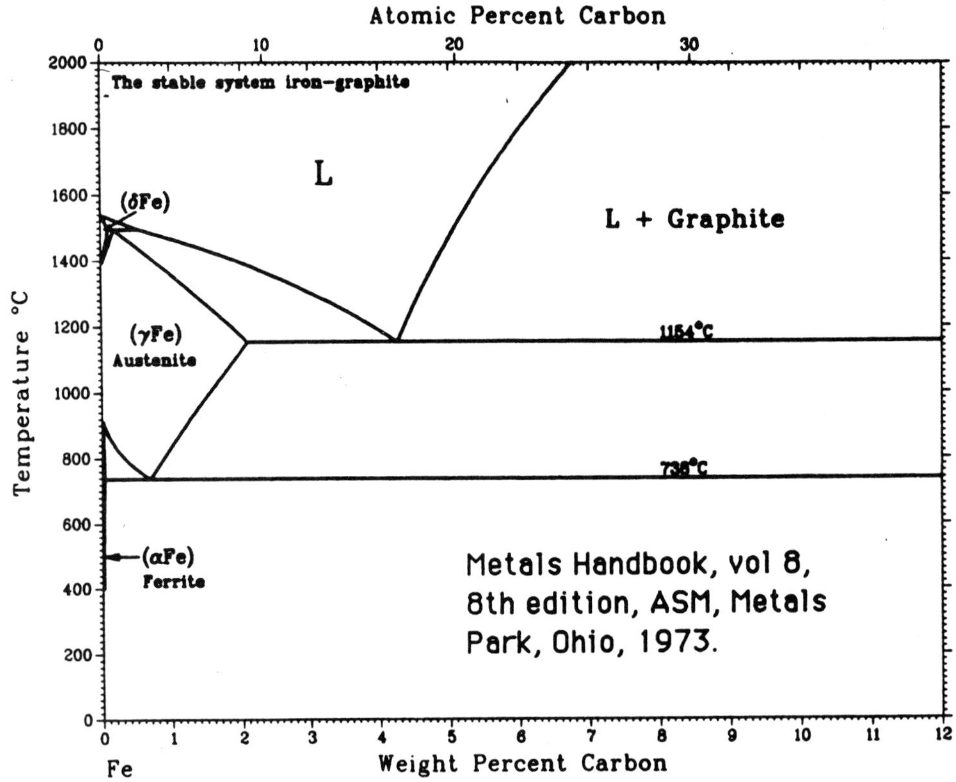
Atomic Percent Carbon
The stable system iron-graphite
L
(δFe)
L + Graphite
(γFe)
Austenite
1154°C
738°C
(αFe)
Ferrite
Metals Handbook, vol 8, 8th edition, ASM, Metals Park, Ohio, 1973.
Temperature °C
Fe
Weight Percent Carbon

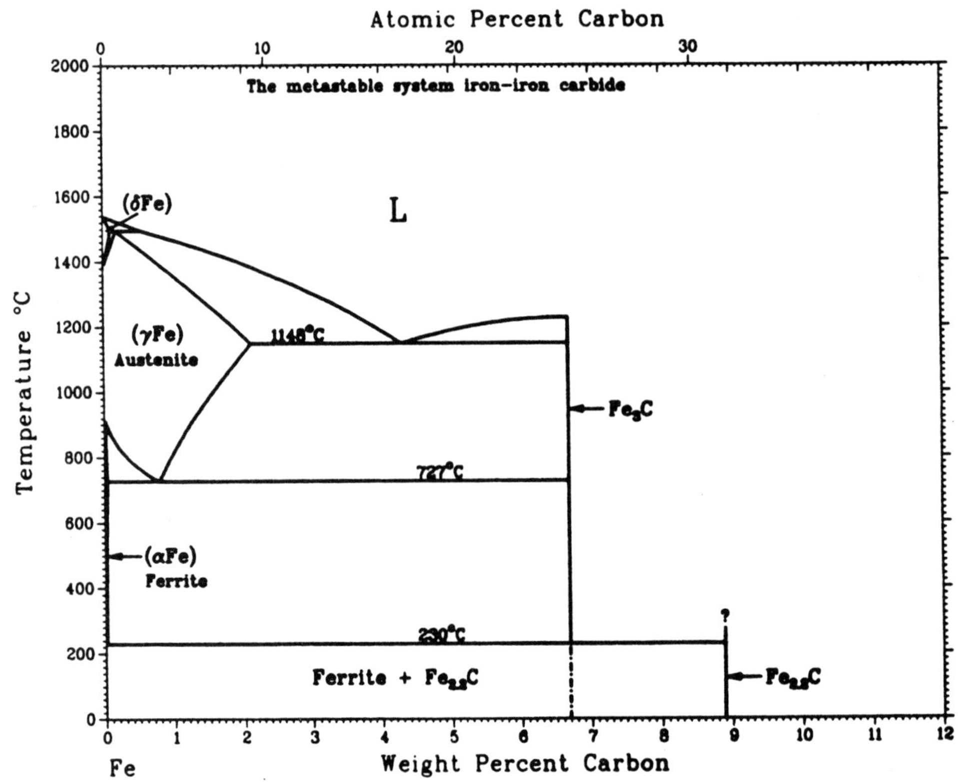
Atomic Percent Carbon
The metastable system iron-iron carbide
L
(δFe)
(γFe)
Austenite
1148°C
Fe3C
727°C
(αFe)
Ferrite
230°C
?
Ferrite + Fe2.5C
Fe2.5C
Temperature °C
Fe
Weight Percent Carbon

55Cu 45Ni

PROPERTY	@ 300K	UNITS	REFERENCES
Density	8.9	g/cm^3	7.1.A
Specific Heat	0.10	cal/g-K	7.2.A-C
Thermal Conductivity	0.05	cal/cm-s-K	7.3.A-D
Thermal Expansion	$16x10^{-6}$	1/K	7.4.A-D
Melting Point	1480	K	7.5.A-B
Resistivity	$50x10^{-6}$	Ohm-cm	7.6.A-C
TCR 1	20	ppm/K	7.7.A-B
Young's Modulus	25,000	ksi	7.8.A-G
Poisson's Ratio	0.32	------	7.9.A-C
Yield Strength	20-30	ksi	7.10.A-C
Ult Tensile Strength	60-70	ksi	7.11.A-C
Elongation at Break	40-50	%	7.12.A-C
Hardness	50	Rockwell B	7.13.A-C

1 The TCR is strongly dependent on small variations in purity and processing.

55Cu 45Ni

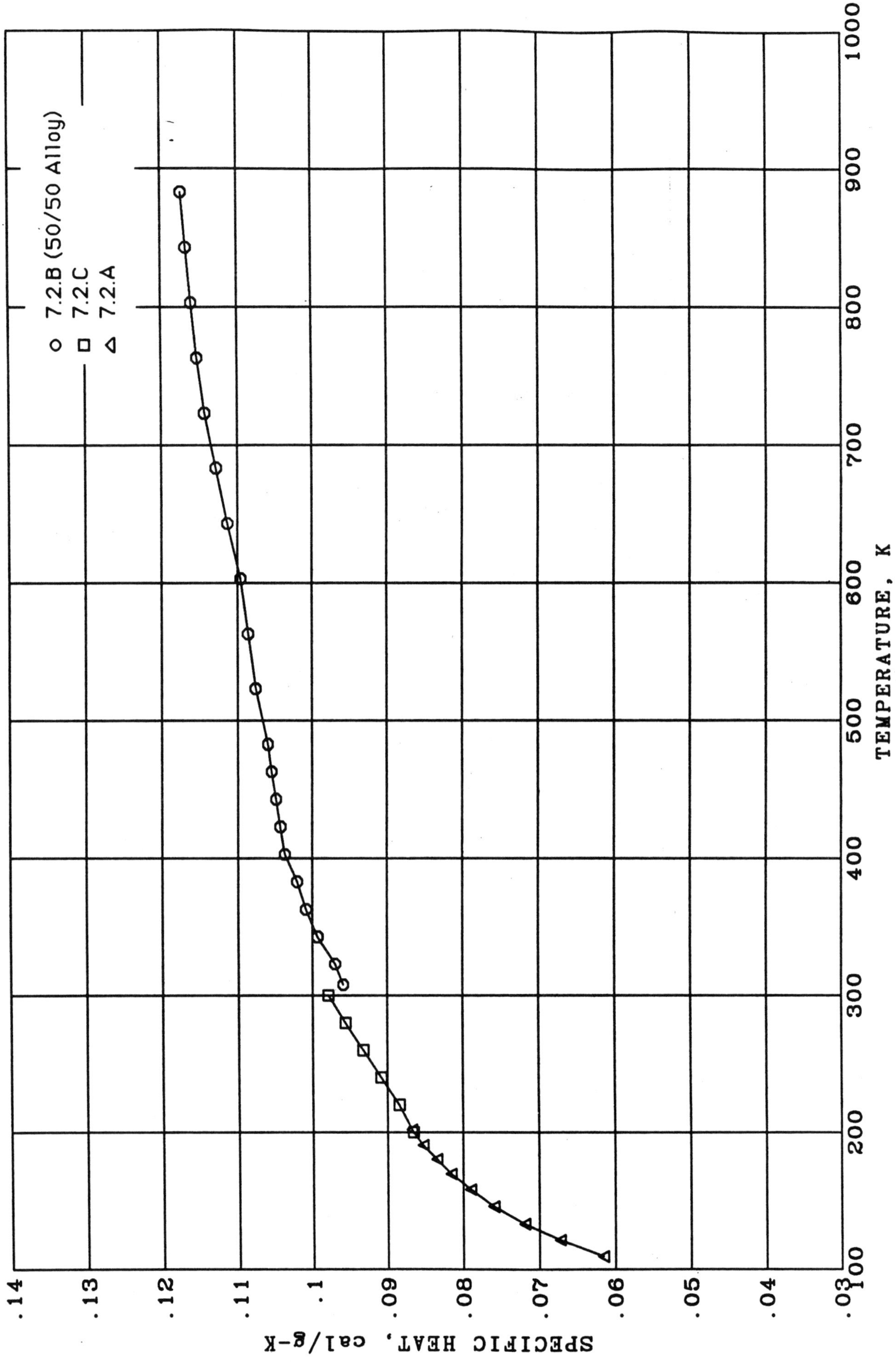

○ 7.2.B (50/50 Alloy)
□ 7.2.C
△ 7.2.A
TEMPERATURE, K
100
200
300
400
500
600
700
800
900
1000
SPECIFIC HEAT, cal/g-K
.03
.04
.05
.06
.07
.08
.09
.1
.11
.12
.13
.14

55Cu 45Ni

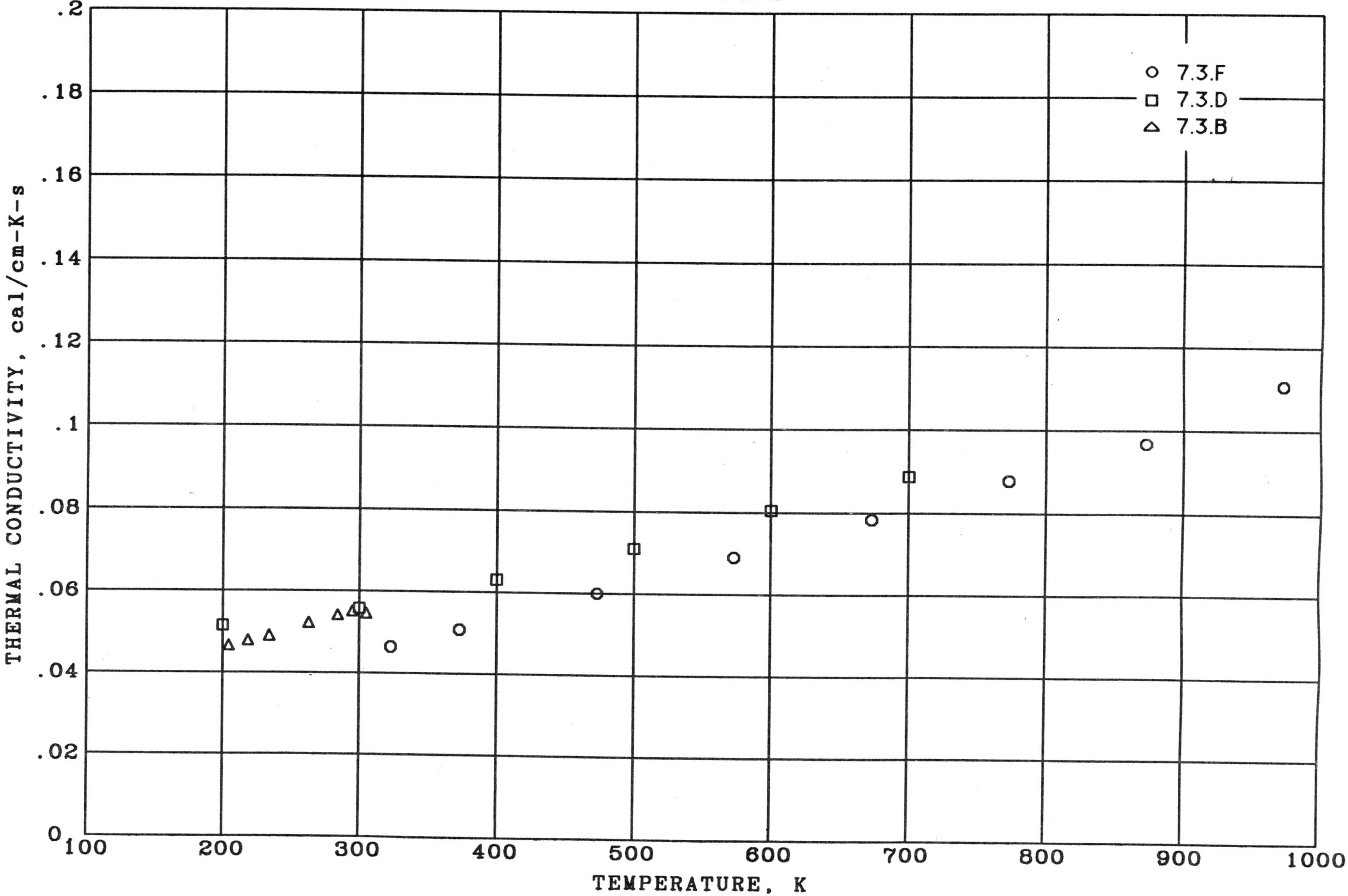

.2
.18
.16
.14
.12
.1
.08
.06
.04
.02
0
100
200
300
400
500
600
700
800
900
1000
THERMAL CONDUCTIVITY, cal/cm-K-s
TEMPERATURE, K
7.3.F
7.3.D
7.3.B

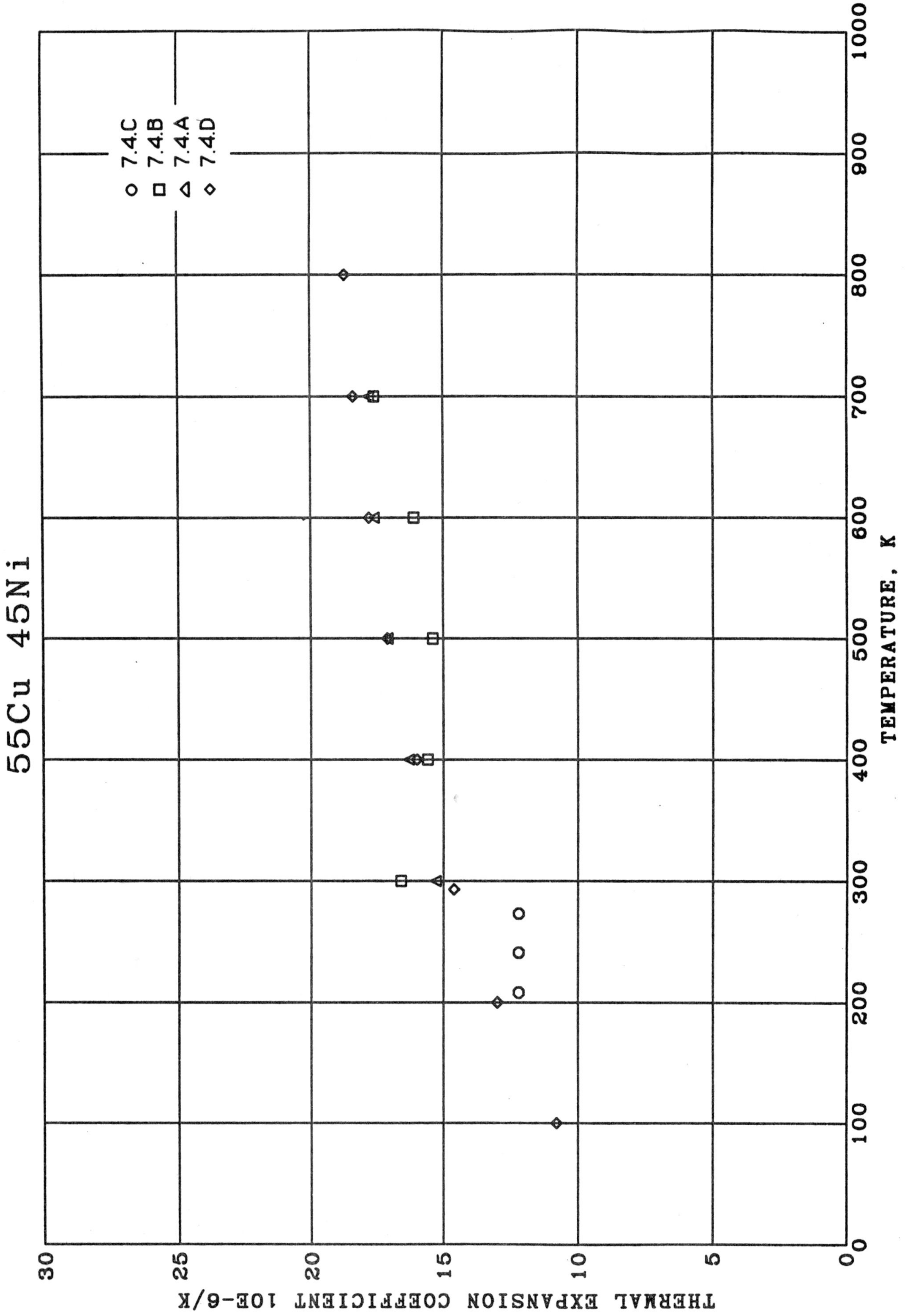
55Cu 45Ni
○ 7.4C
□ 7.4B
△ 7.4A
◇ 7.4D
TEMPERATURE, K
THERMAL EXPANSION COEFFICIENT 10E-6/K
0
100
200
300
400
500
600
700
800
900
1000
0
5
10
15
20
25
30

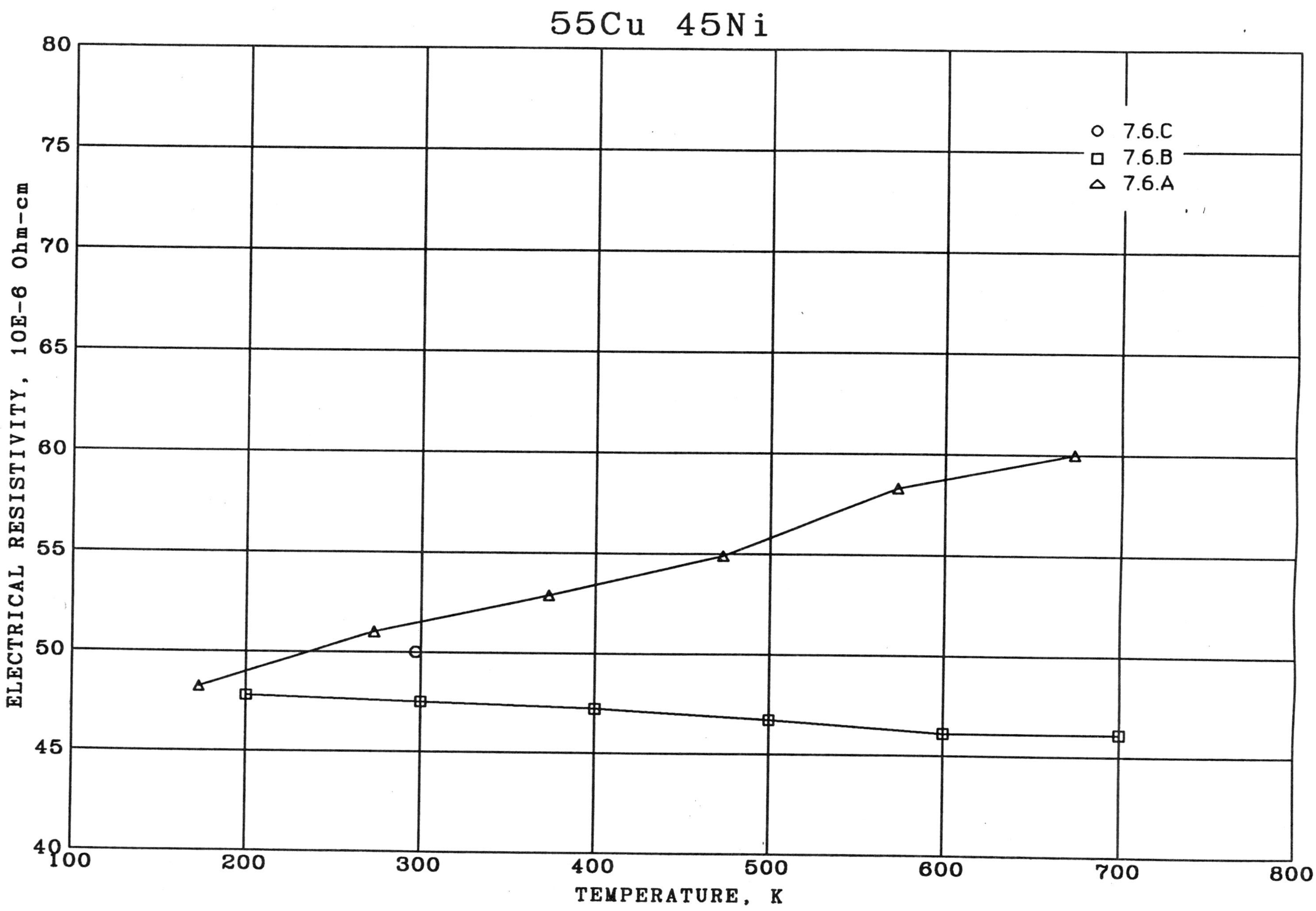
55Cu 45Ni
7.6.C
7.6.B
7.6.A
ELECTRICAL RESISTIVITY, 10E-6 Ohm-cm
TEMPERATURE, K
80
75
70
65
60
55
50
45
40
100
200
300
400
500
600
700
800

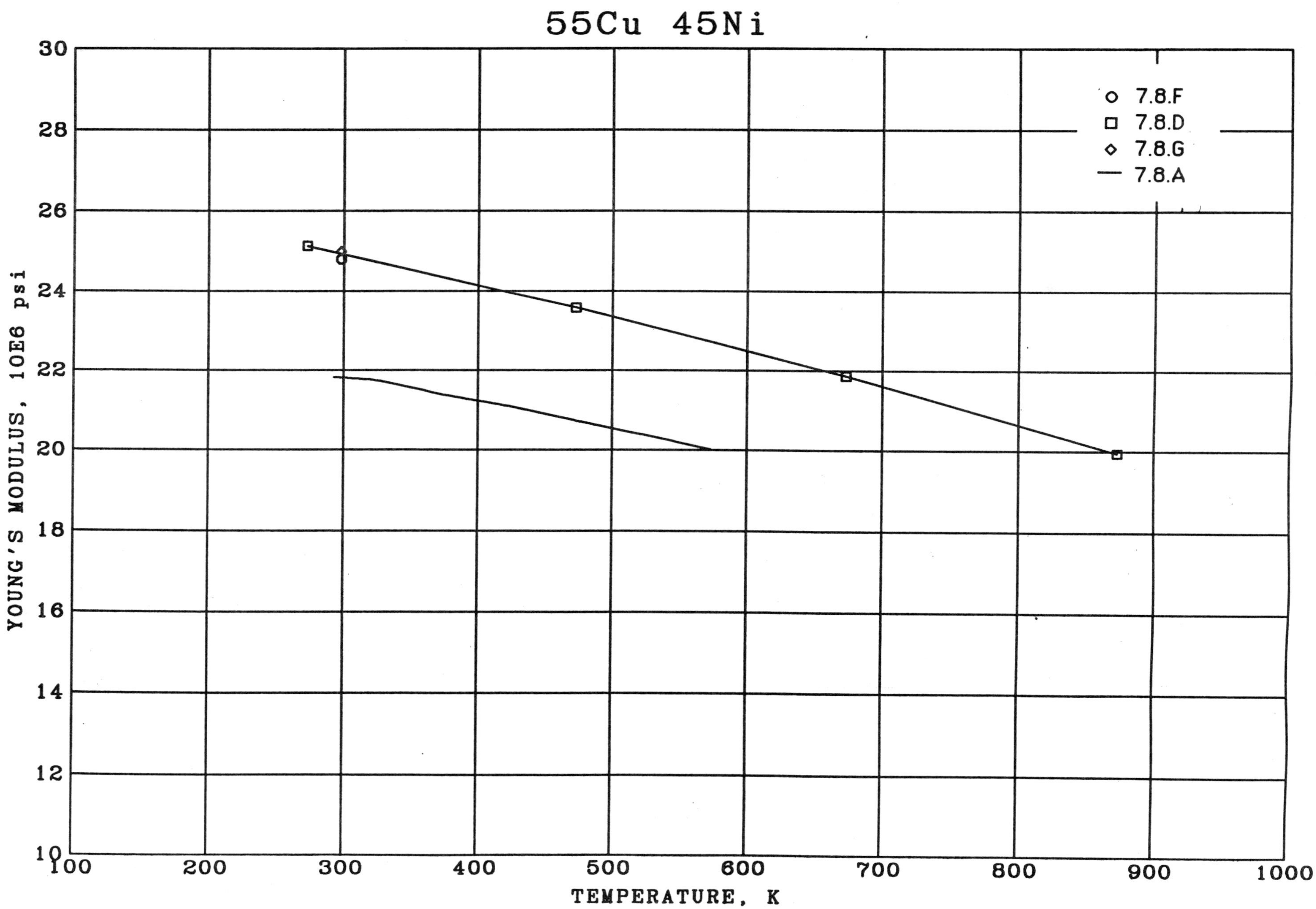

55Cu 45Ni
7.8.F
7.8.D
7.8.G
7.8.A
YOUNG'S MODULUS, 10E6 psi
TEMPERATURE, K
30
28
26
24
22
20
18
16
14
12
10
100
200
300
400
500
600
700
800
900
1000

55Cu 45Ni

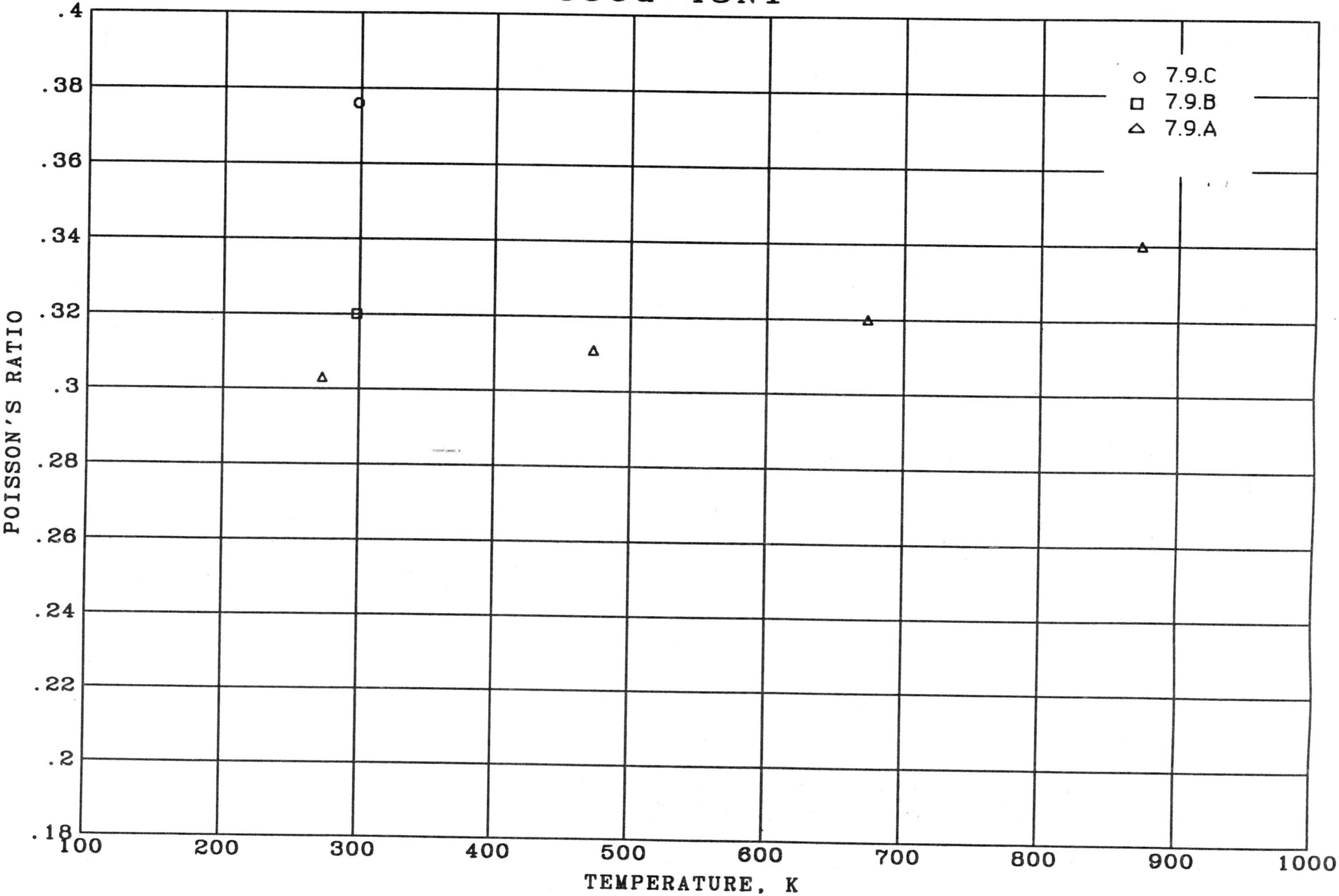

7.9.C
7.9.B
7.9.A
POISSON'S RATIO
.4
.38
.36
.34
.32
.3
.28
.26
.24
.22
.2
.18
100
200
300
400
500
600
700
800
900
1000
TEMPERATURE, K

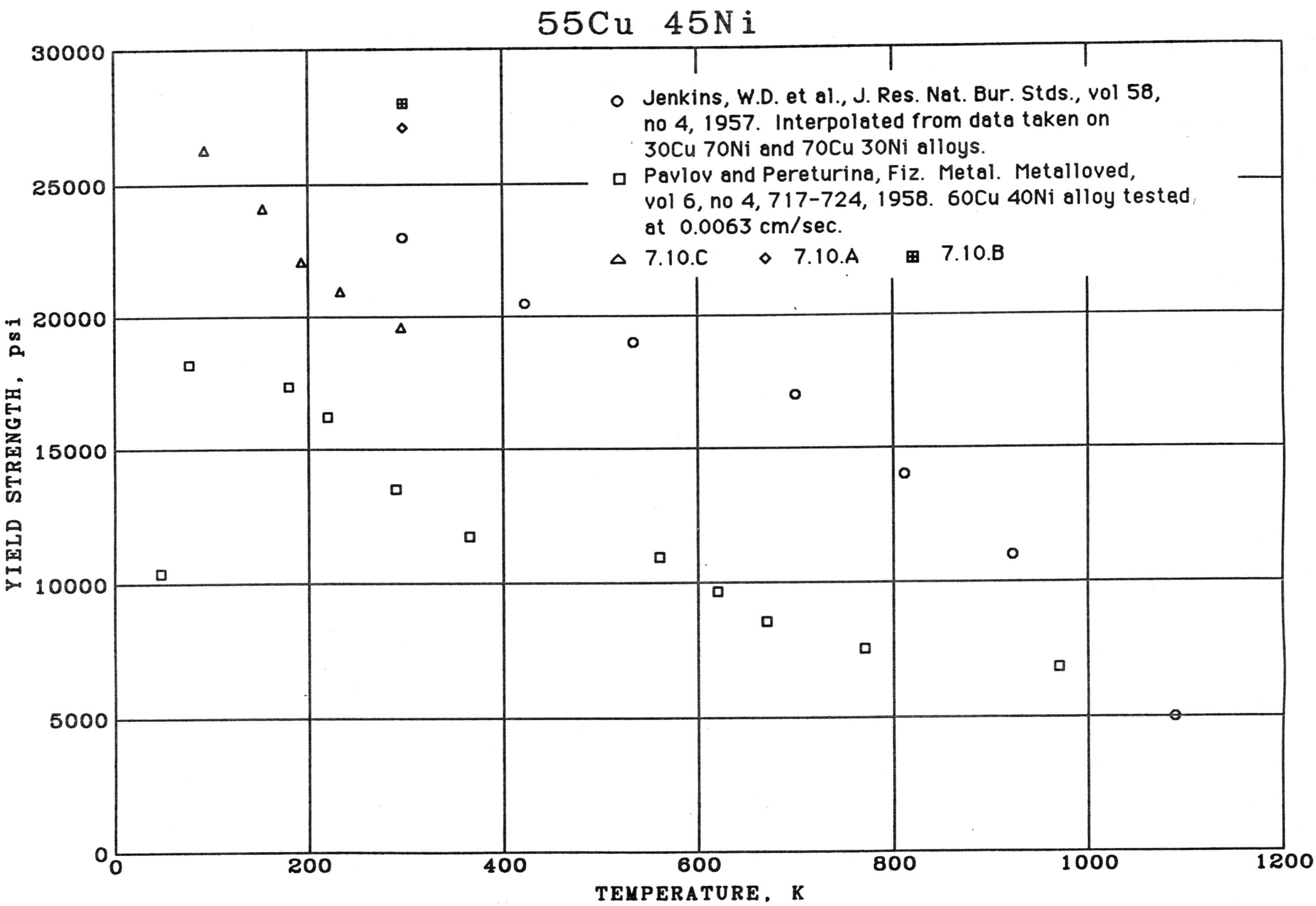
55Cu 45Ni
Jenkins, W.D. et al., J. Res. Nat. Bur. Stds., vol 58, no 4, 1957. Interpolated from data taken on 30Cu 70Ni and 70Cu 30Ni alloys.
Pavlov and Pereturina, Fiz. Metal. Metalloved, vol 6, no 4, 717-724, 1958. 60Cu 40Ni alloy tested at 0.0063 cm/sec.
7.10.C
7.10.A
7.10.B
YIELD STRENGTH, psi
TEMPERATURE, K
30000
25000
20000
15000
10000
5000
0
0
200
400
600
800
1000
1200

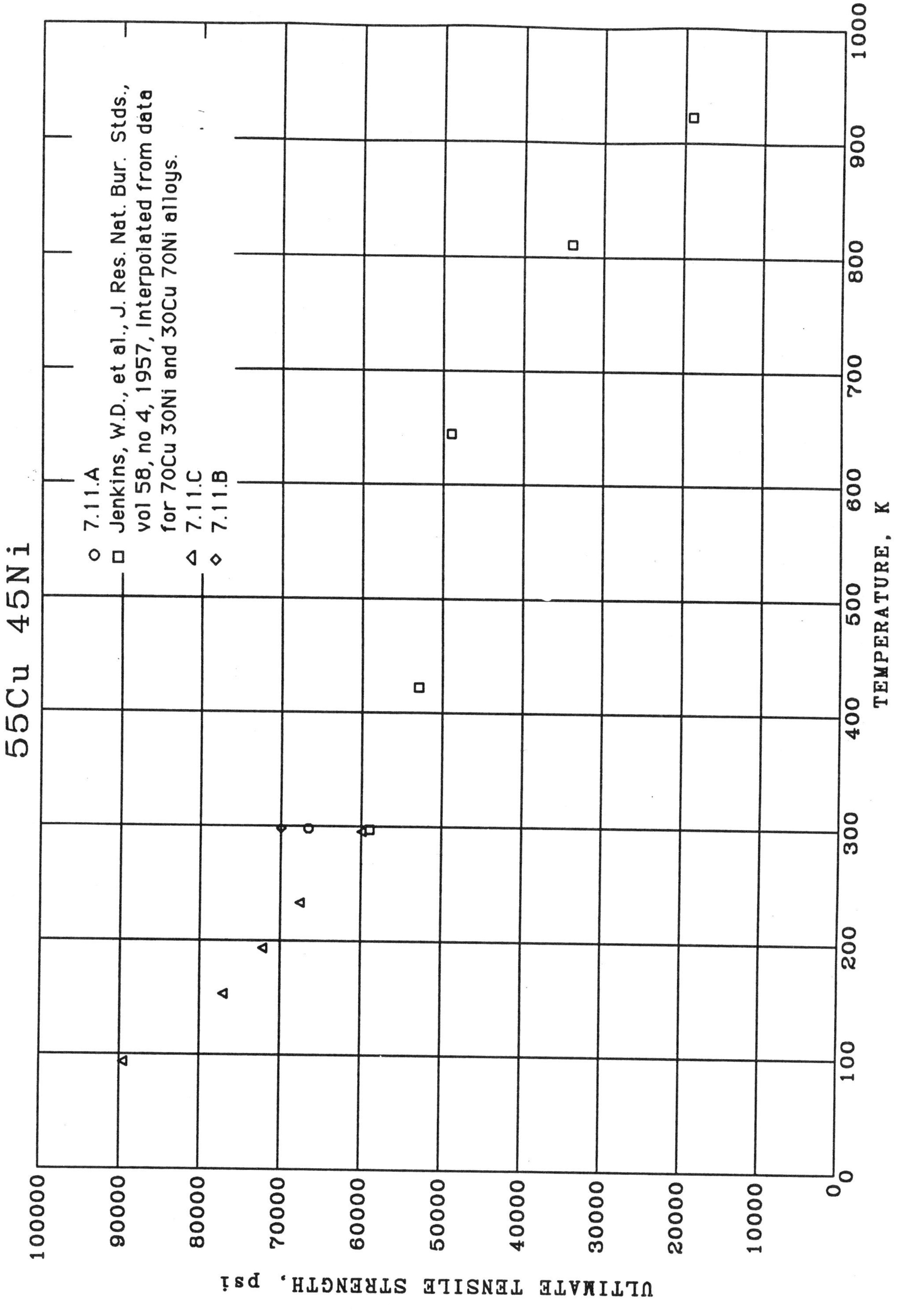
55Cu 45Ni
○ 7.11.A
□ Jenkins, W.D., et al., J. Res. Nat. Bur. Stds., vol 58, no 4, 1957, Interpolated from data for 70Cu 30Ni and 30Cu 70Ni alloys.
△ 7.11.C
◇ 7.11.B
ULTIMATE TENSILE STRENGTH, psi
100000
90000
80000
70000
60000
50000
40000
30000
20000
10000
0
TEMPERATURE, K
0
100
200
300
400
500
600
700
800
900
1000

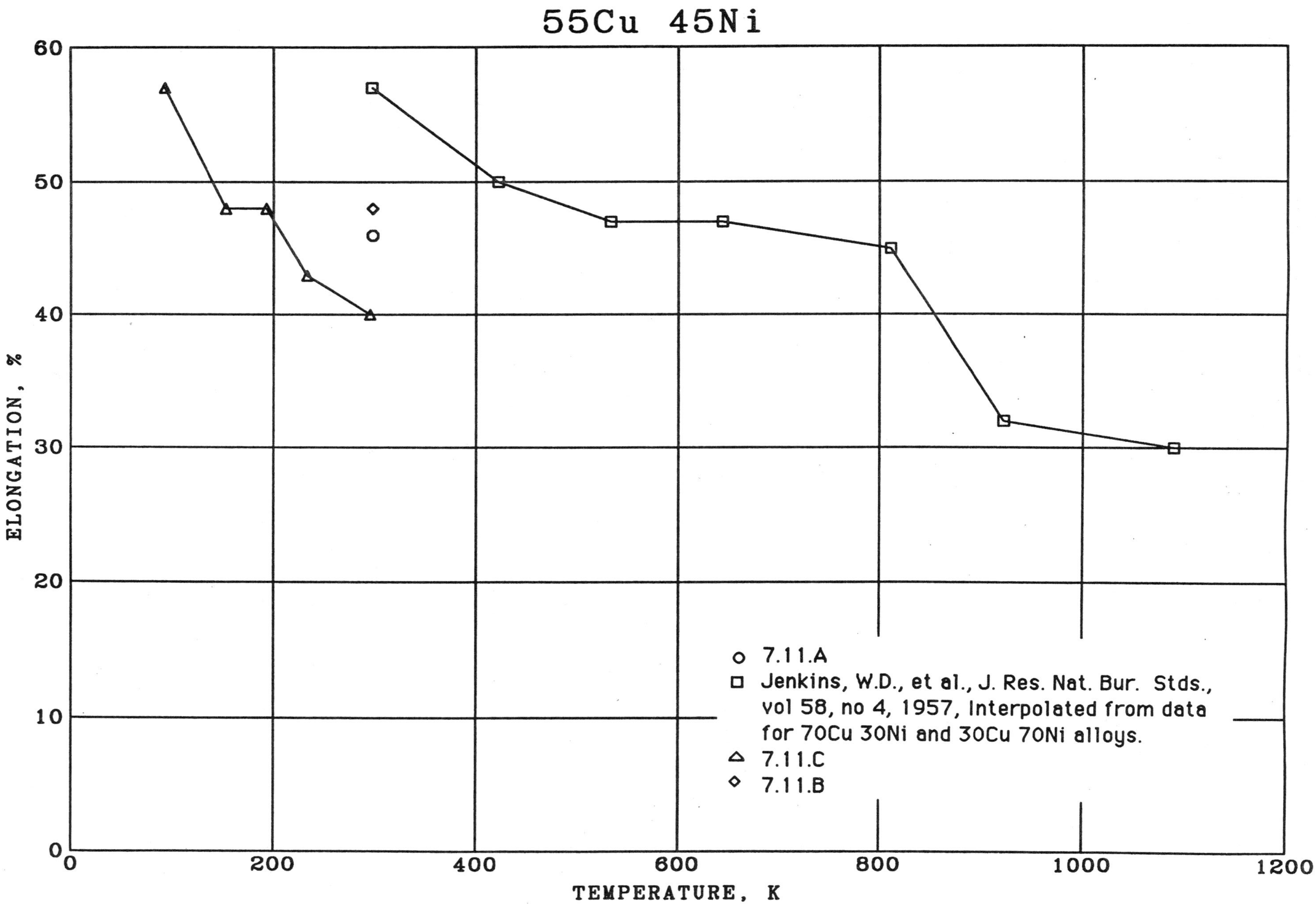
55Cu 45Ni
60
50
40
30
20
10
0
ELONGATION, %
0
200
400
600
800
1000
1200
TEMPERATURE, K
7.11.A
Jenkins, W.D., et al., J. Res. Nat. Bur. Stds., vol 58, no 4, 1957, Interpolated from data for 70Cu 30Ni and 30Cu 70Ni alloys.
7.11.C
7.11.B

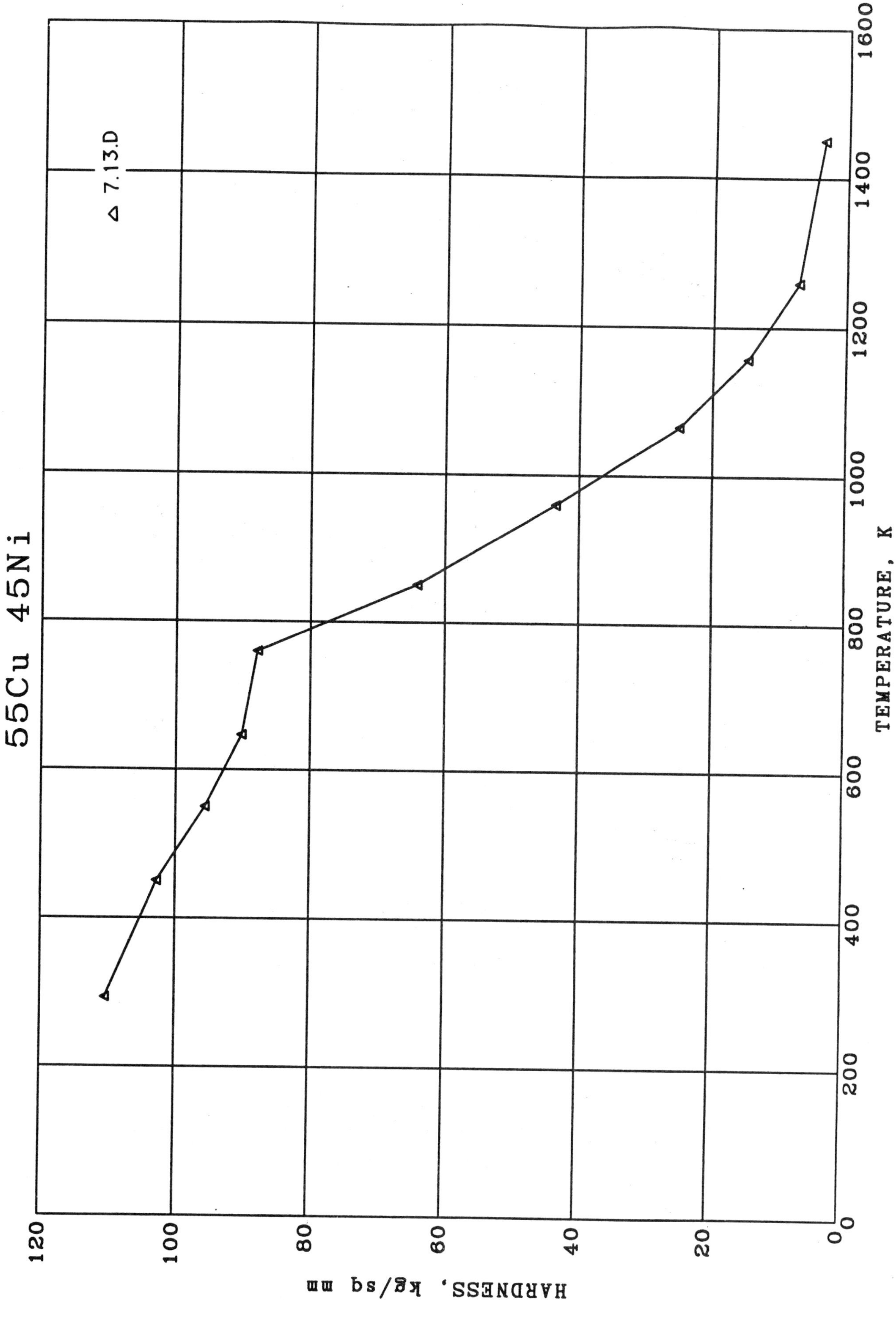
55Cu 45Ni
△ 7.13.D
HARDNESS, kg/sq mm
TEMPERATURE, K
0
20
40
60
80
100
120
0
200
400
600
800
1000
1200
1400
1600

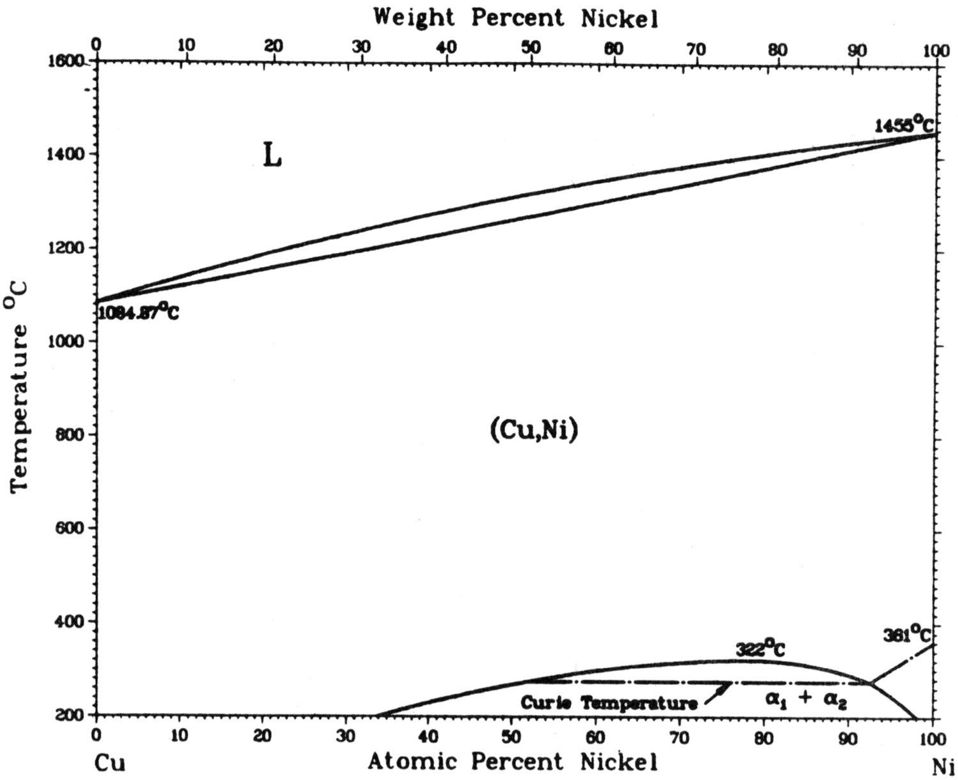
Weight Percent Nickel
Temperature °C
L
1455°C
1084.87°C
(Cu,Ni)
322°C
361°C
Curie Temperature
α1 + α2
Cu
Atomic Percent Nickel
Ni

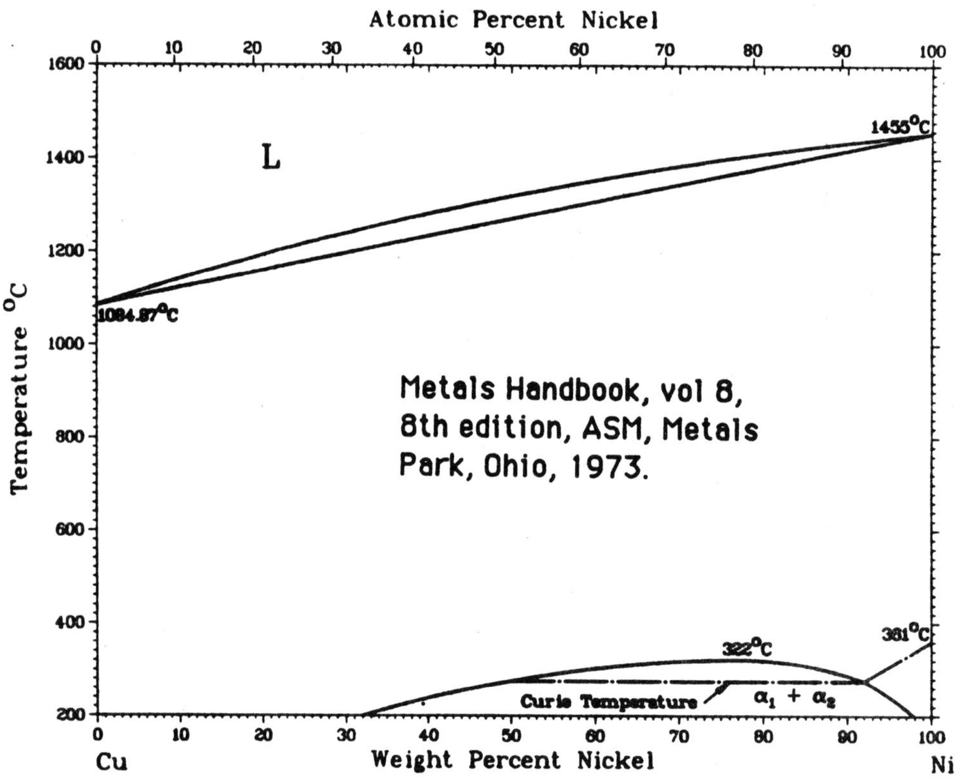
Atomic Percent Nickel
Temperature °C
L
1455°C
1084.87°C
Metals Handbook, vol 8,
8th edition, ASM, Metals
Park, Ohio, 1973.
322°C
361°C
Curie Temperature
α1 + α2
Cu
Weight Percent Nickel
Ni

80Ni 20Cr (Nichrome)

PROPERTY	@ 300K	UNITS	REFERENCES
Density	8.4	g/cm^3	8.1.A-D
Specific Heat	0.105	cal/g-K	8.2.A-C
Thermal Conductivity	0.03	cal/cm-s-K	8.3.D,F
Thermal Expansion	$12x10^{-6}$	1/K	8.4.A-C
Melting Point	1670	K	8.5.A-C
Resistivity	$100x10^{-6}$	Ohm-cm	8.6.A,D,F
TCR	127	ppm/K	8.7.A,D
Young's Modulus	30,000	ksi	8.8.A-D
Poisson's Ratio	0.298	------	8.9.A
Yield Strength	60,000	psi	8.10.A-C
Ult Tensile Strength	100,000	psi	8.11.A-D
Elongation at Break	30	%	8.12.A-C
Hardness	120-300	Brinell	8.13.A-C

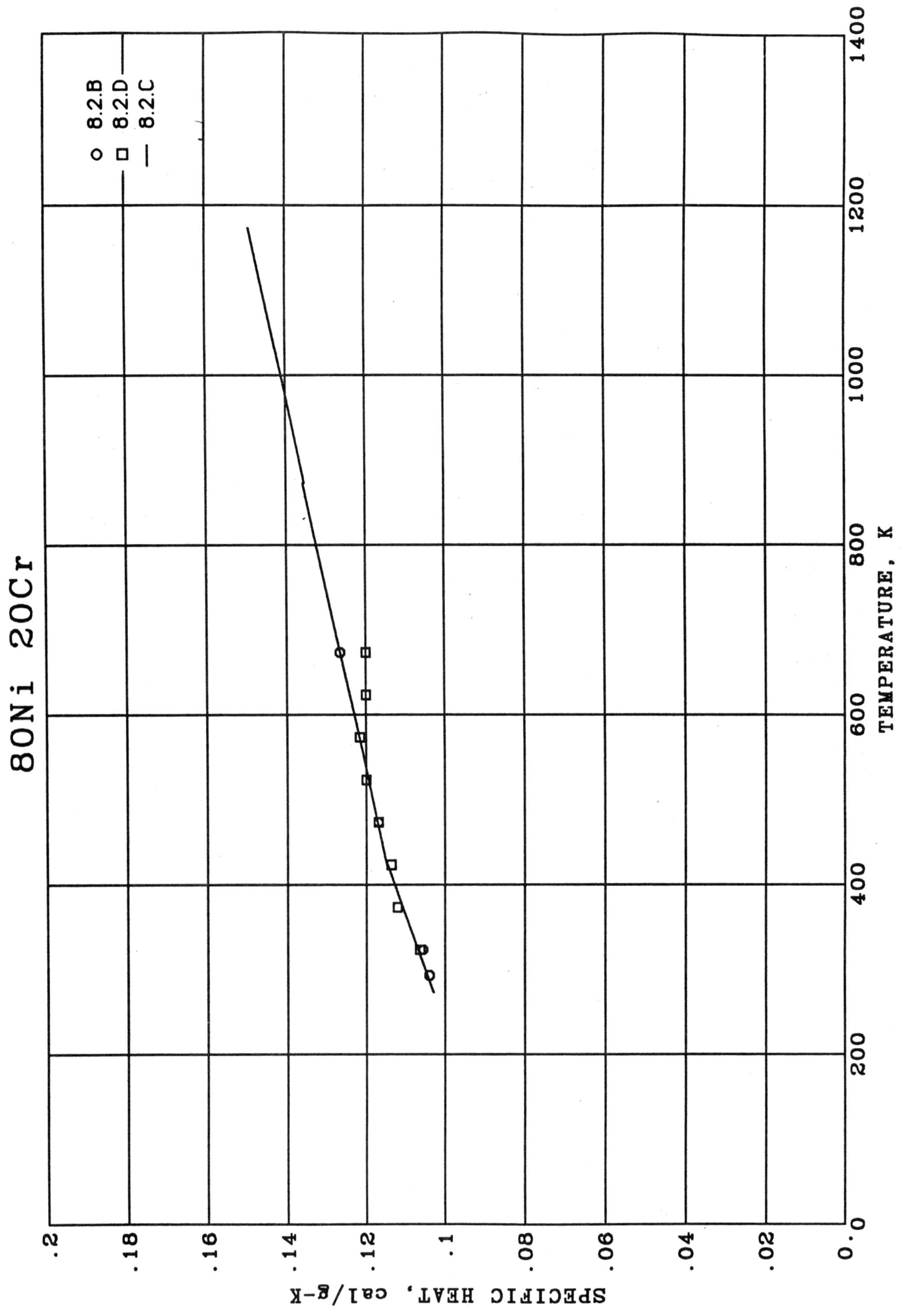
80Ni 20Cr
○ 8.2.B
□ 8.2.D
— 8.2.C
SPECIFIC HEAT, cal/g-K
TEMPERATURE, K
0.
.02
.04
.06
.08
.1
.12
.14
.16
.18
.2
0
200
400
600
800
1000
1200
1400

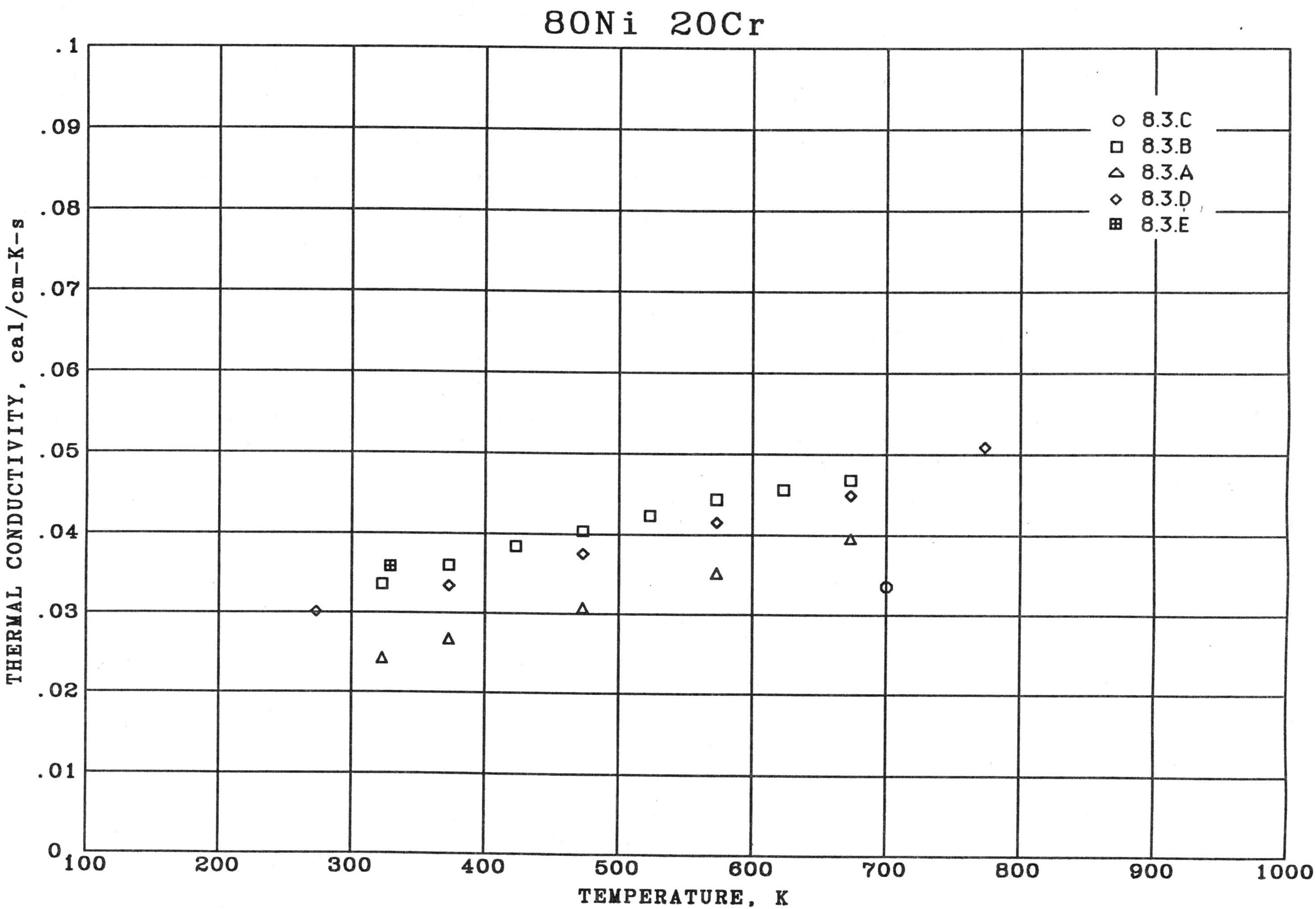
80Ni 20Cr
THERMAL CONDUCTIVITY, cal/cm-K-s
TEMPERATURE, K
8.3.C
8.3.B
8.3.A
8.3.D
8.3.E

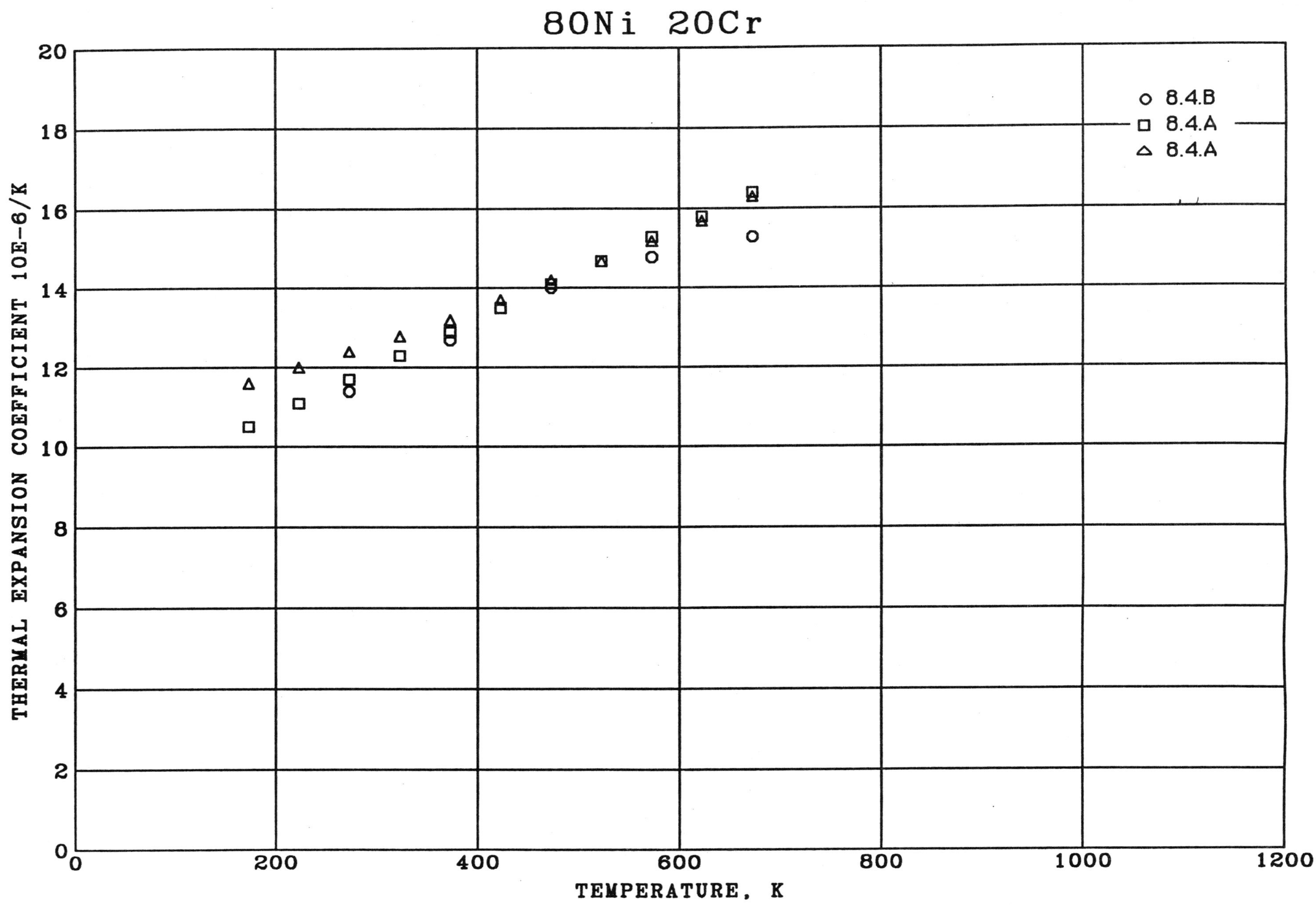
80Ni 20Cr
8.4.B
8.4.A
8.4.A
THERMAL EXPANSION COEFFICIENT 10E-6/K
TEMPERATURE, K
0
200
400
600
800
1000
1200
0
2
4
6
8
10
12
14
16
18
20

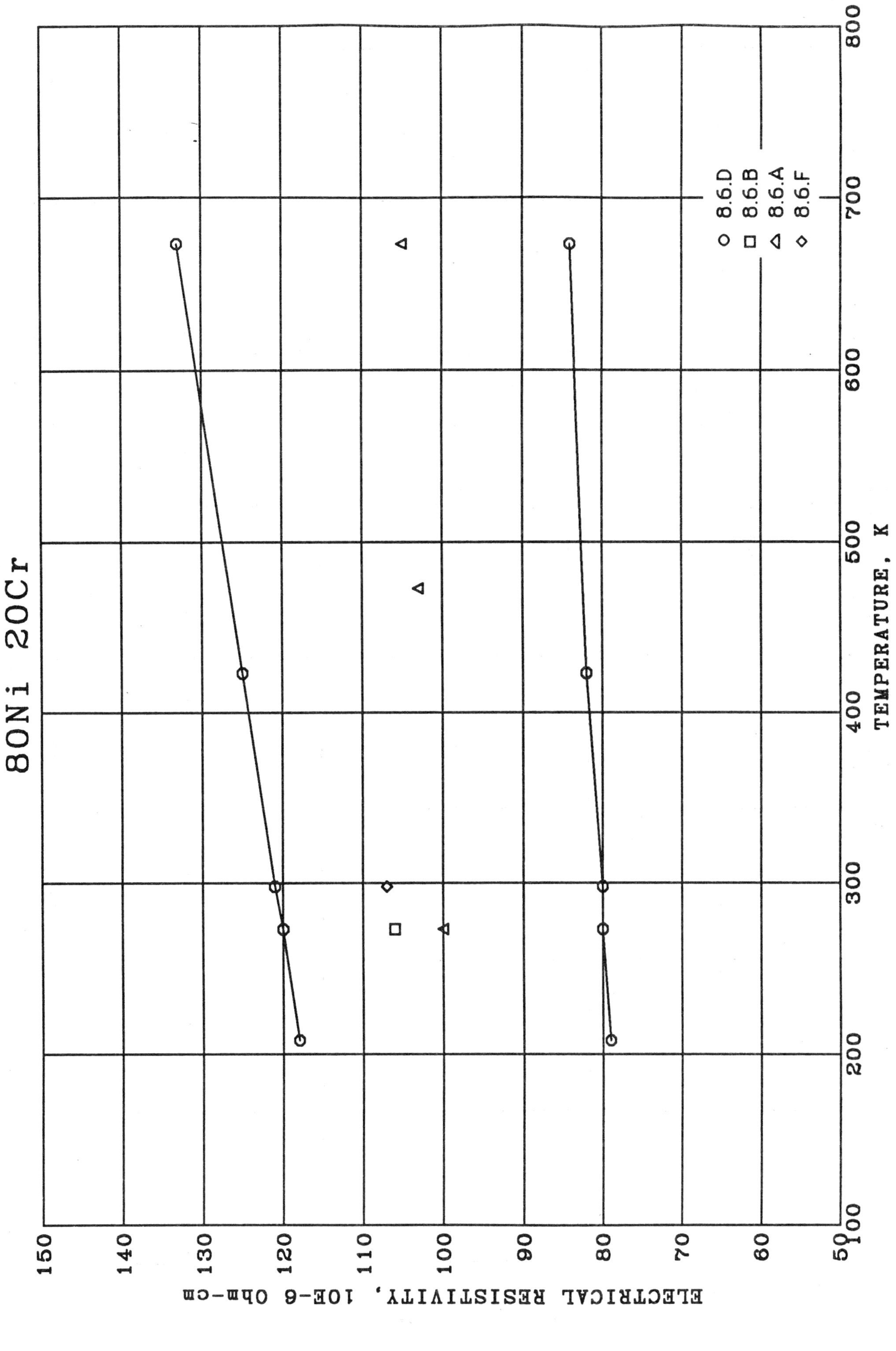
80Ni 20Cr
ELECTRICAL RESISITIVITY, 10E-6 Ohm-cm
TEMPERATURE, K
150
140
130
120
110
100
90
80
70
60
50
100
200
300
400
500
600
700
800
8.6.D
8.6.B
8.6.A
8.6.F

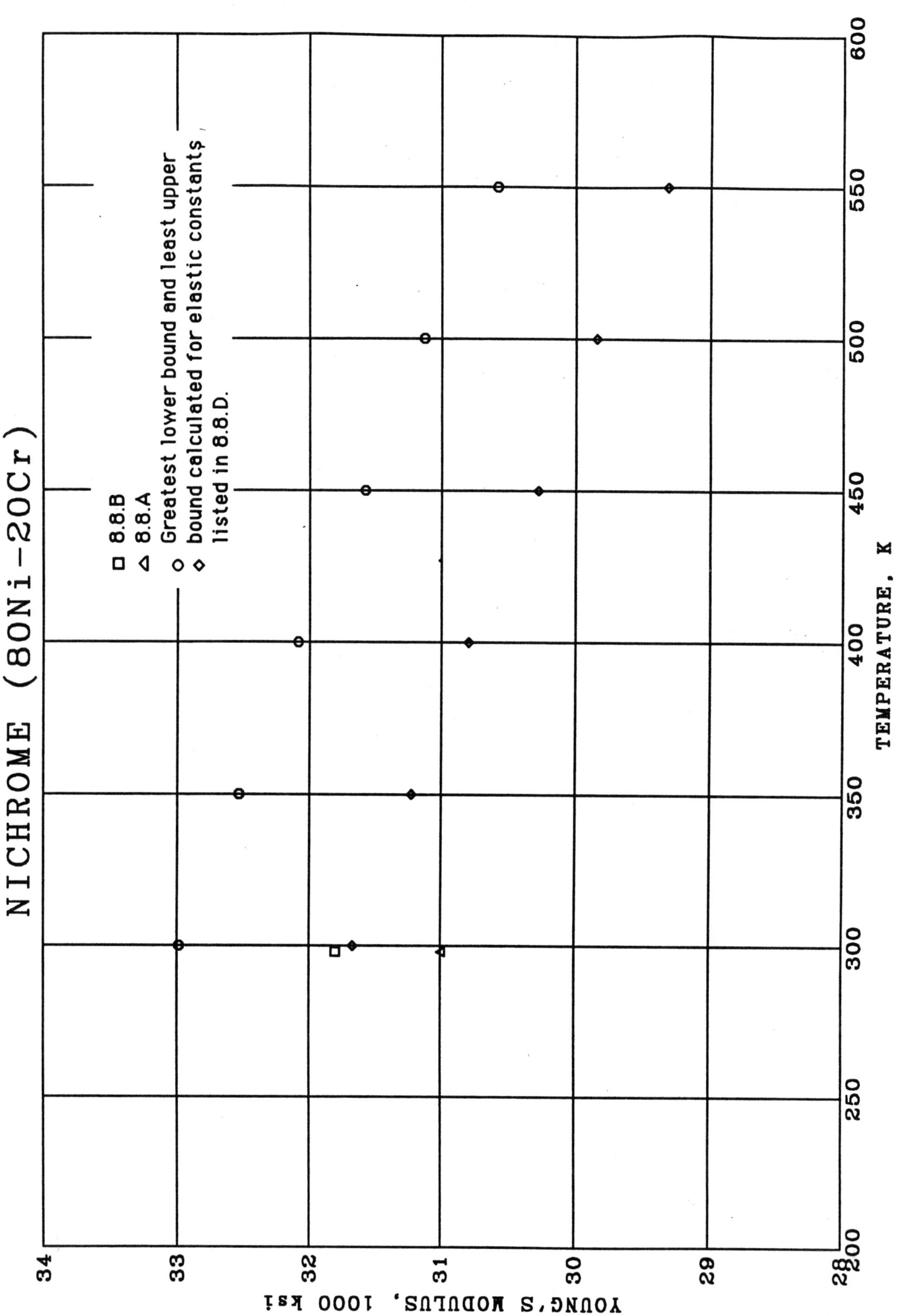
NICHROME (80Ni-20Cr)
□ 8.8.B
△ 8.8.A
○ ◇ Greatest lower bound and least upper bound calculated for elastic constants listed in 8.8.D.
YOUNG'S MODULUS, 1000 ksi
TEMPERATURE, K
34
33
32
31
30
29
28
200
250
300
350
400
450
500
550
600

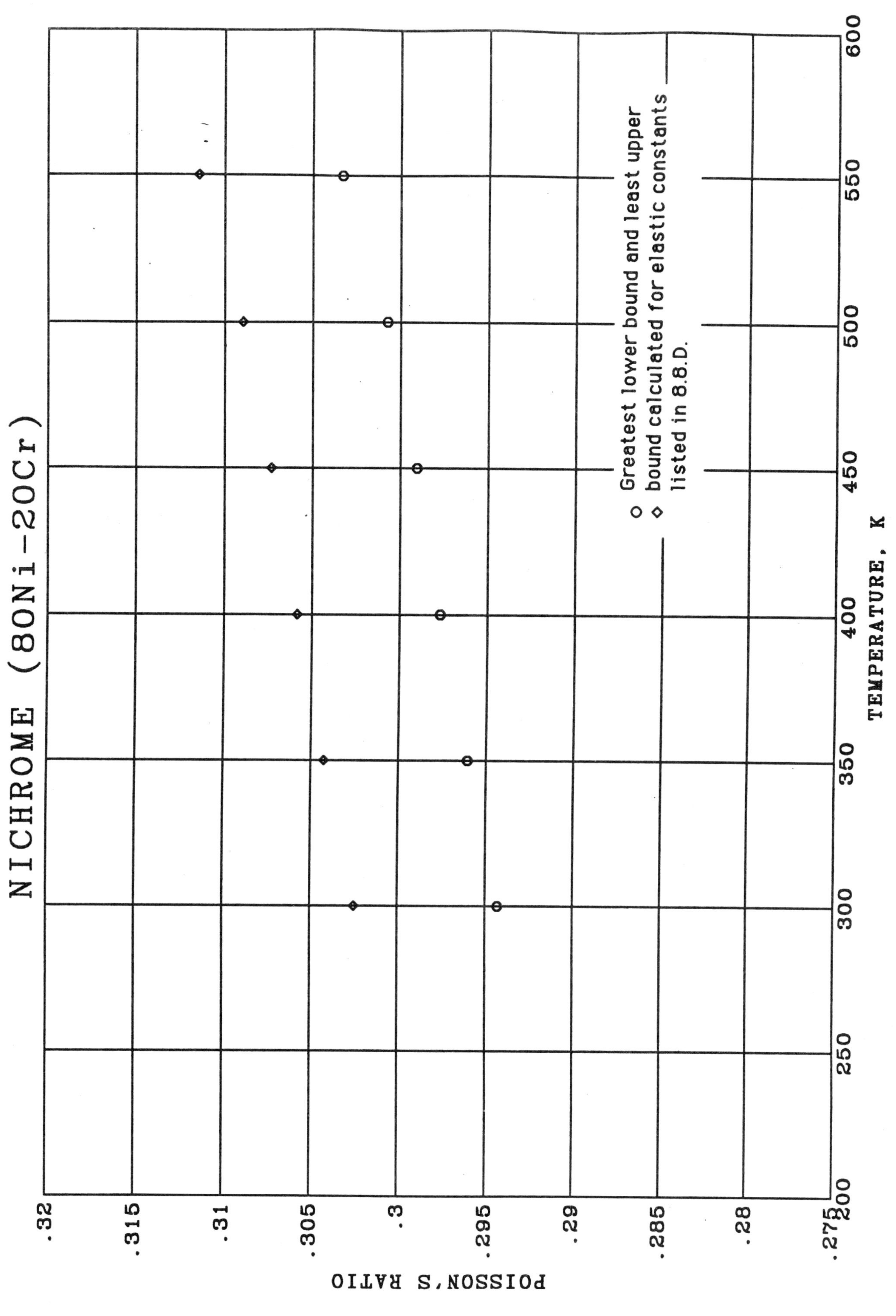
NICHROME (80Ni-20Cr)
Greatest lower bound and least upper bound calculated for elastic constants listed in 8.B.D.
TEMPERATURE, K
POISSON'S RATIO
200
250
300
350
400
450
500
550
600
.32
.315
.31
.305
.3
.295
.29
.285
.28
.275

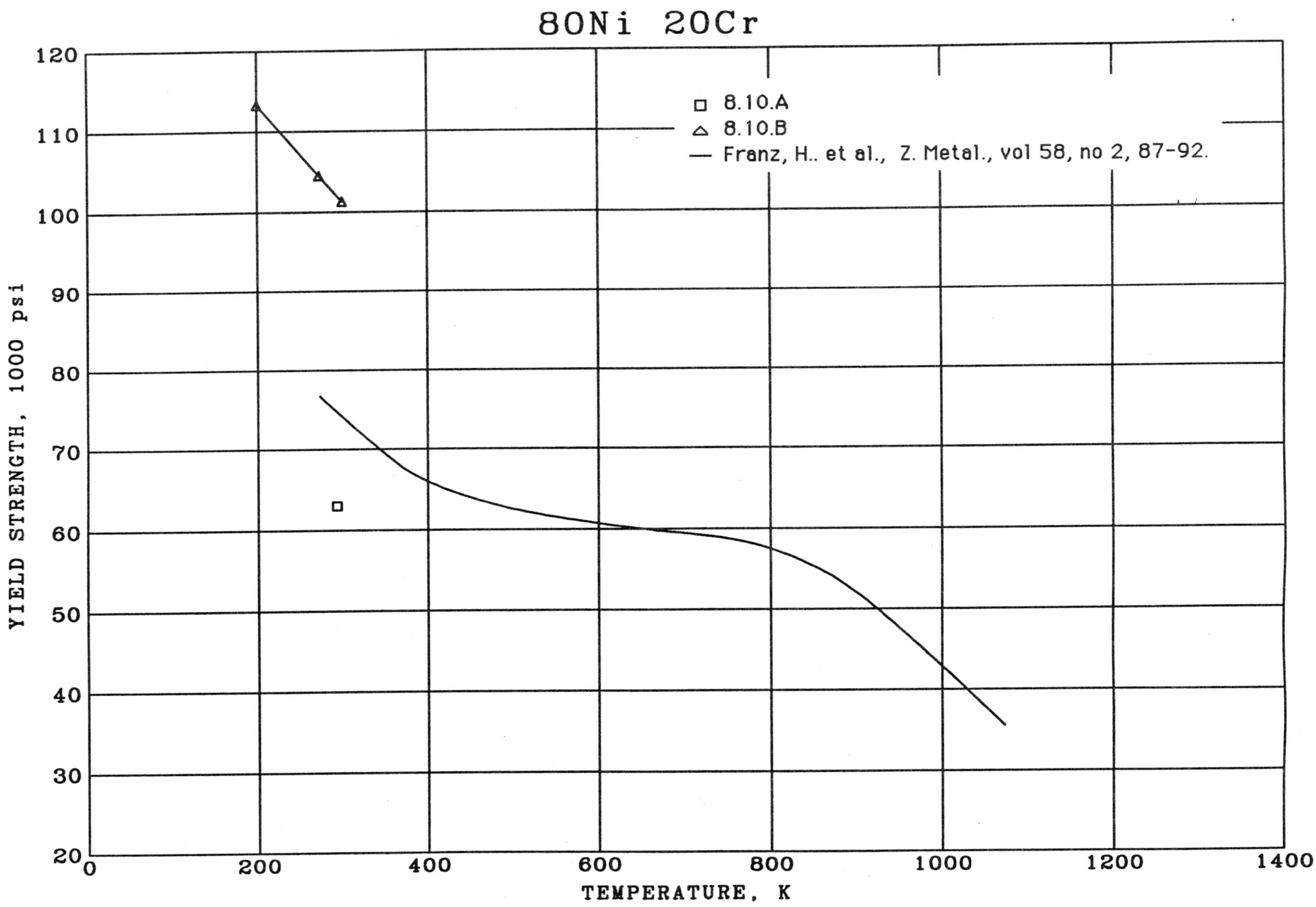
80Ni 20Cr
□ 8.10.A
△ 8.10.B
— Franz, H.. et al., Z. Metal., vol 58, no 2, 87-92.
YIELD STRENGTH, 1000 psi
120
110
100
90
80
70
60
50
40
30
20
0
200
400
600
800
1000
1200
1400
TEMPERATURE, K

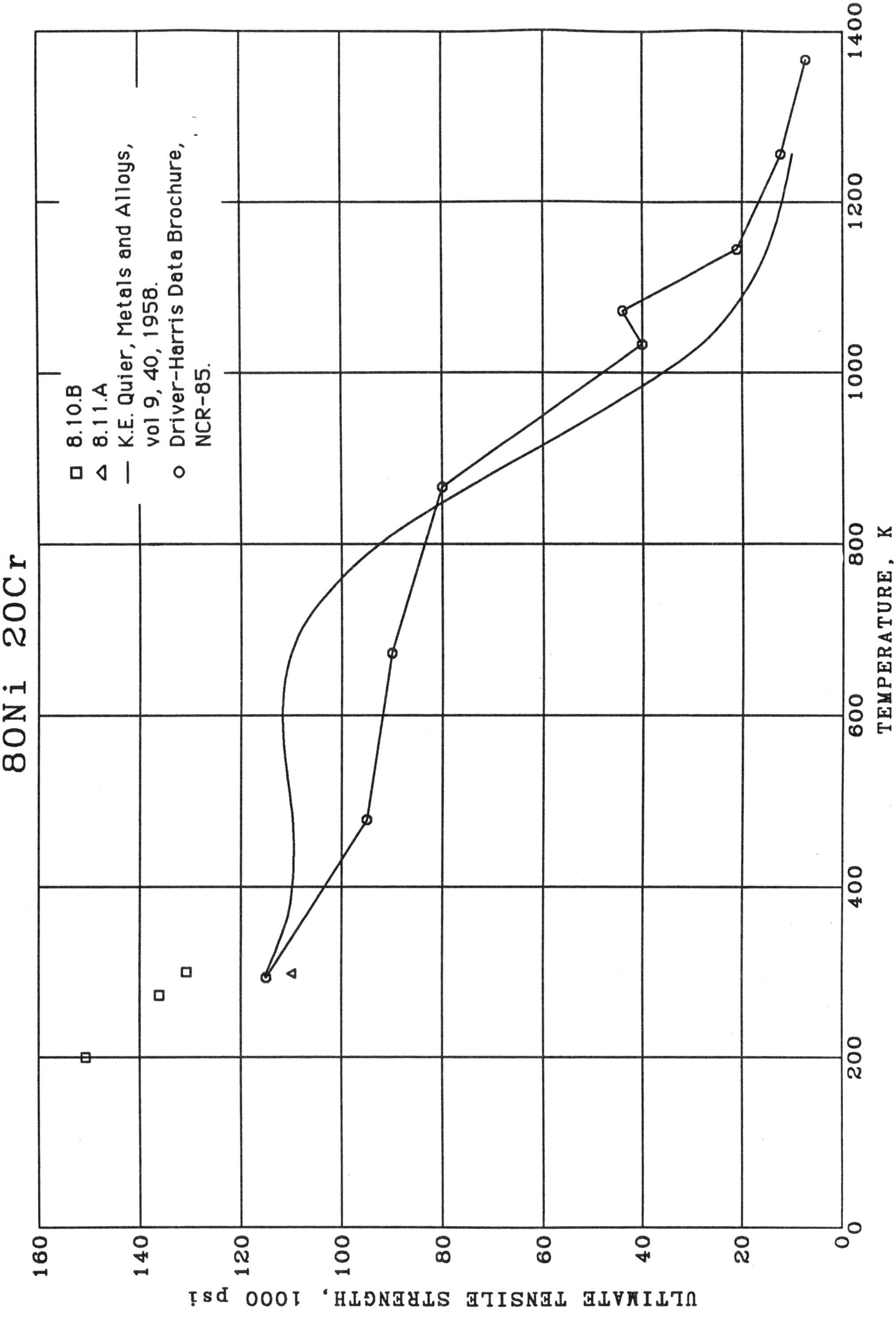
80Ni 20Cr
8.10.B
8.11.A
K.E. Quier, Metals and Alloys, vol 9, 40, 1958.
Driver-Harris Data Brochure, NCR-85.
ULTIMATE TENSILE STRENGTH, 1000 psi
TEMPERATURE, K
0
20
40
60
80
100
120
140
160
200
400
600
800
1000
1200
1400

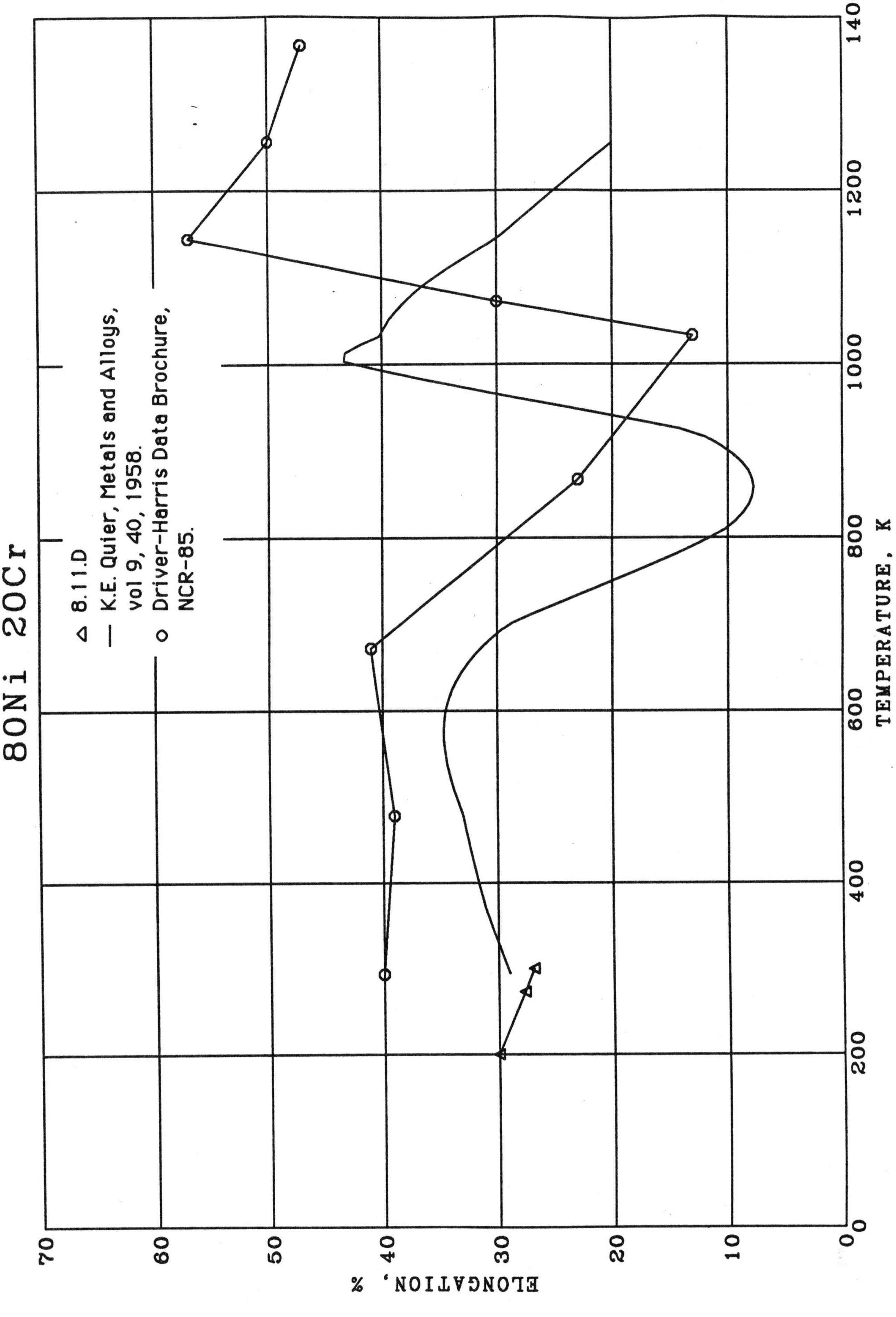
80Ni 20Cr
△ 8.11.D
— K.E. Quier, Metals and Alloys, vol 9, 40, 1958.
○ Driver-Harris Data Brochure, NCR-85.
TEMPERATURE, K
ELONGATION, %
0
200
400
600
800
1000
1200
1400
0
10
20
30
40
50
60
70

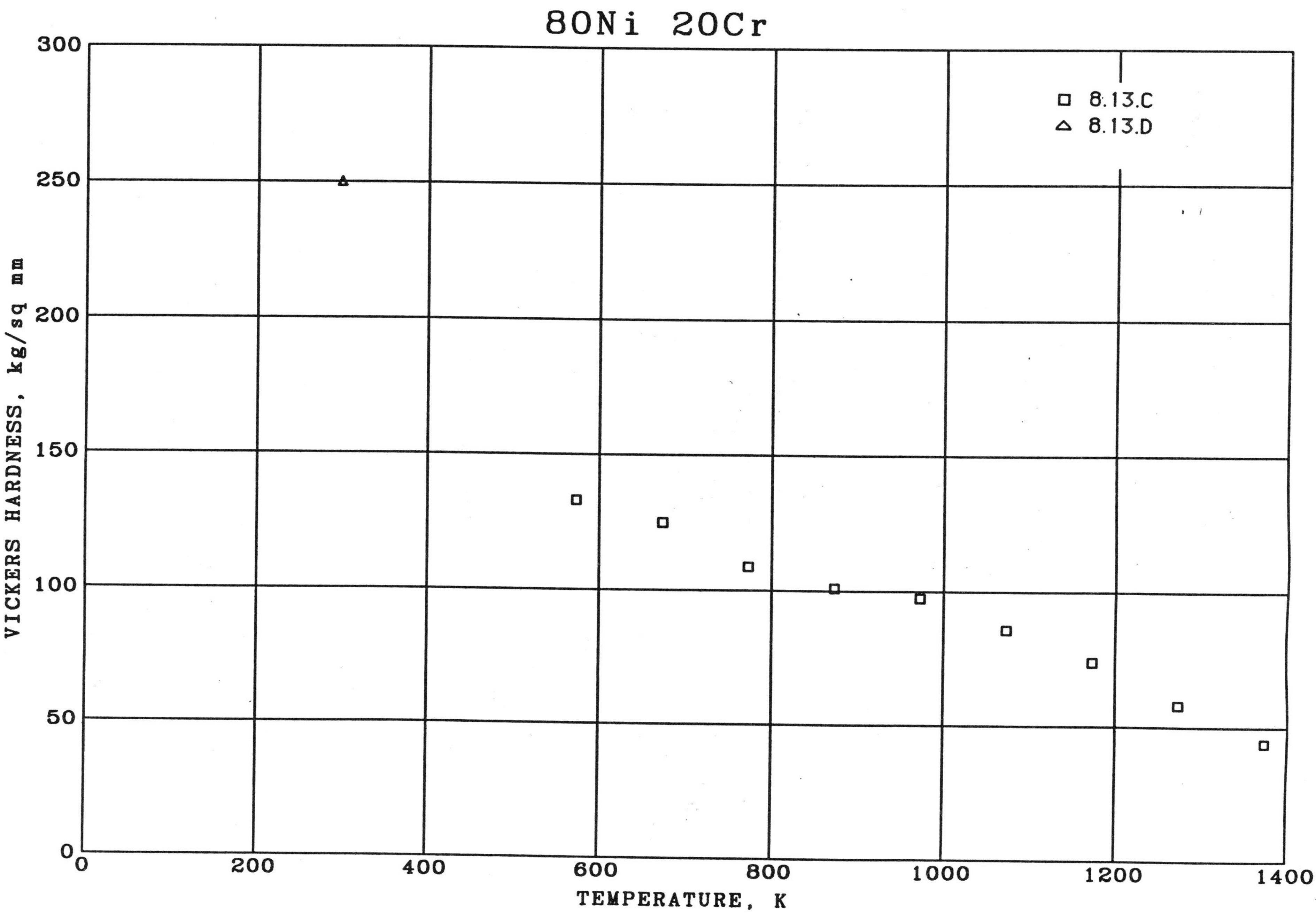
80Ni 20Cr
8.13.C
8.13.D
VICKERS HARDNESS, kg/sq mm
TEMPERATURE, K
0
50
100
150
200
250
300
0
200
400
600
800
1000
1200
1400

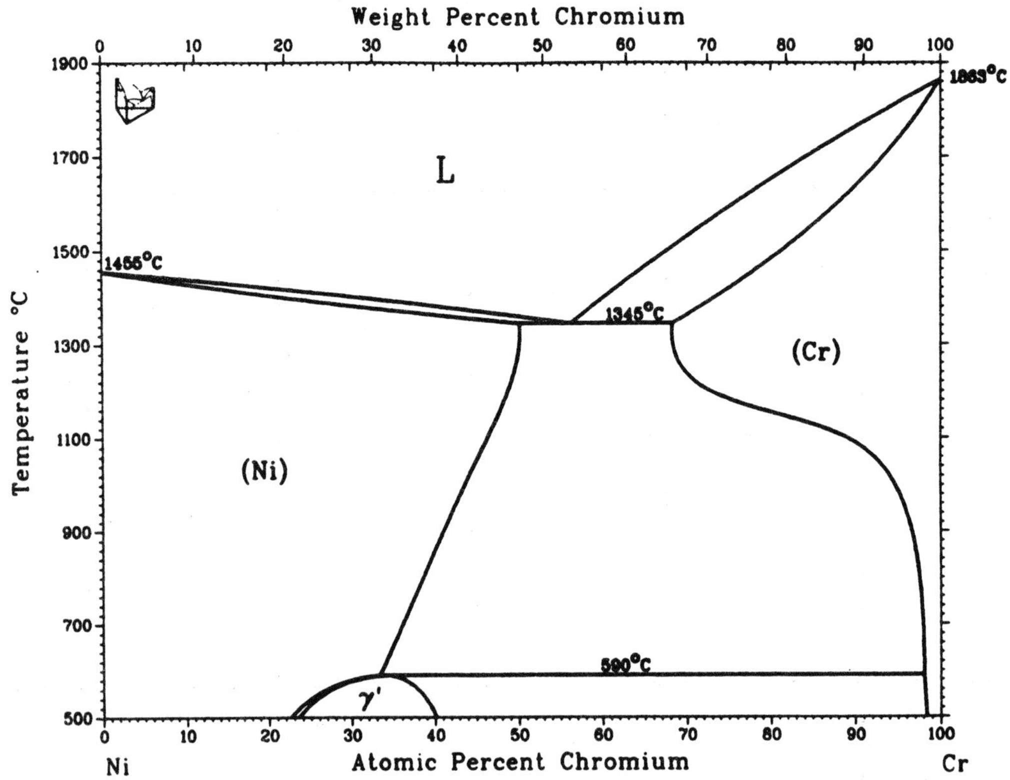
Weight Percent Chromium
Temperature °C
Atomic Percent Chromium
Ni
Cr
L
(Ni)
(Cr)
γ'
1455°C
1345°C
1863°C
590°C

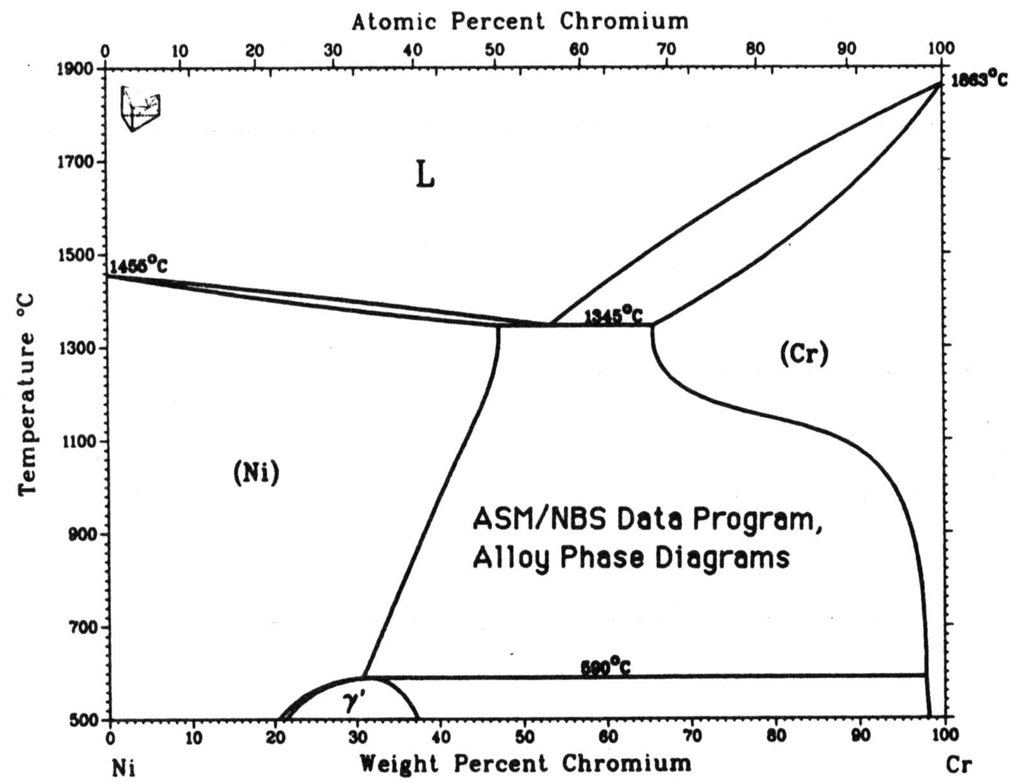
Atomic Percent Chromium
Temperature °C
Weight Percent Chromium
Ni
Cr
L
(Ni)
(Cr)
γ'
1455°C
1345°C
1863°C
590°C
ASM/NBS Data Program,
Alloy Phase Diagrams

42Ni 58Fe

PROPERTY	@ 300K	UNITS	REFERENCES
Density	8.1	g/cm^3	9.1.A-C
Specific Heat	0.121	cal/g-K	9.2.A-B
Thermal Conductivity	0.03	cal/cm-s-K	9.3.A-F
Thermal Expansion	$4x10^{-6}$	1/K	9.4.A-E
Melting Point	1700	K	9.5.A
Resistivity	$60x10^{-6}$	Ohm-cm	9.6.A-G
TCR	2600	ppm/K	9.7.A-C
Young's Modulus	22,600	ksi	9.8.A-C
Poisson's Ratio	0.31	------	9.9.A
Yield Strength	30-50	ksi	9.10.A-C
Ult Tensile Strength	70-80	ksi	9.11.A-C
Elongation at Break	30-40	%	9.12.A-C
Hardness	50	Rockwell B	9.13.A-C

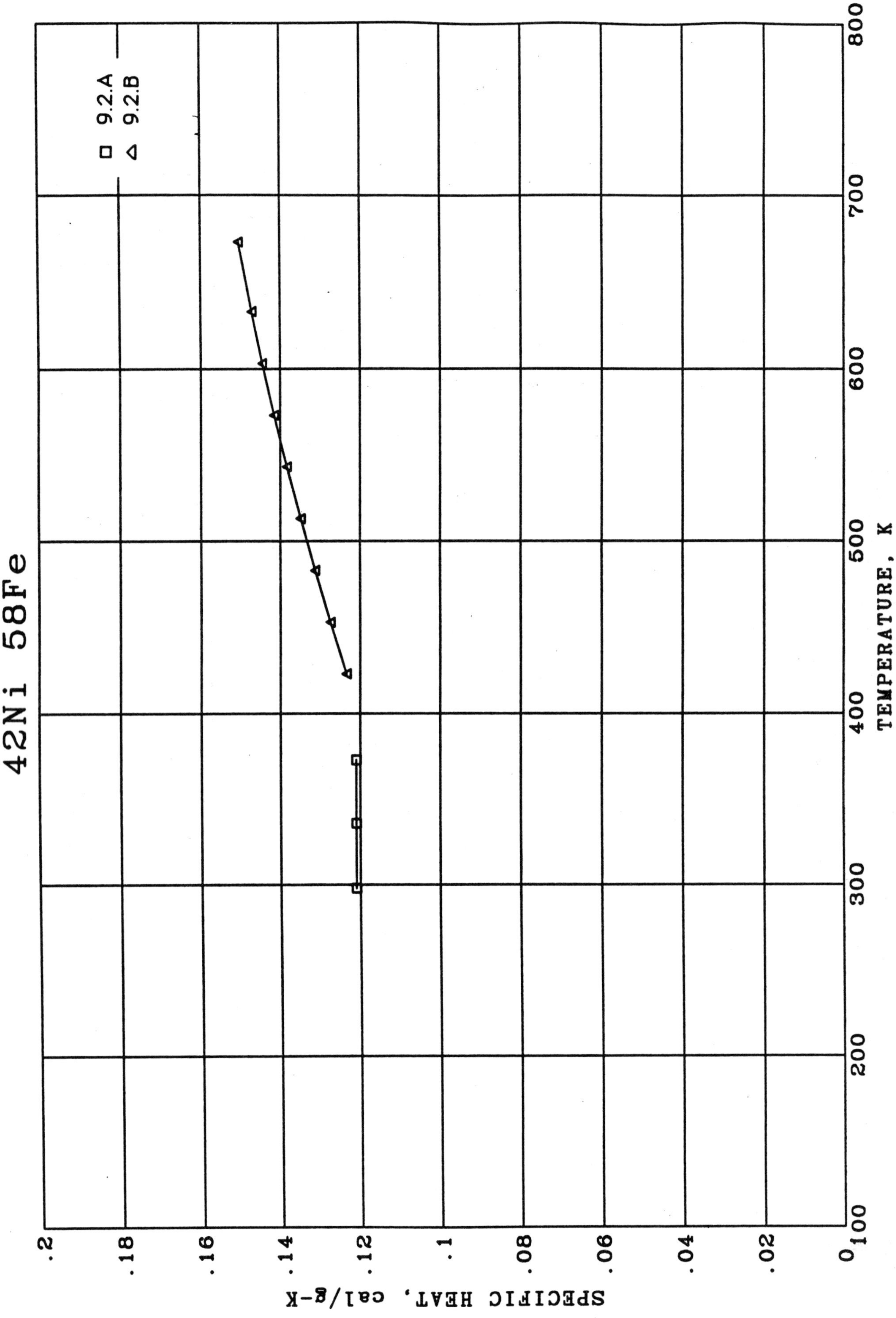
42Ni 58Fe
9.2.A
9.2.B
SPECIFIC HEAT, cal/g-K
TEMPERATURE, K
0
.02
.04
.06
.08
.1
.12
.14
.16
.18
.2
100
200
300
400
500
600
700
800

42Ni 58Fe

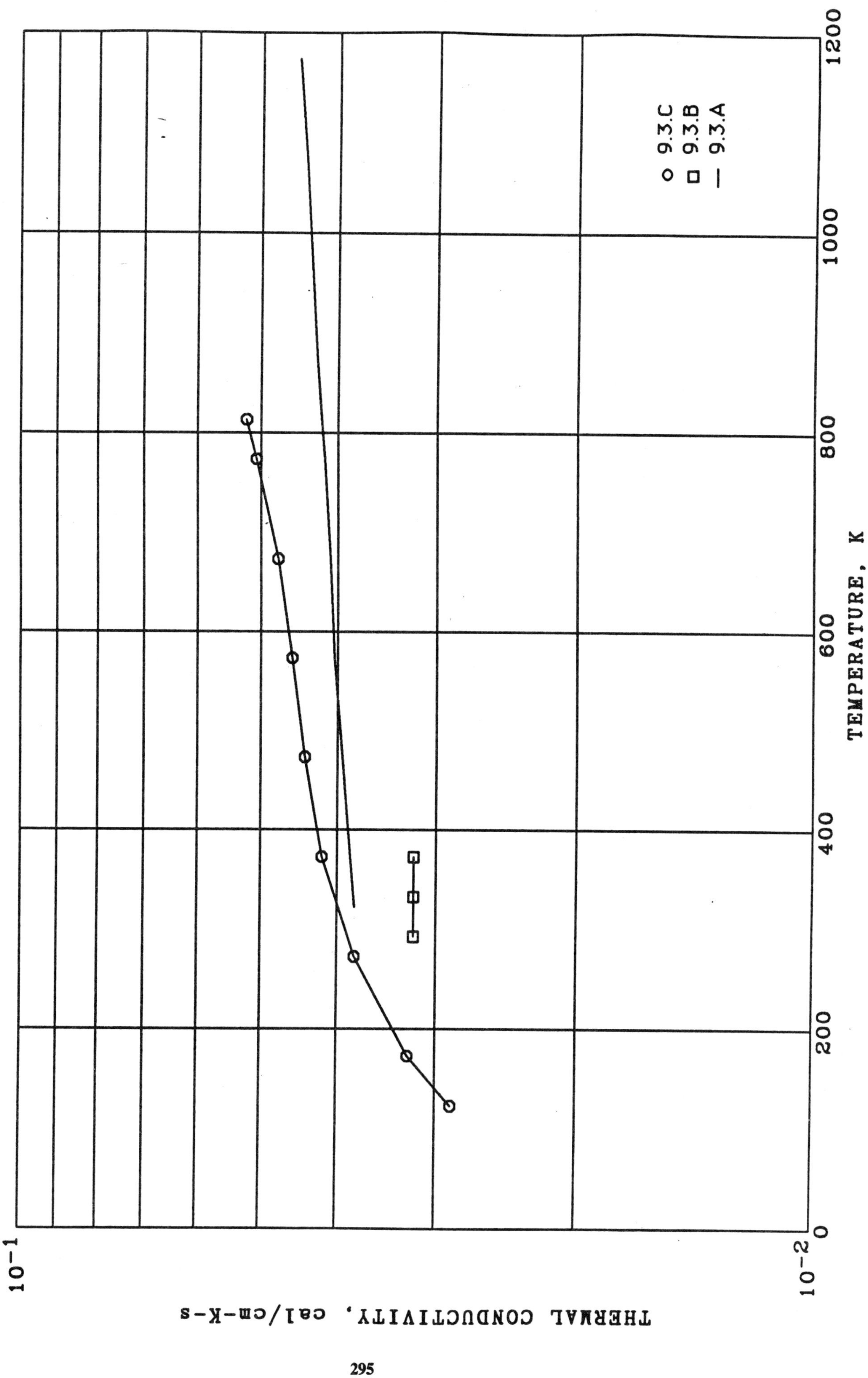

9.3.C
9.3.B
9.3.A
TEMPERATURE, K
0
200
400
600
800
1000
1200
THERMAL CONDUCTIVITY, cal/cm-K-s
10^{-1}
10^{-2}

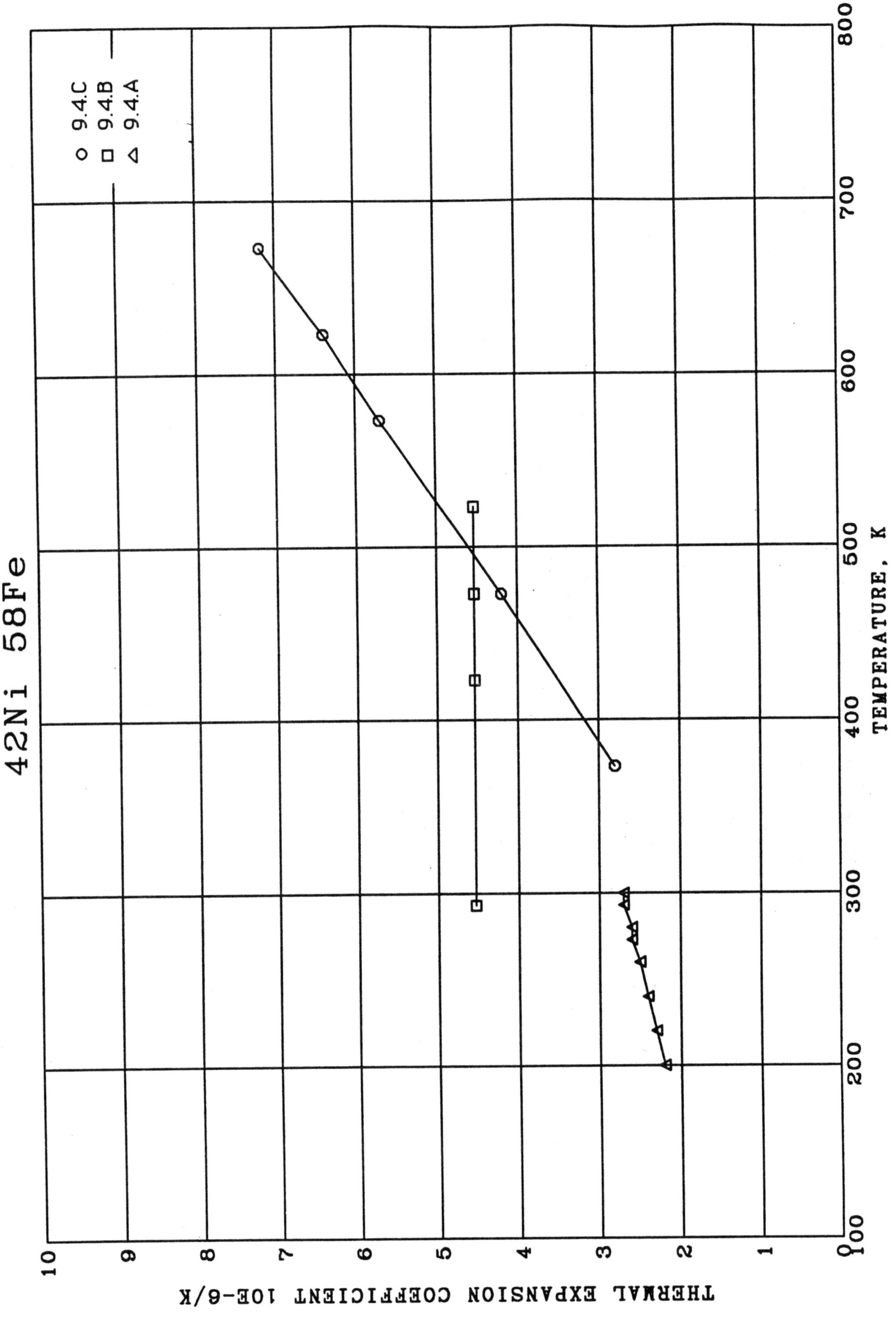

42Ni 58Fe
9.4.C
9.4.B
9.4.A
TEMPERATURE, K
THERMAL EXPANSION COEFFICIENT 10E-6/K

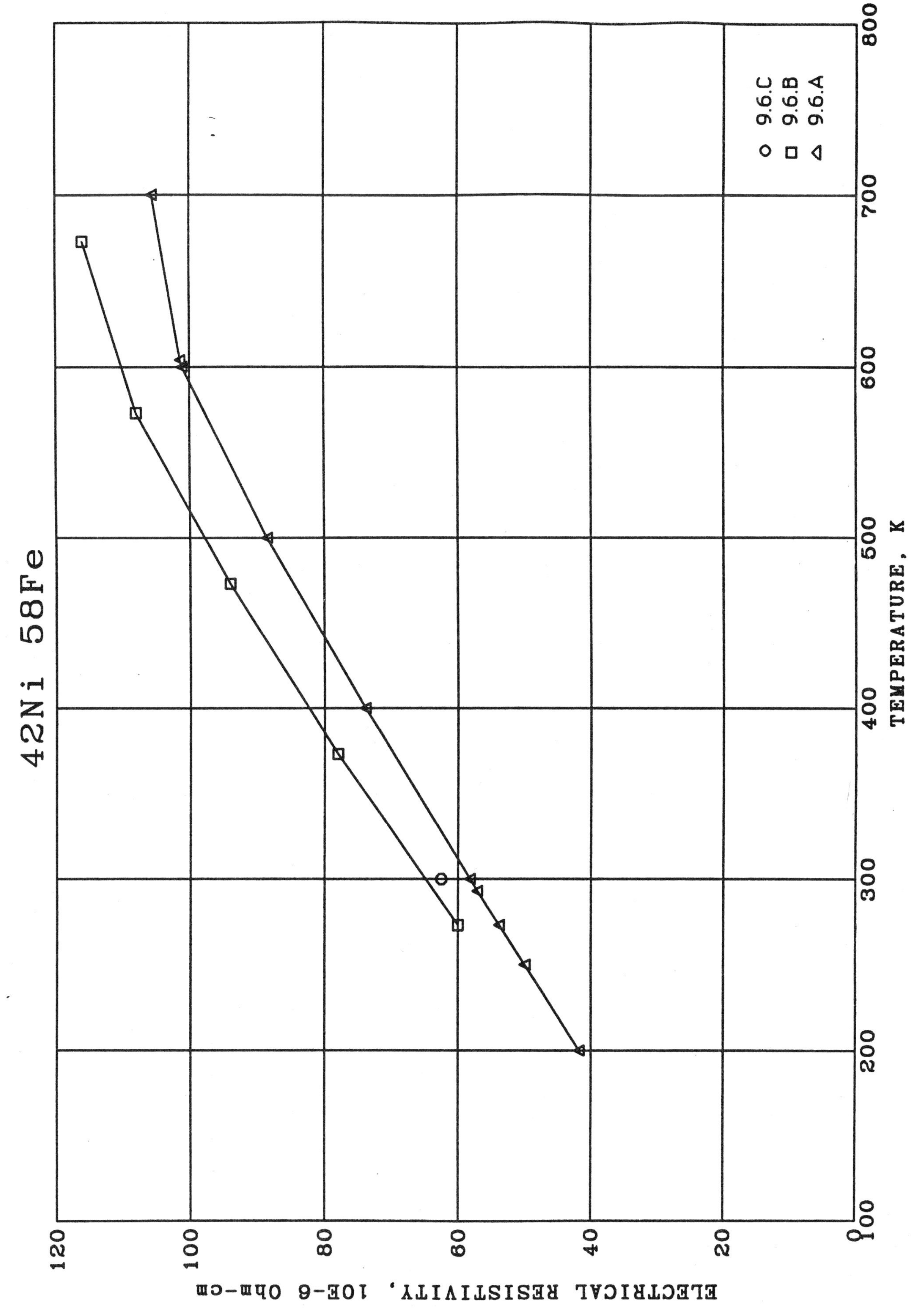
42Ni 58Fe
ELECTRICAL RESISTIVITY, 10E-6 Ohm-cm
TEMPERATURE, K
0
20
40
60
80
100
120
100
200
300
400
500
600
700
800
9.6.C
9.6.B
9.6.A

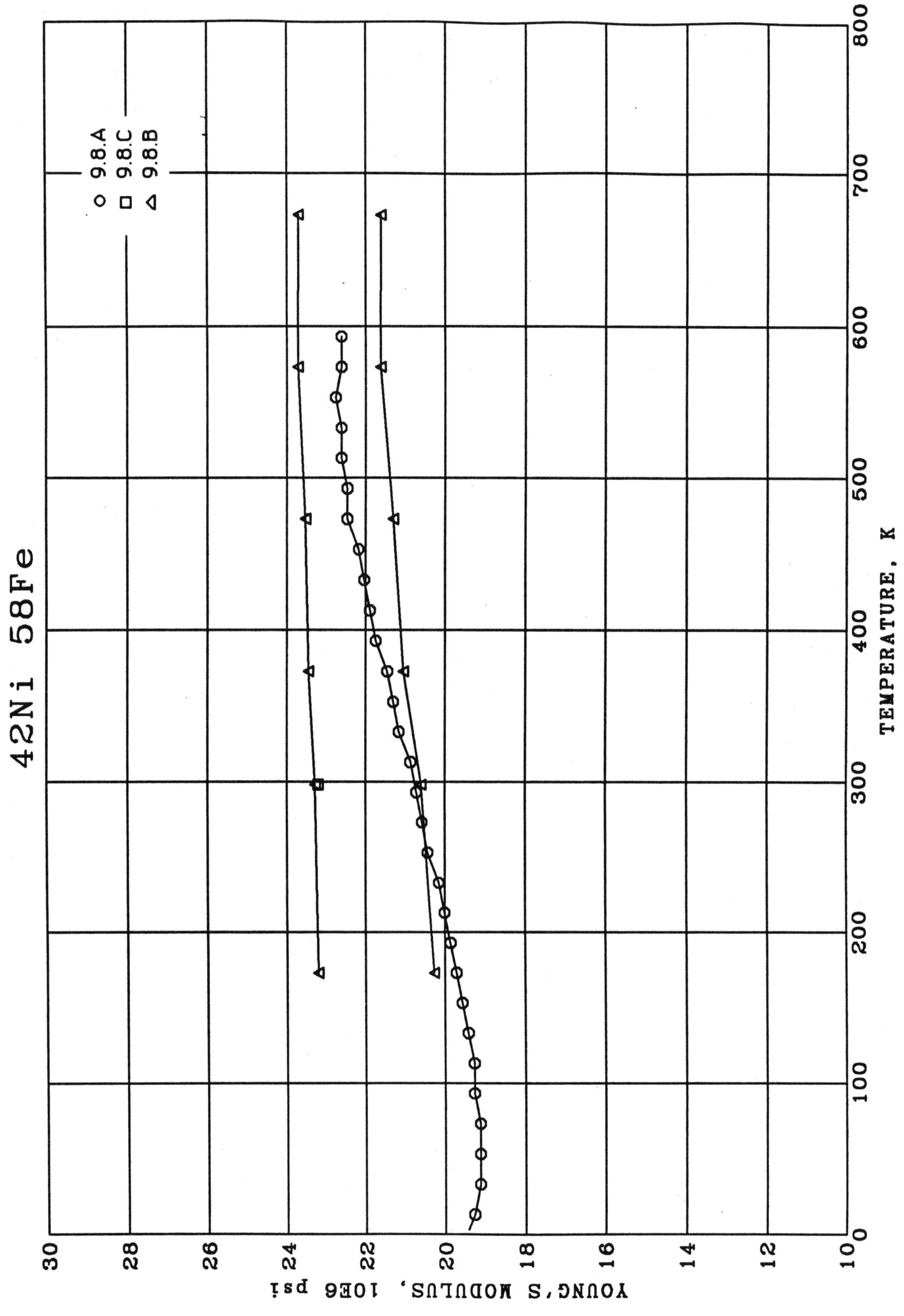
42Ni 58Fe
9.8.A
9.8.C
9.8.B
YOUNG'S MODULUS, 10E6 psi
TEMPERATURE, K
30
28
26
24
22
20
18
16
14
12
10
0
100
200
300
400
500
600
700
800

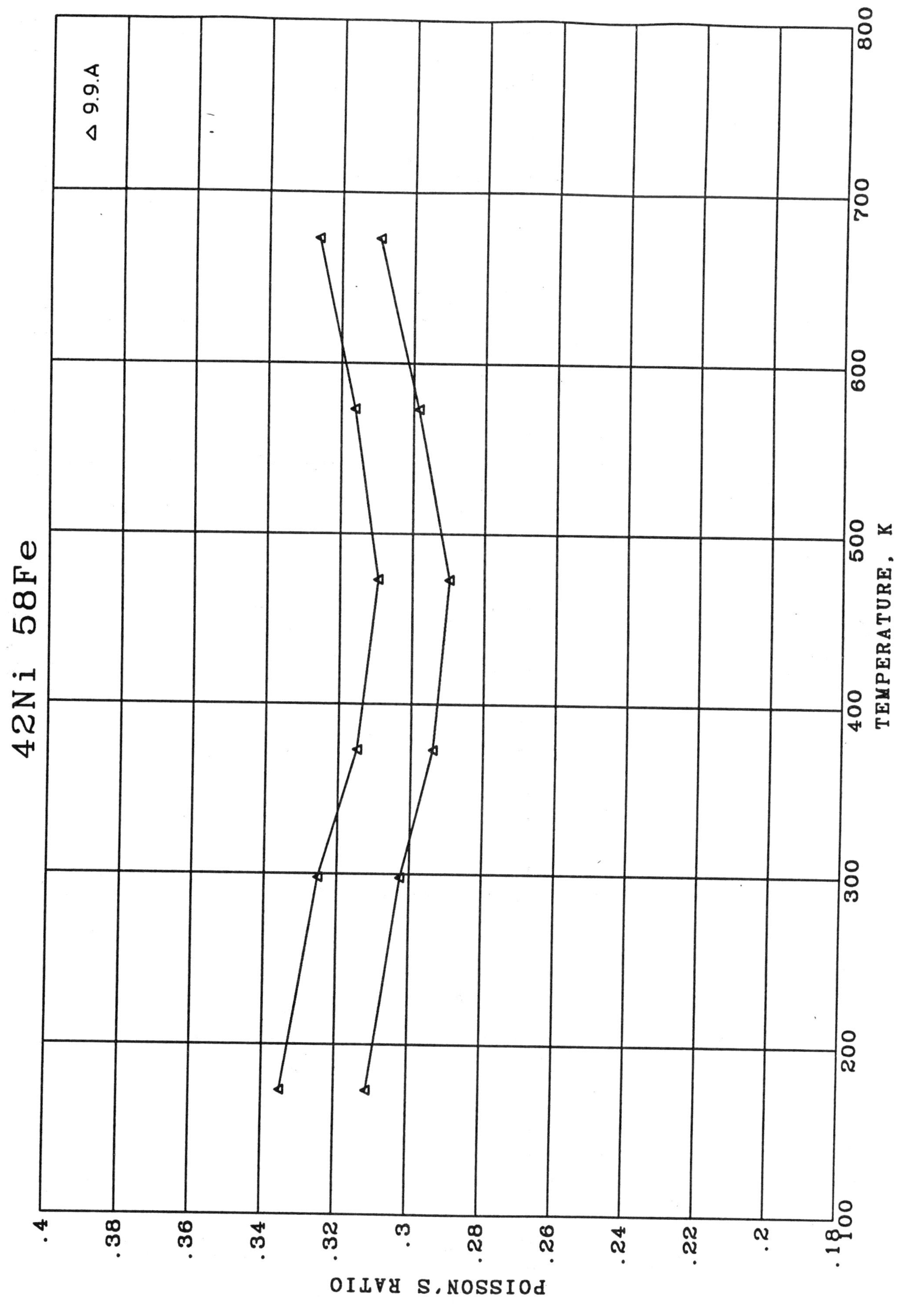
42Ni 58Fe
Δ 9.9.A
POISSON'S RATIO
.4
.38
.36
.34
.32
.3
.28
.26
.24
.22
.2
.18
100
200
300
400
500
600
700
800
TEMPERATURE, K

42Ni 58Fe

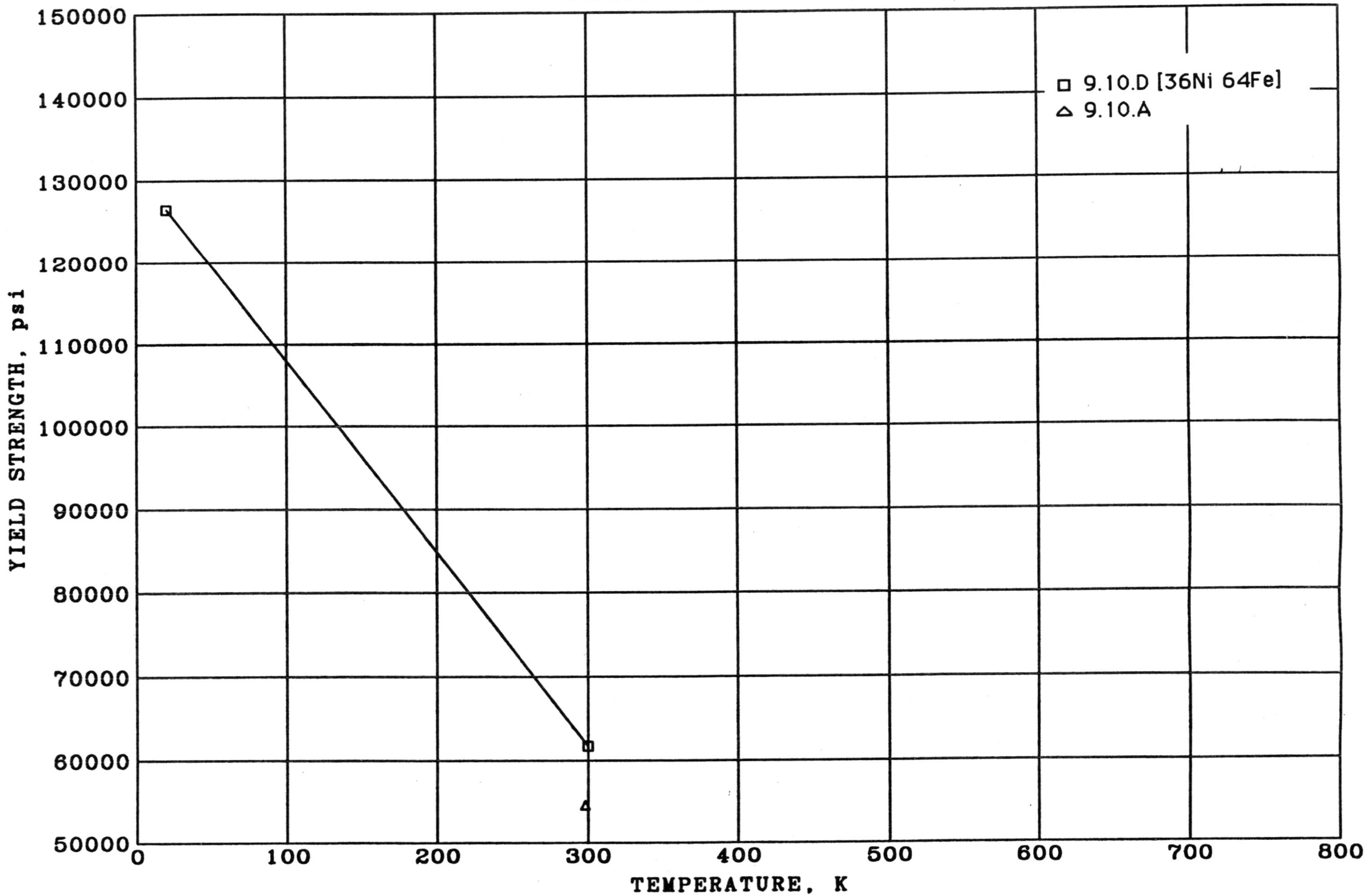

YIELD STRENGTH, psi
150000
140000
130000
120000
110000
100000
90000
80000
70000
60000
50000
0
100
200
300
400
500
600
700
800
TEMPERATURE, K
□ 9.10.D [36Ni 64Fe]
△ 9.10.A

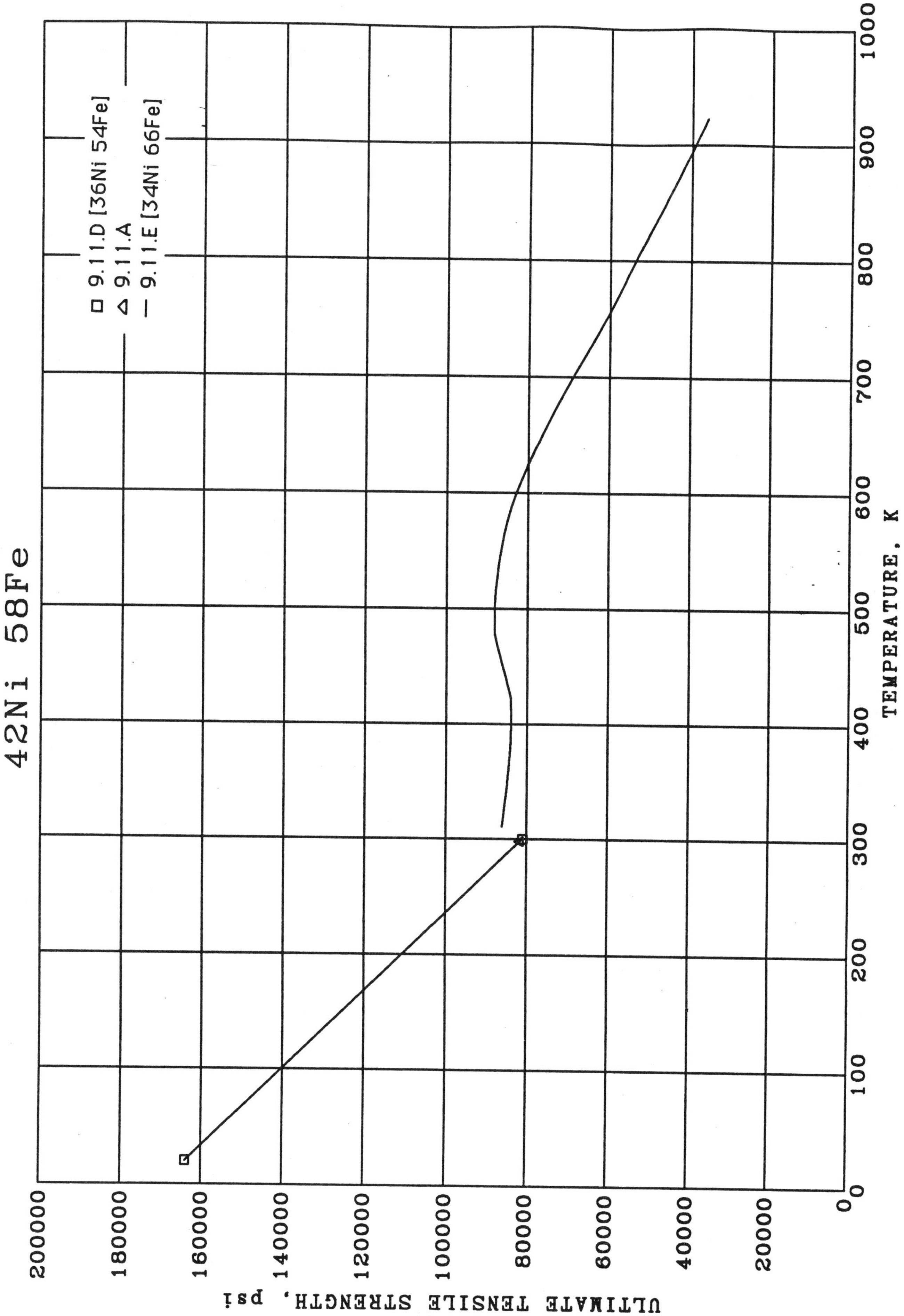

42Ni 58Fe
□ 9.11.D [36Ni 54Fe]
△ 9.11.A
— 9.11.E [34Ni 66Fe]
ULTIMATE TENSILE STRENGTH, psi
200000
180000
160000
140000
120000
100000
80000
60000
40000
20000
0
TEMPERATURE, K
0
100
200
300
400
500
600
700
800
900
1000

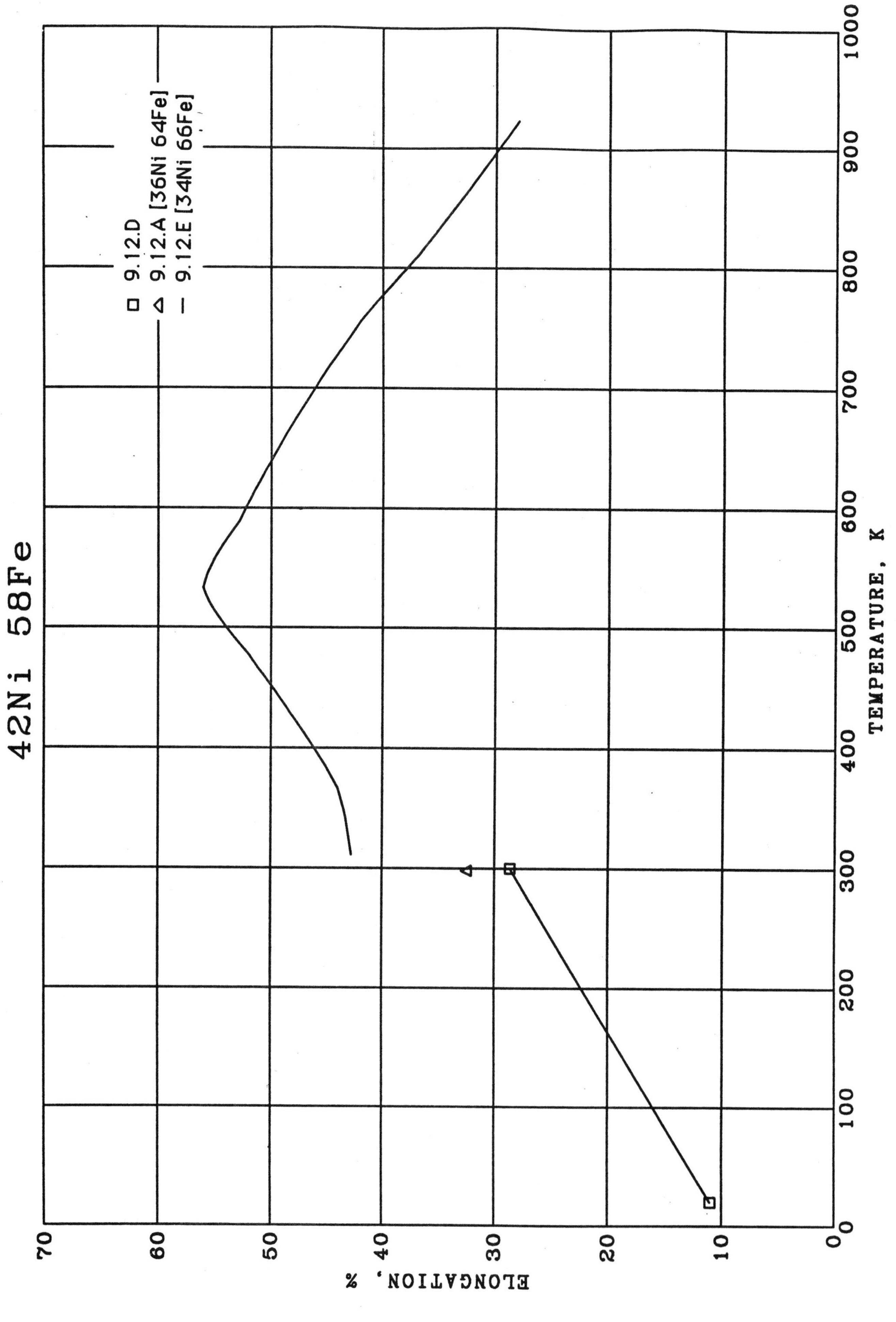
42Ni 58Fe
□ 9.12.D
△ 9.12.A [36Ni 64Fe]
— 9.12.E [34Ni 66Fe]
ELONGATION, %
TEMPERATURE, K
0
10
20
30
40
50
60
70
100
200
300
400
500
600
700
800
900
1000

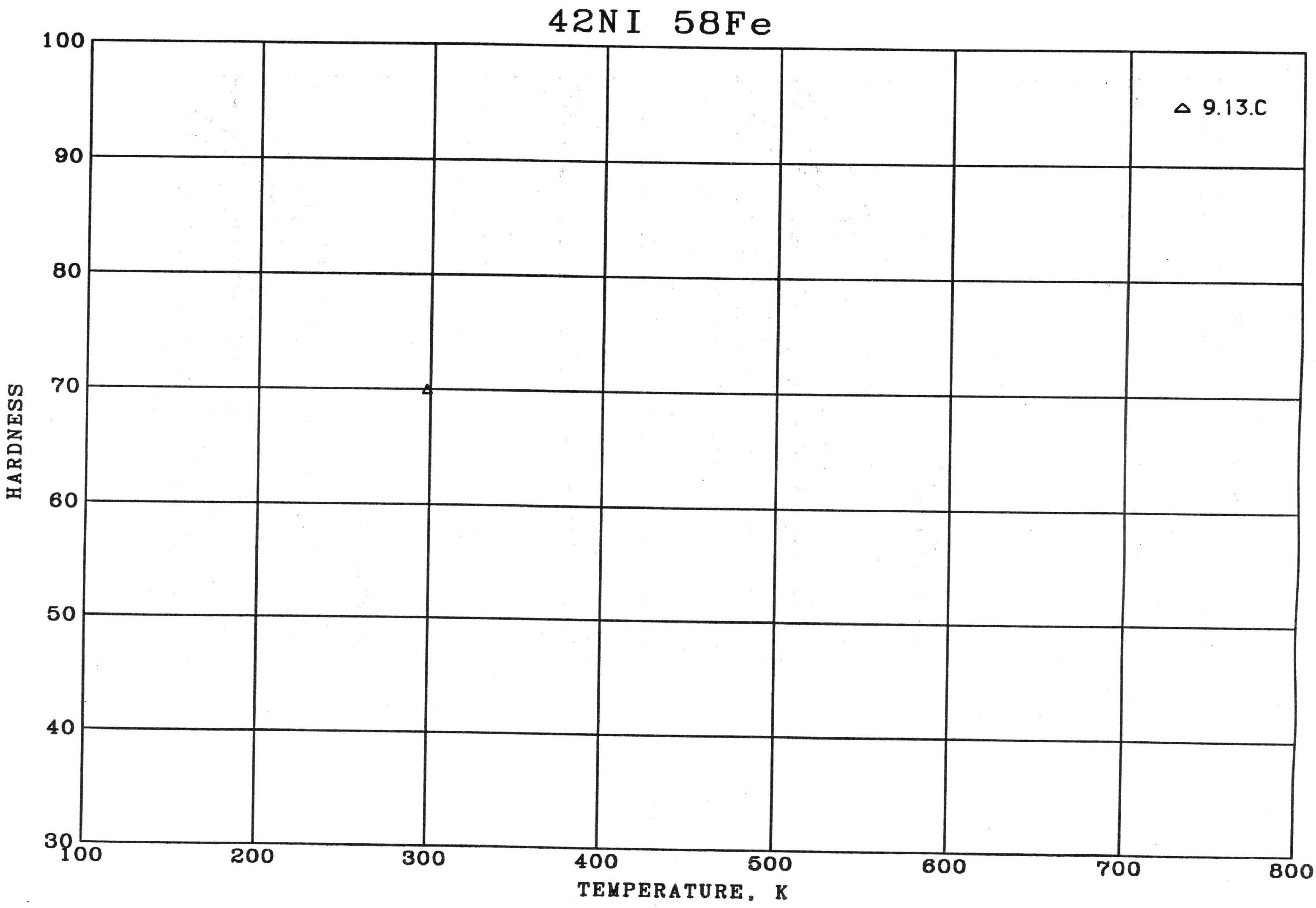
42NI 58Fe
△ 9.13.C
HARDNESS
100
90
80
70
60
50
40
30
100
200
300
400
500
600
700
800
TEMPERATURE, K

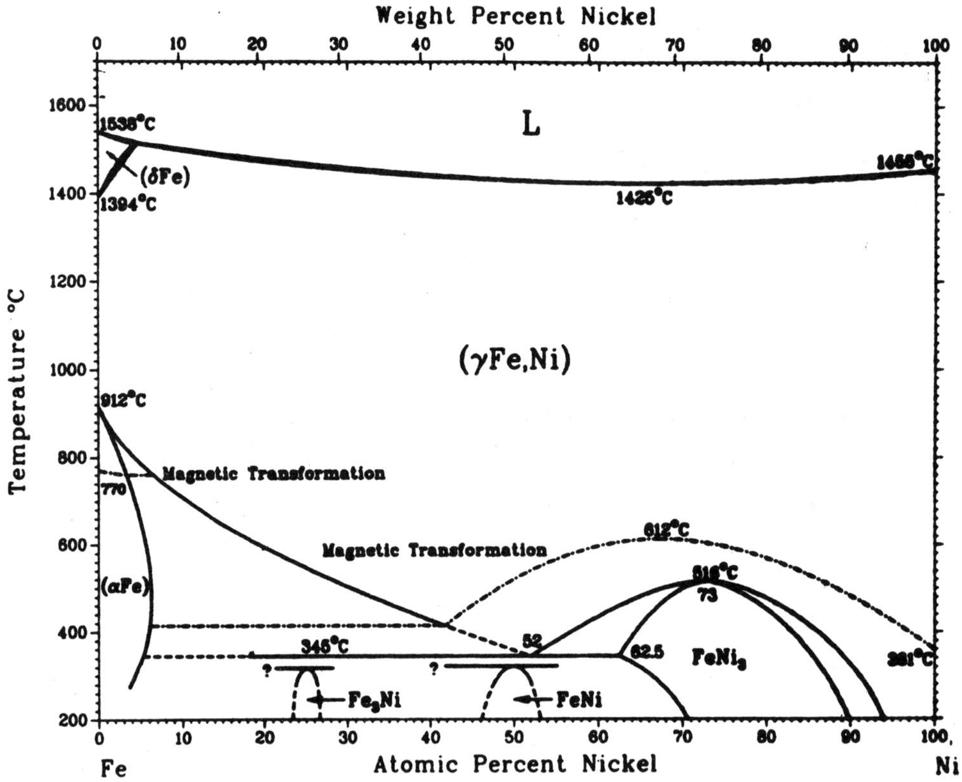

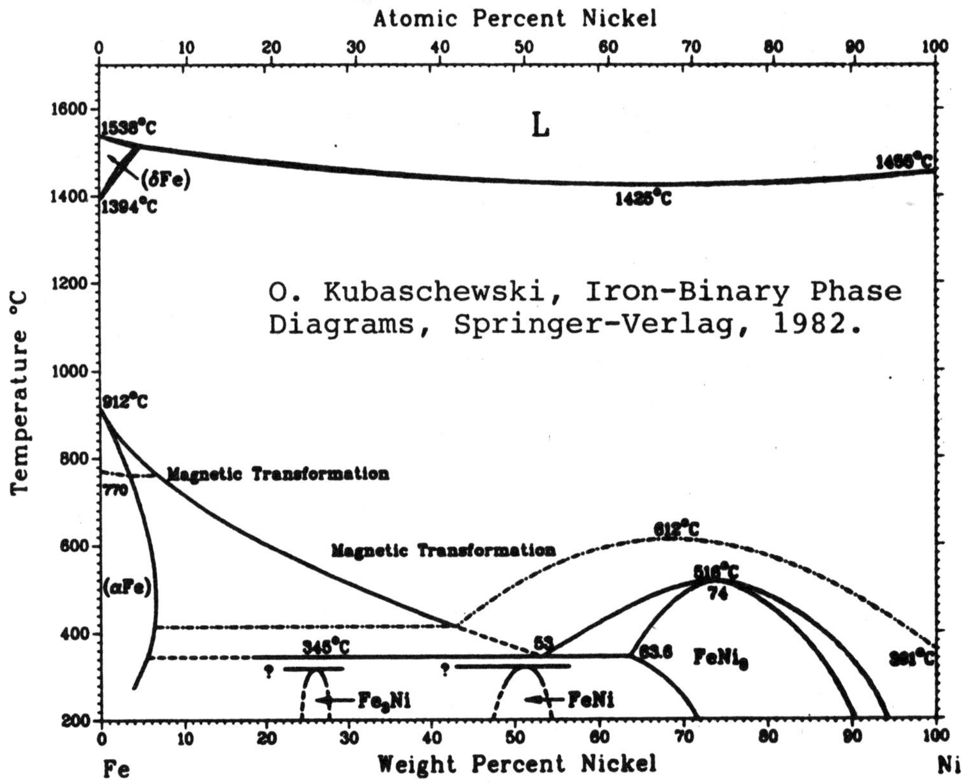

O. Kubaschewski, Iron-Binary Phase Diagrams, Springer-Verlag, 1982.

50Pb 50In

PROPERTY	@ 300K	UNITS	REFERENCES
Density	8.86	g/cm^3	12.1.A-C
Specific Heat	0.047	cal/g-K	12.2.A
Thermal Conductivity	1) 0.0526	cal/cm-s-K	12.3.A
Thermal Expansion	$27x10^{-6}$	1/K	12.4.A
Solidus/Liquidus	453/482	K	12.5.A-C
Resistivity	$28.1x10^{-6}$	Ohm-cm	12.6.A-C
TCR	2000	ppm/K	12.7.A-B
Young's Modulus		ksi	
Poisson's Ratio		------	
Yield Strength		psi	
Ult Tensile Strength	4000	psi	12.11.A-C
Elongation at Break	55	%	12.12.A
Hardness	9.60	Brinell	12.13.A-B

1 Thermal conductivity is for 85C.

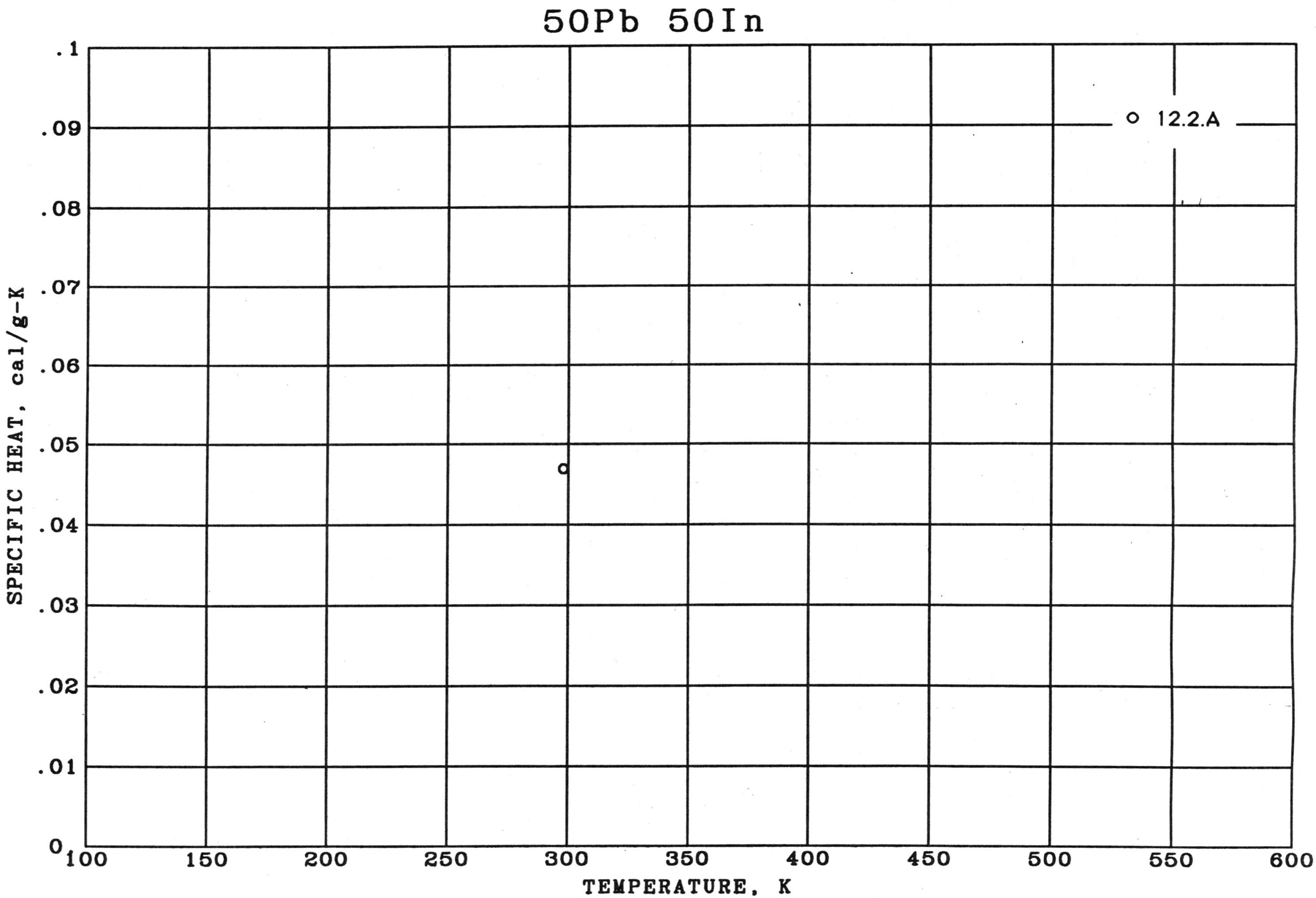
50Pb 50In
SPECIFIC HEAT, cal/g-K
.1
.09
.08
.07
.06
.05
.04
.03
.02
.01
0
100
150
200
250
300
350
400
450
500
550
600
TEMPERATURE, K
12.2.A

50Pb 50In

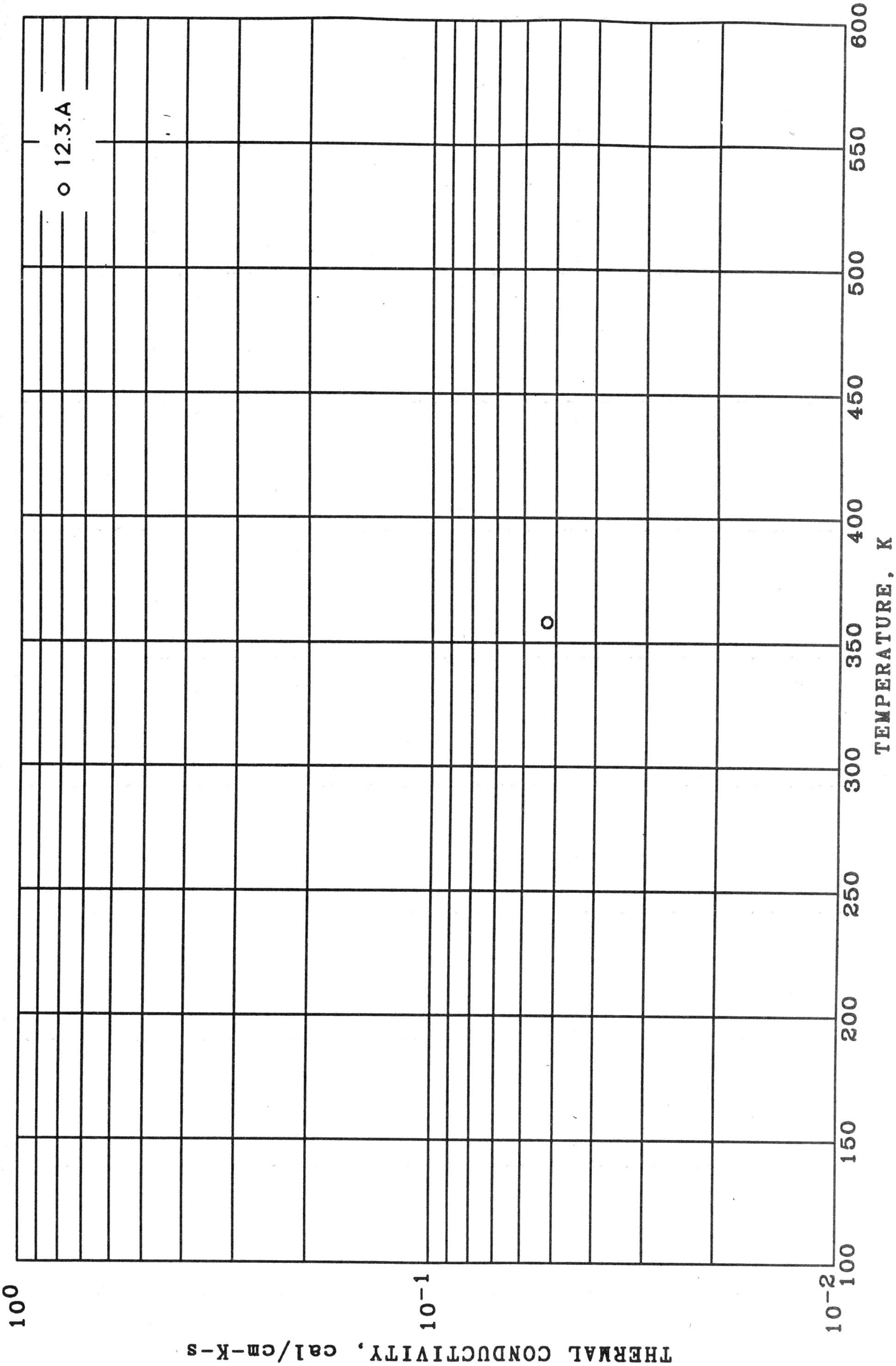

○ 12.3.A
10^0
10^{-1}
10^{-2}
THERMAL CONDUCTIVITY, cal/cm-K-s
100
150
200
250
300
350
400
450
500
550
600
TEMPERATURE, K

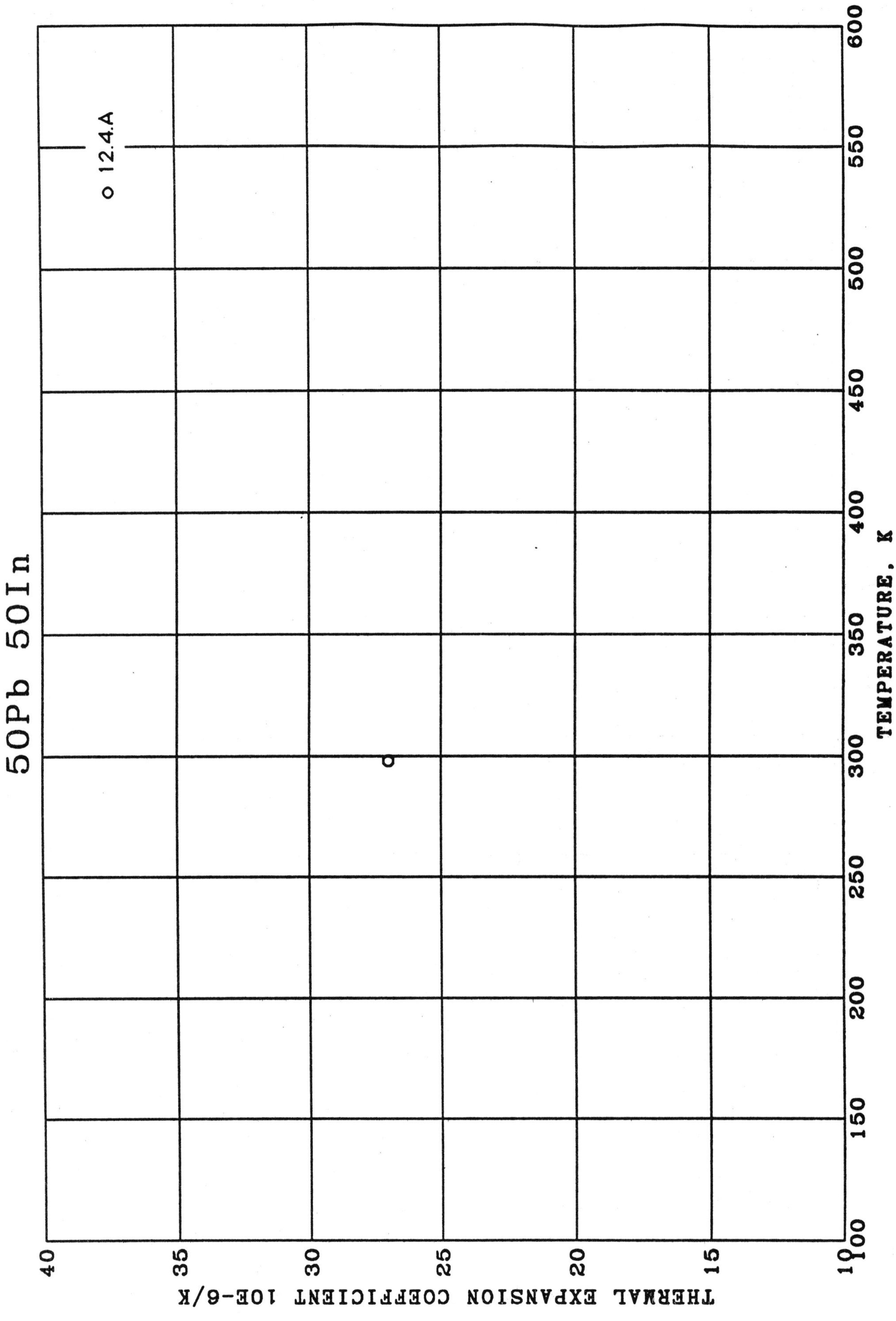
50Pb 50In
o 12.4.A
TEMPERATURE, K
100
150
200
250
300
350
400
450
500
550
600
THERMAL EXPANSION COEFFICIENT 10E-6/K
10
15
20
25
30
35
40

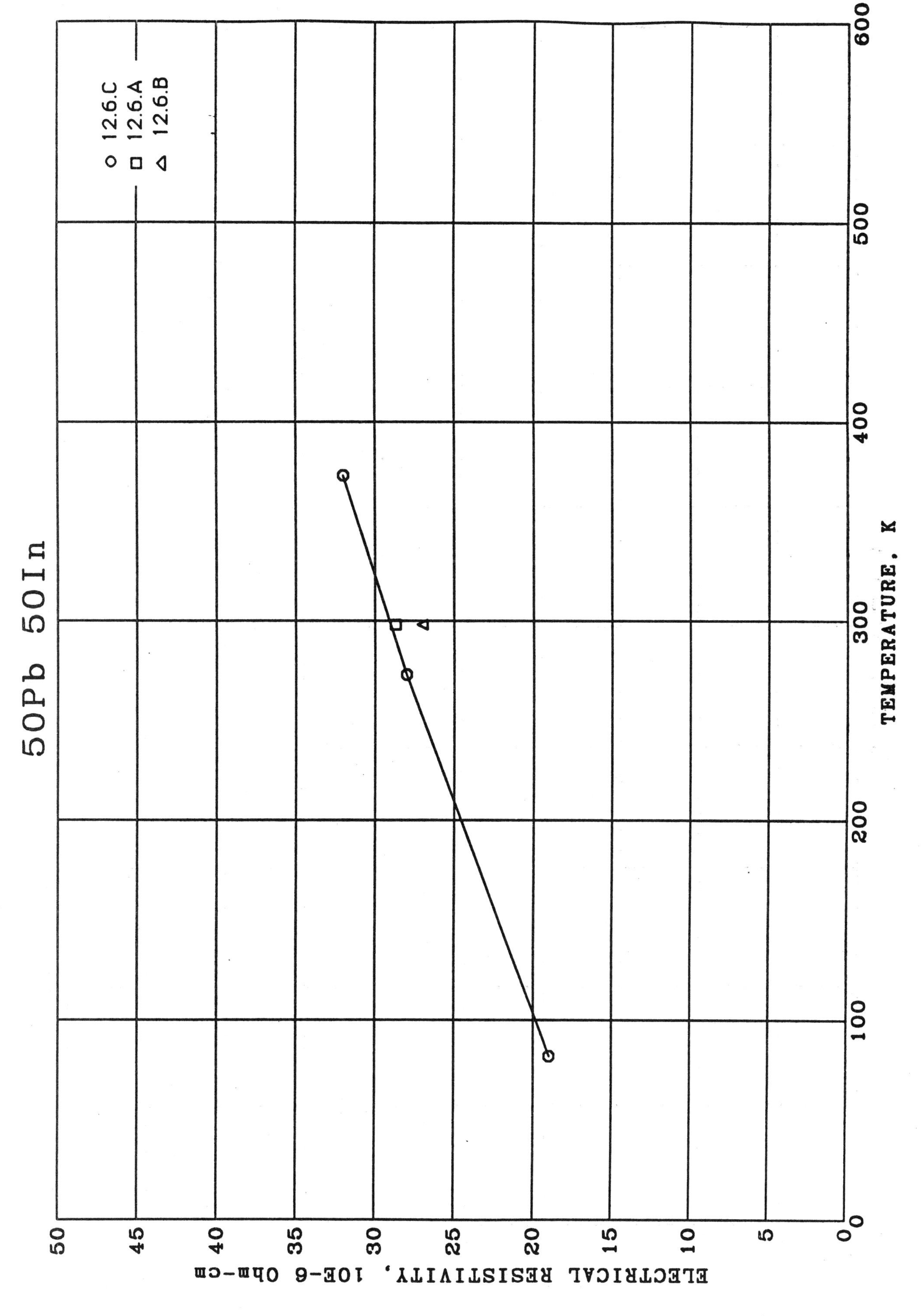
50Pb 50In
○ 12.6.C
□ 12.6.A
△ 12.6.B
ELECTRICAL RESISTIVITY, 10E-6 Ohm-cm
TEMPERATURE, K
0
5
10
15
20
25
30
35
40
45
50
100
200
300
400
500
600

50Pb 50In

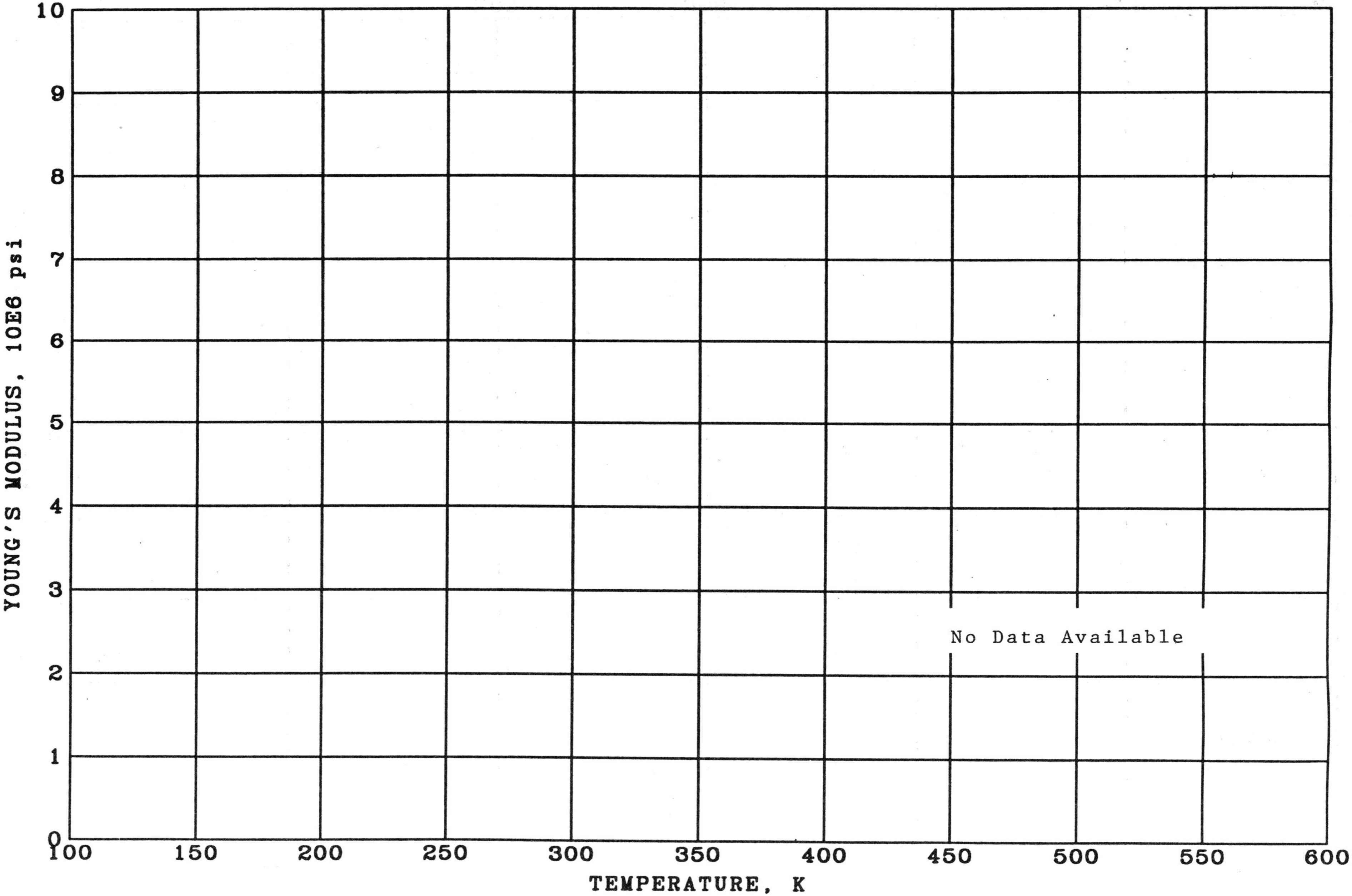
10
9
8
7
6
5
4
3
2
1
0
100
150
200
250
300
350
400
450
500
550
600
YOUNG'S MODULUS, 10E6 psi
TEMPERATURE, K
No Data Available

50Pb 50In

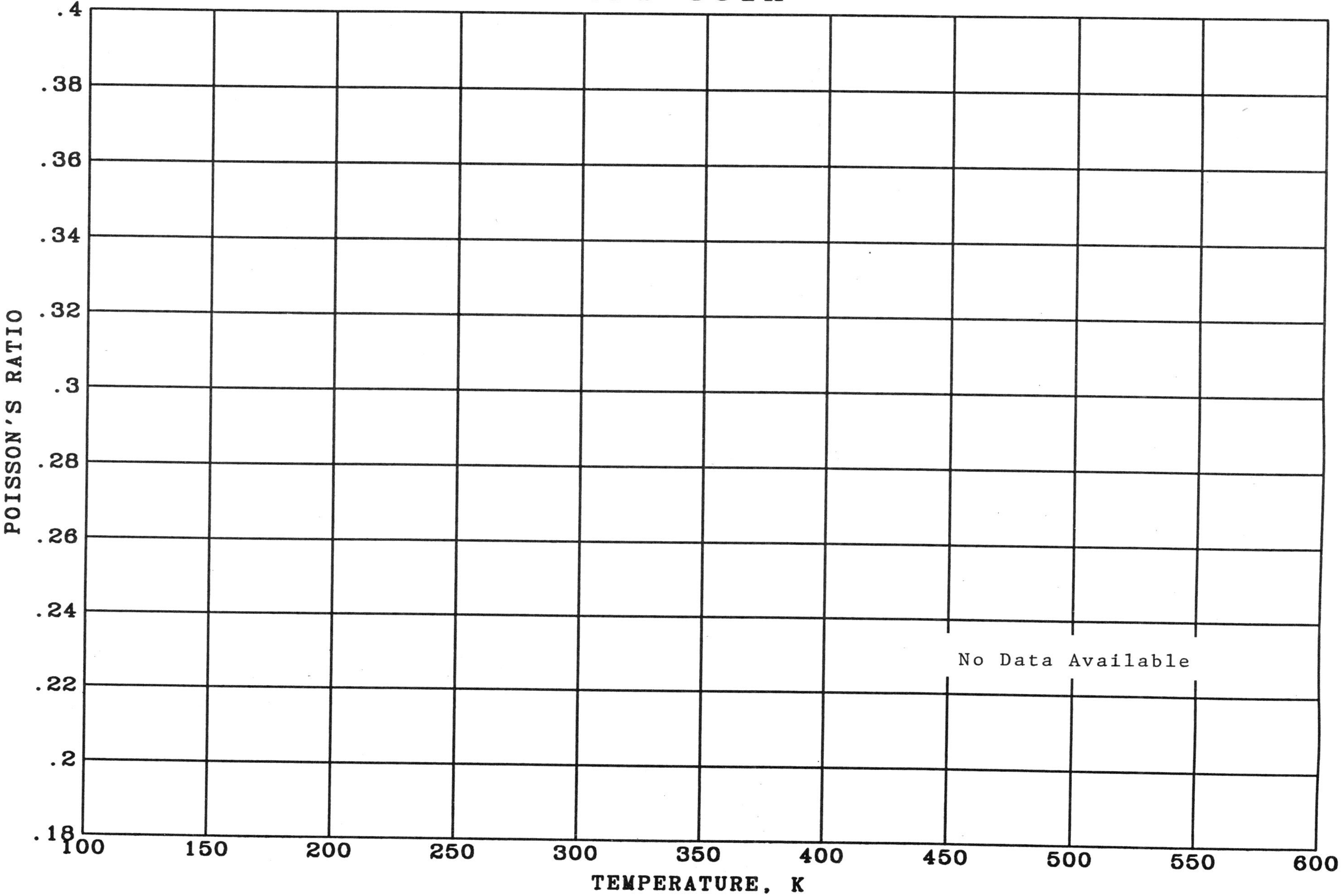
No Data Available
POISSON'S RATIO
.4
.38
.36
.34
.32
.3
.28
.26
.24
.22
.2
.18
100
150
200
250
300
350
400
450
500
550
600
TEMPERATURE, K

50Pb 50In

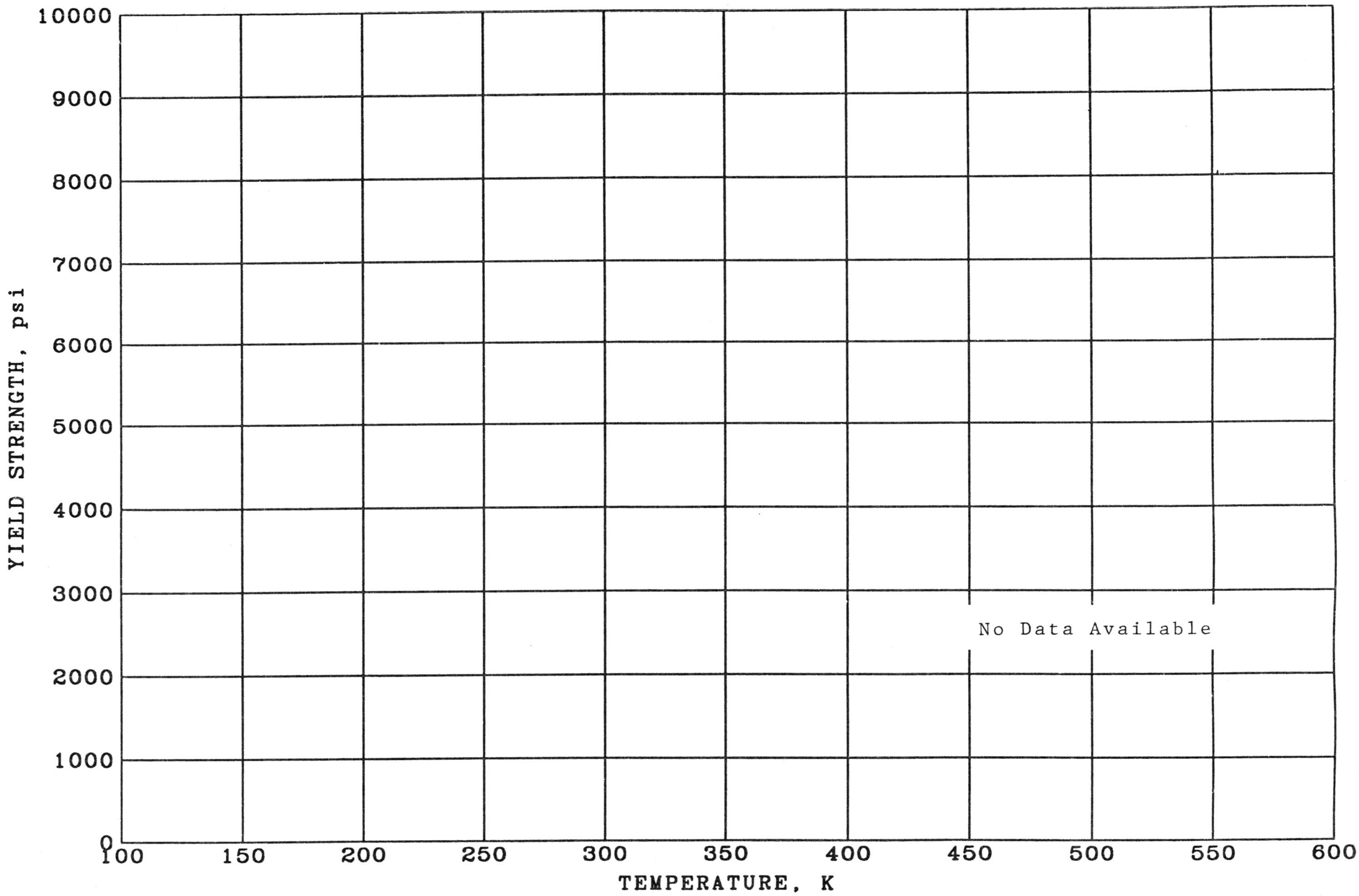
YIELD STRENGTH, psi
10000
9000
8000
7000
6000
5000
4000
3000
2000
1000
0
100
150
200
250
300
350
400
450
500
550
600
TEMPERATURE, K
No Data Available

50Pb 50In

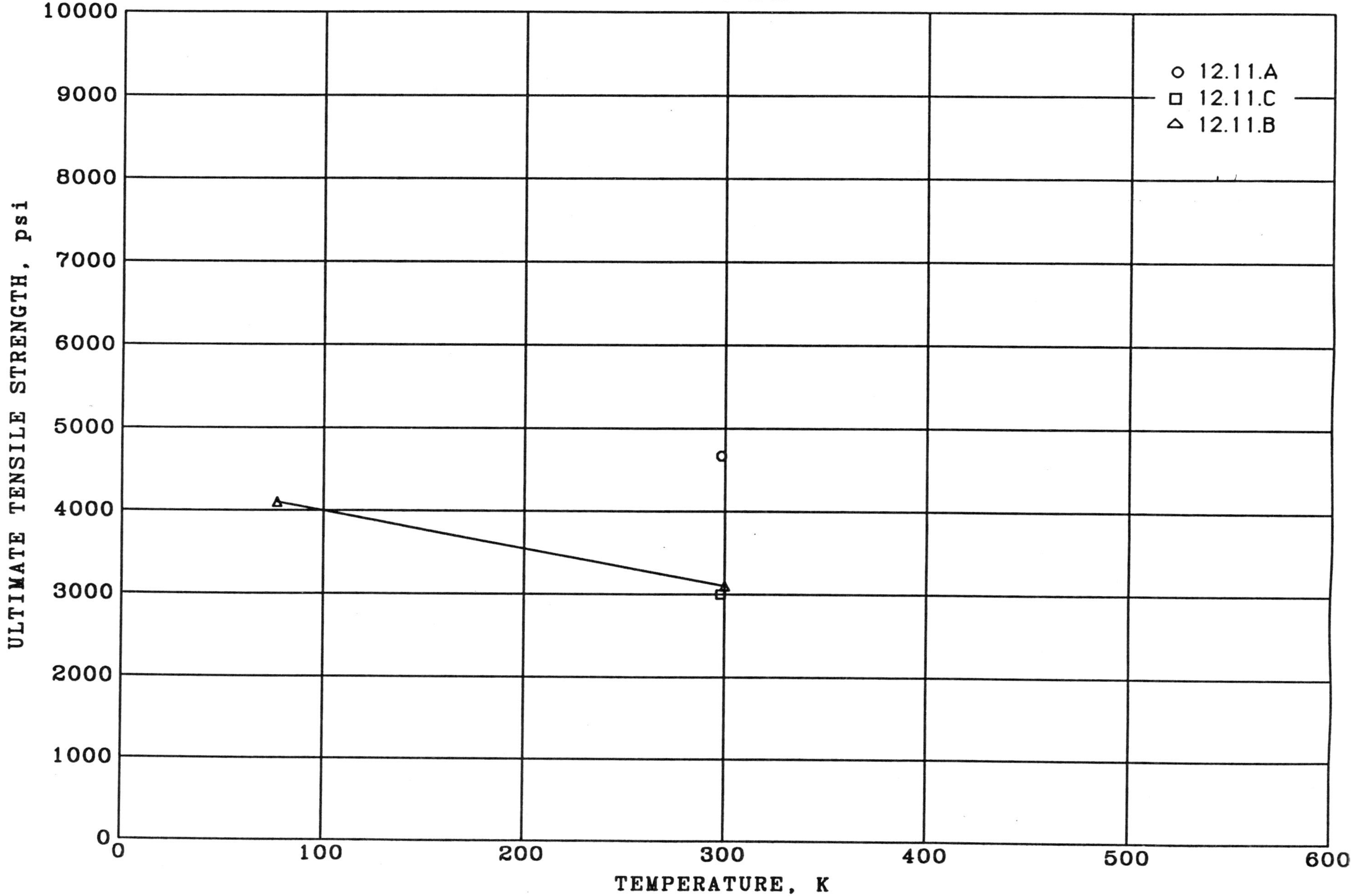
ULTIMATE TENSILE STRENGTH, psi
10000
9000
8000
7000
6000
5000
4000
3000
2000
1000
0
0
100
200
300
400
500
600
TEMPERATURE, K
12.11.A
12.11.C
12.11.B

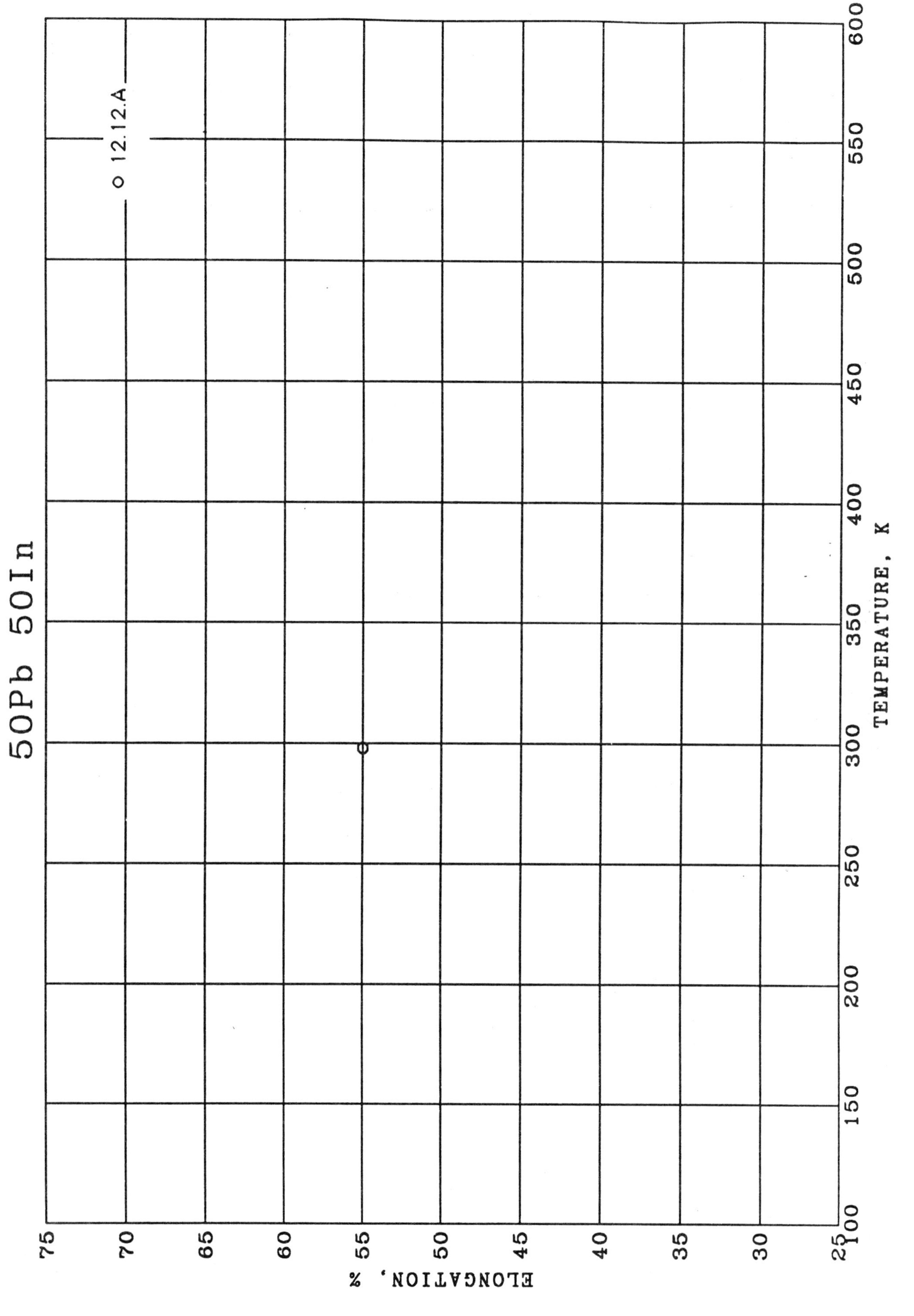
50Pb 50In
○ 12.12.A
ELONGATION, %
TEMPERATURE, K
75
70
65
60
55
50
45
40
35
30
25
100
150
200
250
300
350
400
450
500
550
600

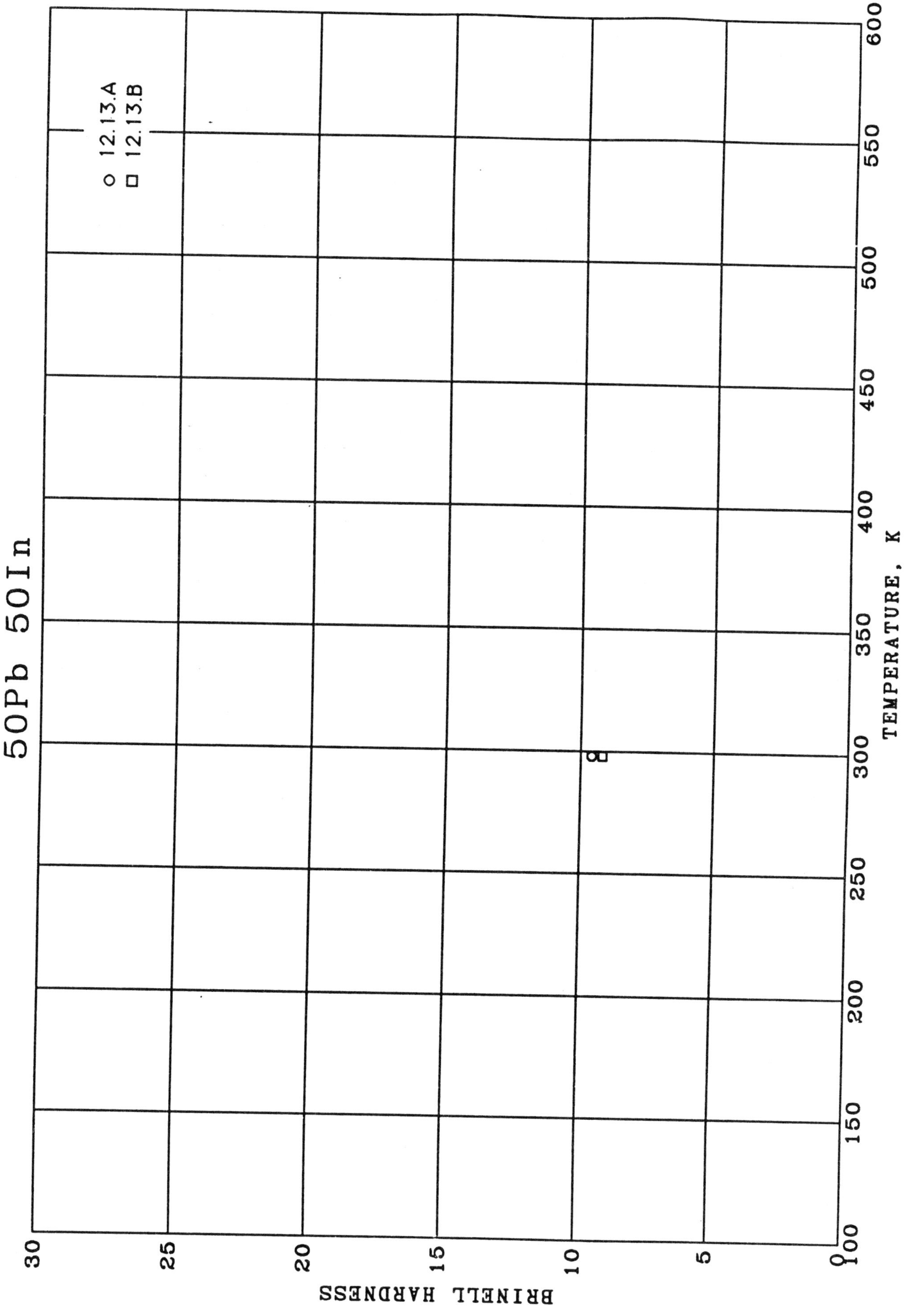

50Pb 50In
○ 12.13.A
□ 12.13.B
BRINELL HARDNESS
0
5
10
15
20
25
30
TEMPERATURE, K
100
150
200
250
300
350
400
450
500
550
600

80Pb 20In

PROPERTY	@ 300K	UNITS	REFERENCES
Density	10.27	g/cm^3	11.1.A-B
Specific Heat	0.0416	cal/g-K	11.2.A
Thermal Conductivity	1) 0.0406	cal/cm-s-K	11.3.A-B
Thermal Expansion	$27x10^{-6}$	1/K	11.4.A
Liquidus/Solidus	553/543	K	11.5.A-C
Resistivity	$35x10^{-6}$	Ohm-cm	11.6.A-D
TCR (20-100C)	2500	ppm/K	11.7.A
Young's Modulus		ksi	
Poisson's Ratio		------	
Yield Strength		psi	
Ult Tensile Strength	5500	psi	11.11.A
Elongation at Break		%	
Hardness			

1 Thermal conductivity listed is for 85C.

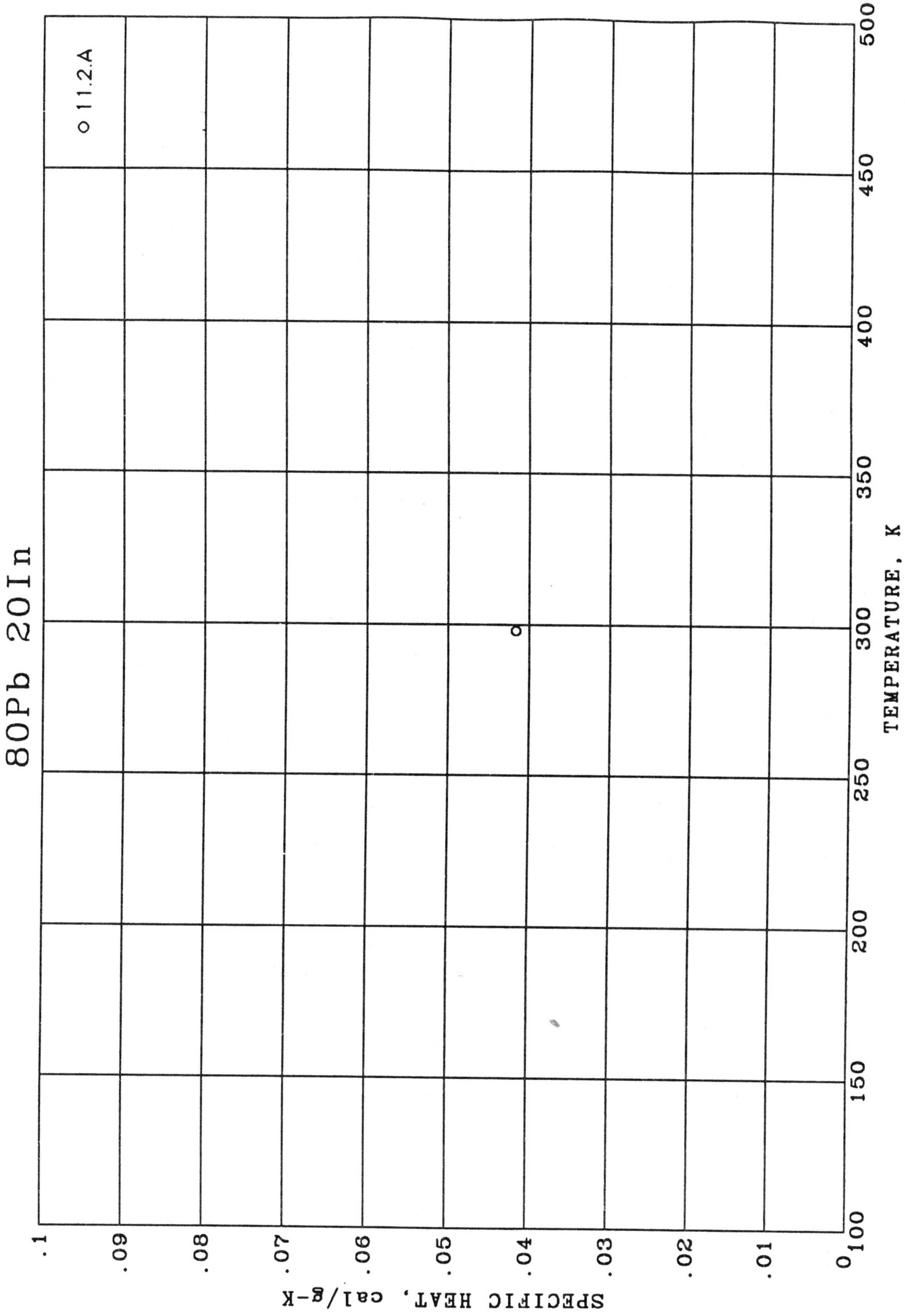
80Pb 20In
○ 11.2.A
SPECIFIC HEAT, cal/g-K
.1
.09
.08
.07
.06
.05
.04
.03
.02
.01
0
TEMPERATURE, K
100
150
200
250
300
350
400
450
500

80Pb 20In

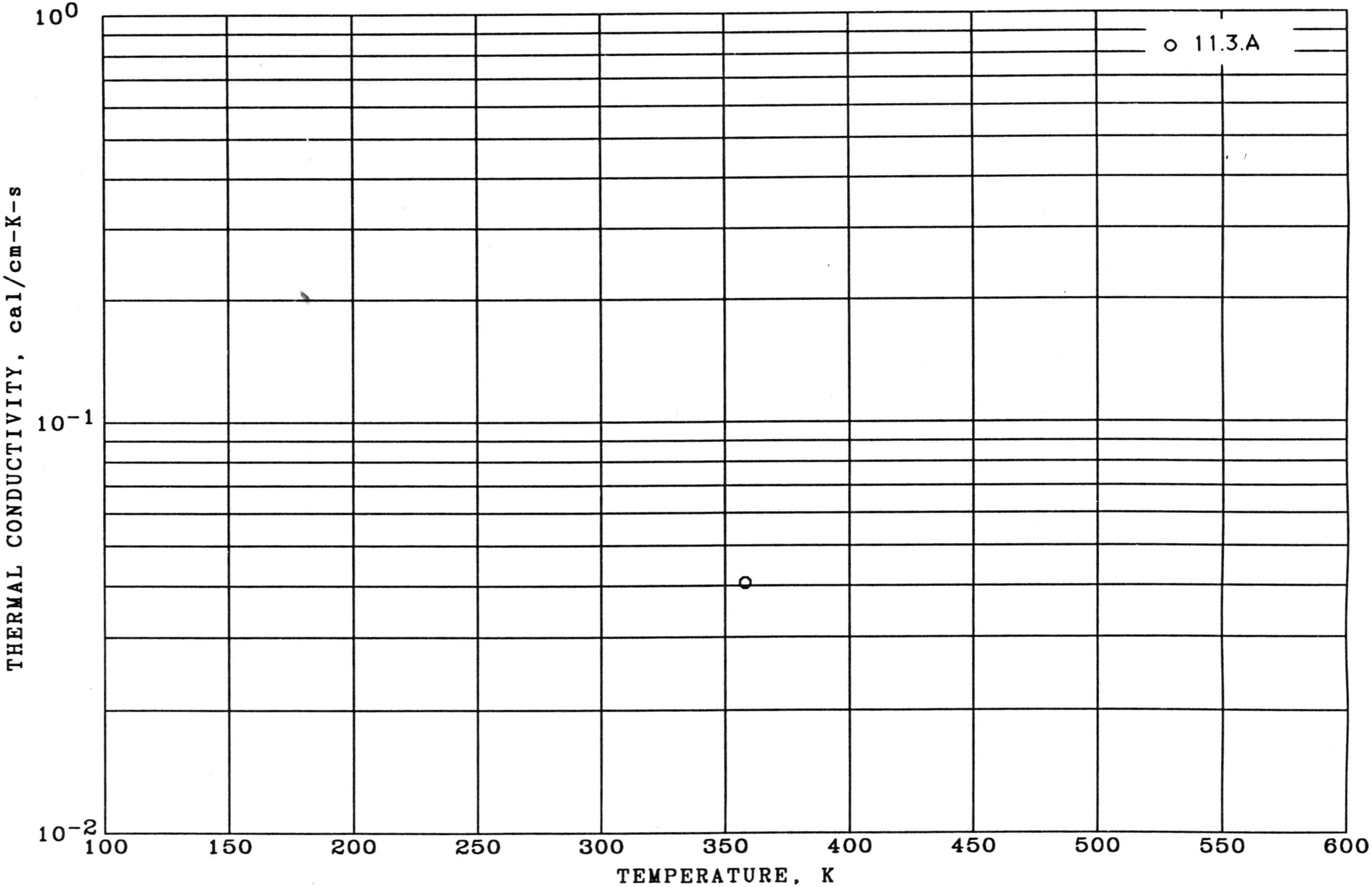

o 11.3.A
10^0
10^{-1}
10^{-2}
THERMAL CONDUCTIVITY, cal/cm-K-s
100
150
200
250
300
350
400
450
500
550
600
TEMPERATURE, K

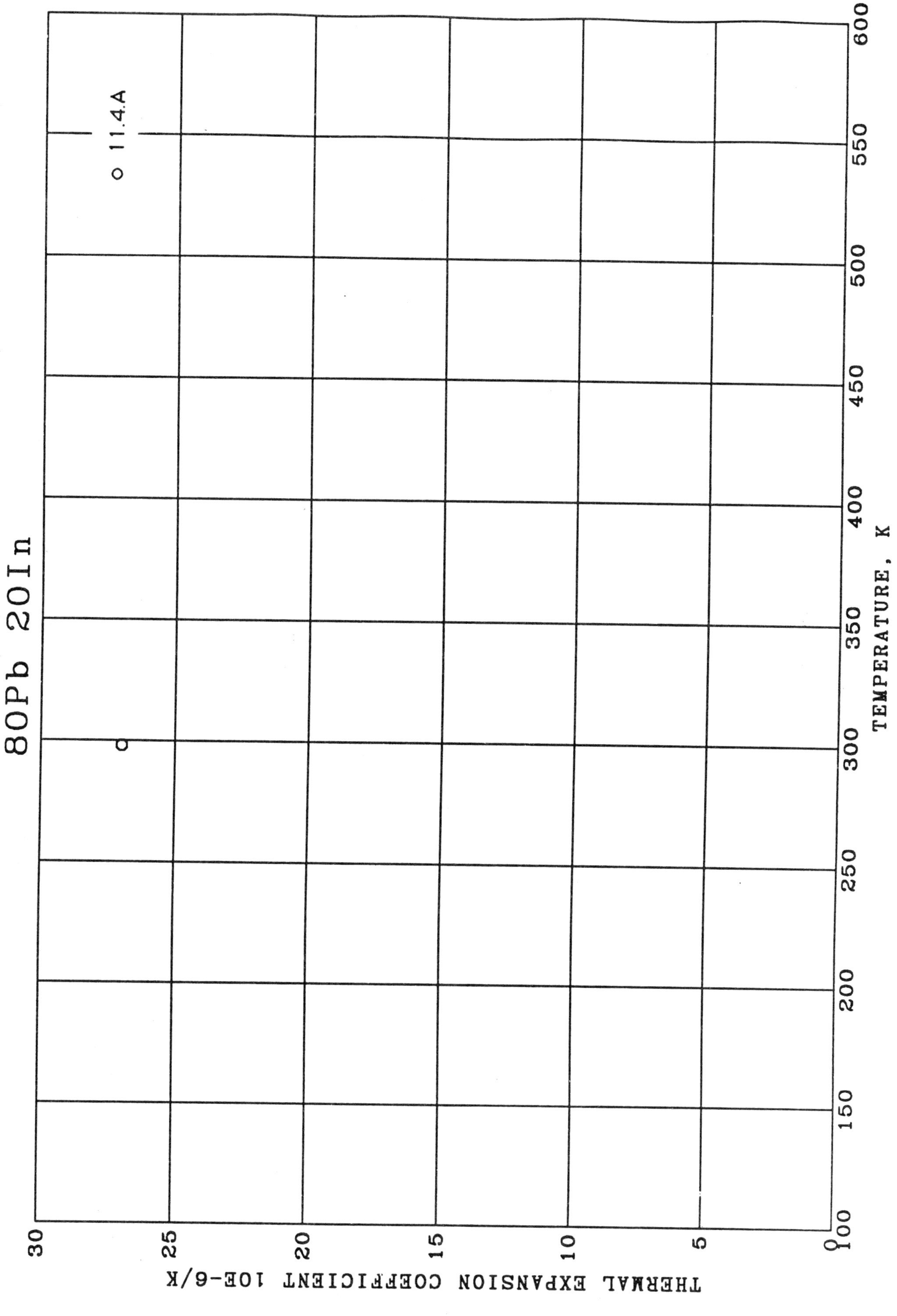
80Pb 20In
o 11.4A
THERMAL EXPANSION COEFFICIENT 10E-6/K
TEMPERATURE, K
0
5
10
15
20
25
30
100
150
200
250
300
350
400
450
500
550
600

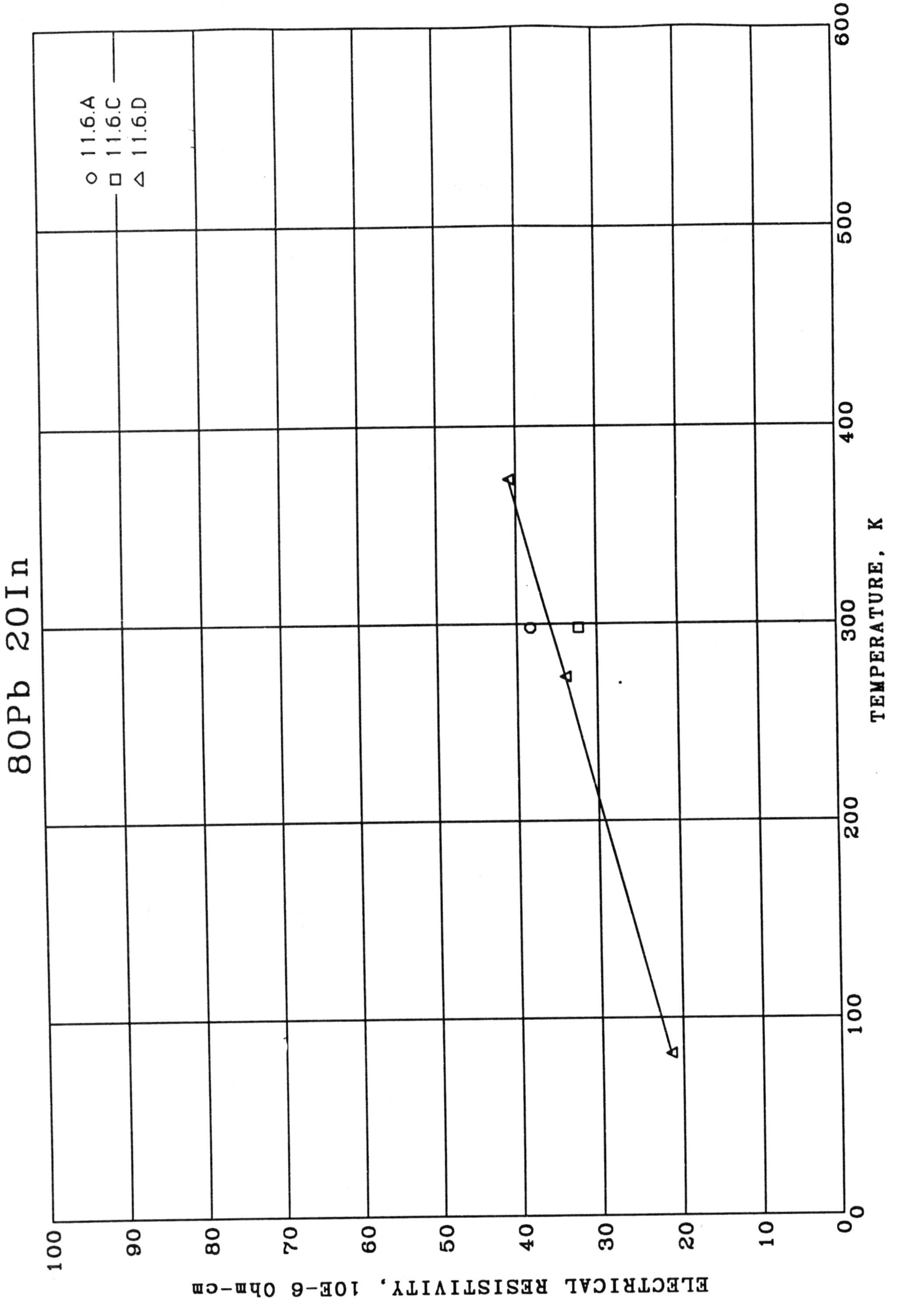
80Pb 20In
○ 11.6.A
□ 11.6.C
△ 11.6.D
ELECTRICAL RESISTIVITY, 10E-6 Ohm-cm
TEMPERATURE, K
0
10
20
30
40
50
60
70
80
90
100
100
200
300
400
500
600

80Pb 20In

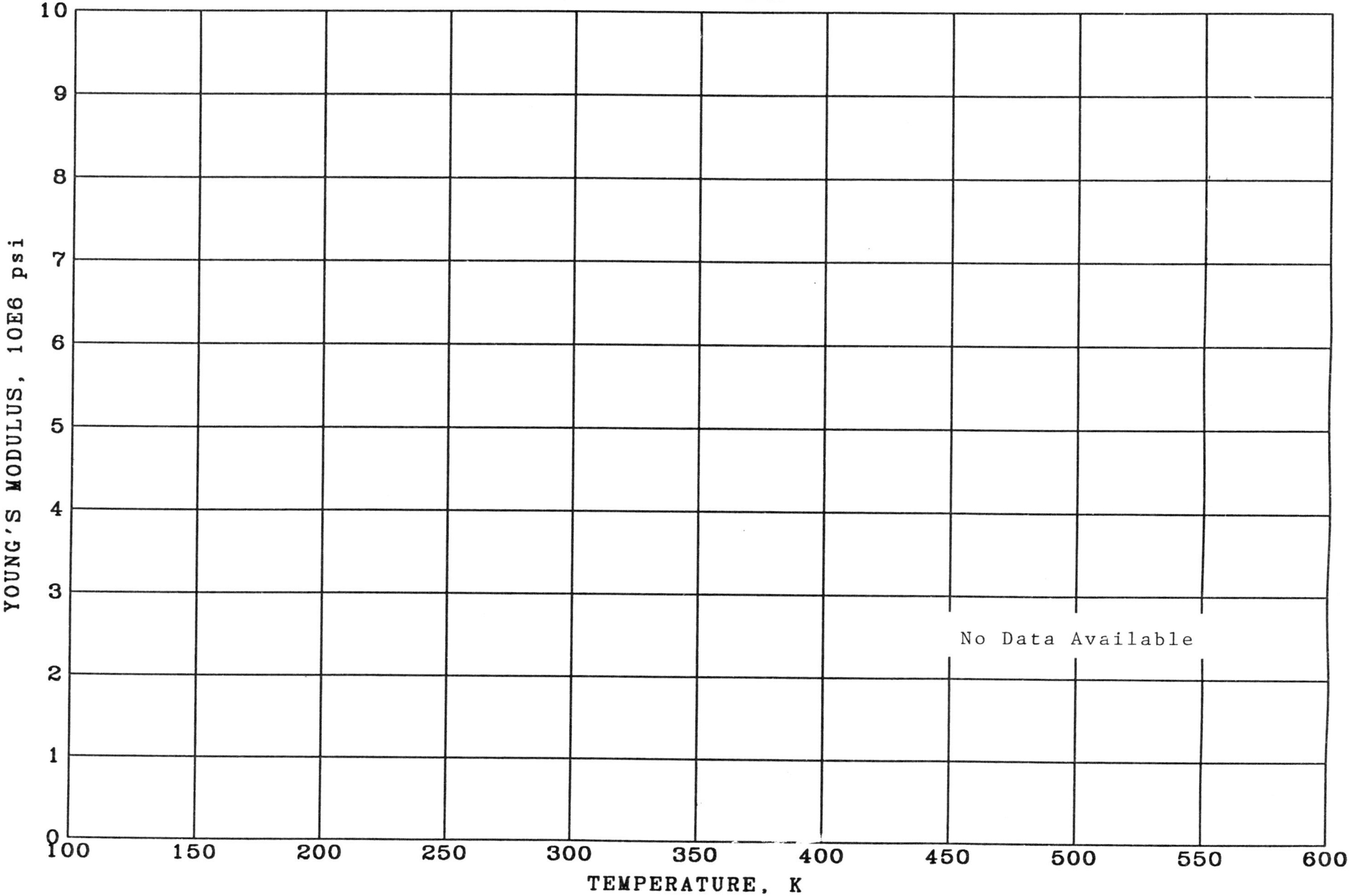

YOUNG'S MODULUS, 10E6 psi
0
1
2
3
4
5
6
7
8
9
10
TEMPERATURE, K
100
150
200
250
300
350
400
450
500
550
600
No Data Available

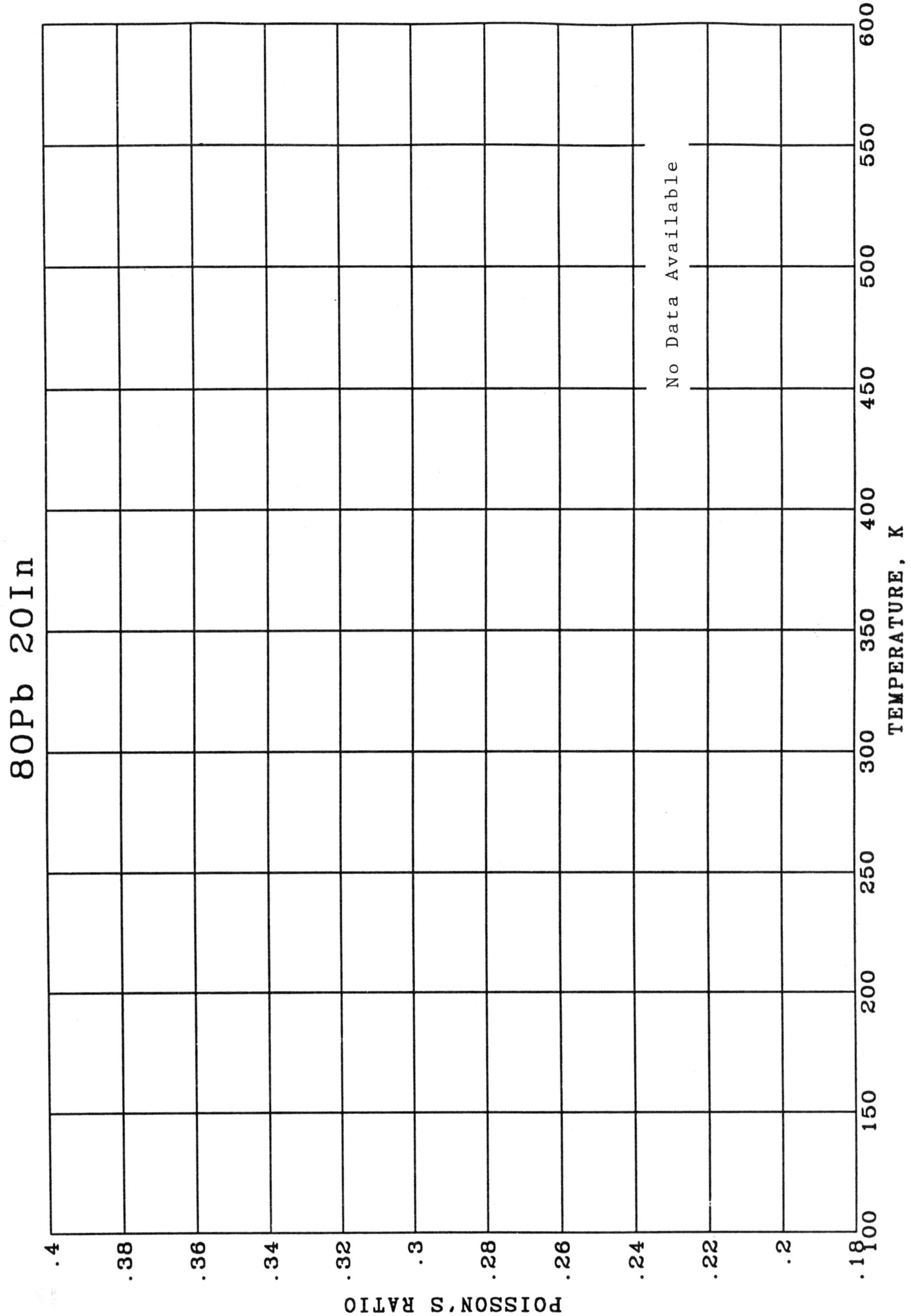
80Pb 20In
No Data Available
POISSON'S RATIO
.4
.38
.36
.34
.32
.3
.28
.26
.24
.22
.2
.18
TEMPERATURE, K
100
150
200
250
300
350
400
450
500
550
600

80Pb 20In

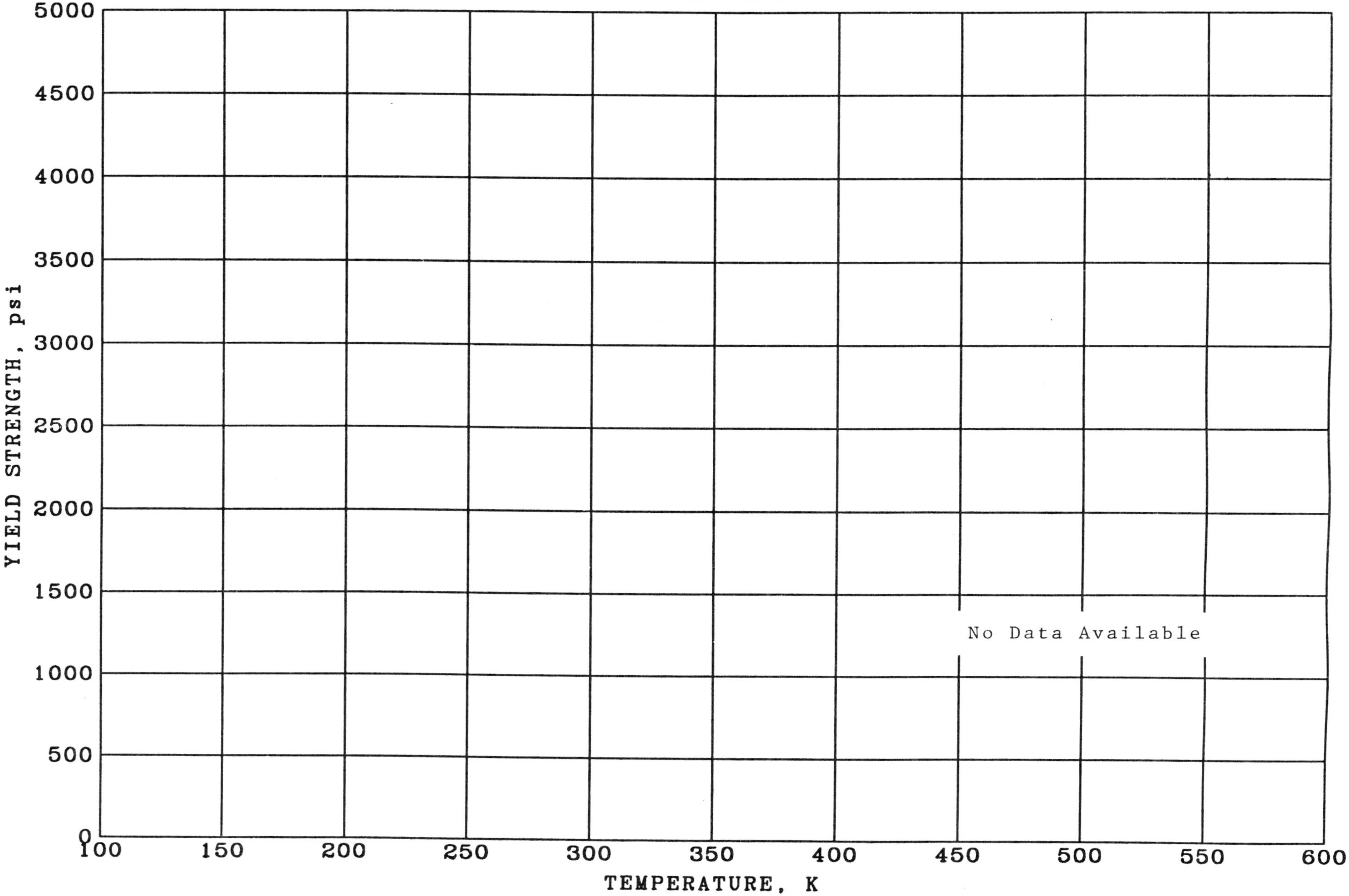
YIELD STRENGTH, psi
0
500
1000
1500
2000
2500
3000
3500
4000
4500
5000
100
150
200
250
300
350
400
450
500
550
600
TEMPERATURE, K
No Data Available

80Pb 20In

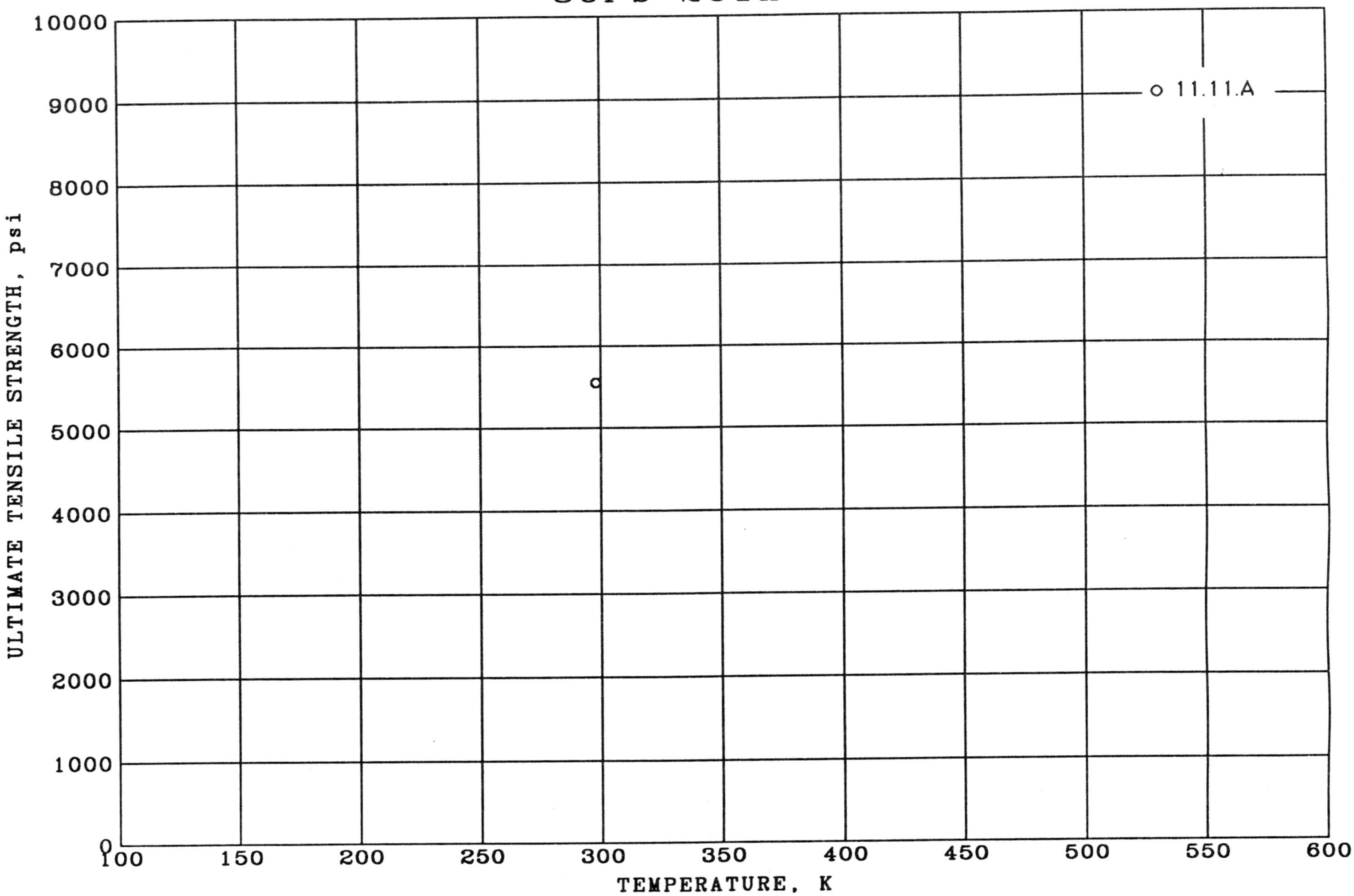

ULTIMATE TENSILE STRENGTH, psi
10000
9000
8000
7000
6000
5000
4000
3000
2000
1000
0
100
150
200
250
300
350
400
450
500
550
600
TEMPERATURE, K
11.11.A

80Pb 20In

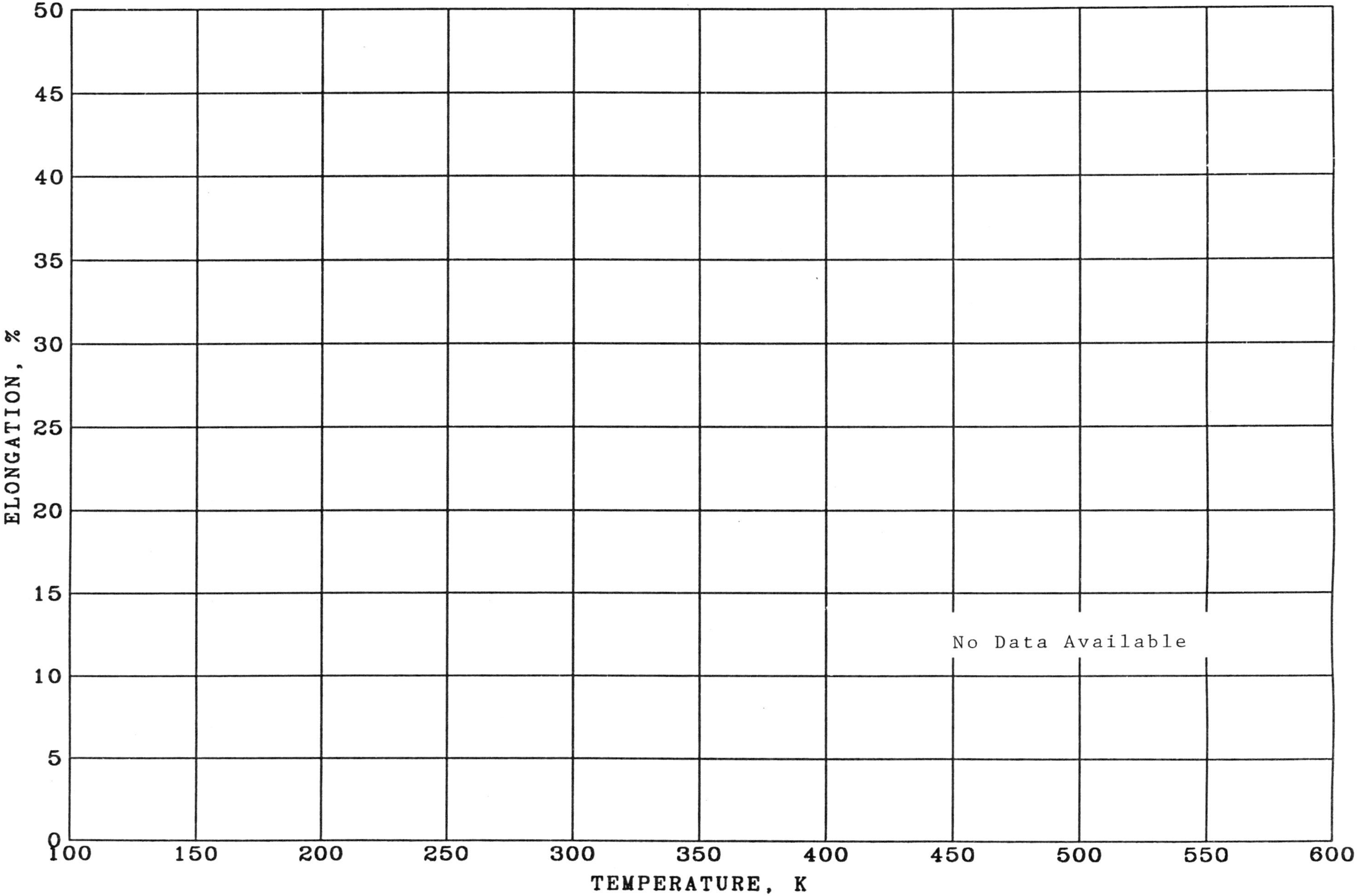

No Data Available
ELONGATION, %
TEMPERATURE, K
0
5
10
15
20
25
30
35
40
45
50
100
150
200
250
300
350
400
450
500
550
600

80Pb 20In

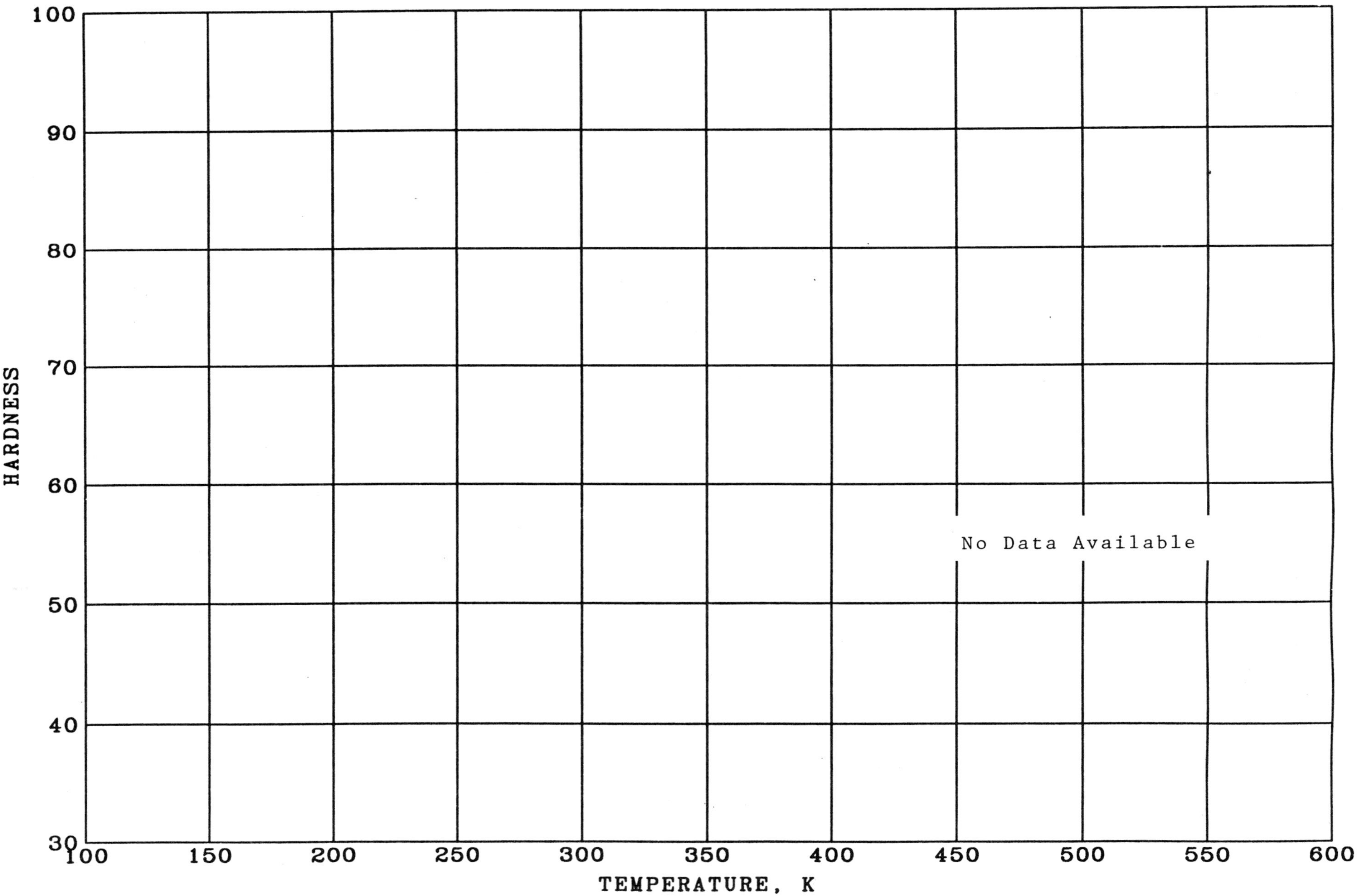

HARDNESS
100
90
80
70
60
50
40
30
100
150
200
250
300
350
400
450
500
550
600
TEMPERATURE, K
No Data Available

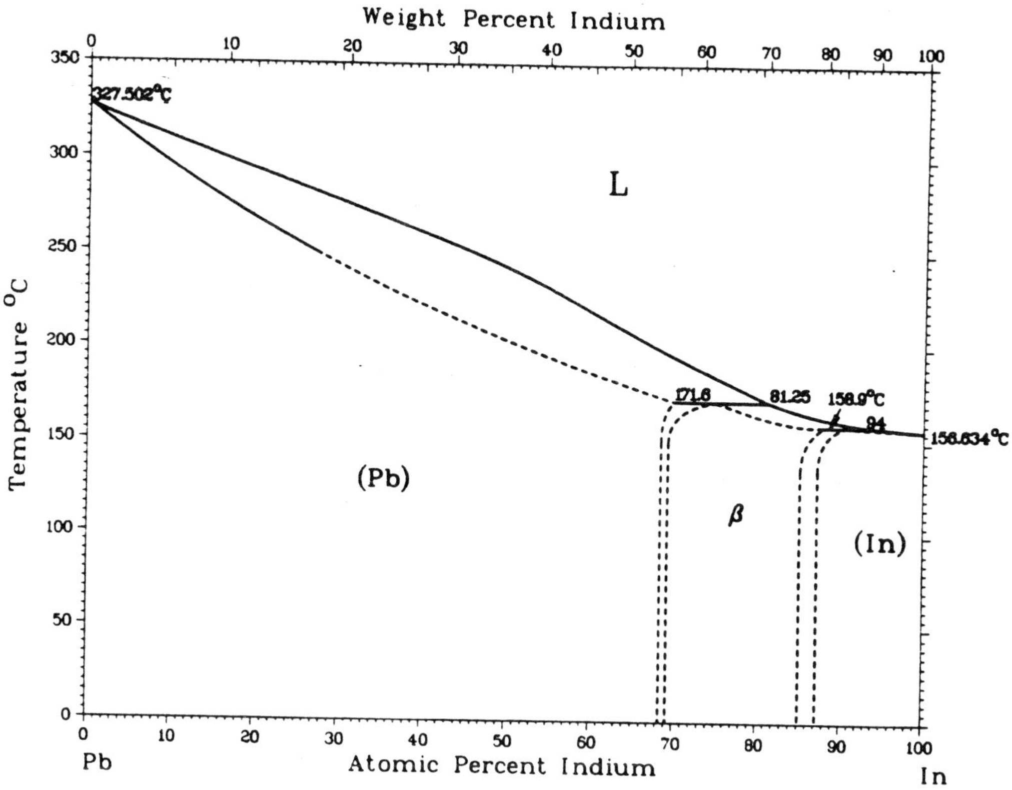
Weight Percent Indium
Atomic Percent Indium
Temperature °C
327.502°C
L
171.6
81.25
158.9°C
94
156.634°C
(Pb)
β
(In)
Pb
In

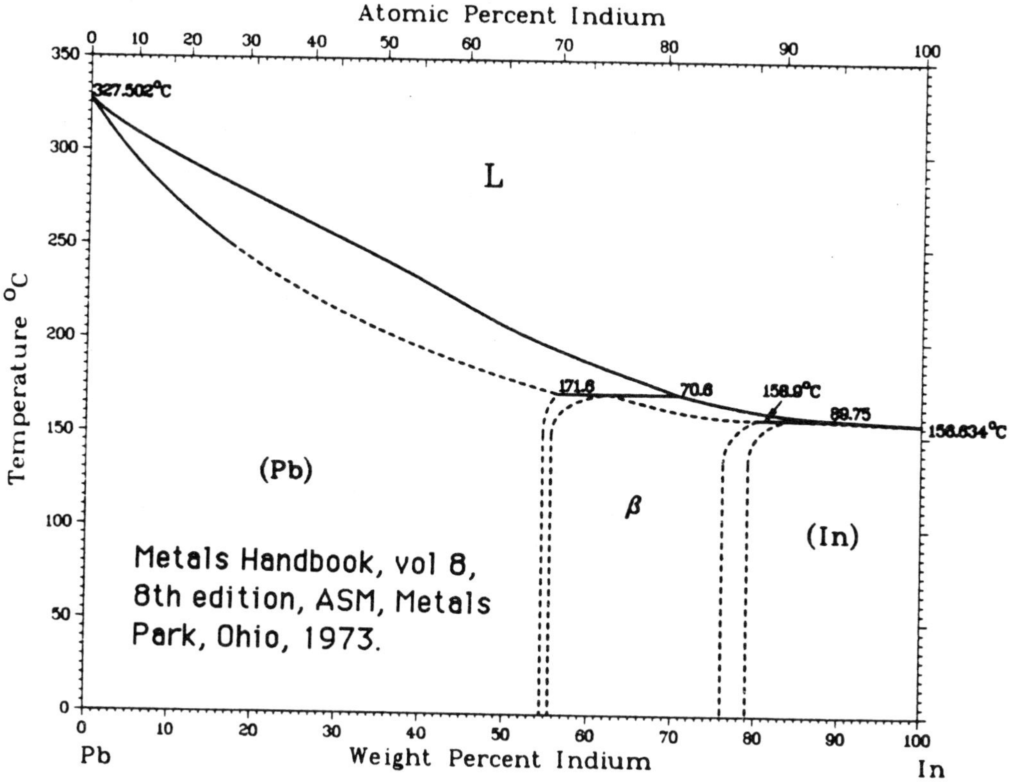
Atomic Percent Indium
Weight Percent Indium
Temperature °C
327.502°C
L
171.6
70.6
158.9°C
89.75
156.634°C
(Pb)
β
(In)
Metals Handbook, vol 8, 8th edition, ASM, Metals Park, Ohio, 1973.
Pb
In

89Ni 11P

PROPERTY	@ 300K	UNITS	REFERENCES
Density	7.92	g/cm3	13.1.A
Specific Heat	0.114	cal/g-K	13.2.A
Thermal Conductivity	0.105-0.013	cal/cm-s-K	13.3.A
Thermal Expansion	13-14.4	ppm/K	13.4.A
Melting Point	1153	K	13.5.A
Resistivity 1	$52\text{-}68\times10^{-6}$	Ohm-cm	13.6.A-C
TCR		ppm/K	
Young's Modulus 2	2840	ksi	13.8.A
Poisson's Ratio		------	
Yield Strength	3	psi	
Ult Tensile Strength	100,000 3	psi	13.11.A-B
Elongation at Break	1.0 3	%	13.12.A-B
Hardness	600	HK100	13.13.A-C

1 Resistivity for 2.2% P material = 30 micro-ohm-cm while a 13% P has a resistivity of 110. Heating a 7% P material deposited at 60C reportedly reduces resistivity from 72 to 20 micro-ohm-cm.

2 Data listed is the reduced Young's modulus from a bulge test.

3 Data is from a bulge test which is a biaxial loading condititon. No permanant deformation is noted and therefore no yield stregth is listed. Heat treatment in the range of 400-800C can significantly improve ductility and strength.

4 Hardness increases to about 1000 when heat treated at 400C for about 30 minutes. Heat treatment at higher temperatures reduces hardness to as-deposited hardness levels.

89Ni 11P

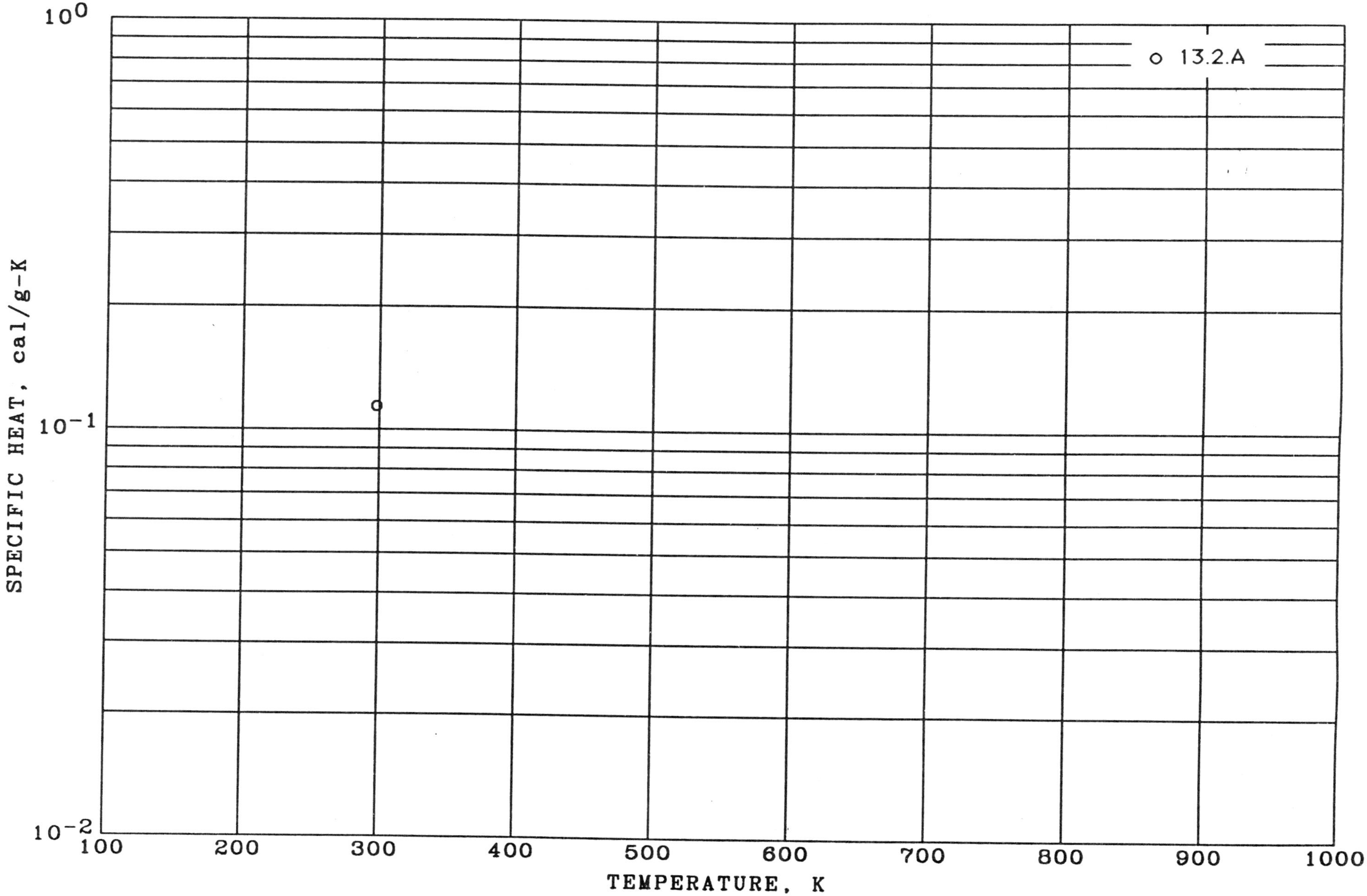

10^0
10^{-1}
10^{-2}
SPECIFIC HEAT, cal/g-K
100
200
300
400
500
600
700
800
900
1000
TEMPERATURE, K
13.2.A

89Ni 11P

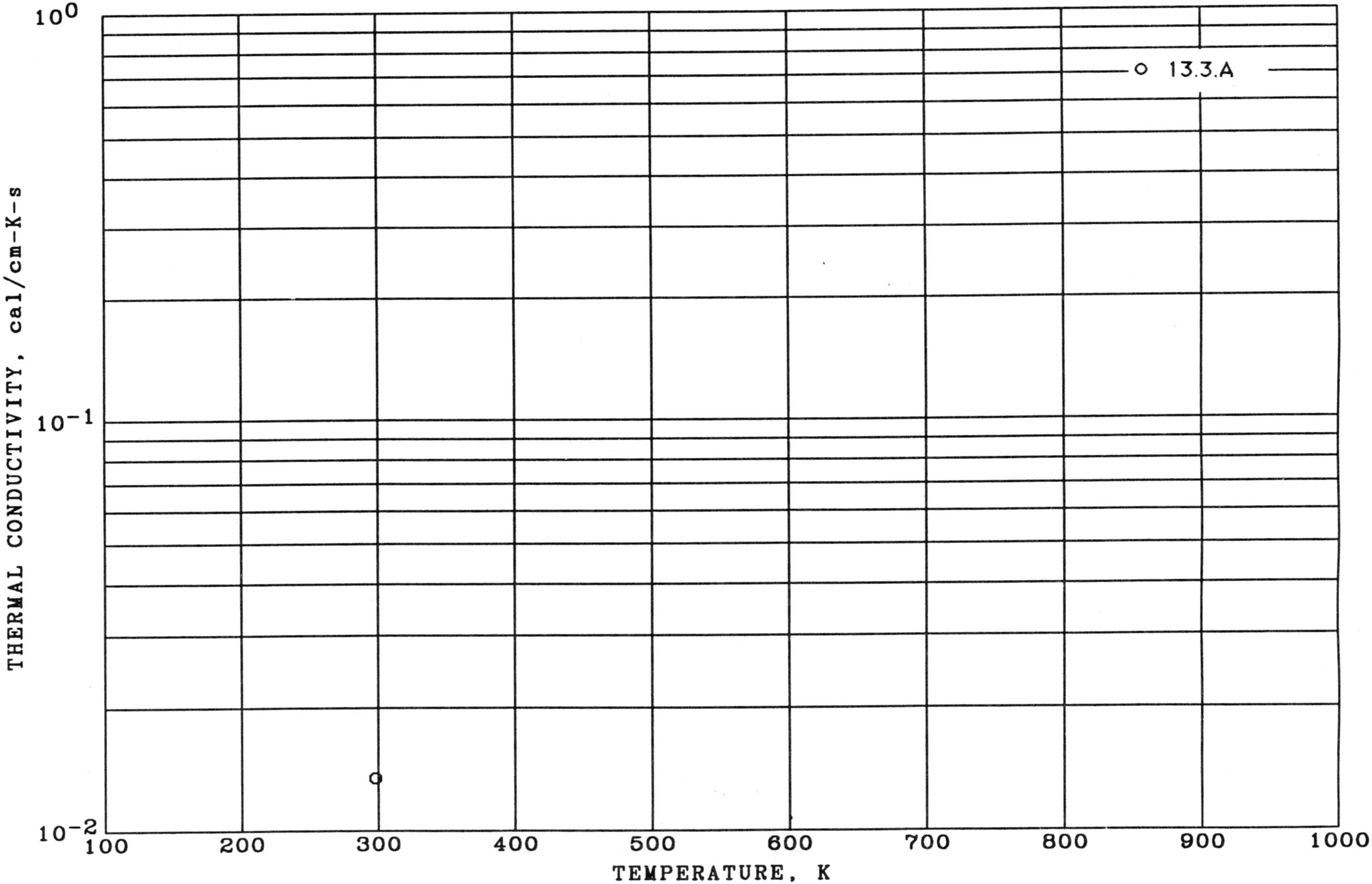

THERMAL CONDUCTIVITY, cal/cm-K-s
10^0
10^-1
10^-2
TEMPERATURE, K
100
200
300
400
500
600
700
800
900
1000
o 13.3.A

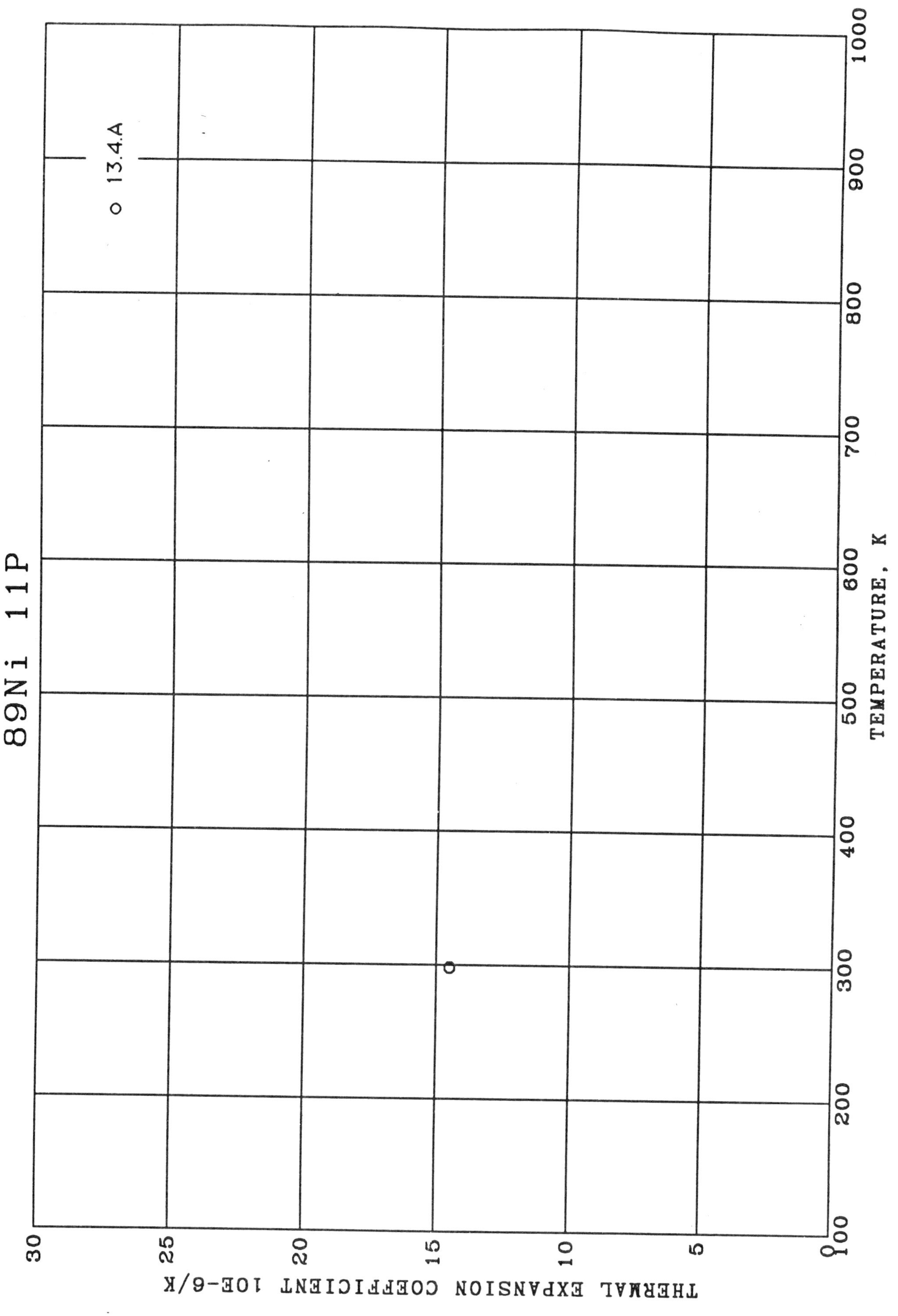
89Ni 11P
13.4A
THERMAL EXPANSION COEFFICIENT 10E-6/K
TEMPERATURE, K
0
5
10
15
20
25
30
100
200
300
400
500
600
700
800
900
1000

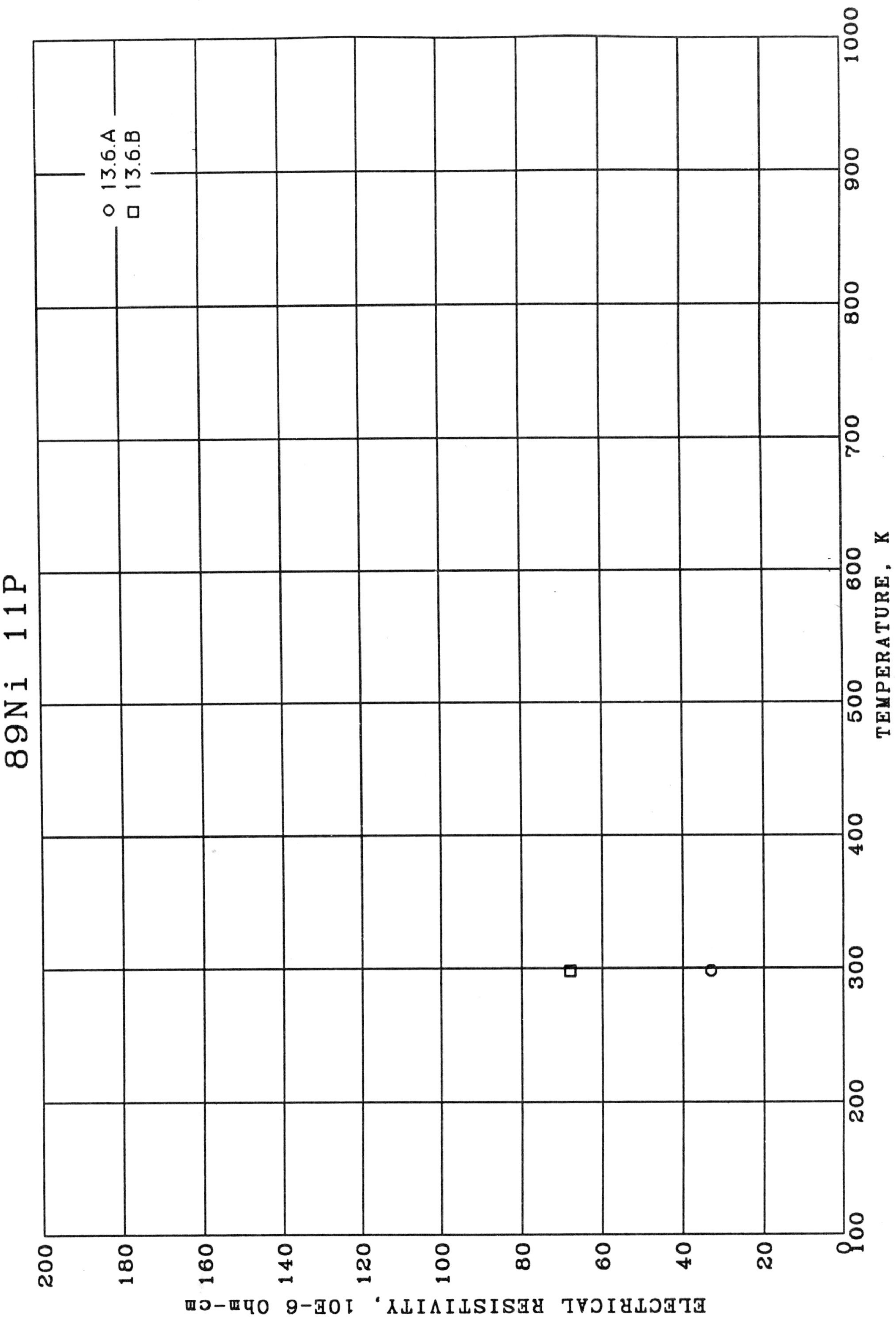
89Ni 11P
○ 13.6.A
□ 13.6.B
ELECTRICAL RESISTIVITY, 10E-6 Ohm-cm
TEMPERATURE, K
0
20
40
60
80
100
120
140
160
180
200
100
200
300
400
500
600
700
800
900
1000

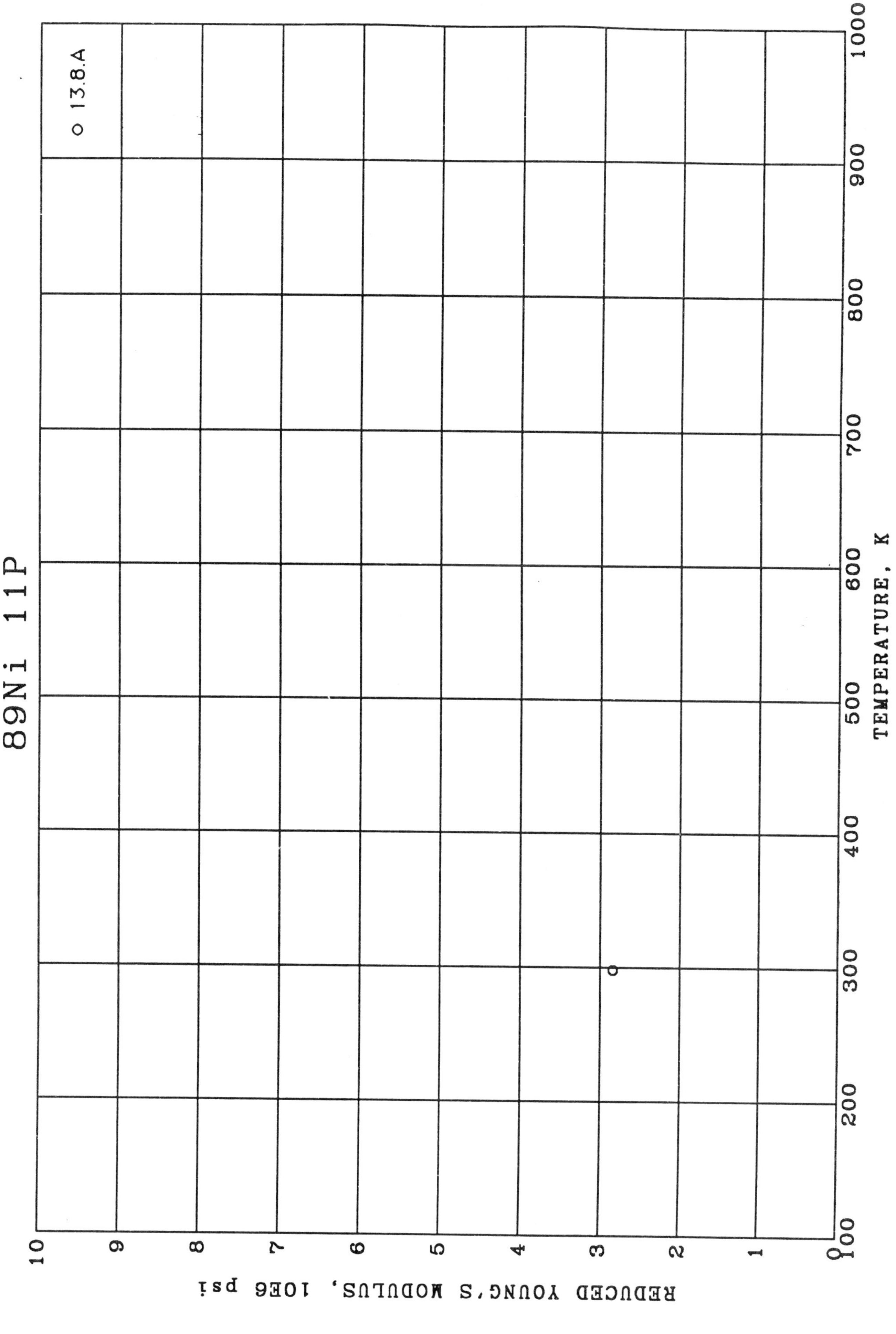
89Ni 11P
○ 13.8.A
REDUCED YOUNG'S MODULUS, 10E6 psi
TEMPERATURE, K
0
1
2
3
4
5
6
7
8
9
10
100
200
300
400
500
600
700
800
900
1000

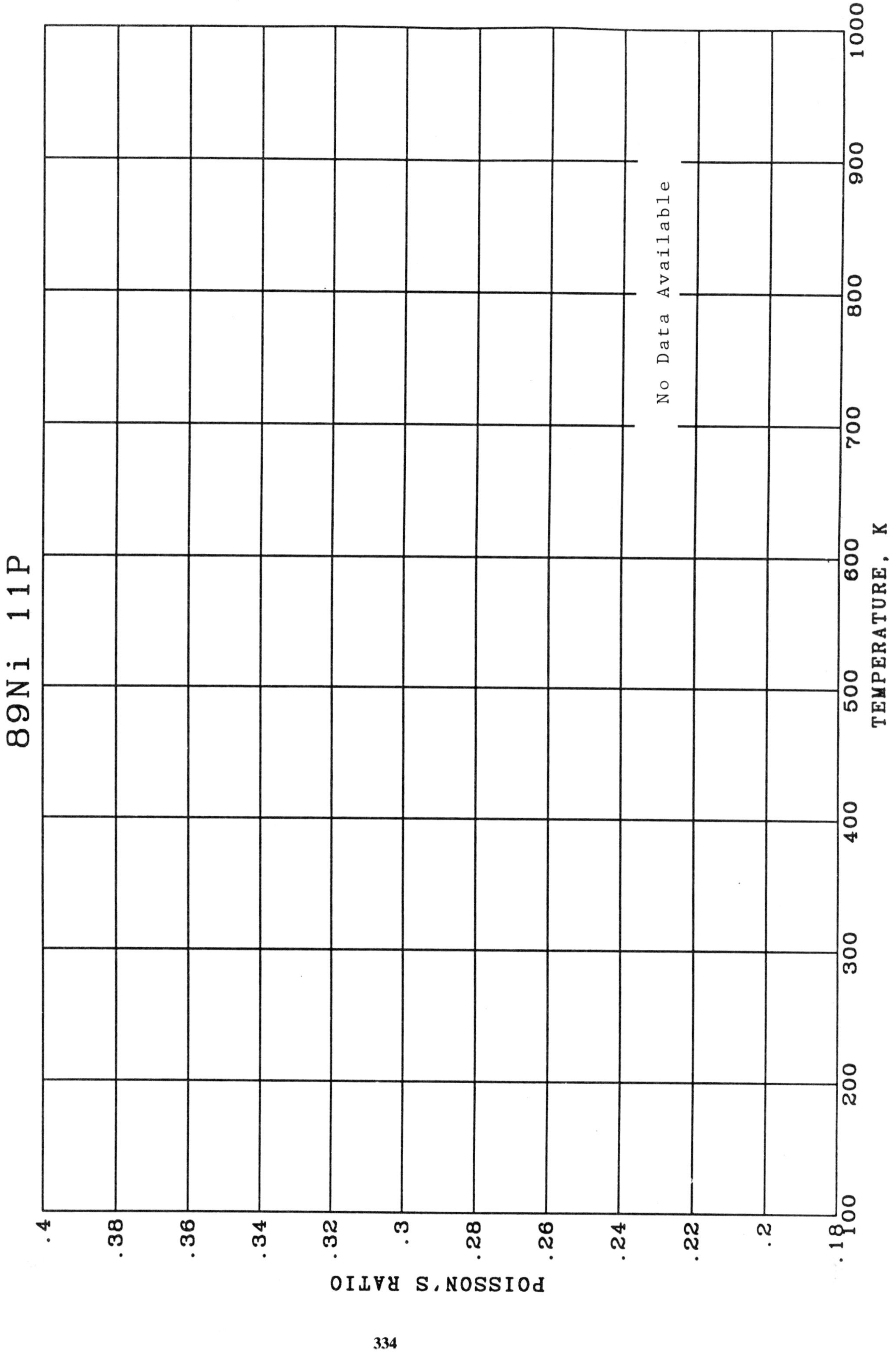
89Ni 11P
No Data Available
POISSON'S RATIO
.4
.38
.36
.34
.32
.3
.28
.26
.24
.22
.2
.18
TEMPERATURE, K
100
200
300
400
500
600
700
800
900
1000

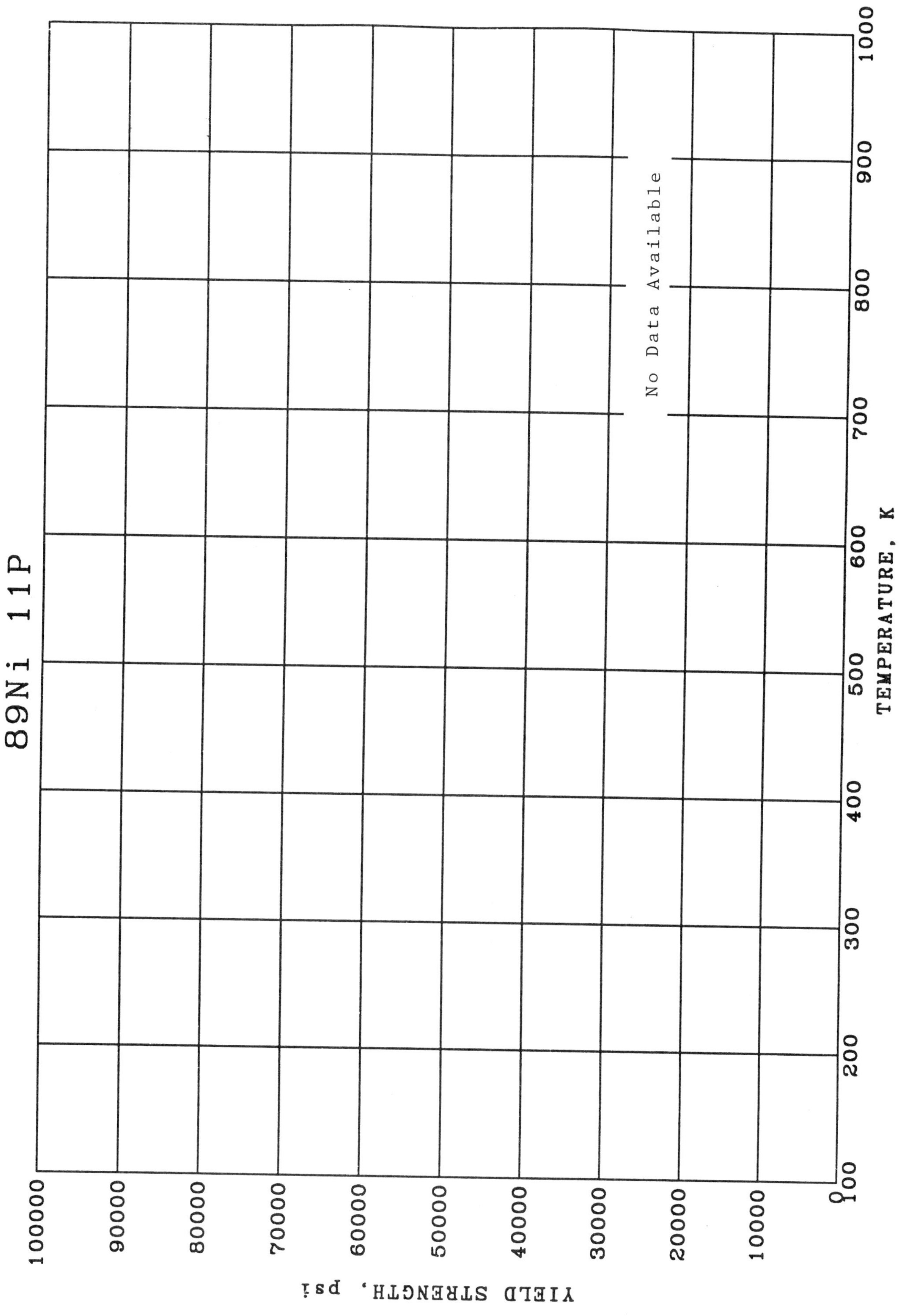
89Ni 11P
No Data Available
YIELD STRENGTH, psi
TEMPERATURE, K
0
10000
20000
30000
40000
50000
60000
70000
80000
90000
100000
100
200
300
400
500
600
700
800
900
1000

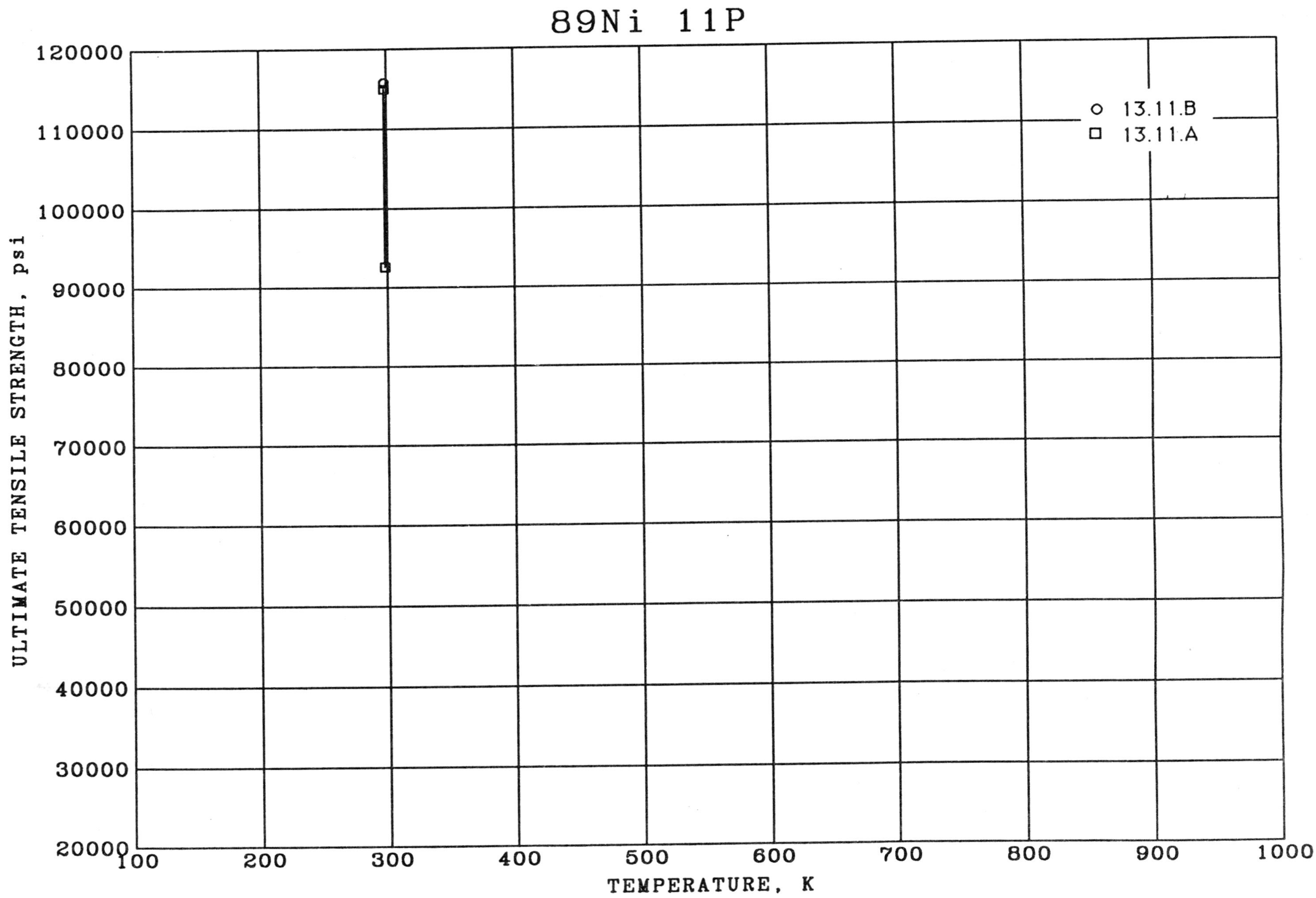
89Ni 11P
13.11.B
13.11.A
ULTIMATE TENSILE STRENGTH, psi
120000
110000
100000
90000
80000
70000
60000
50000
40000
30000
20000
100
200
300
400
500
600
700
800
900
1000
TEMPERATURE, K

89Ni 11P

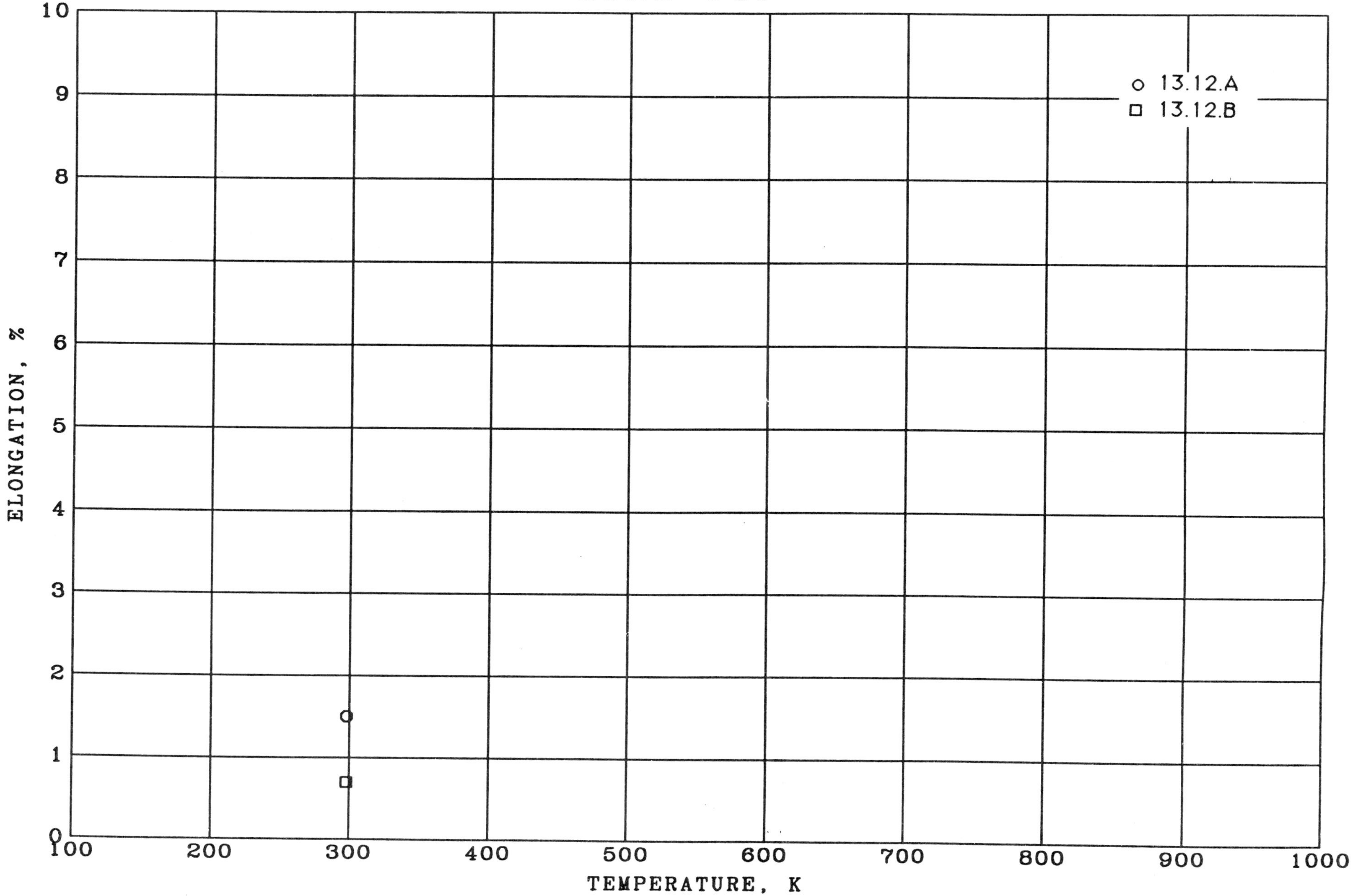

10
9
8
7
6
5
4
3
2
1
0
ELONGATION, %
100
200
300
400
500
600
700
800
900
1000
TEMPERATURE, K
○ 13.12.A
□ 13.12.B

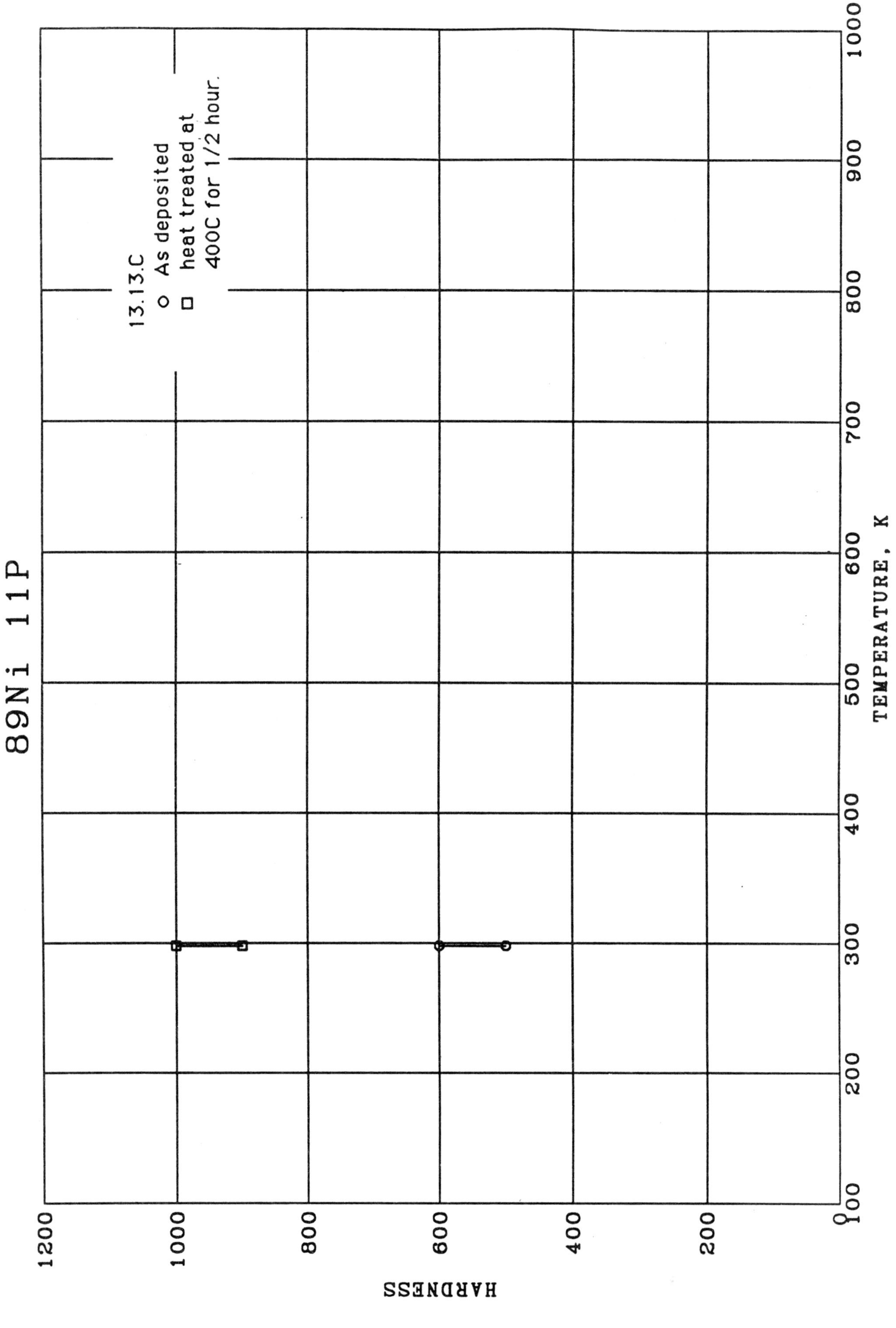
89Ni 11P
13.13.C
○ As deposited
□ heat treated at 400C for 1/2 hour.
HARDNESS
0
200
400
600
800
1000
1200
TEMPERATURE, K
100
200
300
400
500
600
700
800
900
1000

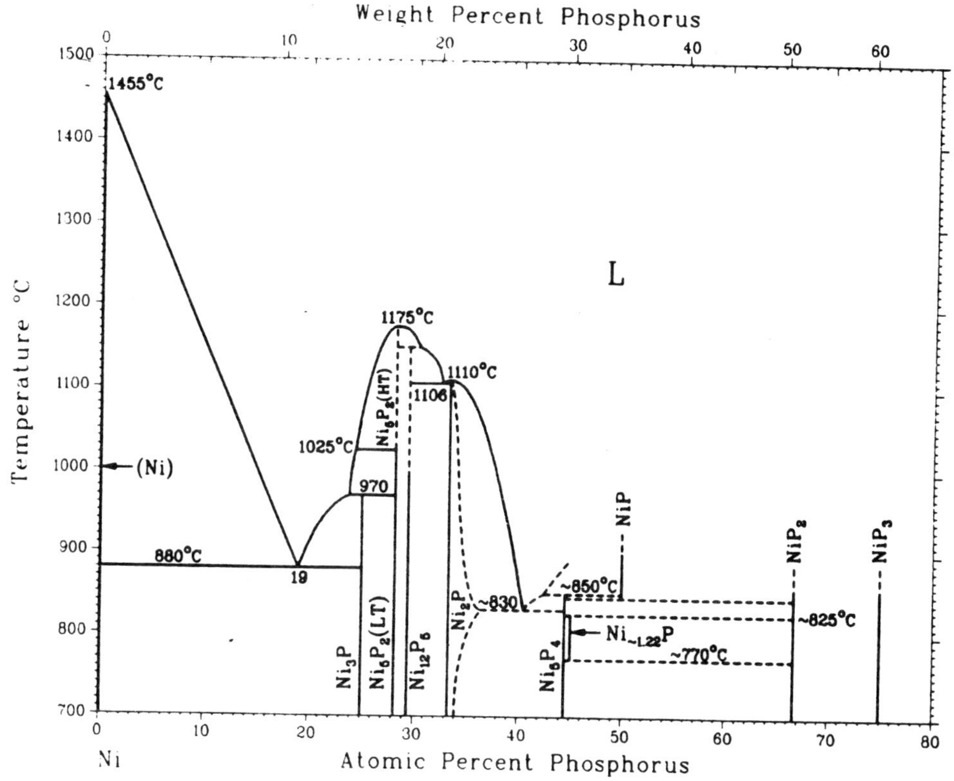
Weight Percent Phosphorus
Temperature °C
Atomic Percent Phosphorus
Ni
1455°C
1175°C
1110°C
1106
1025°C
970
880°C
19
~850°C
~830
~825°C
~770°C
(Ni)
L
Ni_3P
Ni_5P_2(LT)
Ni_5P_2(HT)
$Ni_{12}P_5$
Ni_2P
Ni_5P_4
$Ni_{\sim 1.22}P$
NiP
NiP_2
NiP_3

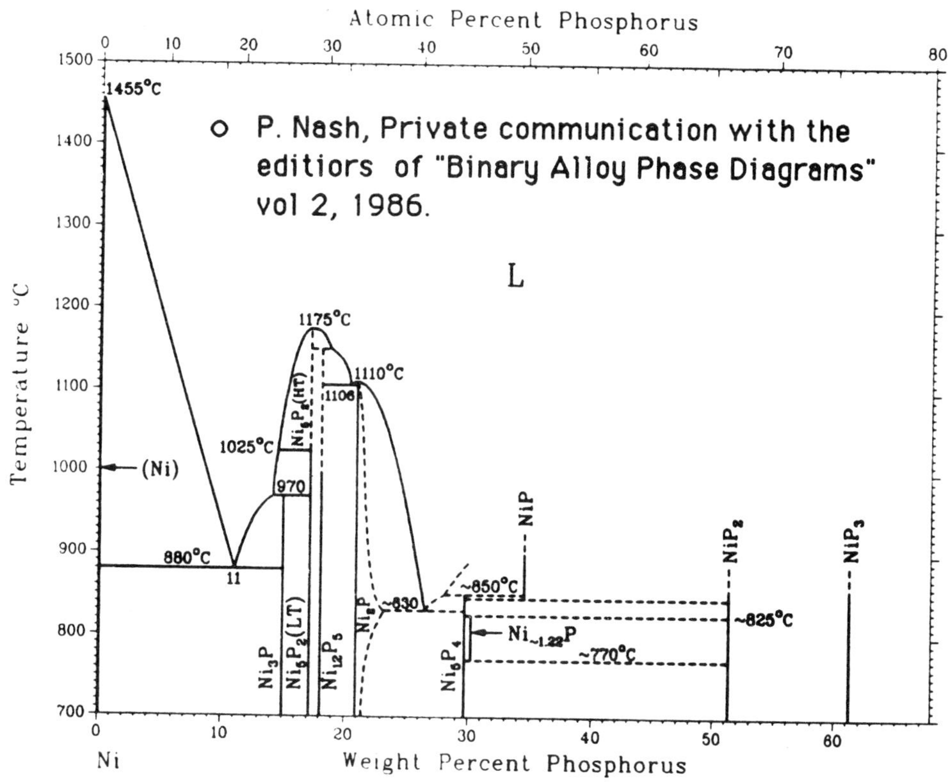
Atomic Percent Phosphorus
Temperature °C
Weight Percent Phosphorus
Ni
o P. Nash, Private communication with the editiors of "Binary Alloy Phase Diagrams" vol 2, 1986.
1455°C
1175°C
1110°C
1106
1025°C
970
880°C
11
~850°C
~830
~825°C
~770°C
(Ni)
L
Ni_3P
Ni_5P_2(LT)
Ni_5P_2(HT)
$Ni_{12}P_5$
Ni_2P
Ni_5P_4
$Ni_{\sim 1.22}P$
NiP
NiP_2
NiP_3

63Sn 37Pb

PROPERTY	@ 300K	UNITS	REFERENCES
Density	8.41	g/cm^3	14.1.A-B
Specific Heat	0.0358	cal/g-K	14.2.A-B
Thermal Conductivity	0.119	cal/cm-s-K	14.3.A-D
Thermal Expansion	$24x10^{-6}$	1/K	14.4.A-C
Melting Point	456	K	14.5.A-C
Resistivity	$14.5x10^{-6}$	Ohm-cm	14.6.A-C
TCR	4500	ppm/K	14.7.A
Young's Modulus	4460	ksi	14.8.A-C
Poisson's Ratio	0.4	------	14.9.A
Yield Strength	4500	psi	14.10.A
Ult Tensile Strength	6360	psi	14.11.A-F
Elongation at Break	50	%	14.12.A-C
Hardness	16	Vickers	14.13.A-B

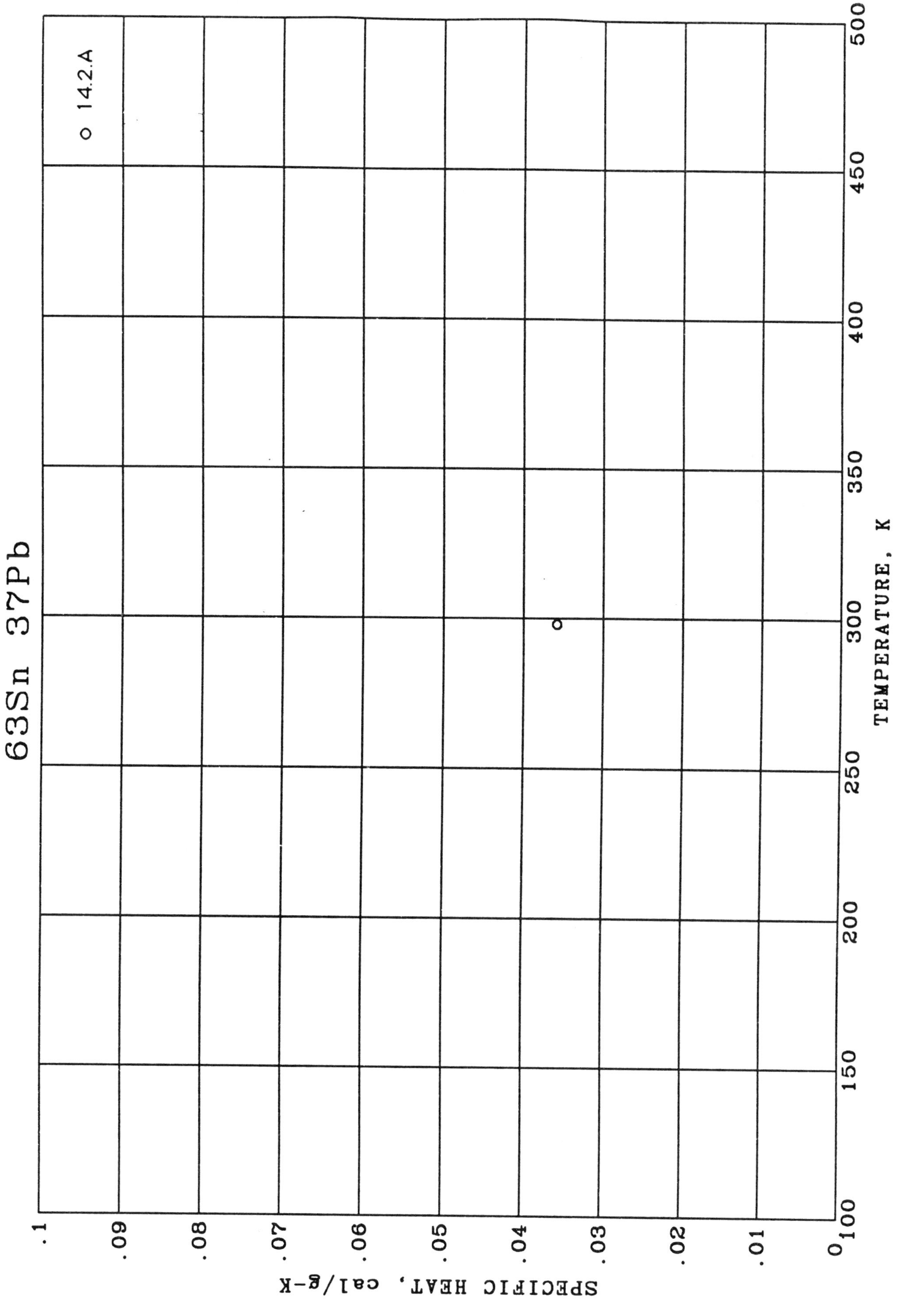
63Sn 37Pb
o 14.2.A
SPECIFIC HEAT, cal/g-K
TEMPERATURE, K
0
.01
.02
.03
.04
.05
.06
.07
.08
.09
.1
100
150
200
250
300
350
400
450
500

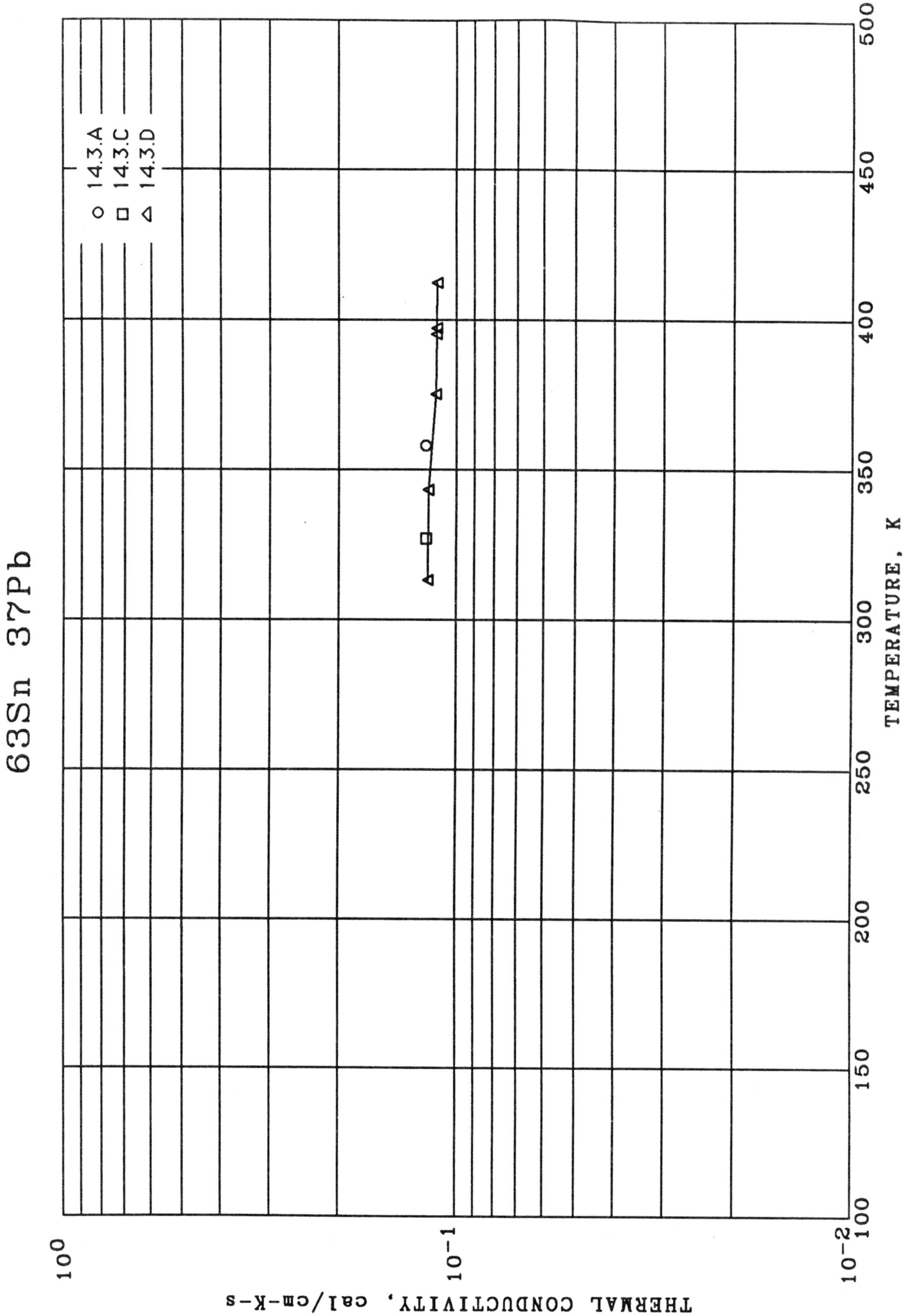
63Sn 37Pb
14.3.A
14.3.C
14.3.D
TEMPERATURE, K
THERMAL CONDUCTIVITY, cal/cm-K-s
100
150
200
250
300
350
400
450
500
10^0
10^-1
10^-2

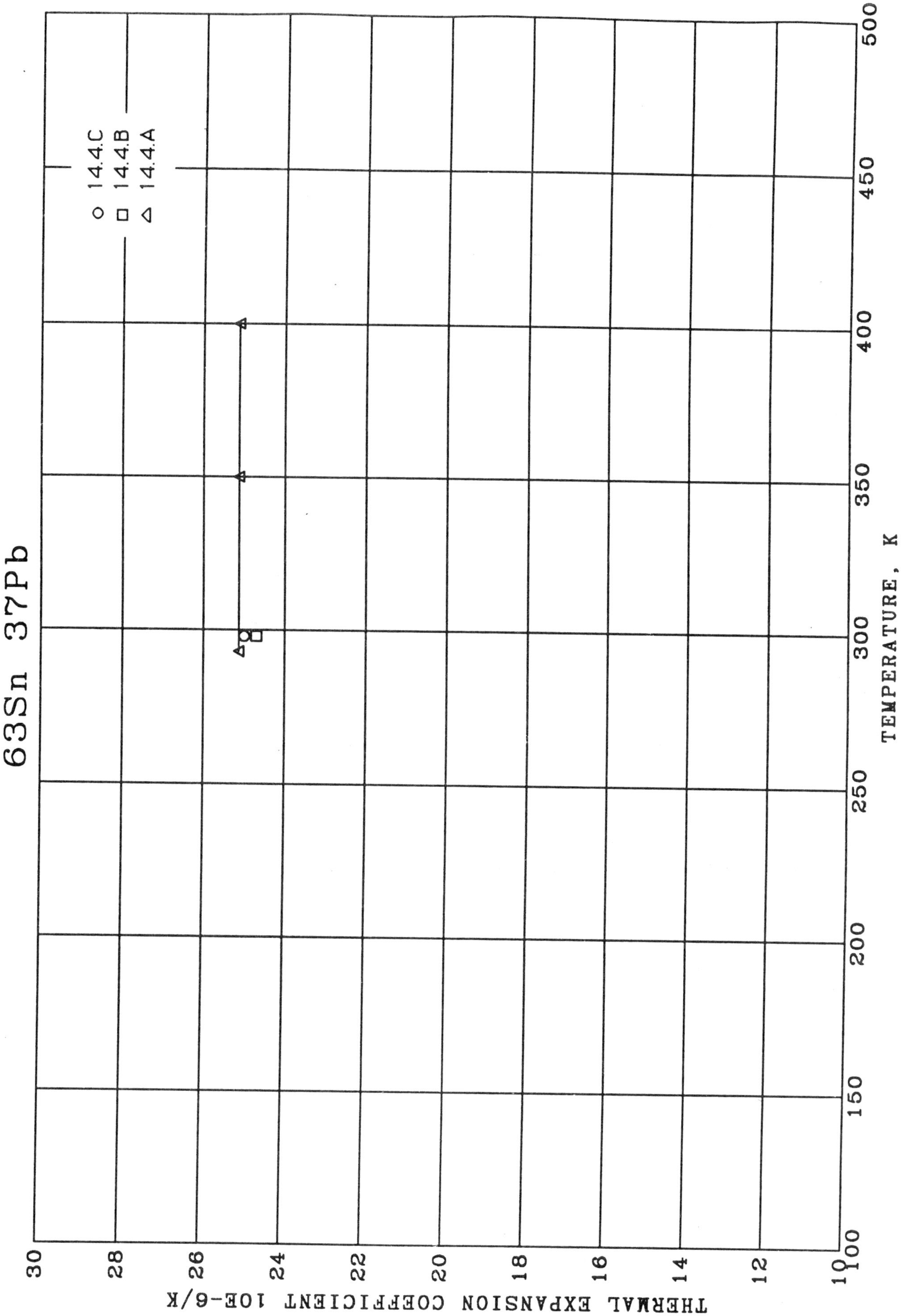
63Sn 37Pb
○ 14.4.C
□ 14.4.B
△ 14.4.A
TEMPERATURE, K
THERMAL EXPANSION COEFFICIENT 10E-6/K
100
150
200
250
300
350
400
450
500
10
12
14
16
18
20
22
24
26
28
30

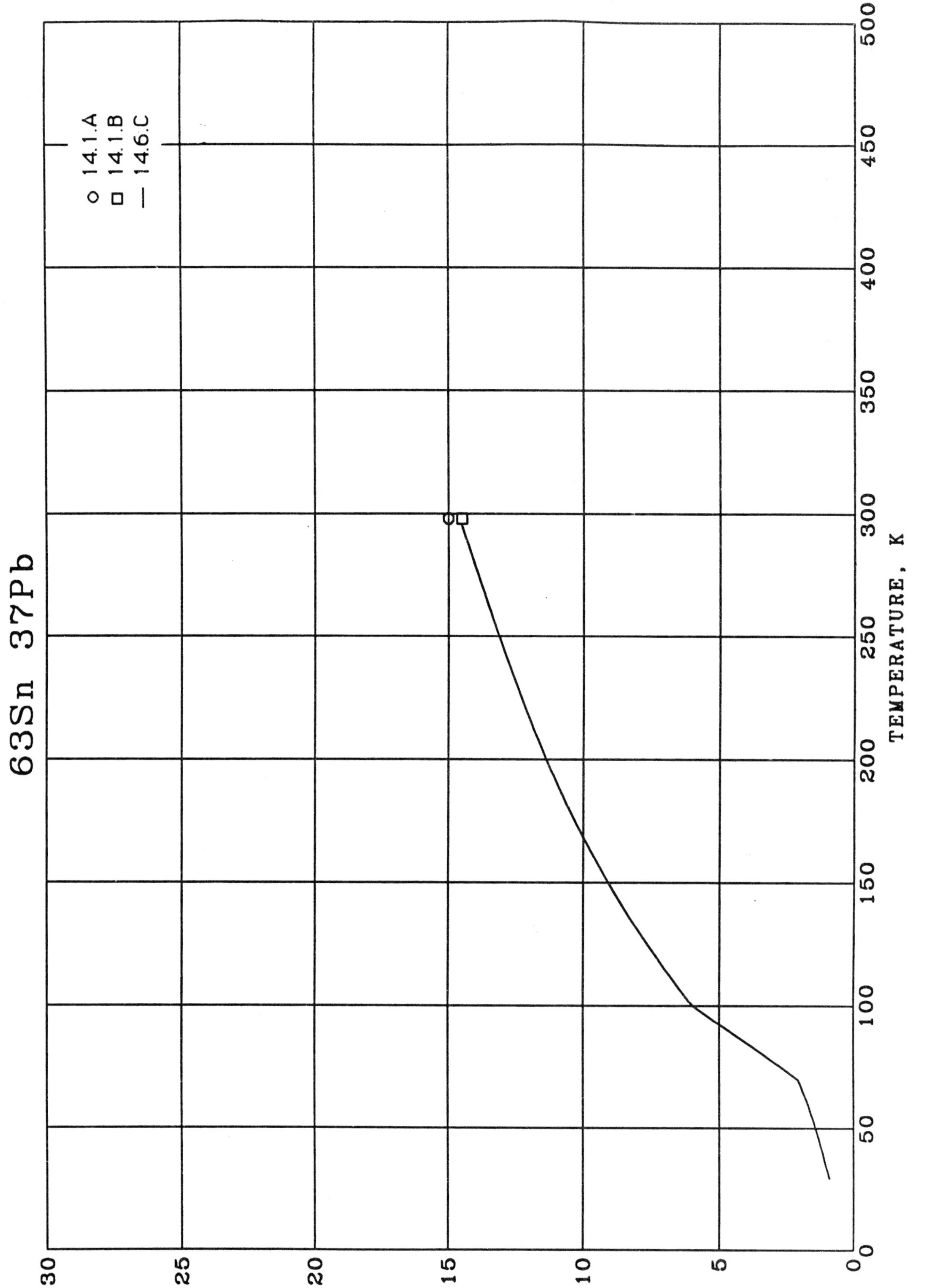

63Sn 37Pb
○ 14.1.A
□ 14.1.B
— 14.6.C
TEMPERATURE, K
ELECTRICAL RESISTIVITY, 10E-6 Ohm-cm

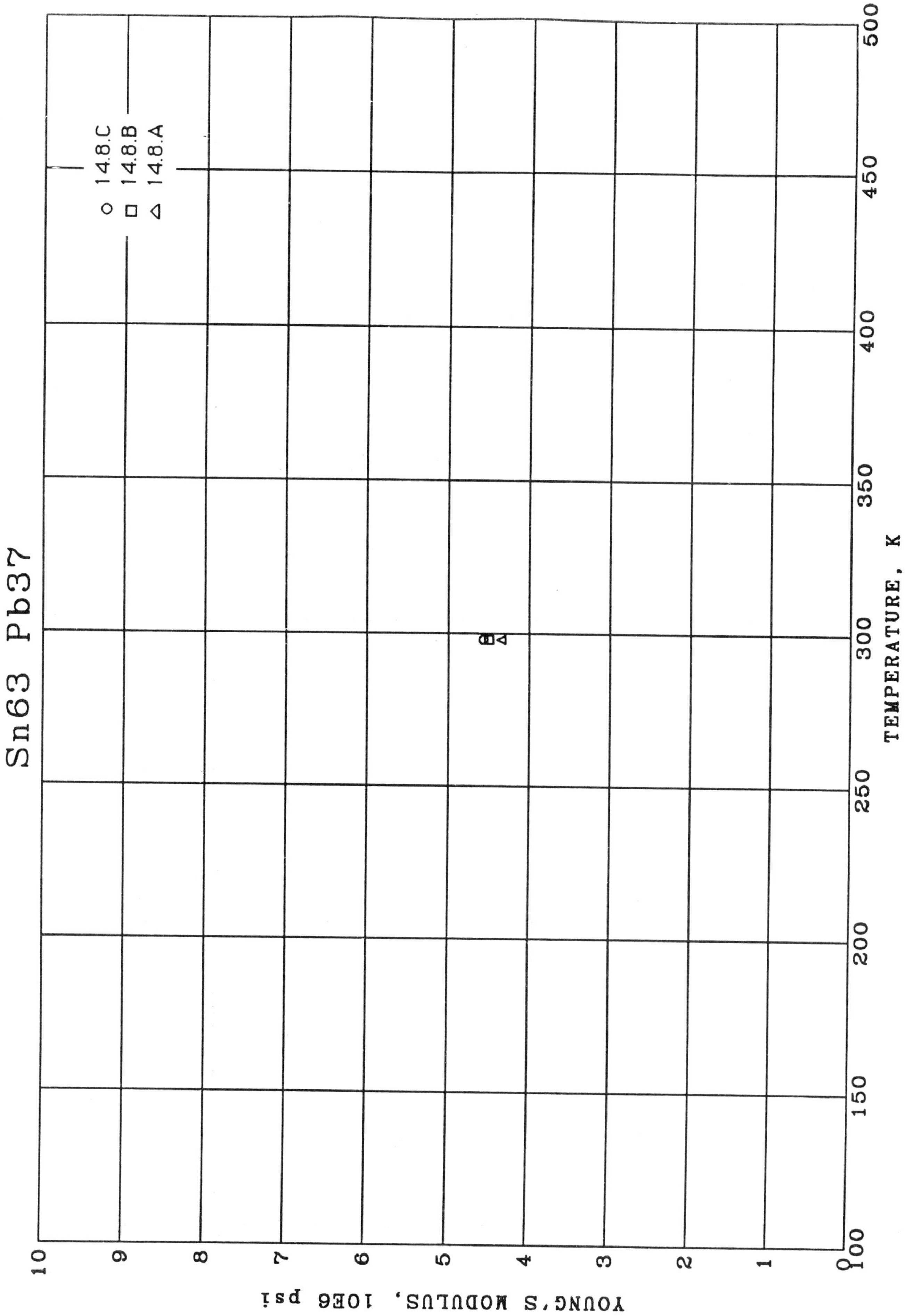
Sn63 Pb37
○ 14.8.C
□ 14.8.B
△ 14.8.A
TEMPERATURE, K
YOUNG'S MODULUS, 10E6 psi
100
150
200
250
300
350
400
450
500
0
1
2
3
4
5
6
7
8
9
10

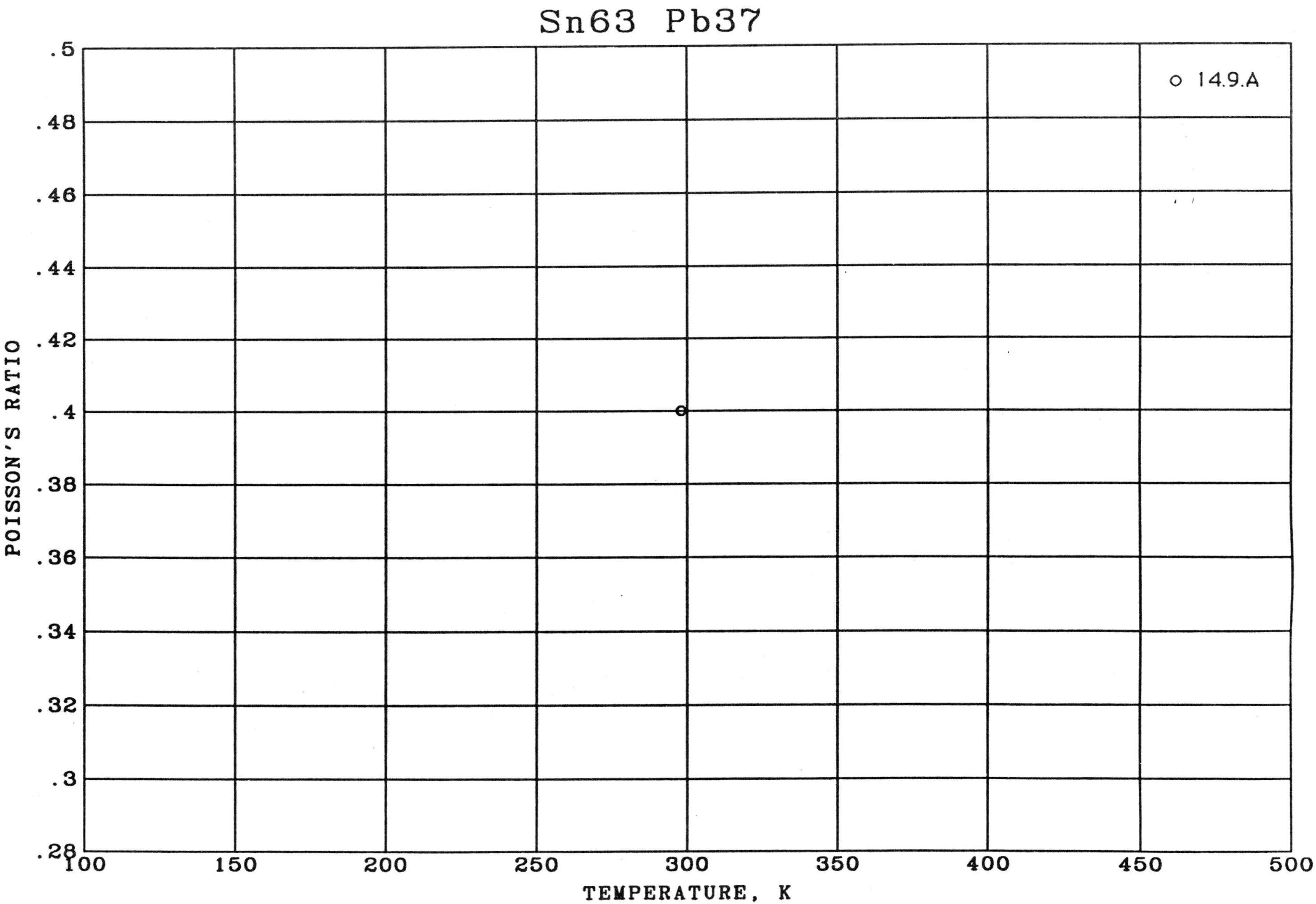
Sn63 Pb37
o 14.9.A
POISSON'S RATIO
.5
.48
.46
.44
.42
.4
.38
.36
.34
.32
.3
.28
100
150
200
250
300
350
400
450
500
TEMPERATURE, K

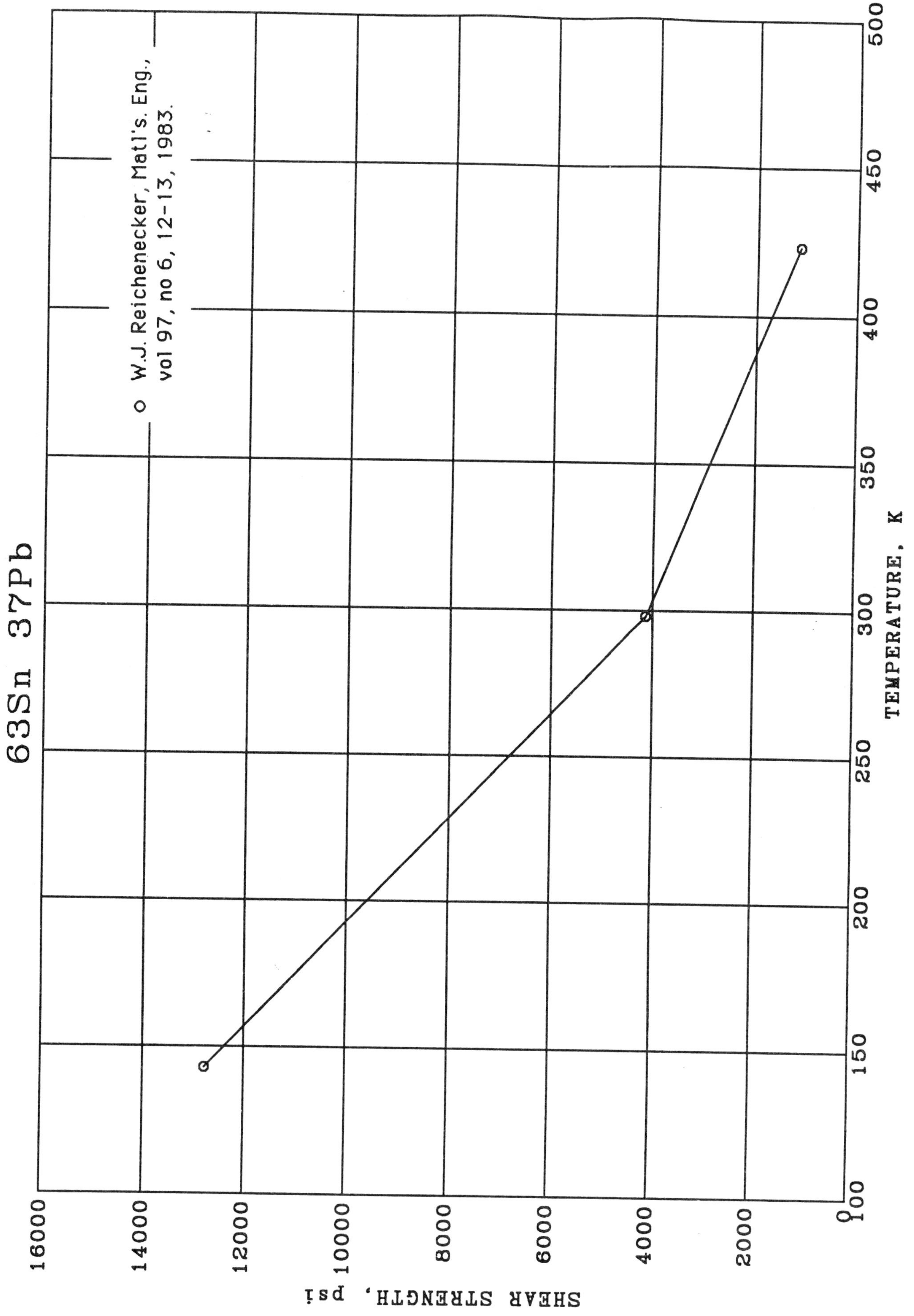
63Sn 37Pb
W.J. Reichenecker, Matl's. Eng., vol 97, no 6, 12-13, 1983.
SHEAR STRENGTH, psi
0
2000
4000
6000
8000
10000
12000
14000
16000
TEMPERATURE, K
100
150
200
250
300
350
400
450
500

63Sn 37Pb

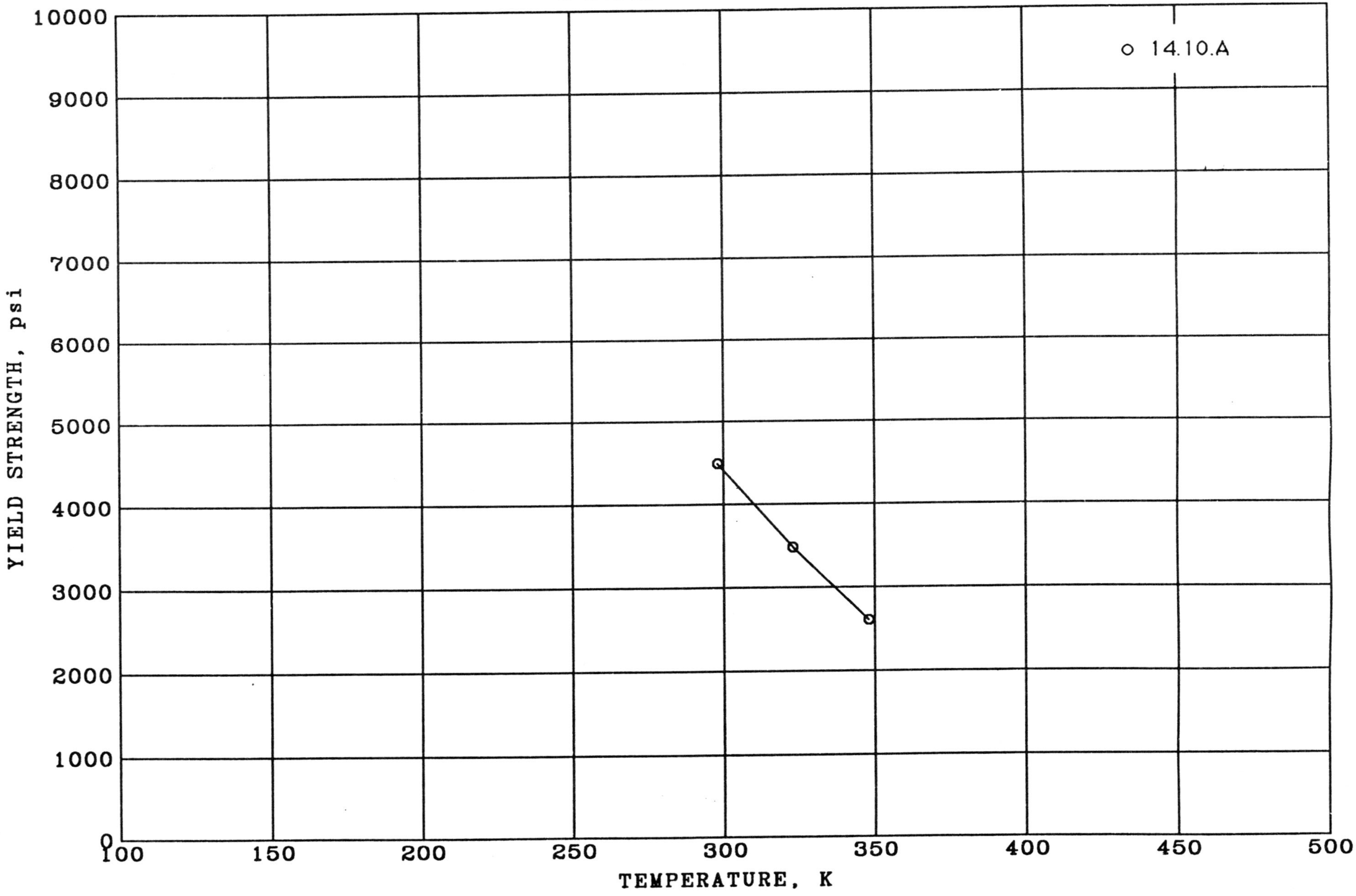

10000
9000
8000
7000
6000
5000
4000
3000
2000
1000
0
YIELD STRENGTH, psi
100
150
200
250
300
350
400
450
500
TEMPERATURE, K
14.10.A

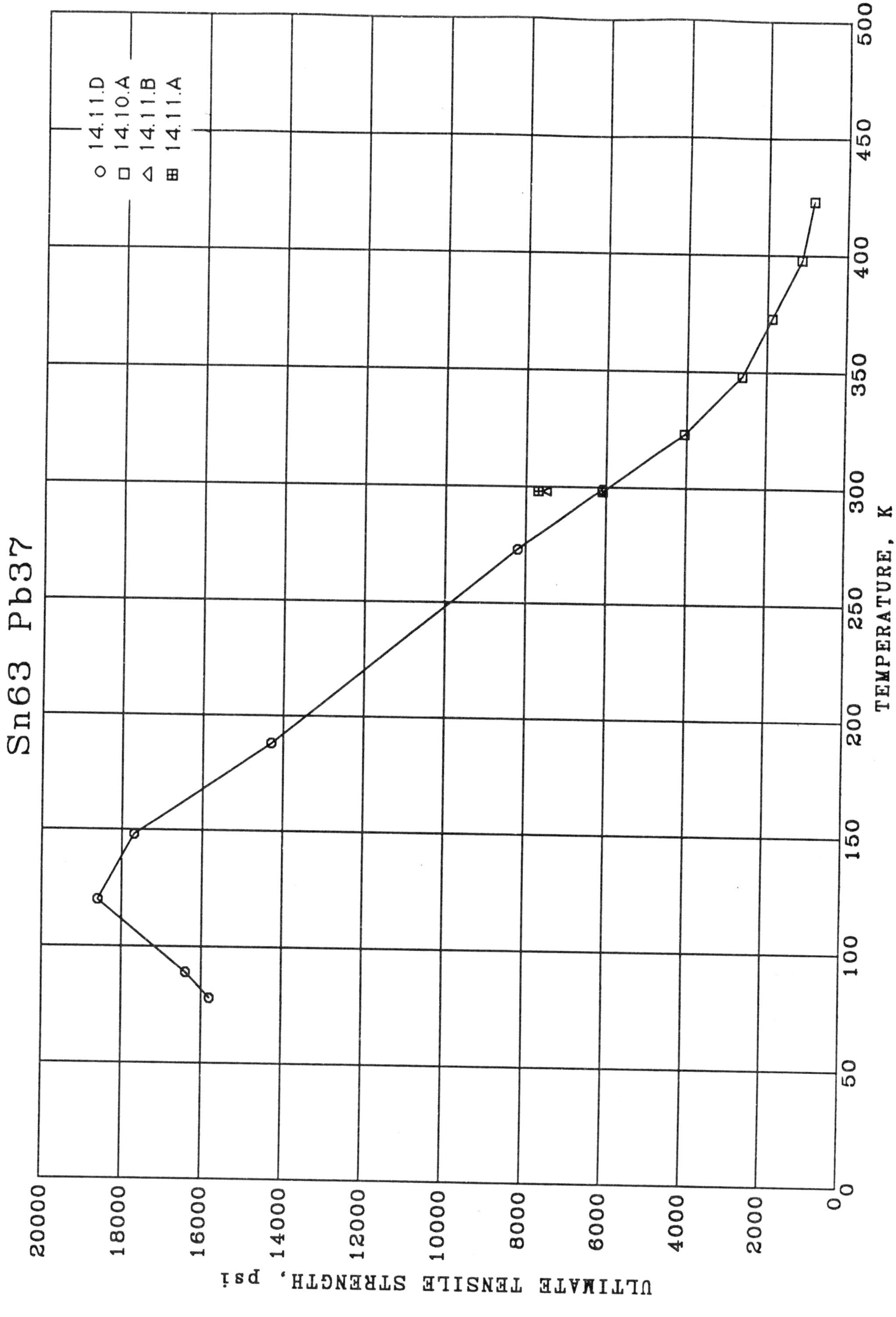
Sn63 Pb37
○ 14.11.D
□ 14.10.A
△ 14.11.B
⊞ 14.11.A
TEMPERATURE, K
ULTIMATE TENSILE STRENGTH, psi
0
50
100
150
200
250
300
350
400
450
500
2000
4000
6000
8000
10000
12000
14000
16000
18000
20000

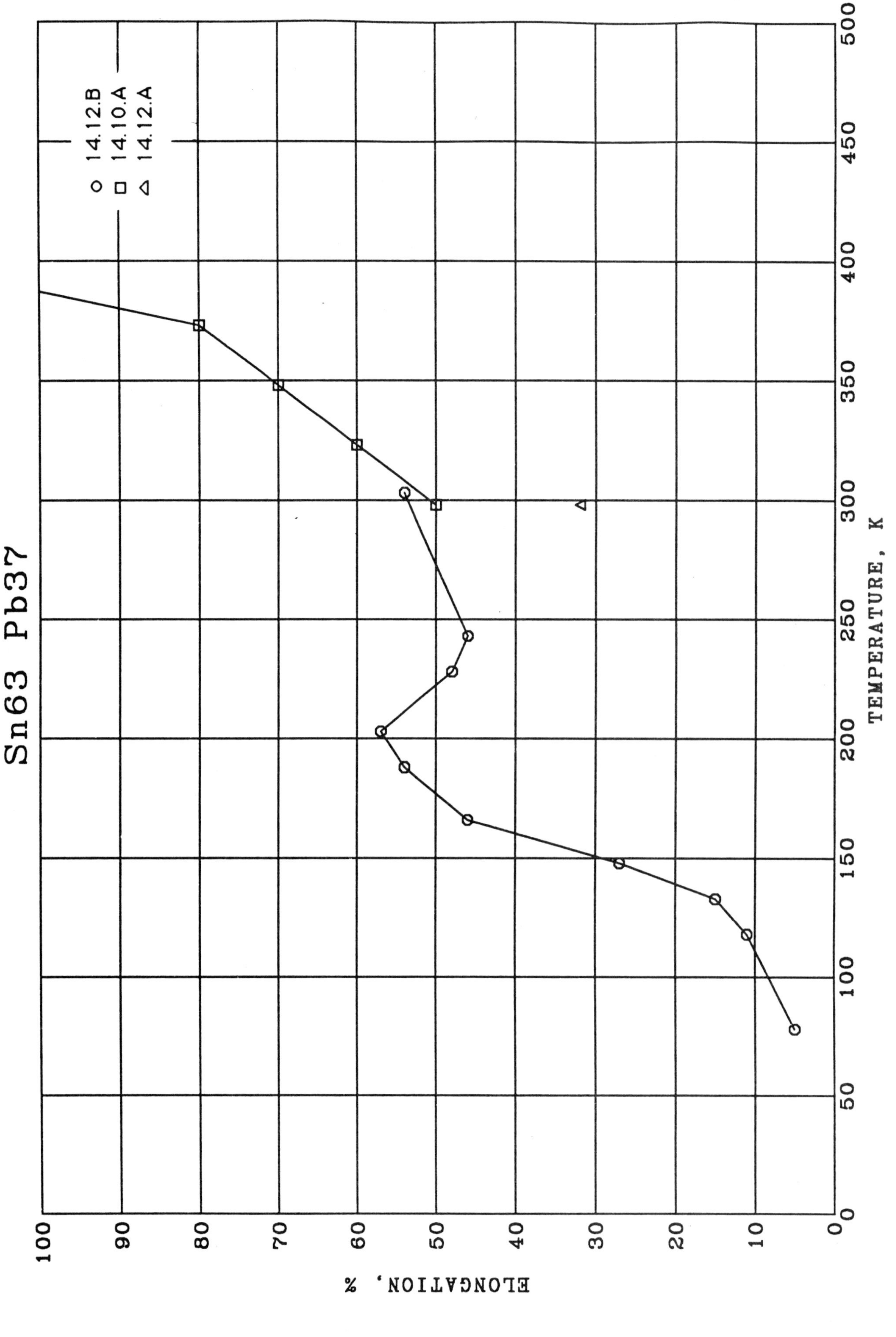

Sn63 Pb37
○ 14.12.B
□ 14.10.A
△ 14.12.A
ELONGATION, %
TEMPERATURE, K
0
10
20
30
40
50
60
70
80
90
100
0
50
100
150
200
250
300
350
400
450
500

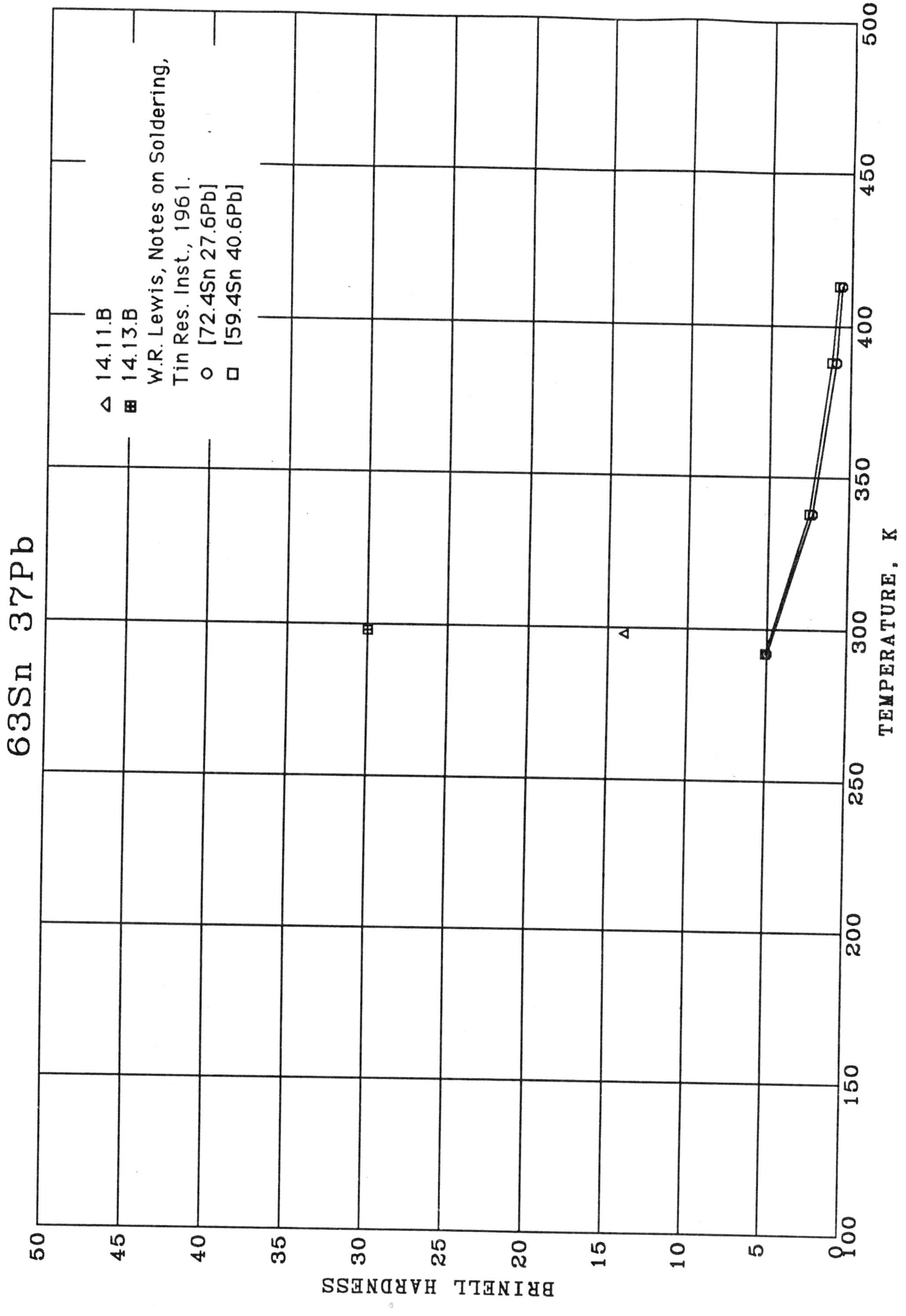

63Sn 37Pb
△ 14.11.B
⊞ 14.13.B
W.R. Lewis, Notes on Soldering,
Tin Res. Inst., 1961.
○ [72.4Sn 27.6Pb]
□ [59.4Sn 40.6Pb]
BRINELL HARDNESS
0
5
10
15
20
25
30
35
40
45
50
TEMPERATURE, K
100
150
200
250
300
350
400
450
500

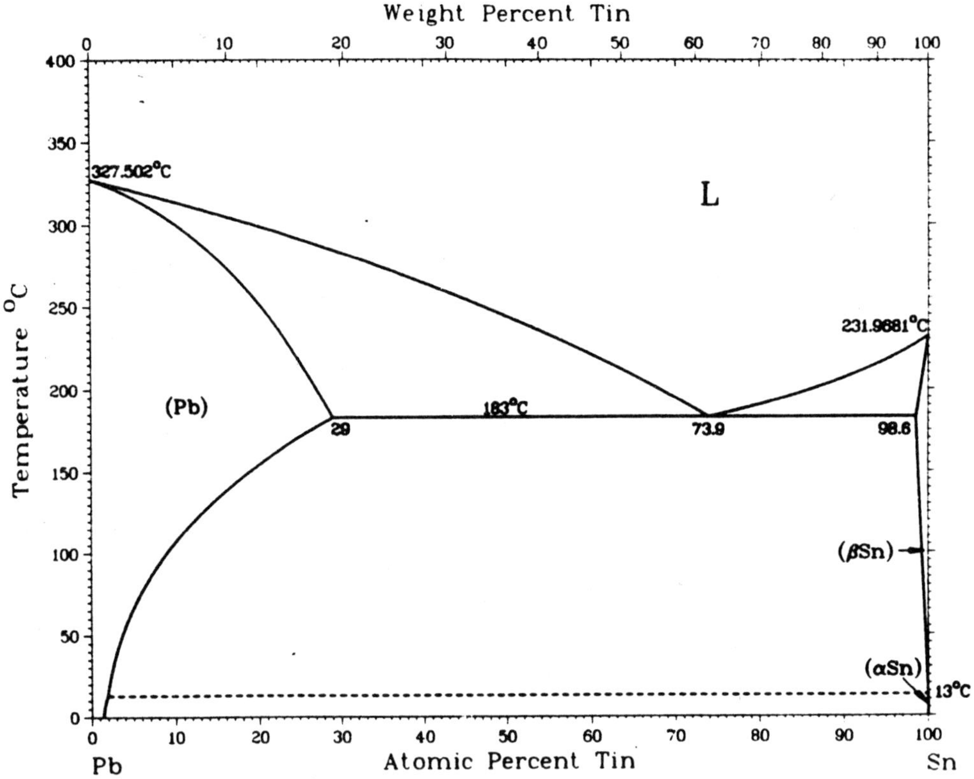
Weight Percent Tin
Temperature °C
327.502°C
L
231.9681°C
(Pb)
183°C
29
73.9
98.6
(βSn)
(αSn)
13°C
Pb
Atomic Percent Tin
Sn

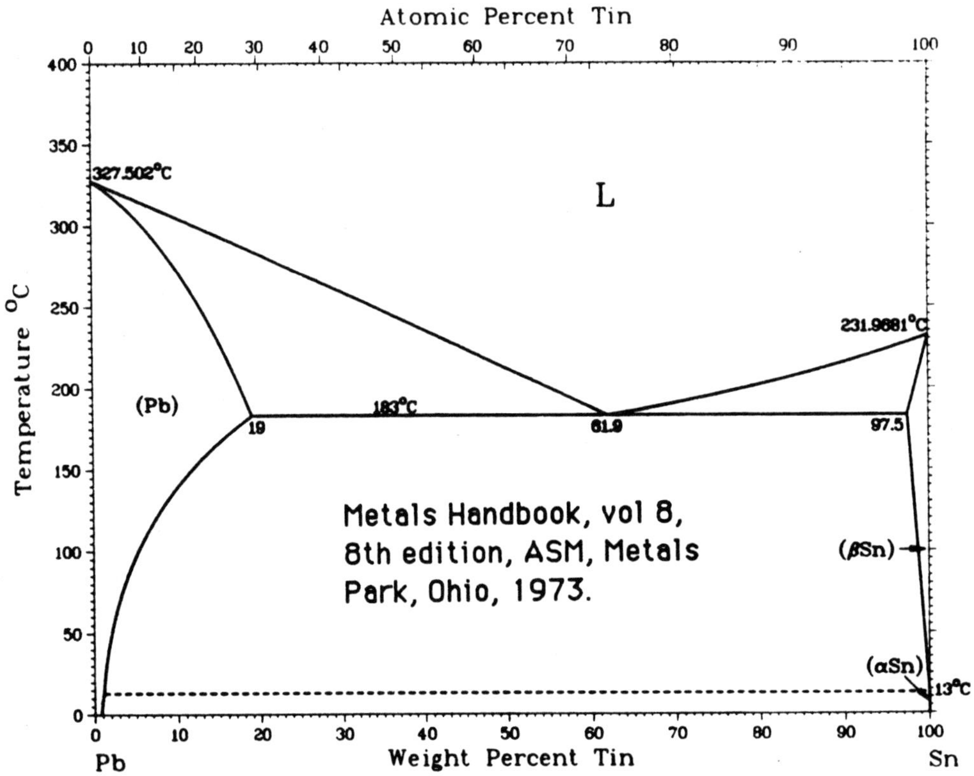
Atomic Percent Tin
Temperature °C
327.502°C
L
231.9681°C
(Pb)
183°C
19
61.9
97.5
Metals Handbook, vol 8,
8th edition, ASM, Metals
Park, Ohio, 1973.
(βSn)
(αSn)
13°C
Pb
Weight Percent Tin
Sn

54Fe 29Ni 17Co

PROPERTY	@ 300K	UNITS	REFERENCES
Density	8.37	g/cm^3	15.1.A-B
Specific Heat	0.105	cal/g-K	15.2.A
Thermal Conductivity	0.0339	cal/cm-s-K	15.3.A-C
Thermal Expansion	$8.4x10^{-6}$	1/K	15.4.A-D
Melting Point	1723	K	15.5.A
Resistivity	$49x10^{-6}$	Ohm-cm	15.6.A-C
TCR	3700	ppm/K	15.7.A
Young's Modulus	20,000	ksi	15.8.A-B
Poisson's Ratio		------	
Yield Strength	55,000	psi	15.10.A-B
Ult Tensile Strength	80,000	psi	15.11.A-B
Elongation at Break	35	%	15.12.A
Hardness	200-250	Brinell	15.13.A

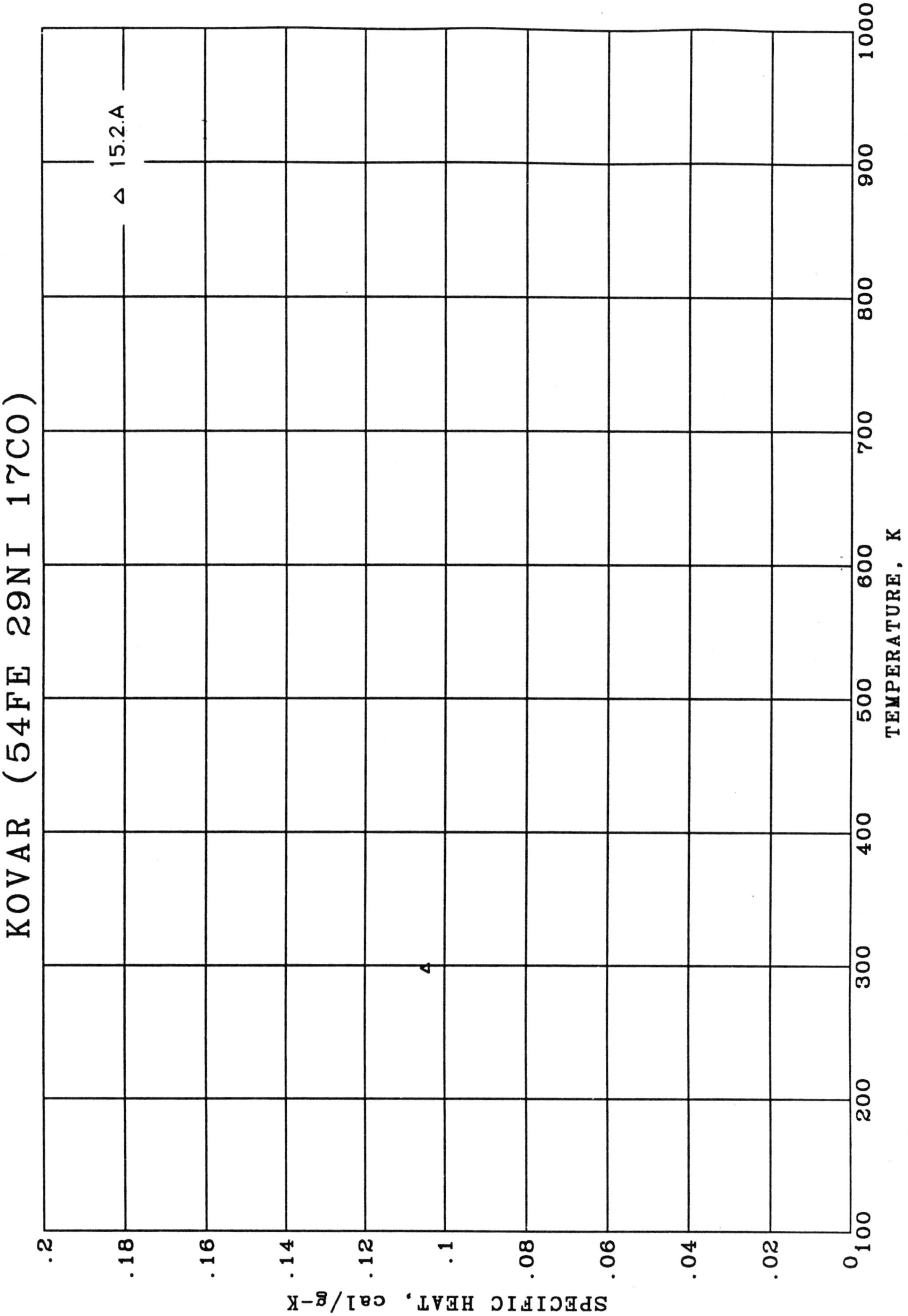

KOVAR (54FE 29NI 17CO)
15.2A
SPECIFIC HEAT, cal/g-K
TEMPERATURE, K
0
.02
.04
.06
.08
.1
.12
.14
.16
.18
.2
100
200
300
400
500
600
700
800
900
1000

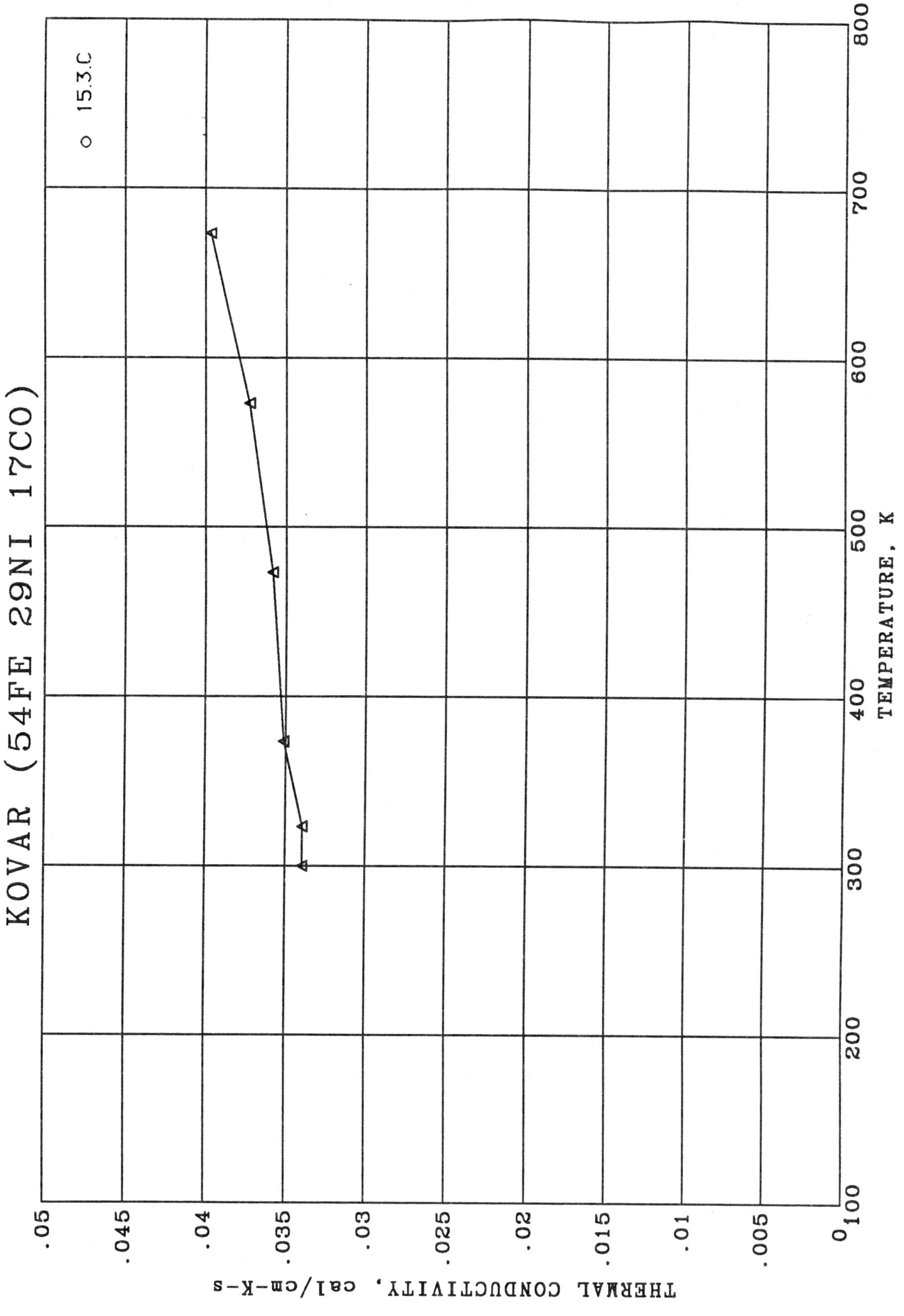
KOVAR (54FE 29NI 17CO)
○ 15.3.C
THERMAL CONDUCTIVITY, cal/cm-K-s
TEMPERATURE, K
0
.005
.01
.015
.02
.025
.03
.035
.04
.045
.05
100
200
300
400
500
600
700
800

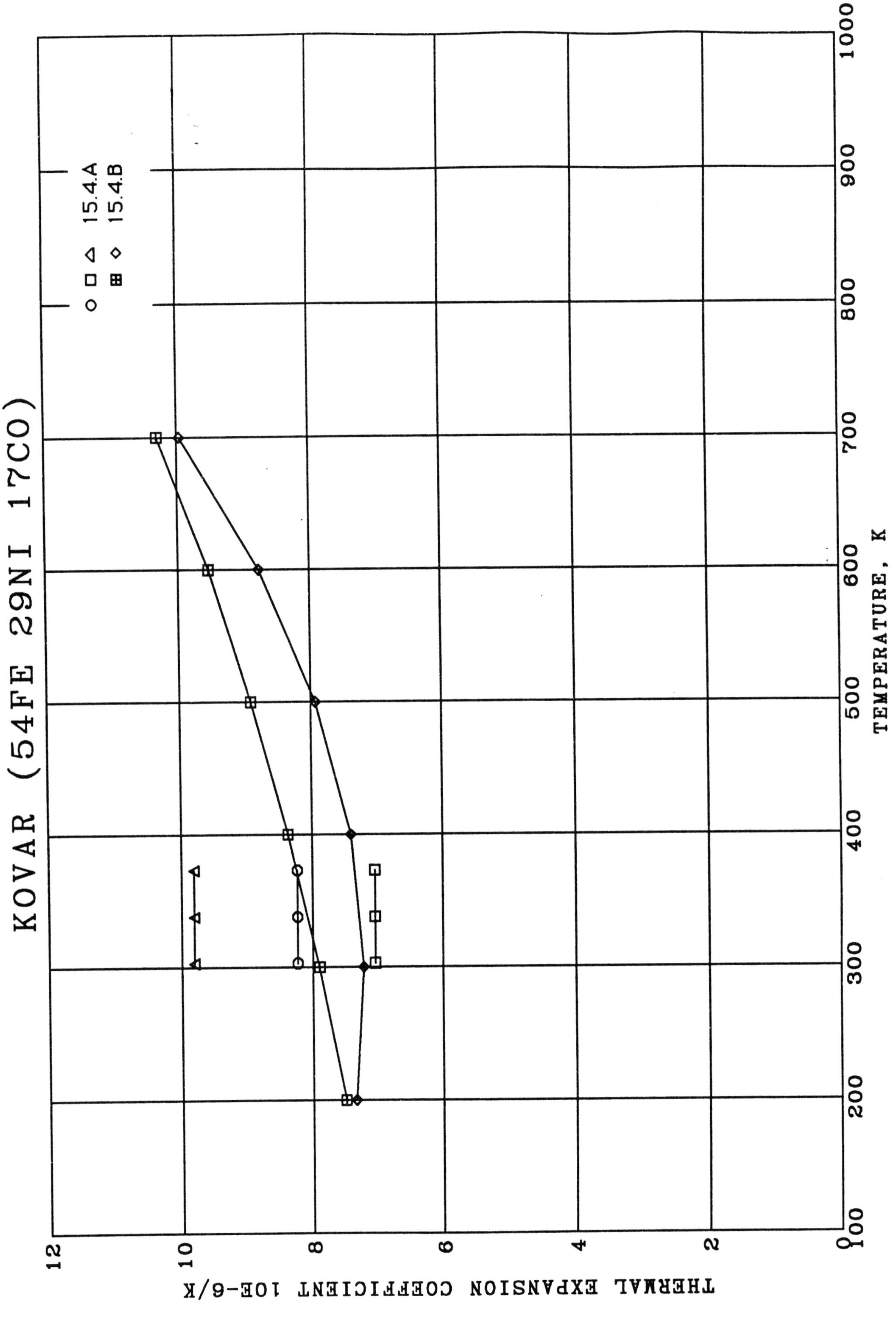
KOVAR (54FE 29NI 17CO)
○ □ △ 15.4.A
⊞ ◇ 15.4.B
THERMAL EXPANSION COEFFICIENT 1OE-6/K
0
2
4
6
8
10
12
TEMPERATURE, K
100
200
300
400
500
600
700
800
900
1000

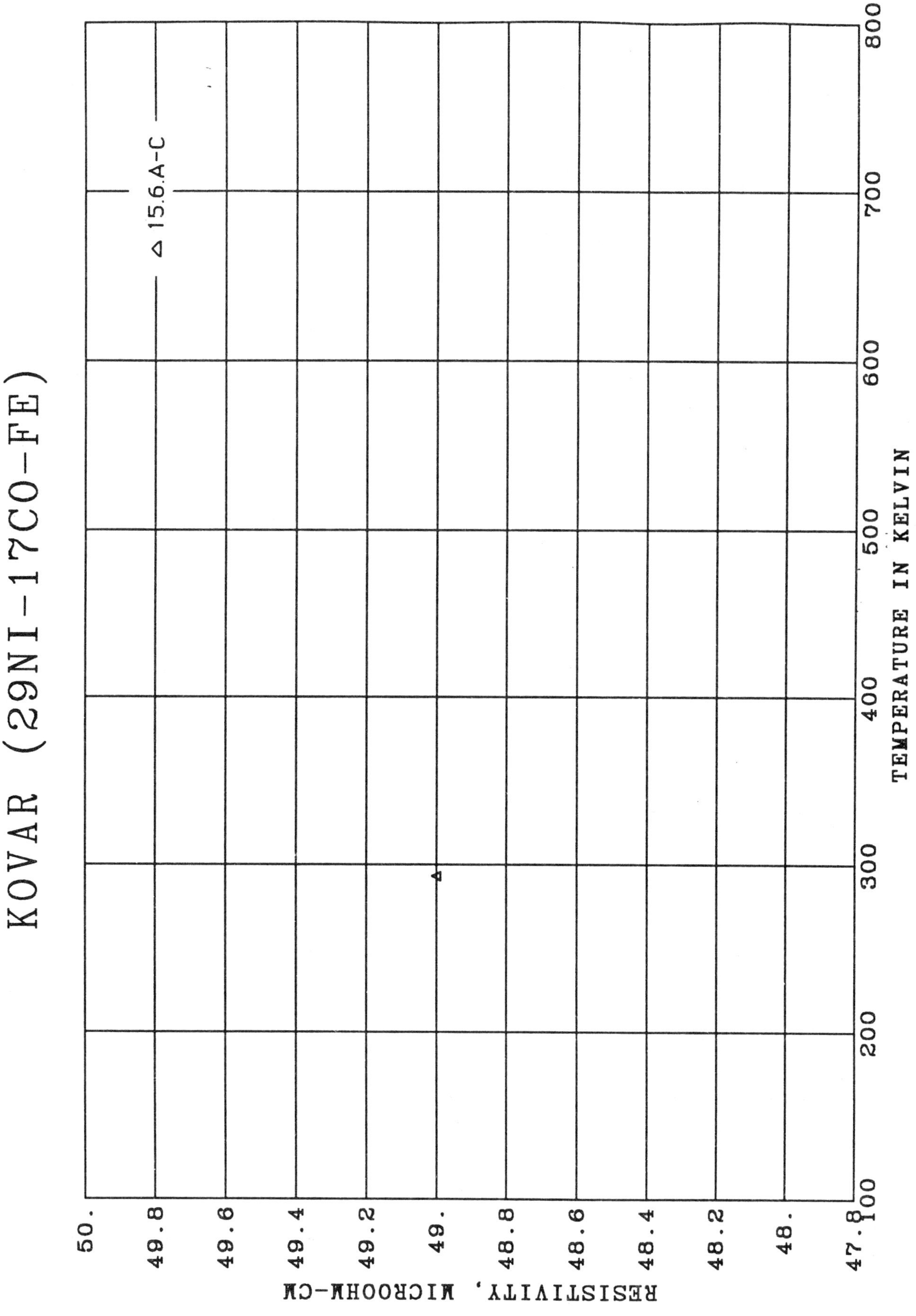
KOVAR (29NI-17CO-FE)
RESISTIVITY, MICROOHM-CM
TEMPERATURE IN KELVIN
50.
49.8
49.6
49.4
49.2
49.
48.8
48.6
48.4
48.2
48.
47.8
100
200
300
400
500
600
700
800
15.6.A-C

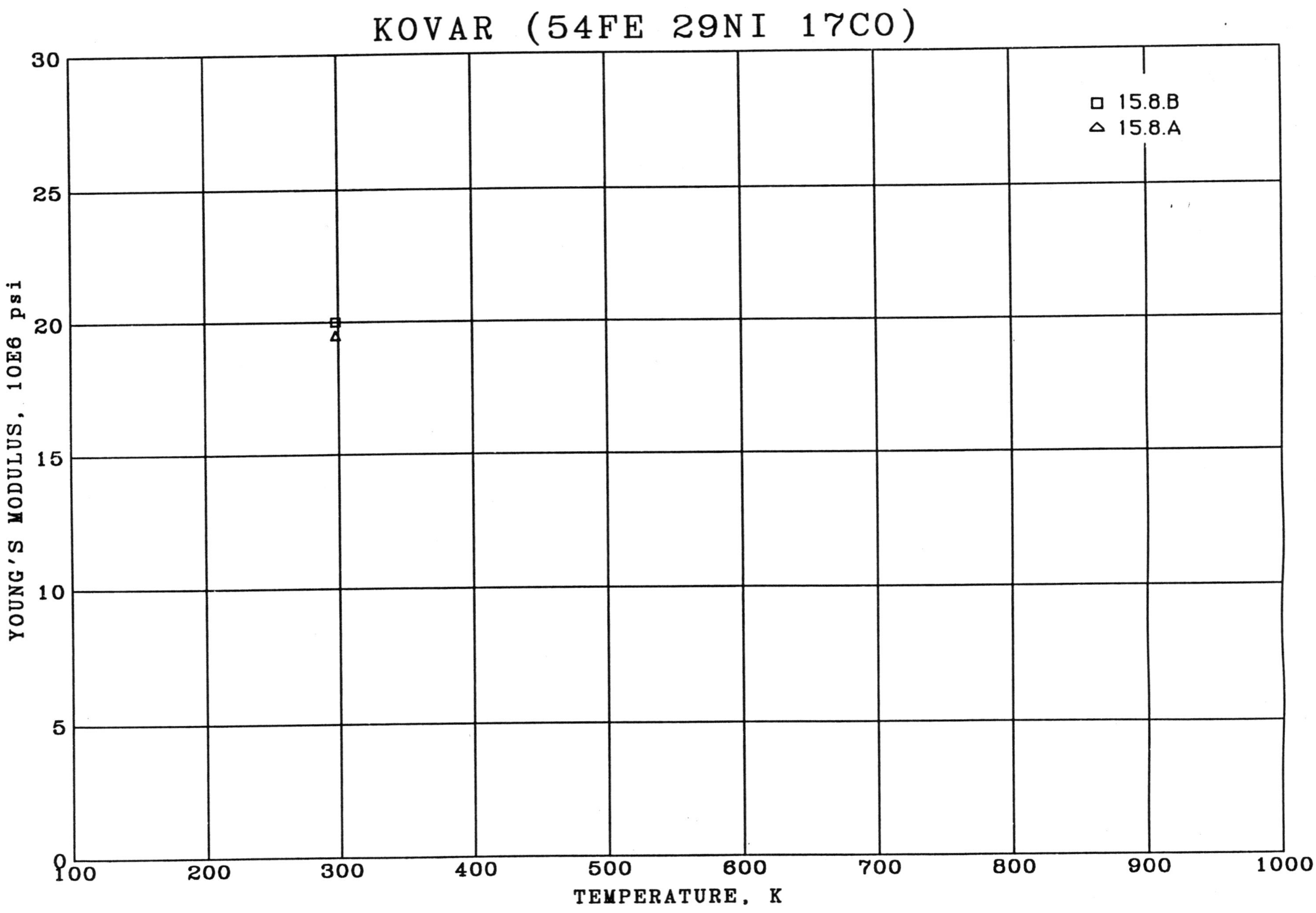
KOVAR (54FE 29NI 17CO)
15.8.B
15.8.A
30
25
20
15
10
5
0
100
200
300
400
500
600
700
800
900
1000
YOUNG'S MODULUS, 10E6 psi
TEMPERATURE, K

KOVAR (54FE 29NI 17CO)

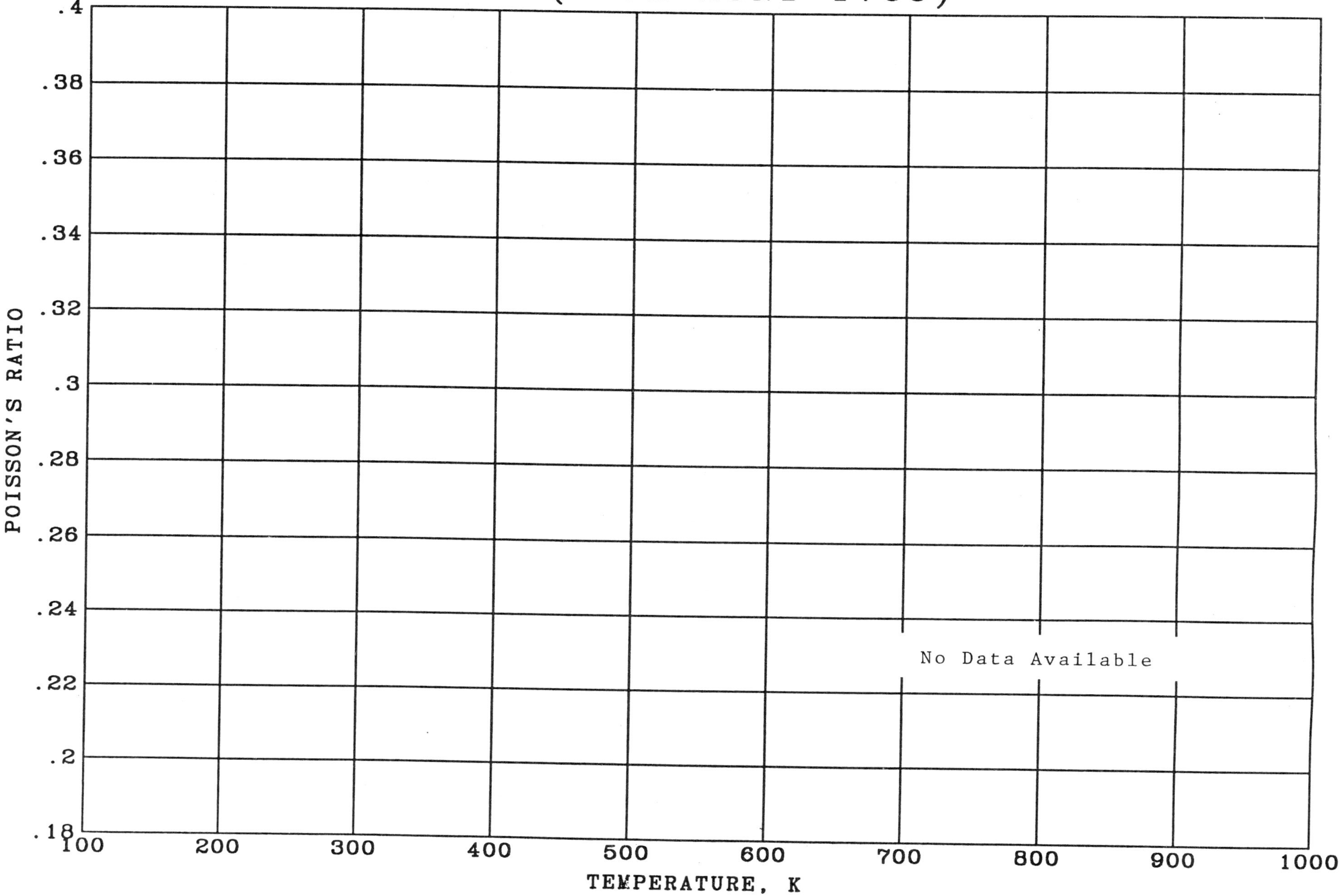

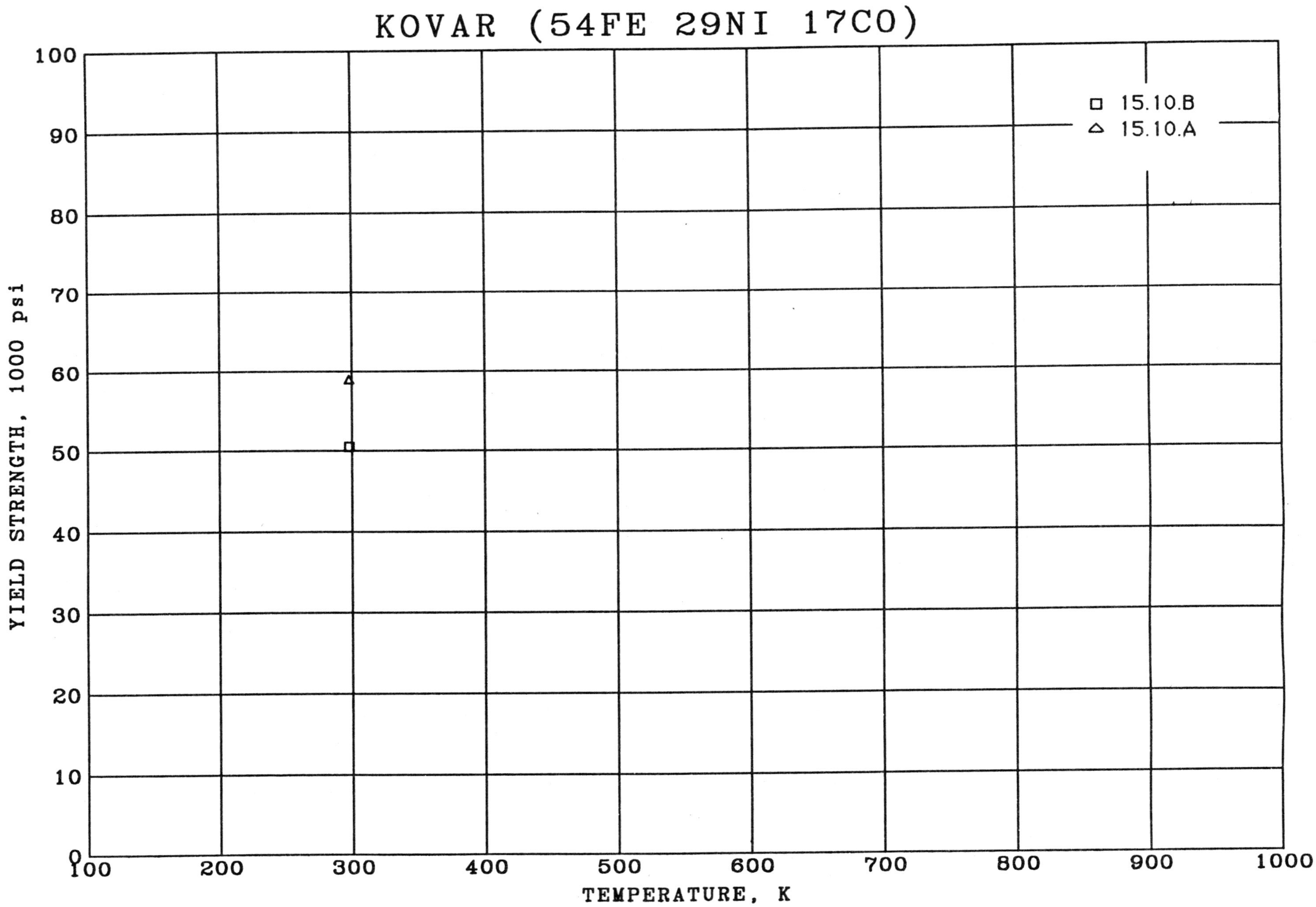
KOVAR (54FE 29NI 17CO)
15.10.B
15.10.A
YIELD STRENGTH, 1000 psi
0
10
20
30
40
50
60
70
80
90
100
TEMPERATURE, K
100
200
300
400
500
600
700
800
900
1000

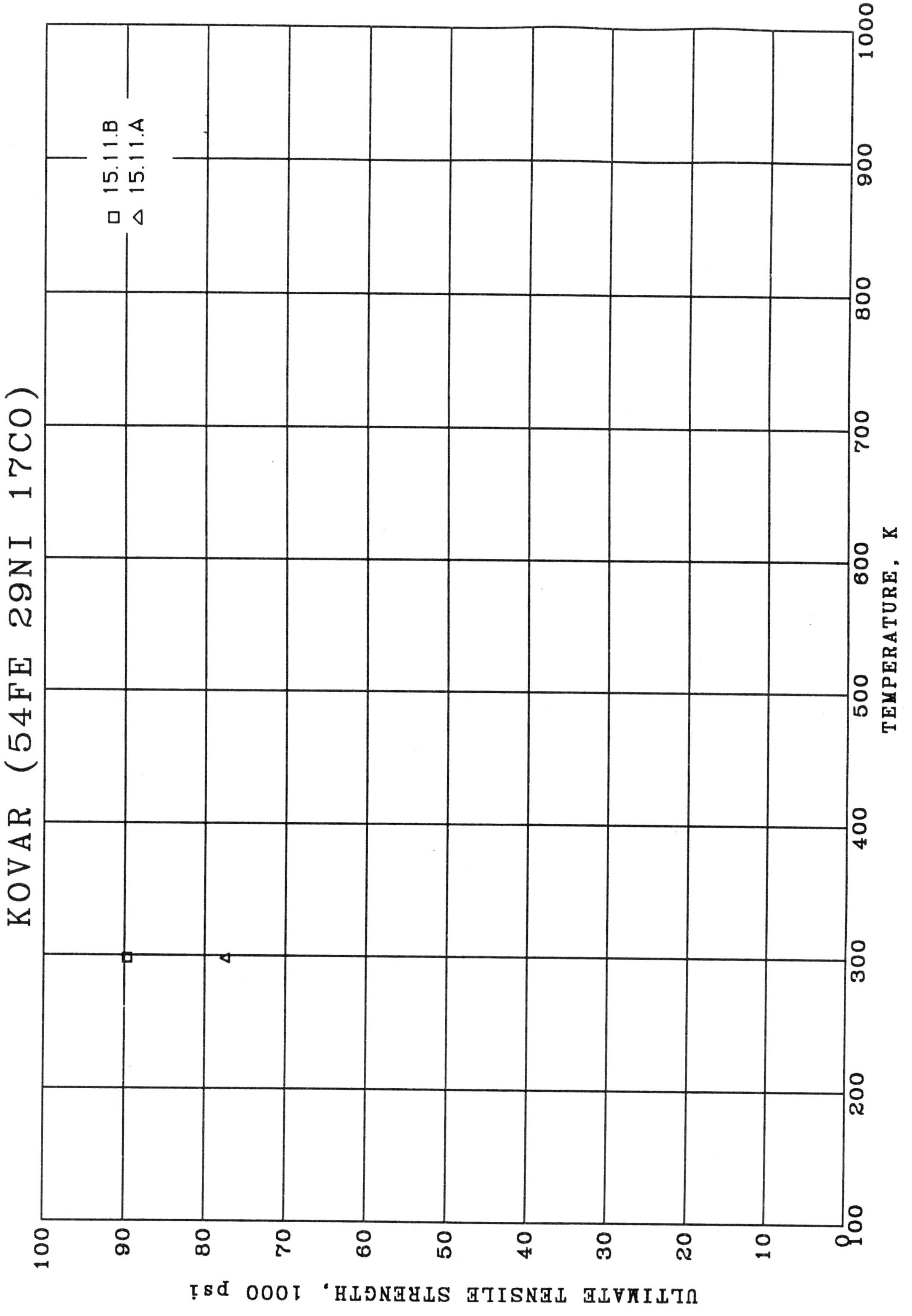

KOVAR (54FE 29NI 17CO)
□ 15.11.B
△ 15.11.A
ULTIMATE TENSILE STRENGTH, 1000 psi
TEMPERATURE, K
0
10
20
30
40
50
60
70
80
90
100
200
300
400
500
600
700
800
900
1000

KOVAR (54FE 29NI 17CO)

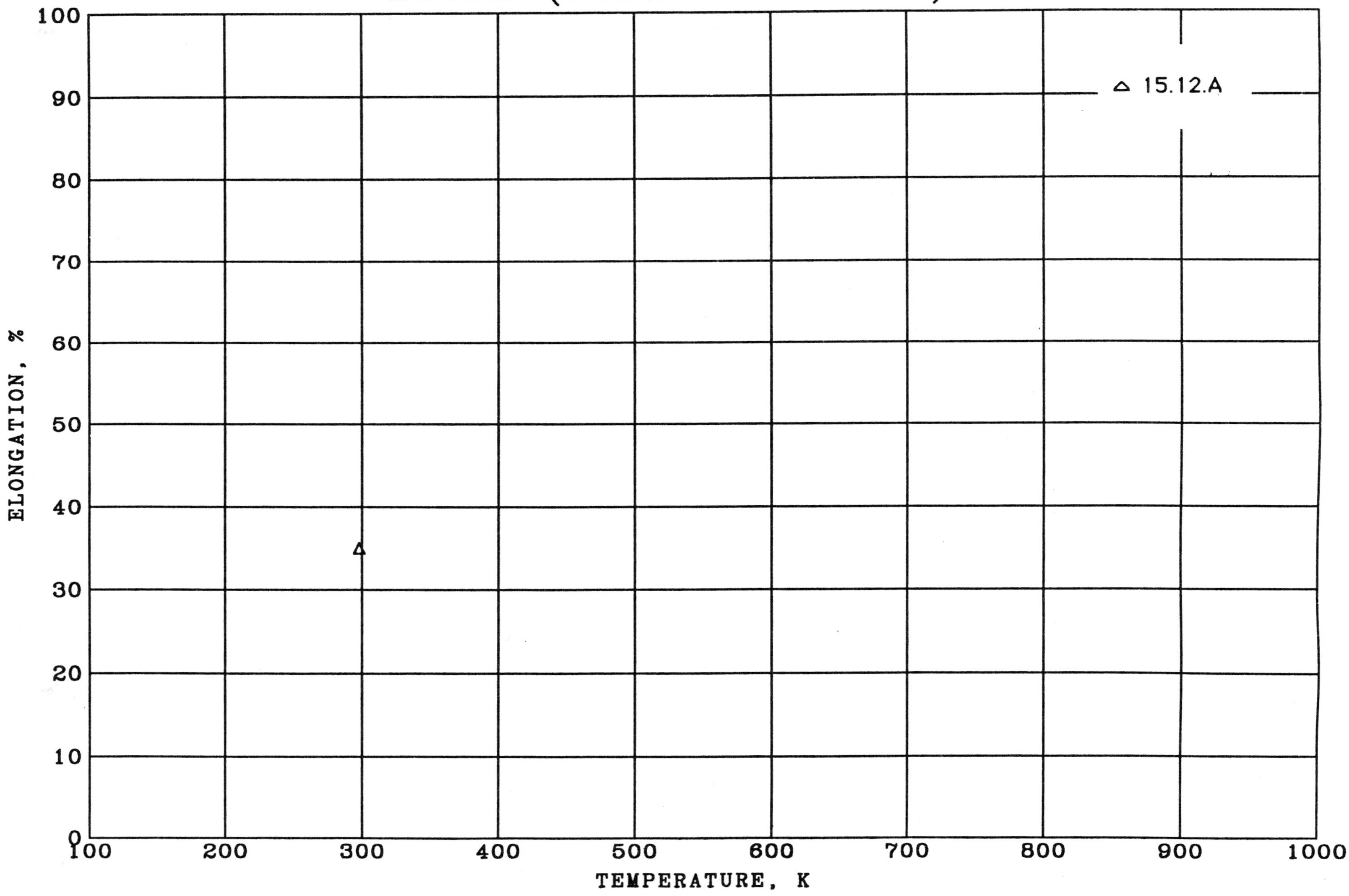

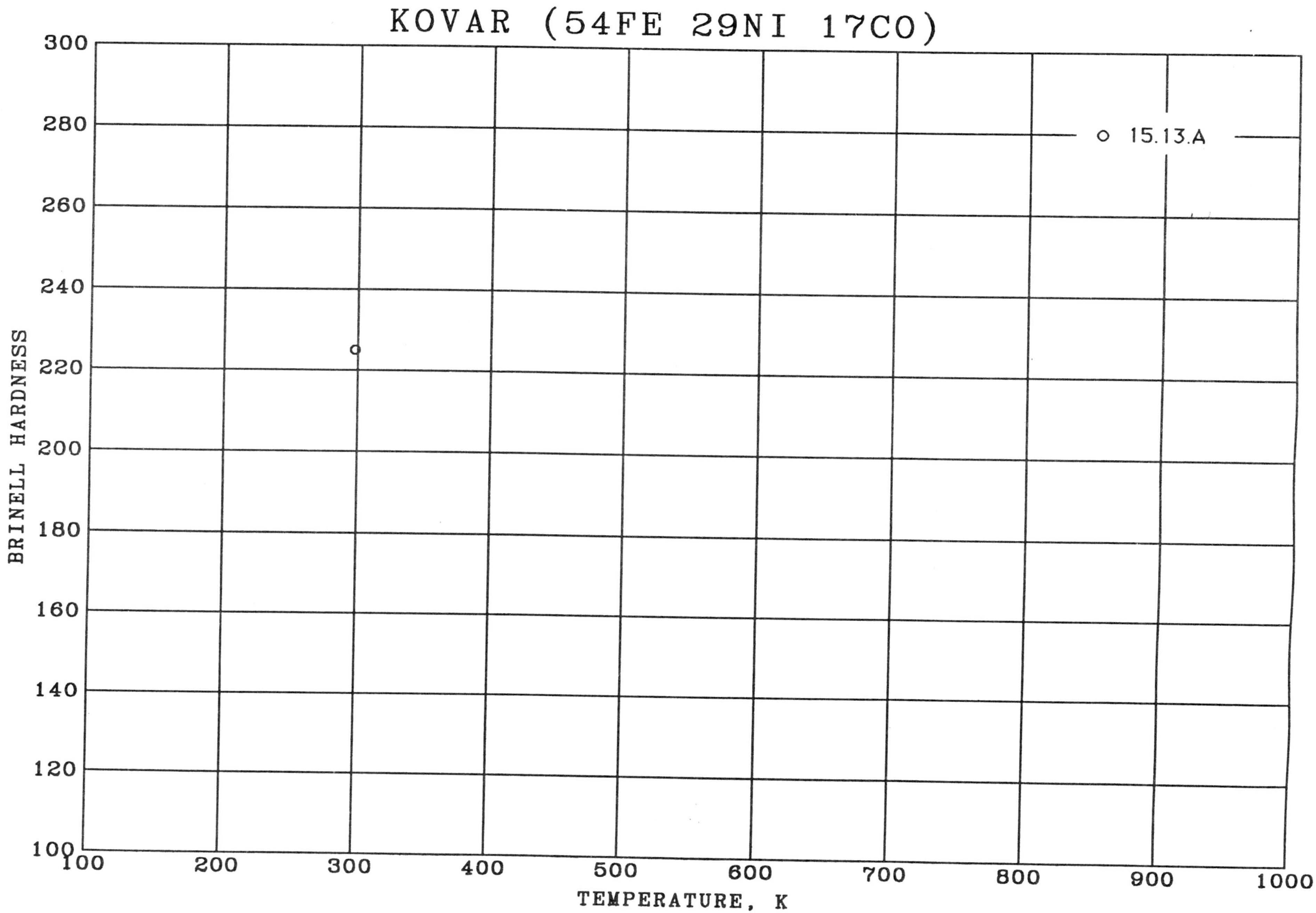
KOVAR (54FE 29NI 17CO)
15.13.A
BRINELL HARDNESS
TEMPERATURE, K
300
280
260
240
220
200
180
160
140
120
100
100
200
300
400
500
600
700
800
900
1000

This page intentionally left blank

88Pb 10Sn 2Ag

PROPERTY	@ 300K	UNITS	REFERENCES
Density	10.8	g/cm^3	16.1.A
Specific Heat	0.034	cal/g-K	16.2.A-B
Thermal Conductivity	0.1	cal/cm-s-K	16.3.A
Thermal Expansion	29.1	1/K	16.4.A
Solidus/Liquidus	540/572	K	16.5.A-C
Resistivity		Ohm-cm	
TCR		ppm/K	
Young's Modulus		psi	
Poisson's Ratio		------	
Yield Strength		psi	
Ult Tensile Strength	5000	psi	16.11.A-C
Elongation at Break	11	%	16.12.A-B
Hardness			

88Pb 10Sn 2Ag

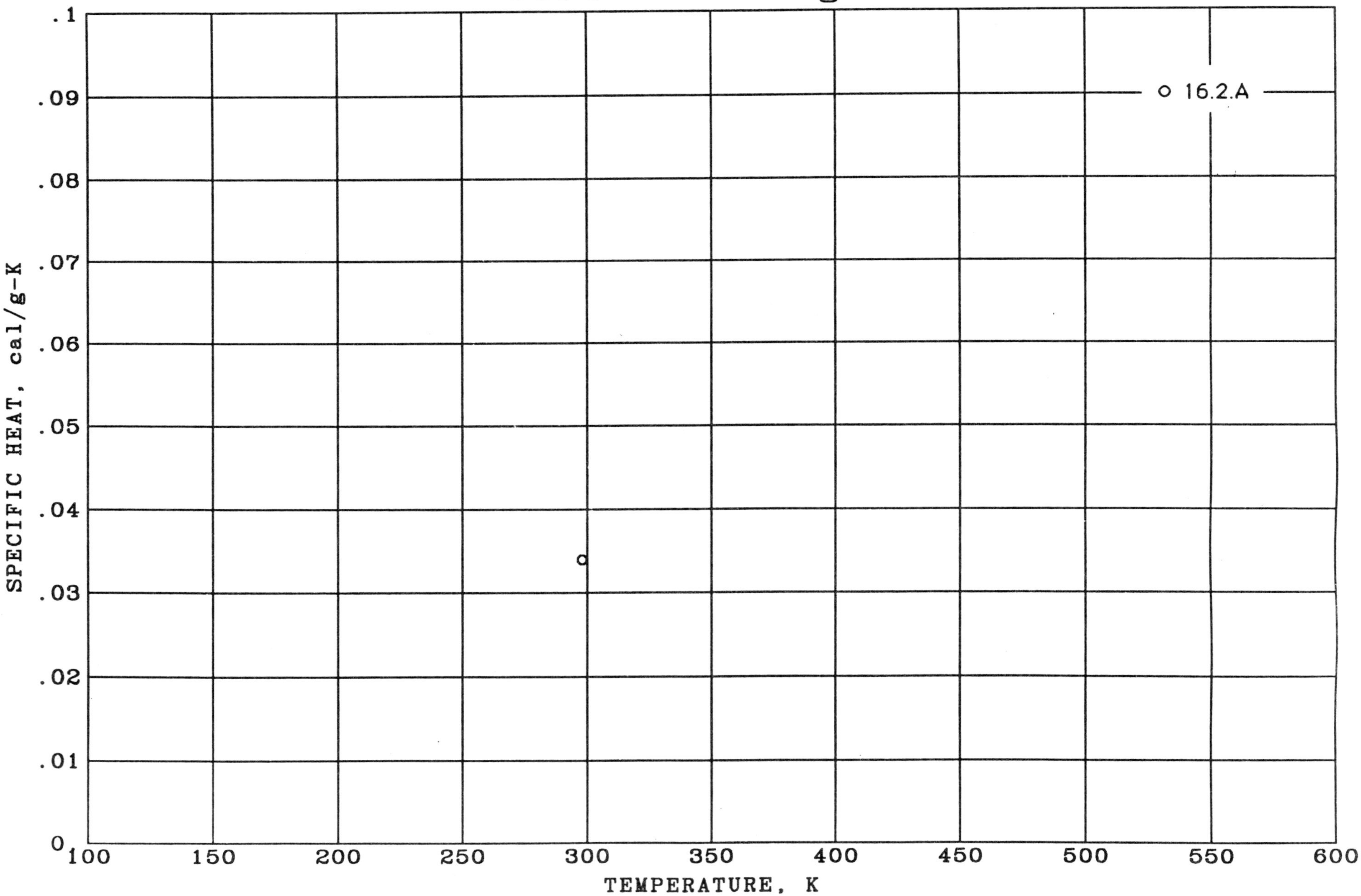

88Pb 10Sn 2Ag

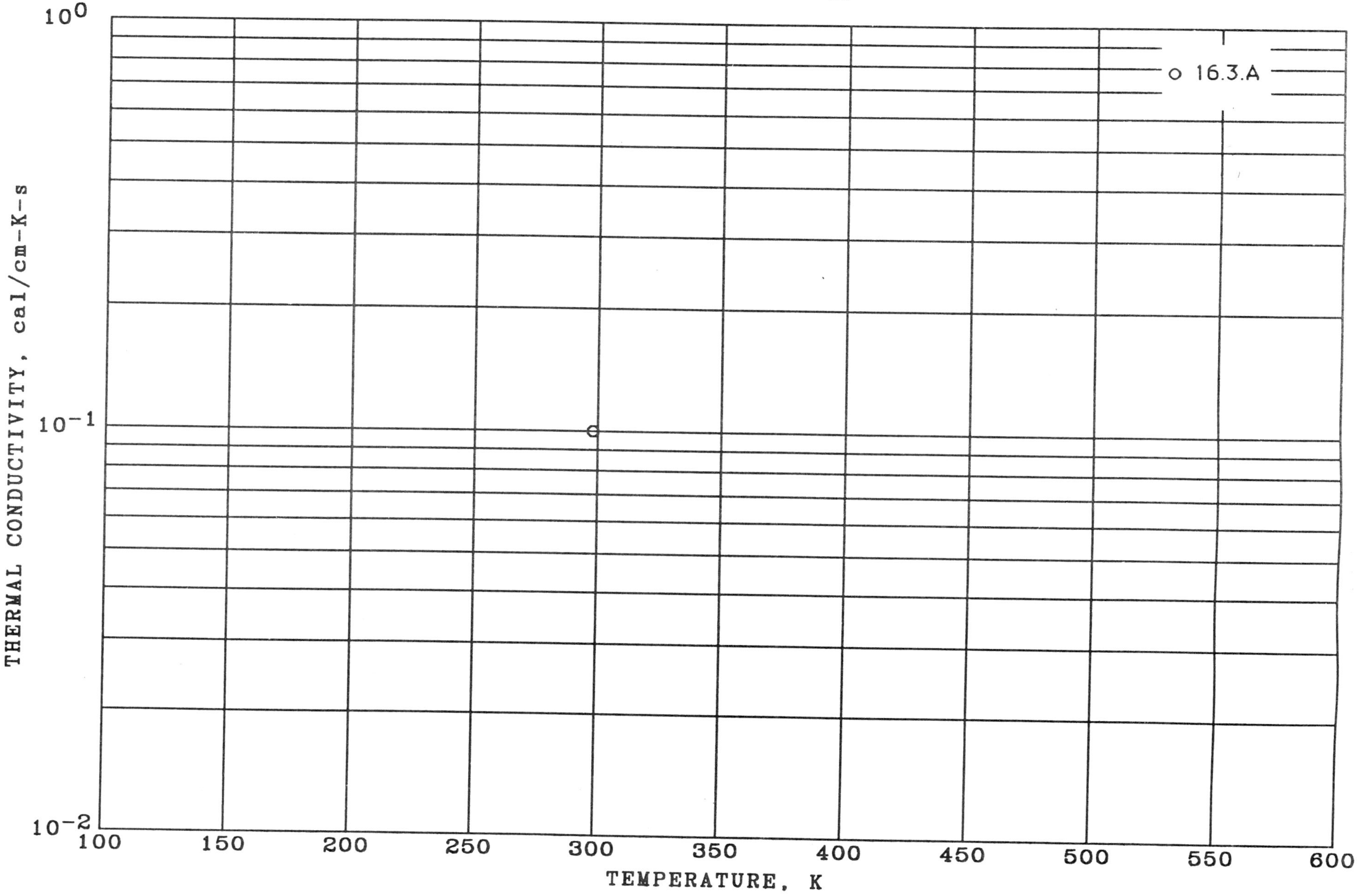

88Pb 10Sn 2Ag

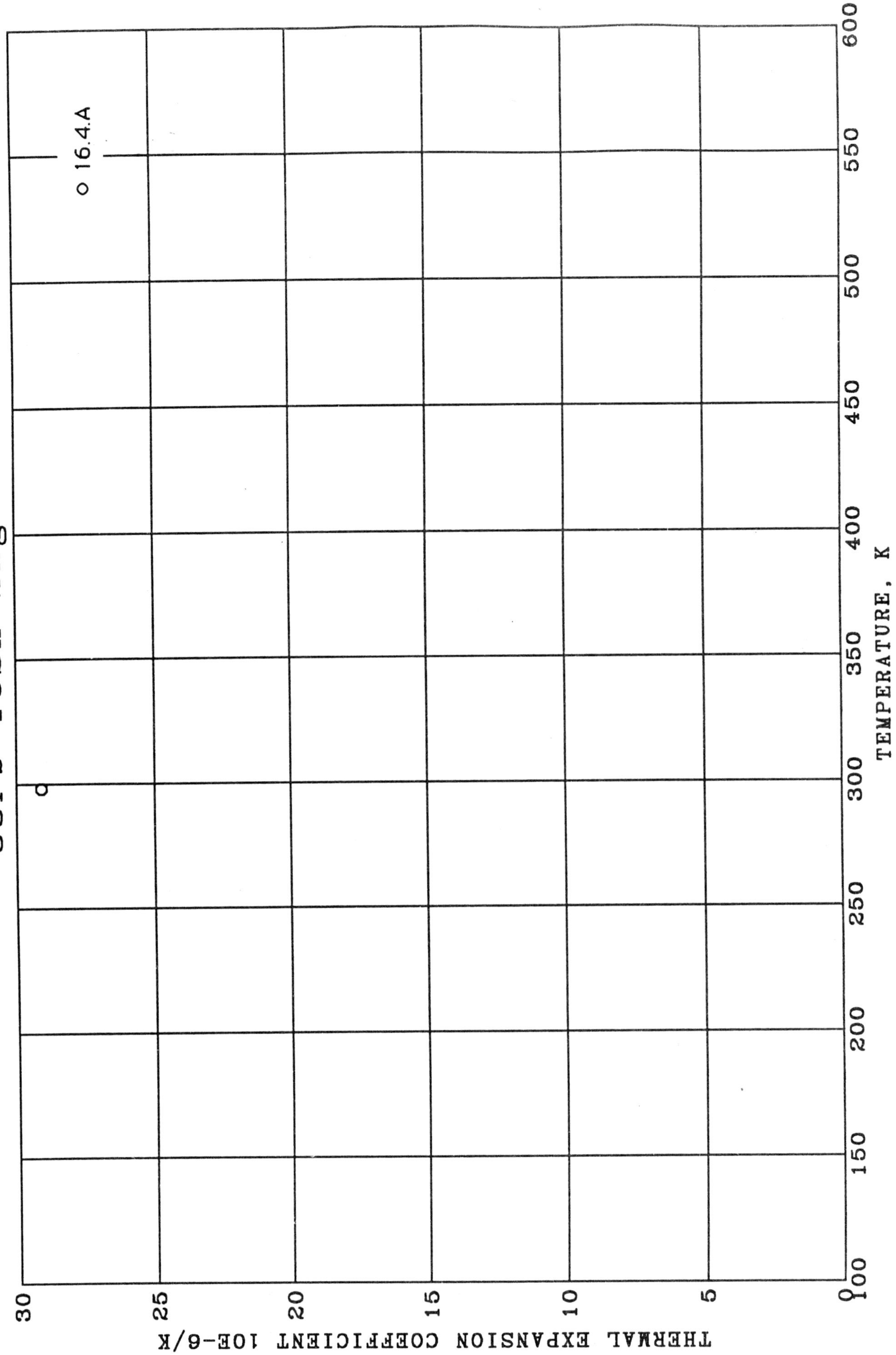

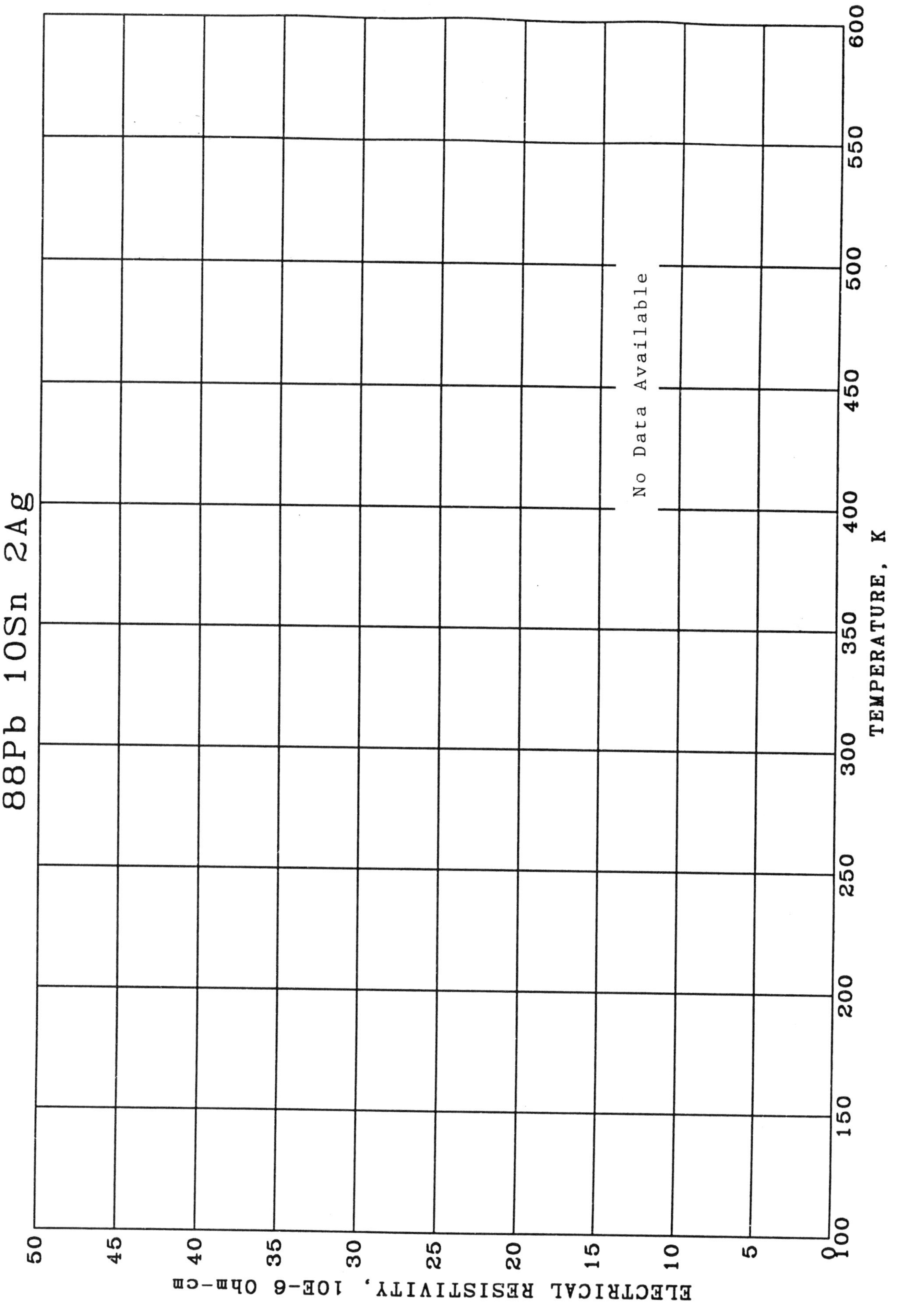
88Pb 10Sn 2Ag
No Data Available
TEMPERATURE, K
100
150
200
250
300
350
400
450
500
550
600
ELECTRICAL RESISTIVITY, 10E-6 Ohm-cm
0
5
10
15
20
25
30
35
40
45
50

88Pb 10Sn 2Ag

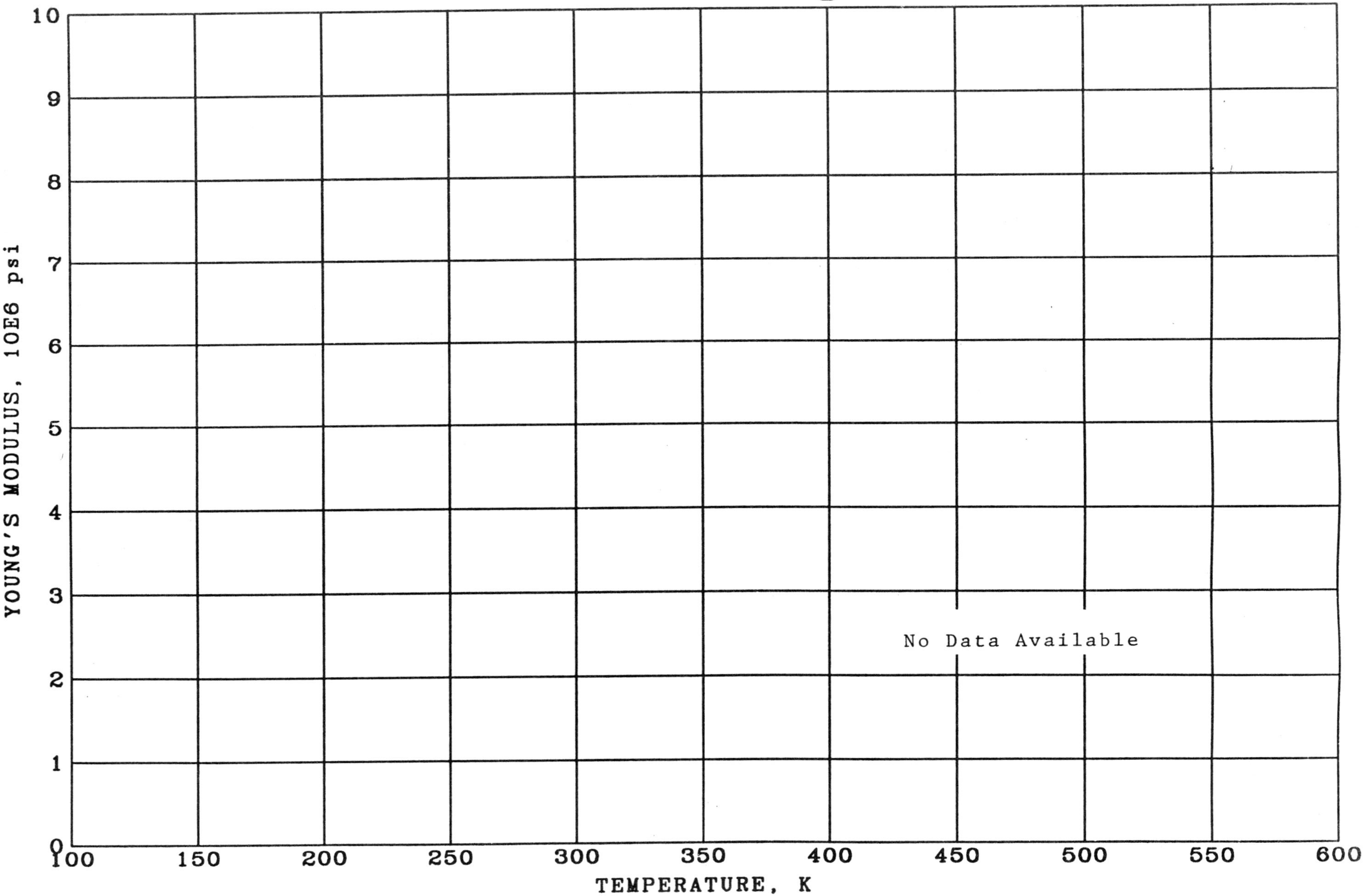

88Pb 10Sn 2Ag

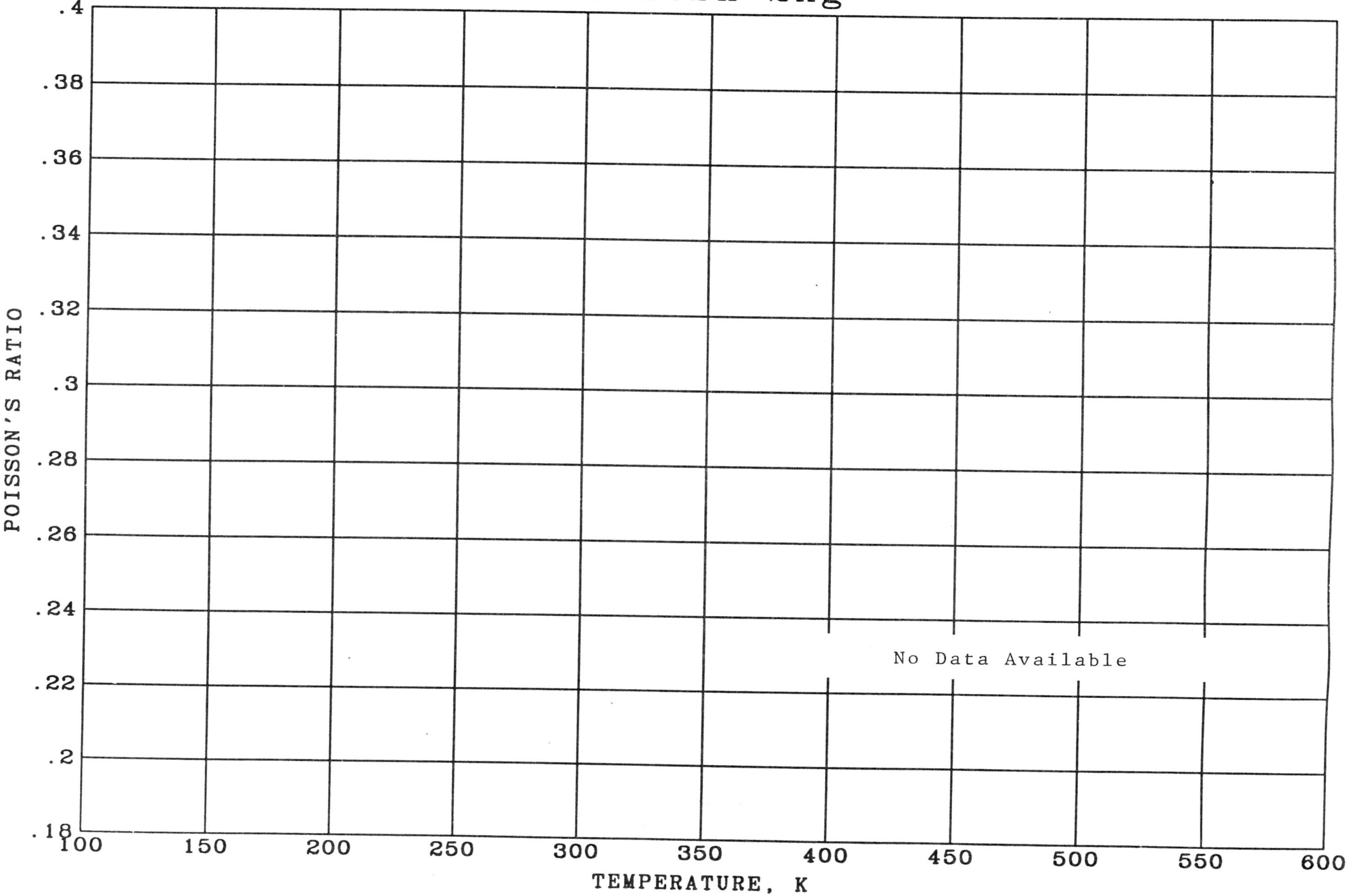

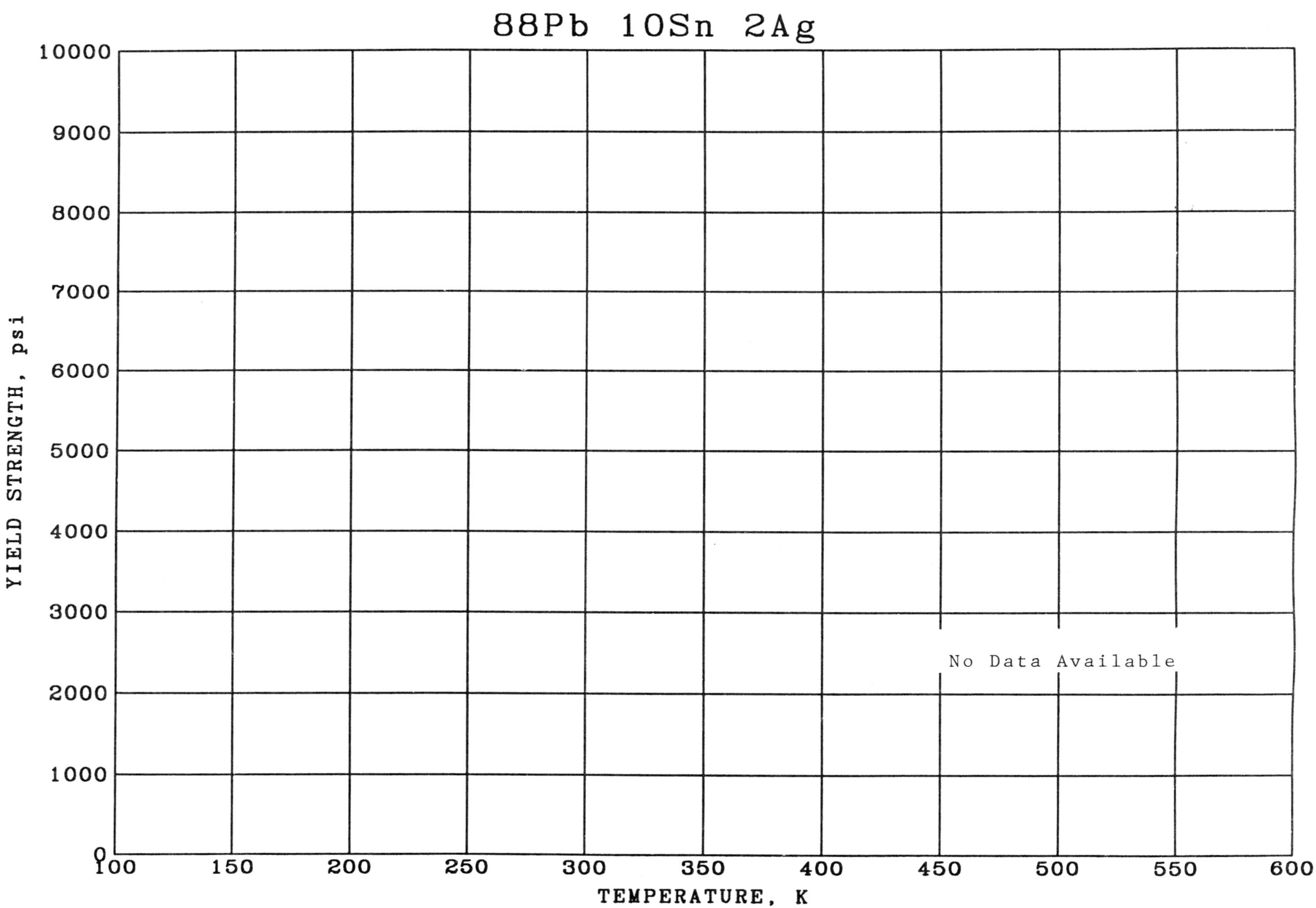
88Pb 10Sn 2Ag
YIELD STRENGTH, psi
0
1000
2000
3000
4000
5000
6000
7000
8000
9000
10000
100
150
200
250
300
350
400
450
500
550
600
TEMPERATURE, K
No Data Available

88Pb 10Sn 2Ag

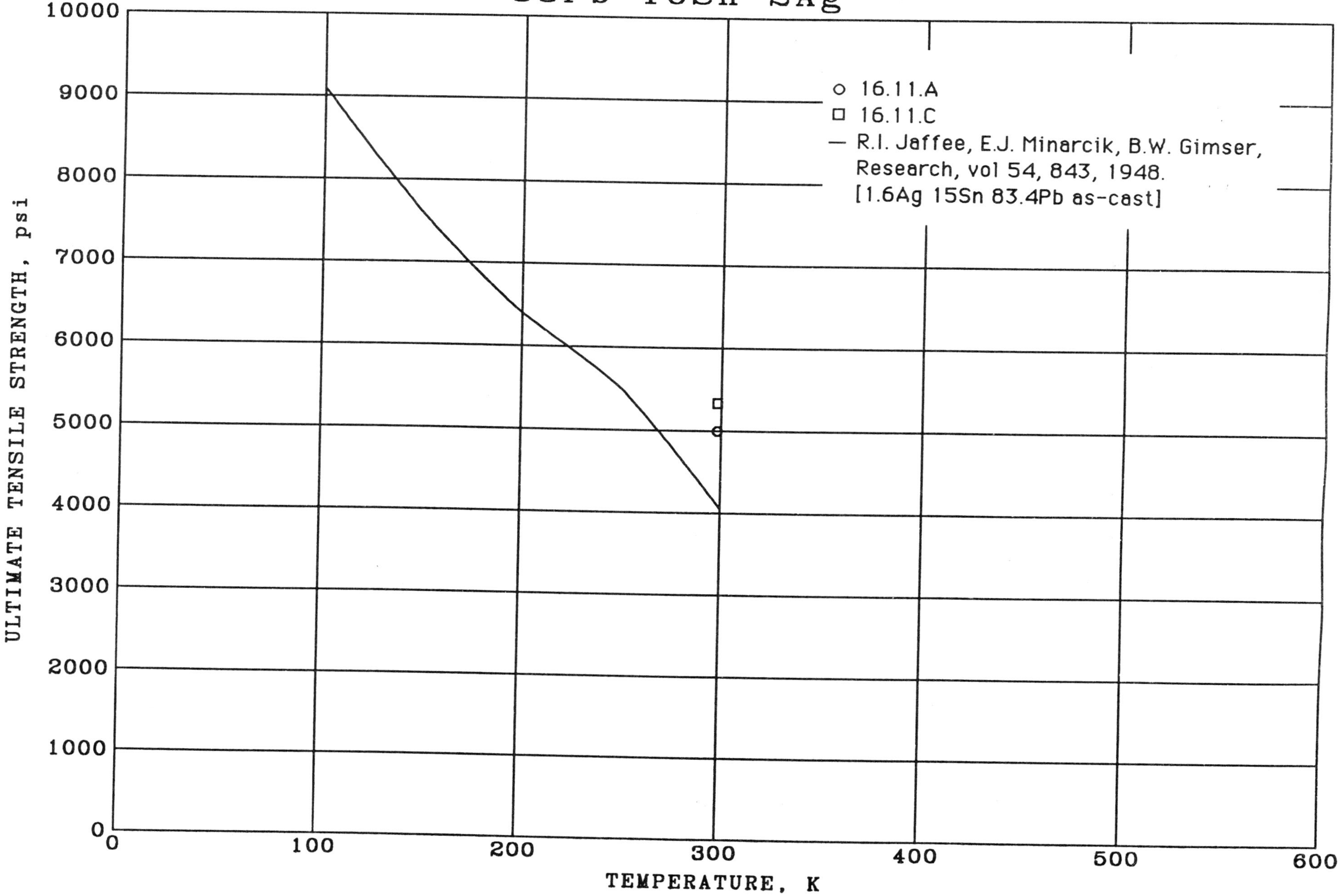

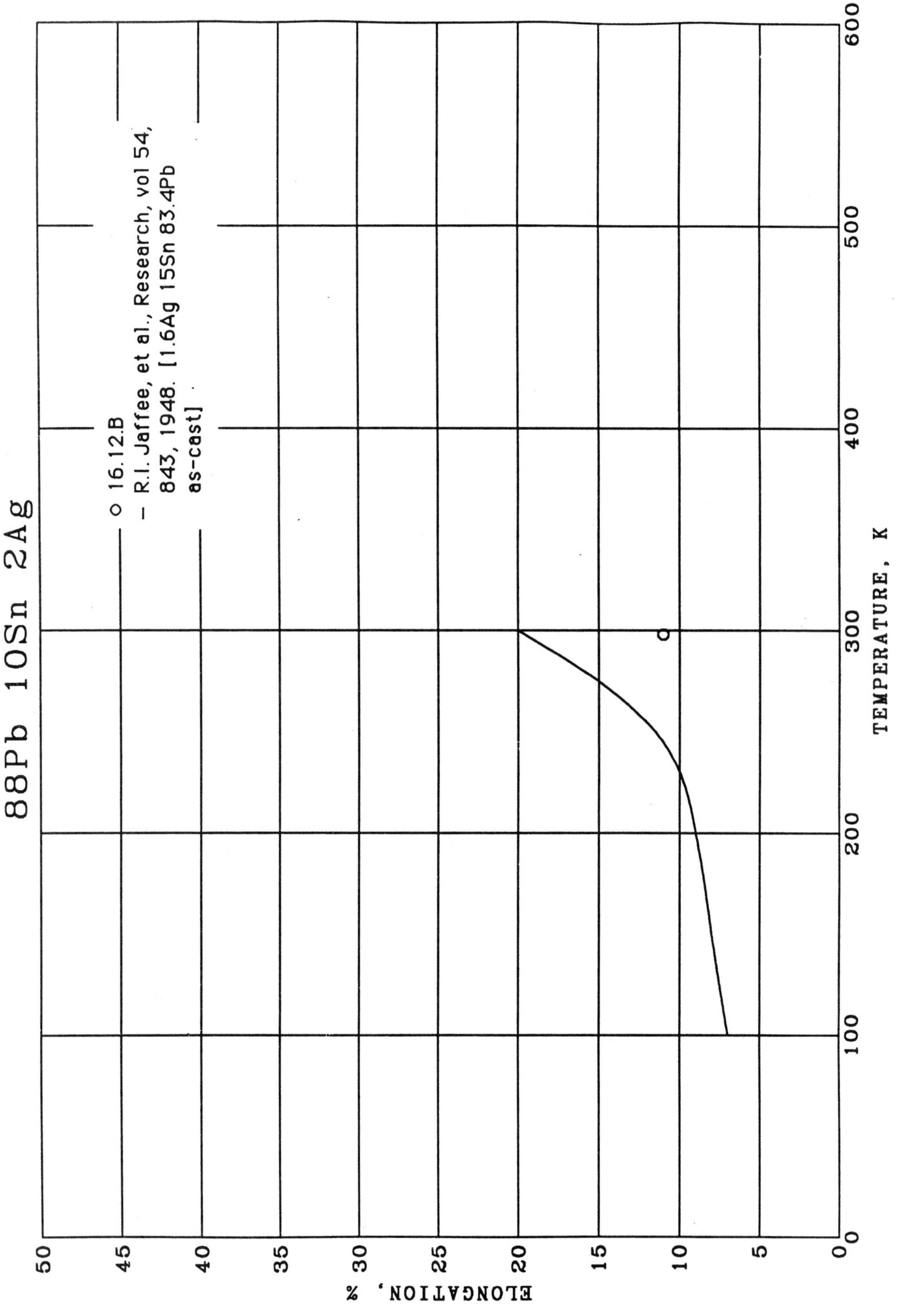

88Pb 10Sn 2Ag
○ 16.12.B
– R.I. Jaffee, et al., Research, vol 54, 843, 1948. [1.6Ag 15Sn 83.4Pb as-cast]
ELONGATION, %
TEMPERATURE, K
0
5
10
15
20
25
30
35
40
45
50
100
200
300
400
500
600

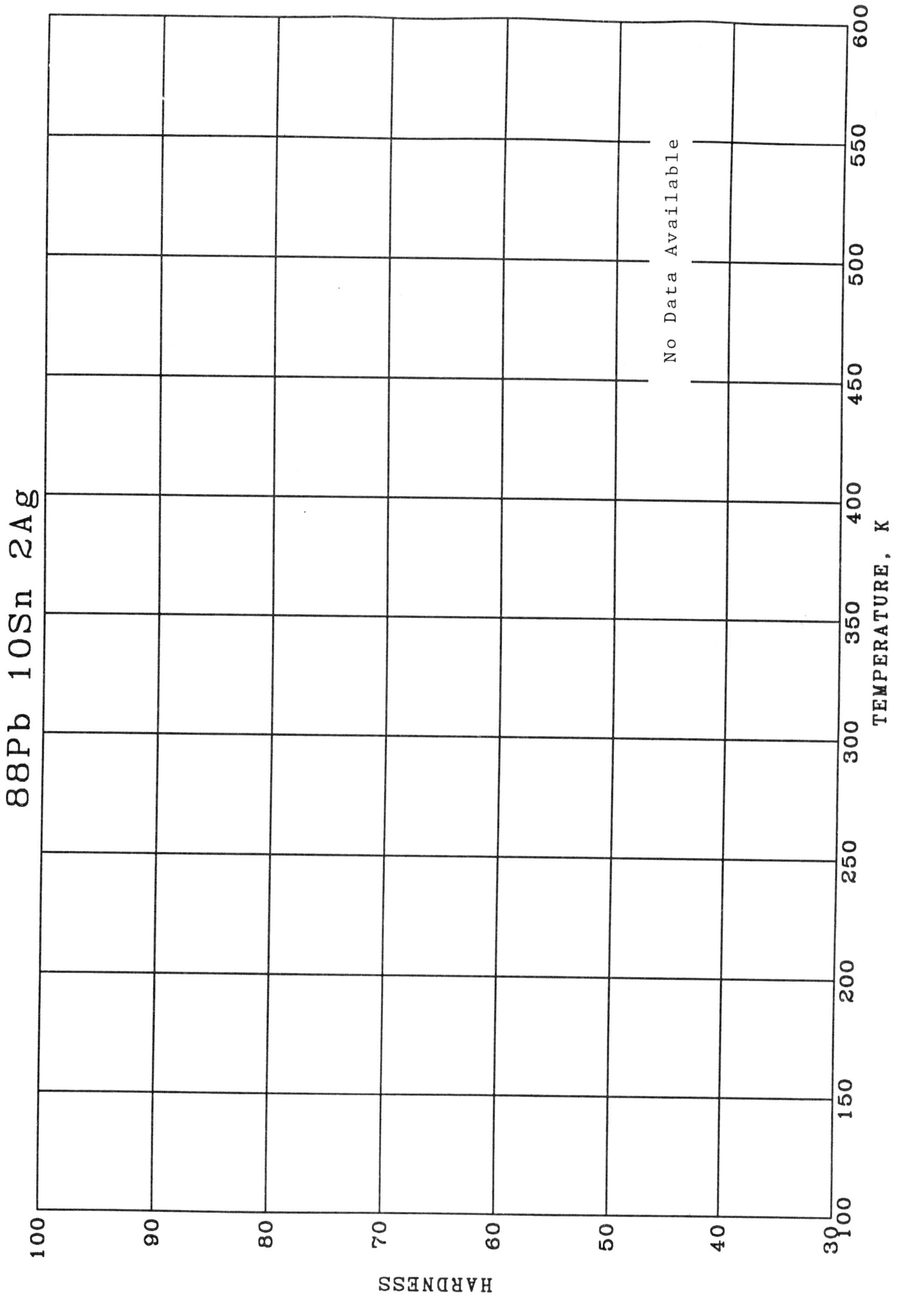
88Pb 10Sn 2Ag
No Data Available
TEMPERATURE, K
100
150
200
250
300
350
400
450
500
550
600
HARDNESS
100
90
80
70
60
50
40
30

This page intentionally left blank

92.5Pb 5Sn 2.5Ag

PROPERTY	@ 300K	UNITS	REFERENCES
Density	11.11	g/cm^3	17.1.A-B
Specific Heat	0.0309	cal/g-K	17.2.A-B
Thermal Conductivity	0.085	cal/cm-s-K	17.3.A-B
Thermal Expansion	27	1/K	17.4.A-B
Solidus/Liquidus	573/583	K	17.5.A-D
Resistivity	$19.5x\ 10^{-6}$	Ohm-cm	17.6.A
TCR	<3300	ppm/K	17.7.A
Young's Modulus	2000	ksi	17.8.A
Poisson's Ratio		------	
Yield Strength	2240	psi	17.10.A
Ult Tensile Strength	5000	psi	17.11.A-C
Elongation at Break	42	%	17.12.A-B
Hardness			

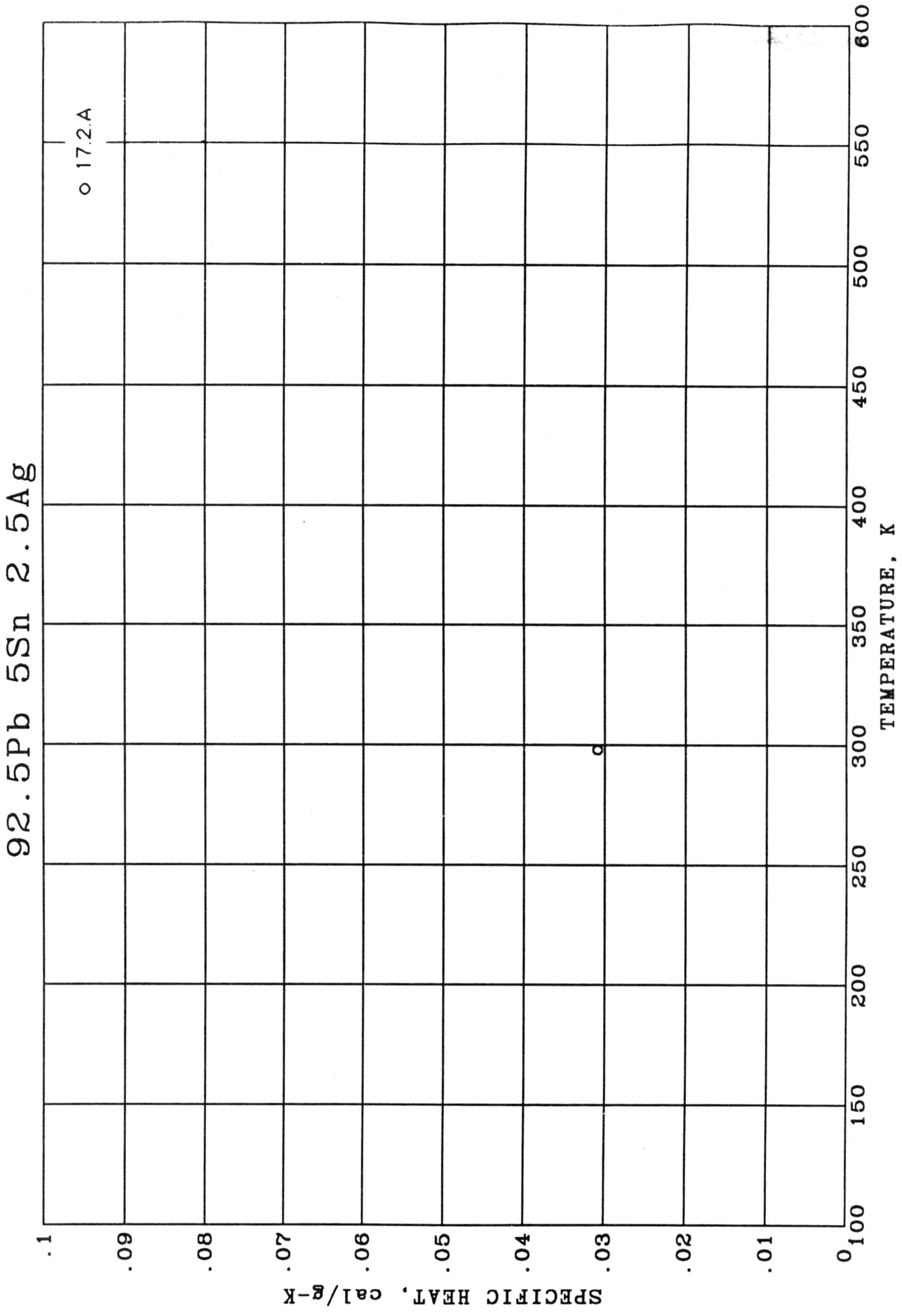
92.5Pb 5Sn 2.5Ag
○ 17.2A
SPECIFIC HEAT, cal/g-K
0
.01
.02
.03
.04
.05
.06
.07
.08
.09
.1
TEMPERATURE, K
100
150
200
250
300
350
400
450
500
550
600

92.5Pb 5Sn 2.5Ag

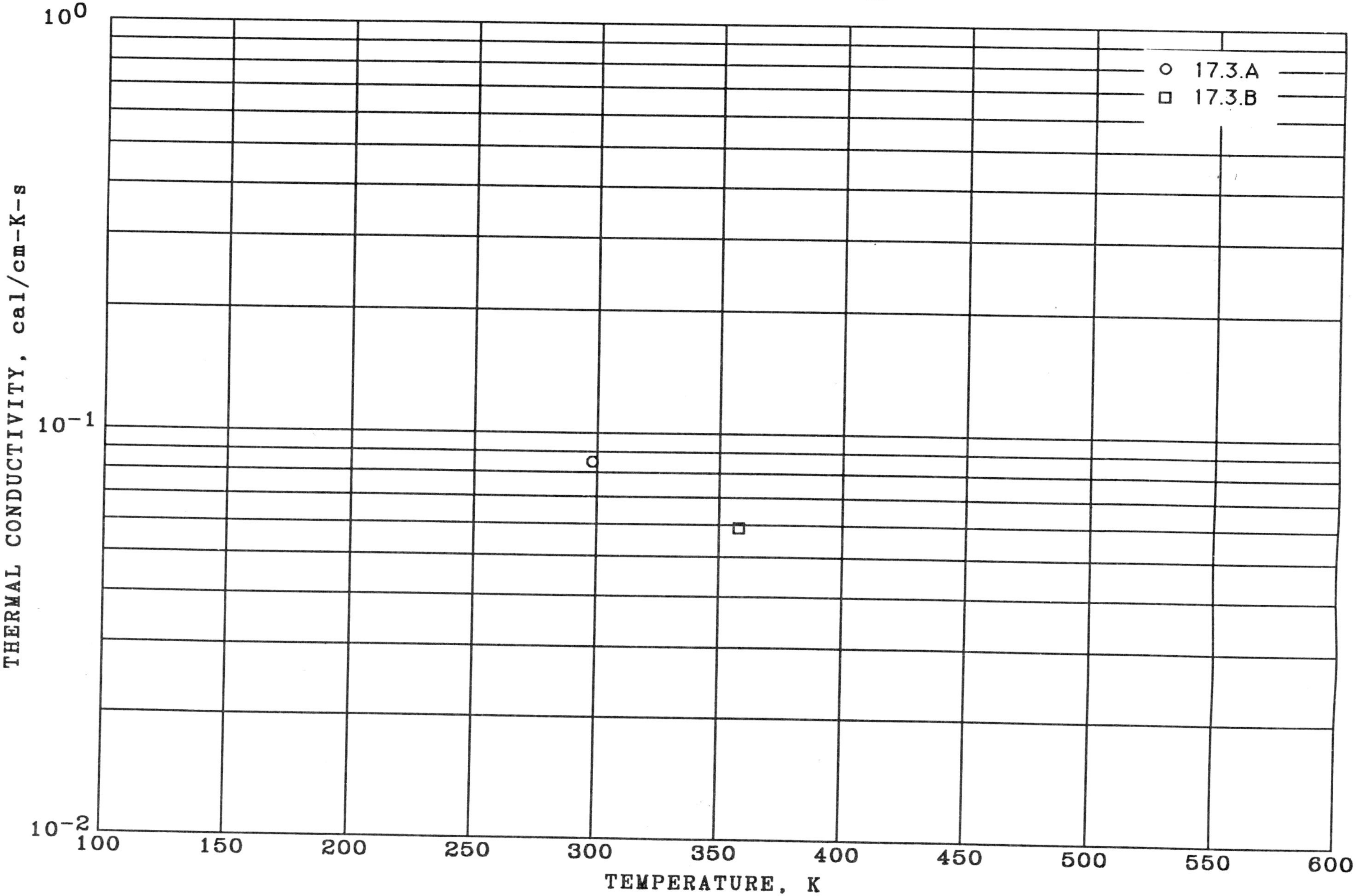

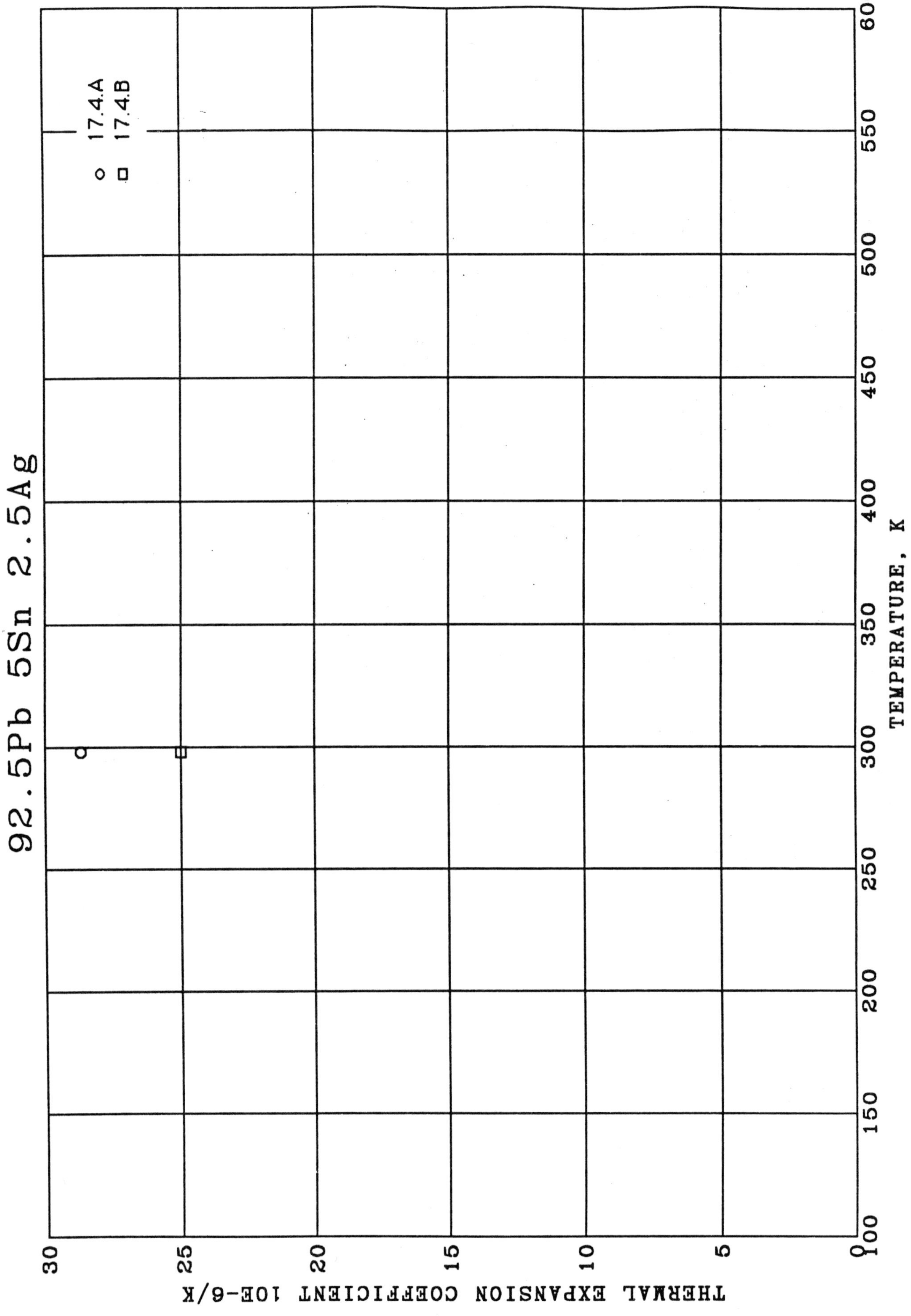
92.5Pb 5Sn 2.5Ag
○ 17.4.A
□ 17.4.B
TEMPERATURE, K
THERMAL EXPANSION COEFFICIENT 10E-6/K
0
5
10
15
20
25
30
100
150
200
250
300
350
400
450
500
550
600

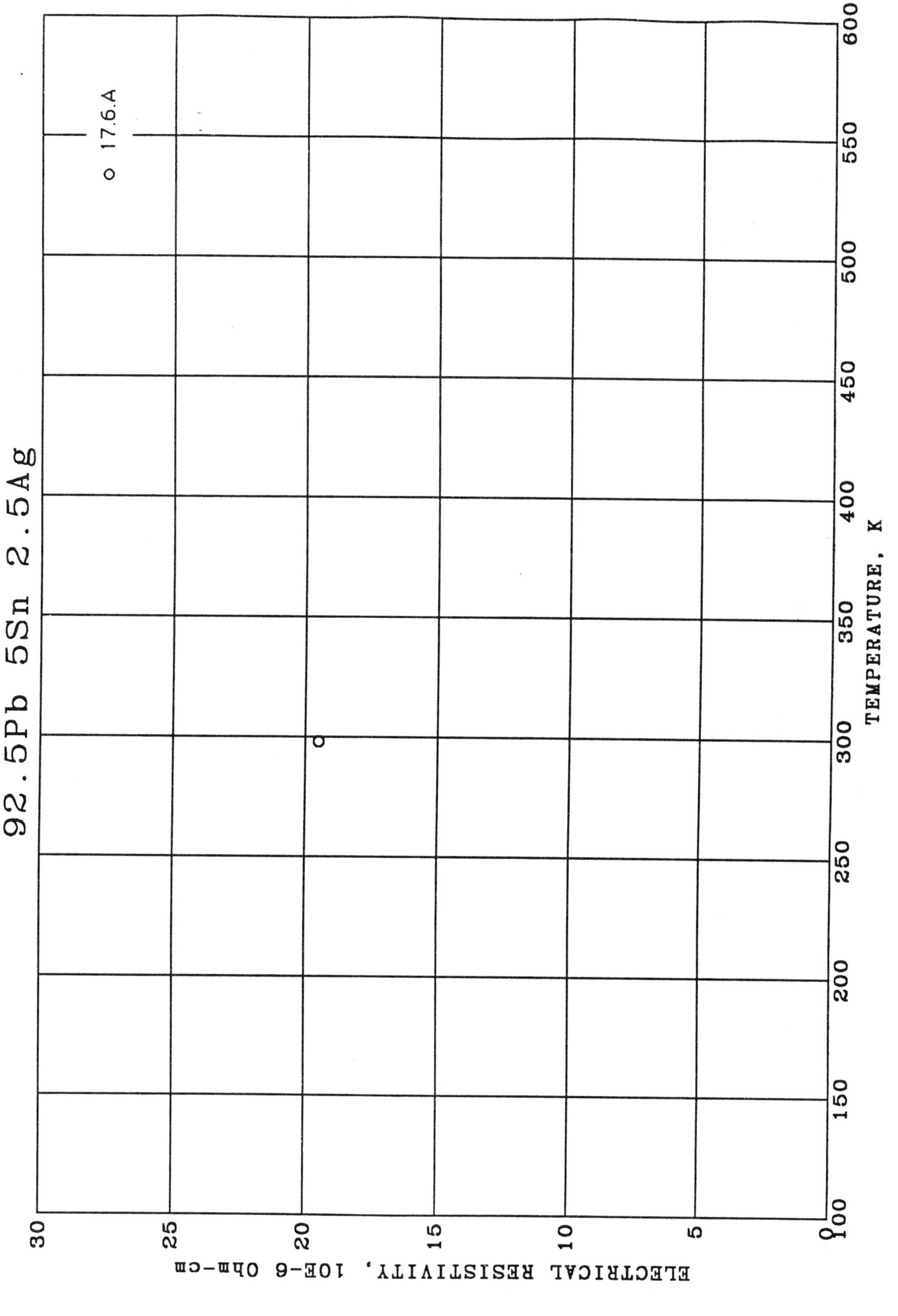
92.5Pb 5Sn 2.5Ag
17.6.A
ELECTRICAL RESISTIVITY, 10E-6 Ohm-cm
TEMPERATURE, K
0
5
10
15
20
25
30
100
150
200
250
300
350
400
450
500
550
600

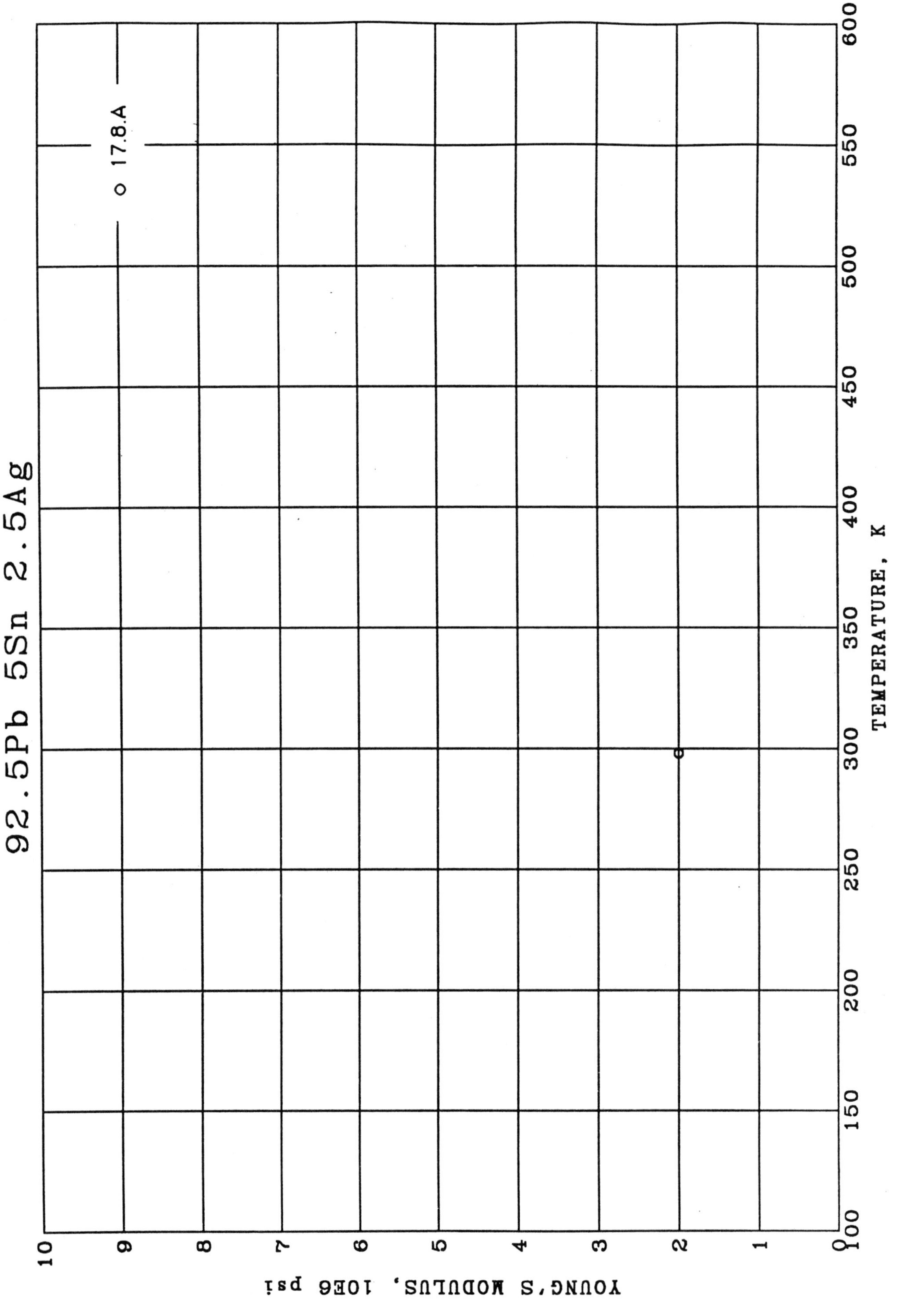
92.5Pb 5Sn 2.5Ag
○ 17.8.A
TEMPERATURE, K
100
150
200
250
300
350
400
450
500
550
600
YOUNG'S MODULUS, 10E6 psi
0
1
2
3
4
5
6
7
8
9
10

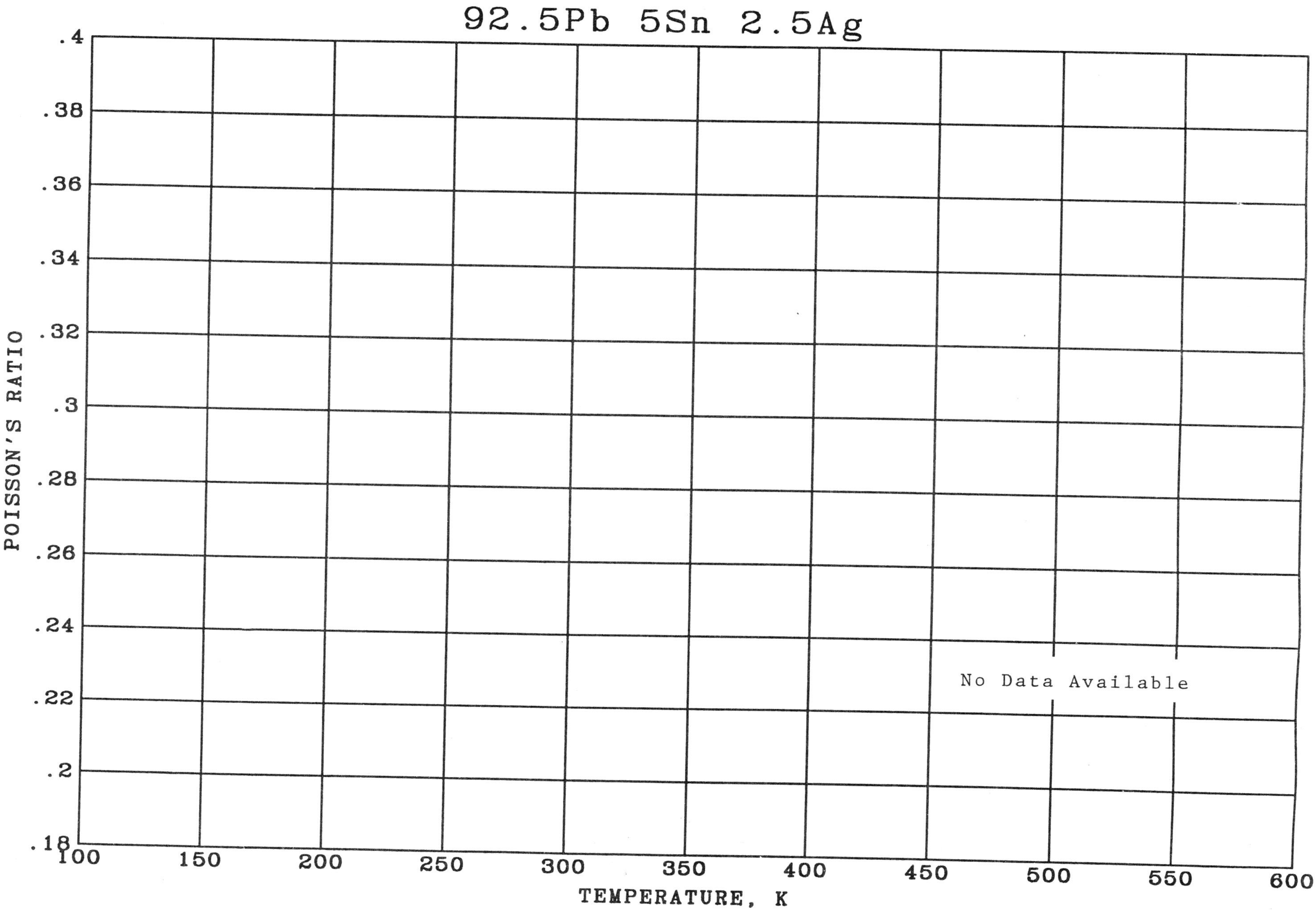
92.5Pb 5Sn 2.5Ag
POISSON'S RATIO
.4
.38
.36
.34
.32
.3
.28
.26
.24
.22
.2
.18
100
150
200
250
300
350
400
450
500
550
600
TEMPERATURE, K
No Data Available

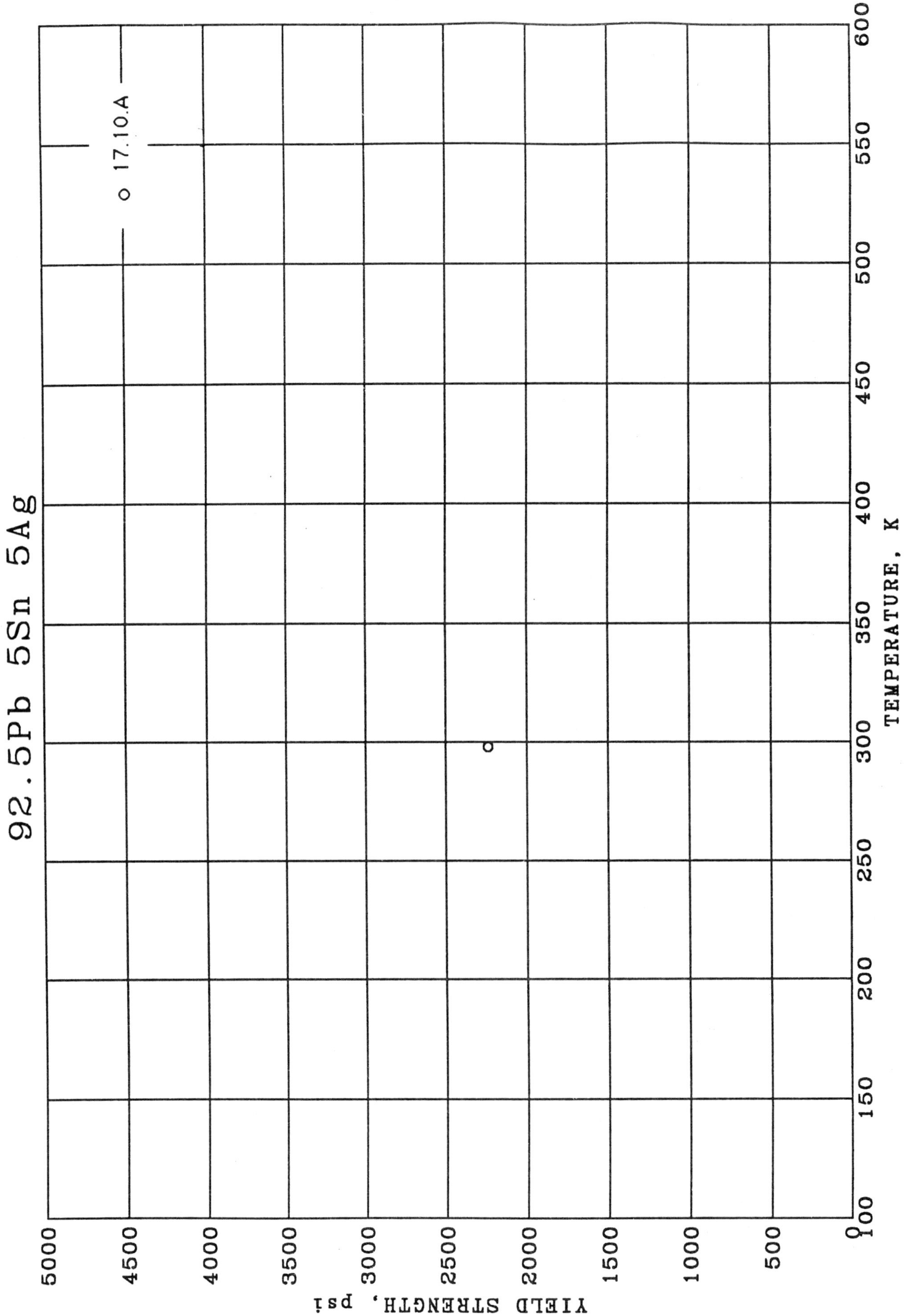
92.5Pb 5Sn 5Ag
○ 17.10.A
YIELD STRENGTH, psi
0
500
1000
1500
2000
2500
3000
3500
4000
4500
5000
TEMPERATURE, K
100
150
200
250
300
350
400
450
500
550
600

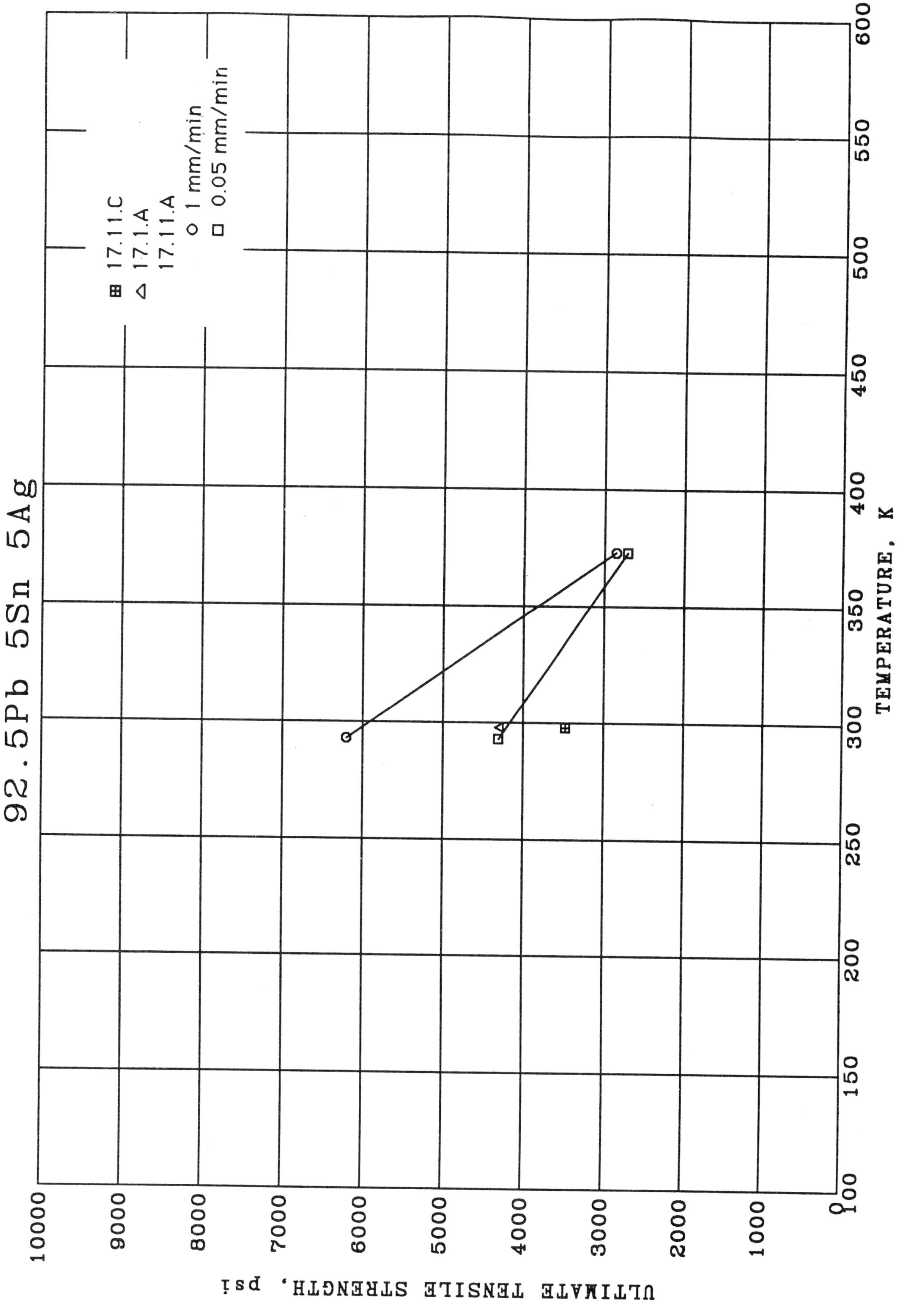
92.5Pb 5Sn 5Ag
⊞ 17.11.C
△ 17.1.A
17.11.A
○ 1 mm/min
□ 0.05 mm/min
ULTIMATE TENSILE STRENGTH, psi
TEMPERATURE, K
10000
9000
8000
7000
6000
5000
4000
3000
2000
1000
0
100
150
200
250
300
350
400
450
500
550
600

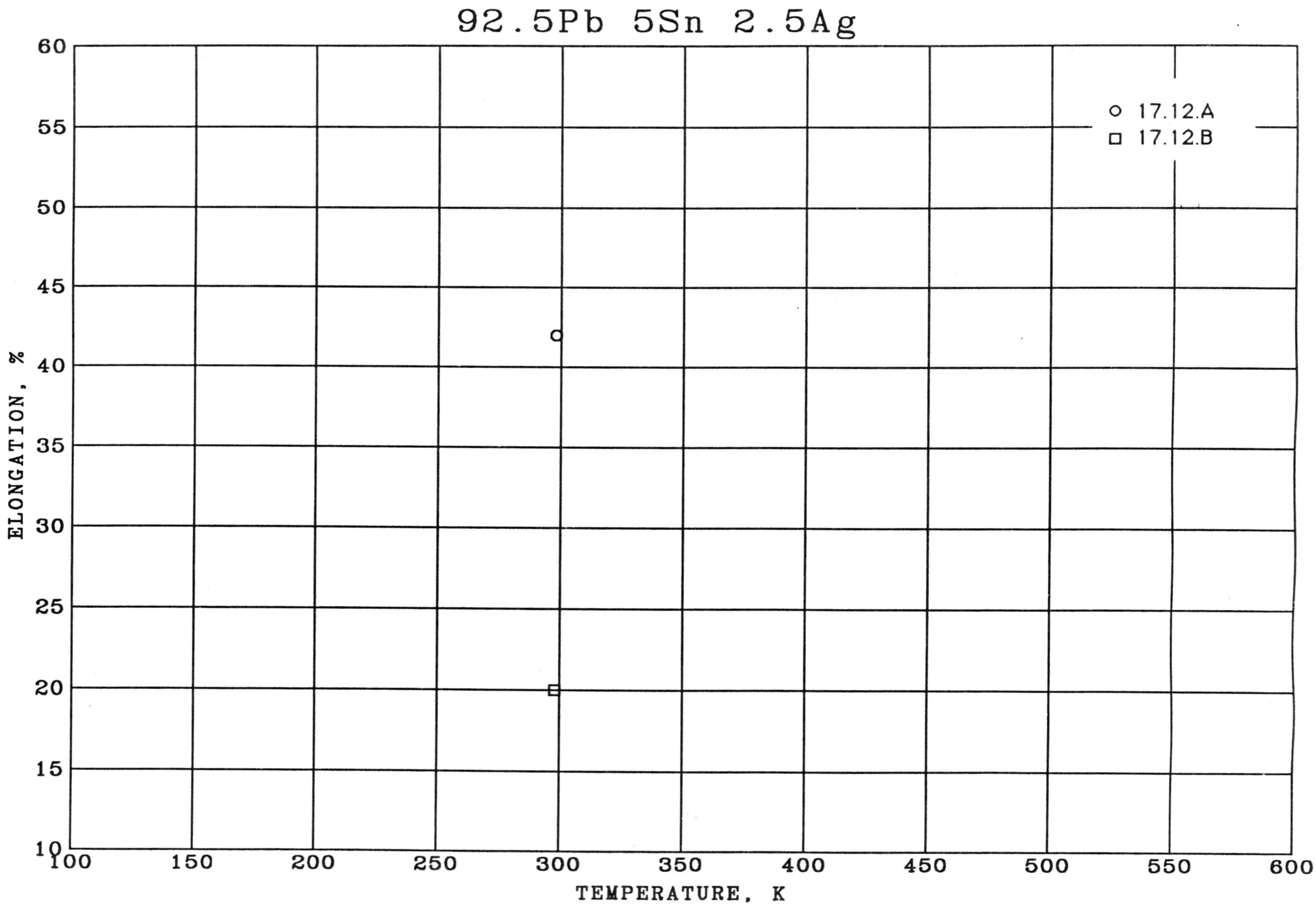
92.5Pb 5Sn 2.5Ag
ELONGATION, %
TEMPERATURE, K
17.12.A
17.12.B
60
55
50
45
40
35
30
25
20
15
10
100
150
200
250
300
350
400
450
500
550
600

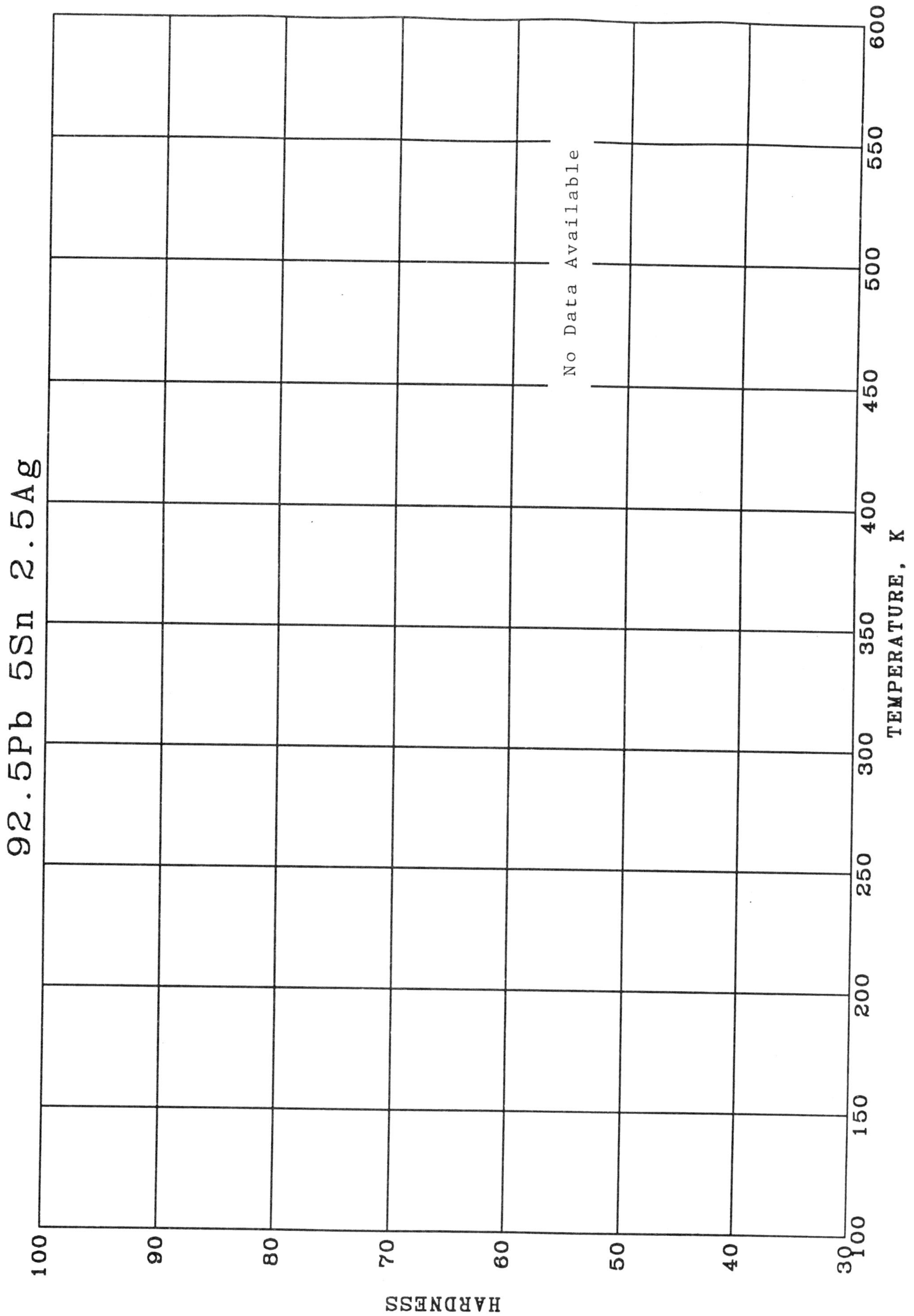
92.5Pb 5Sn 2.5Ag
No Data Available
HARDNESS
100
90
80
70
60
50
40
30
TEMPERATURE, K
100
150
200
250
300
350
400
450
500
550
600

62Sn 36Pb 2Ag

PROPERTY	@ 300K	UNITS	REFERENCES
Density	8.50	g/c^{m3}	18.1.A-C
Specific Heat	0.045	cal/g-K	18.2.A-B
Thermal Conductivity	0.13	cal/cm-s-K	18.3.A-B
Thermal Expansion	26×10^{-6}	1/K	18.4.A-C
Melting Point	451	K	18.5.A-C
Resistivity	14.7×10^{-6}	Ohm-cm	18.6.A-B
TCR		ppm/K	
Young's Modulus	3330	ksi	18.8.A
Poisson's Ratio		------	
Yield Strength		psi	
Ult Tensile Strength	7000	psi	18.11.A-D
Elongation at Break	7	%	18.12.A
Hardness	33.4	Brinell	18.13.A

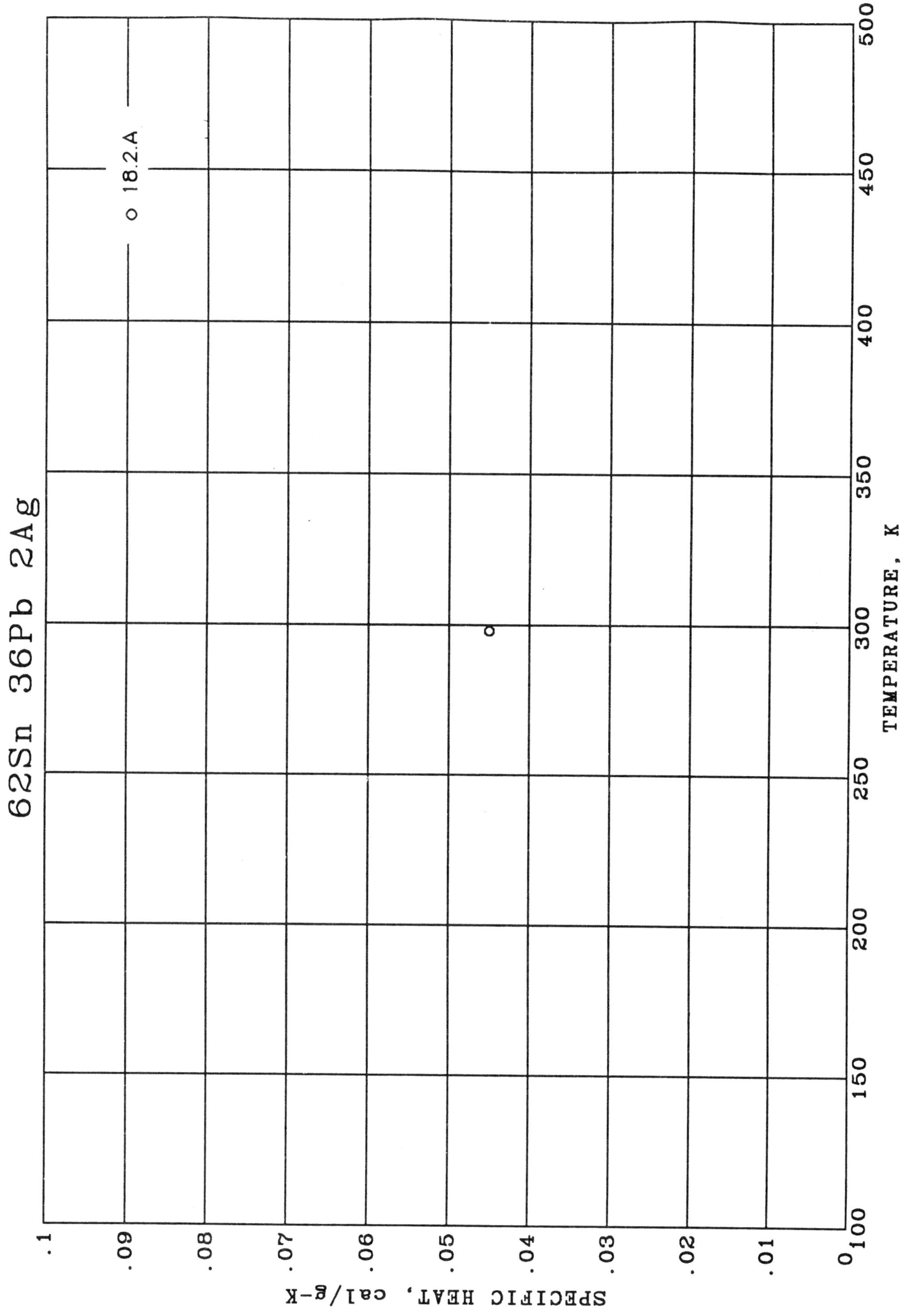

62Sn 36Pb 2Ag
18.2.A
SPECIFIC HEAT, cal/g-K
TEMPERATURE, K
.1
.09
.08
.07
.06
.05
.04
.03
.02
.01
0
100
150
200
250
300
350
400
450
500

62Sn 36Pb 2Ag

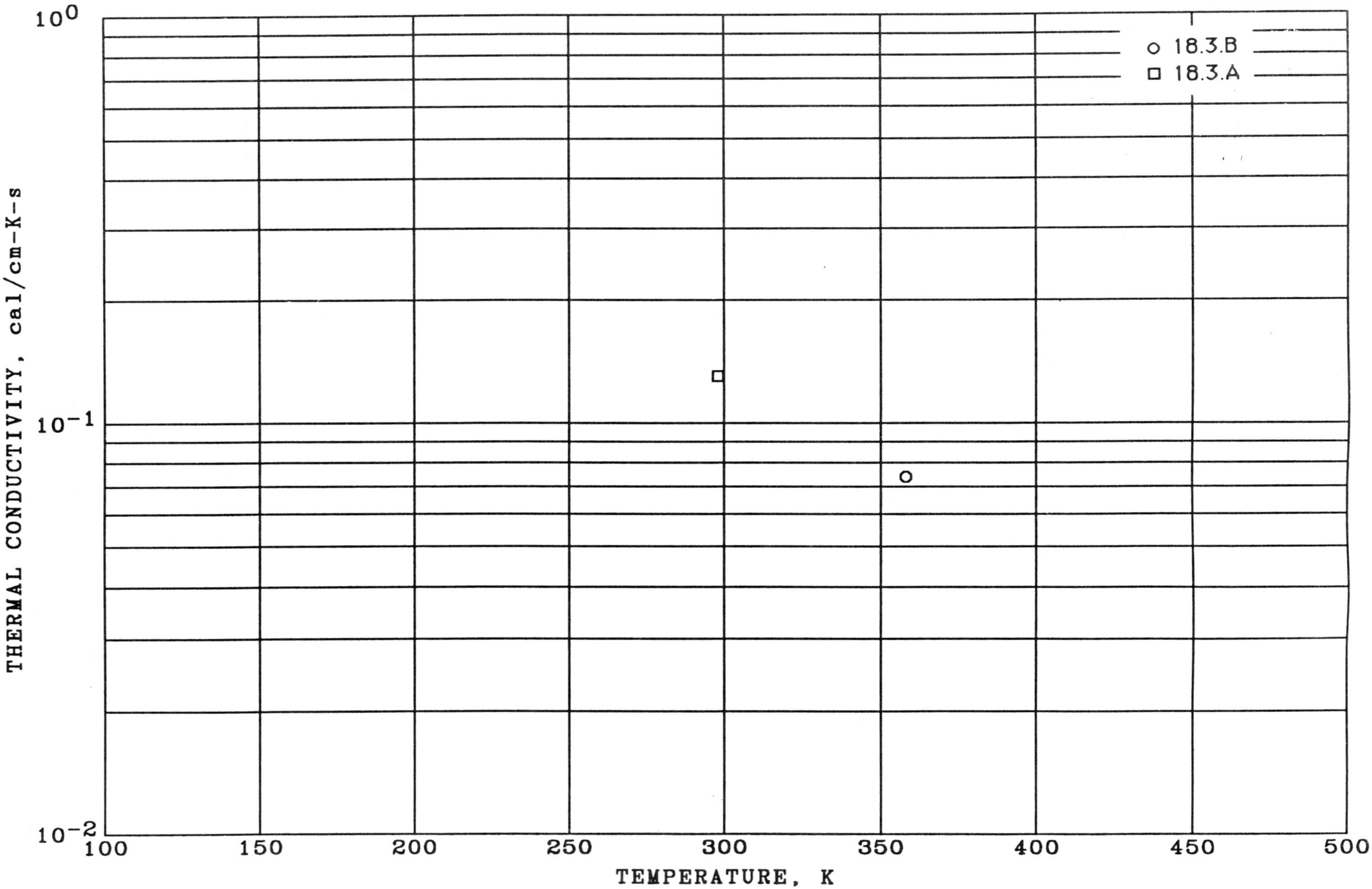

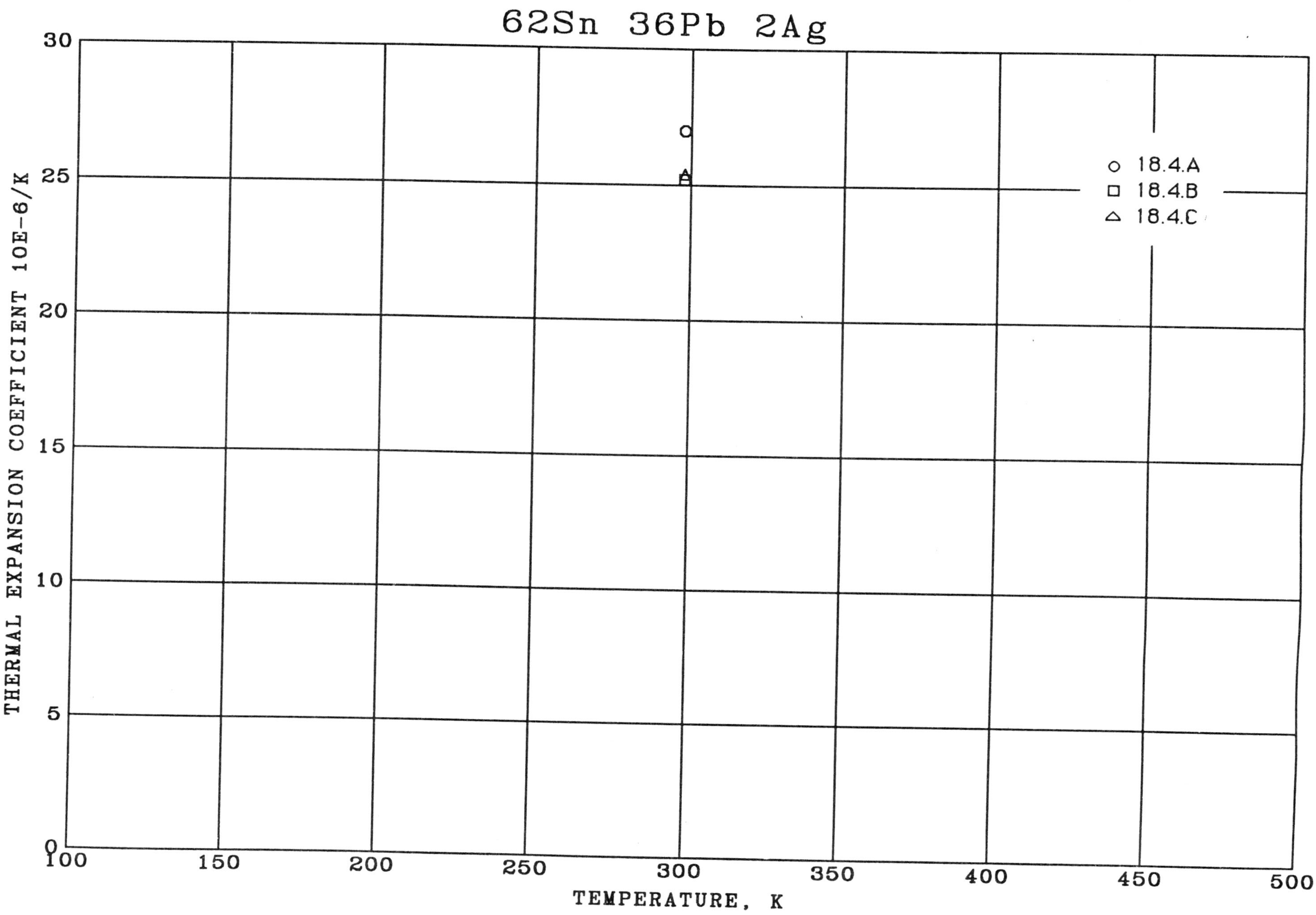
62Sn 36Pb 2Ag
THERMAL EXPANSION COEFFICIENT 10E-6/K
TEMPERATURE, K
0
5
10
15
20
25
30
100
150
200
250
300
350
400
450
500
○ 18.4.A
□ 18.4.B
△ 18.4.C

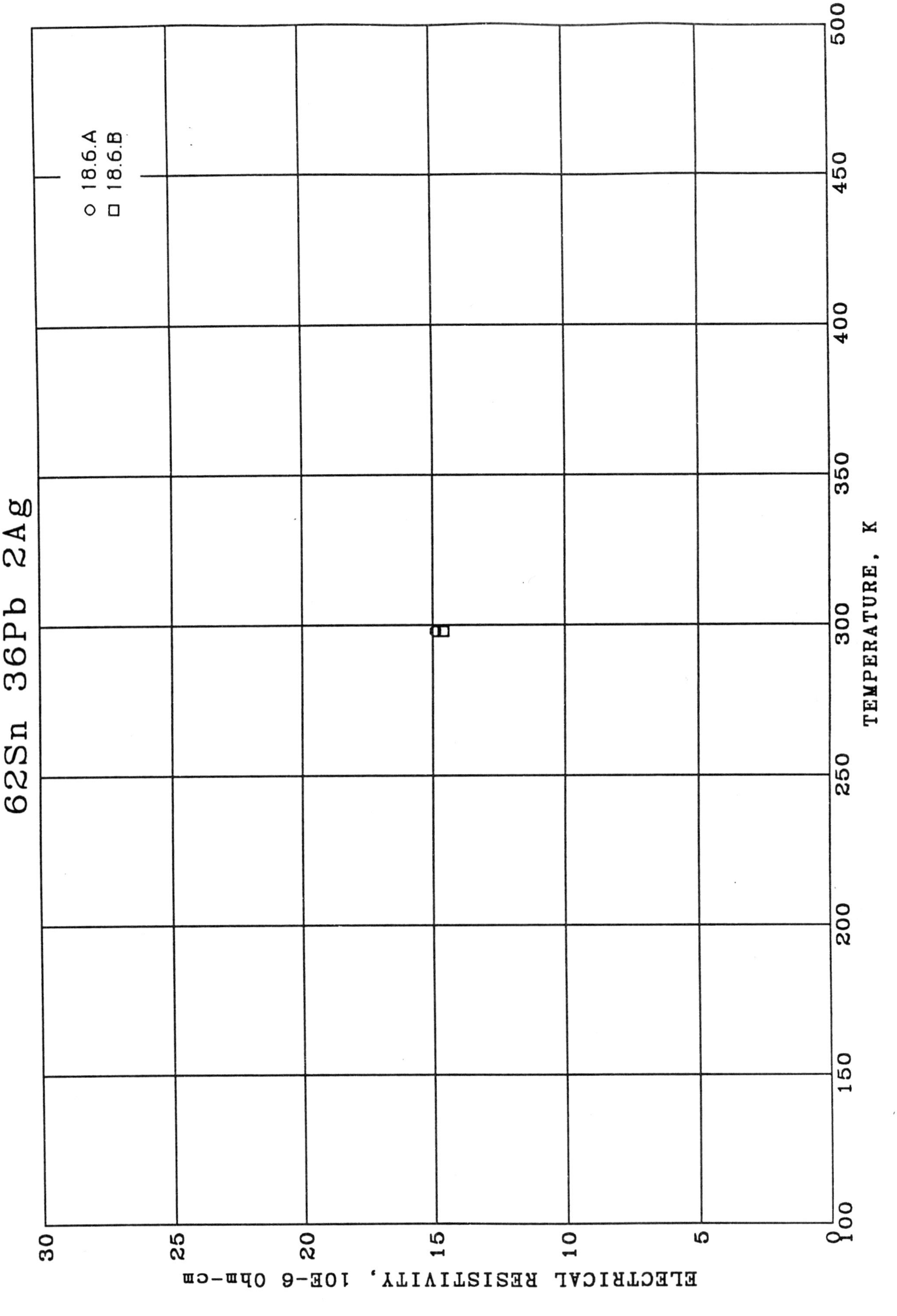
62Sn 36Pb 2Ag
○ 18.6.A
□ 18.6.B
ELECTRICAL RESISTIVITY, 10E-6 Ohm-cm
TEMPERATURE, K
0
5
10
15
20
25
30
100
150
200
250
300
350
400
450
500

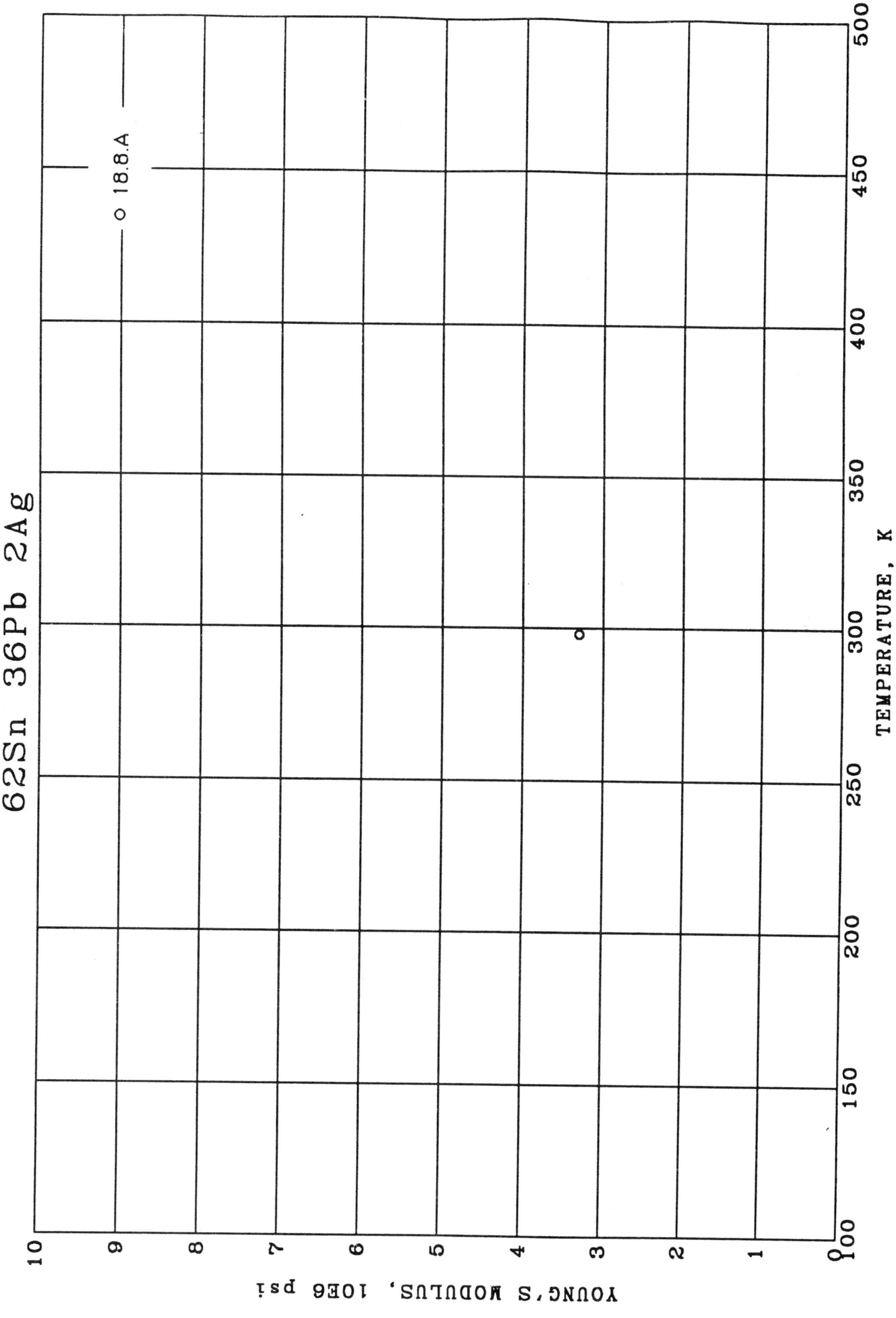

62Sn 36Pb 2Ag
18.8.A
TEMPERATURE, K
YOUNG'S MODULUS, 10E6 psi
0
1
2
3
4
5
6
7
8
9
10
100
150
200
250
300
350
400
450
500

62Sn 36Pb 2Ag

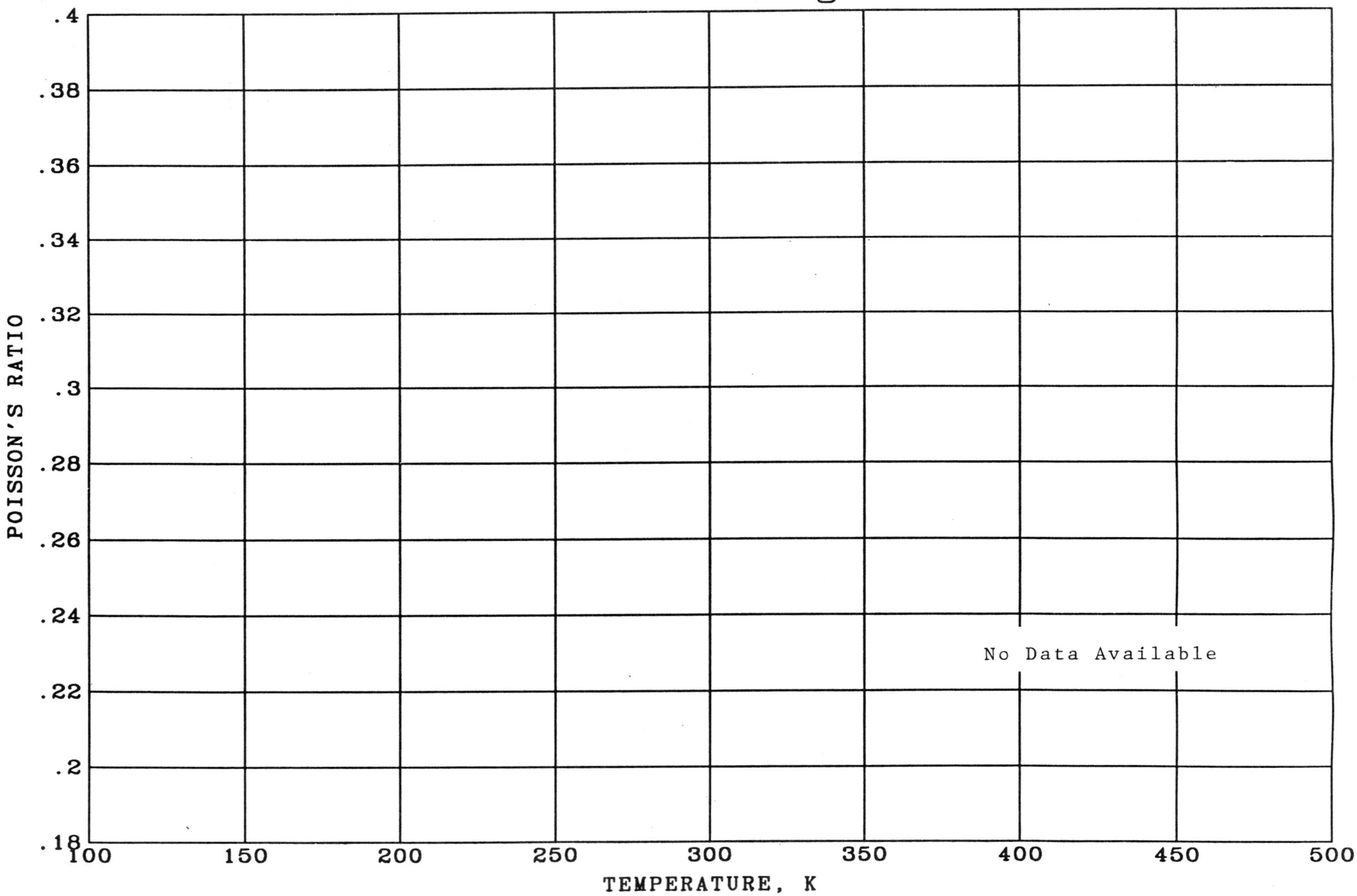

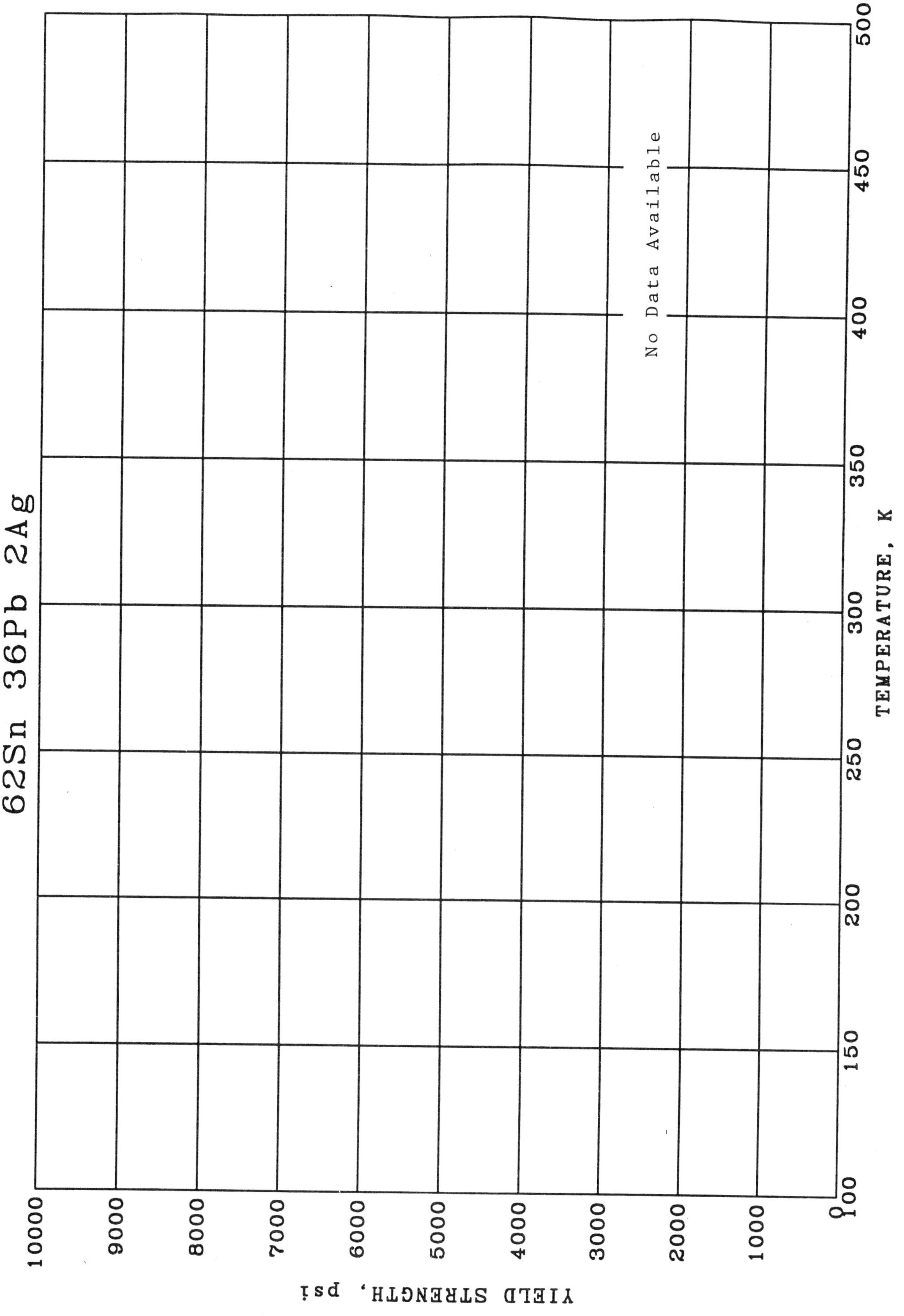
62Sn 36Pb 2Ag
No Data Available
TEMPERATURE, K
0 50 100 150 200 250 300 350 400 450 500
YIELD STRENGTH, psi
0 1000 2000 3000 4000 5000 6000 7000 8000 9000 10000

62Sn 36Pb 2Ag

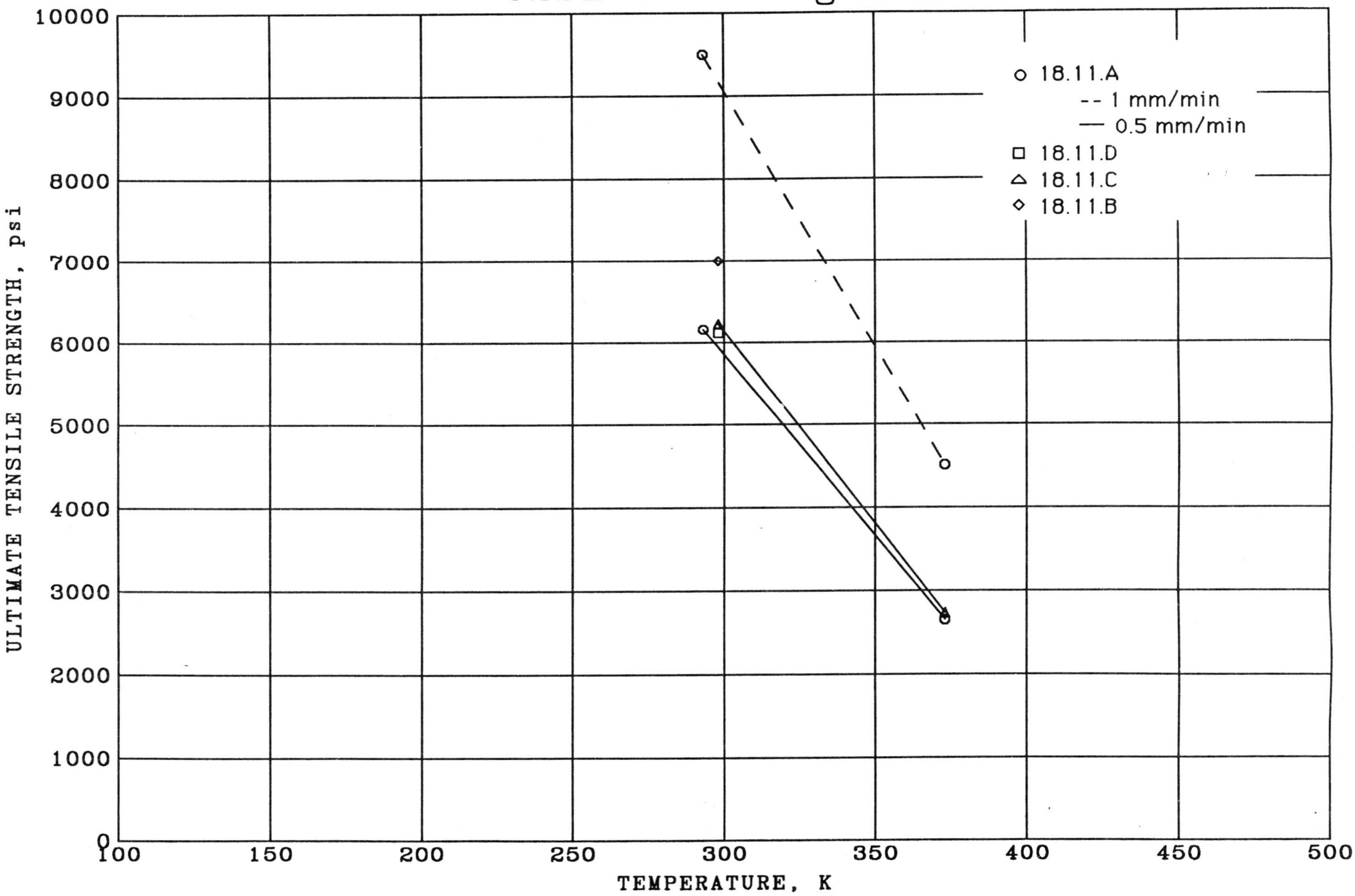

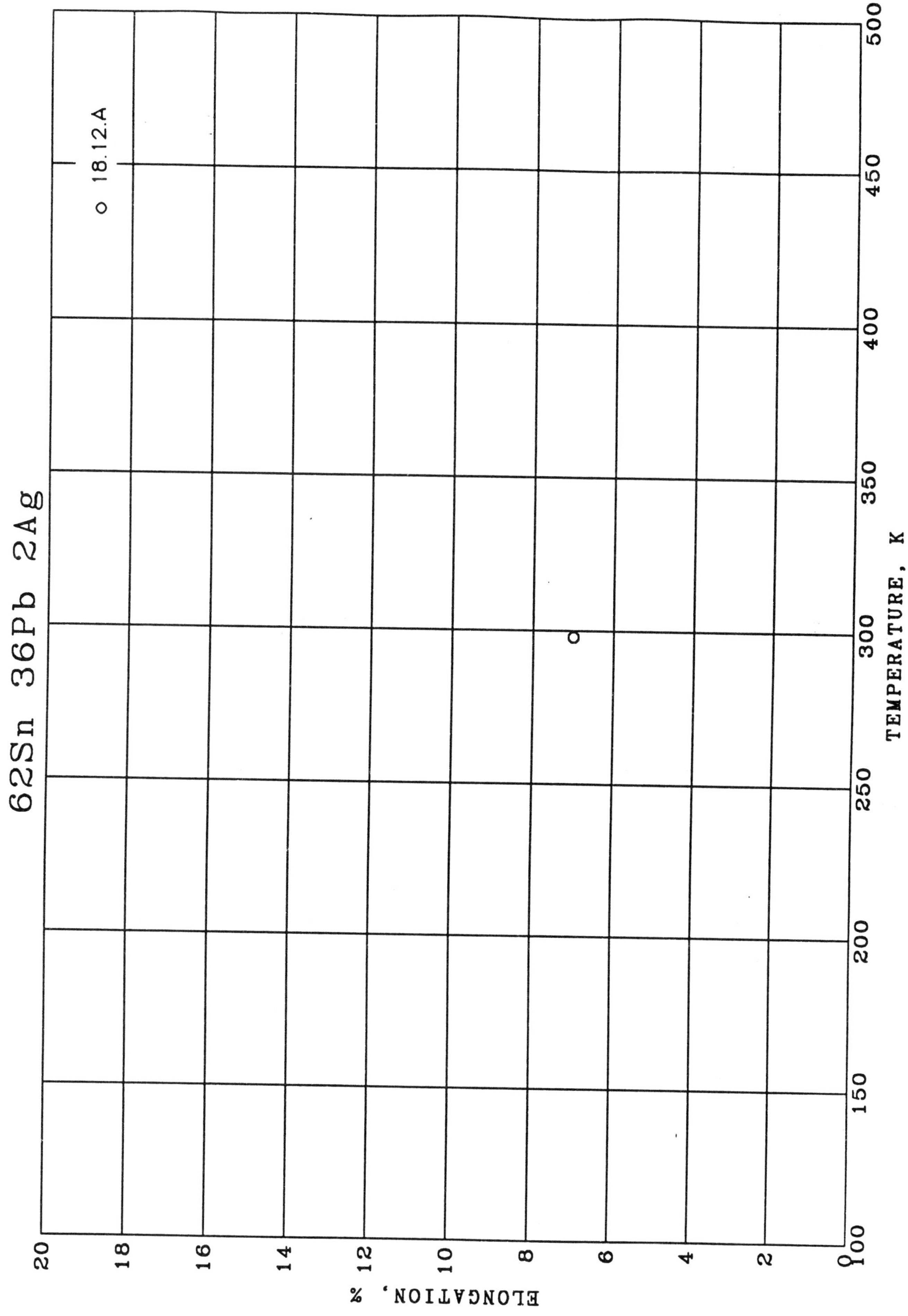
62Sn 36Pb 2Ag
○ 18.12.A
ELONGATION, %
TEMPERATURE, K
0
2
4
6
8
10
12
14
16
18
20
100
150
200
250
300
350
400
450
500

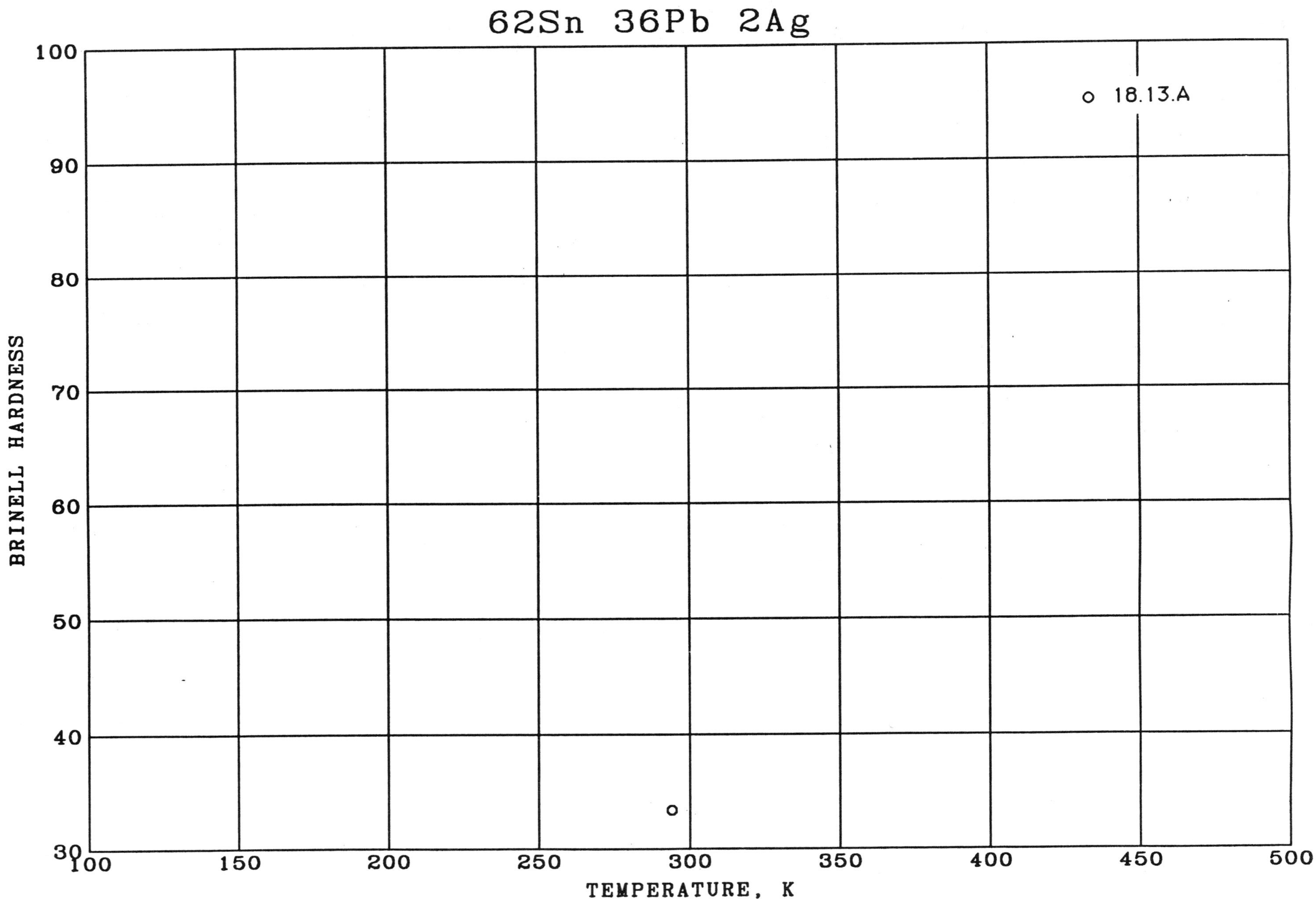
62Sn 36Pb 2Ag
18.13.A
BRINELL HARDNESS
100
90
80
70
60
50
40
30
TEMPERATURE, K
100
150
200
250
300
350
400
450
500

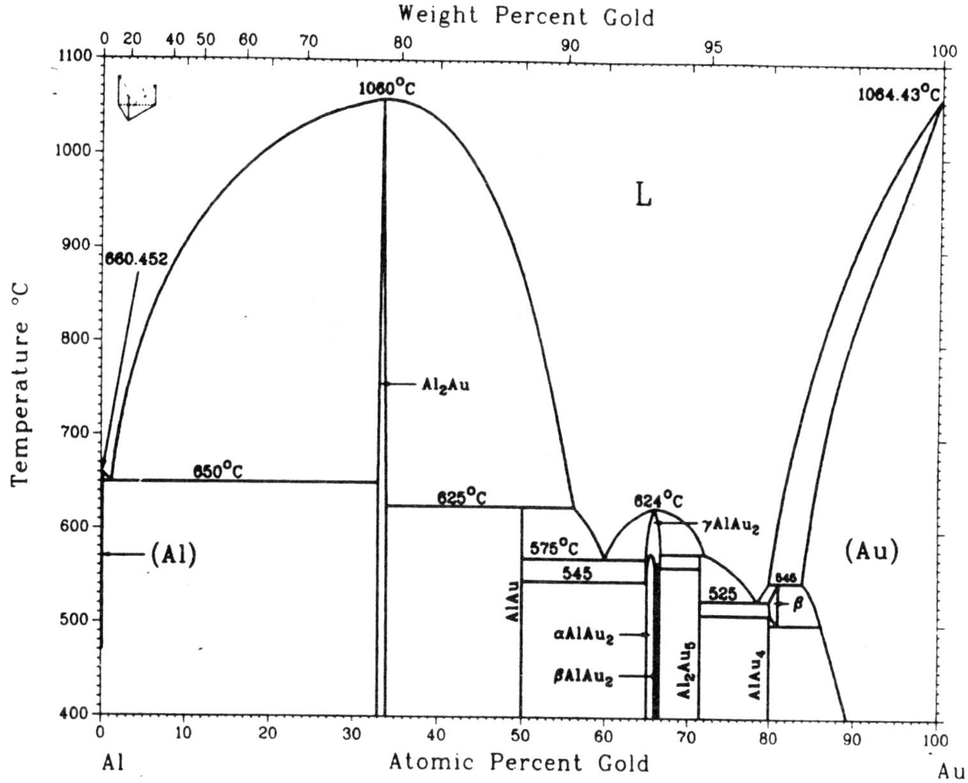
Weight Percent Gold
0 20 40 50 60 70 80 90 95 100
1100
1000
900
800
700
600
500
400
Temperature °C
1060°C
1064.43°C
L
660.452
Al2Au
650°C
625°C
624°C
γAlAu2
(Al)
575°C
545
(Au)
525
546
β
AlAu
αAlAu2
βAlAu2
Al2Au5
AlAu4
0 10 20 30 40 50 60 70 80 90 100
Al
Atomic Percent Gold
Au

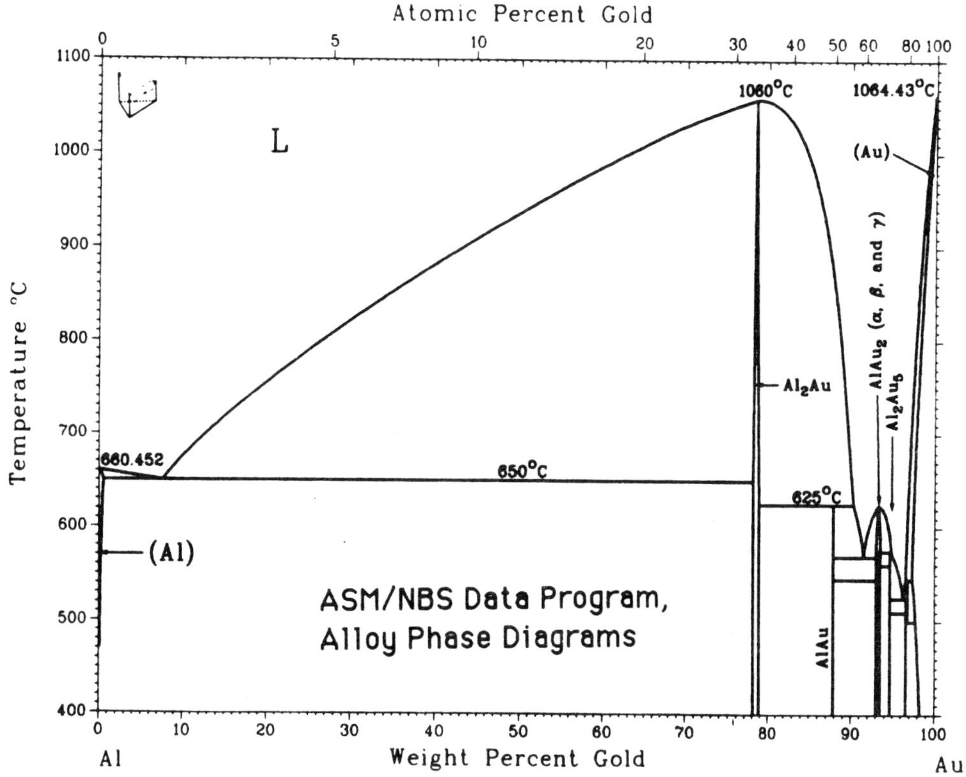
Atomic Percent Gold
0 5 10 20 30 40 50 60 80 100
1100
1000
900
800
700
600
500
400
Temperature °C
1060°C
1064.43°C
L
(Au)
AlAu2 (α, β, and γ)
Al2Au
Al2Au5
660.452
650°C
625°C
(Al)
ASM/NBS Data Program,
Alloy Phase Diagrams
AlAu
0 10 20 30 40 50 60 70 80 90 100
Al
Weight Percent Gold
Au

REFERENCES

95Sn 5Ag

1.1.A Allied Metglas Products, Parsippany, NJ. Density at room temperature = 7.39 g/cc. Solidus/liquidus = 494/513 K.

1.1.B John, W.G., E.J. Evans, Phil Mag, Series 7, vol 23, 1033-48. Density = 7.44 g/cc. Resistivity at 273K = 10.45 micro-ohm-cm. TCR = 4230 ppm/K between 0 and 100C for chill cast and annealed material with 5.62% silver.

1.1.C Teledyne Microelectronics Handbook, Appendix A, Material Properties. Density = 7.39 g/cc. Specific heat = 0.055 cal/gm-K. Thermal conductivity = 0.079 cal/cm-s-K. Yield strength = 3540 psi. Tensile strength = 4280 psi. Elongation = 49%. Electrical resistivity = 10.4 micro-ohm-cm. Solidus/Liquidus = 221/240.

1.2.A See 1.1.C

1.3.A Smith, A. W., Ohio State University, Engineering Experimental Station Bulletin, vol 30, 1925. Thermal conductivity = 0.147 cal/cm-sec-K at 333K (60C).

1.3.B See 1.1.C

1.4.A See 1.1.C

1.5.A See 1.1.A

1.5.B Smith, J. F., Materials Engineering, vol 80, no 3, 76-78, 1974. Solidus/liquidus = 494/518K.

1.5.C See 1.1.C

1.6.A Smith, A. W., Ohio State University, Engineering Experimental Station Bulletin, vol 30, 1925. Resistivity at room temperature = 11.7 $\mu\Omega$cm.

1.6.B Howard H. Manko, "Solders and Soldering", 125. McGraw-Hill, 1979. Resistivity at room temperature = 13.70 $\mu\Omega$cm.

1.6.C Data is summary of 2 sources: a. Matthiessen, Philosophical Transactions of the Royal Society of London, Series A, vol 150, 161, 1960, and b. same author, journal and series, vol 154, 167, 1965. Resistivity at 293.3 K = 13.43 $\mu\Omega$cm.

1.6.D See 1.1.B

1.7.A See 1.1.B

1.8.A Subrahmanyam, B. and B. Krishnamurty, Ind J Pure Appl

Phys, vol 2, 139-140, 1964. Young's modulus = 8.2 x 10^6 psi. Poisson's ratio = 0.37.

1.9.A See 1.8.A

1.10.A See 1.1.C

1.10.B Addicks, L. (Editor), Silver in Industry, Reinhold Publishing Corp, New York, New York, 1940. Yield strength = 3600 psi. Tensile strength = 4600 psi. Elongation in 2 inches = 49%. Properties are for 0.040 inch thick sheet aged 14 days at room temperature.

1.11.A Howard H. Manko, "Solders and Soldering", McGraw-Hill, 1979, 125. Ultimate tensile strength at room temperature = 8000 psi. Elongation at break at room temperature = 30 %. Brinell hardness at room temperature = 13.7.

1.11.B See 1.10.B

1.11.C Turkus, S. and A.A. Smith, Metals and Alloys, March, 1942, 412. Tensile strength = 4915 psi.

1.22.D Rhines, F.N. and W.A. Anderson, Metals and Alloys, vol 14, 704, 1941. Tensile strength = 14,000 psi. Shear strength = 10,600 psi.

1.12.A See 1.11.A

1.12.B See 1.1.C

1.12.C See 1.10.B

1.13.A See 1.11.A

1.13.B Cole, S.S. and A.R. Kroehs, 1969 Electronic Components Conference Proceedings, 298-302. Hardness at 21C = 22.3 Brinell.

96.5Sn 3.5Ag

2.1.A John, W.G. and E.J. Evans, Phil Mag, Series 7, vol 23, no 158, 1033-48, 1937. Density at room temperature = 7.48 g/cc. TCR = 4300 ppm/K between 273 and 373K. Resistivity = 10.5 micro-ohm-cm.

2.1.B Indium Company of America, Indalloy #121, Density = 7.36 g/cc. Thermal conductivity at 358K(85C) = 0.079 cal/cm-sec-K. Thermal coefficient of expansion at 293K = 30.2 x 10^{-6} /K. Tensile strength = 2860 psi.

2.1.C Teledyne Microelectronics Handbook, Appendix A, Material Properties. Density = 7.36 g/cc. Specific heat = 0.054 cal/g-K. Thermal conductivity = 0.079 cal/cm-sec-K. Thermal Expansion = 23.4 x 10^{-6}/K. Melting point = 494K. Resistivity = 10.63 micro-ohm-cm. Yield strength = 3600 psi. Ultimate tensile strength = 4600 psi. Elongation at break = 49%.

2.2.A Lee, H.J. and Taylor, R.E., 14th Thermal Conductivity Conference. Specific heat at 300 K = 0.0729 cal/g-K. Thermal conductivity at 300 K = 0.187 cal/s-cm-K.

2.2.B See 2.1.C

2.3.A See 2.2.A

2.3.B Smith, A. W., Ohio State University, Engineering Experimental Station Bulletin, vol 30, 1925. Thermal conductivity at 60 C = 0.149 cal/s-cm-K.

2.3.C See 2.1.B

2.4.A See 2.1.B

2.4.B See 2.1.C

2.5.A Hanson, D., Sandford, E. J. and Stevens, H., J. Inst. Metals, 55, 115-31, 1934. Melting point = 221 C.

2.5.B Smithells Metals Reference Book, 6th Edition, Butterworths. Melting point = 221 C.

2.5.C Smith, J. F., Materials Engineering, vol 80, no 3, 76-78, 1974. Melting point = 221 C.

2.5.D SCM Metal Products, Inc. Melting point = 221 C.

2.6.A Manko, Howard H., Product Engineering, 32, 10, 39-46, 1961. Resisitvity at room temperature = 12.31$\mu\Omega$cm.

2.6.B M. C. Canfield and Sons. Resistivity at room temperature = 11.95 $\mu\Omega$cm.

2.6.C Data is summary of two sources: a. Matthiessen, Transactions of the Royal Society of London, Series A, 150, 161, 1960, and b. same author, journal and series, 154, 167, 1965. Resistivity at 293.3 K = 13.43 $\mu\Omega$cm.

2.6.D See 2.1.A

2.7.A See 2.1.A

2.8.A Subrahmanyam, B. and B. Krishnamurty, Ind J Pure Appl Phys, 2, 1964, 139-140. Young's modulus = 7.8 x 10^6 psi. Poisson's ratio = 0.36.

2.9.A See 2.8.A

2.10.A Addicks, L., (Editor), Silver in Industry, Reinhold Publishing Co. New York, New York, 1940. Yield strength = 4500 psi. Tensile strength = 5300 psi. Elongation in 2 inches = 37%. Properties are for 0.040 inch thick sheet aged 14 days at room temperature.

2.10.B See 2.1.C

2.11.A International Tin Research Institute Pub. no. 656, "Solder Alloy Data", 24. Ultimate tensile strength at 293 K = 6240 psi. Hardness at room temperture = 14.8 HV.

2.11.B Turkus, S. and A. Smith, Metals and Alloys, March, 1942, 412. Tensile strength = 4980 psi.

2.11.C London, J. and D. W. Ashall, Brazing and Soldering, no 10, Spring 1986, 17-23. Tensile strength = 7250 psi for bulk solder. Shear strength = 2430 psi.

2.12.A See 2.10.A

2.13.A See 2.11.A

2.13.B Manko, H. H., Solders and Soldering, McGraw Hill, 1979. Brinnell hardness = 40 at room temperature.

2.13.C Davies, R.L., Welding Journal, Oct, 1976. Bulk tensile strength = 8922 psi. Brinell hardness = 13.8.

88Au 12Ge

3.1.A Indium Corp. of America. Density at room temperature = 14.673 g/cc. Melting point = 356C (629K). Thermal expansion coefficient = 13.35 x 10^{-6}/K. Thermal conductivity = 0.105 cal/cm-sec-K. Young's modulus = 10.55 x 10^6 psi. Yield strength = tensile strength = 26,800 psi. Elongation < 2%.

3.1.B GTE Wesco, Technical Ceramics and Metals. Density = 14.7 g/cm3. Melting point = 356C (629K). Tensile strength = 29,000 psi.

3.1.C Teledyne Microelectronics Handbook, Appendix A, Material Properties. Density = 14.7 g/cc. Thermal expansion coefficient = 12.6 x 10^{-6}/K. Thermal conductivity = 0.145 cal/cm-sec-K. Melting point = 356C.

3.2.A Calculated using Kopp's law and elemental specific heats listed by Kelly in 1961 Bureau of Mines Bulletin 592.

3.3.A See 3.1.A

3.3.B Motorola Inc., NASA contractor report 96283, 1968. Thermal conductivity at 100C = 0.060 cal/cm-sec-K. Thermal expansion coefficient = 14.1 x 10^{-6}/K from 20 to 200C.

3.3.C See 3.1.C

3.4.A See 3.1.A

3.4.B See 3.3.B

3.4.C See 3.1.C

3.4.D Bernstein, L., Semiconductor Products, August, 35-39, 1961. Thermal expansion coefficient between 20 and 220C = 16.5 x 10^{-6} /K.

3.5.A See 3.1.A

3.5.B See 3.1.B

3.5.C See 3.1.C

3.6.A Johnson, B.C., D.H. Aspin and W.H.Waldin, J Vac Sci Tech, vol 16, no 2, 145-6, 1979. Resistivity of 0.3 vol fraction Ge in Au film 10 angstrom thick deposited on SiO_2 is 30 micro-ohm-cm.

3.8.A See 3.1.A

3.10.A See 3.1.A

3.11.A See 3.1.A

3.11.B See 3.1.B

3.12.A See 3.1.A

3.13.A Jaffee, R.I., E.M. Smith and B.W. Gunser, Trans AIMME, Metals Div, vol 161, 366-72, 1945. Hardness = 39.09 kg/mm^2 or 206 Vickers.

3.13.B Cole, S.S. and A.R. Kroehs, Proceeding 1969 Electronics Components Conference, 298-302. Hardness = 10.7 Brinell at 21C.

97Au 3Si

4.1.A Indium Company of America, Indalloy #184, [96.76Au-3.24Si] Density = 15.4 g/cc. Thermal conductivity = 0.065 cal/cm-sec-K. Thermal expansion coefficient = 12.33 x 10^{-6} /K. Melting point = 636K (363C). Young's modulus = 12.04 x 10^{6} psi. Yield strength = 31,900 psi. Tensile strength = 36,975 psi.

4.1.B Teledyne Microelectronics Handbook, Appenedix A, Material Properties. Density = 15.4 g/cc. Thermal conductivity = 0.157 cal/cm-sec-K. Thermal expansion coefficient = 13.7 x 10^{-6}/K. Melting point = 370C (653K).

4.2.A Chen, H.S. and D. Turnbull, J Appl Phys, vol 38, no 9, 3646-50, 1967. Heat capacity = 0.0350 cal/g-cm.

4.3.A See 4.1.A

4.3.B See 4.1.B

4.3.C Motorola Inc, Semiconductor Division, Contract 952022 (for JPL), Fourth Quarterly Report, July, 1968. Thermal Conductivity for 97Au-3Si alloy = 0.031 cal/cm-sec-K at 100C (373K). Thermal expansion coefficient = 14.2 x 10^{-6} between 30 and 160C (Fifth Quarterly Report).

4.4.A See 4.1.A

4.4.B Misik, A. M., Novikov, E. N., and Ivankina, M. S., Soviet Physics Journal 14, 5, 668-70, 1971. Thermal linear coeff. of expansion at room temperature = 10.1 ppm/K. Young's modulus at room temperature = 11.67 x 10^{6}.

4.4.C See 4.3.C

4.4.D Bernstein, L., Semiconductor Products, August, 35-39, 1961. Thermal expansion coefficient = 13.3 x 10^{-6} /K between 20 and 200C.

4.5.A See 4.1.A

4.5.B See 4.1.B

4.6.A Hauser, J. J. and J. Tauc, Phys Rev B, vol 17, no 8, 3371-3380. Electrical resistivity for amorphous, sputtered films at 77K = 117 micro-ohm-cm. TCR over the range 4K to 77K = 150 ppm for films annealed at room temperature.

4.7.A See 4.6.A

4.8.A See 4.1.A

4.8.B See 4.4.B

4.10.A See 4.1.A

4.11.A See 4.1.A

4.13.A Cole, S.S. and A.R. Kroehs, 1969 Proceedings Electronics Components Conference, 298-302. Hardness of 98Au 2Si = 56 Brinell at 21C.

80Au 20Sn

5.1.A Indium Corp. of America. Density at room temperature = 14.52 g/cc. Specific heat at room temperature = 0.036 cal/K-g. Thermal conductivity at room temperature = 0.14 cal/s-cm-K. Thermal linear coeff. of expansion = 15.93 ppm/K. Melting point = 280 C. Young's modulus at room = 8570 ksi. Yield strength at room = 40000 psi. Ultimate tensile strength at room = 40000 psi. Elongation at break at room temperature = 2 % in 1 inch.

5.1.B Mattiessen, A. and M. Holzman, Pogg Ann, vol 110, 21-37, 1860. Density at 14.2C = 14.243 g/cc for 22.8% tin alloy. Interpolated value gives density = 14.84 g/cc.

5.1.C Teledyne Microelectronics Handbook, Appendix A, Material Properties. Density = 12.5 g/cc. Thermal Expansion coefficient = 15.93 x 10^{-6}/K. Specific heat = 0.036 cal/gm-K. Thermal conductivity = 0.137 cal/cm-sec-K. Young's modulus = 8.57 x 10^{6} psi. Melting point = 280C. Yield strength = 40,000 psi. Tensile strength = 40,000 psi. Elongation <2%.

5.2.A See 5.1.A

5.2.B See 5.1.C

5.3.A See 5.1.A

5.3.B Motorola Inc., NASA contractor report 96283, 1968. Thermal conductivity at 100C = 0.069 cal/cm-sec-K. Thermal expansion coefficient in the ranges 34-74C and 88-130C = 17.1 x 10^{-6}. Some anomolies exist at intermediate temperatures.

5.3.C See 5.1.C

5.3.D Kadota T., et al., Rev Elec Comm Lab, vol 25, no 9-10, 1052-1061. Thermal conductivity = 0.067 cal/cm-sec-K.

5.4.A See 5.1.A

5.4.B See 5.3.B

5.4.C See 5.1.C

5.4.D Bernstein, L., Semiconductor Products, August, 35-39, 1961. Thermal Expansion coefficient = 16.1 between 20 and 120C.

5.5.A Smithells Metal Reference Book, 6th Edition, Butterworths. Melting point = 280 C.

5.5.B See 5.1.A

5.5.C See 5.1.C

5.6.A Data is summary from 2 sources: a. Matthiessen, Transactions of the Royal Society of London, Series A, vol 150, 161, 1960, and b. same author, journal and series, vol 154, 167, 1965. Resistivity at 289 K = 17.35 $\mu\Omega$cm.

5.8.A See 5.1.A

5.8.B See 5.1.C

5.10.A See 5.1.A

5.10.B See 5.1.C

5.11.A See 5.1.A

5.11.B See 5.1.C

5.12.A See 5.1.A

5.12.B See 5.1.C

5.13.A Cole, S.S. and A.R. Kroehs, Proceedings 1969 Elec Comp Conf, 298-302. Hardness at 21C = 115 Brinell.

0.1C-Fe (Low Carbon Steel)

6.1.A Lucks, C. F. and Deem, H. W., ASTM Special Tech Pub, no. 227, 1-29. Density at room temperature = 7.858 g/cc.

6.1.B Lucks, C.F., et al, USAF Tech Rept 6145, I, 7-16. [AD 117 715] Density at room temperature = 7.85 g/cc.

6.2.A Lucks, C.F., et al, USAF TR 6145, 1-71, 1954. [AD 95 406] Specific heat at 298 K = 0.105 cal/g-K (interpolated).

6.2.B Pallister, P.R., J Iron Steel Inst, vol 185, 474-82, 1957. Specific heat at 293 K = 0.1061 cal/g-K.

6.2.D Hatfield, Brown-Firth Research Laboratories. Specific heat at room temperture = about 0.113 cal/g-K.

6.2.E Honda, Japanese J of Physics, vol 1, 71, 1922. Specific heat at room temperature = about 0.113 cal/g-K.

6.3.A Gruneisen, E., Ann Physik, vol 4, no 3, 43-74, 1900. Thermal conductivity at 291 K = 0.171 cal/cm-sec-K.

6.3.B Lucks, C.F., et al, USAF TR 6145, I, 1-127, 1951. [ATI 117 715] Thermal conductivity at 300 K = 0.164 cal/cm-sec-K (interpolated).

6.3.C Bollenrath, F. and Bungardt, W., Arch. Eisenhuttenw., vol 9, no 5, 253-262, 1935. Thermal conductivity at 293 K = 0.127 cal/cm-sec-K.

6.3.D Raezer, S. D., Office Ordn. Research TR 1, 1-36, 1954. [AD 49 544] Thermal conductivity at 298 K = 0.136 cal/cm-sec-K.

6.3.F Pozdnyak, N.Z., Proc of the 8th Thermal Conductivity Conference, 457-64, 1969.

6.4.A Lucks, C. F., et al, U. S. Air Force Rept. AF-TR-6145 (Pt. I), 1951. See 6.4.C below for data at room temperature.

6.4.B Johnston, H.L., et al, J Chem Eng Data vol 10, no 3, 241-2, 1965. See 6.4.C below for data at room temperature.

6.4.C Beenakker, J.J.M. and Swenson, C.A., Rev Sci Instr, vol 26, 1204-5, 1955. Summary of data from 6.4.A-C: thermal linear coeff. of expansion at 293 K = 10.9 ppm/K.

6.5.A Data from phase diagram.

6.6.A Raezer, S.D., Irst Res Lehigh U., Bethelhem, Pa. Tech Report 1, 1-48. [AD49-544]. Resistivity at room temperature = 10.4 micro-ohm-cm.

6.6.B Yensen, T.D., Trans Amer Inst Elec Eng, 43, 145-75, 1969. Resistivity at room temperature = 11.7$\mu\Omega$cm.

6.6.C Sawyer, R. B., USAEC Publ. NP-5866, 1-32. Resistivity at 300 K = 13.3 $\mu\Omega$cm (interpolated)

6.6.D Hogan, C.L. and Sawyer, R.B., J Appl Phys, 23, 177-80.

6.7.A Calculated from data provided by reference 6.6.A. TCR from 298-405K = 7000 ppm/K on the basis of a reference resistivity of 13.4 micro-ohm-cm.

6.7.B Calculated from data provided by reference 6.6.C. TCR from 273-473K = 4100 ppm/K on the basis of a reference resistivity of 13.4 micro-ohm-cm.

6.7.C Calculated from data provided by reference 6.6.D. TCR from 373-473K = 11000 ppm/K on the basis of a reference resistivity of 13.4 micro-ohm-cm.

6.8.A Benedicks, Thesis, Upsala, 1904. See 6.8.C below for data.

6.8.B Crawford, Proceedings of the Royal Society of Edinburgh, 32, 348, 1912. See 6.8.C below for data.

6.8.C Wahlberg, J. of the Iron and Steel Institute (London), 60, 234, 1901. Summary of data from 6.8.A-C: Young's modulus at room temperature = 29,300 to 30,000 ksi.

6.8.D Hoyt, S.L., METAL DATA, Reinhold, New York, 1952. Young's modulus at 21C = 29,500 ksi.

6.8.E Kimball, A.L. and D.E. Lovell, Phys Rev, 26, 121, 1925.

6.9.A Bhagavantam, S. and B. Ramachandra, Nature, vol 157, 624, 1946. Poisson's ratio calculated from Young's modulus and shear modulus = 0.292. Data is from ultrasonic methods.

6.9.B Calculated from shear modulus and Young's modulus given in reference 6.8.D. At 21C, Poisson's ratio = 0.25.

6.10.A Timken Steel and Tube Co. [0.10-0.20% C, 0.30-0.60% Mn, 0.25% Si, 0.45-0.65% Mo] Yield strength = 33,000 psi. Tensile strength = 64,000 psi. Elongation = 37% in 2 inches.

6.10.B ASME and ASTM, Compilation of Available High Temperature Data, ASTM, 1938. [0.15% C, 0.46% Mn, 0.28% Si, one inch diameter wrought bar annealed 1550F, grain size = 5-6] Yield strength = 34,000 psi. Tensile strength = 57,300 psi. Elongation = 40%. Hardness = 111 Brinell.

6.10.C Lander, C.H. and J.V. Howard, Proc Royal Soc Lon A, vol 156, 411-426, 1936. [Mild steel] Yield strength = 50,000 psi. Ultimate tensile strength = 60,000 psi.

6.10.D Gillett, H.W., ASTM Spec Pub No 47, 1941. [1020 steel, cold rolled] Yield strength = 52,000 psi. Ultimate tensile strength = 72,000 psi.

6.11.A See 6.10.A

6.11.B See 6.10.B

6.11.C See 6.10.C

6.11.D See 6.10.D

6.12.A See 6.10.A

6.12.B See 6.10.B

6.12.C Hatfield, W.H., Metallurgia, vol 20, 107, 1939. [0.1% C, 0.72% Mn, 0.06% Si, Normalized at 1650F] Yield strength = 36,300 psi. Ultimate tensile strength = 63,800 psi. Elongation = 37%.

6.13.A Guillet and Cournot, Comptes rendus hebdomadaires des s'eances de l'acad'emie des sciences, de l'institut de France, vol 174, 384, 1922. For data see 6.13.B below.

6.13.B Ito, Science Reports of the Tohoku Imperial University, Series I: Mathematics, Physics and Chemistry, vol 12, 137, 1923. Summary of data in 6.13.A-B: Brinell hardness at room temperature = 110 to 112.

6.13.C See 6.10.B

55Cu 45Ni

7.1.A Guettel, C.L., Driver Harris Co. Density (wrought) = 8.9 g/cc. Specific heat = 0.094 cal/g-K. Thermal conductivity = 0.051 cal/cm-sec-K. Thermal expansion coefficient (20-100C) = 14.9 ppm/K. Melting temperature = 1480K. TCR (20-150 C) = 20 ppm/K. Tensile strength = 135,000 psi (cold worked) and 60,000 psi (annealed). Elongation = 32%. Hardness = 48-54 Rockwell B.

7.1.B Grahm, W.A. Metals Handbook, ASM, 1418, 1939. Density of cast material = 8.6 g/cc. Melting range = 1225 to 1275C.

7.2.A Eucken, Von A. and Werth, H., Z Anorg Allgem Chem, vol 188, 152-72, 1930. Specific heat at 300 K = 0.1108 cal/g-K. (extrapolated).

7.2.B Pawel, R.E. and E.E. Stansbury, J Phys Chem Solid, vol 26, 607-13, 1965. Specific heat for 50/50 alloy = 0.094 cal/g-K.

7.2.C Corruccini, R.J. and Gniewek, J.J., National Bureau of Standards Monograph 21, 15, 1960. Specific heat at 300 K = 0.0979 cal/g-K.

7.3.A Gruneisen, E., Ann Physik, vol 4, 3, 43-74, 1900. Thermal conductivity at 291 K = 0.049 cal/cm-sec-K.

7.3.B Powers, R.W., et al, USAF TR 264-8, 1-14, 1951. [ATI 105 925] Thermal conductivity at 295 K = 0.055 cal/cm-sec-K.

7.3.C Ellis, W. C., et al, Rensselaer Polytech Inst Eng and Sci Series, Bull 21, 1-23, 1928. Thermal conductivity at 305 K = 0.049 cal/cm-sec-K.

7.3.D Ho, C.Y., et al, 14th Thermoconductivity Conference, 233. Thermal conductivity at 300 K = 0.056 cal/cm-sec-K.

7.3.E Smith, A.W., Ohio State Univ. Engineering Experimental Station Bulletin, vol 30, 1925.

7.3.F Silverman, L., J Metals, vol 5, 631-2, 1953.

7.4.A Kantola, M. and Tokola, E., Ann Acad Sci Fenn, A VI(233), 1967. Thermal linear coeff. of expansion at 300 K = 16.0 ppm/K. (summary of two sets of data)

7.4.B NBS distillation of Data from: 1) Aoyama, S., and T. Ito, Tohoku Univ Sci Repts, vol 27, 348-364, 1939 and 2) Krupkowski, M.A. and W.J. DeHaas, Lieden Comm no 194b, 1928. Thermal expansion of 50/50 alloy = 13.8 ppm/K.

7.4.C Henning, Annalen Der Physik, vol 22, 631, 1907.

7.4.D Interpolation of Thermophysical Properties of Matter provisional values for 35Cu 65Ni and 65Cu 35Ni. Thermal expansion coefficient = 14.6 ppm/K at 293K.

7.5.A See 7.1.A

7.5.B See 7.1.B

7.6.A Chevenard, P., J. Inst. Metals, vol 36, 39-62, 1926. Resistivity at room temperature = 48.9 μΩcm. (interpolated)

7.6.B Ho, C. Y., et al, Thermal Conductivity Conference, 14th, 233. Resistivity at 300 K = 47.5 μΩcm.

7.6.C Smith, A.W., Ohio State Univ. Engineering Experimental Station Bulletin, vol 30, 1925. Resistivity at 297 K = 50 μΩcm. TCR at room temperature = -53 ppm/K, defined with respect to resistivity at 25 C.

7.7.A See 7.6.C

7.7.B See 7.1.A

7.8.A Iguchi, E. and K. Udagawa, J Phys F, vol 15, 214-226. Young's modulus at 293K = 21,820 ksi.

7.8.B NBS Circular 58, Second Edition, 1923. Young's modulus for 44% Ni alloy = 23,200 ksi.

7.8.C Yousef, Y.L., Proc Phys Soc Lon, vol 57, 412-18, 1945. Young's modulus = 24,000 ksi.

7.8.D. Orlov, A.F. and Fedotov, Fiz Metal Metaloved, vol 22, no 1, 137-8, 1966. Young's modulus = 24,600 ksi. Poisson's ratio = 0.291.

7.8.E Faninger, G. Z. MetallK., vol 60, no 7, 601-5, 1979. Young's modulus = 24,000 ksi. Poisson's ratio = 0.307 to 0.352.

7.8.F Mudge, W.A. and L.W. Luff. Proc ASTM, vol 28, part 2, 278, 1928. Young's modulus = 24,800 ksi. Poisson's ratio = 0.376. Yield strength = 27,100 psi. Tensile strength = 66,600 psi. Elongation = 46% in 2 inches. Hardness = 54 Rockwell B. (annealed 4 hours at 1400F)

7.8.G McAdam, D.J., Trans Am Soc Steel Treat, vol 7, 54, 1925. Young's modulus = 25,000 ksi. Yield strength = 28,000 psi. Tensile strength = 70,000 psi. Elongation = 48%. Hardness = 100 Brinell.

7.9.A See 7.8.D

7.9.B See 7.8.E

7.9.C See 7.8.F

7.10.A See 7.8.F

7.10.B See 7.8.G

7.10.C Colbeck, E.W. and MacGillivray, Trans Inst Chem Eng II, 107, 1933. Yield strength (0.1%) at 300 K = 19,600 psi. Ultimate tensile strength at 300 K = 60,000 psi. Elongation at break at 300 K = 40 % in 2 inches.

7.11.A See 7.8.F

7.11.B See See 7.8.G

7.11.C See 7.10.C

7.12.A See 7.8.F

7.12.B See 7.8.G

7.12.C See 7.10.C

7.13.A See 7.1.A

7.13.B See 7.8.F

7.13.C See 7.8.G

7.13.D Engl, J. and J. Fowler, Z Metallk, 702, 1936.

7.13.E Comptes Rendus (Doklady) URSS, vol 53, no 9, 1946. Brinell Hardness = 86.5 with 10 mm ball and 500 kg load.

80Ni 20Cr (Nichrome)

8.1.A Gatward, W.A., Hoskins Mfg Co data sheet. Density = 8.4 g/cc. Specific heat = 0.107 cal/g-K. Thermal conductivity = 0.033 cal/cm-sec-K. Thermal expansion coefficient (0-1700C = 17.96 ppm/K. Resistivity = 107.9 micro-ohm-cm. Melting point = 1400C. Young's modulus = 31 x 10^6 psi. Yield strength = 63,000 psi Ultimate tensile strength = 110,000 psi. Elongation = 25%. Hardness = 85 to 90 Rockwell B.

8.1.B Harrison Alloys, Inc, NCR-85. Density = 8.41 g/cc. Thermal conductivity = 0.026 cal/cm-sec-K at 100C. Coefficient of thermal expansion = 17 ppm/K. Melting point = 1400C. Ultimate tensile strenth at 20C = 100,000 psi for fully annealed material and 200,000 psi for hard drawn material. Elongation = 40% in 2 inches.

8.1.C Betty, B.B. and W.A. Mudge, Mechanical Engr, Feb, 123-129, 1945 (International Nickel Company). Density = 8.42 g/cc. Thermal conductivity = 0.0358 cal/cm-sec-K. Thermal expansion coefficient = 7.2 ppm/K from 0 to 100C. Melting point = 1232C. Yield strength = 50-190 ksi. Ultimate tensile strength = 100-200 ksi. Elongation = 35-0.5%. Brinell hardness = 120-300.

8.1.D Seibel, R.D. and Mason, G.L., Denver University, Denver Res. Inst., Colo., WADC TR-57-468, 1-58. (AD 155 605). Density at 298 K = 8.35 g/cc.

8.2.A See 8.1.A

8.2.B Douglas, T.B. and Victor, A.C., J Res NBS, vol 65C, no 1, 65-9. Specific heat at 293 K = 0.104 cal/g-K.

8.2.C Douglas, T.B. and J.L. Denver, USAEC Rept, 1953, (NBS-2302). Specific heat = 0.105 cal/g-K.

8.2.D Neimark, B.E., et al, High Temperature, vol 1, no 1, 9-12.

8.3.A Silverman, L., J Metals, 5, 631-2.

8.3.B See 8.2.D

8.3.C Seibel, R.D. and Mason, G.L., Denver University, Denver Res. Inst., Colo., WADC TR-57-468, 1-58. (AD 155 605).

8.3.D Powell, R.W., Proc Phys Soc (London), 48(3), 381-92, 1936. Thermal Conductivity at room temperature = 0.130 cal/s-cm-K. (interpolated)

8.3.E Smith, A.W., Ohio State Univ. Engineering Experimental Station Bulletin, vol 30, 1925.

8.3.F References 8.1.A-C list thermal conductivity for commercially available material

8.4.A Joint Research Committee on the Effect of Temperature on the Properties of Metals, American Soc. Testing Materials Proc, vol 27, no 1, 153-62, 1927. Coeff. of thermal linear expansion at room temperature = 12.3 ppm/K (interpolated).

8.4.B Chevenard, P., J Inst Metals, vol 36, 39-62, 1926. Coeff. of thermal linear expansion at room temperature = 11.7 ppm/K.

8.4.C References 8.1.A-8.1.C list thermal expansion coefficient for commercially available material.

8.5.A Hitchcock, R.C. and Zemansky, M.W., Amer J Physics, vol 13, 329, 1945. Melting point = 1670K.

8.5.B References 8.1.A-8.1.C list melting point.

8.6.A Thomas, H.Z., Physik, vol 129, 219-32. Resisitivity at room temperature = 100 μΩcm. TCR at room temperature = 150 ppm/K, defined with respect to resistivity at 0 C.

8.6.B Koster, W. and Rocholl, P., Z Metallk, vol 48, 485-95.

8.6.C Swanson, J.G. and D.S. Campbell, Thin Solid Films, vol 1, 183-20, 1967. Resistivity varies with thickness.

8.6.D Belu-Marian, A., et al, Thin Solid Films, vol 39, 15-24, 1986. Resisitivity at 298 K = 80-121 μΩcm.

8.6.E Nenadovic, T.M., et al, Thin Solid Films, vol 10, 45-56, 1972. Resisitivity varies with thickness.

8.6.F Smith, A.W., Ohio State Univ. Engineering Experimental Station Bulletin, vol 30, 1925. Resistivity at 298 K = 107 μΩcm. TCR at room temperature = 104 ppm/K, defined with respect to resistivity at 25 C.

8.7.A See 8.6.A above.

8.7.B Swanson, J.G. and Campbell, D.S., Thin Solid Films, vol 1, 183-202, 1967. TCR varies with thickness.

8.7.C Belu-Marian, A., et al, Thin Solid Films, vol 39, 15-24, 1986. TCR varies with thickness.

8.7.D See 8.6.F

8.8.A See 8.1.A

8.8.B MeKeown, J. and E.D. Ward, BNFMRA, Series #43, 1938. Young's modulus = 31,800 psi. Proportional limit = 49,300 psi. Tensile strength = 113,000 psi (hot rolled).

8.8.C Nickel Rev, vol 11, 33, 1938. Young's modulus = 31,000 ksi. Yield strength = 63,000 psi. Tensile strength = 110 psi. Elongation = 30 % in 10 inches.

8.8.D Landolt-Bornstein Numerical Data and Functional Relationships in Science and Technology, New Series, III/18, 5, 69. Young's modulus at 300 K = 31,600 ksi to 33,000 ksi. Poisson's ratio at 300 K = 0.295 to 0.303

8.9.A See 8.8.D

8.10.A See 8.1.A

8.10.B Dehaas, W.J. and Hadfield, R., Trans Roy Soc (London), 232A, 297, 1933. Yield strength at 300 K = 101,300 psi. Elongation at break at 300 K = 26.9 %. Ultimate tensile strength at 300 K = 130,700 psi. Material is water quenched from 1832F.

8.10.C See 8.1.C

8.11.A See 8.1.A

8.11.B See 8.1.B

8.11.C See 8.1.C

8.11.D See 8.10.B

8.12.A See 8.1.A

8.12.A See 8.1.B

8.12.D See 8.10.B

8.13.A See 8.1.A

8.13.B See 8.1.C

8.13.C Osipov, E.M. and E.M. Miroshkima, Tr Inst Met Akad Nauk SSSR, No 3, 160-164, 1958.

8.13.D Koster, W. and P. Rocholl, Z Metallk, vol 48, 485-95, 1957. Hardness of fully recrystallized material = 250 kg/mm^2. Hardness of 50% cold worked material = 328. Hardness of 50% cold work plus 450C anneal = 385.

42Ni 58Fe

9.1.A Jette, E.R. and F. Foote, Trans AIMME, vol 120, 259-272, 1936. Density = 8.07 g/cc.

9.1.B Huntington Alloys, Nilo Alloy 42. Density = 8.13 g/cc. Thermal conductivity = 0.0251 cal/cm-sec-K. Thermal expansion coefficient (30-200C) = 4.2 ppm/K. Electrical resistivity = 64 micro-ohm-cm. Yield strength = 37,000 psi. Tensile strength = 72,000 psi. Elongation = 43%. Hardness = 80 Rockwell B.

9.1.C Harrison Alloys (Driver Harris) Technical Information Bulletin for Alloys 142, 146 and 152. Density = 8.13 g/cc. Thermal conductivity = 0.0257 cal/cm-sec-K. Melting point = 1425C. Coefficient of thermal expansion (30-300C) = 4.4 ppm/K. Yield strength = 24,000 psi. Tensile strength = 70,000 psi. Elongation = 20%. Hardness of strip = 70 Rockwell B.

9.2.A Zuithoff, A.J., Rec. Trav. Chim. 59, 131-60, 1940. Specific heat = 0.1154 cal/g-K at 120C.

9.2.B Ingersoll, L.R., et al, Phys Rev 16, 126-32, 1920. Specific heat at 298 K = 0.121 cal/g-K. Thermal conductivity at 293 K = 0.032 cal/sec-cm-K.

9.3.A Silverman, L., J. Metals, 5, 631-2. Thermal conductivity at 50C = 0.033 cal/cm-sec-K.

9.3.B See 9.2.B

9.3.C Watson, T.W. and H.E. Robinson, J Heat Trans, Nov 1961, 403-408. Thermal conductivity = 0.038 cal/cm-sec-K at 273K.

9.3.D See 9.1.B

9.3.E See 9.1.C

9.3.F Inco Alloys, Int. Quick Reference Guide to High Temperature Alloys. Thermal conductivity = 0.0251 cal/cm-sec-K. Yield strength = 37,000 psi. Tensile strength = 72,000 psi. Elongation = 43%. Hardness = 139 Brinell.

9.4.A Beenakker, J.J.M. and C.A. Swenson, Rev Sci Instr vol 26, 1204-5, 1955. Coeff. of thermal linear expansion at 300 K = 2.7 ppm/K.

9.4.B Brookhaven National Lab., USAEC Publ BNL-67, 1-37. Coeff. of thermal linear expansion at 293 K = 4.54 ppm/K.

9.4.C Millner, T. and Welesz, R., Acta Tech Acad Sci Hung, vol 14, 279-91, 1956.

9.4.D See 9.1.B

9.4.E See 9.1.C

9.6.A Ho, C.Y., et al, J Phys Chem Ref Data, vol 12, no 2, 287, 1983. Resistivity at 300 K = 58.2 μΩcm. TCR (derived from resistivity) at 300 = 3054 ppm/K (with respect to 273 K).

9.6.B Ingersoll, L.R., et al, Phys Rev, vol 16, 126-32, 1920. Resistivity at room temperature 64.5 μΩcm. TCR at 300K = 2200 ppm/K, defined with respect to resistivity at 0C.

9.6.C Window, B., J. Appl. Phys. 44(6), 2853-65, 1973. Resistivity at 300K = 62.5 μΩcm.

9.7.A See 9.6.A

9.7.B See 9.6.B

9.7.C MWS Wire Company. TCR = 3000 ppm/K.

9.8.A Ledbetter, H.M., Cryogenics, Dec, 1982, 653-656. Young's modulus = 20,500 ksi.

9.8.B Landolt-Bornstein Numerical Data and Functional Relationships in Science and Technology, New Series, III/11, 16, 25, III/18, 6, 74. Young's modulus at 298K = 20,615 ksi to 23,278 ksi. Poisson's ratio at 298K = 0.302 to 0.325.

9.8.C Data is a summary of: a. Colby, Proc ASTM, no 3, 141, 1903, and b. Guillaume, Congr'es international des methods d'essai des mat'eriaur de construction, Paris, 1900. Young's modulus at room temperature = 23,200 ksi.

9.9.A See 9.8.B

9.10.A Data is summary of 2 sources: a. Burgess and Aston, Bull, Univ Wisconsin, no 346, 1910, and b. Roush, Chemical and Metallurgical Engineering, vol 8, 468, 1910.
Yield strength at room temperature = 54600 psi. Ultimate tensile strength at room temperature = 81,900 psi.
Elongation at break at room temperature = 32.5 %.
Hardness at room temperature = 17 ScH.

9.10.B See 9.1.B

9.10.C See 9.1.C

9.10.D DeHaas, W.J. and R. Hadfield, Trans Roy Soc Lon A, vol 232, 297, 1933

9.11.A See 9.10.A

9.11.B See 9.1.B

9.11.C See 9.1.C

9.11.D See 9.10.D

9.11.E MacPherran, R.S., Proc ASTM, vol 21, 852, 1921.

9.12.A See 9.10.A

9.12.B See 9.1.B

9.12.C See 9.1.C

9.12.D See 9.10.D

9.13.A See 9.10.A

9.13.B See 9.1.B

9.13.C See 9.1.C

9.13.D See 9.10.D

80Pb 20In

11.1.A Indium Corp. of America, Indalloy No 150 (81Pb 19In). Density at room temperature = 10.28 g/cc. Thermal conductivity at 85C =0.0406 cal/cm-sec-K. Resistivity at room temperature = 38.3 μΩcm. Coeff. of thermal linear expansion = 27 ppm/K. Solidus = 244C. Liquidus = 275C. Tensile strength = 5,500 psi.

11.1.B Lead Industries Assoc, Lead and Lead Alloys [81Pb 19In]. Density = 10.27 g/cc. Solidus = 270C. Liquidus = 280C. Thermal conductivity = 0.0406 cal/cm-sec-K. Resistivity = 38.3 micro-ohm-cm. For 80Pb 20In: Density = 10.16 g/cc. Solidus = 180C. Liquidus = 270C.

11.2.A Calculated from Kopp's Law.

11.3.A See 11.1.A

11.3.B See 11.1.B

11.4.B See 11.1.B

11.5.A See 11.1.A

11.5.B See 11.1.B

11.5.C Alpha Metals, Indium Alloy Materials, 1985. [81Pb 19In] Solidus = 270C. Liquidus = 280C.

11.6.A See 11.1.A

11.6.B See 11.1.B

11.6.C Kurnakov and Zhemchuzhnui, Z anorg allgem Chemie, vol 64, 149, 1909. Resistivity at room temperature = 32.3 μΩcm. TCR at 298K = 2500 ppm/K, defined with respect to resistivity at 25C.

11.6.D Valentine, S., Z Physik, vol 133, 1-20, 1938. Resistivity at 0C = 34 micro-ohm-cm. TCR = from -191 to 100C based on resistivity at 0C.

11.7.A See 11.6.C

11.7.B See 11.6.D

11.11.A See 11.1.A

50Pb 50In

12.1.A Indium Corp. of America. Density at room temperature = 8.87 g/cc. Solidus = 180C. Liquidus = 209C. Resistivity at room temperature: 28.7 μΩcm. Thermal conductivity at 85C = 0.0526 cal/cm-sec-K. Coefficient of thermal linear expansion at room temperature = 27 ppm/K. Ultimate tensile strength at room temperature = 4670 psi. Hardness = 9.60 Brinell (100 Kg load for 30 seconds).

12.1.B Lead Industries Assoc., Inc. Density at room temperature = 8.86 g/cc. Ultimate tensile strength at room temperature = 4575 psi. Solidus = 180C. Liquidus = 209C.

12.1.C Smith, J.F., Materials Engr, vol 80, no 3, 76-78, 1974. Density = 9.13 g/cc.

12.2.A Calculated with Kopp's Law.

12.3.A See 12.1.A

12.4.A See 12.1.A

12.5.A See 12.1.A

12.5.B See 12.1.B

12.5.C Alpha Metals, Inc., Indium Solders. Solidus = 180C. Liquidus = 209C.

12.6.A See 12.1.A

12.6.B Kurnakov and Zhemchuzhnui, Z anorg allgem Chemie, vol 64, 149, 1909. Resistivity at 298K = 27 μΩcm. TCR at 298K = 2300 ppm/K, defined with respect to resistivity at 298K.

12.6.C Valentine, S., Z Physik, vol 115, 11-16, 1940. Resistivity at 0C = 28 micro-ohm-cm. TCR from -191 to 100C = 1600 ppm/K.

12.7.A See 12.6.B

12.7.B See 12.6.C

12.11.A See 12.1.A

12.11.B Indium Company of America, INDIUM, 138, 1959. Tensile strength = 4670 psi. Elongation = 55% in 2 inches. Hardness = 9.60 Brinell (100 kg load for 30 seconds).

12.11.C Kaufman, A.B., Materials in Design Eng, vol 48, 114, 1958. Ultimate tensile strength at room temperature = 3000 psi.

12.12.A See 12.1.A

12.13.A See 12.11.B

12.13.B Cole, S.S. Cole and A.R. Kroehs, 1969 Elec Comp Conf Proc, 298-302. Hardness = 9.2 Brinell at 21C.

<u>89Ni 11P</u>

13.1.A Metal Finishing Suppliers Association, Quality Metal Finishing Guide, Electroless Nickel Plating, vol 1, no 2. Density with 7.5% P = 7.92 g/cc. Thermal expansion coefficient for 8 to 9% P = 13 to 14.5 ppm/K. Thermal conductivity for 8 to 9% P = 0.0105 to 0.0135 cal/cm-sec-K. Melting point = 880C for 7 to 9% P. Electrical resistivity = 52 to 68 micro-ohm-cm.

13.2.A Calculated from Kopp's Law.

13.3.A See 13.1.A

13.4.A See 13.1.A

13.5.A See 13.1.A

13.5.B Thwaites, C.J., Capillary Joining-Brazing and Soft Soldering, Research Studies Press, Wiley, 81. Melting point = 875C.

13.6.A See 13.1.A

13.6.B See ASTM B656-84. Resistivity of 10% P material as deposited is about 120 micro-pohm-cm. Resistivity drops to 33 when heated to 800C.

13.6.C Flechm, J., and G. Ormancey, J Physique et le Radium, Phys Applique, Suppl 12, 201, Dec, 1962. Resisitivity of 0.054 micron thick film with 10.1% P = 24,000 micro-ohm-cm.

13.8.A Jovanovic and C.S. Smith, J Appl Phys, vol 32, 121, 1961. Reduced modulus from bulge test = 2.84 x 10^6 psi for amorphous coatings 1-10 microns thick with 7-10% P. Biaxial tensile strength = 115,800 psi. Elongation = 0.7%.

13.11.A Graham, A.H., et al, J. Electrochem. Soc., 112(4), 401-13, 1965. Ultimate tensile strength at room temperature = 92700-115000 psi. Hardness at room temperature = 526 KHN. Ductility in the bulge test = 1.5%

13.11.B See 13.8.A

13.12.A See 13.11.A

13.12.B See 13.8.A

13.13.A See 13.11.A

13.13.B Johnson, C.E. and Ogbrun, F., Surface Technology, 4, 161-72, 1976. Hardness = 515-916 (HK100 scale).

13.13.C Higgs, C.E., Electrodeposition and Surface Treatment, vol 2, 315-326, 1973/74. As deposited hardness is about 500 VPN. Hardness increases to about 1000 when heat treated 2 hours at about 400C.

63Sn 37Pb

14.1.A Indium Company of America, Indalloy #106. Density = 8.41 g/cc. Thermal conductivity = 0.1195 cal/cm-sec-K at 85C. Thermal expansion coefficient = 25 ppm/K. Melting point = 183C(456K). Resistivity = 15.0 micro-ohm-cm. Tensile strength = 7500 psi.

14.1.B Lead Association, Inc., Lead and Lead Alloys, Density = 8.42 g/cc. Thermal expansion coefficient = 24.7 ppm/K. Melting point = 183C(456K). Resistivity = 14.5 micro-ohm-cm. Tensile strength = 7500 psi.

14.1.C Subrahmanyam, B., et al., Ind J Pure Appl Phys, vol 12, 367-8, 1964. Density = 8.60 g/cc for 67% tin alloy.

14.2.A Metals Handbook, 9th Edition, ASM, 1979. Specific heat at room temperature = 0.0358 cal/g-K. Thermal conductivity at room temperature = 0.19 cal/cm-sec-K.

14.2.B Calculated from Kopp's Law. Specific heat=0.0462 cal/g-K.

14.3.A See 14.1.A

14.3.B See 14.1.B

14.3.C Smith, A.W. Ohio State University, Engineering Experimental Station Bulletin, vol 30, 1925. Thermal conductivity at 54C = 0.119 cal/cm-sec-K.

14.3.D Brown, W.B., Phys Rev, vol 22, 171, 1923. Thermal conductivity of 62Sn 38Pb = 0.1125 cal/cm-sec-K at 102C.

14.4.A See 14.1.A

14.4.B See 14.1.B

14.4.C Balasundaram, L.J. and A.N. Sinha, J Appl Phys, vol 42, no 12, 5207, 1971. Coefficient of thermal expansion at 293 = 24 ppm/K.

14.5.A See 14.1.A

14.5.B See 14.2.B

14.5.6 Reichenecker, W.J., Matls Engr, vol 97, no 6, 12-13, 1983. Melting point = 183C.

14.6.A See 14.1.A

14.6.B See 14.1.B

14.6.C Chauk, G.S., et al, J low temp Phys, vol 31, no 1/2 153, 1978. Resistivity of $Sn_{70}Pb_{30}$ = 14.5 micro-ohm-cm.

14.7.A TCR calculated for 14.6.C from 29-295C = 3500 ppm/K

14.8.A Allied Metglass Products. Young's modulus at room temperature = 4570 ksi.

14.8.B Harper, C.A., Handbook of Electrical Packaging, McGraw Hill, 3-12. Young's modulus = 4500 psi.

14.8.C Greenfield, L.T. and P.G. Forrester, ITRI Publ no 155, 1947. Young's modulus = 4348 ksi for 60/40 alloy at room temperature.

14.9.A Westinghouse Defense and Space Center, 1968, August, 414pp. Poisson's ratio = 0.4 at 298K.

14.10.A Baker, E., IEEE Trans Parts, Hybrids & Packaging, vol PHP-8, no 4, 1972. Yield strength at 298 K = 4800 psi.

14.11.A See 14.1.A

14.11.B See 14.1.B

14.11.C Rhines, F.N. and W.A. Anderson, Metals and Alloys, vol 14, 704, 1941. Tensile strength = 29,000 psi. Shear strength = 8800 psi. Elongation = 60%. BHN = 14.

14.11.D Kalish, H.S. and F.J. Dunkerley, Trans AIMME, vol 180, 637-656, 1949. Tensile strength = 6100 psi at 295K. Elongatio = 54%.

14.11.E McNaughten, D.J., E.S. Hodges and W.R. Lewis, Int Tin Res Devlp Council Bul, no 2, 1935. Tensile strength = 7500 psi. Elongation = 32% in 4 inches. Hardness = 14. Shear strength = 6900 psi.

14.11.F Alpha Metals, Inc., Jersey City, N.J. Technical Data Bulletin A103B. Tensile strength = 7733 psi.

14.12.A See 14.11.E

14.12.B See 14.11.D

14.12.C See 14.11.C

14.13.A See 14.11.C

14.13.B Cole, S.S. and A.R. Kroehs, 1969 Elec Comp Conf. Hardness = 30 Brinell at 21C.

Kovar (54Fe 29Ni 17Co)

15.1.A Teledyne Microelectronics Handbook, Appendix A - Material Properties. Density = 8.36 g/cc. Specific heat = 0.105 cal/g-K. Thermal conductivity = 0.04 cal/cm-sec-K. Thermal expansion coefficient = 5.2 ppm/K. Resistivity = 49 micro-ohm-cm. Young's modulus = 19.5 x 10^6 psi. Yield strength = 59,500 psi. Tensile strength = 77,500 psi. Elongation = 35%.

15.1.B. Harrison Alloys Inc. Technical Information Bulletin. Density = 8.38 g/cc. Thermal conductivity = 0.0339 cal/cm-sec-K. Thermal Expansion coefficient (30 to 400C) = 4.6 to 5.2 ppm/K. Resistivity = 49 micro-ohm-cm. Modulus of elasticity = 20 x 10^6 psi. Melting point = 1450C. Yield strength = 50,500 psi. Tensile strnegth = 89,700 psi. Brinell hardness = 200 to 250.

15.2.A See 15.1.A

15.3.A See 15.1.A

15.3.B See 15.1.B

15.3.C Silverman, L., J Metals, 5, 631-2, 1953. Thermal conductivity at 50C = 0.0339 cal/cm-sec-K.

15.4.A Masumoto, H., Sci. Reports of the Tohoku Univ., 20, 101-23, 1931. Coeff. of thermal linear expansion at 303 K = 8.36 ppm/K. (average of three sets of data)

15.4.B Shevlin, T.S., et al, U.S. Air Force Rept. WADC-TR-54-33 (Pt. 2), 127 pp., 1956. [AD 90 041] Coeff. of thermal linear expansion at room temperature = 8.7 ppm/K. (average of 3 sets of data)

15.4.C See 15.1.A

15.4.D See 15.1.B

15.5.A See 15.1.B

15.6.A See 15.1.A

15.6.B See 15.1.B

15.6.C Smithells Metals Reference Book, 6th Edition, Butterworths, 19-5. Resistivity at room temperature = 49 $\mu\Omega$cm. TCR at room temperature = 3700 ppm/K, defined with respect to 0 C.

15.7.A See 15.6.C

15.8.A See 15.1.A

15.8.B See 15.1.B

15.10.A See 15.1.A

15.10.B See 15.1.B

15.11.A See 15.1.A

15.11.B See 15.1.B

15.12.A See 15.1.A

15.13.A See 15.1.B

88Pb 10Sn 2Ag

16.1.A Allied Metglas Products. Density at room temperature = 10.75 g/cc. Solidus = 267C. Liquidus = 299C.

16.1.B Teledyne Microelectronics Handbook, Appendis A, Material Properties. Density = 10.9 g/cc. Specific heat = 0.034 cal/gm-K. Thermal conductivity = 0.10 cal/cm-sec-K. Thermal expansion coefficient = 29.1 ppm/K. Solidus = 268C. Liquidus = 290C

16.2.A See 16.1.B

16.2.B Specific heat = 0.0397 cal/g-K when calculated using Kopp's law.

16.3.A See 16.1.B

16.4.A See 16.1.B

16.5.A See 16.1.A

16.5.B See 16.1.B

16.5.C Lead Industries Assoc., Inc. Solidus = 268C. Liquidus = 299C.

16.11.A Turkis, S. and A.A. Smith, Jr. Metals and Alloys, vol 15, 412, 1942. Tensile strength of chill cast trip = 5000 psi.

16.11.B Jaffee, R.I., E.J. Minarcik and B.W. Gonser, Res, vol 54, 843, 1948. [83Pb 15Sn 1.6Ag] Tensile strength of as-cast material at 300K = 4000 psi. Elongation = 20%.

16.11.C Federated Metals Division, American Smelting and Refining Co. Tensile strength of 0.5Sb 1.5Ag 10Sn bal Pb = 5335 psi. Elongation in 2 inches = 11.5%.

16.12.A See 16.11.A

16.12.B See 16.11.C

92.5Pb 5Sn 2.5Ag

17.1.A Teledyne Microelectronics Handbook, Appendix A, Material Properties. Density = 11.0 g/cc. Specific heat = 0.0309 cal/cm-sec-K. Thermal conductivity = 0.085 cal/cm-sec-K. Thermal expansion coefficient = 28.7 ppm/K. Melting point = 280C. Electrical resistivity = 19.5 micro-ohm-cm. TCR <3360 ppm/K. Young's modulus = 2 x 10^6 psi. Yield strength = 2240 psi. Tensile strength = 3480 psi. Elongation = 42% .

17.1.B Allied Metglas Products. Density at room temperature = 11.02 g/cc.

17.1.C Indium Company of America, Indalloy #164. Density = 11.02 g/cc. Thermal conductivity = 0.059 cal/cm-sec-K. Thermal expansion coefficient = 25 ppm/K. Tensile strength = 4560 psi.

17.2.A See 17.1.A

17.2.B Specific heat = 0.0391 cal/g-K using Kopp's law.

17.3.A See 17.1.A

17.3.B See 17.1.C

17.4.A See 17.1.A

17.4.B See 17.1.C

17.5.A See 17.1.A

17.5.B See 17.1.B

17.5.C See 17.1.C

17.5.D Lead Assoc, Inc., Properties of Lead and Lead Alloys. [93.5Pb 5.0Sn 1.5Ag] Solidus = 296C. Liquidus = 301C. Tensile strength = 4350 psi. Elongation = 20%.

17.6.A See 17.1.A

17.7.A See 17.1.A

17.8.A See 17.1.A

17.10.A See 17.1.A

17.11.A International Tin Research Institute Publication No. 656, p. 96. Ultimate tensile strength at 293 K = 6200 psi for a testing speed of 1.0 mm/min.

17.11.B See 17.5.C

17.11.C See 17.1.A

17.12.A See 17.1.A

17.12.B See 17.5.C

62Sn 36Pb 2Ag

18.1.A Allen, B.M., "Soldering Handbook", Iliffe Books, London, 1969. Density at room temperature = 8.50 g/cc.

18.1.B Indium company of America, Indalloy #104 [62.5Sn 36.1Pb 1.4Ag] Density = 8.42 g/cc. Thermal conductivity = 0.074 cal/cm-sec-K at 85C. Thermal expansion coefficient = 25.2 ppm/K. Melting point = 179C (452K). Resistivity = 14.9 micro-ohm-cm. Tensile strength = 7000 psi.

18.1.C Teledyne Microelectronics Handbook, Appendix A, Material Properties. Density = 8.41 g/cc. Specific heat = 0.045 cal/g-K. Thermal conductivity = 0.13 cal/cm-sec-K. Thermal expansion coefficient = 25.4 ppm/K. Melting point = 179C (452K). Electrical resistivity = 14.6 micro-ohm-cm. Tensile strength = 6120 psi.

18.2.A See 18.1.C

18.2.B Specific heat = 0.057 cal/g-K when using Kopp's Law.

18.3.A See 18.1.B

18.3.B See 18.1.C

18.4.A Wild, R.N., IBM No 69-825-2451, Oswego, NY, 1969. Coeff. of thermal linear expansion at room temperature = 27.0 ppm/K. Young's modulus at room temperature = 3329 ksi.

18.4.B See 18.1.B

18.4.C See 18.1.C

18.5.A Mackay, Proc Internepcon, U.K., Brighton, 61-67, 1980. Melting point = 450 K.

18.5.B SCM Metal Products, Inc. Melting point = 452 K.

18.5.C Lead Industries Assoc., Inc. Melting point = 452 K.

18.6.A See 18.1.B

18.6.B See 18.1.C

18.8.A See 18.4.A

18.11.A Internaional Tin Research Institute Publication No. 656. Ultimate tensile strength at room temperature = 6160-11,700 psi. Tensile strength is strongly dependent on strain rate.

18.11.B See 18.1.B

18.11.C See 18.5.C

18.11.D See 18.1.C

18.12.A See 18.5.C

18.13.A Cole, S.S., A.r. Kroehs, IEEE 1969 Proc Elec Comp Conf, 298-302. Hardness at 21C = 33.4 Brinell.

INSULATORS

ALUMINUM NITRIDE

PROPERTY	@ 300K	UNITS	REFERENCES
Density	3.26	g/cm^3	AN1 AN2 AN3
Specific Heat	0.173	cal/g-K	AN4 AN5 AN6
Thermal Conductivity[1]	0.16	cal/cm-s-K	AN7 AN8 AN9
Thermal Exp Coef	$4.0x10^{-6}$	1/K	AN2 AN6 AN10
Dielectric Constant	9.0	------	AN2 AN6 AN12
Dissipation Factor	0.005	------	AN2 AN6 AN9
Dielectric Strength 2	380	V/mil	AN9 AN11 AN18
Young's Modulus	50,000	ksi	AN2 AN6 AN14
Poisson's Ratio		------	
Flexural Strength	50,000	psi	AN2 AN9 AN15
Ult Tensile Strength		psi	
Compressive Strength	300,000	psi	AN2 AN9 AN16
Hardness	1200	Knoop	AN2 AN9 AN17

[1] Data is for polycrystalline material. Theoretical maximum = 0.76 cal/cm-s-K. [AN7] Recent data for ceramic material gives values as high as 0.478 cal/cm-s-k. [AN19]

2 Dielectric strength for thin films can be as high as 5000 volts/mil. Dielectric strength decreases with thickness and temperature. [AN11]

ALUMINUM NITRIDE

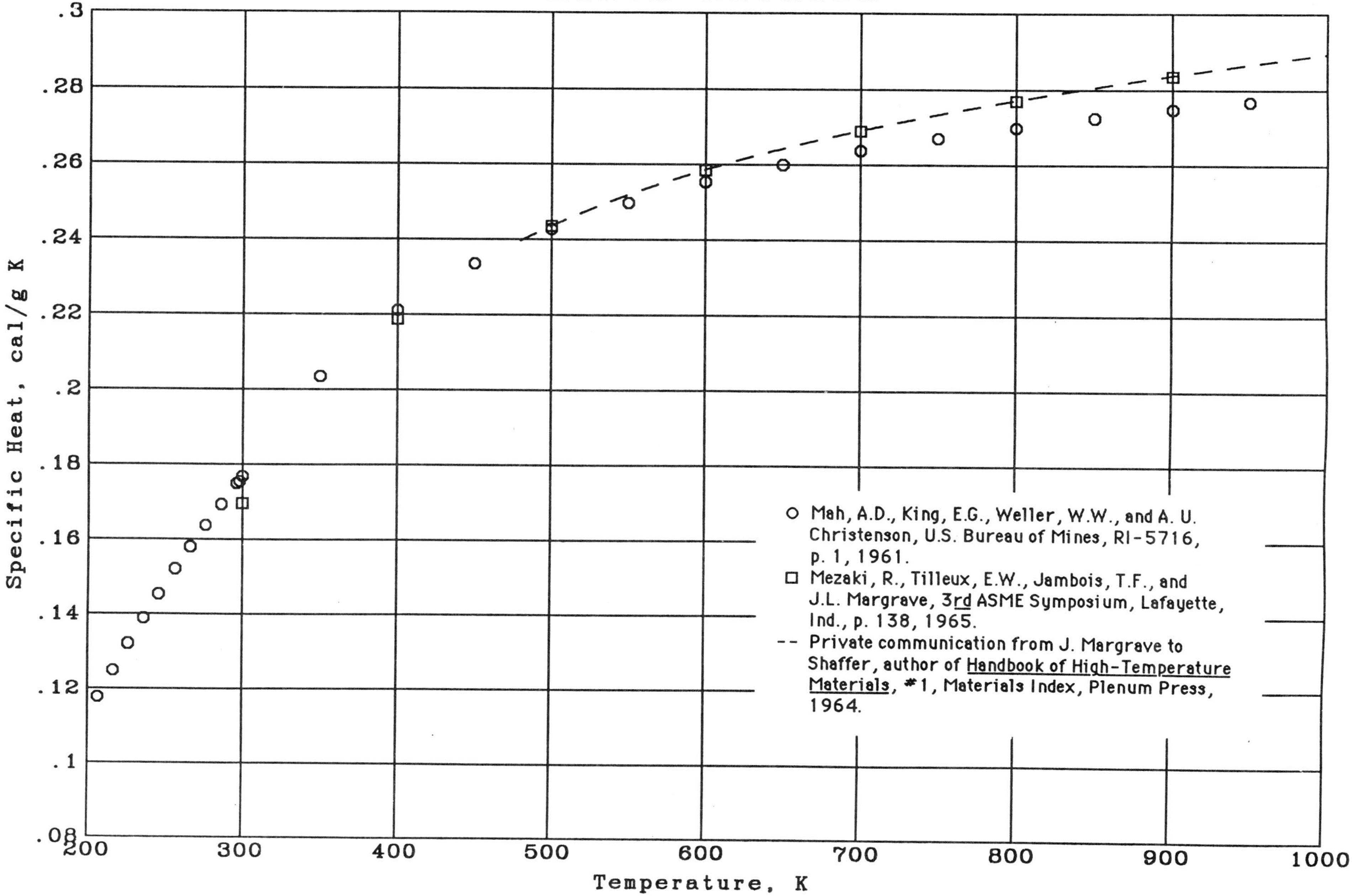

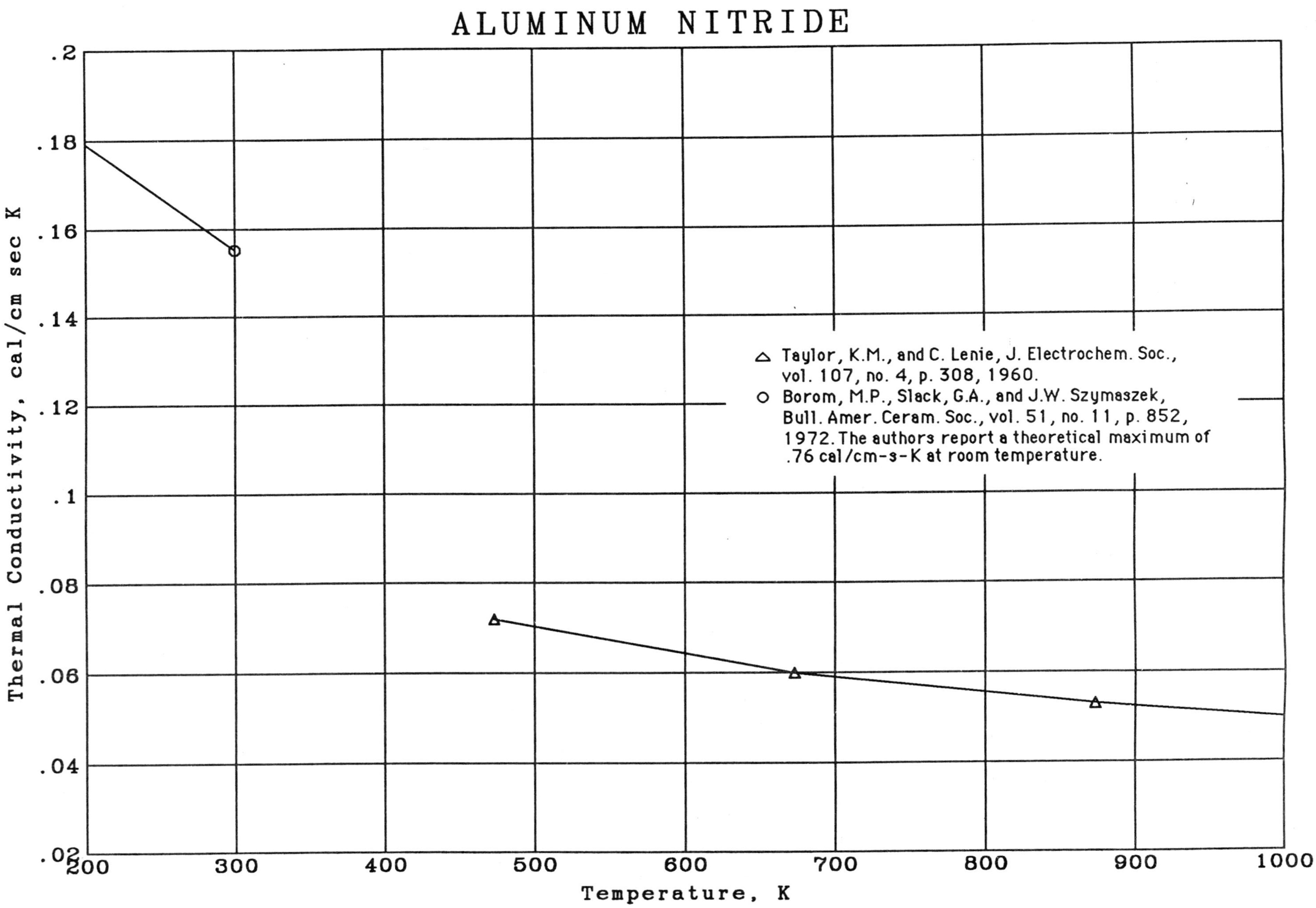
ALUMINUM NITRIDE
Thermal Conductivity, cal/cm sec K
.2
.18
.16
.14
.12
.1
.08
.06
.04
.02
200
300
400
500
600
700
800
900
1000
Temperature, K
△ Taylor, K.M., and C. Lenie, J. Electrochem. Soc., vol. 107, no. 4, p. 308, 1960.
○ Borom, M.P., Slack, G.A., and J.W. Szymaszek, Bull. Amer. Ceram. Soc., vol. 51, no. 11, p. 852, 1972. The authors report a theoretical maximum of .76 cal/cm-s-K at room temperature.

Aluminum Nitride

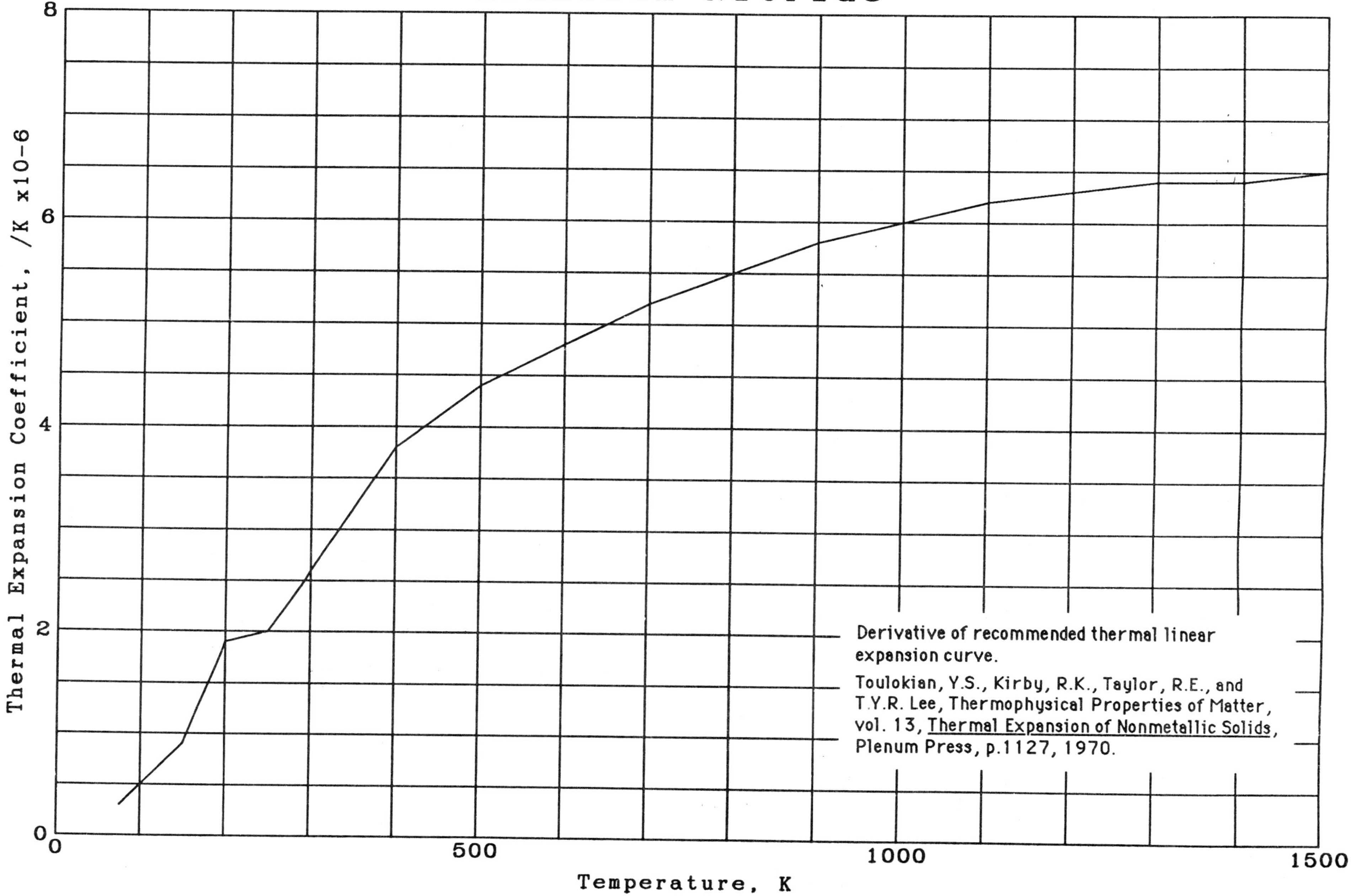

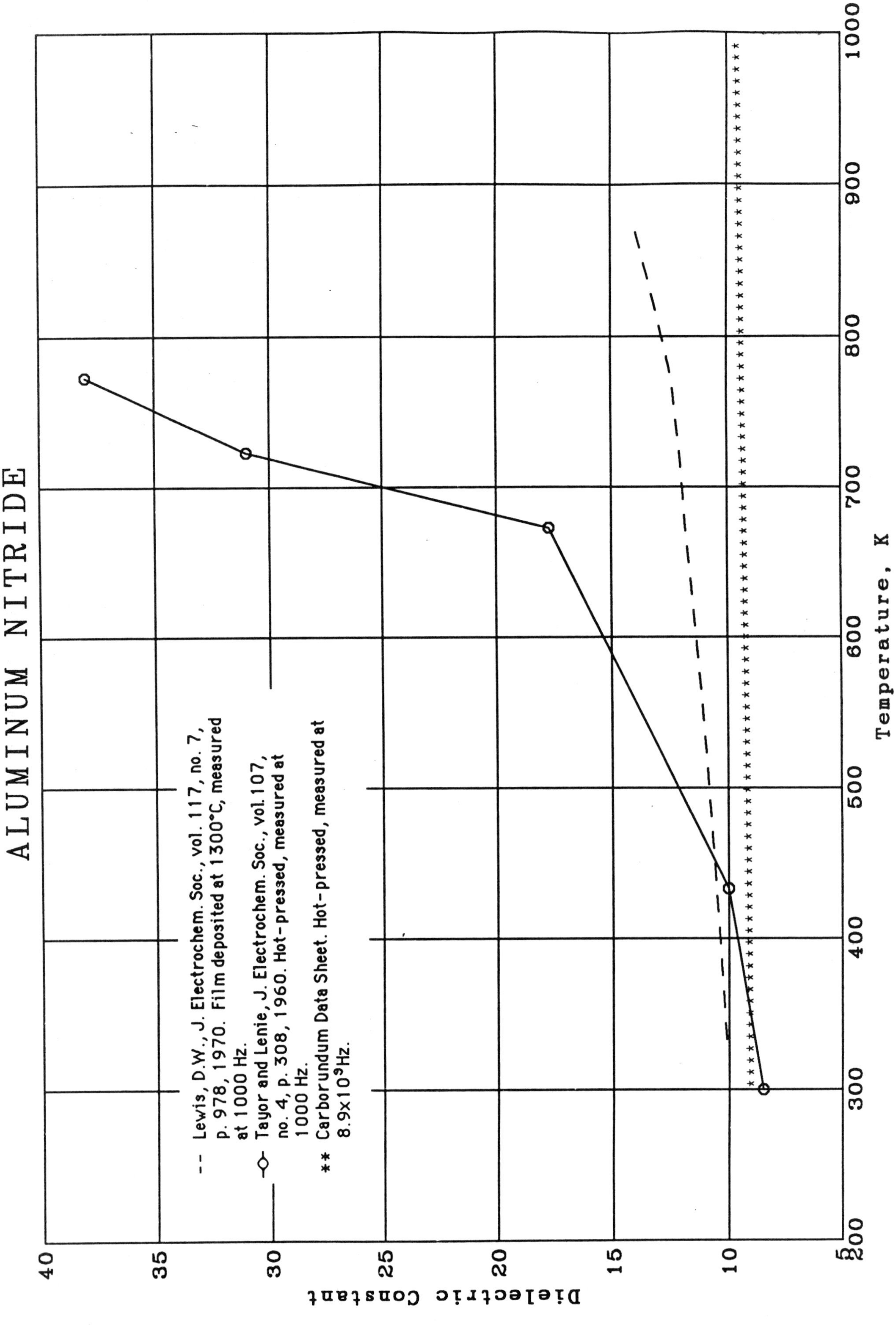
ALUMINUM NITRIDE
Dielectric Constant
Temperature, K
40
35
30
25
20
15
10
5
200
300
400
500
600
700
800
900
1000
-- Lewis, D.W., J. Electrochem. Soc., vol. 117, no. 7, p. 978, 1970. Film deposited at 1300°C, measured at 1000 Hz.
-o- Tayor and Lenie, J. Electrochem. Soc., vol.107, no. 4, p. 308, 1960. Hot-pressed, measured at 1000 Hz.
** Carborundum Data Sheet. Hot-pressed, measured at 8.9x10⁹Hz.

ALUMINUM NITRIDE

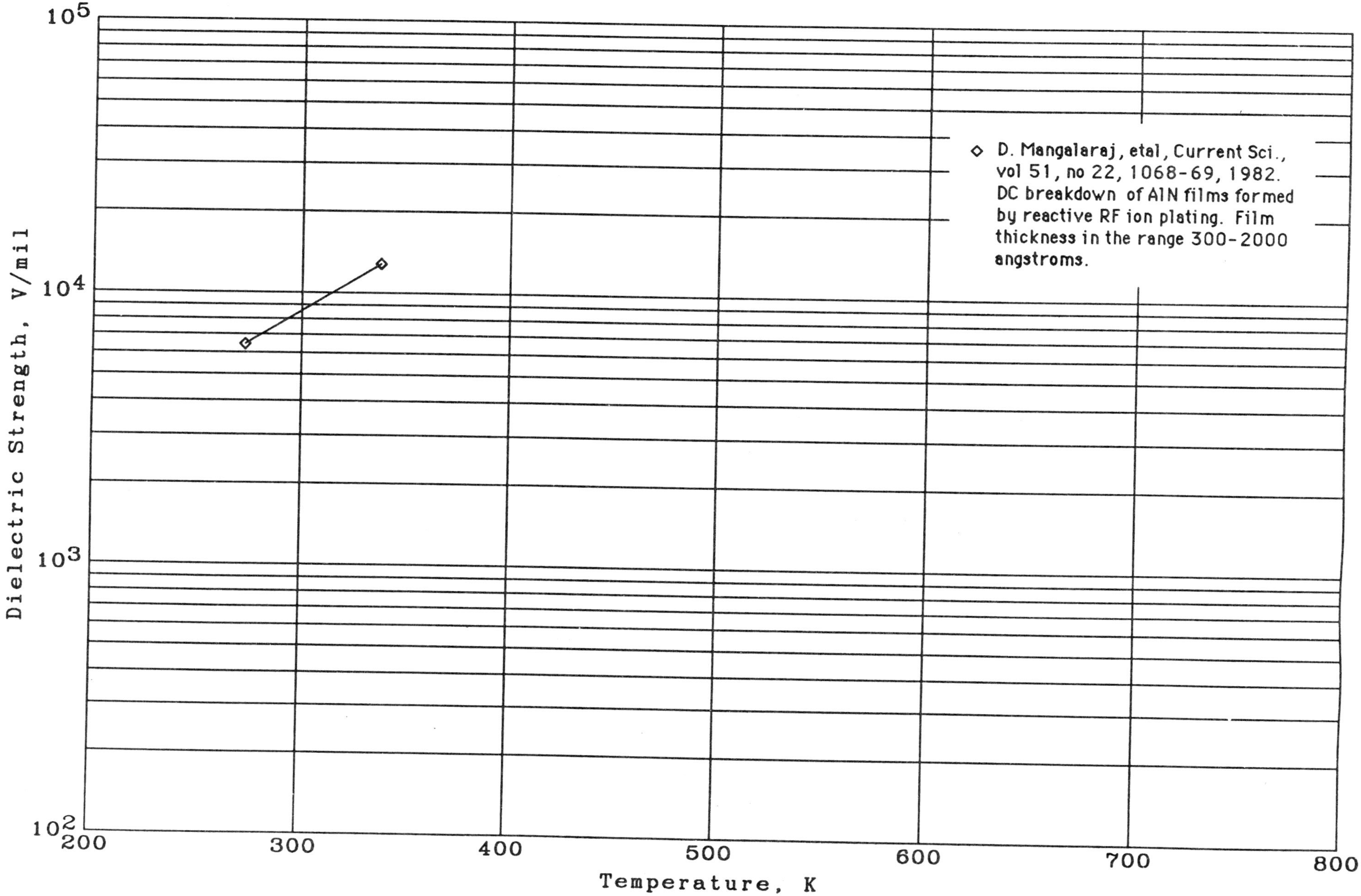

ALUMINUM NITRIDE

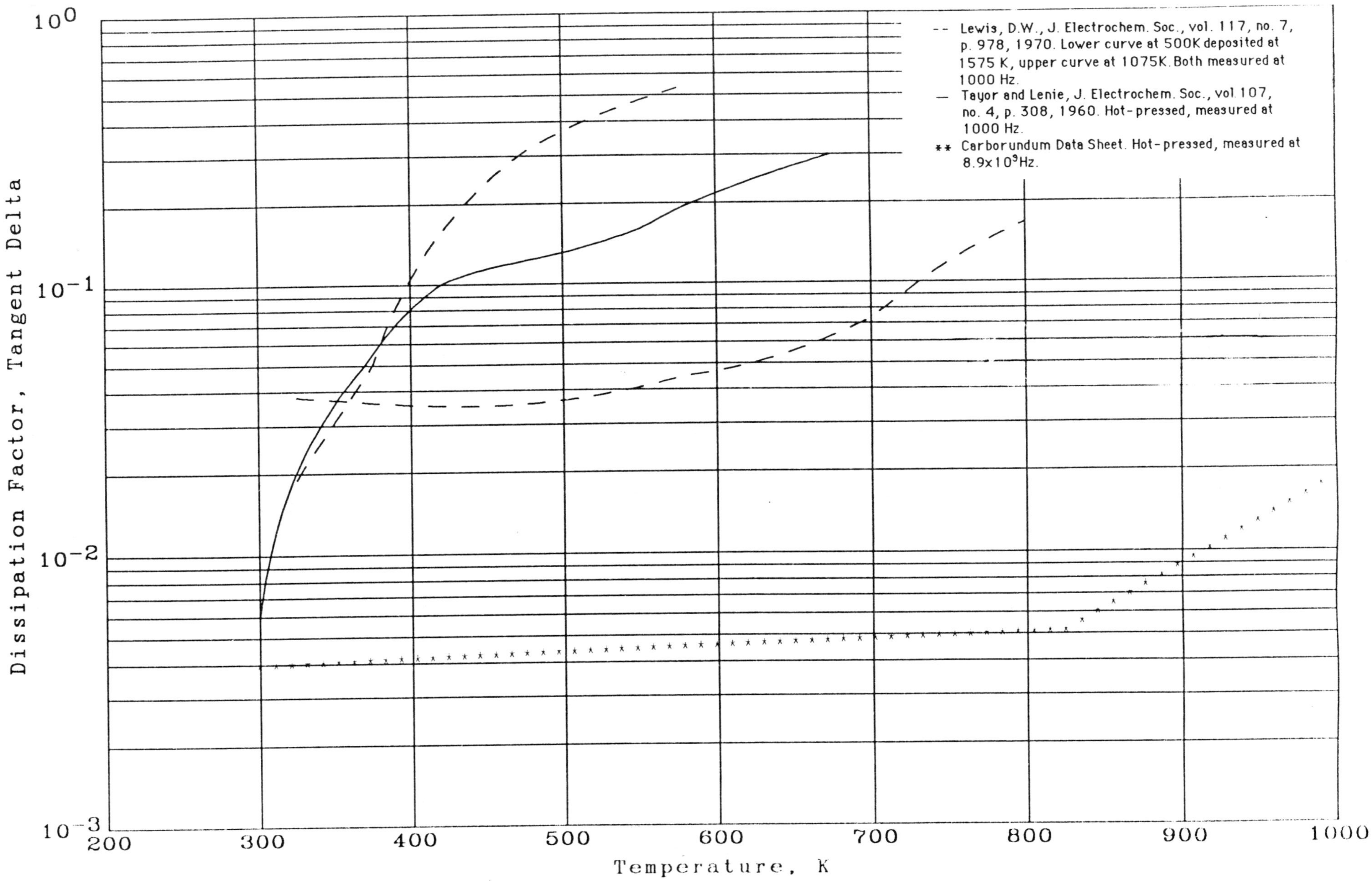

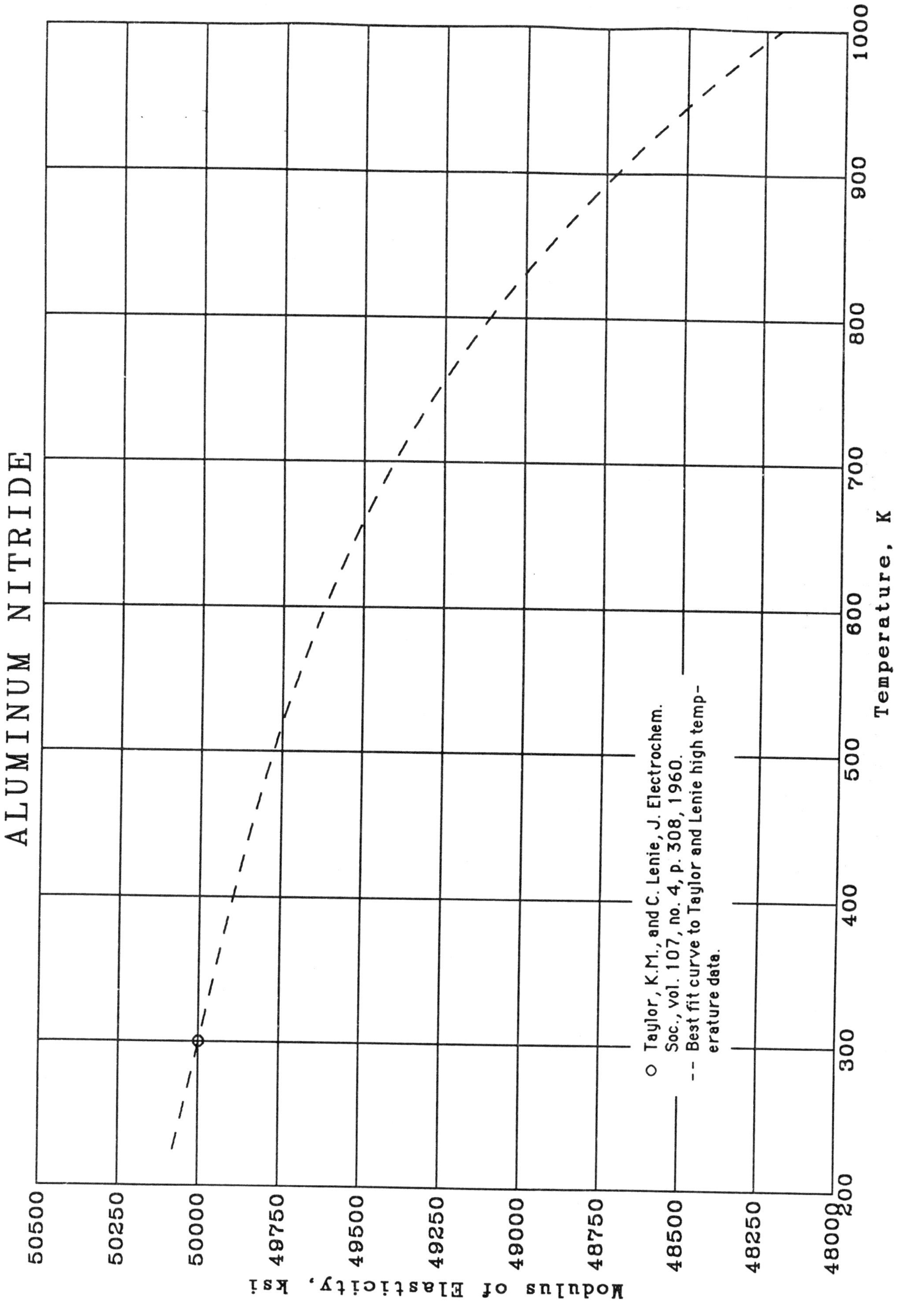

ALUMINUM NITRIDE
Modulus of Elasticity, ksi
50500
50250
50000
49750
49500
49250
49000
48750
48500
48250
48000
Temperature, K
200
300
400
500
600
700
800
900
1000
○ Taylor, K.M., and C. Lenie, J. Electrochem. Soc., vol. 107, no. 4, p. 308, 1960.
-- Best fit curve to Taylor and Lenie high temperature data.

ALUMINUM NITRIDE

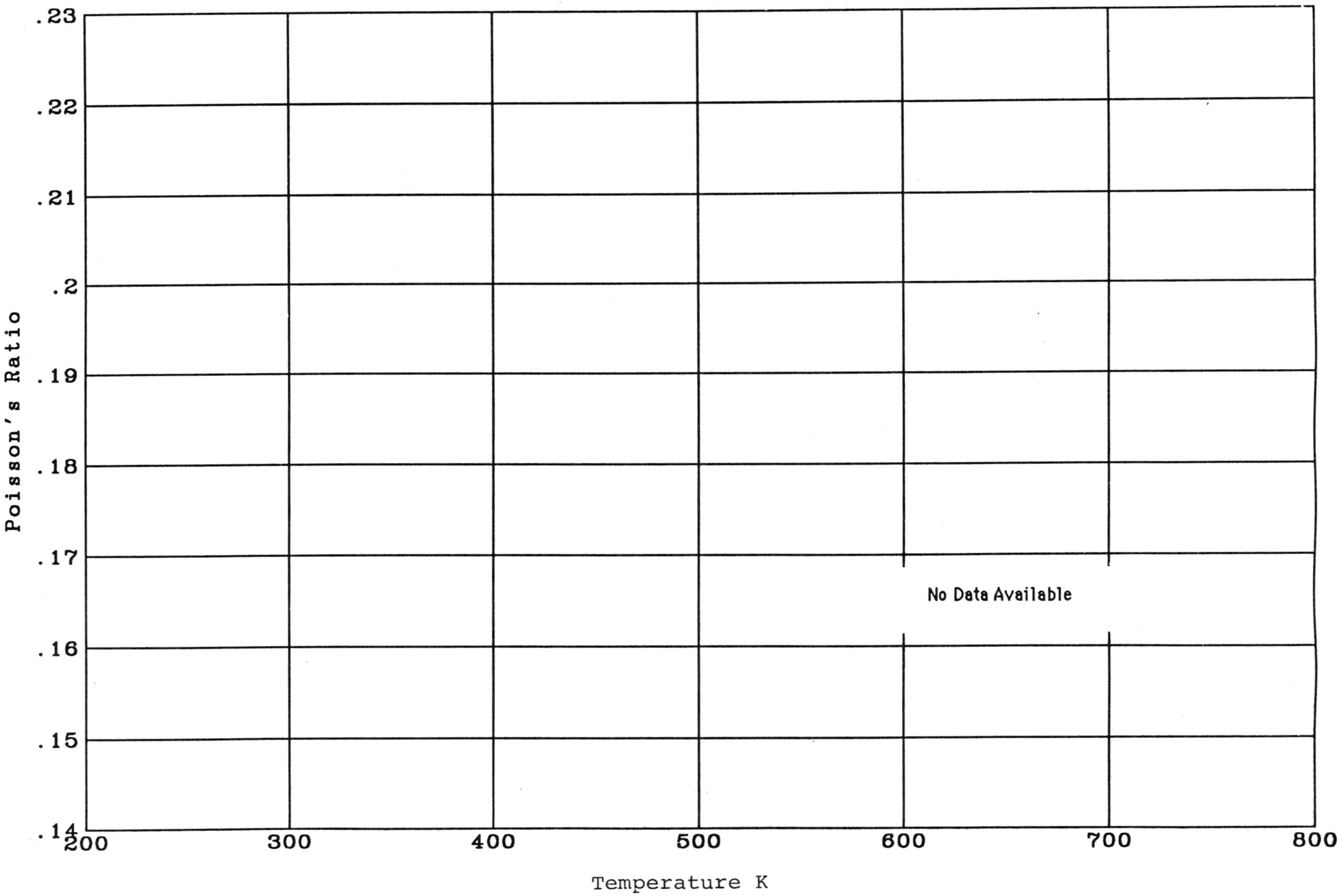

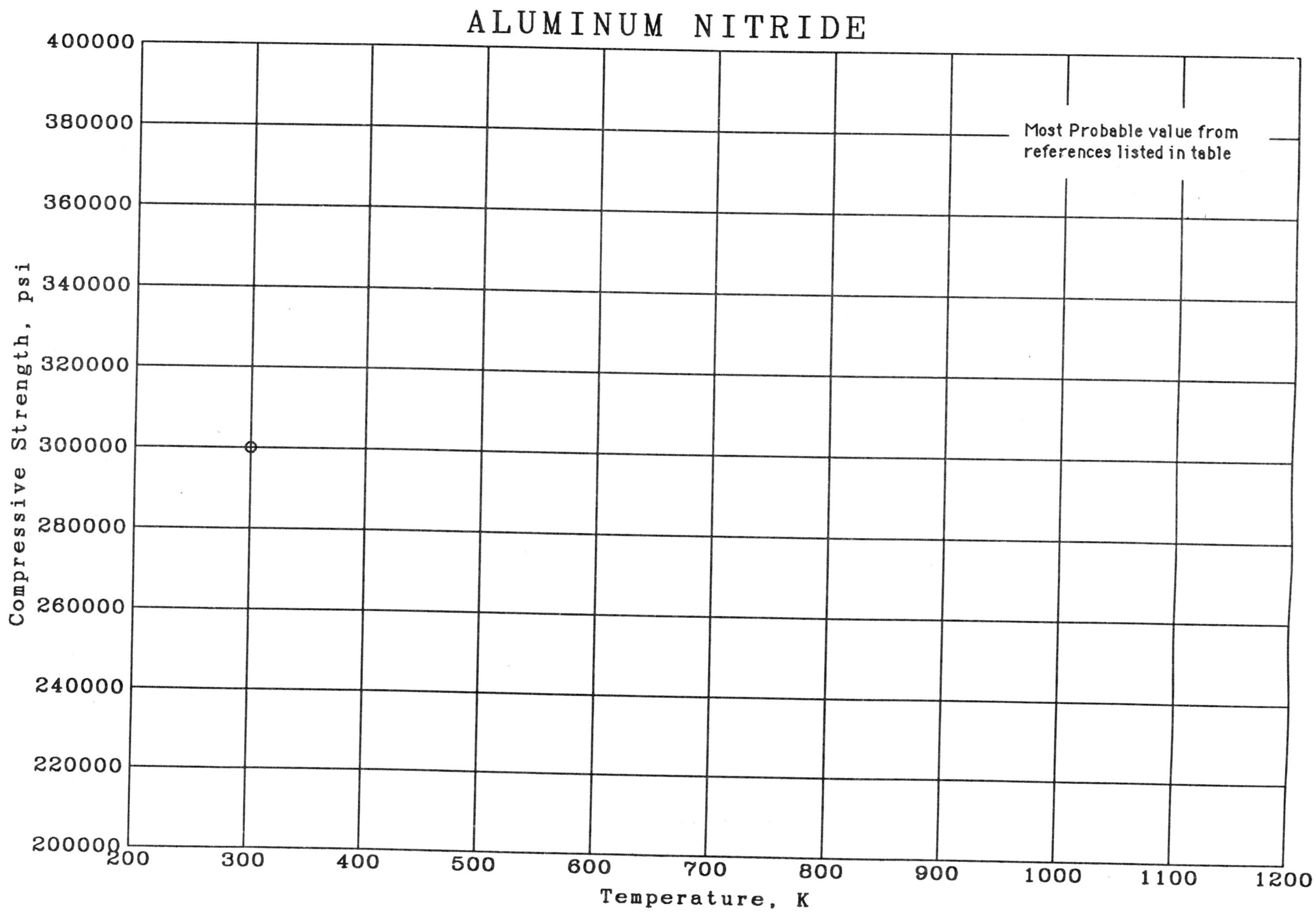
ALUMINUM NITRIDE
Most Probable value from references listed in table
Compressive Strength, psi
400000
380000
360000
340000
320000
300000
280000
260000
240000
220000
200000
200
300
400
500
600
700
800
900
1000
1100
1200
Temperature, K

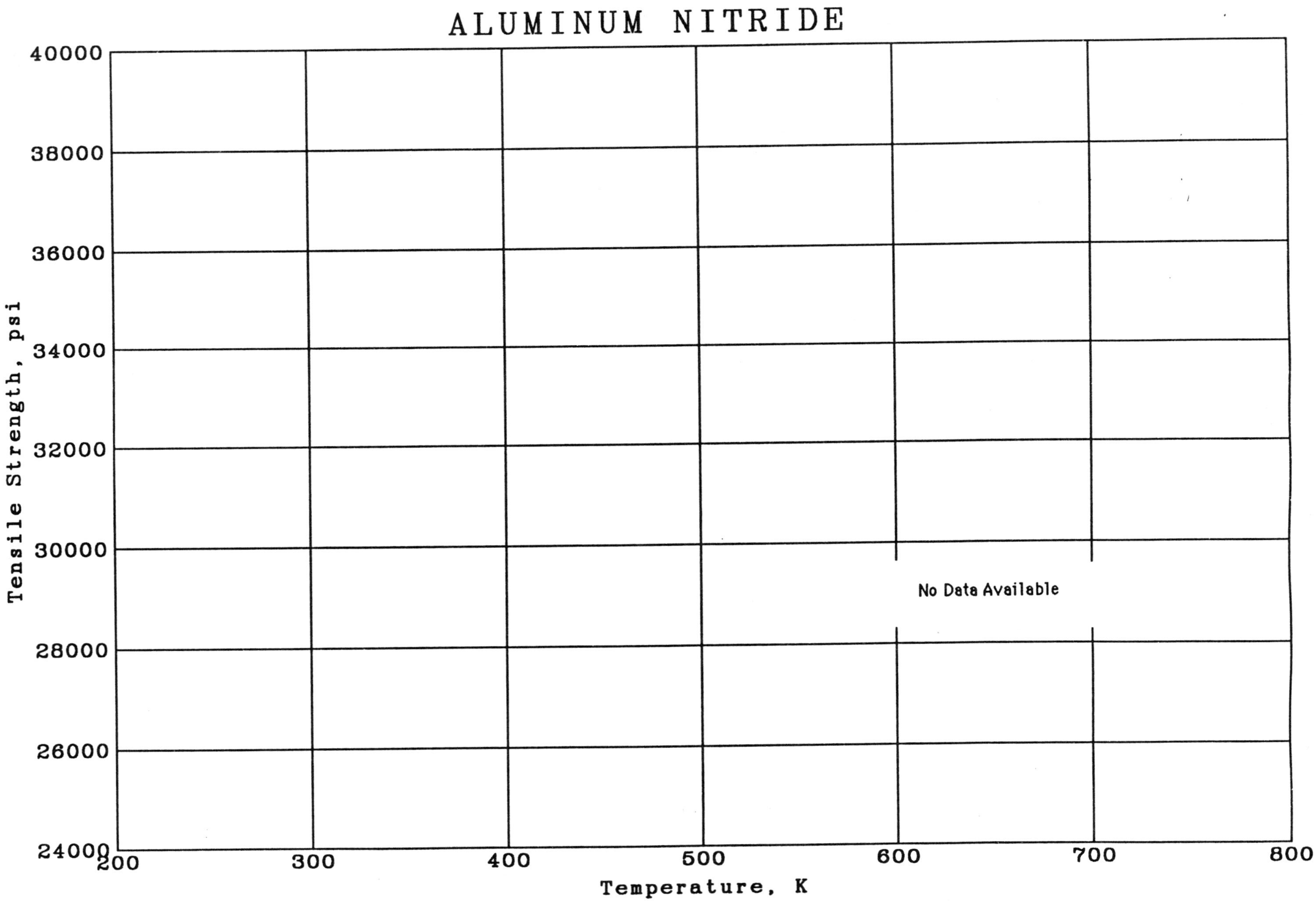
ALUMINUM NITRIDE
Tensile Strength, psi
40000
38000
36000
34000
32000
30000
28000
26000
24000
200
300
400
500
600
700
800
Temperature, K
No Data Available

ALUMINUM NITRIDE

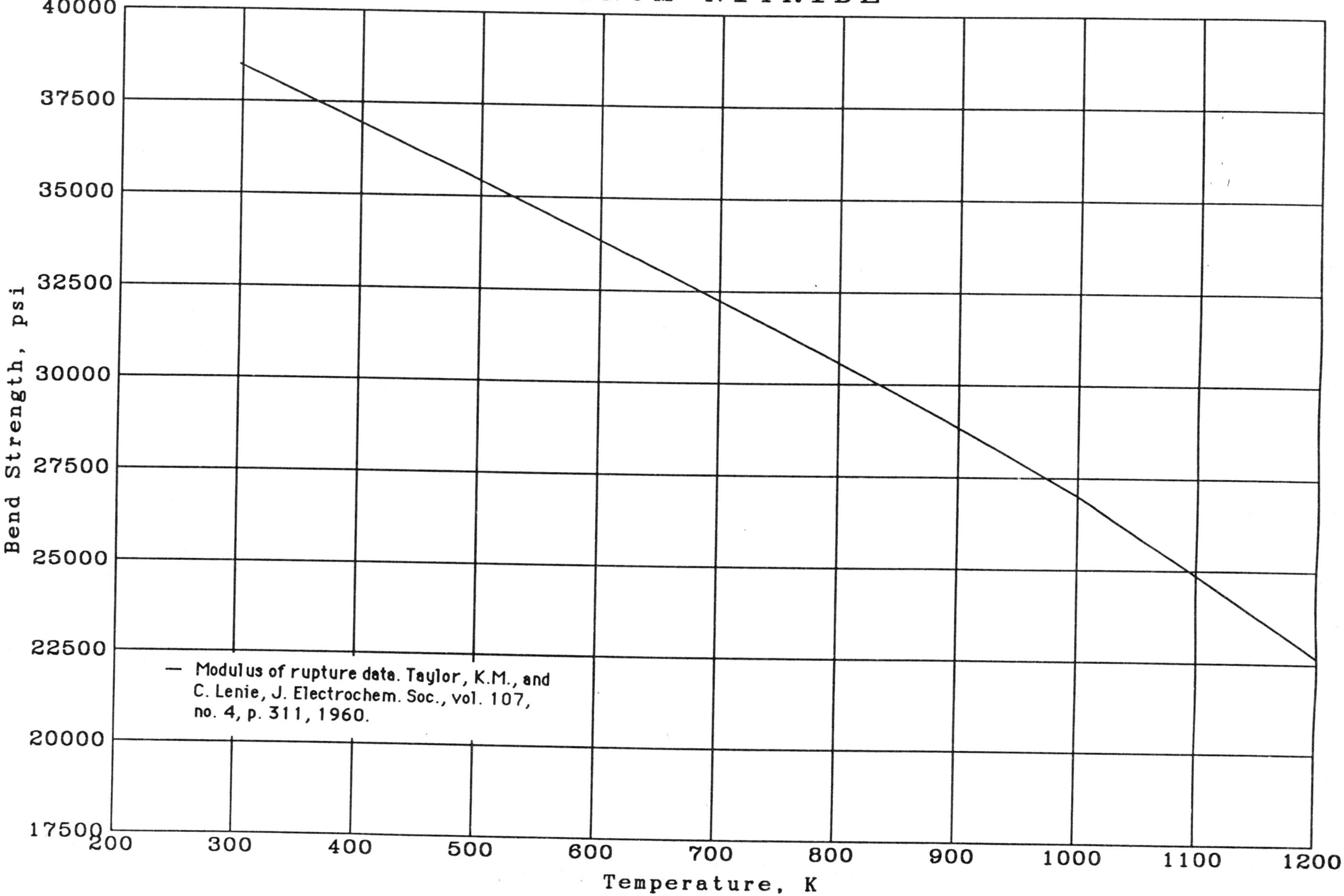

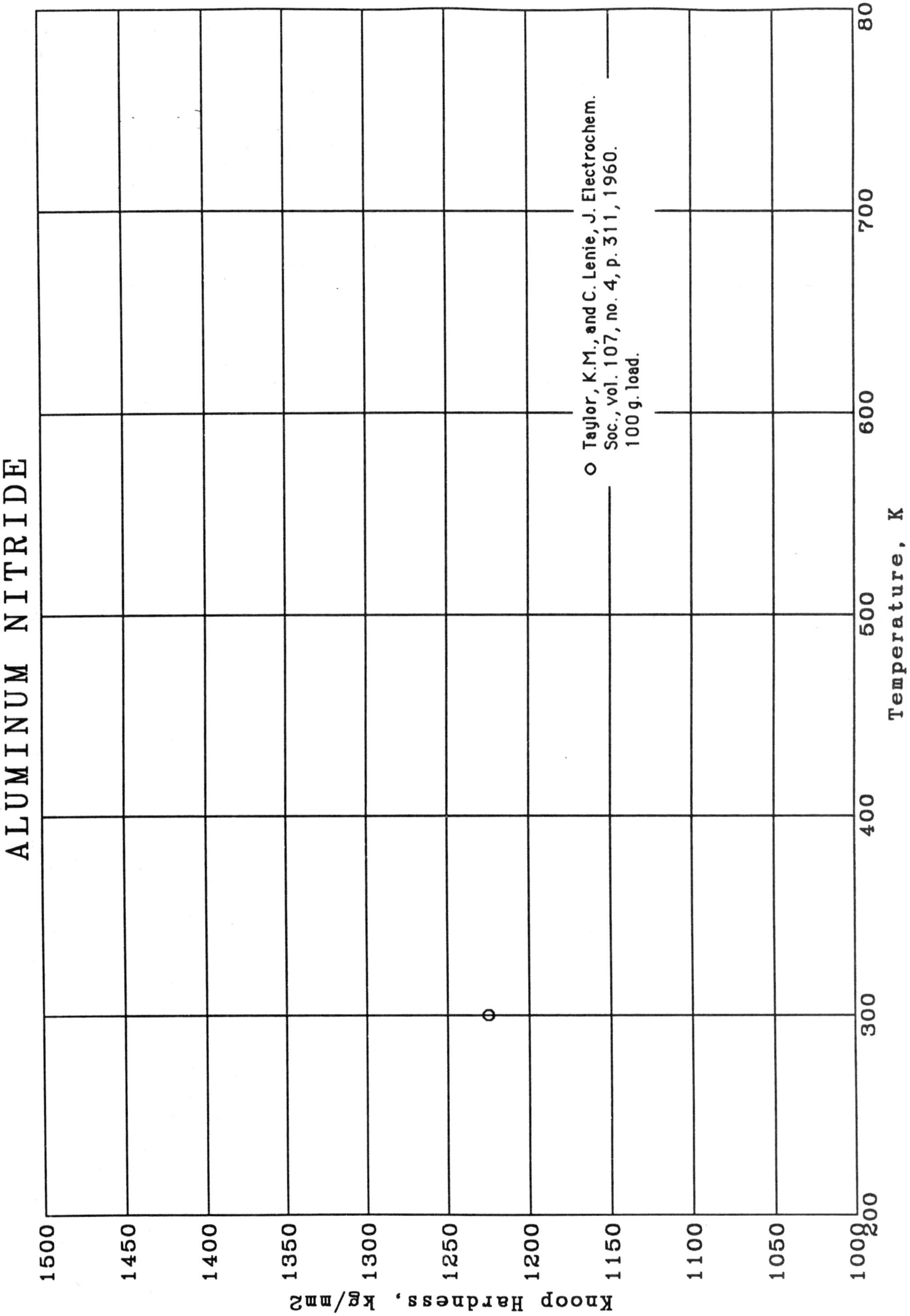
ALUMINUM NITRIDE
Knoop Hardness, kg/mm2
1500
1450
1400
1350
1300
1250
1200
1150
1100
1050
1000
Temperature, K
200
300
400
500
600
700
800
o Taylor, K.M., and C. Lenie, J. Electrochem. Soc., vol. 107, no. 4, p. 311, 1960. 100 g. load.

96% ALUMINA

PROPERTY	@ 300K	UNITS	REFERENCES
Density	3.72	g/cm^3	96A1 96A2 96A3
Specific Heat	0.185	cal/g-K	96A4 96A5 96A6
Thermal Conductivity	0.06	cal/cm-s-K	96A1 96A2 96A7
Thermal Exp Coef	6.2×10^{-6}	1/K	96A1 96A2 96A3
Dielectric Constant	9.0	------	96A1 96A2 96A3
Dissipation Factor	0.0011	------	96A1 96A2 96A3
Dielectric Strength	210	V/mil	96A1 96A2 96A3
Young's Modulus	44,000	ksi	96A1 96A3 96A8
Poisson's Ratio	0.21	------	96A1 96A3
Flexural Strength	50,000	psi	96A1 96A2 96A3
Ult Tensile Strength	26,000	psi	96A1 96A3 98A8
Compressive Strength	300,000	psi	96A1 96A2 96A3
Hardness	78	Rock 45N	96A1 96A2 96A8

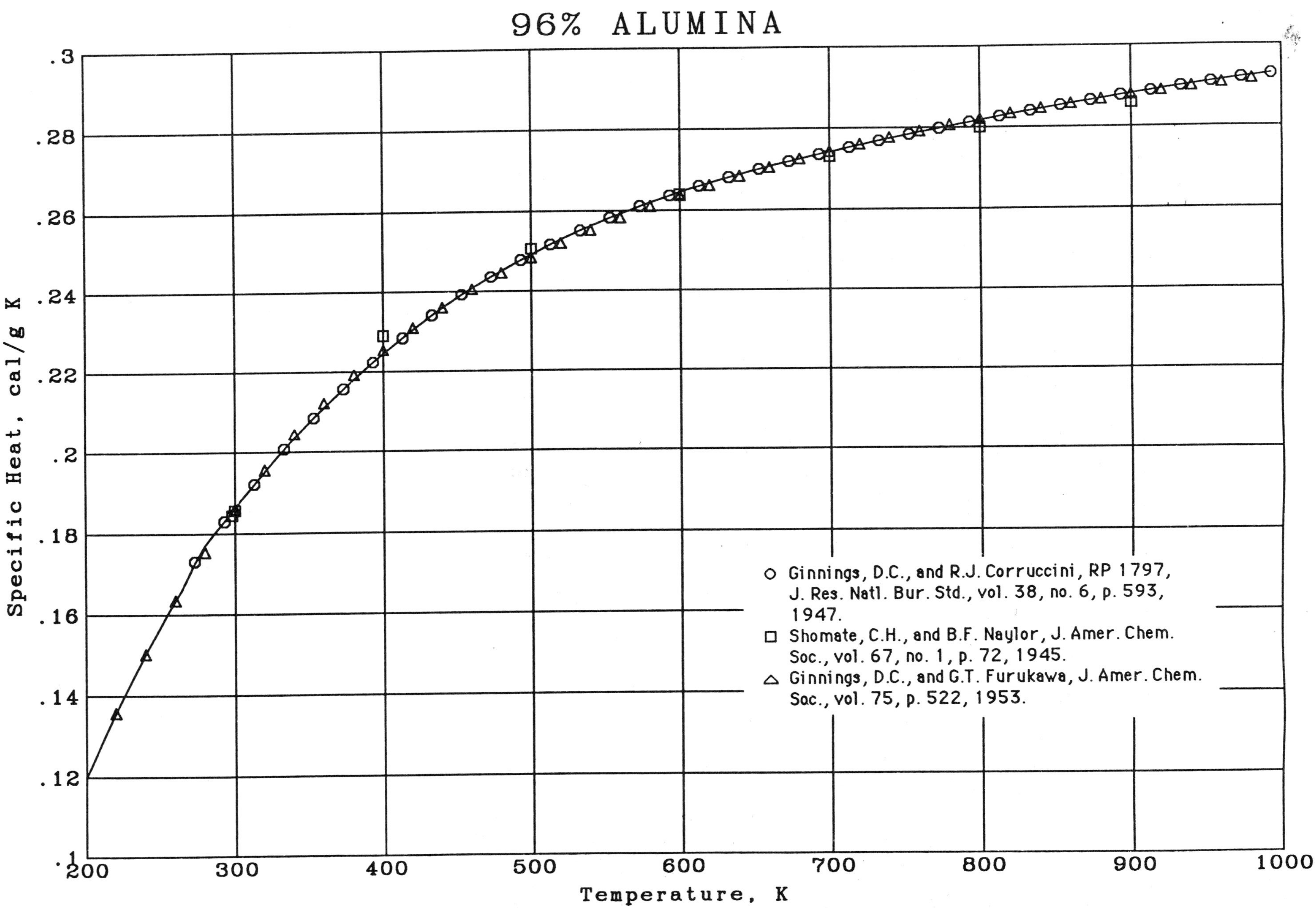
96% ALUMINA
Specific Heat, cal/g K
Temperature, K
.1
.12
.14
.16
.18
.2
.22
.24
.26
.28
.3
200
300
400
500
600
700
800
900
1000
○ Ginnings, D.C., and R.J. Corruccini, RP 1797, J. Res. Natl. Bur. Std., vol. 38, no. 6, p. 593, 1947.
□ Shomate, C.H., and B.F. Naylor, J. Amer. Chem. Soc., vol. 67, no. 1, p. 72, 1945.
△ Ginnings, D.C., and G.T. Furukawa, J. Amer. Chem. Soc., vol. 75, p. 522, 1953.

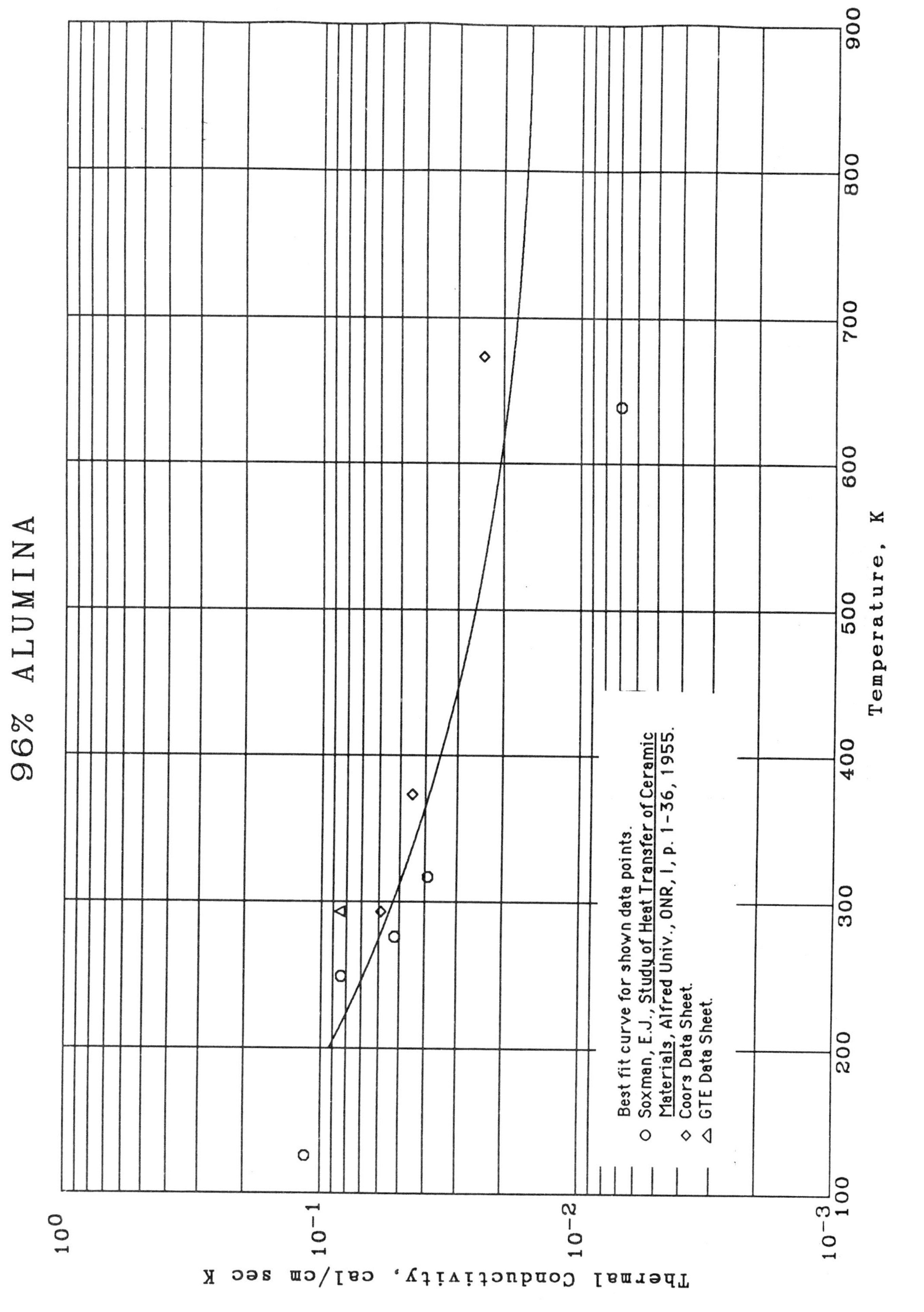
96% ALUMINA
Thermal Conductivity, cal/cm sec K
10^0
10^{-1}
10^{-2}
10^{-3}
Temperature, K
100
200
300
400
500
600
700
800
900
Best fit curve for shown data points.
○ Soxman, E.J., Study of Heat Transfer of Ceramic Materials, Alfred Univ., ONR, I, p. 1-36, 1955.
◇ Coors Data Sheet.
△ GTE Data Sheet.

96% ALUMINA

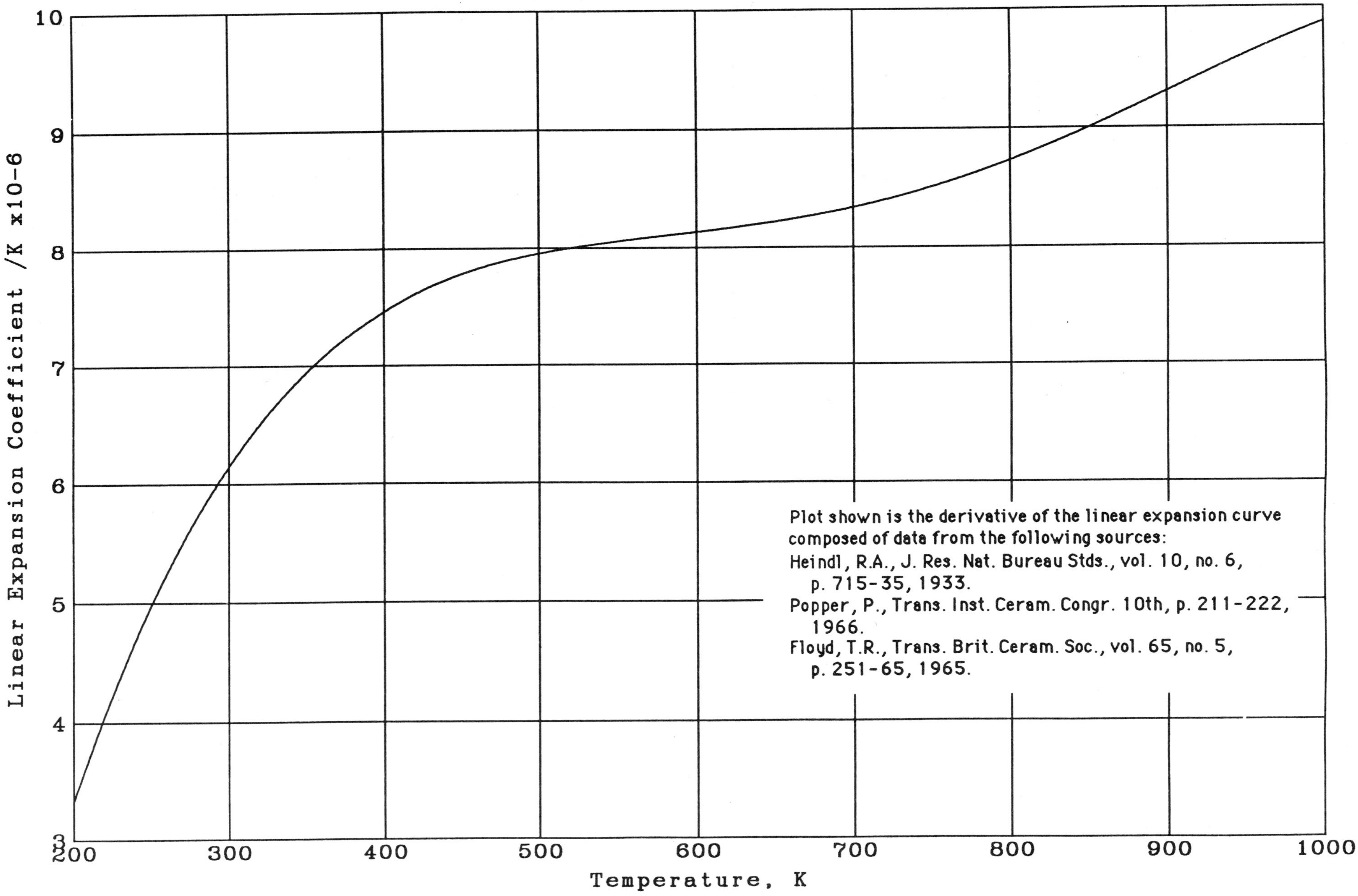

Linear Expansion Coefficient /K x10-6
Temperature, K
10
9
8
7
6
5
4
3
200
300
400
500
600
700
800
900
1000
Plot shown is the derivative of the linear expansion curve composed of data from the following sources:
Heindl, R.A., J. Res. Nat. Bureau Stds., vol. 10, no. 6, p. 715-35, 1933.
Popper, P., Trans. Inst. Ceram. Congr. 10th, p. 211-222, 1966.
Floyd, T.R., Trans. Brit. Ceram. Soc., vol. 65, no. 5, p. 251-65, 1965.

96% ALUMINA

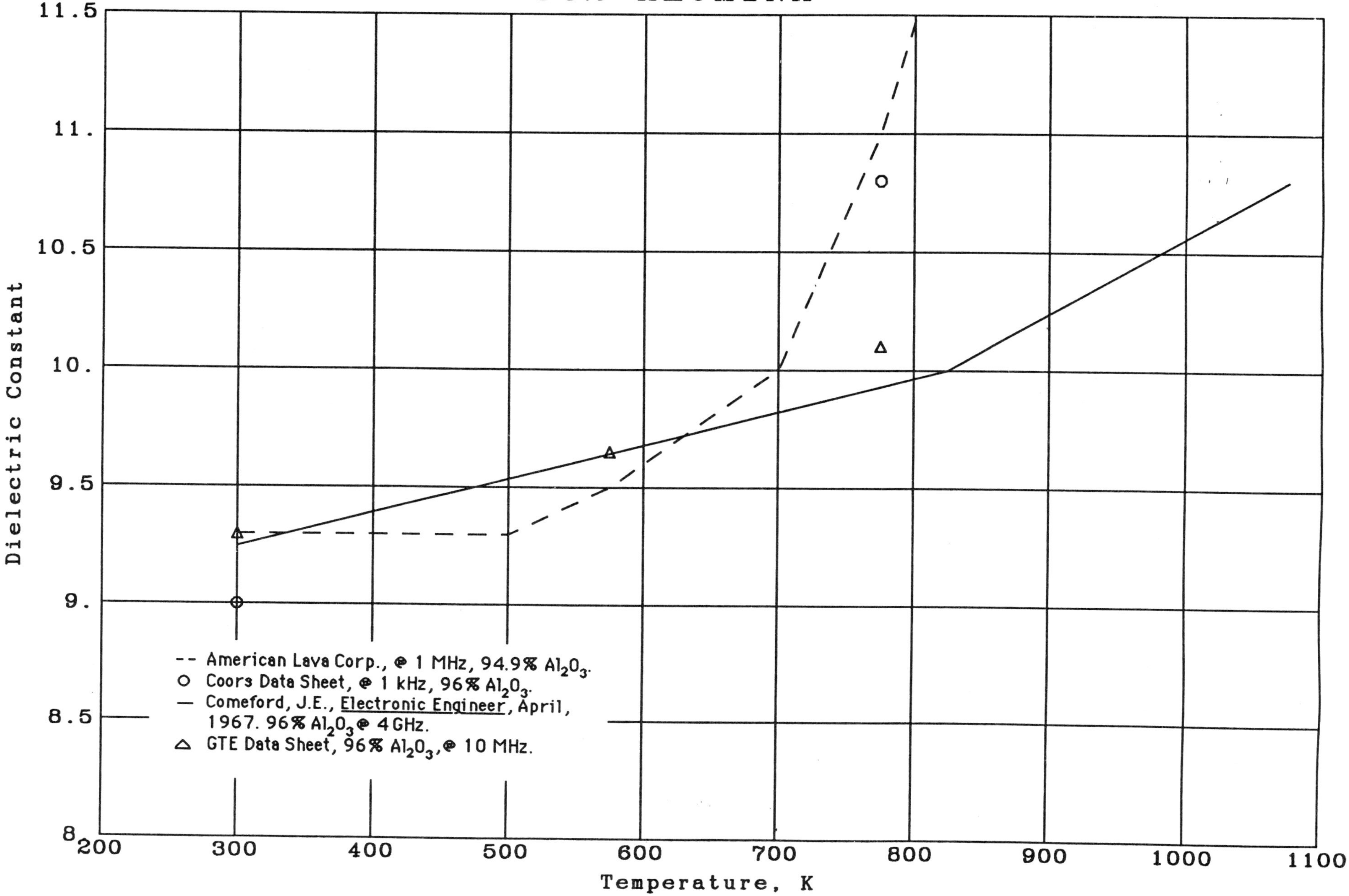

11.5
11.
10.5
10.
9.5
9.
8.5
8
Dielectric Constant
200
300
400
500
600
700
800
900
1000
1100
Temperature, K
-- American Lava Corp., @ 1 MHz, 94.9% Al_2O_3.
○ Coors Data Sheet, @ 1 kHz, 96% Al_2O_3.
— Comeford, J.E., Electronic Engineer, April, 1967. 96% Al_2O_3 @ 4 GHz.
△ GTE Data Sheet, 96% Al_2O_3, @ 10 MHz.

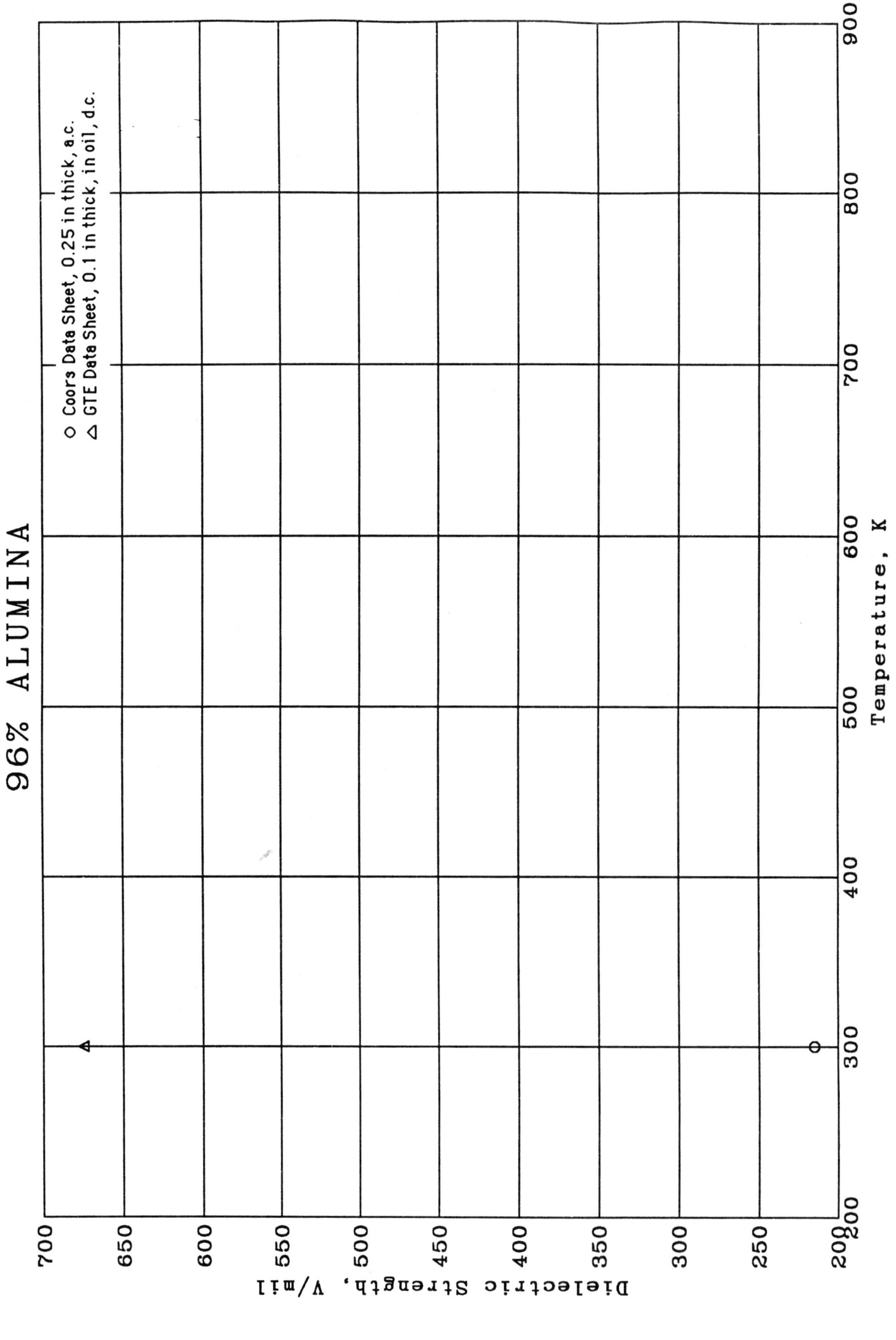
96% ALUMINA
○ Coors Data Sheet, 0.25 in thick, a.c.
△ GTE Data Sheet, 0.1 in thick, in oil, d.c.
Temperature, K
Dielectric Strength, V/mil
200
300
400
500
600
700
800
900
250
350
450
550
650

96% ALUMINA

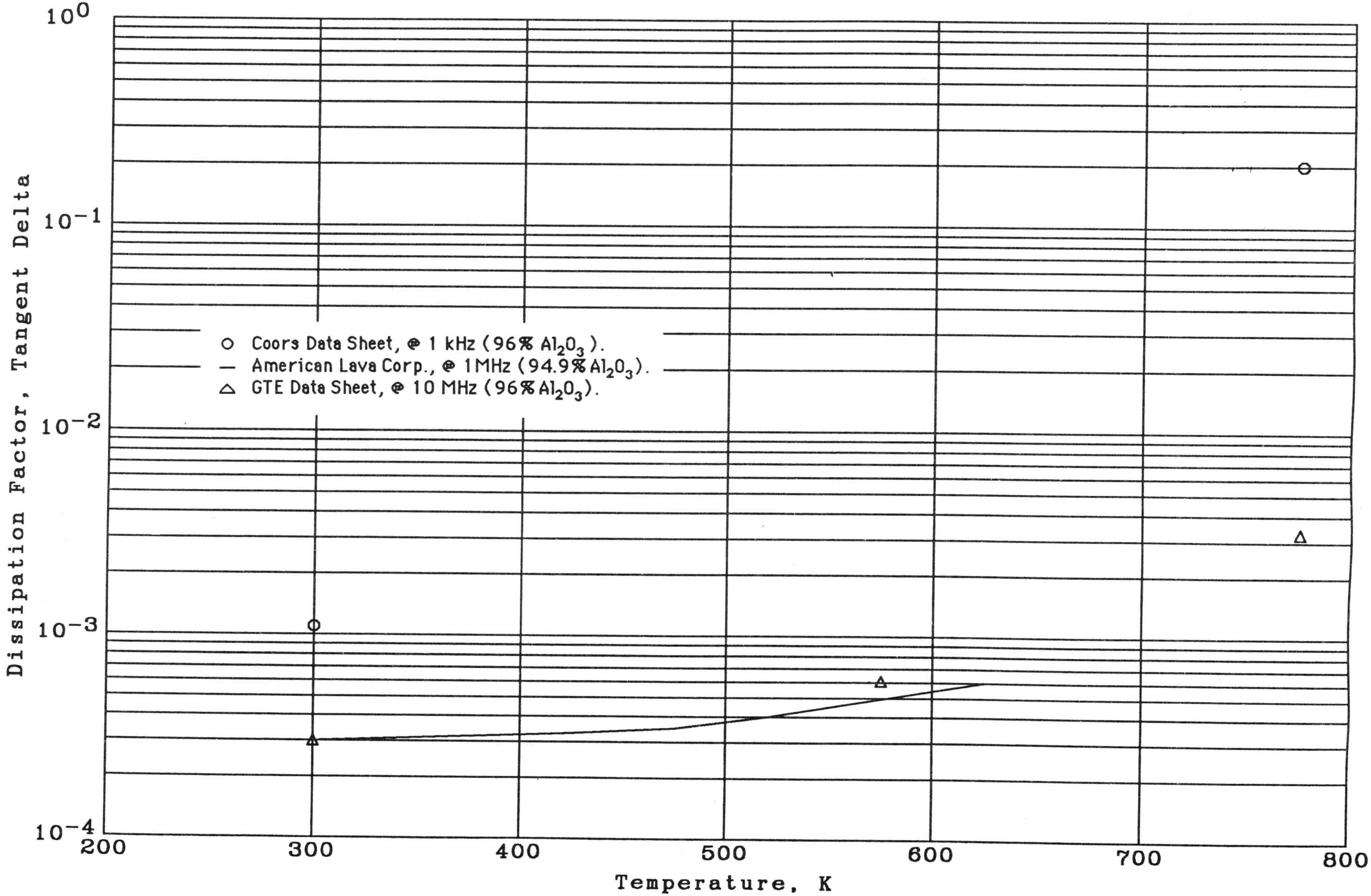

10^0
10^{-1}
10^{-2}
10^{-3}
10^{-4}
Dissipation Factor, Tangent Delta
200
300
400
500
600
700
800
Temperature, K
○ Coors Data Sheet, @ 1 kHz (96% Al_2O_3).
— American Lava Corp., @ 1MHz (94.9% Al_2O_3).
△ GTE Data Sheet, @ 10 MHz (96% Al_2O_3).

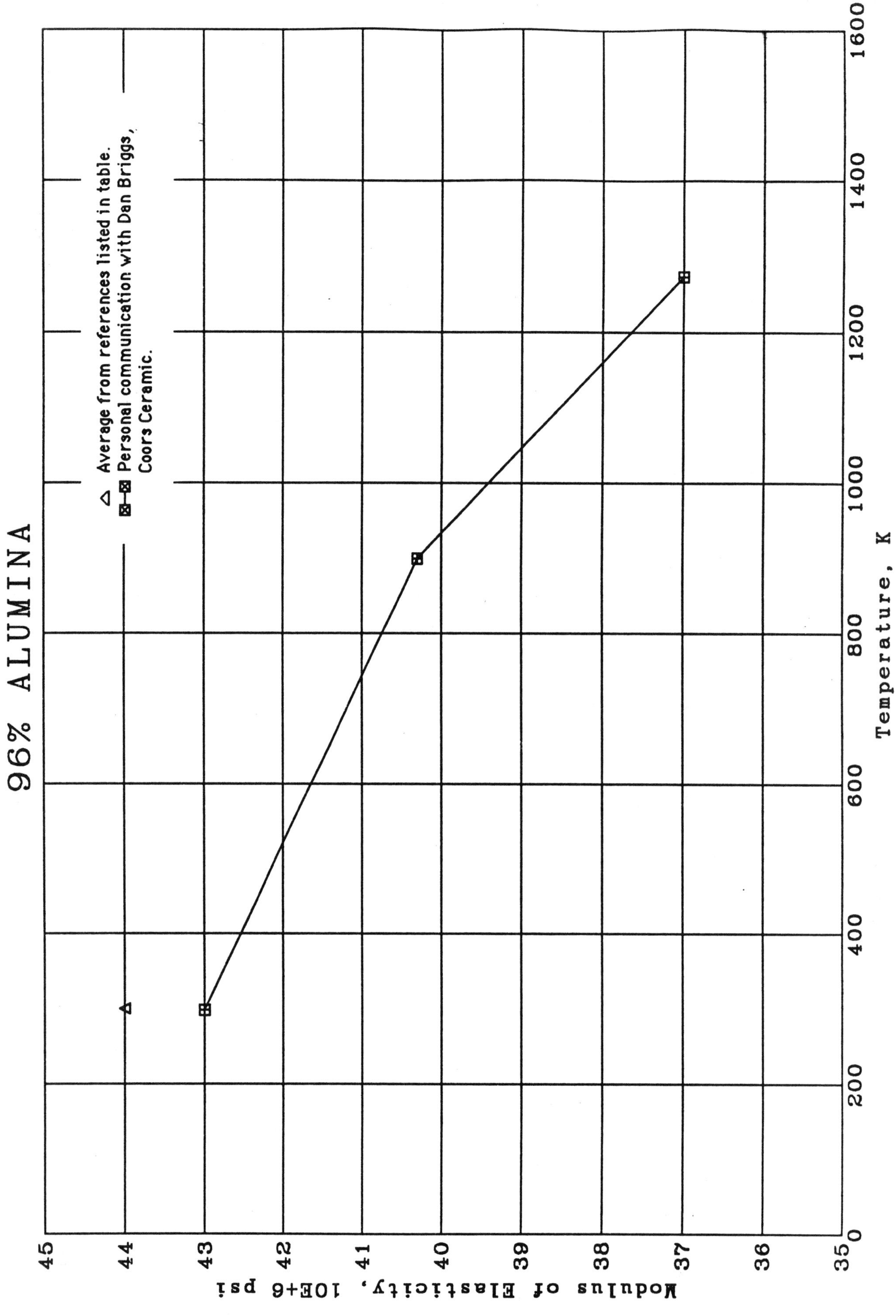

96% ALUMINA
Average from references listed in table.
Personal communication with Dan Briggs, Coors Ceramic.
Modulus of Elasticity, 10E+6 psi
Temperature, K
45
44
43
42
41
40
39
38
37
36
35
0
200
400
600
800
1000
1200
1400
1600

96% ALUMINA

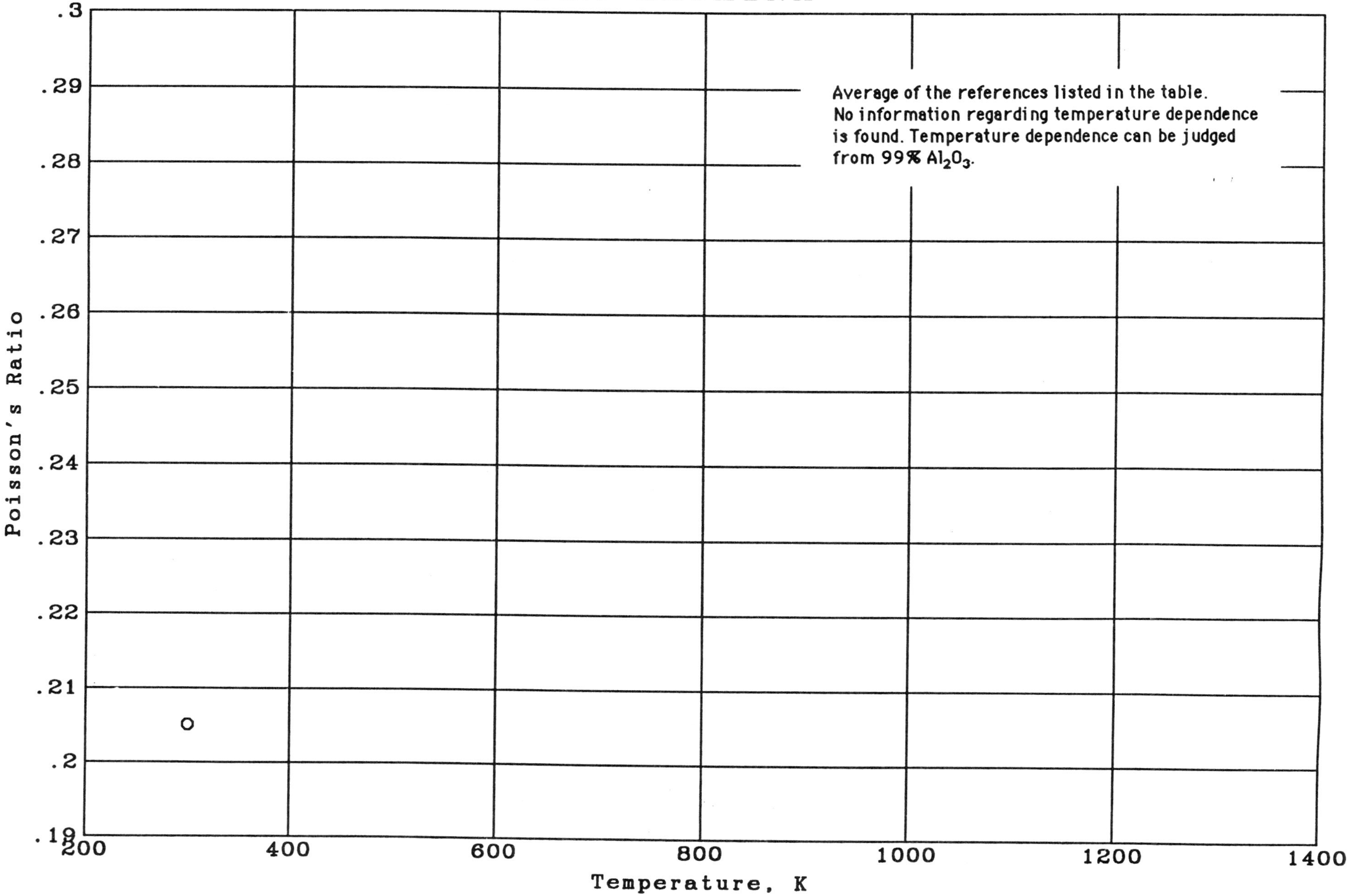
.3
.29
.28
.27
.26
.25
.24
.23
.22
.21
.2
.19
Poisson's Ratio
200
400
600
800
1000
1200
1400
Temperature, K
Average of the references listed in the table.
No information regarding temperature dependence
is found. Temperature dependence can be judged
from 99% Al_2O_3.

96% ALUMINA

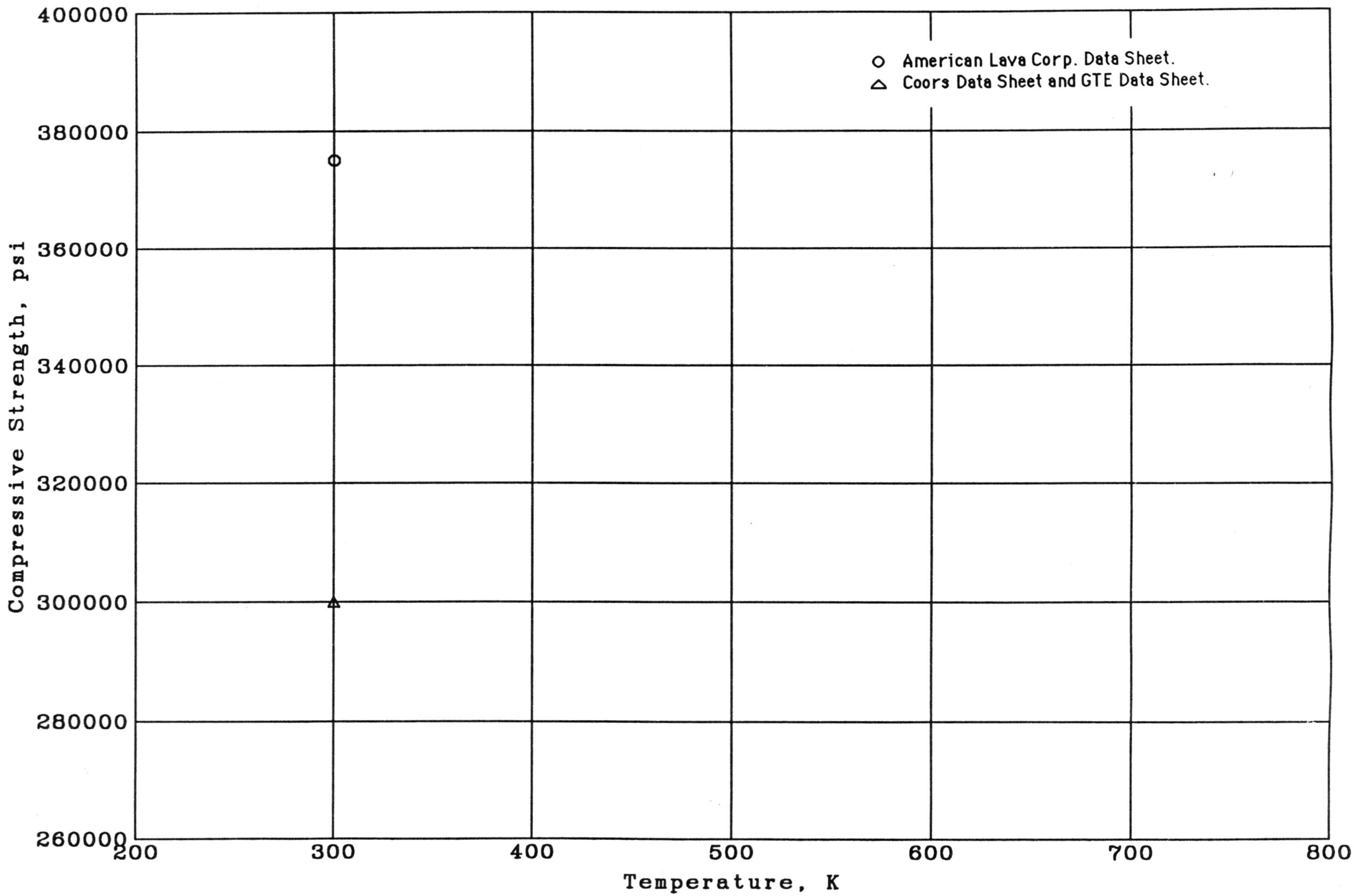

American Lava Corp. Data Sheet.
Coors Data Sheet and GTE Data Sheet.
400000
380000
360000
340000
320000
300000
280000
260000
Compressive Strength, psi
200
300
400
500
600
700
800
Temperature, K

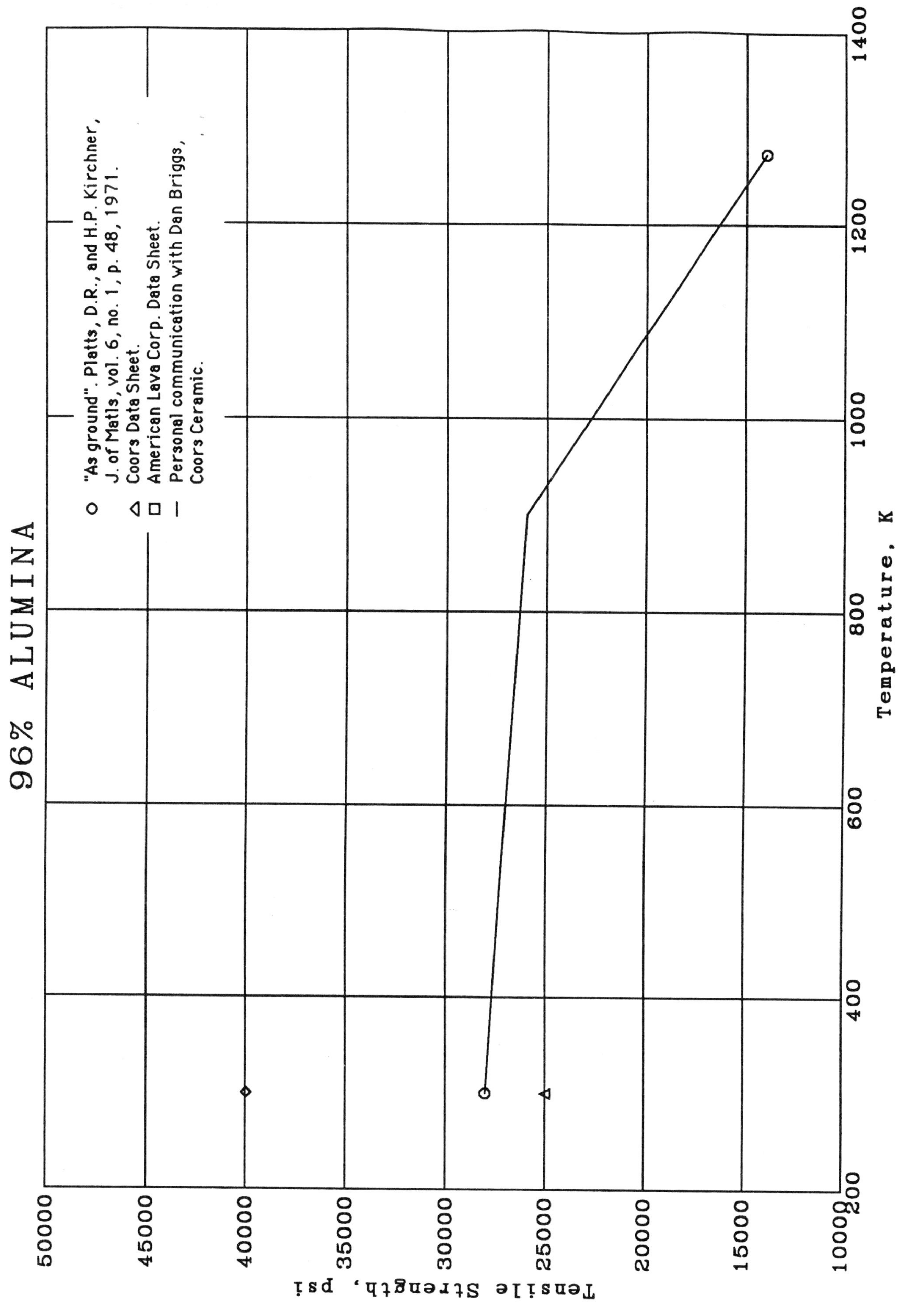
96% ALUMINA
○ "As ground". Platts, D.R., and H.P. Kirchner, J. of Matls, vol. 6, no. 1, p. 48, 1971.
△ Coors Data Sheet.
□ American Lava Corp. Data Sheet.
— Personal communication with Dan Briggs, Coors Ceramic.
Tensile Strength, psi
Temperature, K
50000
45000
40000
35000
30000
25000
20000
15000
10000
200
400
600
800
1000
1200
1400

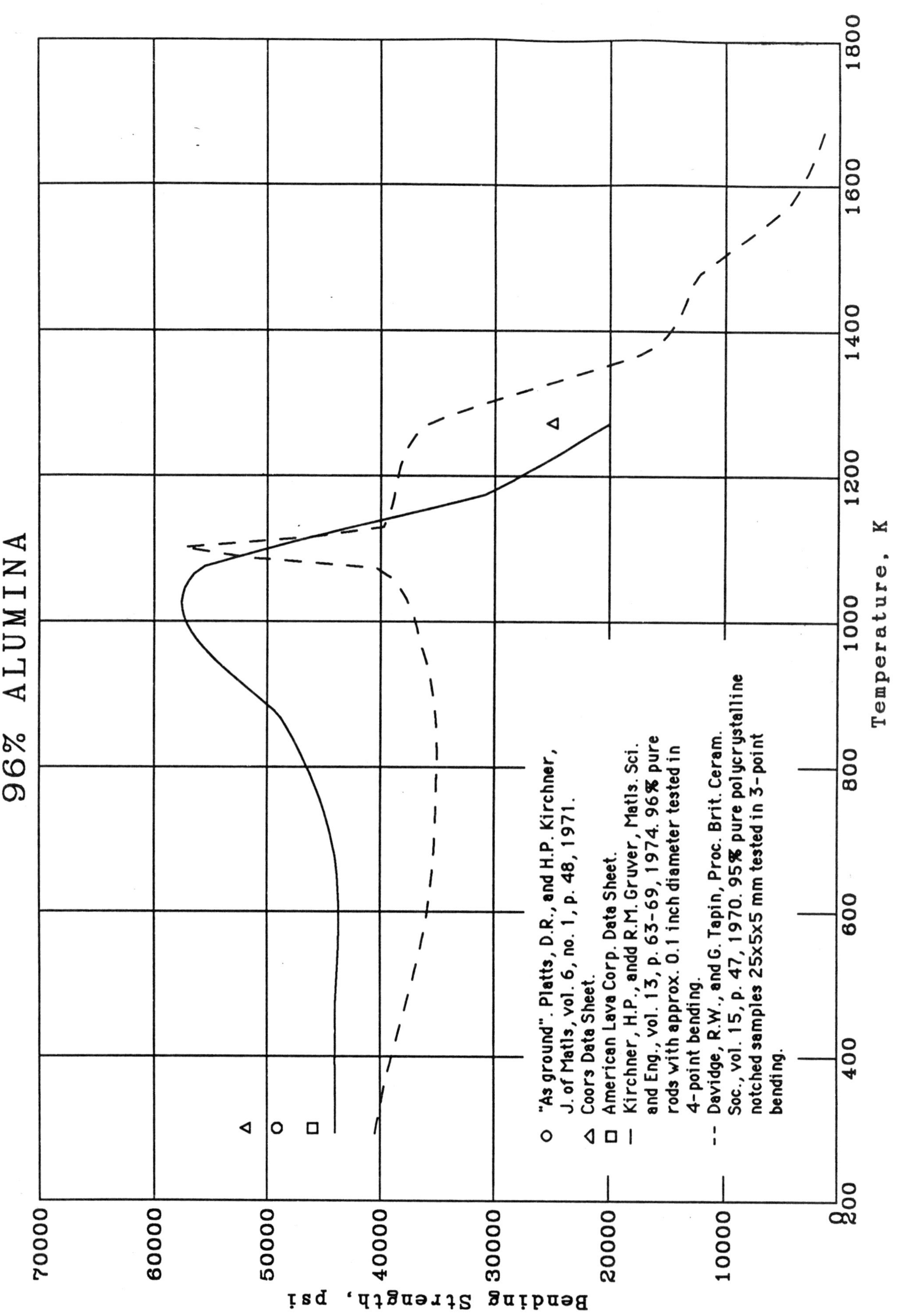

96% ALUMINA
70000
60000
50000
40000
30000
20000
10000
0
Bending Strength, psi
200
400
600
800
1000
1200
1400
1600
1800
Temperature, K
○ "As ground". Platts, D.R., and H.P. Kirchner, J. of Matls, vol. 6, no. 1, p. 48, 1971.
△ Coors Data Sheet.
□ American Lava Corp. Data Sheet.
— Kirchner, H.P., andd R.M. Gruver, Matls. Sci. and Eng., vol. 13, p. 63-69, 1974. 96% pure rods with approx. 0.1 inch diameter tested in 4-point bending.
-- Davidge, R.W., and G. Tapin, Proc. Brit. Ceram. Soc., vol. 15, p. 47, 1970. 95% pure polycrystalline notched samples 25x5x5 mm tested in 3-point bending.

96% ALUMINA

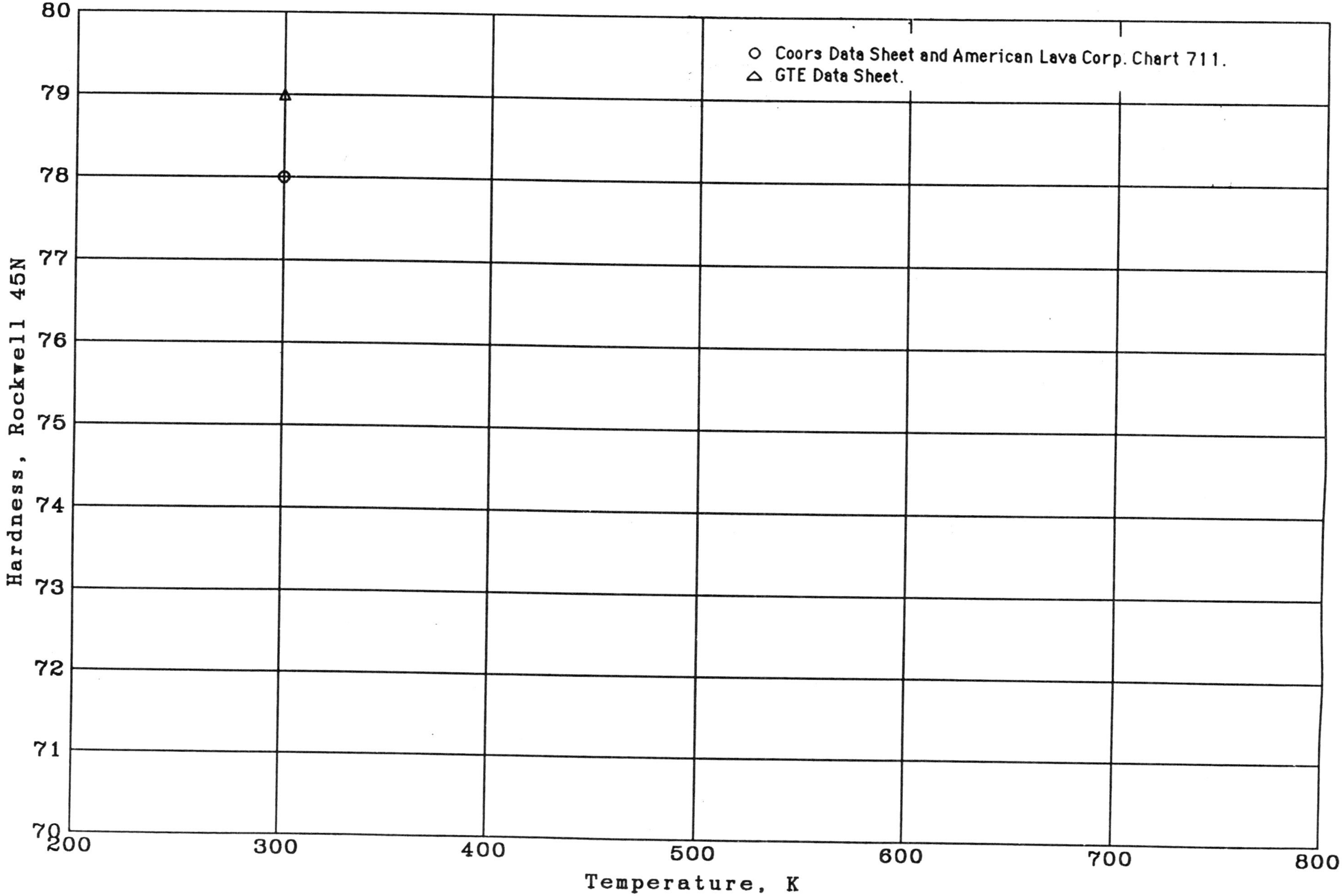

Coors Data Sheet and American Lava Corp. Chart 711.
GTE Data Sheet.
80
79
78
77
76
75
74
73
72
71
70
Hardness, Rockwell 45N
200
300
400
500
600
700
800
Temperature, K

99% ALUMINA

PROPERTY	@ 300K	UNITS	REFERENCES
Density	3.83	g/cm^3	99A1 99A2 99A3
Specific Heat	0.185	cal/g-K	99A4 99A5 99A6
Thermal Conductivity	0.070	cal/cm-s-K	99A1 99A2 99A7
Thermal Exp Coef	6.6×10^{-6}	1/K	99A1 99A2 99A10
Dielectric Constant	9.4	------	99A1 99A2 99A10
Dissipation Factor	0.004	------	99A1 99A2 99A10
Dielectric Strength	215	V/mil	99A1 99A2 99A9
Young's Modulus 1	54,000	ksi	99A1 99A3 99A8
Poisson's Ratio 2	0.24	------	99A1 99A3 99A8
Flexural Strength	48,000	psi	99A1 99A2 99A3
Ult Tensile Strength	31,000	psi	99A1 99A3 99A9
Compressive Strength	345,000	psi	99A1 99A2 99A3
Hardness	80	Rock 45N	99A1 99A2 99A9

1 Young's modulus is given by $59.49 \times 10^6 \exp\{-3.95P\}$ where P = fractional pore volume.[A11]
2 Poisson's ratio is given by 0.257 - 0.35P where P = fractional pore volume.[A12]

99% ALUMINA

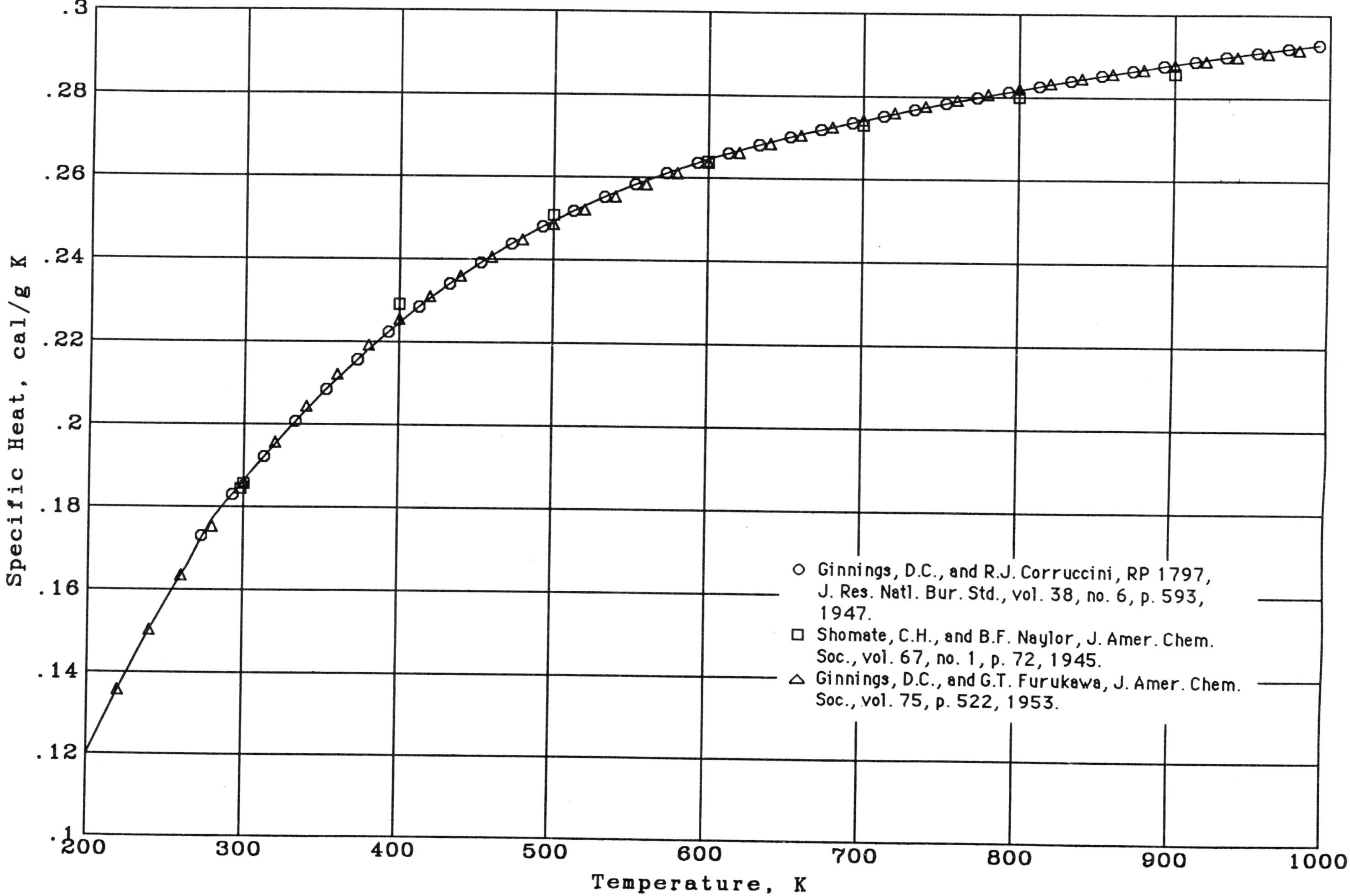

Specific Heat, cal/g K
.3
.28
.26
.24
.22
.2
.18
.16
.14
.12
.1
200
300
400
500
600
700
800
900
1000
Temperature, K
○ Ginnings, D.C., and R.J. Corruccini, RP 1797, J. Res. Natl. Bur. Std., vol. 38, no. 6, p. 593, 1947.
□ Shomate, C.H., and B.F. Naylor, J. Amer. Chem. Soc., vol. 67, no. 1, p. 72, 1945.
△ Ginnings, D.C., and G.T. Furukawa, J. Amer. Chem. Soc., vol. 75, p. 522, 1953.

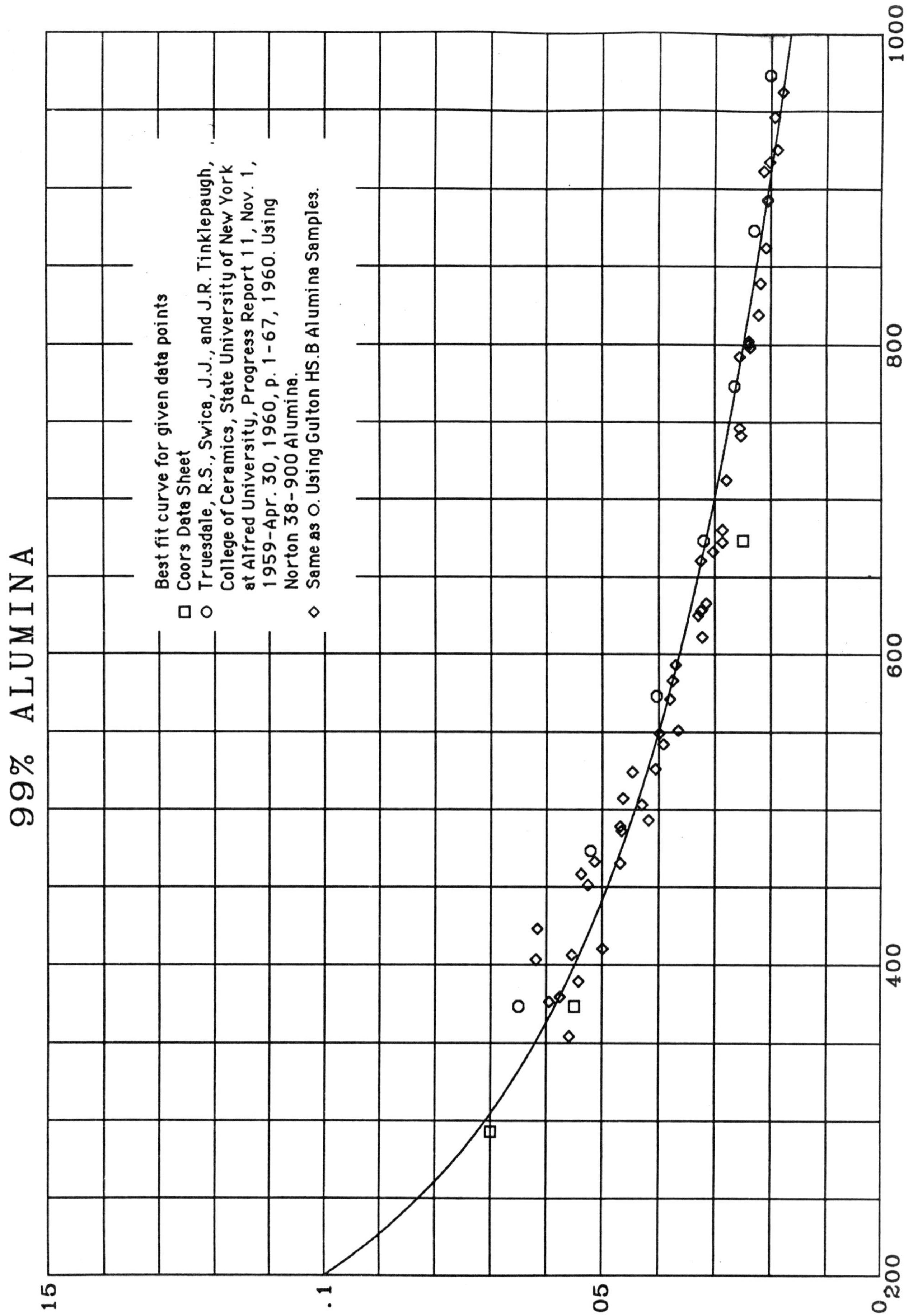
99% ALUMINA
Best fit curve for given data points
□ Coors Data Sheet
○ Truesdale, R.S., Swica, J.J., and J.R. Tinklepaugh, College of Ceramics, State University of New York at Alfred University, Progress Report 11, Nov. 1, 1959-Apr. 30, 1960, p. 1-67, 1960. Using Norton 38-900 Alumina.
◇ Same as ○. Using Gulton HS.B Alumina Samples.
Thermal Conductivity, cal/sec cm K
Temperature, K
.15
.1
.05
0
200
400
600
800
1000

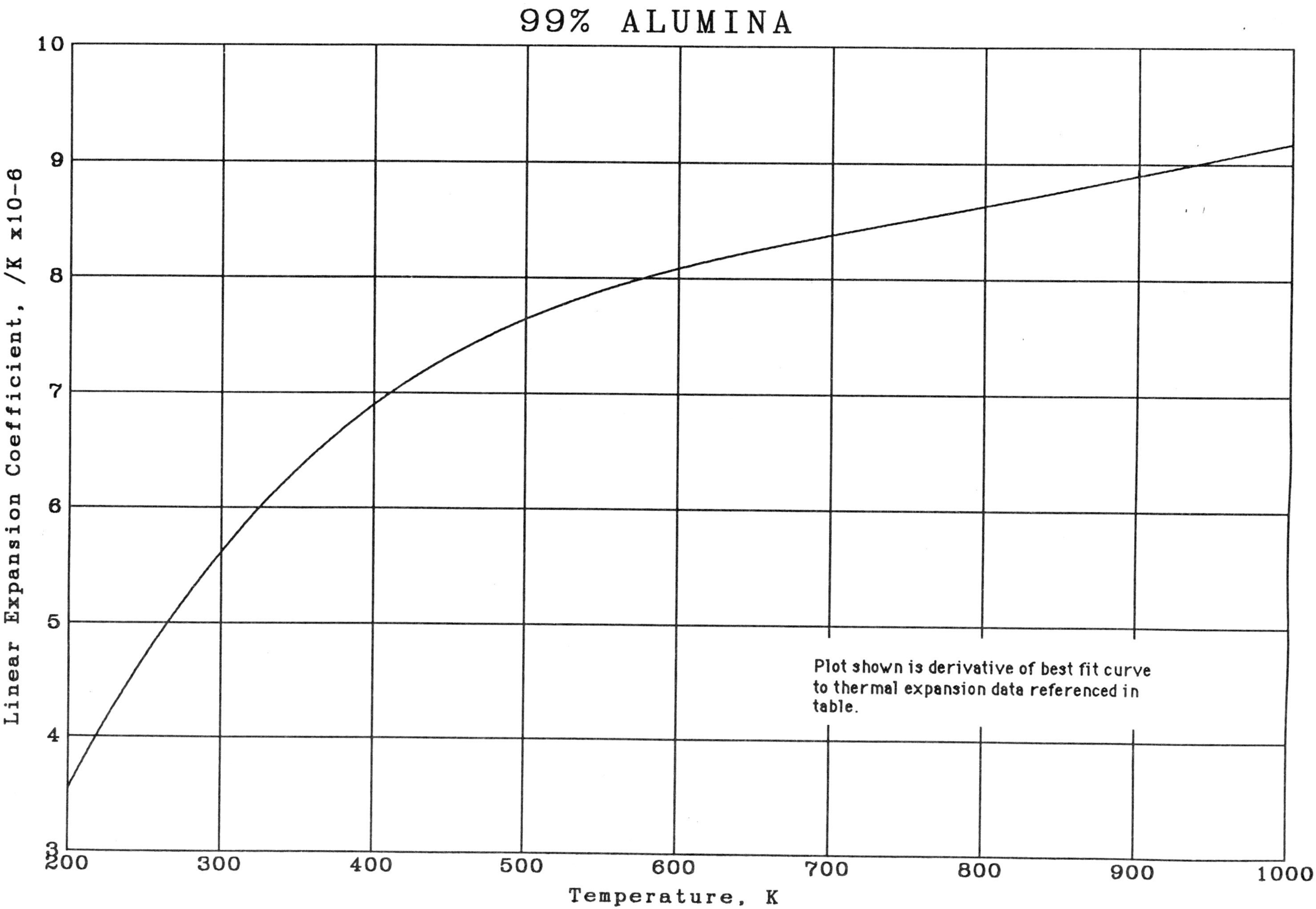
99% ALUMINA
Linear Expansion Coefficient, /K x10-6
10
9
8
7
6
5
4
3
200
300
400
500
600
700
800
900
1000
Temperature, K
Plot shown is derivative of best fit curve to thermal expansion data referenced in table.

99% ALUMINA

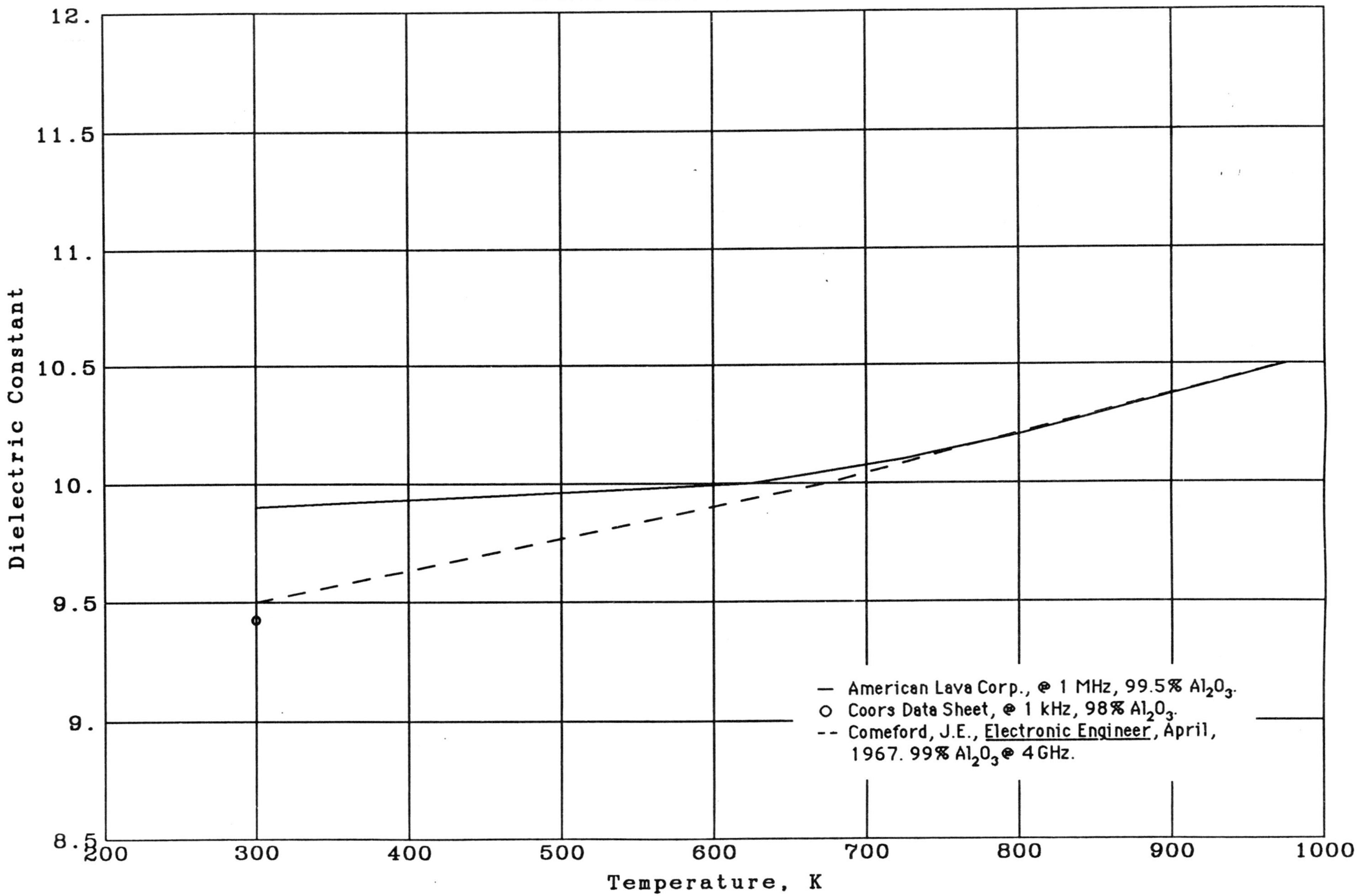
Dielectric Constant
12.
11.5
11.
10.5
10.
9.5
9.
8.5
200
300
400
500
600
700
800
900
1000
Temperature, K
— American Lava Corp., @ 1 MHz, 99.5% Al_2O_3.
○ Coors Data Sheet, @ 1 kHz, 98% Al_2O_3.
-- Comeford, J.E., Electronic Engineer, April, 1967. 99% Al_2O_3 @ 4 GHz.

99% ALUMINA

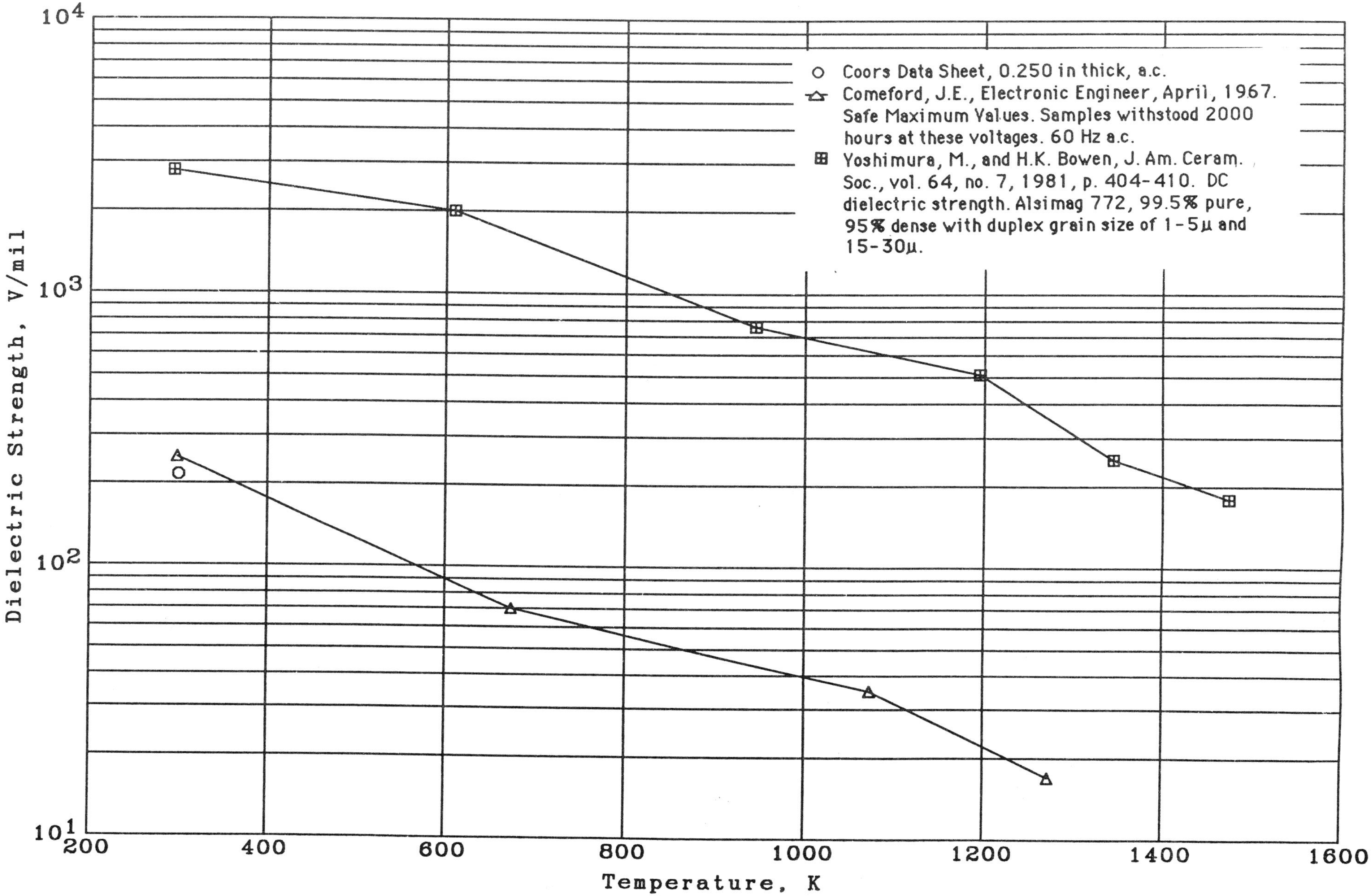

Coors Data Sheet, 0.250 in thick, a.c.
Comeford, J.E., Electronic Engineer, April, 1967. Safe Maximum Values. Samples withstood 2000 hours at these voltages. 60 Hz a.c.
Yoshimura, M., and H.K. Bowen, J. Am. Ceram. Soc., vol. 64, no. 7, 1981, p. 404-410. DC dielectric strength. Alsimag 772, 99.5% pure, 95% dense with duplex grain size of 1-5μ and 15-30μ.
Dielectric Strength, V/mil
10^4
10^3
10^2
10^1
200
400
600
800
1000
1200
1400
1600
Temperature, K

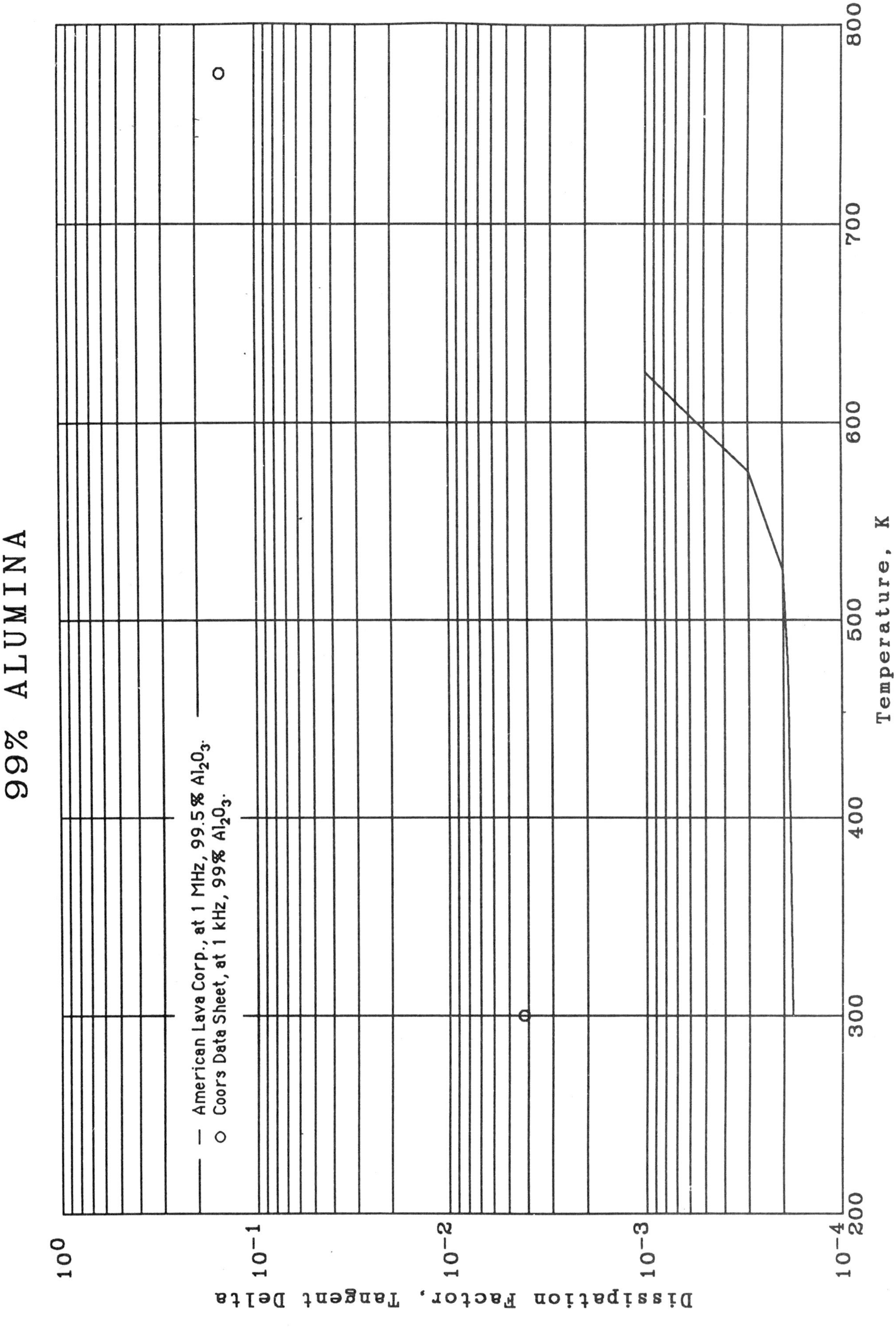
99% ALUMINA
— American Lava Corp., at 1 MHz, 99.5% Al_2O_3.
○ Coors Data Sheet, at 1 kHz, 99% Al_2O_3.
Dissipation Factor, Tangent Delta
10^0
10^{-1}
10^{-2}
10^{-3}
10^{-4}
Temperature, K
200
300
400
500
600
700
800

99% ALUMINA

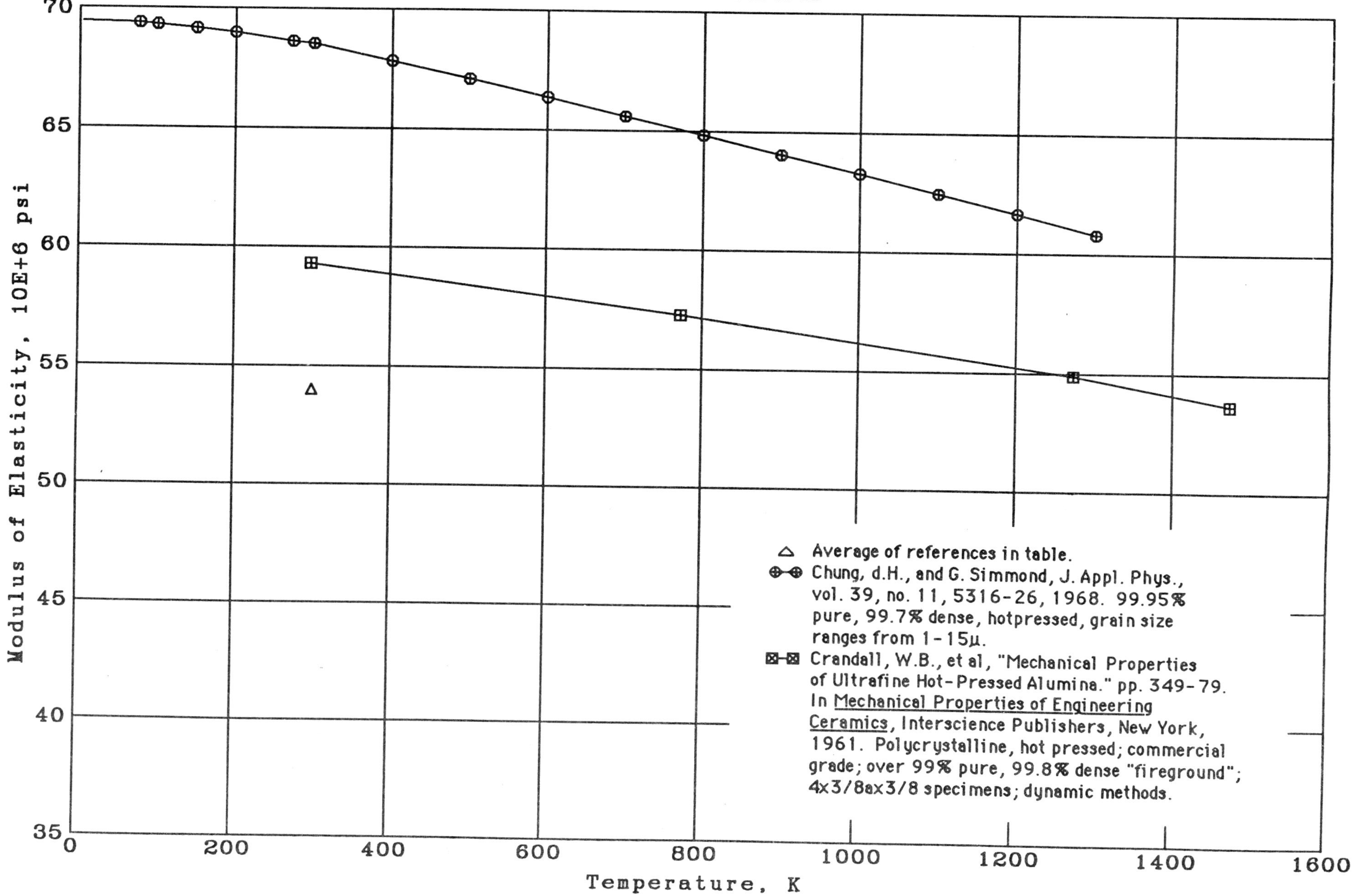

Modulus of Elasticity, 10E+6 psi
70
65
60
55
50
45
40
35
0
200
400
600
800
1000
1200
1400
1600
Temperature, K
△ Average of references in table.
Chung, d.H., and G. Simmond, J. Appl. Phys., vol. 39, no. 11, 5316-26, 1968. 99.95% pure, 99.7% dense, hotpressed, grain size ranges from 1-15μ.
Crandall, W.B., et al, "Mechanical Properties of Ultrafine Hot-Pressed Alumina." pp. 349-79. In Mechanical Properties of Engineering Ceramics, Interscience Publishers, New York, 1961. Polycrystalline, hot pressed; commercial grade; over 99% pure, 99.8% dense "fireground"; 4x3/8ax3/8 specimens; dynamic methods.

99% ALUMINA

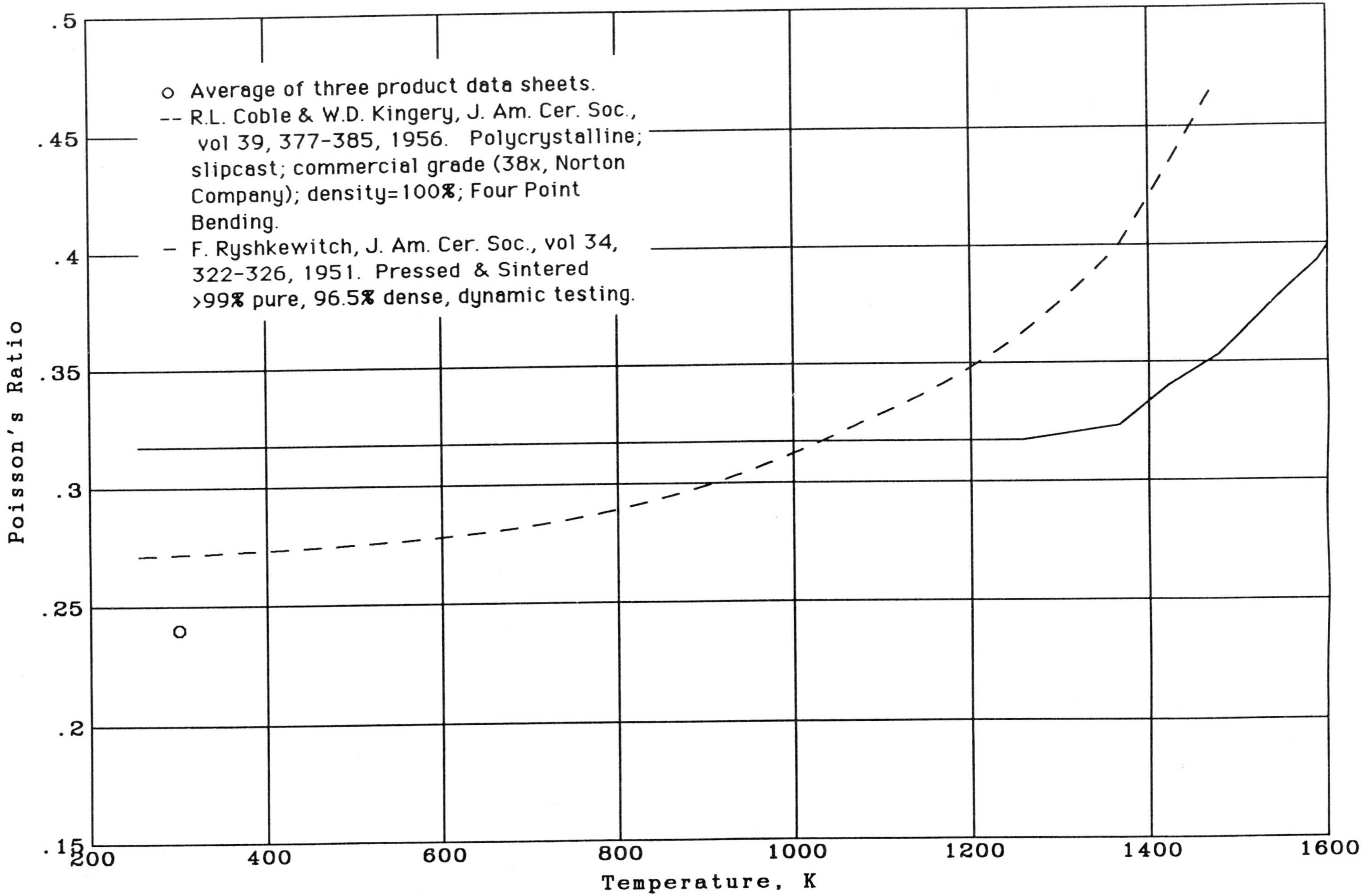

Poisson's Ratio
.5
.45
.4
.35
.3
.25
.2
.15
200
400
600
800
1000
1200
1400
1600
Temperature, K
o Average of three product data sheets.
-- R.L. Coble & W.D. Kingery, J. Am. Cer. Soc., vol 39, 377-385, 1956. Polycrystalline; slipcast; commercial grade (38x, Norton Company); density=100%; Four Point Bending.
— F. Ryshkewitch, J. Am. Cer. Soc., vol 34, 322-326, 1951. Pressed & Sintered >99% pure, 96.5% dense, dynamic testing.

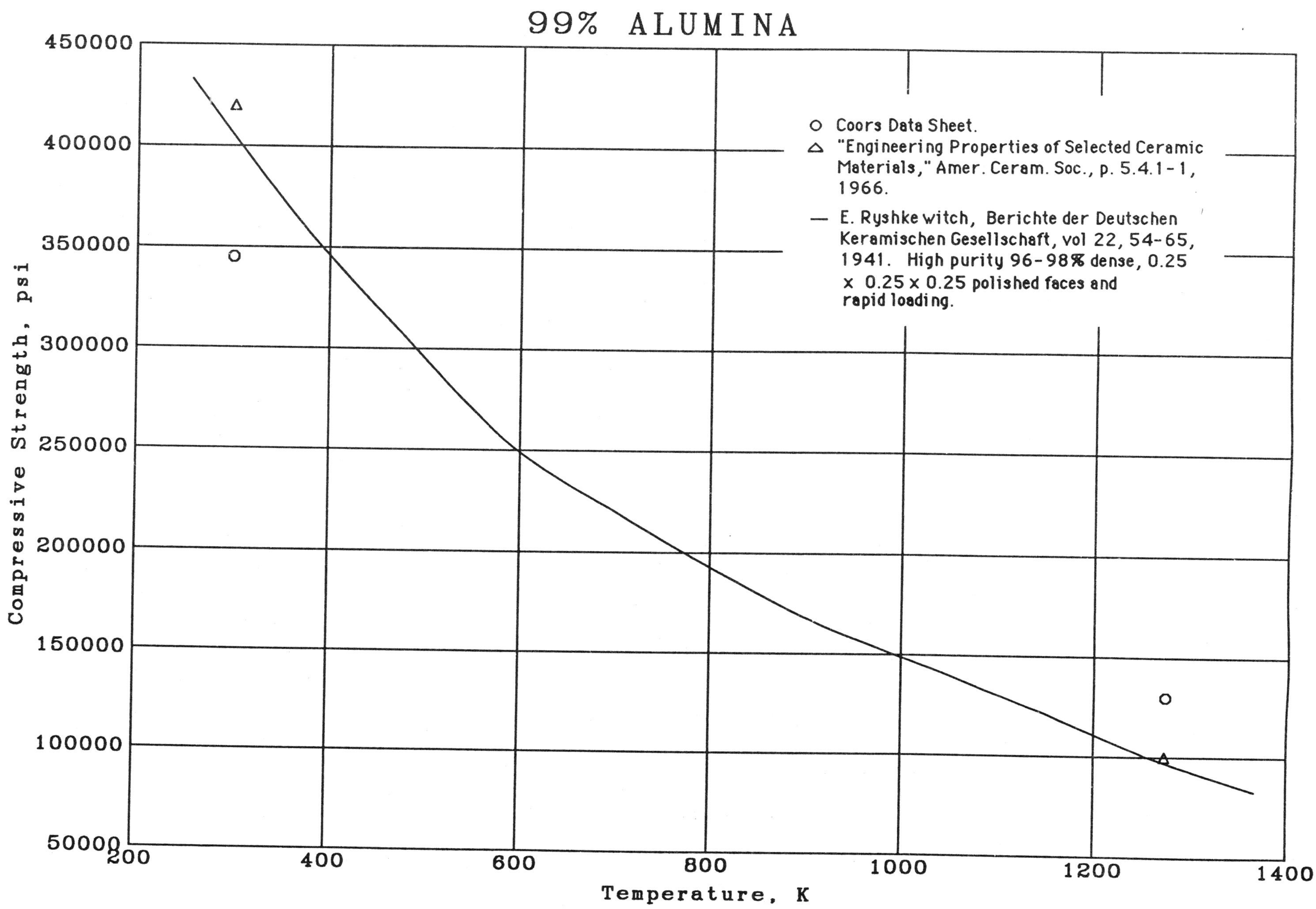
99% ALUMINA
450000
400000
350000
300000
250000
200000
150000
100000
50000
Compressive Strength, psi
200
400
600
800
1000
1200
1400
Temperature, K
Coors Data Sheet.
"Engineering Properties of Selected Ceramic Materials," Amer. Ceram. Soc., p. 5.4.1-1, 1966.
E. Ryshke witch, Berichte der Deutschen Keramischen Gesellschaft, vol 22, 54-65, 1941. High purity 96-98% dense, 0.25 x 0.25 x 0.25 polished faces and rapid loading.

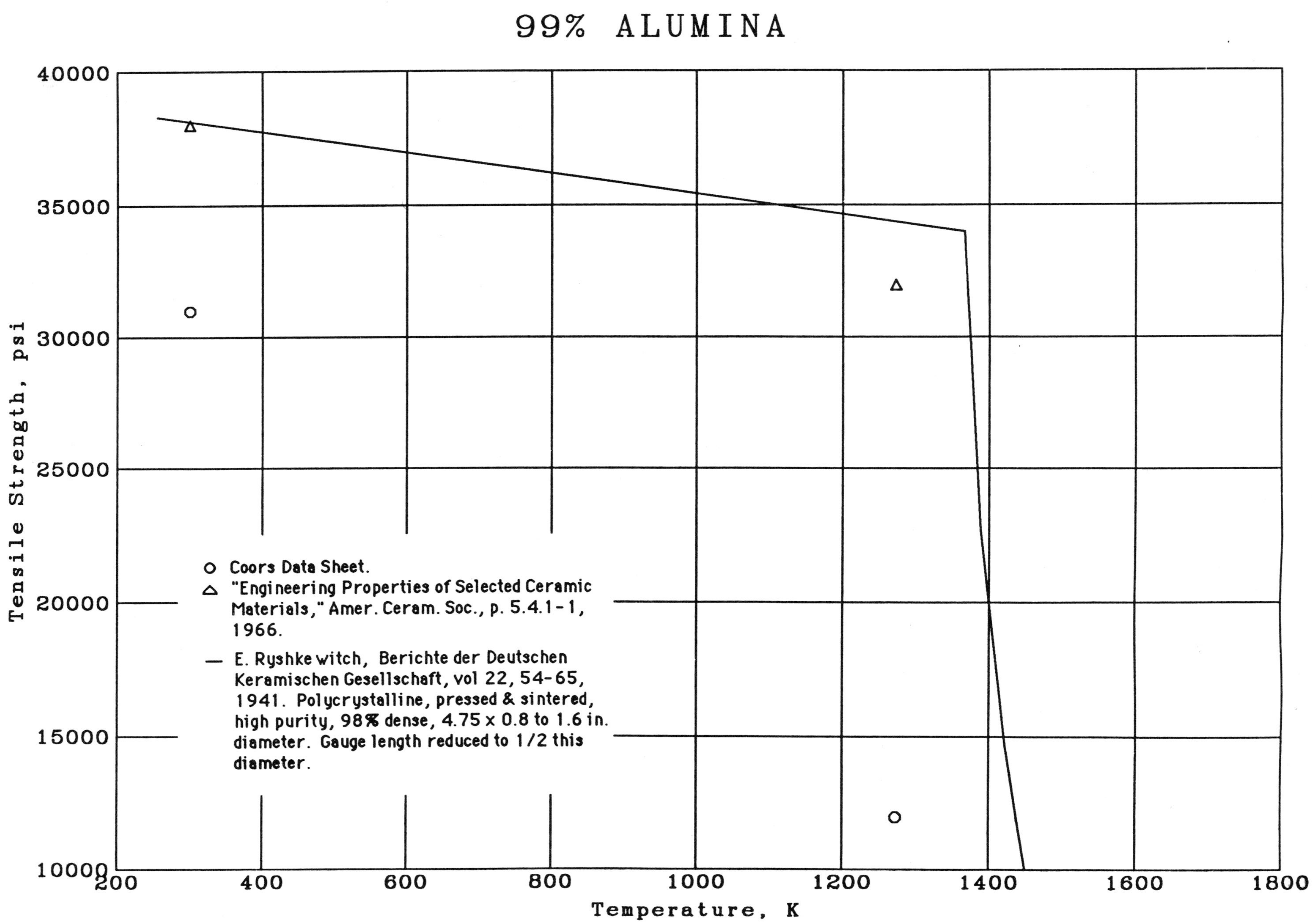
99% ALUMINA
Tensile Strength, psi
40000
35000
30000
25000
20000
15000
10000
200
400
600
800
1000
1200
1400
1600
1800
Temperature, K
○ Coors Data Sheet.
△ "Engineering Properties of Selected Ceramic Materials," Amer. Ceram. Soc., p. 5.4.1-1, 1966.
— E. Ryshkewitch, Berichte der Deutschen Keramischen Gesellschaft, vol 22, 54-65, 1941. Polycrystalline, pressed & sintered, high purity, 98% dense, 4.75 x 0.8 to 1.6 in. diameter. Gauge length reduced to 1/2 this diameter.

99% ALUMINA

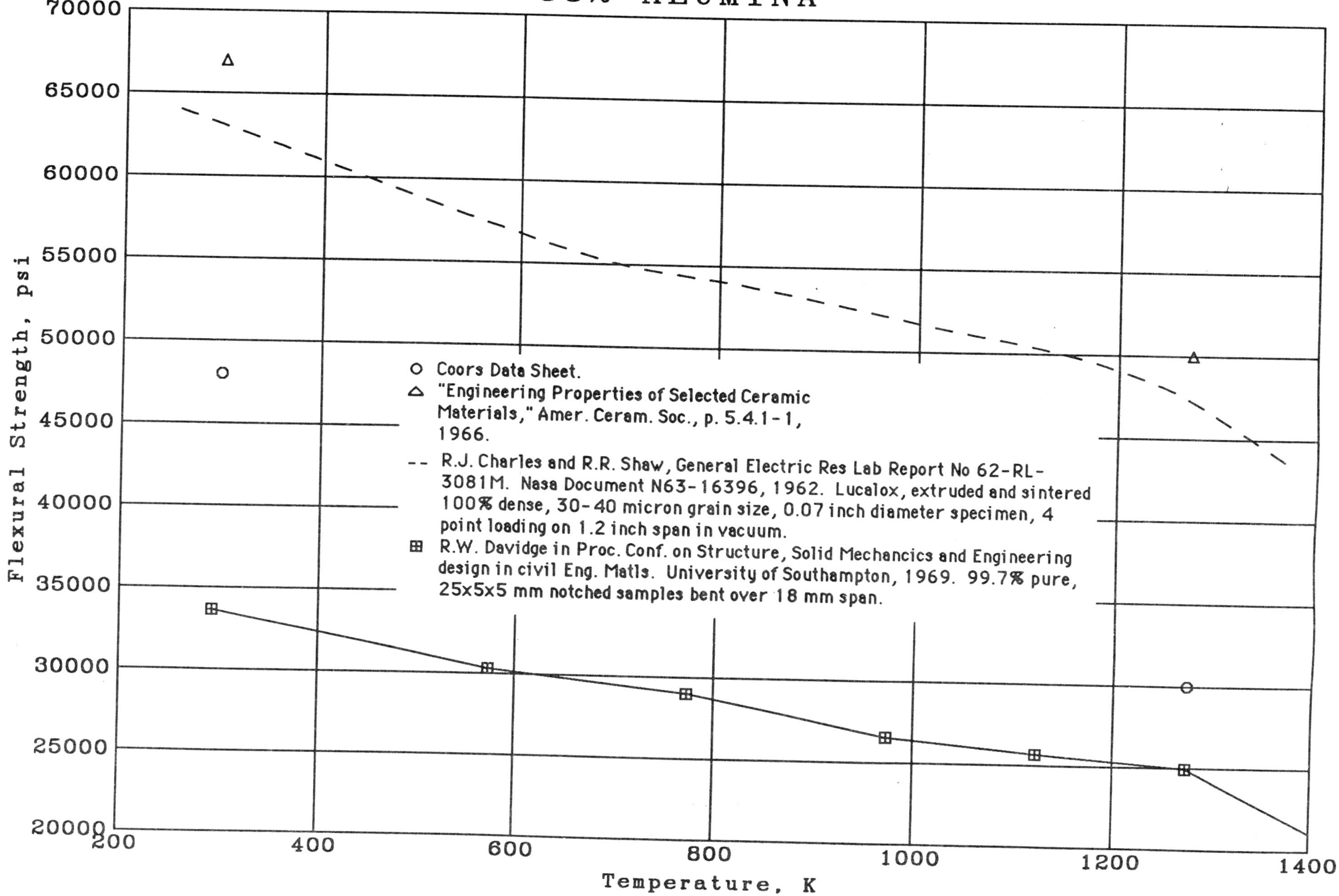
70000
65000
60000
55000
50000
45000
40000
35000
30000
25000
20000
Flexural Strength, psi
200
400
600
800
1000
1200
1400
Temperature, K
Coors Data Sheet.
"Engineering Properties of Selected Ceramic Materials," Amer. Ceram. Soc., p. 5.4.1-1, 1966.
R.J. Charles and R.R. Shaw, General Electric Res Lab Report No 62-RL-3081M. Nasa Document N63-16396, 1962. Lucalox, extruded and sintered 100% dense, 30-40 micron grain size, 0.07 inch diameter specimen, 4 point loading on 1.2 inch span in vacuum.
R.W. Davidge in Proc. Conf. on Structure, Solid Mechancics and Engineering design in civil Eng. Matls. University of Southampton, 1969. 99.7% pure, 25x5x5 mm notched samples bent over 18 mm span.

99% ALUMINA

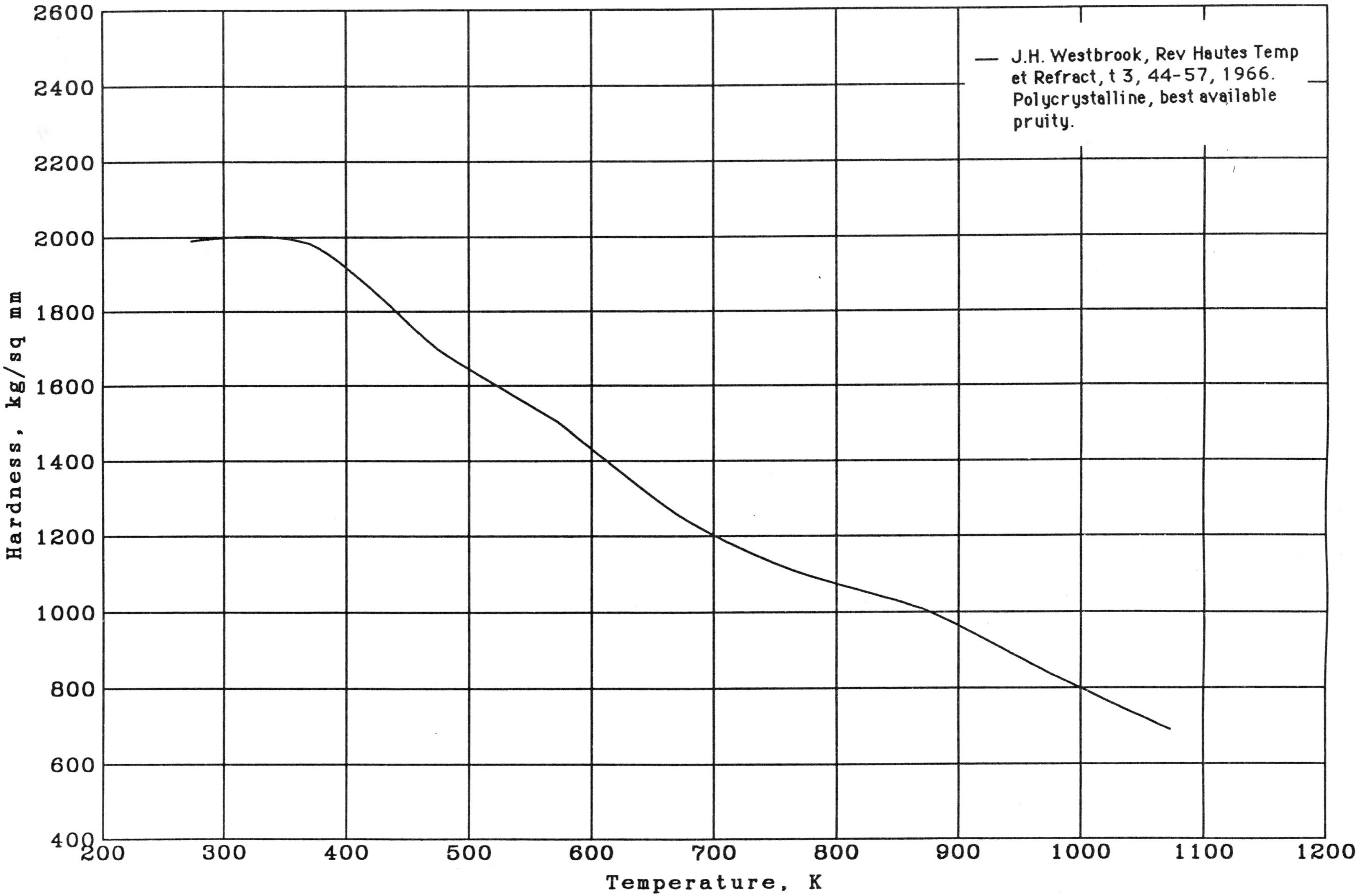

2600
2400
2200
2000
1800
1600
1400
1200
1000
800
600
400
Hardness, kg/sq mm
200
300
400
500
600
700
800
900
1000
1100
1200
Temperature, K
J.H. Westbrook, Rev Hautes Temp et Refract, t 3, 44-57, 1966. Polycrystalline, best available pruity.

ALUMINA (SAPPHIRE)

PROPERTY	@ 300K	UNITS	REFERENCES
Density	3.99	g/cm^3	S1 S2 S3
Specific Heat	0.185	cal/g-K	S4 S5 S6
Thermal Conductivity	0.110	cal/cm-s-K	S7 S8 S9
Thermal Exp Coef	6.0×10^{-6}	1/K	S1 S10 S11
Dielectric Constant	10.3	------	S1 S12 S13
Dissipation Factor	0.002	------	S1 S12 S11
Dielectric Strength	1200	V/mil	S11 S21 S23
Young's Modulus	60,000	ksi	S1 S16 S17
Poisson's Ratio	0.23	------	S17 S18 S19
Flexural Strength	92,000	psi	S16 S11 S22
Ult Tensile Strength[1]	370,000	psi	S20 S24 S26
Compressive Strength	300,000	psi	S1 S24
Hardness	9	Mohs	S1 S11 S25

[1] Average of 22 samples with 2.5-mm diam., tensile axis oriented along the c-axis. Samples were tested in water with a strain rate of 0.132/min. 20 samples tested with a strain rate of 0.00066/min averaged 330,000 psi. Strength will decrease as the size of the sample increases. The tensile strength of polycrystalline machined samples is typically in the range of 20 to 70 ksi.

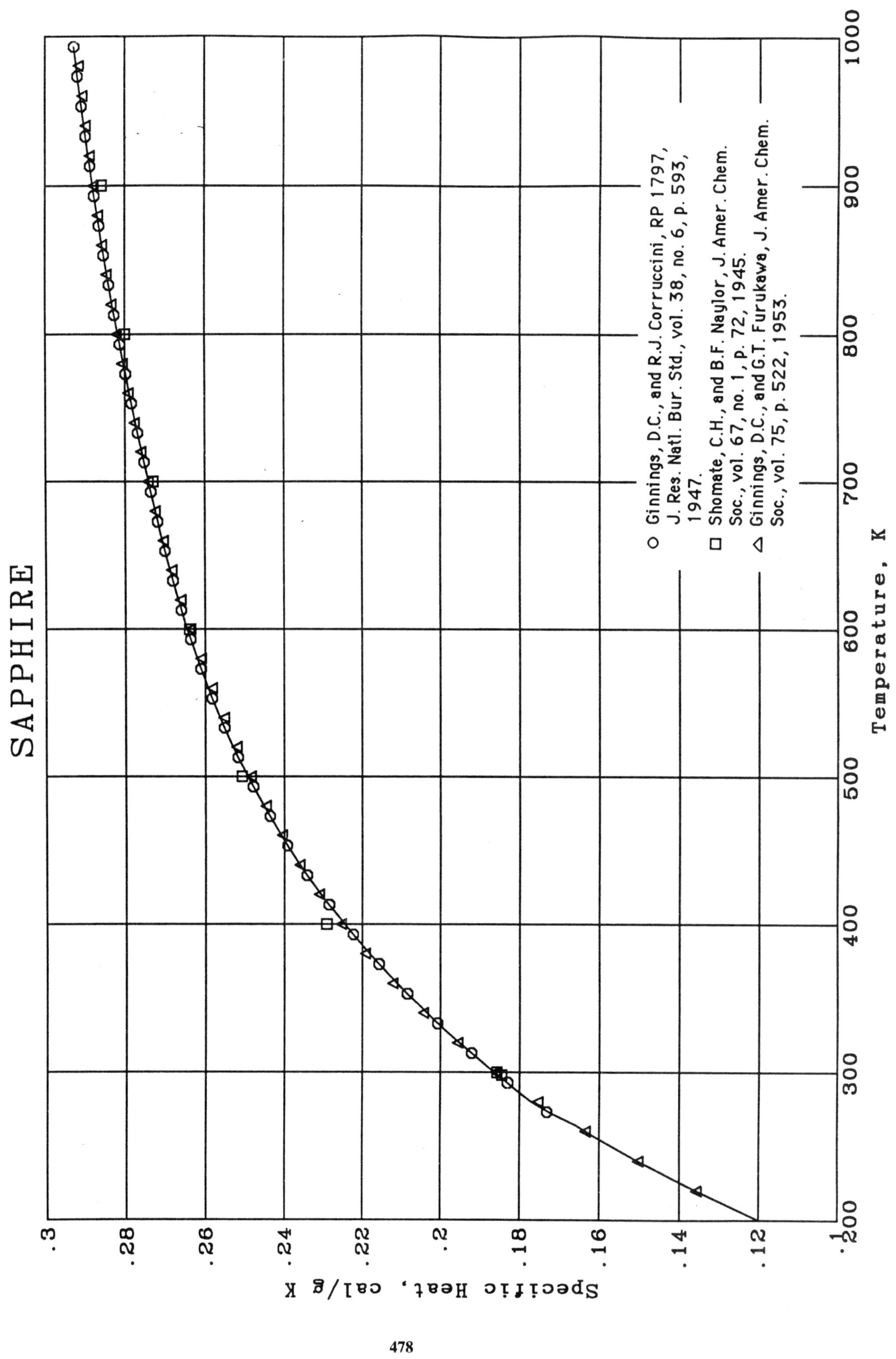
SAPPHIRE
Specific Heat, cal/g K
Temperature, K
.1
.12
.14
.16
.18
.2
.22
.24
.26
.28
.3
200
300
400
500
600
700
800
900
1000
○ Ginnings, D.C., and R.J. Corruccini, RP 1797, J. Res. Natl. Bur. Std., vol. 38, no. 6, p. 593, 1947.
□ Shomate, C.H., and B.F. Naylor, J. Amer. Chem. Soc., vol. 67, no. 1, p. 72, 1945.
△ Ginnings, D.C., and G.T. Furukawa, J. Amer. Chem. Soc., vol. 75, p. 522, 1953.

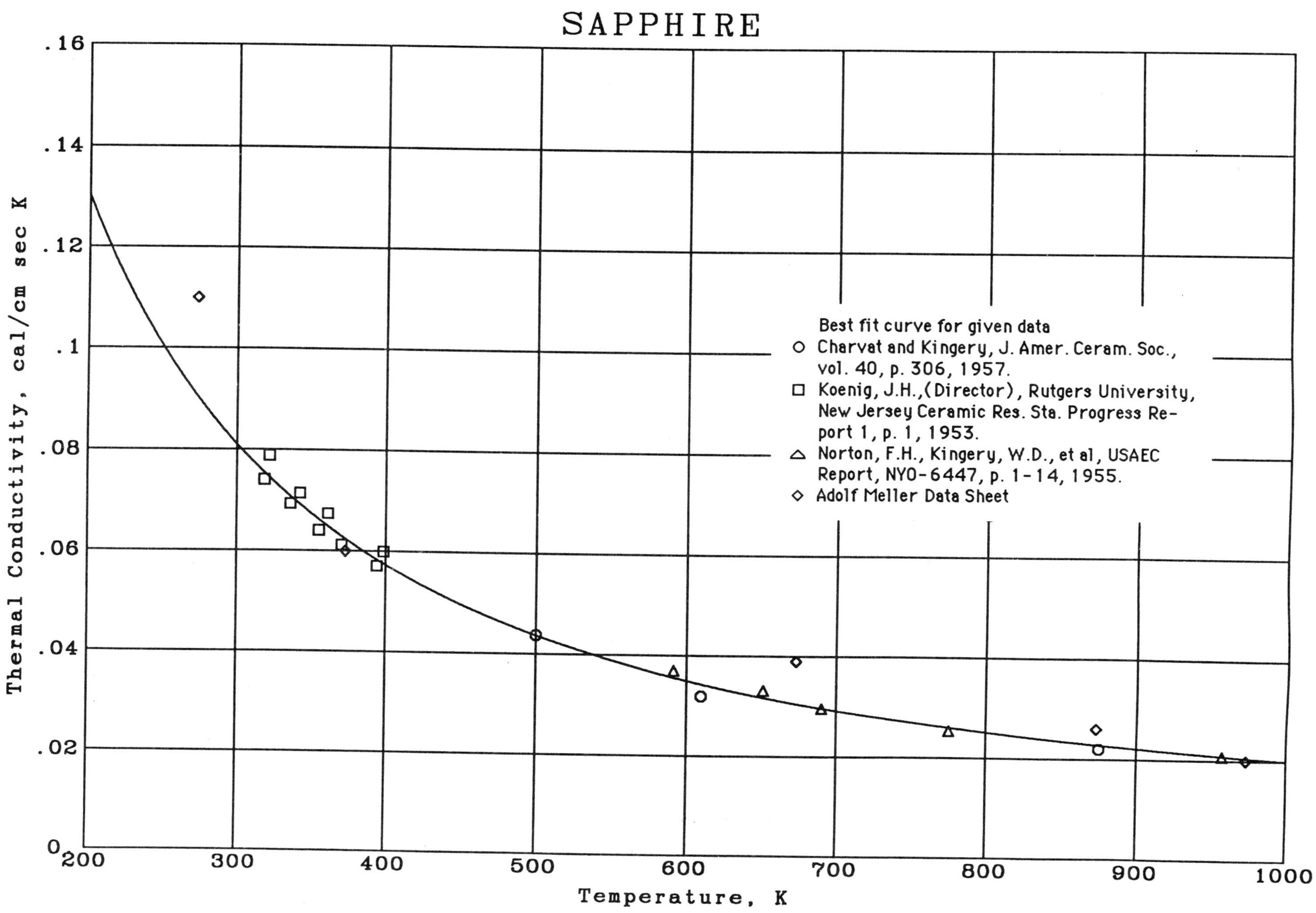
SAPPHIRE
Thermal Conductivity, cal/cm sec K
Temperature, K
0
.02
.04
.06
.08
.1
.12
.14
.16
200
300
400
500
600
700
800
900
1000
Best fit curve for given data
○ Charvat and Kingery, J. Amer. Ceram. Soc., vol. 40, p. 306, 1957.
□ Koenig, J.H.,(Director), Rutgers University, New Jersey Ceramic Res. Sta. Progress Report 1, p. 1, 1953.
△ Norton, F.H., Kingery, W.D., et al, USAEC Report, NYO-6447, p. 1-14, 1955.
◇ Adolf Meller Data Sheet

SAPPHIRE

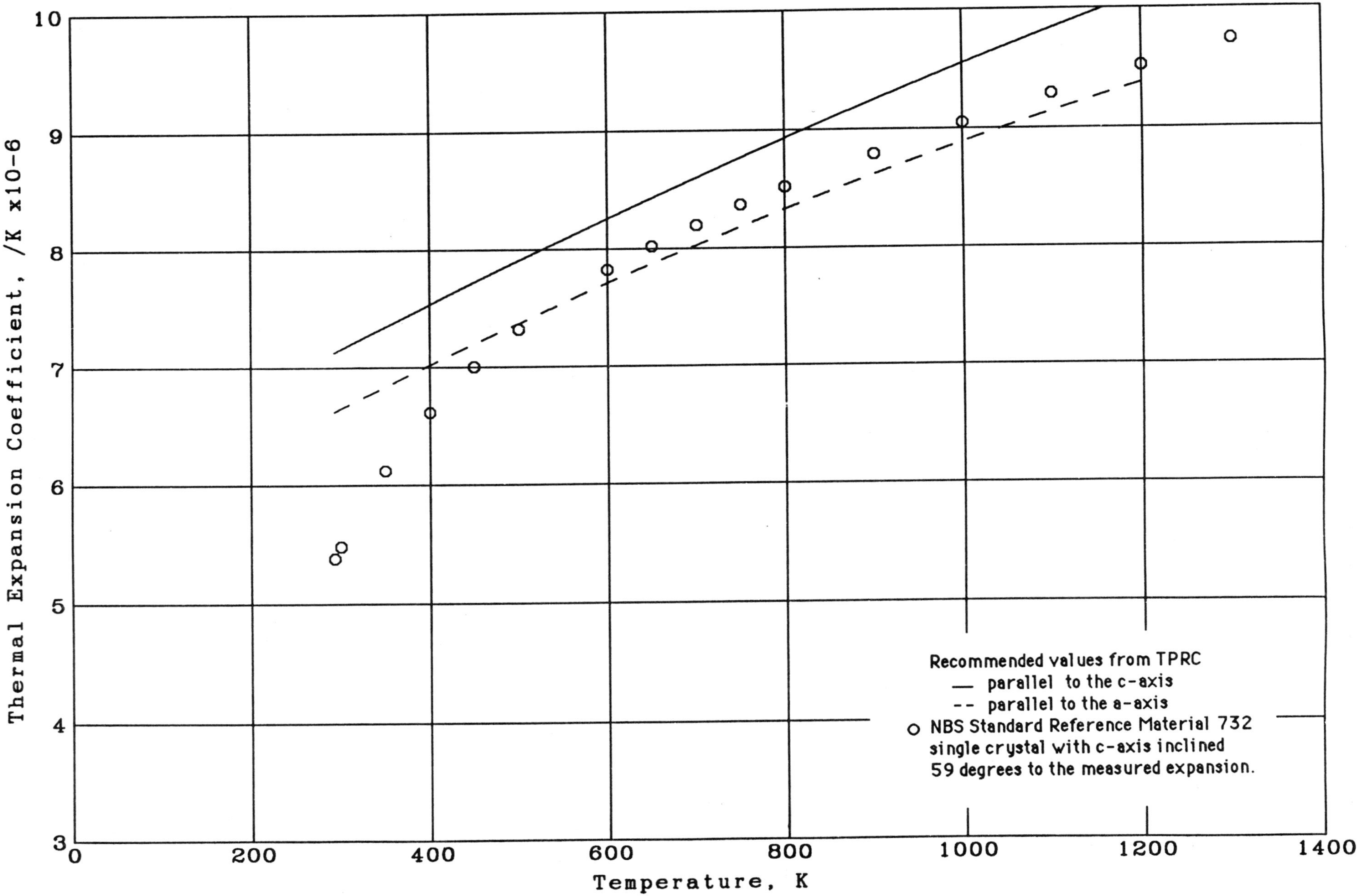

Thermal Expansion Coefficient, /K x10-6
10
9
8
7
6
5
4
3
0
200
400
600
800
1000
1200
1400
Temperature, K
Recommended values from TPRC
— parallel to the c-axis
-- parallel to the a-axis
o NBS Standard Reference Material 732
single crystal with c-axis inclined
59 degrees to the measured expansion.

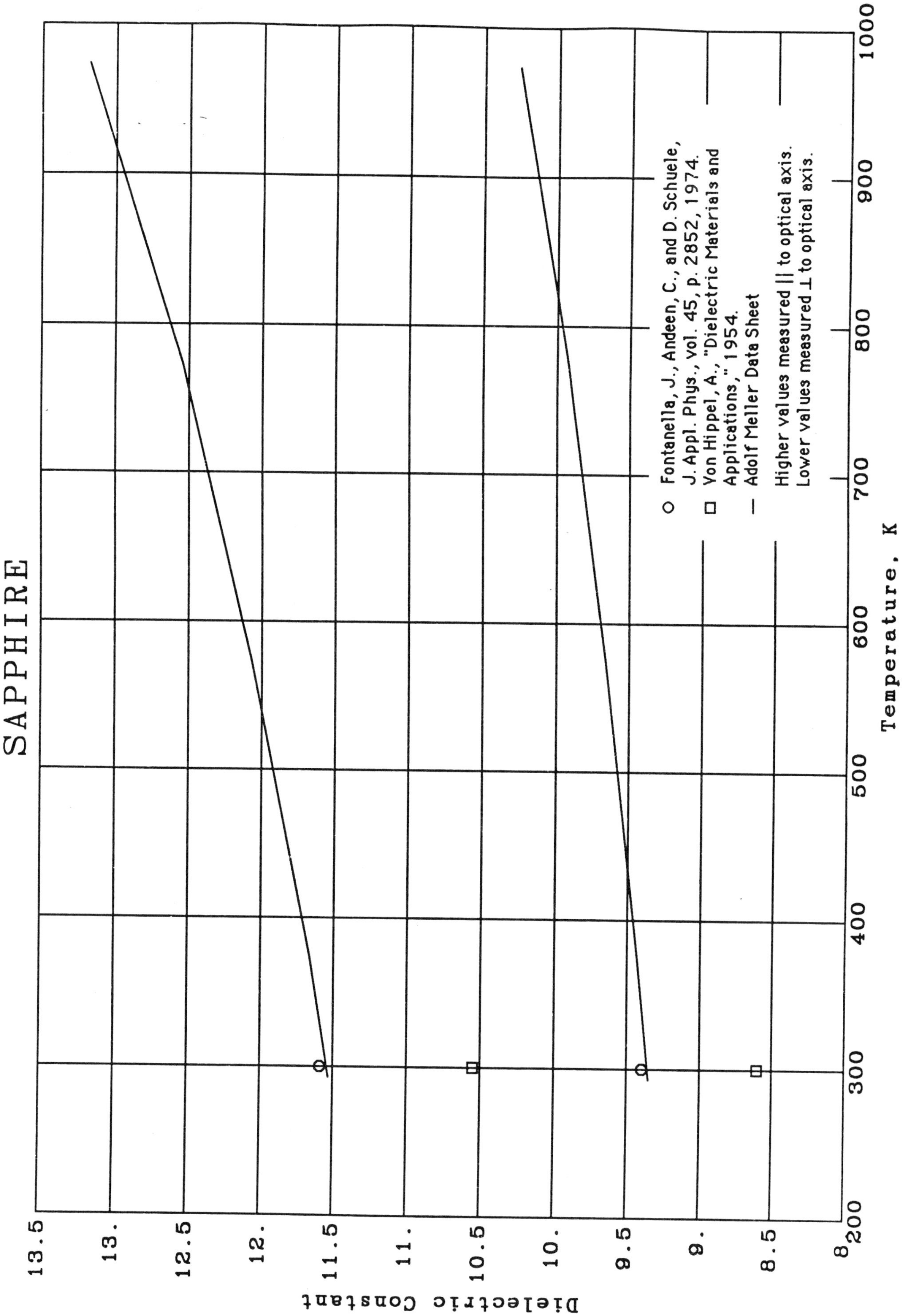
SAPPHIRE
○ Fontanella, J., Andeen, C., and D. Schuele, J. Appl. Phys., vol. 45, p. 2852, 1974.
□ Von Hippel, A., "Dielectric Materials and Applications," 1954.
— Adolf Meller Data Sheet
Higher values measured || to optical axis.
Lower values measured ⊥ to optical axis.
Dielectric Constant
Temperature, K
13.5
13.
12.5
12.
11.5
11.
10.5
10.
9.5
9.
8.5
8
200
300
400
500
600
700
800
900
1000

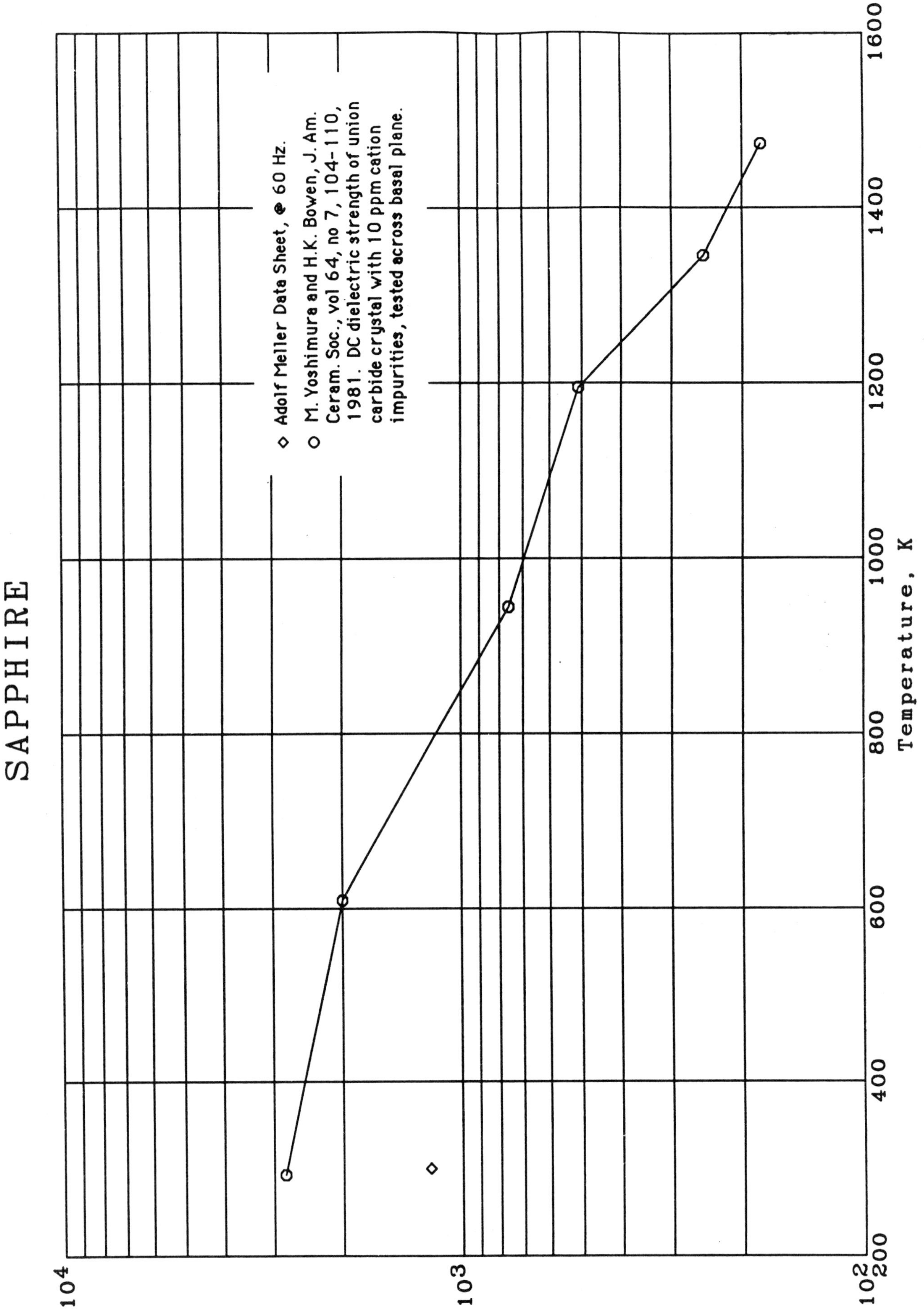

SAPPHIRE
◇ Adolf Meller Data Sheet, @ 60 Hz.
○ M. Yoshimura and H.K. Bowen, J. Am. Ceram. Soc., vol 64, no 7, 104-110, 1981. DC dielectric strength of union carbide crystal with 10 ppm cation impurities, tested across basal plane.
Dielectric Strength, V/mil
10^4
10^3
10^2
200
400
600
800
1000
1200
1400
1600
Temperature, K

SAPPHIRE

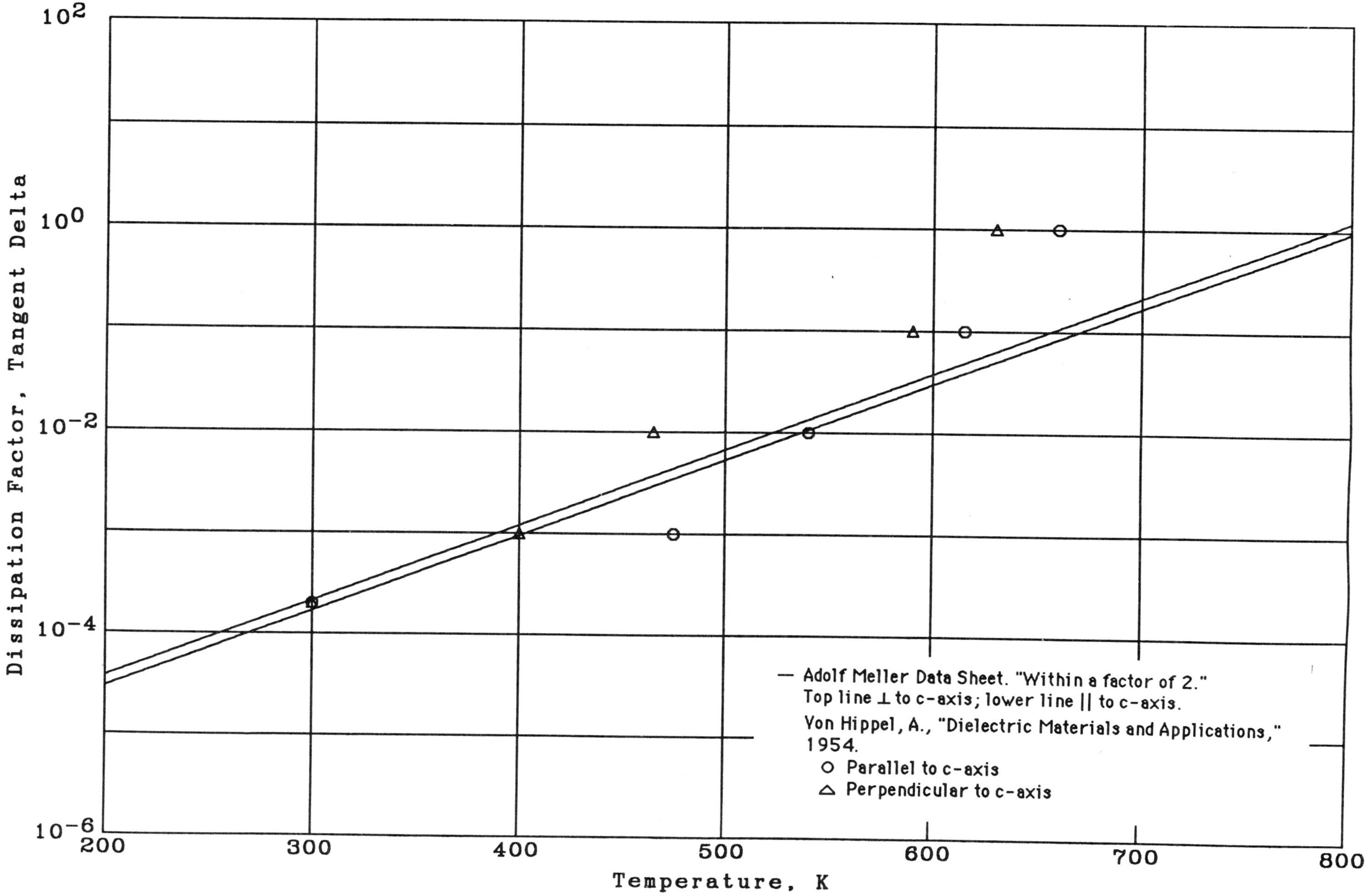

10^2
10^0
10^{-2}
10^{-4}
10^{-6}
Dissipation Factor, Tangent Delta
200
300
400
500
600
700
800
Temperature, K
— Adolf Meller Data Sheet. "Within a factor of 2."
Top line ⊥ to c-axis; lower line || to c-axis.
Von Hippel, A., "Dielectric Materials and Applications," 1954.
○ Parallel to c-axis
△ Perpendicular to c-axis

SAPPHIRE

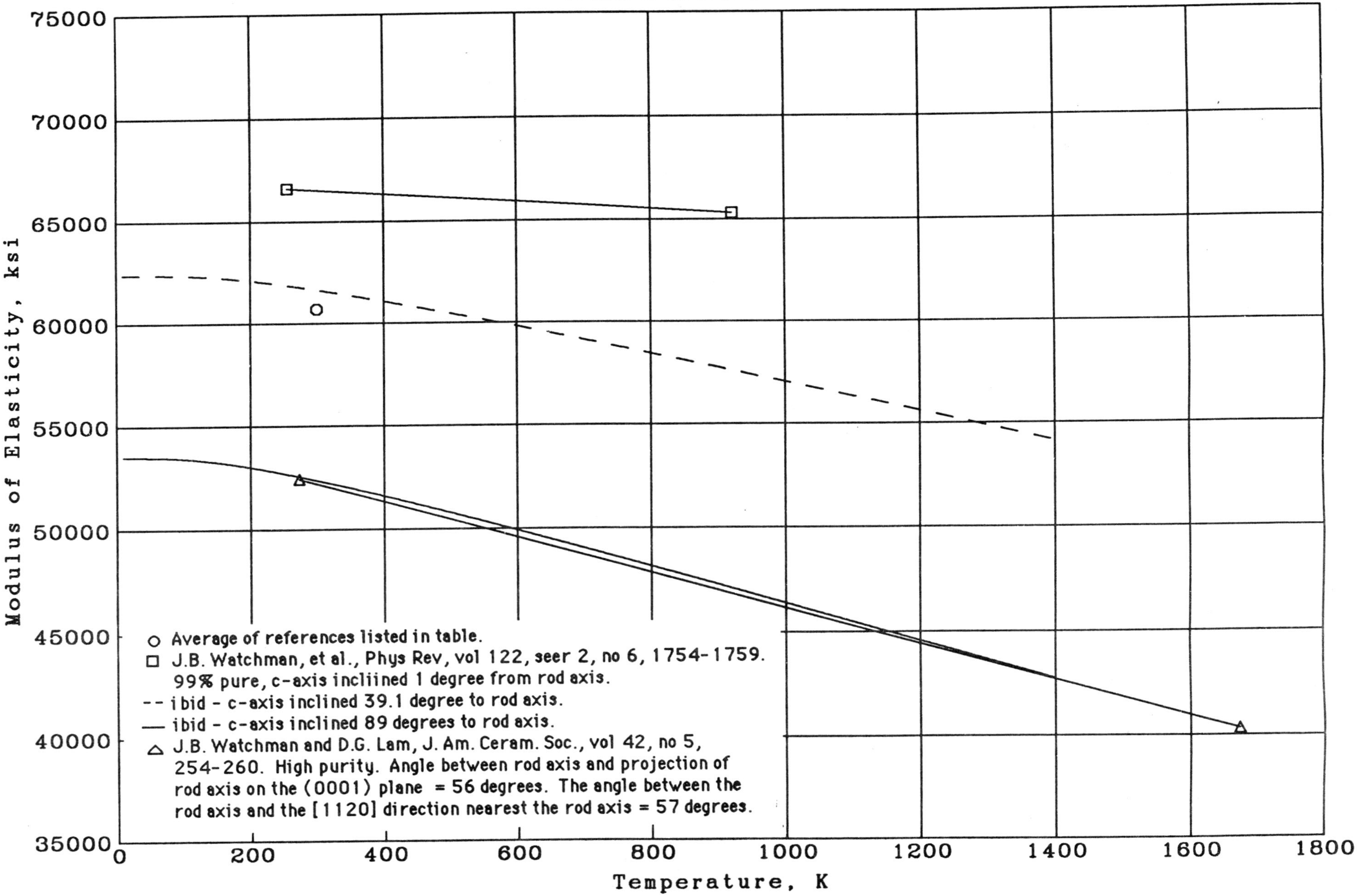

75000
70000
65000
60000
55000
50000
45000
40000
35000
Modulus of Elasticity, ksi
0
200
400
600
800
1000
1200
1400
1600
1800
Temperature, K
○ Average of references listed in table.
□ J.B. Watchman, et al., Phys Rev, vol 122, seer 2, no 6, 1754-1759. 99% pure, c-axis incliined 1 degree from rod axis.
-- ibid - c-axis inclined 39.1 degree to rod axis.
— ibid - c-axis inclined 89 degrees to rod axis.
△ J.B. Watchman and D.G. Lam, J. Am. Ceram. Soc., vol 42, no 5, 254-260. High purity. Angle between rod axis and projection of rod axis on the (0001) plane = 56 degrees. The angle between the rod axis and the [1120] direction nearest the rod axis = 57 degrees.

SAPPHIRE

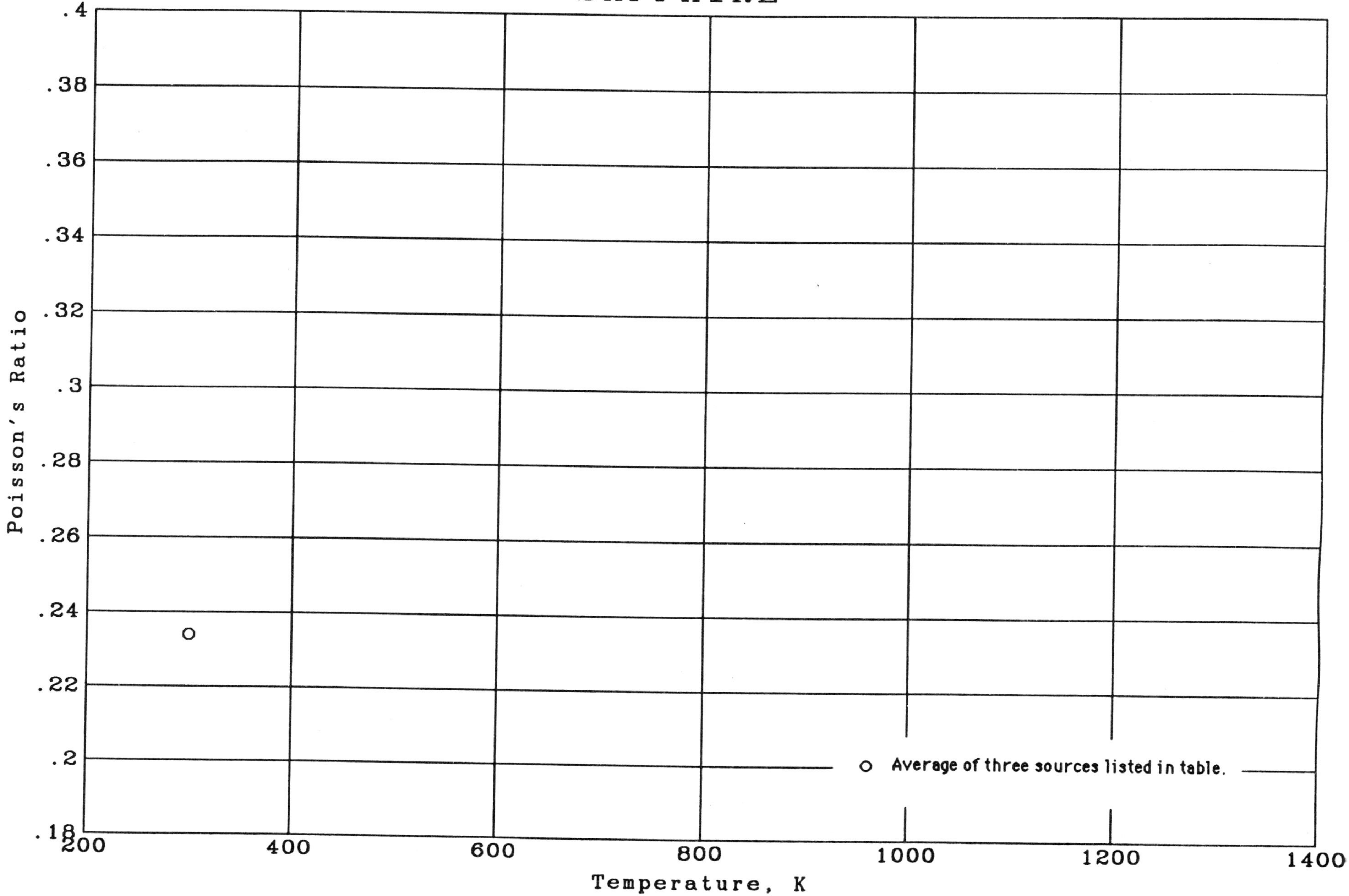

.4
.38
.36
.34
.32
.3
.28
.26
.24
.22
.2
.18
Poisson's Ratio
200
400
600
800
1000
1200
1400
Temperature, K
o Average of three sources listed in table.

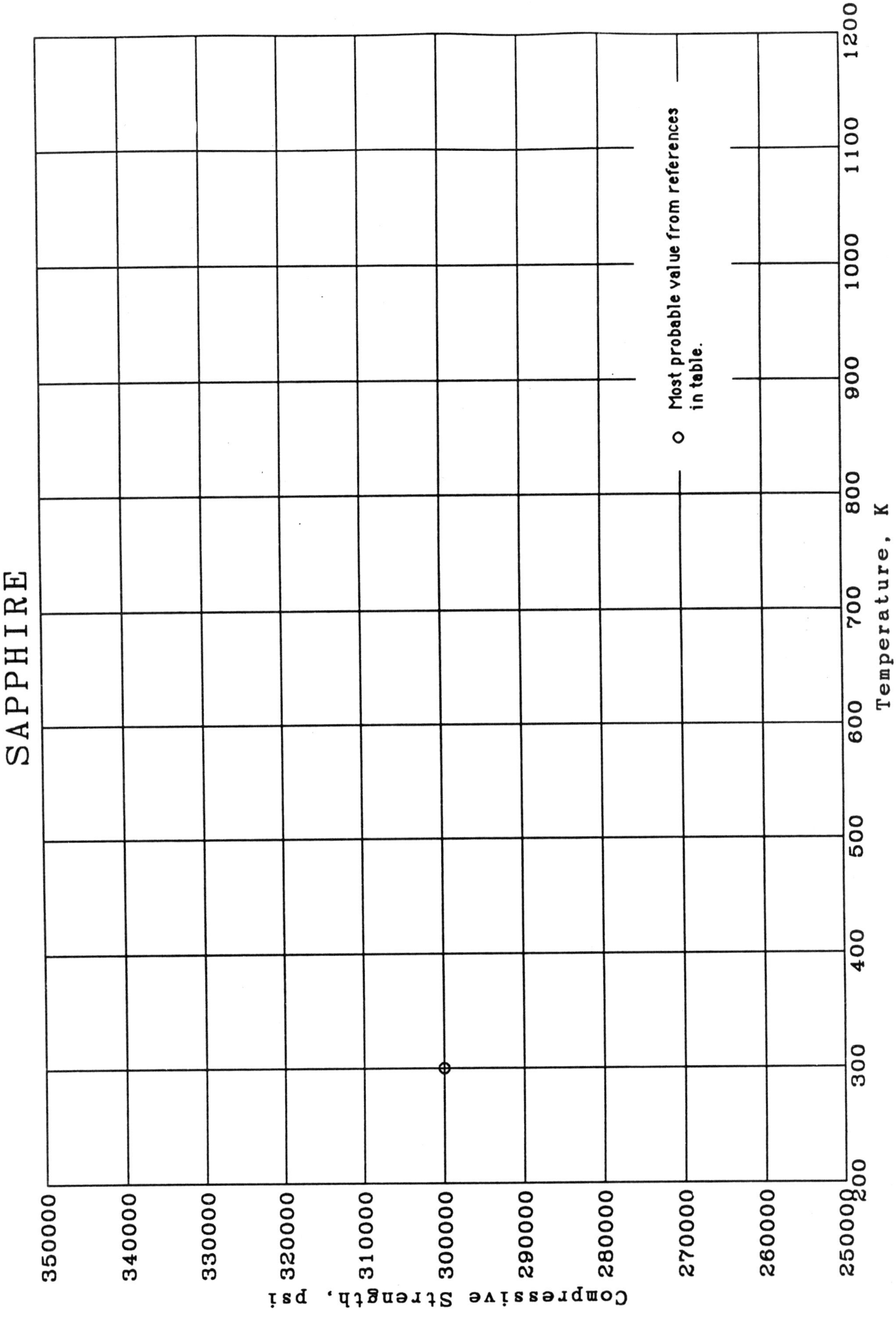
SAPPHIRE
Most probable value from references in table.
Compressive Strength, psi
350000
340000
330000
320000
310000
300000
290000
280000
270000
260000
250000
Temperature, K
200
300
400
500
600
700
800
900
1000
1100
1200

SAPPHIRE

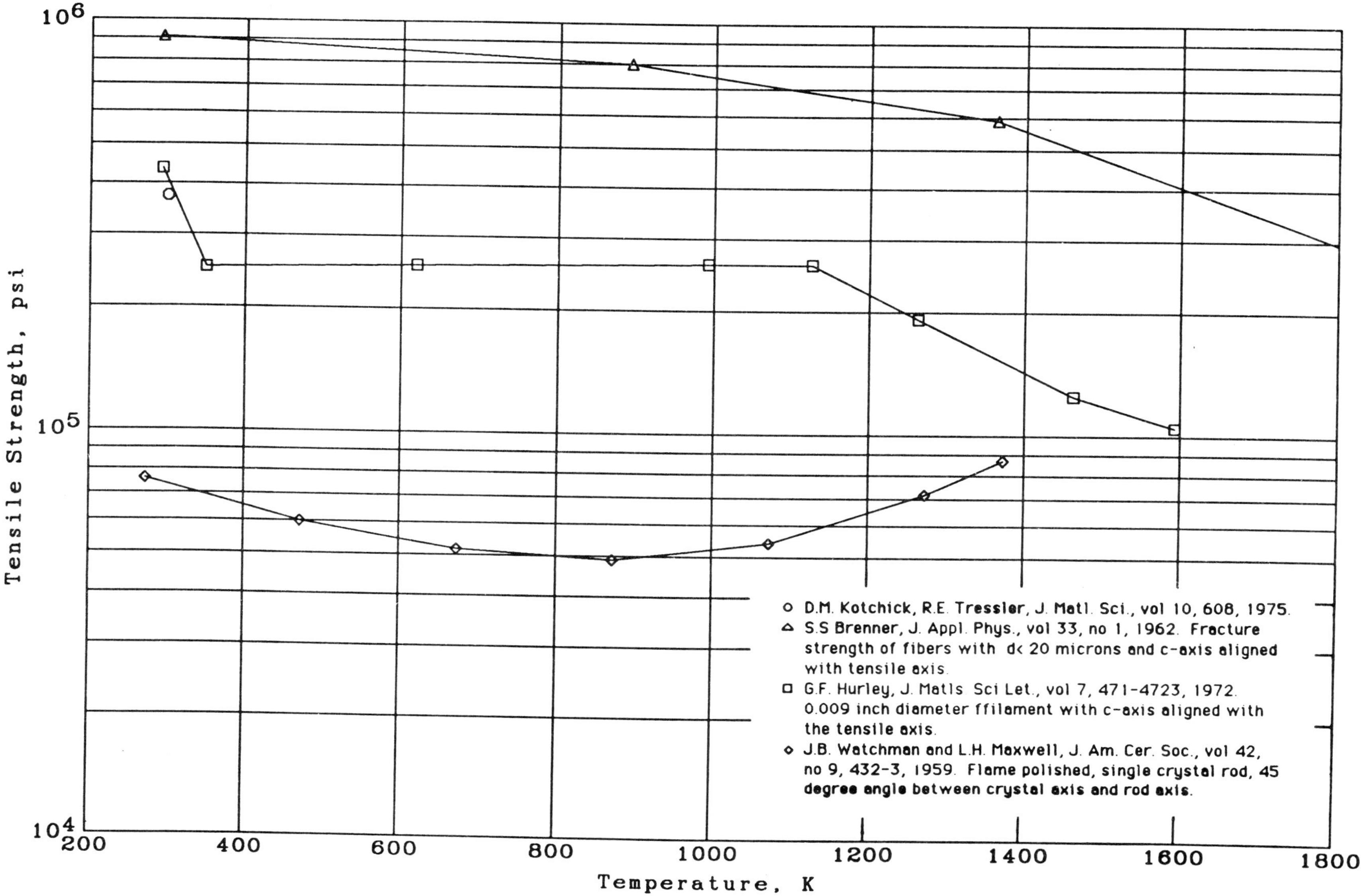

10^6
10^5
10^4
Tensile Strength, psi
200
400
600
800
1000
1200
1400
1600
1800
Temperature, K
○ D.M. Kotchick, R.E. Tressler, J. Matl. Sci., vol 10, 608, 1975.
△ S.S Brenner, J. Appl. Phys., vol 33, no 1, 1962. Fracture strength of fibers with d< 20 microns and c-axis aligned with tensile axis.
□ G.F. Hurley, J. Matls. Sci Let., vol 7, 471-4723, 1972. 0.009 inch diameter ffilament with c-axis aligned with the tensile axis.
◇ J.B. Watchman and L.H. Maxwell, J. Am. Cer. Soc., vol 42, no 9, 432-3, 1959. Flame polished, single crystal rod, 45 degree angle between crystal axis and rod axis.

SAPPHIRE

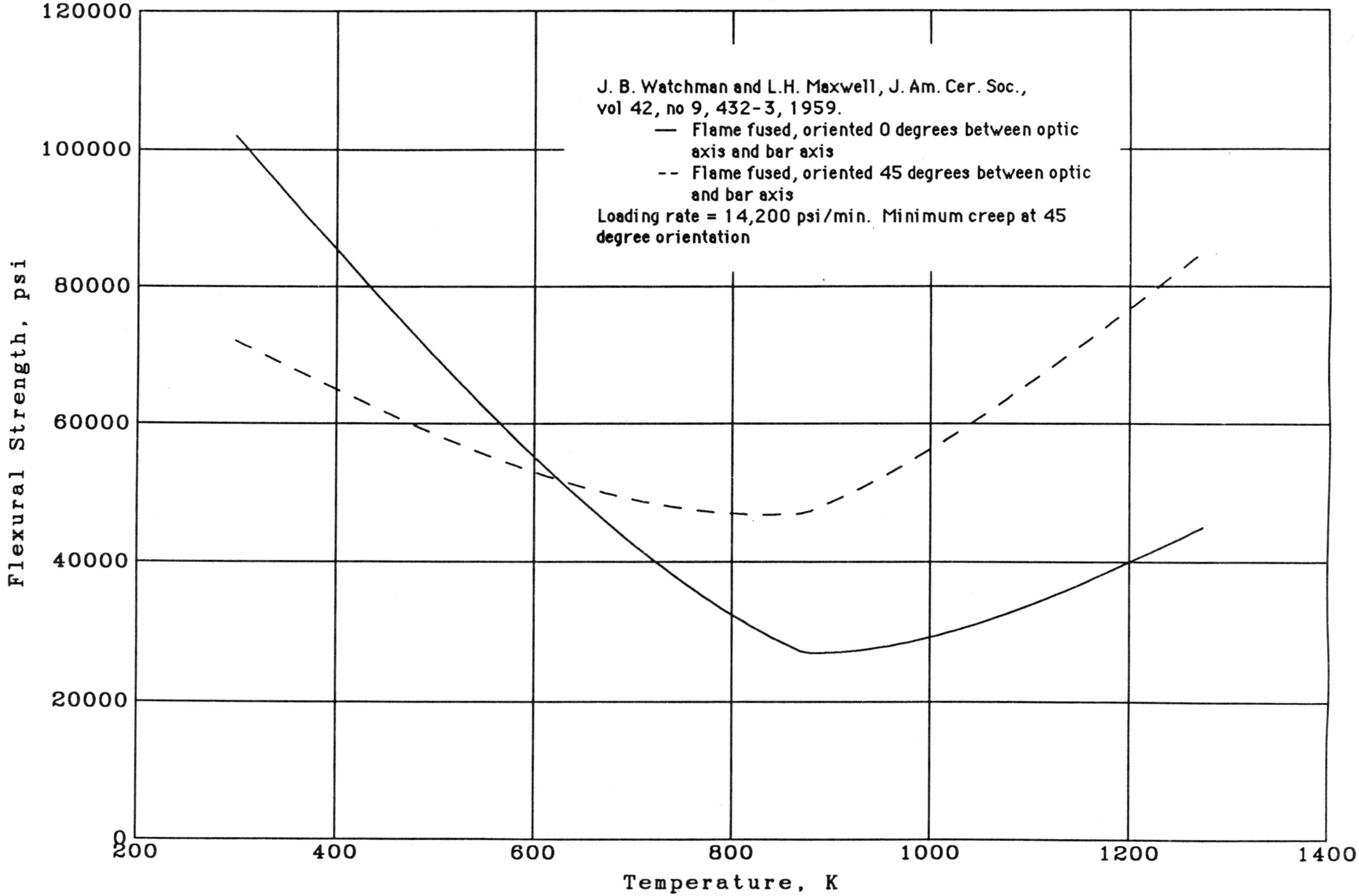

J. B. Watchman and L.H. Maxwell, J. Am. Cer. Soc., vol 42, no 9, 432-3, 1959.
— Flame fused, oriented 0 degrees between optic axis and bar axis
-- Flame fused, oriented 45 degrees between optic and bar axis
Loading rate = 14,200 psi/min. Minimum creep at 45 degree orientation
120000
100000
80000
60000
40000
20000
0
Flexural Strength, psi
200
400
600
800
1000
1200
1400
Temperature, K

SAPPHIRE

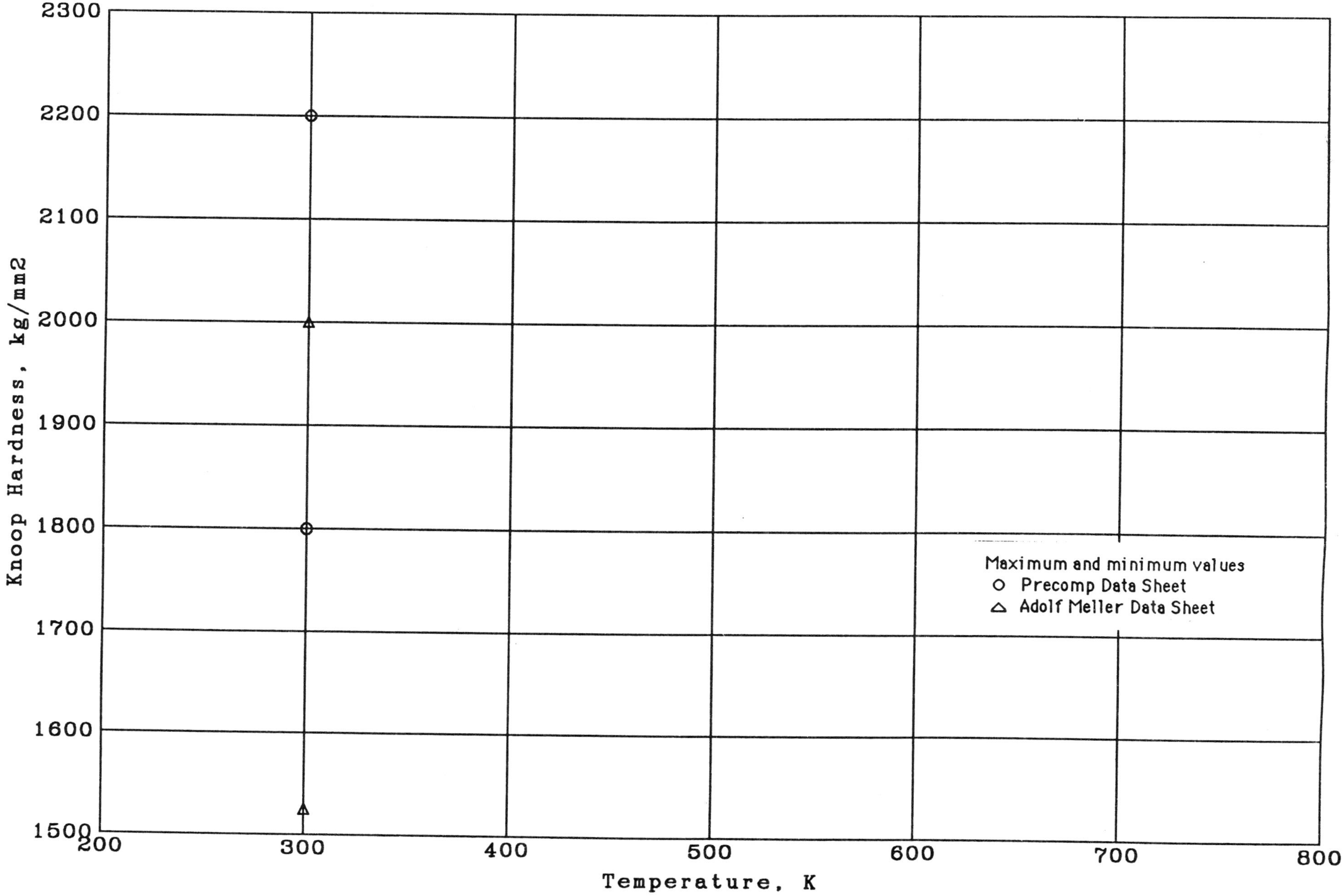

Knoop Hardness, kg/mm2
2300
2200
2100
2000
1900
1800
1700
1600
1500
200
300
400
500
600
700
800
Temperature, K
Maximum and minimum values
Precomp Data Sheet
Adolf Meller Data Sheet

BERYLLIA

PROPERTY	@ 300K	UNITS	REFERENCES
Density	2.90	g/cm^3	B1 B2 B3
Specific Heat	0.243	cal/g-K	B4 B5 B6
Thermal Conductivity	0.650	cal/cm-s-K	B7 B8 B9
Thermal Exp Coef	6.3×10^{-6}	1/K	B1 B2 B10
Dielectric Constant	6.8	------	B1 B2 B3
Dissipation Factor	.0004-.001	------	B1 B2 B3
Dielectric Strength	260	V/mil	B1 B2 B3
Young's Modulus	50,000	ksi	B1 B11 B12
Poisson's Ratio	0.21	------	B1 B11 B12
Flexural Strength	40,000	psi	B1 B2 B3
Ult Tensile Strength	20,000	psi	B1 B2 B3
Compressive Strength	250,000	psi	B1 B2 B3
Hardness	67	Rock 45N	B1 B2 B3

BERYLLIA

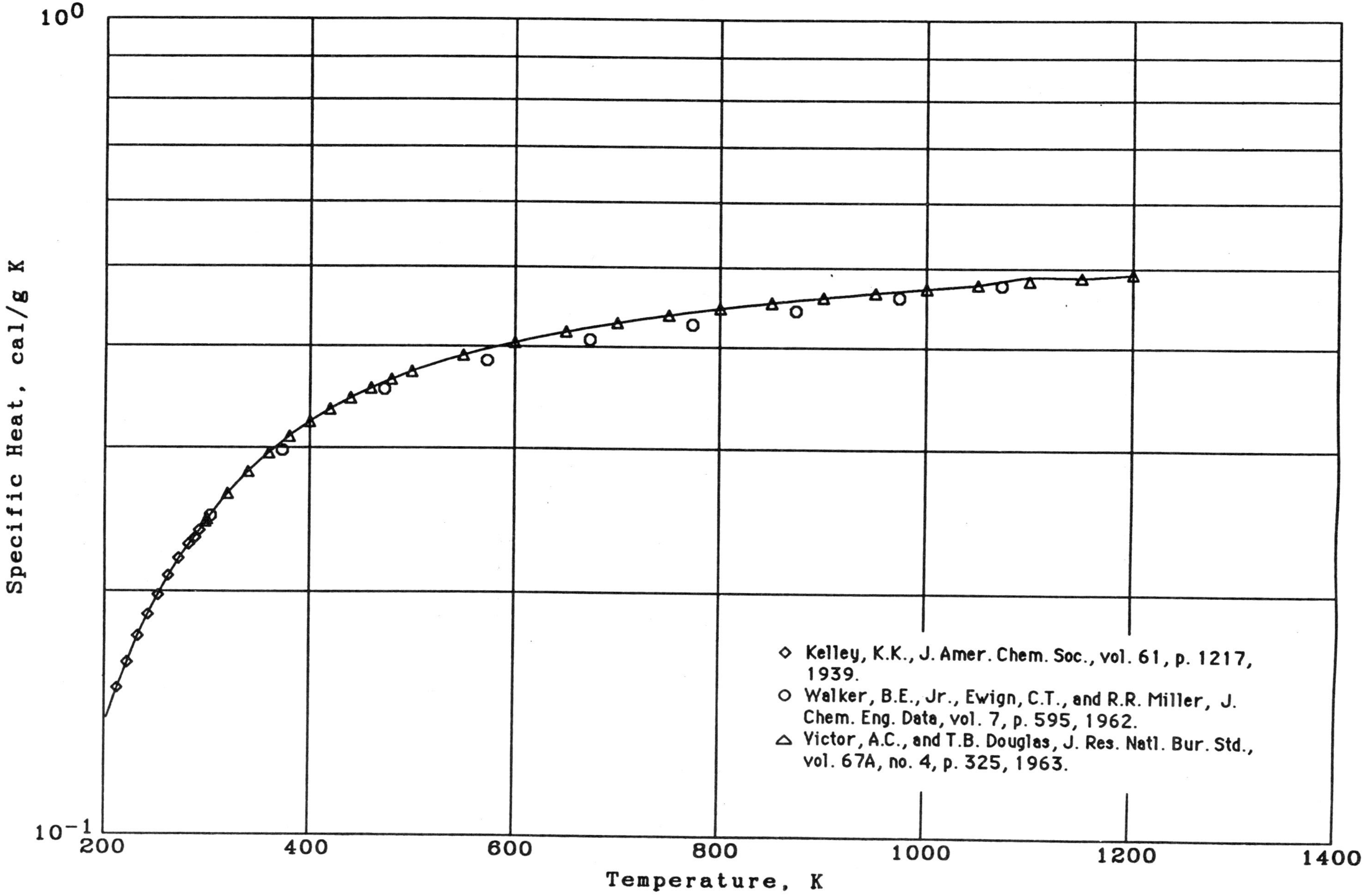

Specific Heat, cal/g K
10^0
10^-1
200
400
600
800
1000
1200
1400
Temperature, K
◇ Kelley, K.K., J. Amer. Chem. Soc., vol. 61, p. 1217, 1939.
○ Walker, B.E., Jr., Ewign, C.T., and R.R. Miller, J. Chem. Eng. Data, vol. 7, p. 595, 1962.
△ Victor, A.C., and T.B. Douglas, J. Res. Natl. Bur. Std., vol. 67A, no. 4, p. 325, 1963.

BERYLLIA

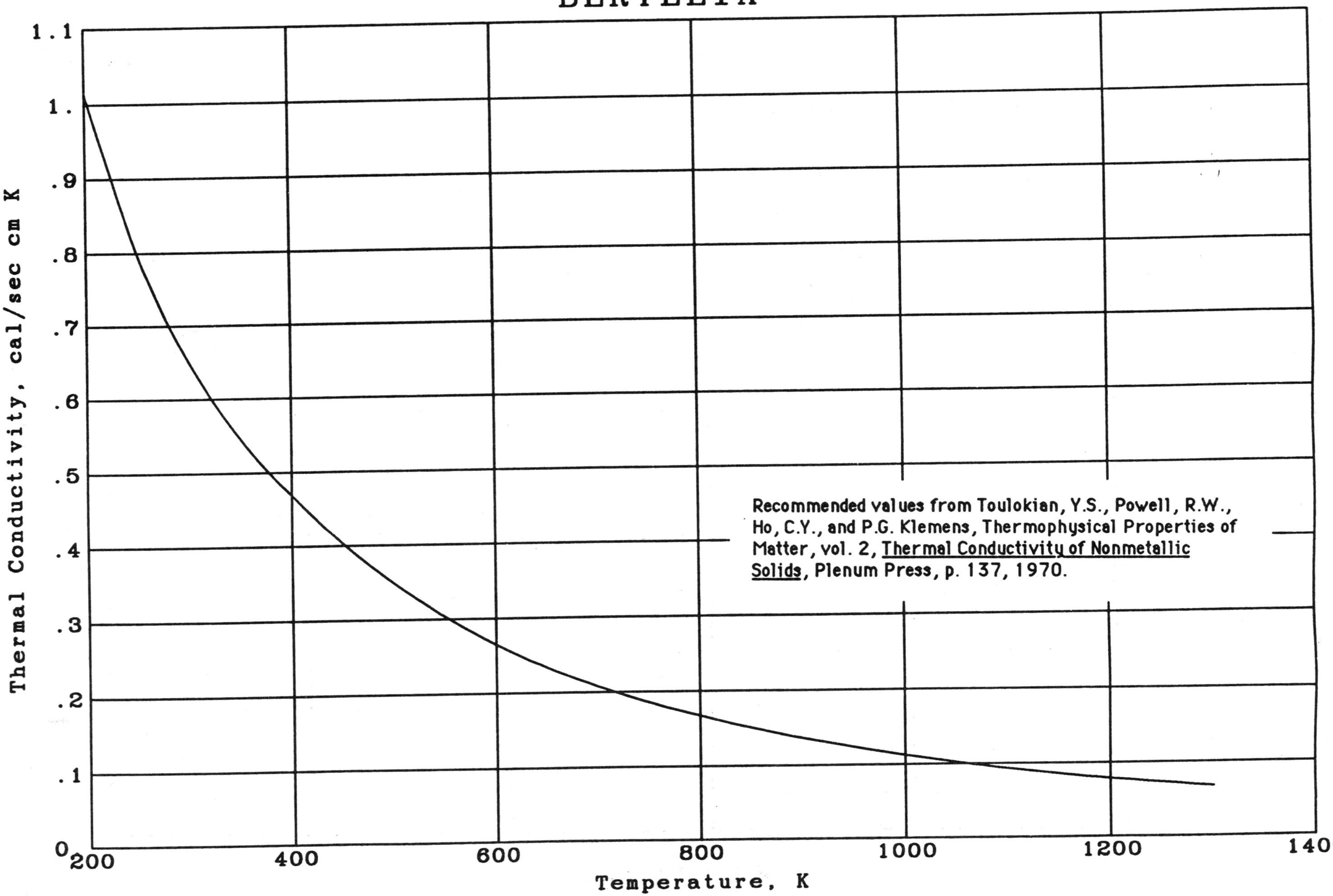
1.1
1.
.9
.8
.7
.6
.5
.4
.3
.2
.1
0
Thermal Conductivity, cal/sec cm K
200
400
600
800
1000
1200
1400
Temperature, K
Recommended values from Touloukian, Y.S., Powell, R.W., Ho, C.Y., and P.G. Klemens, Thermophysical Properties of Matter, vol. 2, Thermal Conductivity of Nonmetallic Solids, Plenum Press, p. 137, 1970.

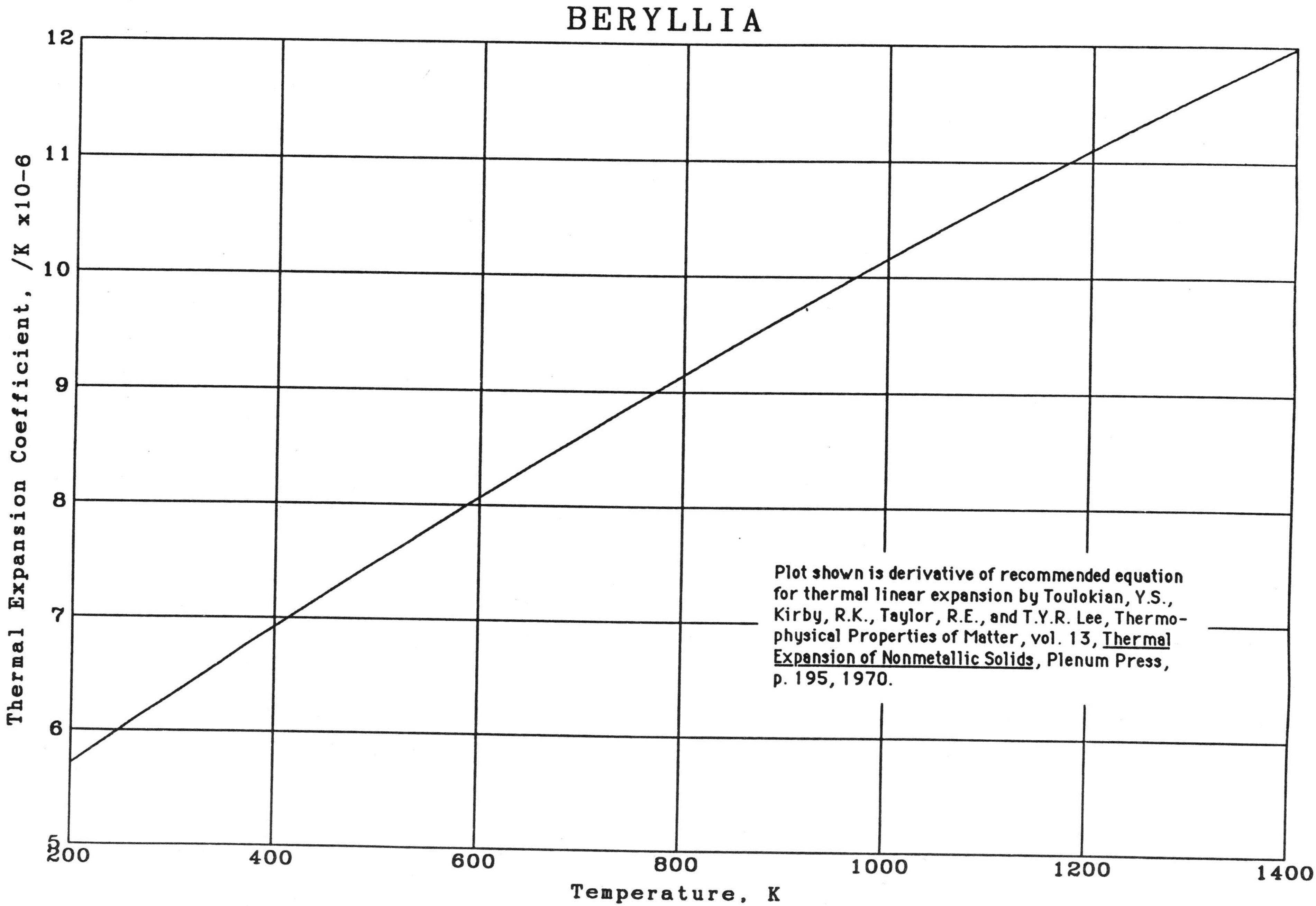
BERYLLIA
Thermal Expansion Coefficient, /K x10-6
Temperature, K
12
11
10
9
8
7
6
5
200
400
600
800
1000
1200
1400
Plot shown is derivative of recommended equation for thermal linear expansion by Toulokian, Y.S., Kirby, R.K., Taylor, R.E., and T.Y.R. Lee, Thermophysical Properties of Matter, vol. 13, Thermal Expansion of Nonmetallic Solids, Plenum Press, p. 195, 1970.

BERYLLIA

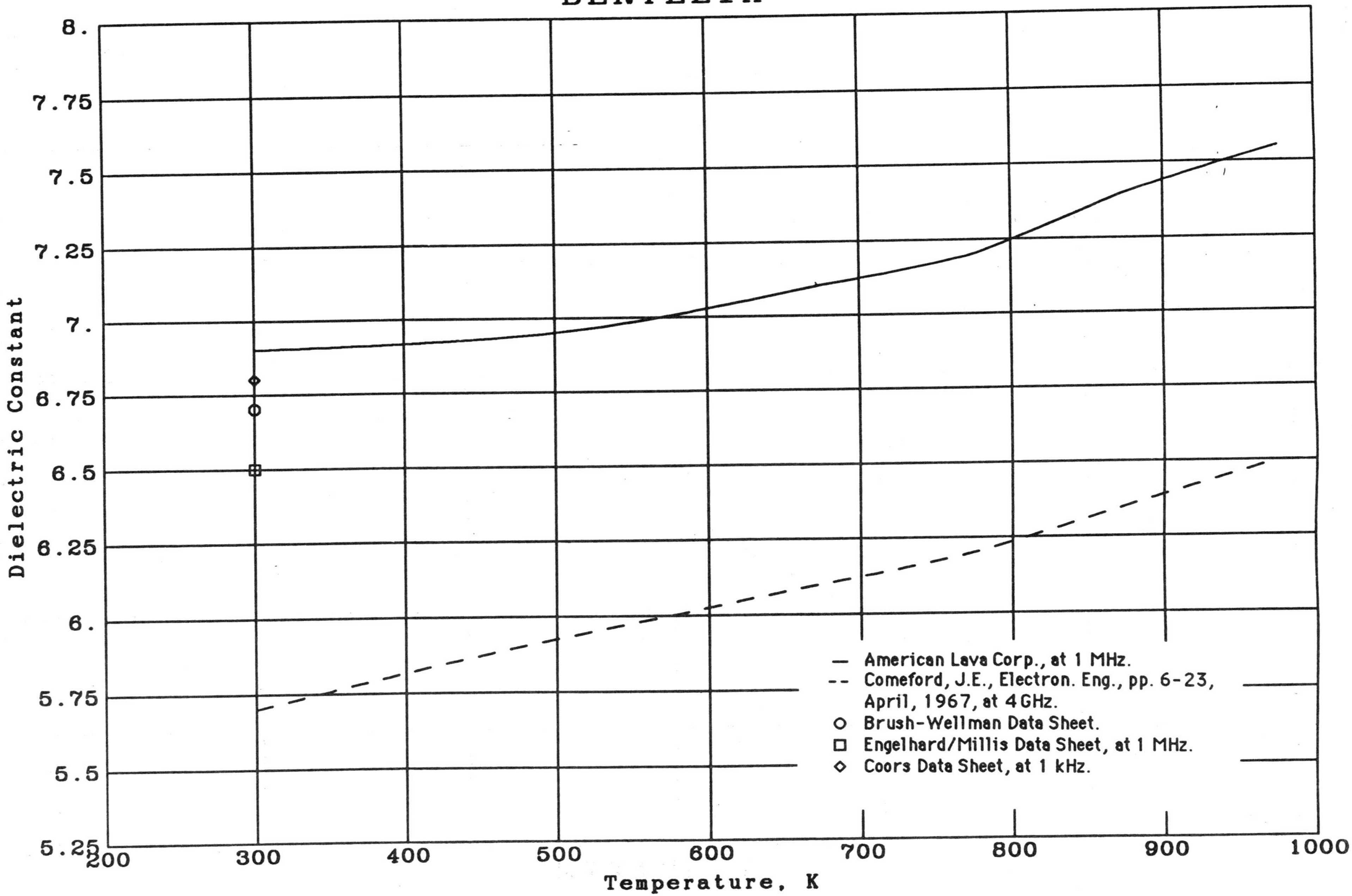

8.
7.75
7.5
7.25
7.
6.75
6.5
6.25
6.
5.75
5.5
5.25
Dielectric Constant
200
300
400
500
600
700
800
900
1000
Temperature, K
— American Lava Corp., at 1 MHz.
-- Comeford, J.E., Electron. Eng., pp. 6-23, April, 1967, at 4 GHz.
○ Brush-Wellman Data Sheet.
□ Engelhard/Millis Data Sheet, at 1 MHz.
◇ Coors Data Sheet, at 1 kHz.

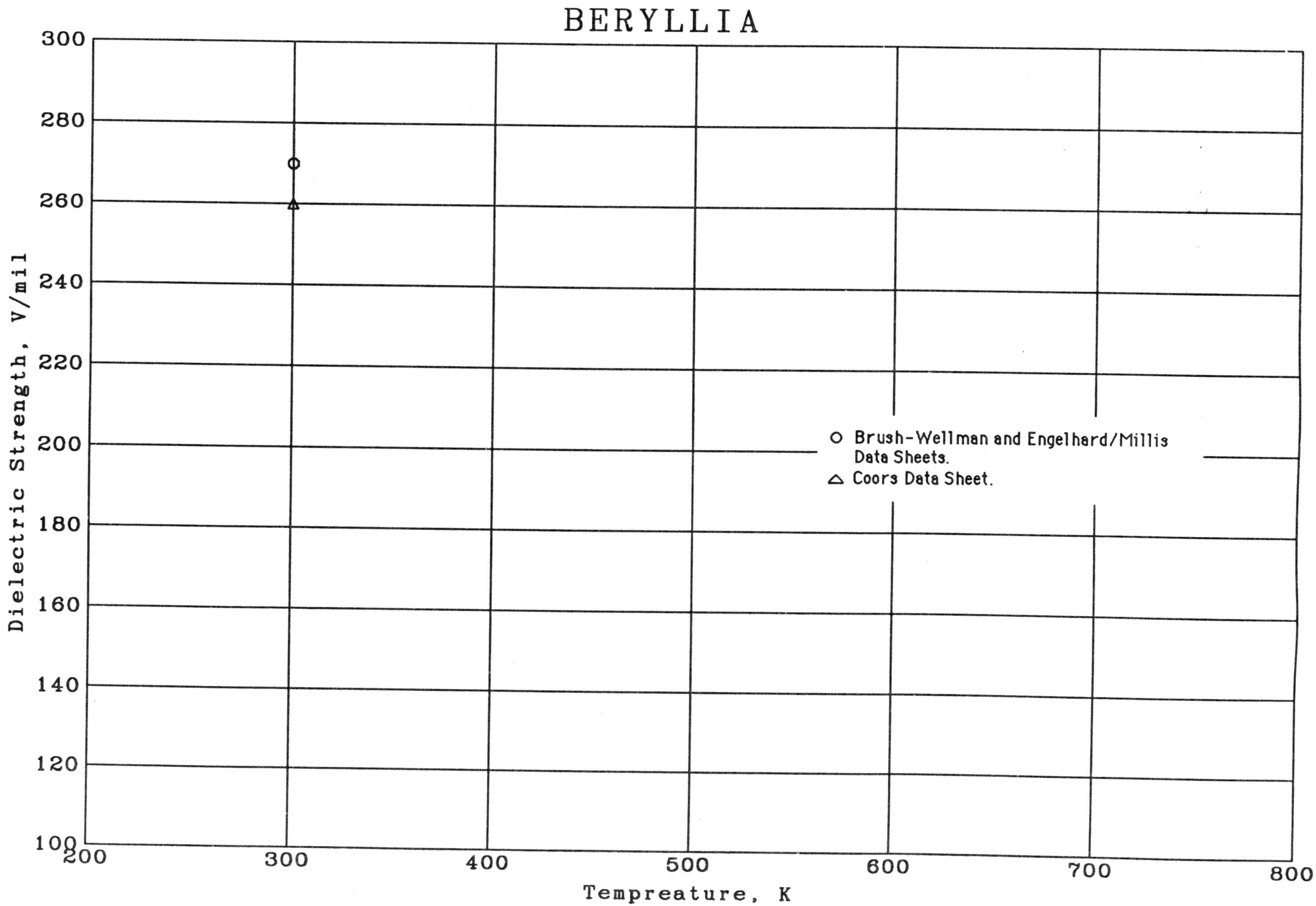
BERYLLIA
Dielectric Strength, V/mil
Tempreature, K
300
280
260
240
220
200
180
160
140
120
100
200
300
400
500
600
700
800
○ Brush-Wellman and Engelhard/Millis Data Sheets.
△ Coors Data Sheet.

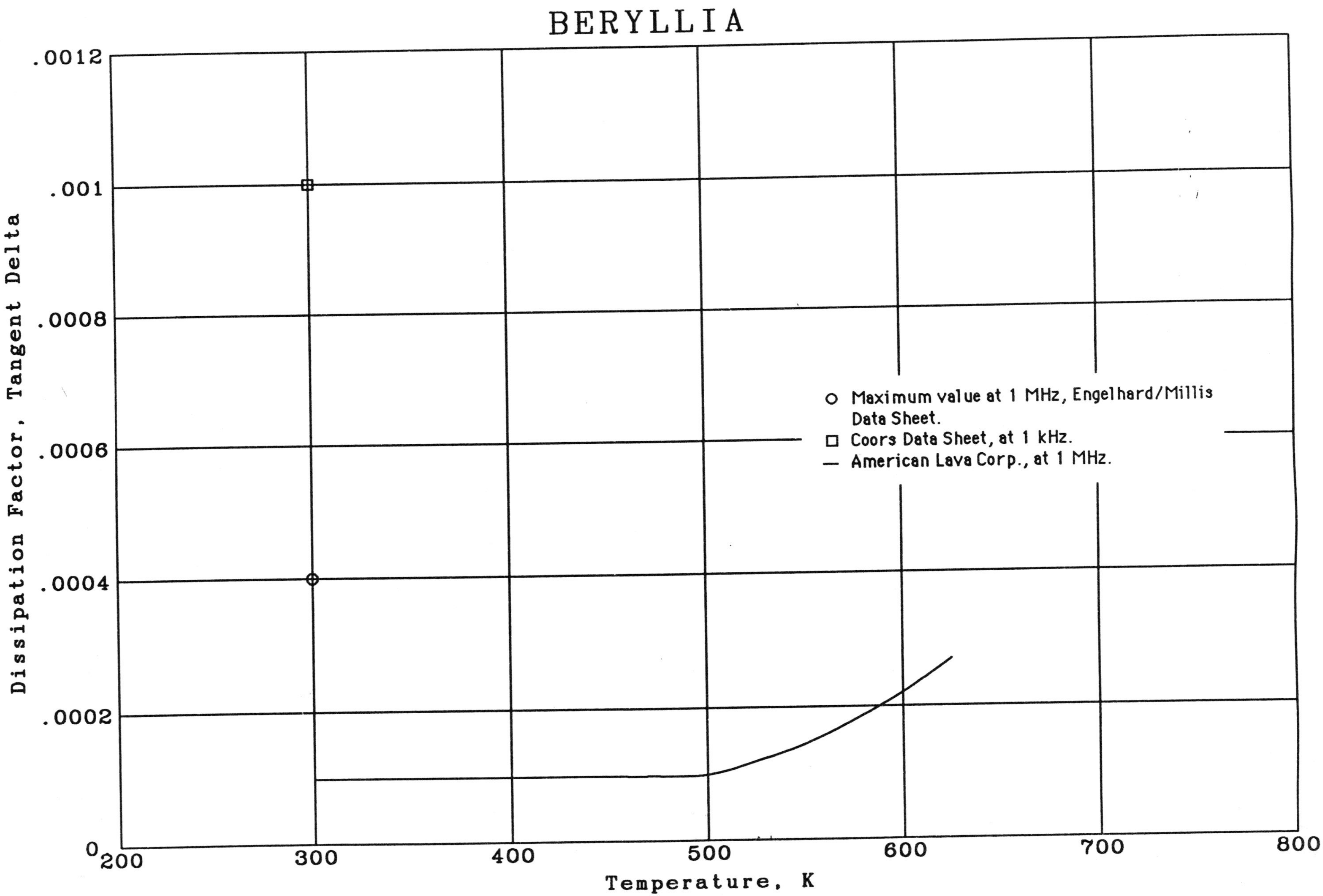

BERYLLIA
.0012
.001
.0008
.0006
.0004
.0002
0
200
300
400
500
600
700
800
Temperature, K
Dissipation Factor, Tangent Delta
○ Maximum value at 1 MHz, Engelhard/Millis Data Sheet.
□ Coors Data Sheet, at 1 kHz.
— American Lava Corp., at 1 MHz.

BERYLLIA

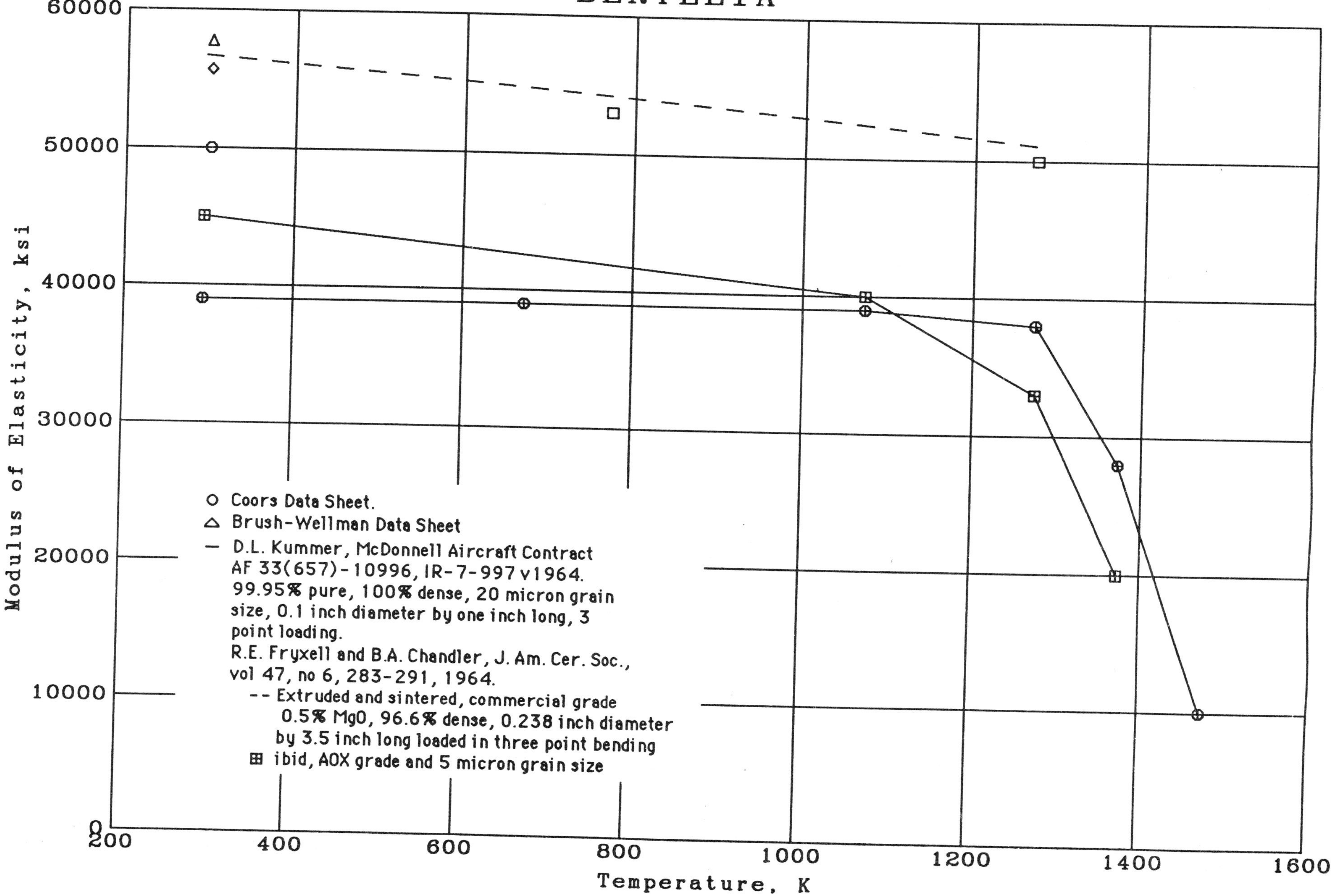

60000
50000
40000
30000
20000
10000
0
Modulus of Elasticity, ksi
200
400
600
800
1000
1200
1400
1600
Temperature, K
○ Coors Data Sheet.
△ Brush-Wellman Data Sheet
— D.L. Kummer, McDonnell Aircraft Contract AF 33(657)-10996, IR-7-997 v1964. 99.95% pure, 100% dense, 20 micron grain size, 0.1 inch diameter by one inch long, 3 point loading.
R.E. Fryxell and B.A. Chandler, J. Am. Cer. Soc., vol 47, no 6, 283-291, 1964.
-- Extruded and sintered, commercial grade 0.5% MgO, 96.6% dense, 0.238 inch diameter by 3.5 inch long loaded in three point bending
⊞ ibid, AOX grade and 5 micron grain size

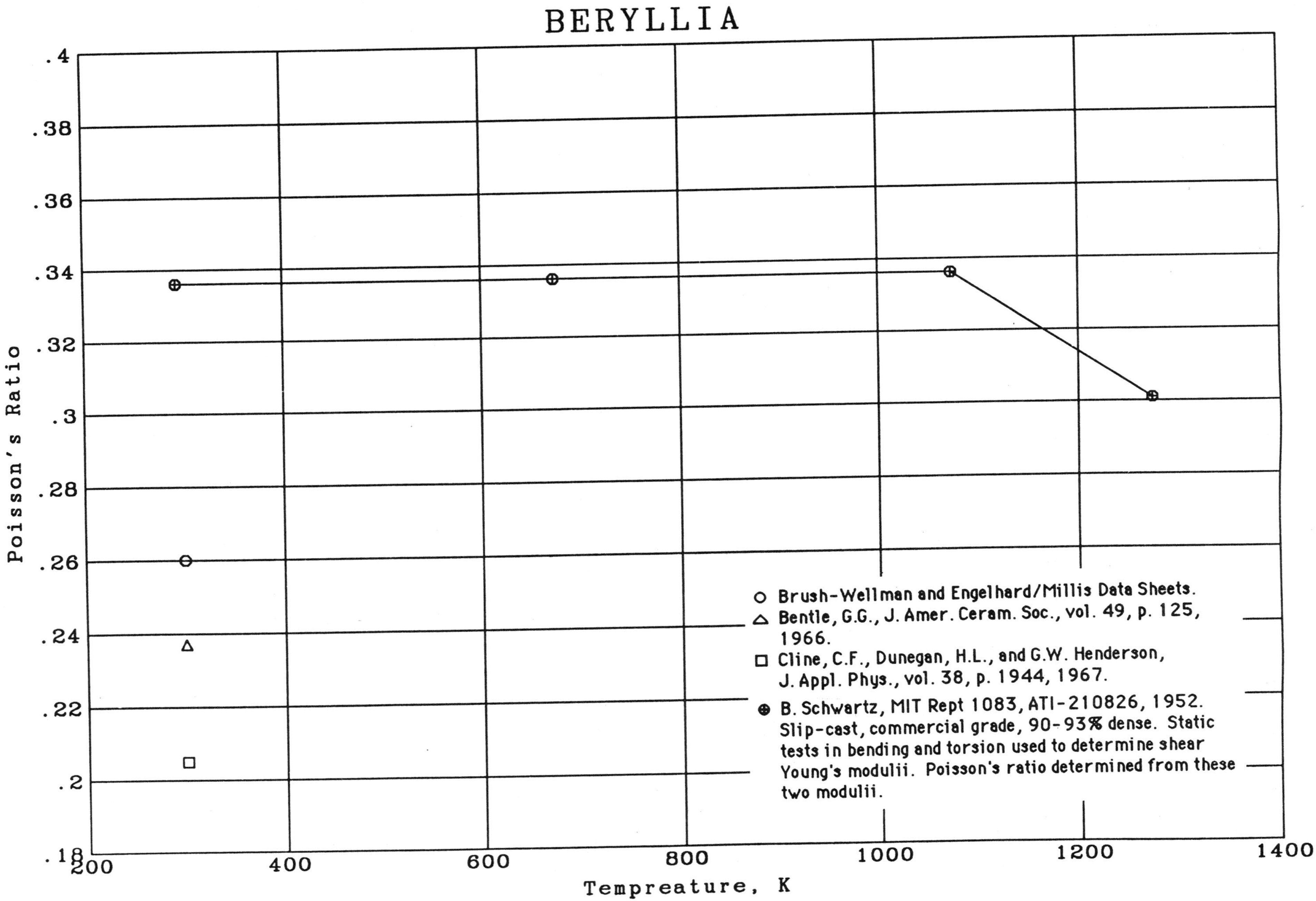
BERYLLIA
Poisson's Ratio
.4
.38
.36
.34
.32
.3
.28
.26
.24
.22
.2
.18
200
400
600
800
1000
1200
1400
Tempreature, K
○ Brush-Wellman and Engelhard/Millis Data Sheets.
△ Bentle, G.G., J. Amer. Ceram. Soc., vol. 49, p. 125, 1966.
□ Cline, C.F., Dunegan, H.L., and G.W. Henderson, J. Appl. Phys., vol. 38, p. 1944, 1967.
⊕ B. Schwartz, MIT Rept 1083, ATI-210826, 1952. Slip-cast, commercial grade, 90-93% dense. Static tests in bending and torsion used to determine shear Young's modulii. Poisson's ratio determined from these two modulii.

BERYLLIA

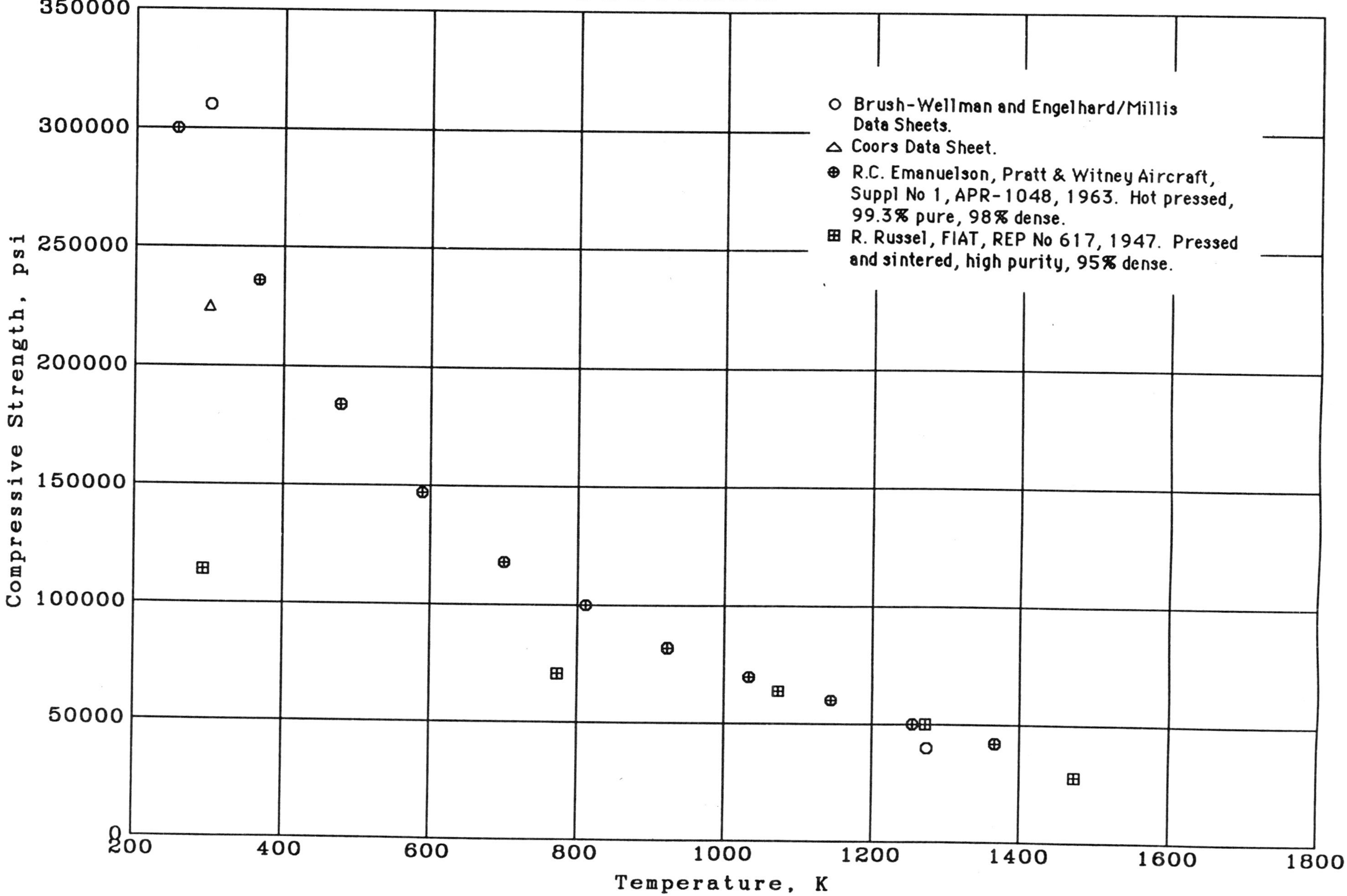

Brush-Wellman and Engelhard/Millis Data Sheets.
Coors Data Sheet.
R.C. Emanuelson, Pratt & Witney Aircraft, Suppl No 1, APR-1048, 1963. Hot pressed, 99.3% pure, 98% dense.
R. Russel, FIAT, REP No 617, 1947. Pressed and sintered, high purity, 95% dense.
Compressive Strength, psi
Temperature, K
350000
300000
250000
200000
150000
100000
50000
0
200
400
600
800
1000
1200
1400
1600
1800

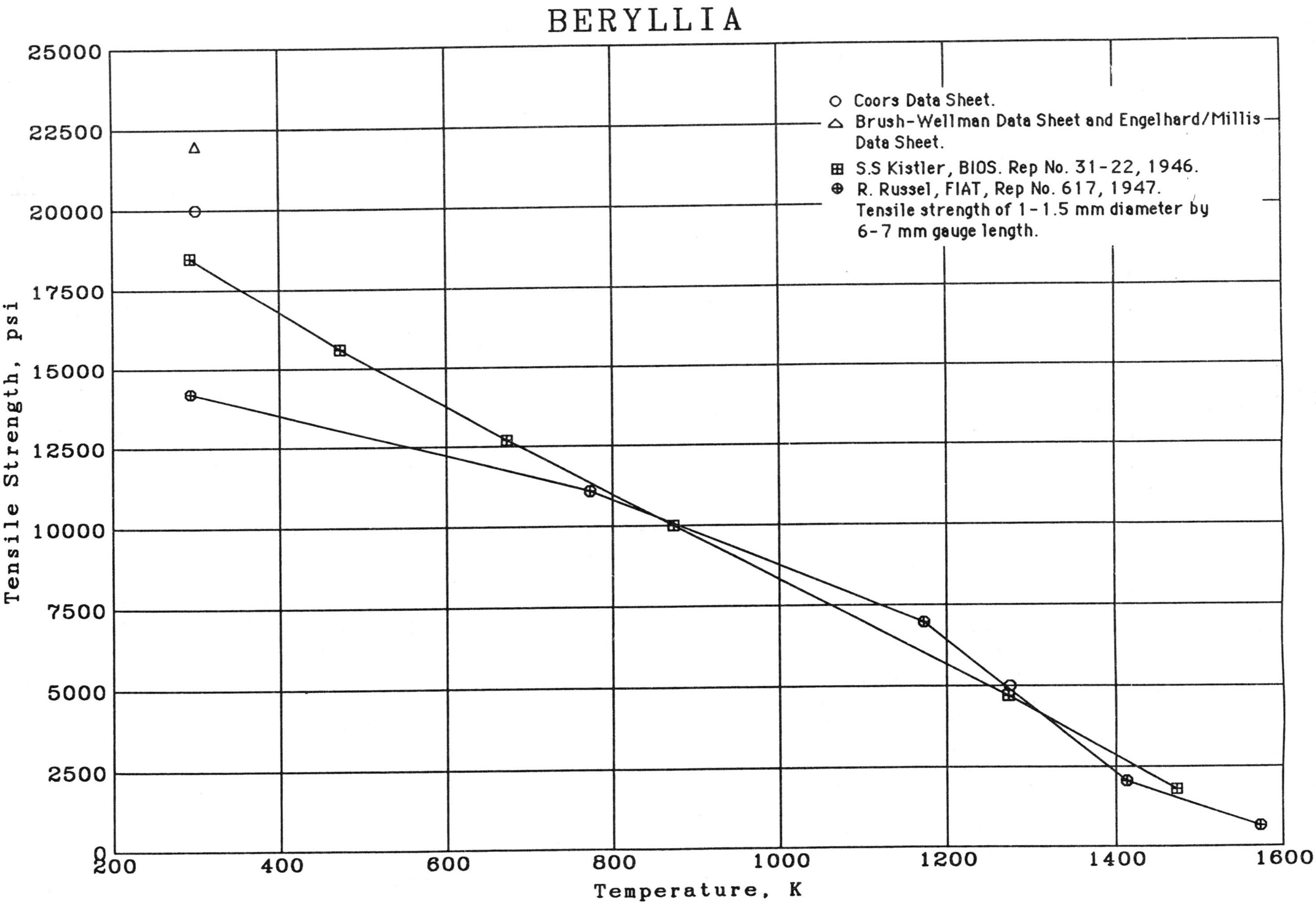
BERYLLIA
25000
22500
20000
17500
15000
12500
10000
7500
5000
2500
0
200
400
600
800
1000
1200
1400
1600
Tensile Strength, psi
Temperature, K
○ Coors Data Sheet.
△ Brush-Wellman Data Sheet and Engelhard/Millis Data Sheet.
⊞ S.S Kistler, BIOS. Rep No. 31-22, 1946.
⊕ R. Russel, FIAT, Rep No. 617, 1947. Tensile strength of 1-1.5 mm diameter by 6-7 mm gauge length.

BERYLLIA

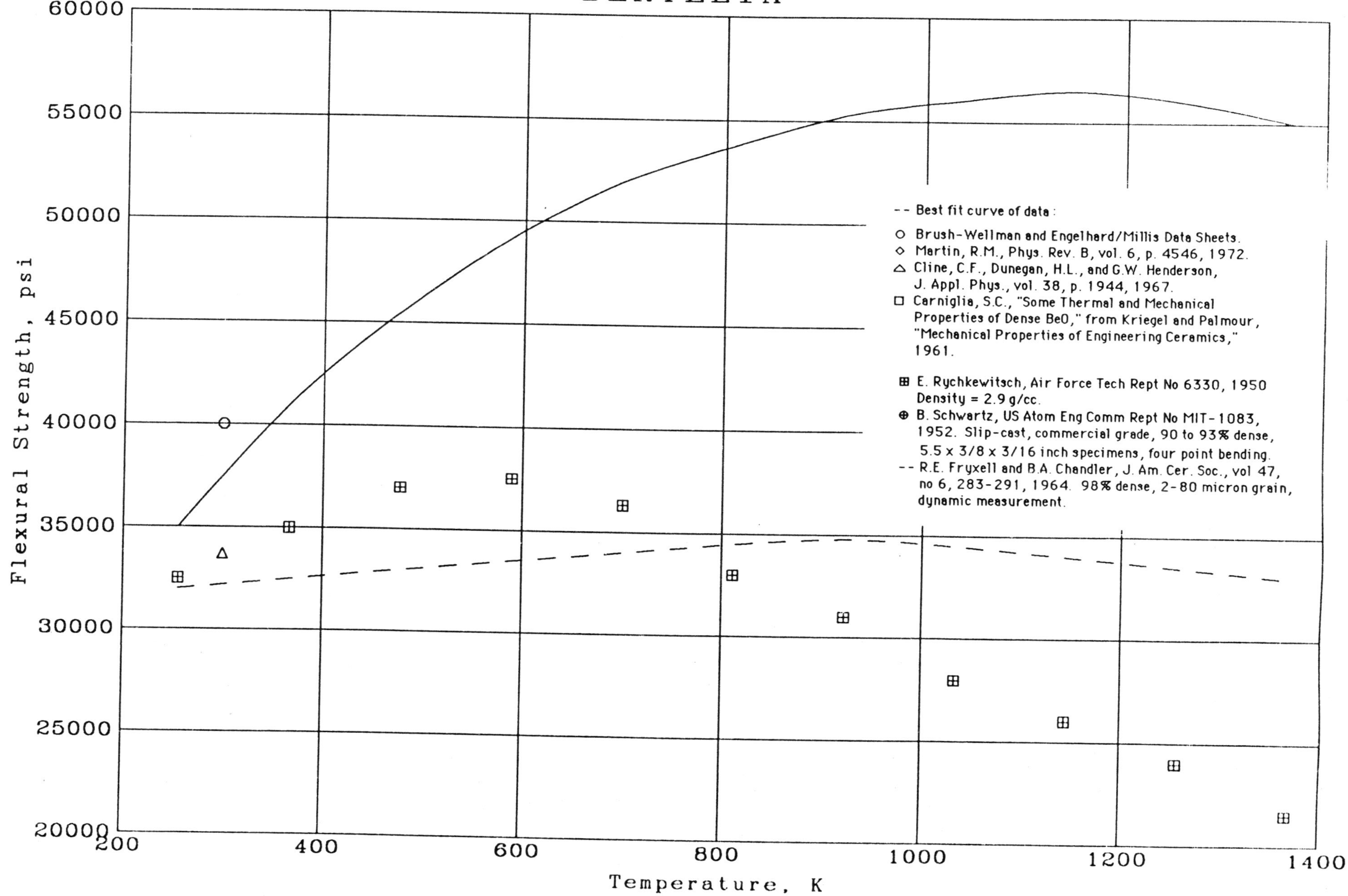
60000
55000
50000
45000
40000
35000
30000
25000
20000
Flexural Strength, psi
200
400
600
800
1000
1200
1400
Temperature, K
-- Best fit curve of data :
○ Brush-Wellman and Engelhard/Millis Data Sheets.
◇ Martin, R.M., Phys. Rev. B, vol. 6, p. 4546, 1972.
△ Cline, C.F., Dunegan, H.L., and G.W. Henderson, J. Appl. Phys., vol. 38, p. 1944, 1967.
□ Carniglia, S.C., "Some Thermal and Mechanical Properties of Dense BeO," from Kriegel and Palmour, "Mechanical Properties of Engineering Ceramics," 1961.
⊞ E. Rychkewitsch, Air Force Tech Rept No 6330, 1950 Density = 2.9 g/cc.
⊕ B. Schwartz, US Atom Eng Comm Rept No MIT-1083, 1952. Slip-cast, commercial grade, 90 to 93% dense, 5.5 x 3/8 x 3/16 inch specimens, four point bending.
-- R.E. Fryxell and B.A. Chandler, J. Am. Cer. Soc., vol 47, no 6, 283-291, 1964. 98% dense, 2-80 micron grain, dynamic measurement.

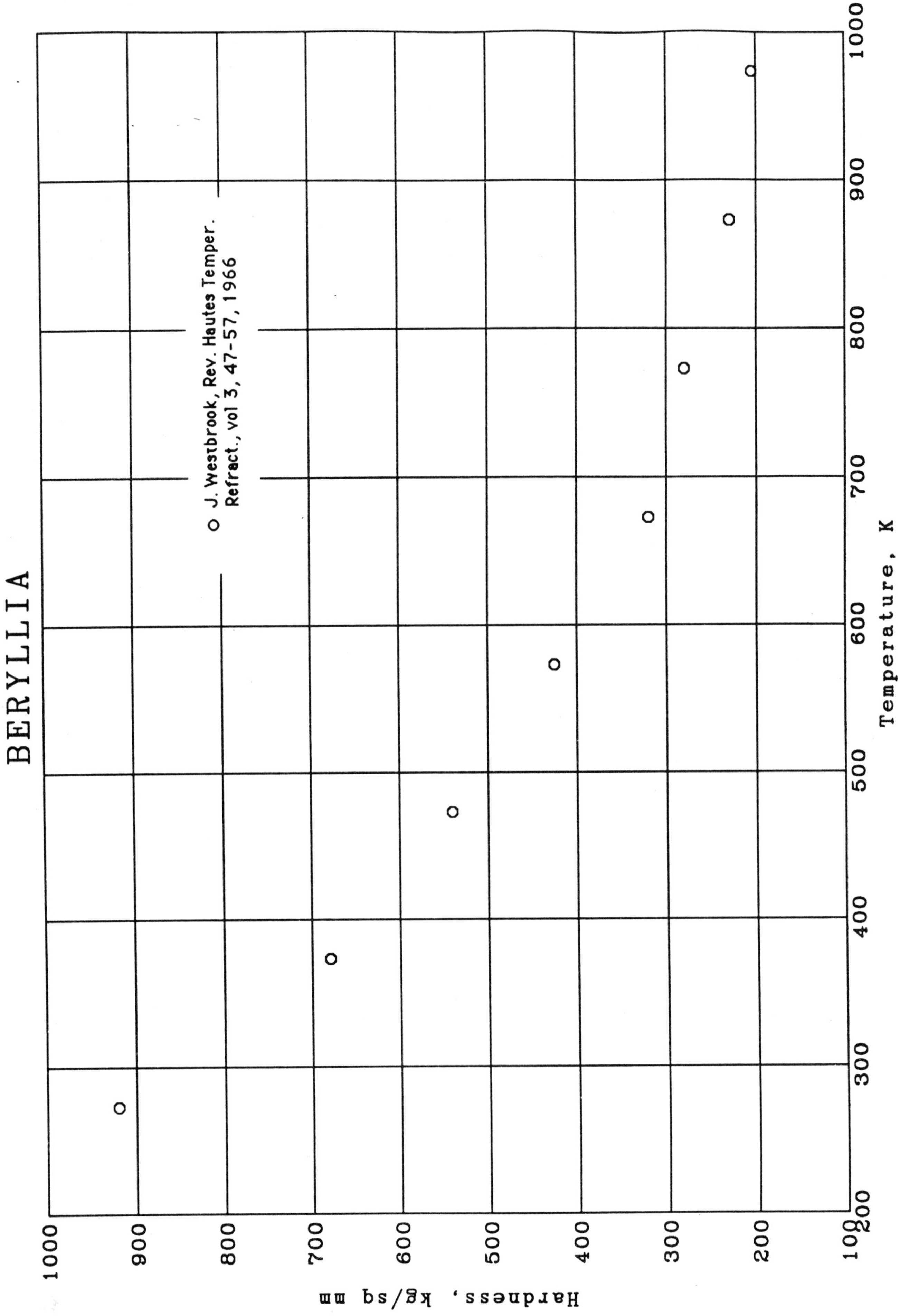
BERYLLIA
○ J. Westbrook, Rev. Hautes Temper. Refract., vol 3, 47-57, 1966
Temperature, K
Hardness, kg/sq mm
200
300
400
500
600
700
800
900
1000
100
200
300
400
500
600
700
800
900
1000

DIAMOND

PROPERTY	@ 300K	UNITS	REFERENCES
Density	3.515	g/cm^3	D1 D2 D3
Specific Heat	0.1233	cal/g-K	D4 D5 D6
Thermal Conductivity[1]	4.78	cal/cm-s-K	D7 D8 D9
Thermal Exp Coef	1.02×10^{-6}	1/K	D10 D11 D12
Dielectric Constant	5.70	------	D13 D14 D15
Dissipation Factor		------	
Dielectric Strength	25 - 3000	V/mil	D22 D23 D24
Young's Modulus	152,000	ksi	D16 D17 D18
Poisson's Ratio	0.104	------	D16 D17 D18
Flexural Strength	40,000	psi	D25 D26
Ult Tensile Strength	400-500,000	psi	D19 D20
Compressive Strength	1,260,000	psi	D21 D25 D26
Hardness	6-10,000	Knoop	D8 D15 D20

1 Thermal Conductivity is provided for type Ia diamonds. For type IIa the thermal conductivity = 5.521. For type IIb the thermal conductivity = 3.226.

DIAMOND

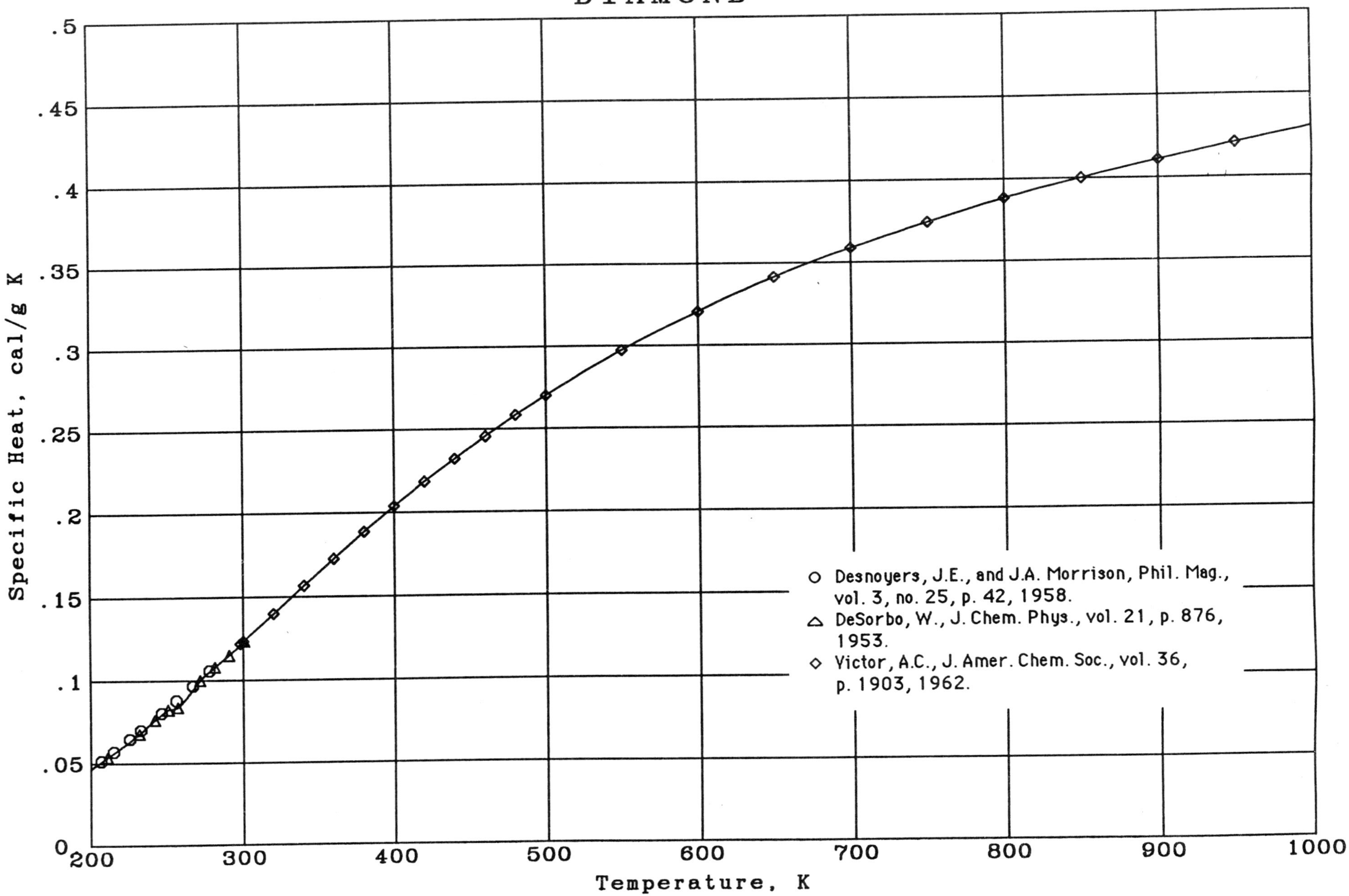
Specific Heat, cal/g K
Temperature, K
.5
.45
.4
.35
.3
.25
.2
.15
.1
.05
0
200
300
400
500
600
700
800
900
1000
○ Desnoyers, J.E., and J.A. Morrison, Phil. Mag., vol. 3, no. 25, p. 42, 1958.
△ DeSorbo, W., J. Chem. Phys., vol. 21, p. 876, 1953.
◇ Victor, A.C., J. Amer. Chem. Soc., vol. 36, p. 1903, 1962.

DIAMOND

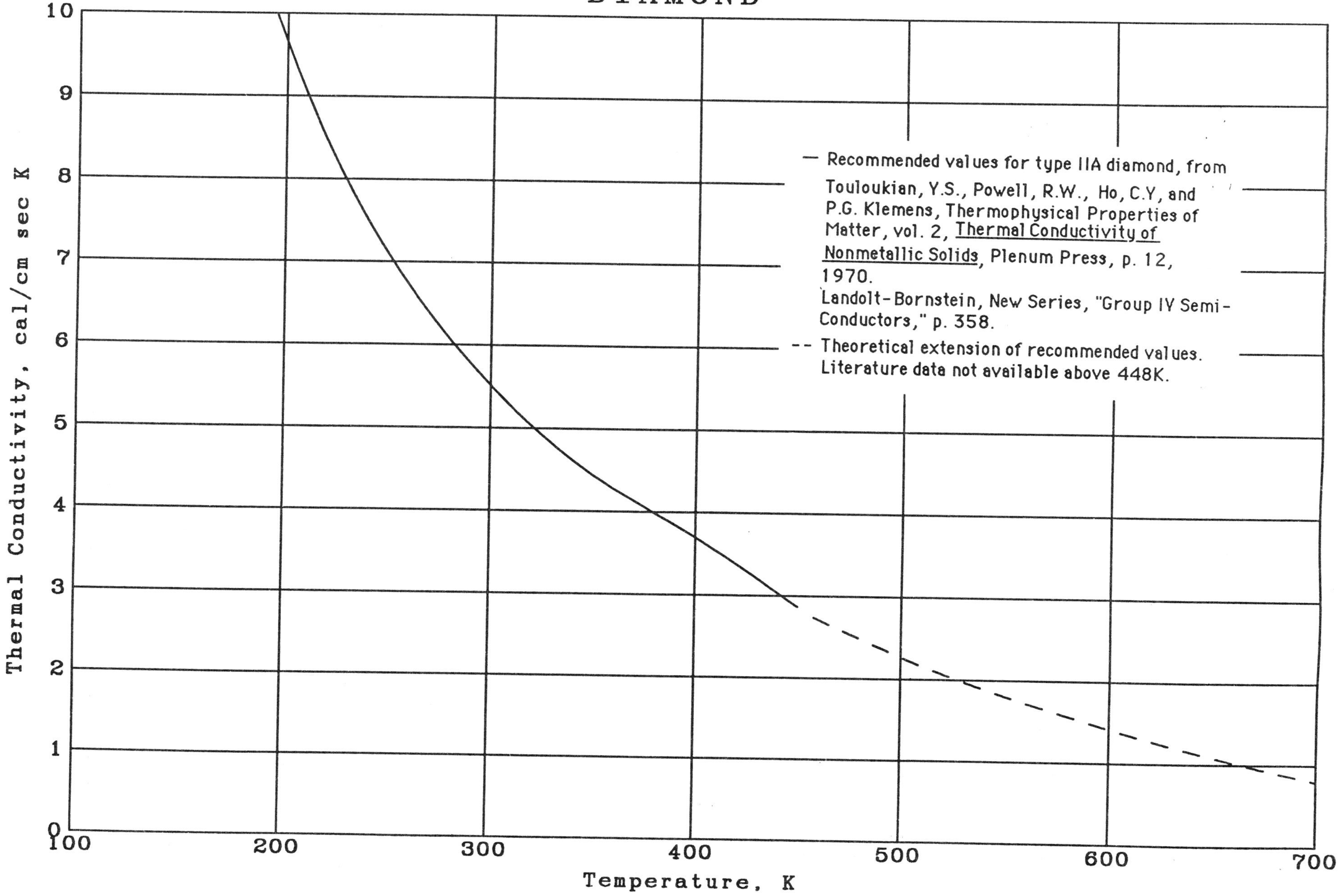

— Recommended values for type IIA diamond, from Touloukian, Y.S., Powell, R.W., Ho, C.Y, and P.G. Klemens, Thermophysical Properties of Matter, vol. 2, Thermal Conductivity of Nonmetallic Solids, Plenum Press, p. 12, 1970.
Landolt-Bornstein, New Series, "Group IV Semi-Conductors," p. 358.
-- Theoretical extension of recommended values. Literature data not available above 448K.
Thermal Conductivity, cal/cm sec K
Temperature, K
0
1
2
3
4
5
6
7
8
9
10
100
200
300
400
500
600
700

DIAMOND

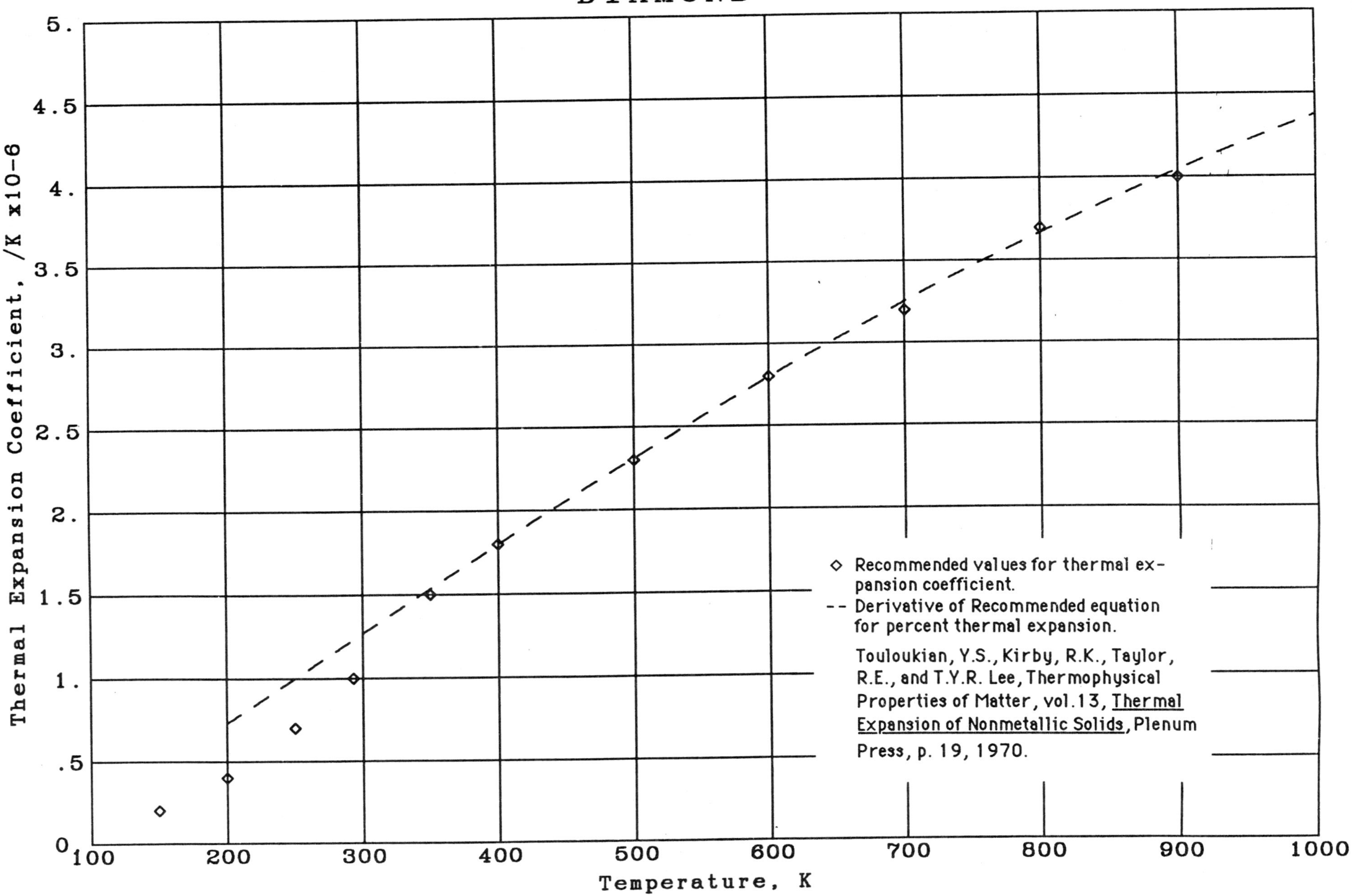

Thermal Expansion Coefficient, /K x10-6
5.
4.5
4.
3.5
3.
2.5
2.
1.5
1.
.5
0
100
200
300
400
500
600
700
800
900
1000
Temperature, K
◇ Recommended values for thermal expansion coefficient.
-- Derivative of Recommended equation for percent thermal expansion.
Touloukian, Y.S., Kirby, R.K., Taylor, R.E., and T.Y.R. Lee, Thermophysical Properties of Matter, vol.13, Thermal Expansion of Nonmetallic Solids, Plenum Press, p. 19, 1970.

DIAMOND

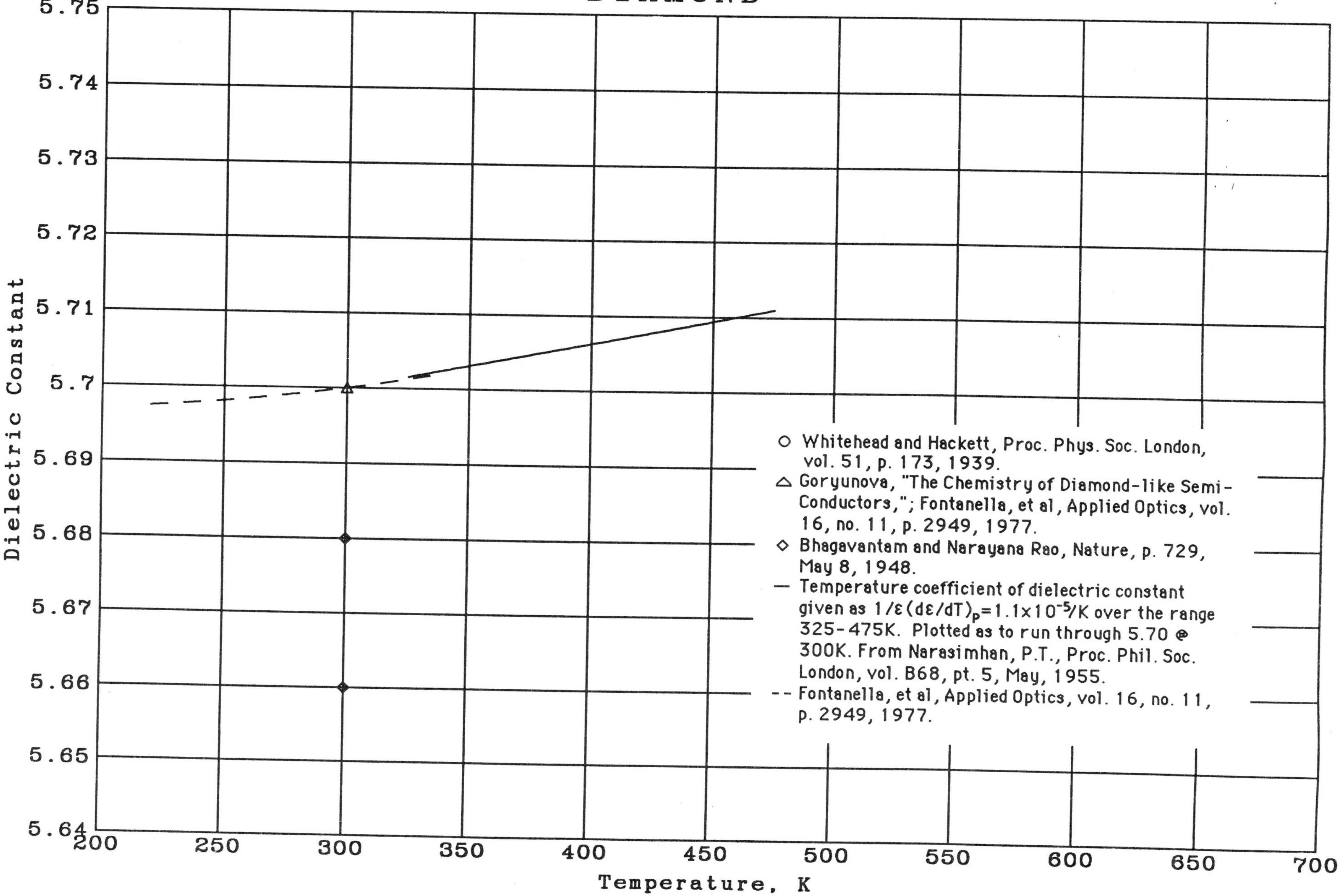
Dielectric Constant
Temperature, K
5.75
5.74
5.73
5.72
5.71
5.7
5.69
5.68
5.67
5.66
5.65
5.64
200
250
300
350
400
450
500
550
600
650
700
○ Whitehead and Hackett, Proc. Phys. Soc. London, vol. 51, p. 173, 1939.
△ Goryunova, "The Chemistry of Diamond-like Semi-Conductors,"; Fontanella, et al, Applied Optics, vol. 16, no. 11, p. 2949, 1977.
◇ Bhagavantam and Narayana Rao, Nature, p. 729, May 8, 1948.
— Temperature coefficient of dielectric constant given as $1/\varepsilon(d\varepsilon/dT)_p=1.1\times10^{-5}$/K over the range 325-475K. Plotted as to run through 5.70 @ 300K. From Narasimhan, P.T., Proc. Phil. Soc. London, vol. B68, pt. 5, May, 1955.
-- Fontanella, et al, Applied Optics, vol. 16, no. 11, p. 2949, 1977.

DIAMOND

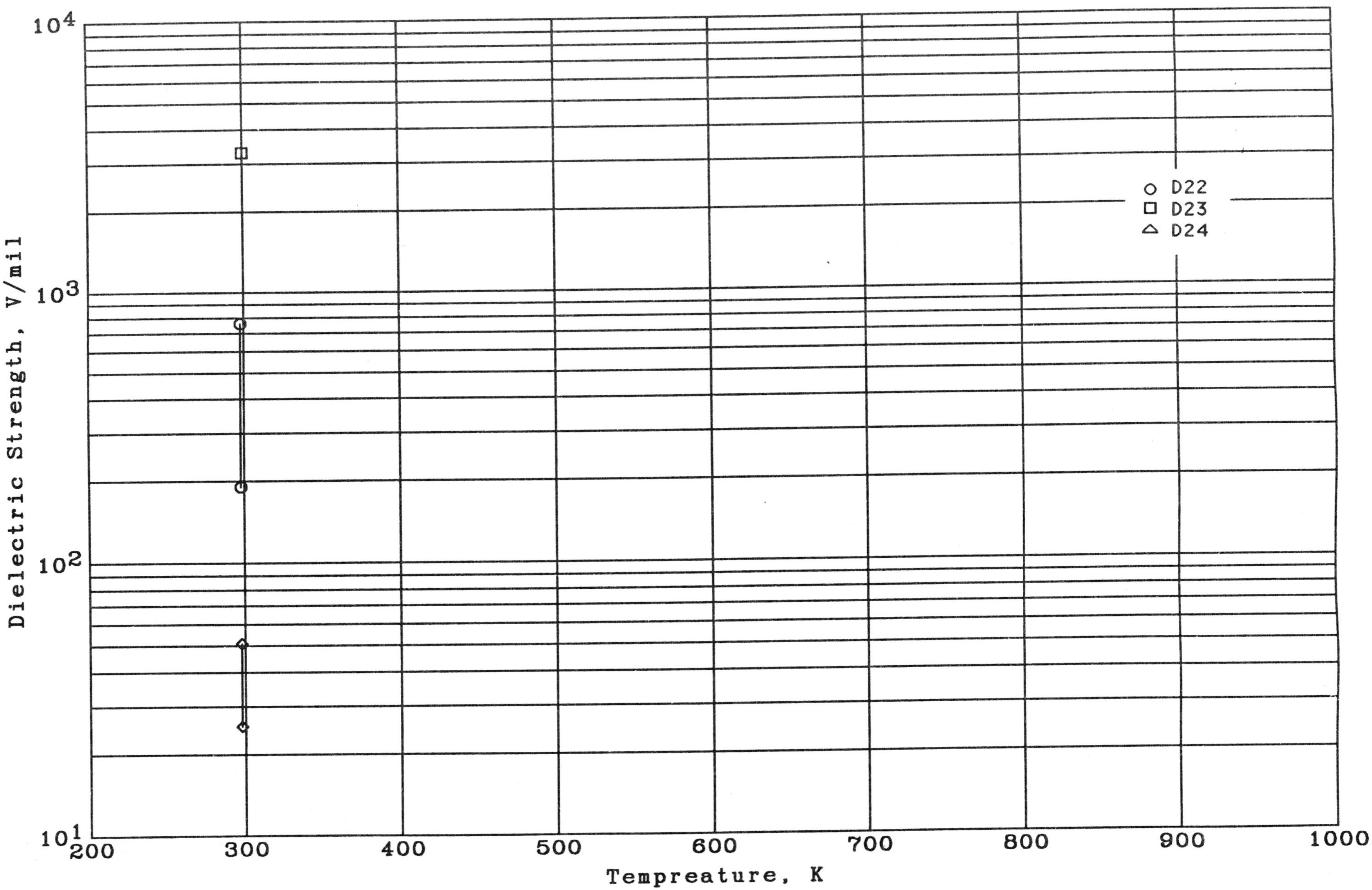

D22
D23
D24
10^4
10^3
10^2
10^1
200
300
400
500
600
700
800
900
1000
Dielectric Strength, V/mil
Tempreature, K

DIAMOND

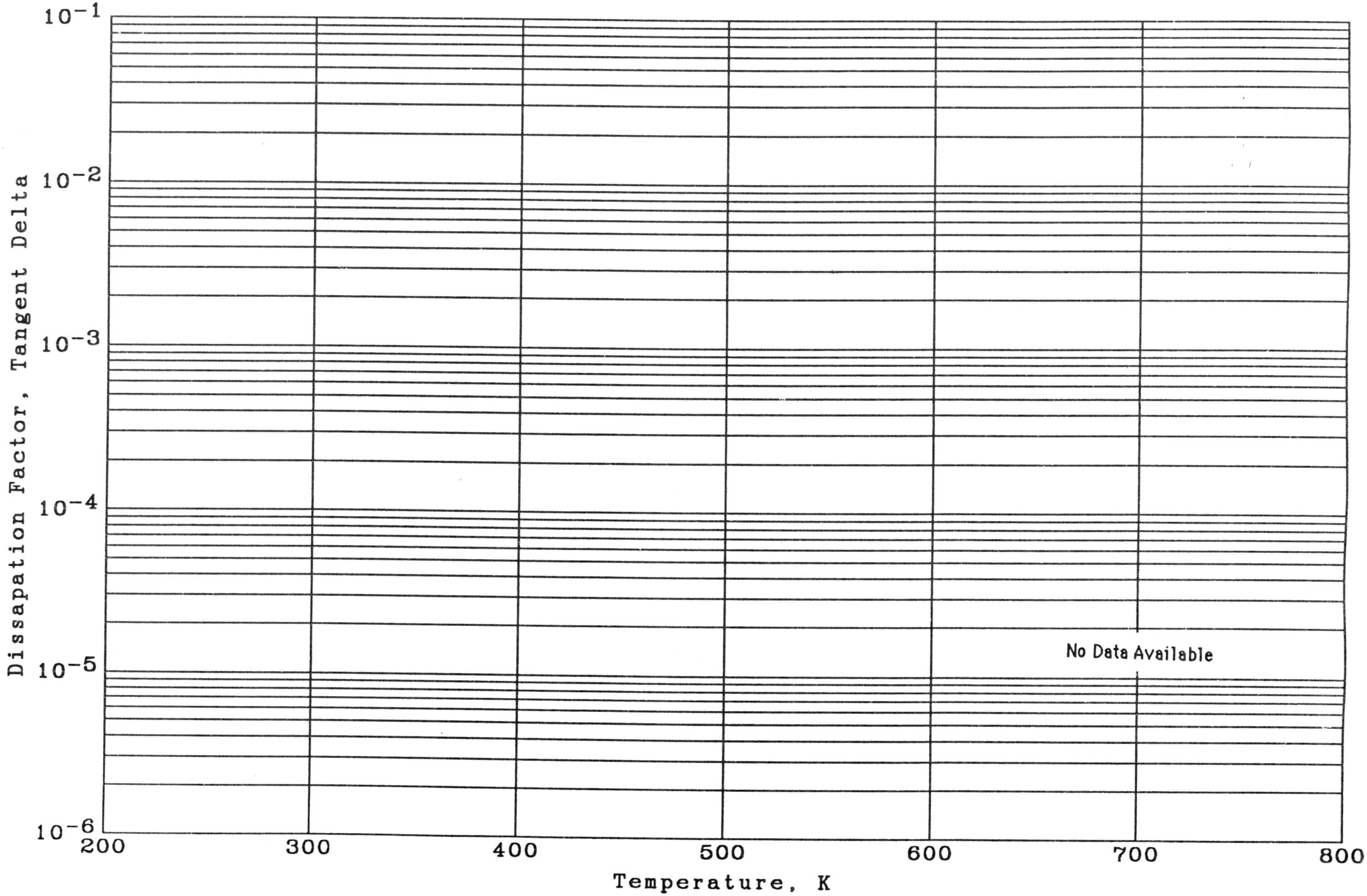
No Data Available
Dissapation Factor, Tangent Delta
10⁻¹
10⁻²
10⁻³
10⁻⁴
10⁻⁵
10⁻⁶
Temperature, K
200
300
400
500
600
700
800

DIAMOND

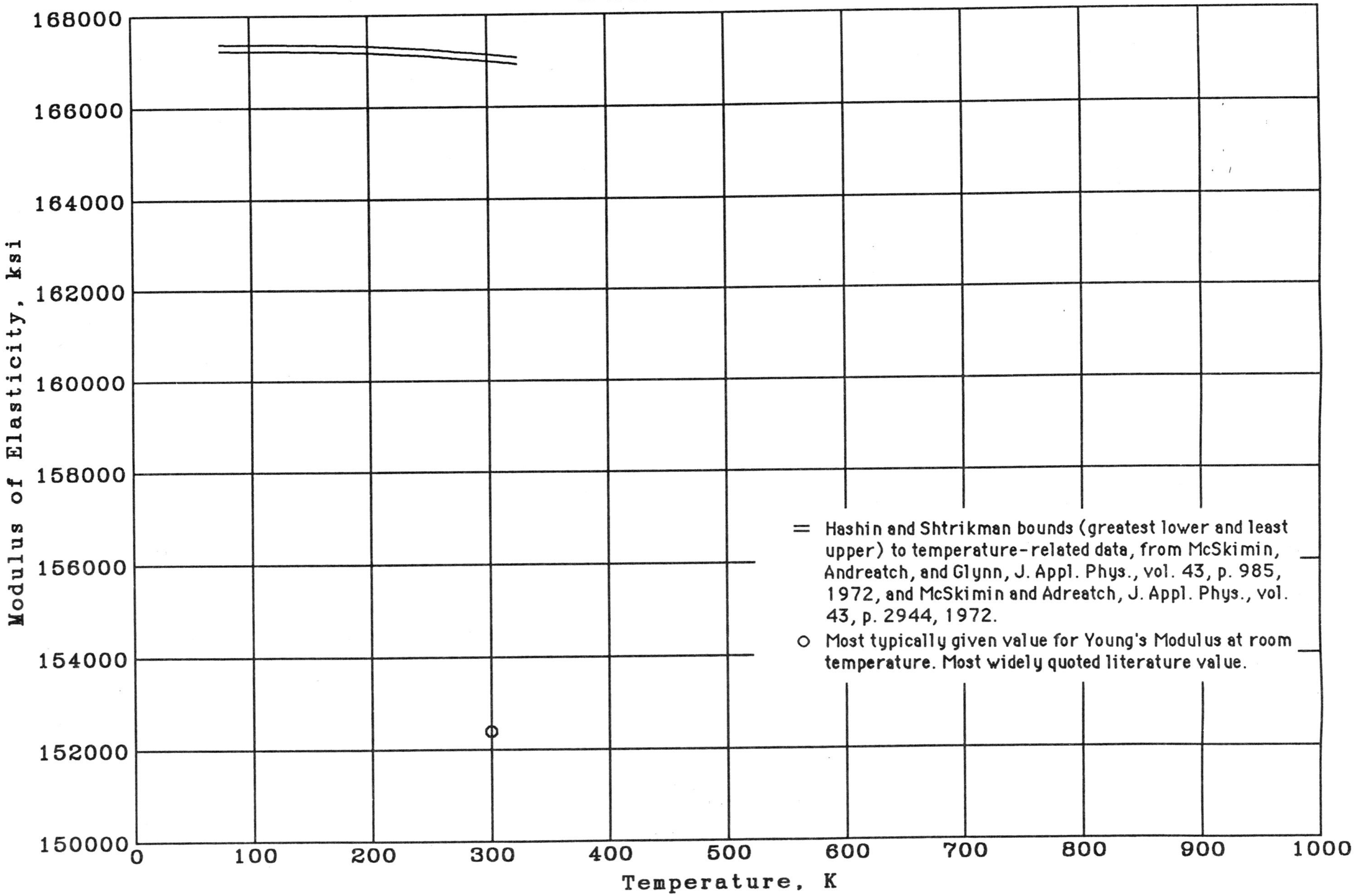

168000
166000
164000
162000
160000
158000
156000
154000
152000
150000
Modulus of Elasticity, ksi
0
100
200
300
400
500
600
700
800
900
1000
Temperature, K
= Hashin and Shtrikman bounds (greatest lower and least upper) to temperature-related data, from McSkimin, Andreatch, and Glynn, J. Appl. Phys., vol. 43, p. 985, 1972, and McSkimin and Adreatch, J. Appl. Phys., vol. 43, p. 2944, 1972.
o Most typically given value for Young's Modulus at room temperature. Most widely quoted literature value.

DIAMOND

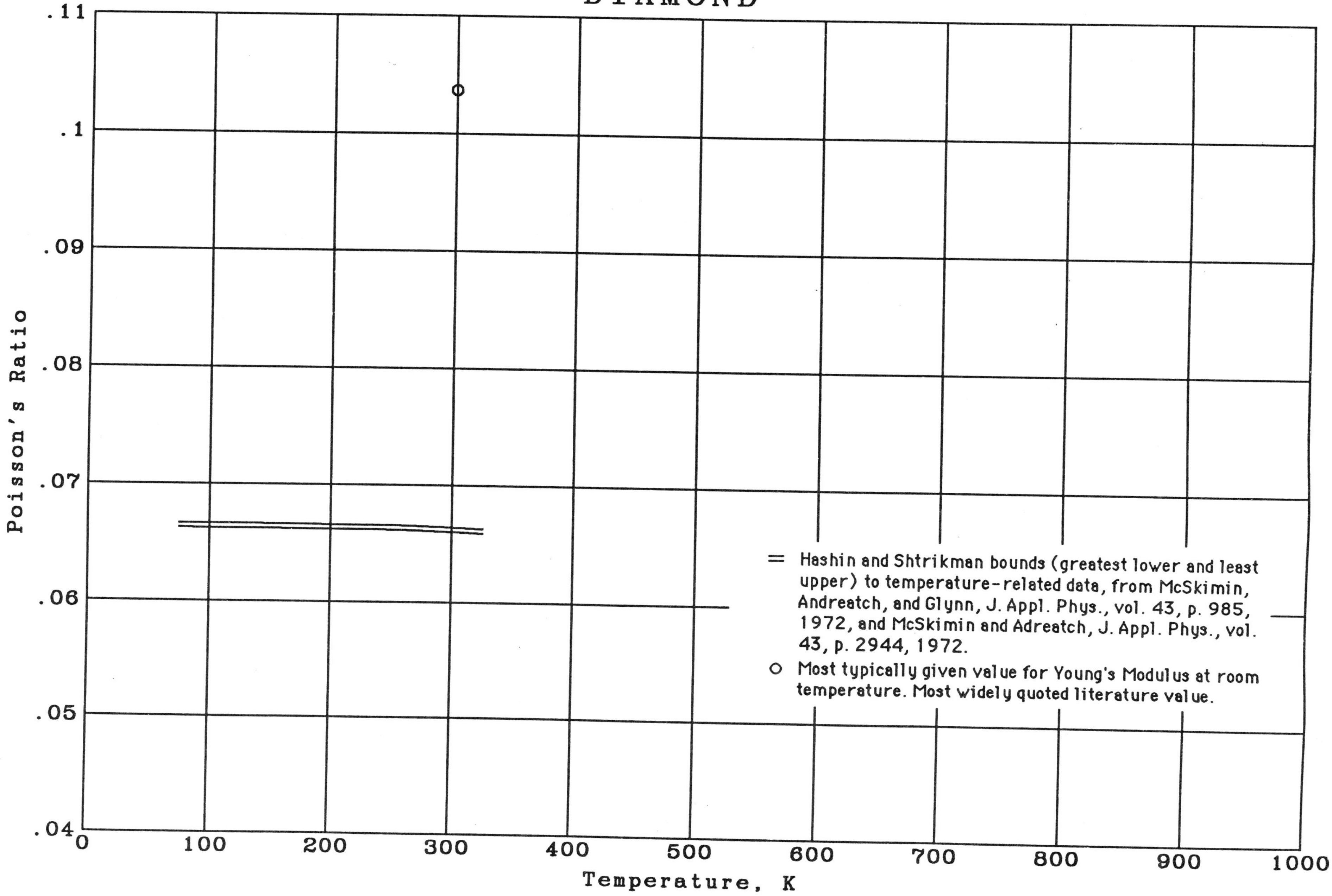
.11
.1
.09
.08
.07
.06
.05
.04
Poisson's Ratio
0
100
200
300
400
500
600
700
800
900
1000
Temperature, K
= Hashin and Shtrikman bounds (greatest lower and least upper) to temperature-related data, from McSkimin, Andreatch, and Glynn, J. Appl. Phys., vol. 43, p. 985, 1972, and McSkimin and Adreatch, J. Appl. Phys., vol. 43, p. 2944, 1972.
○ Most typically given value for Young's Modulus at room temperature. Most widely quoted literature value.

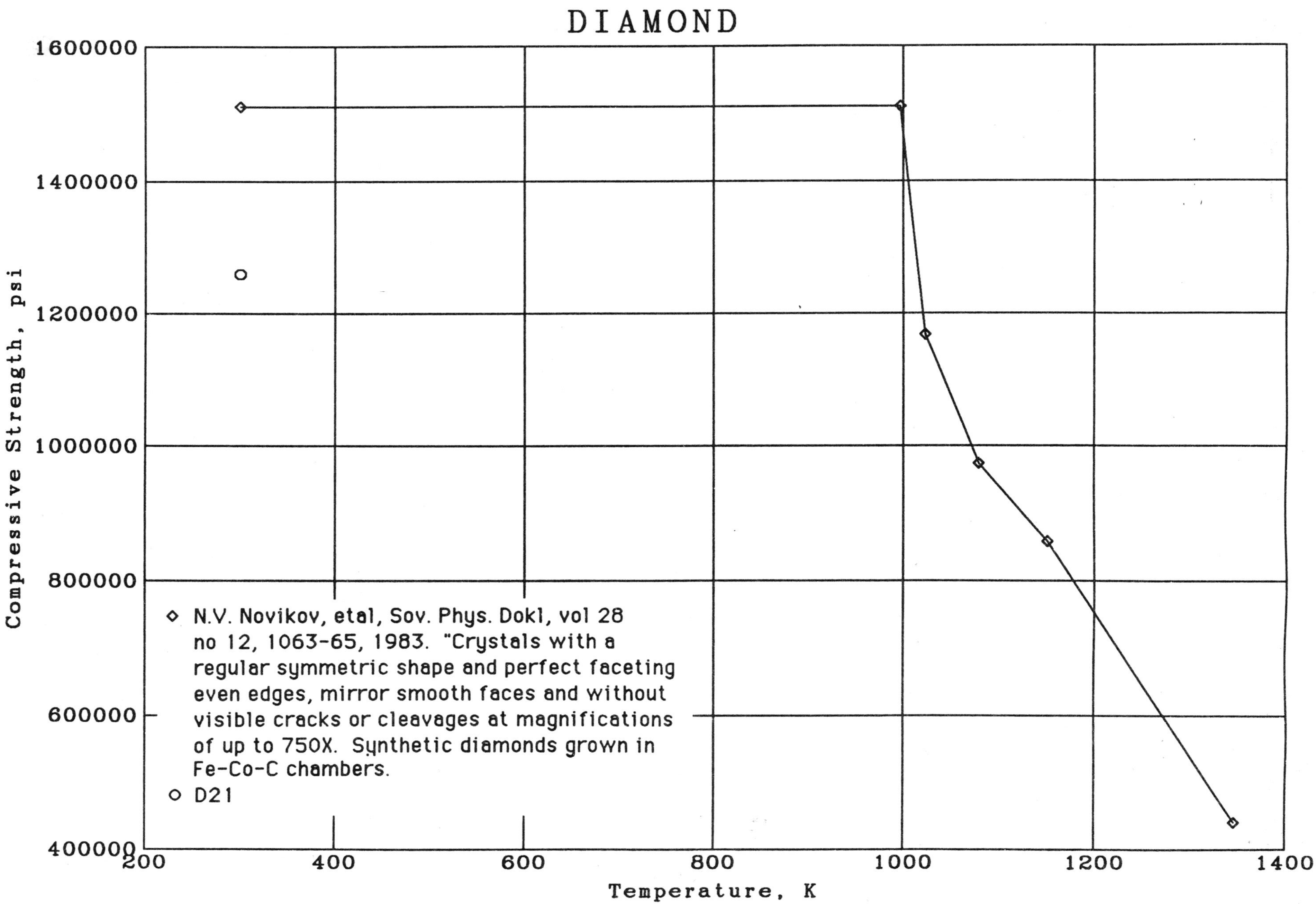
DIAMOND
Compressive Strength, psi
1600000
1400000
1200000
1000000
800000
600000
400000
200
400
600
800
1000
1200
1400
Temperature, K
◇ N.V. Novikov, etal, Sov. Phys. Dokl, vol 28 no 12, 1063-65, 1983. "Crystals with a regular symmetric shape and perfect faceting even edges, mirror smooth faces and without visible cracks or cleavages at magnifications of up to 750X. Synthetic diamonds grown in Fe-Co-C chambers.
○ D21

DIAMOND

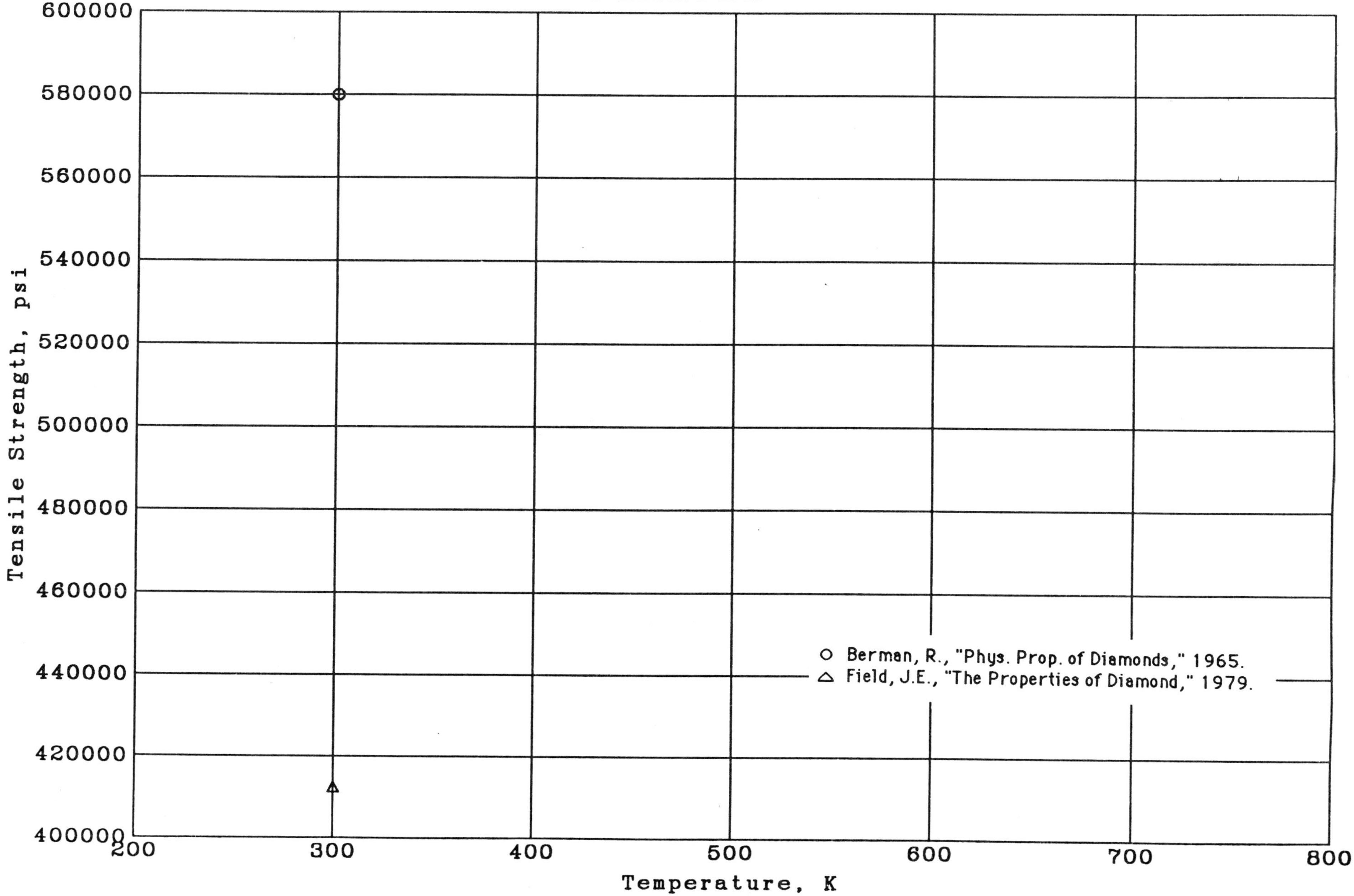

600000
580000
560000
540000
520000
500000
480000
460000
440000
420000
400000
Tensile Strength, psi
200
300
400
500
600
700
800
Temperature, K
○ Berman, R., "Phys. Prop. of Diamonds," 1965.
△ Field, J.E., "The Properties of Diamond," 1979.

DIAMOND

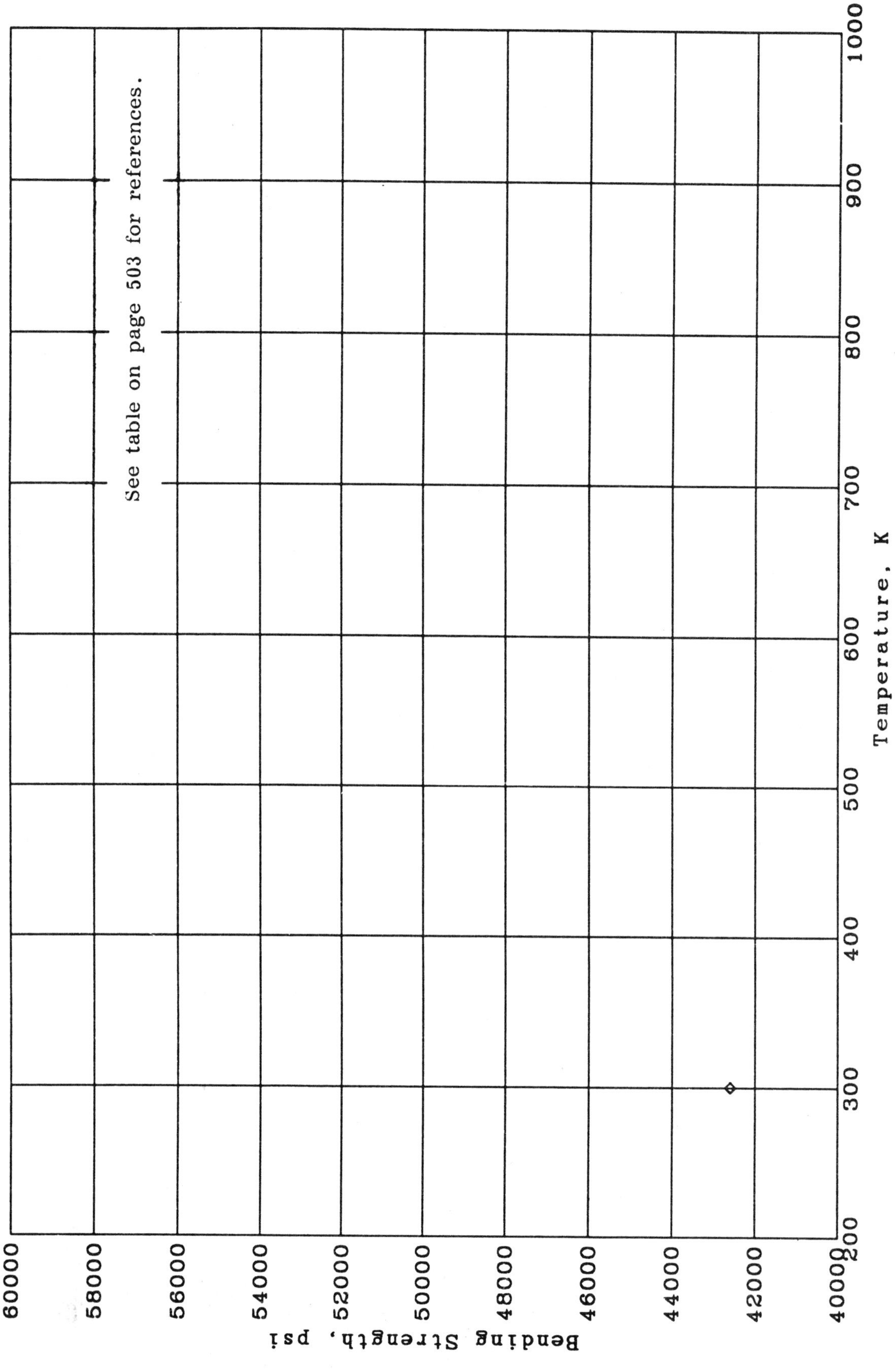

See table on page 503 for references.
Bending Strength, psi
60000
58000
56000
54000
52000
50000
48000
46000
44000
42000
40000
Temperature, K
200
300
400
500
600
700
800
900
1000

DIAMOND

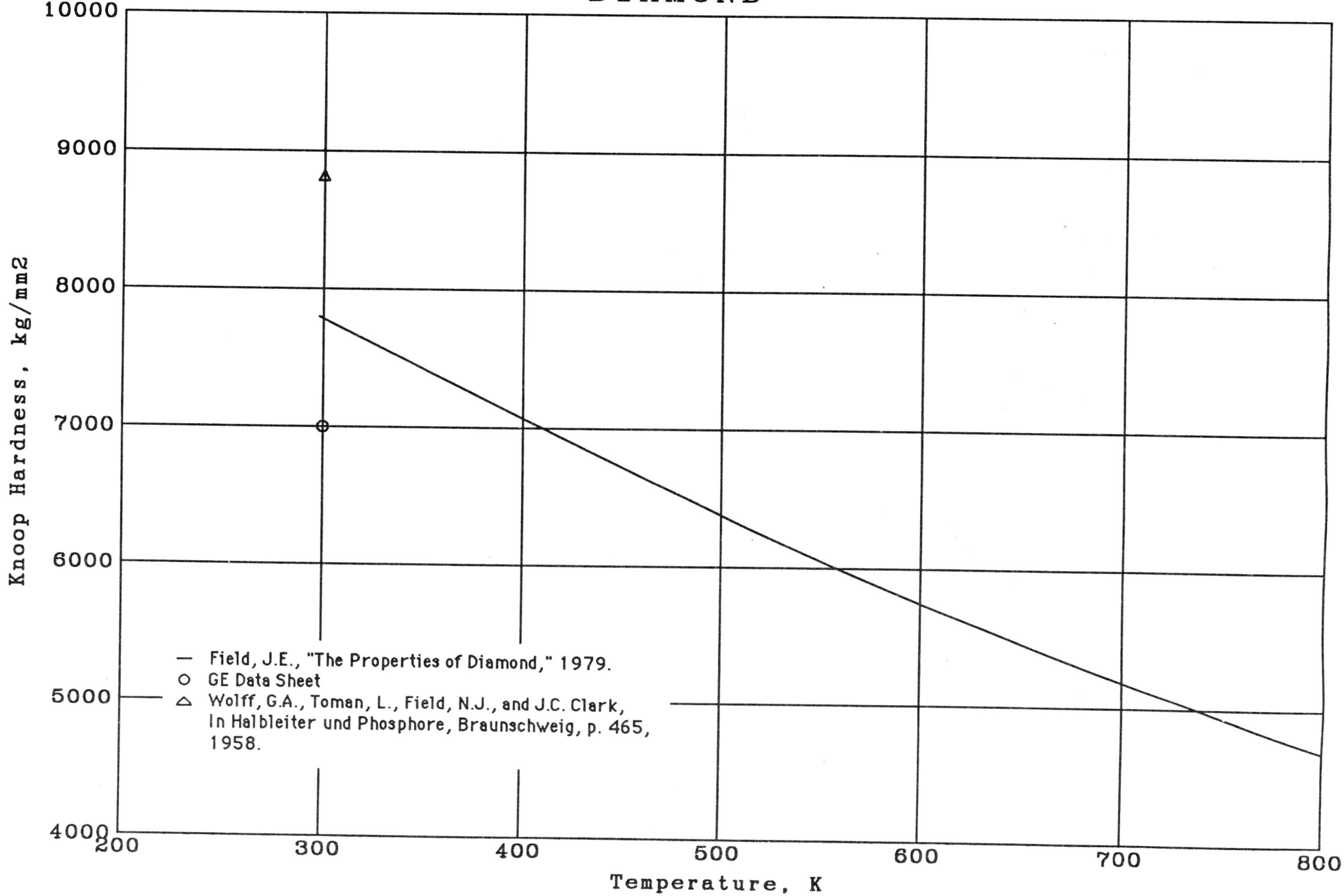
10000
9000
8000
7000
6000
5000
4000
Knoop Hardness, kg/mm2
200
300
400
500
600
700
800
Temperature, K
— Field, J.E., "The Properties of Diamond," 1979.
○ GE Data Sheet
△ Wolff, G.A., Toman, L., Field, N.J., and J.C. Clark,
In Halbleiter und Phosphore, Braunschweig, p. 465,
1958.

FUSED QUARTZ

PROPERTY	@ 300K	UNITS	REFERENCES
Density	2.203	g/cm^3	FQ1 FQ2 FQ3
Specific Heat	0.1775	cal/g-K	FQ4 FQ5 FQ6
Thermal Conductivity	0.0033	cal/cm-s-K	FQ7 FQ8 FQ9
Thermal Exp Coef	0.5×10^{-6}	1/K	FQ10 FQ11 FQ12
Dielectric Constant	3.75	------	FQ13 FQ14 FQ15
Dissipation Factor	0.0015	------	FQ15 FQ16 FQ17
Dielectric Strength	700-1000	V/mil	FQ15 FQ17 FQ20
Young's Modulus	10,500	ksi	FQ2 FQ15 FQ16
Poisson's Ratio	0.14	------	FQ2 FQ15 FQ16
Flexural Strength	9700	psi	FQ13 FQ15 FQ17
Ult Tensile Strength[1]	4000-11,000	psi	FQ15 FQ17 FQ18
Compressive Strength	161,000	psi	FQ15 FQ17 FQ19
Hardness	500	Knoop	FQ15 FQ16 FQ17

[1] The ultimate strength of glass is determined by the size and distribution of flaws present in the material. Smaller samples have a narrow distribution of smaller flaws which makes them inherently stronger.

FUSED QUARTZ

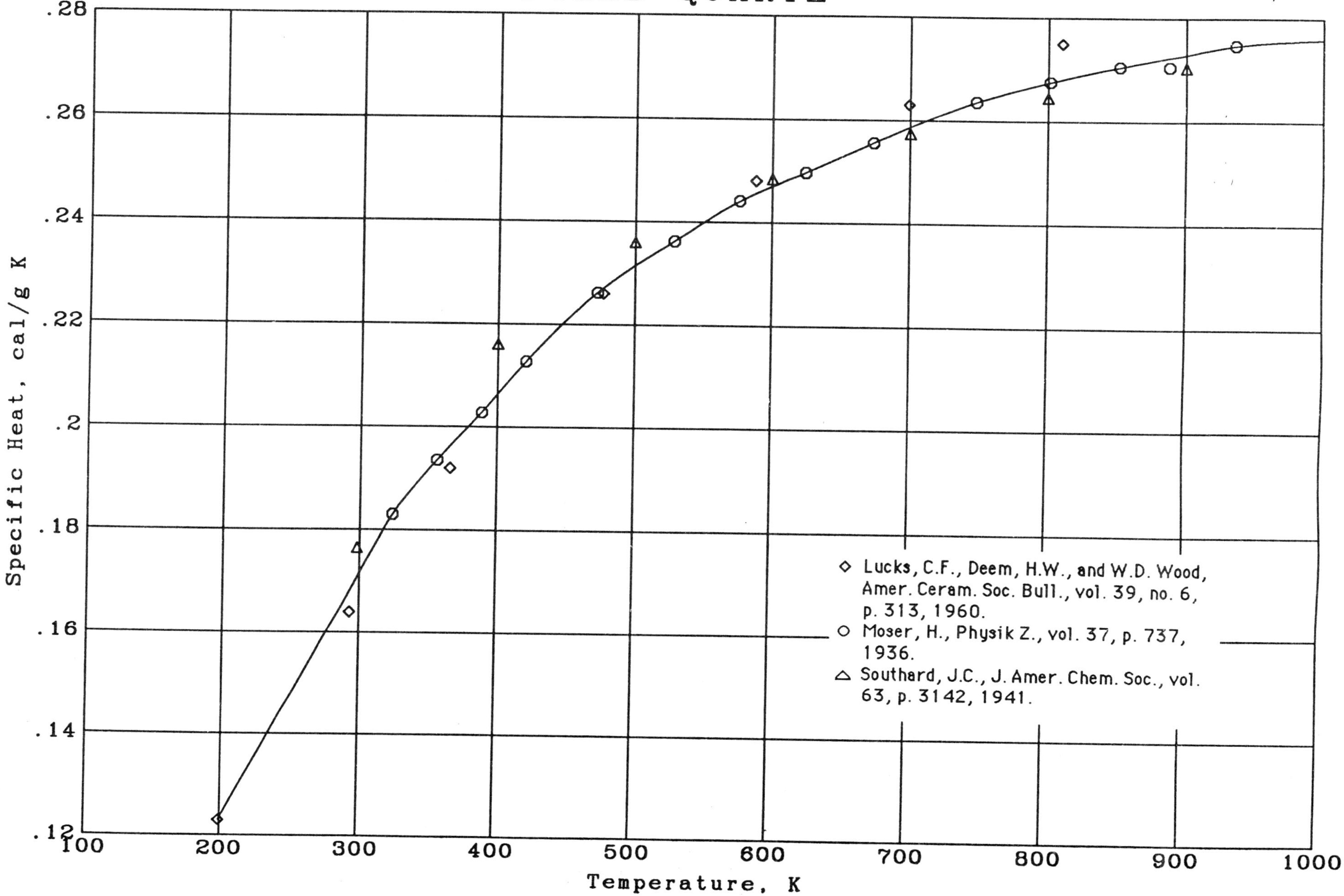

FUSED QUARTZ

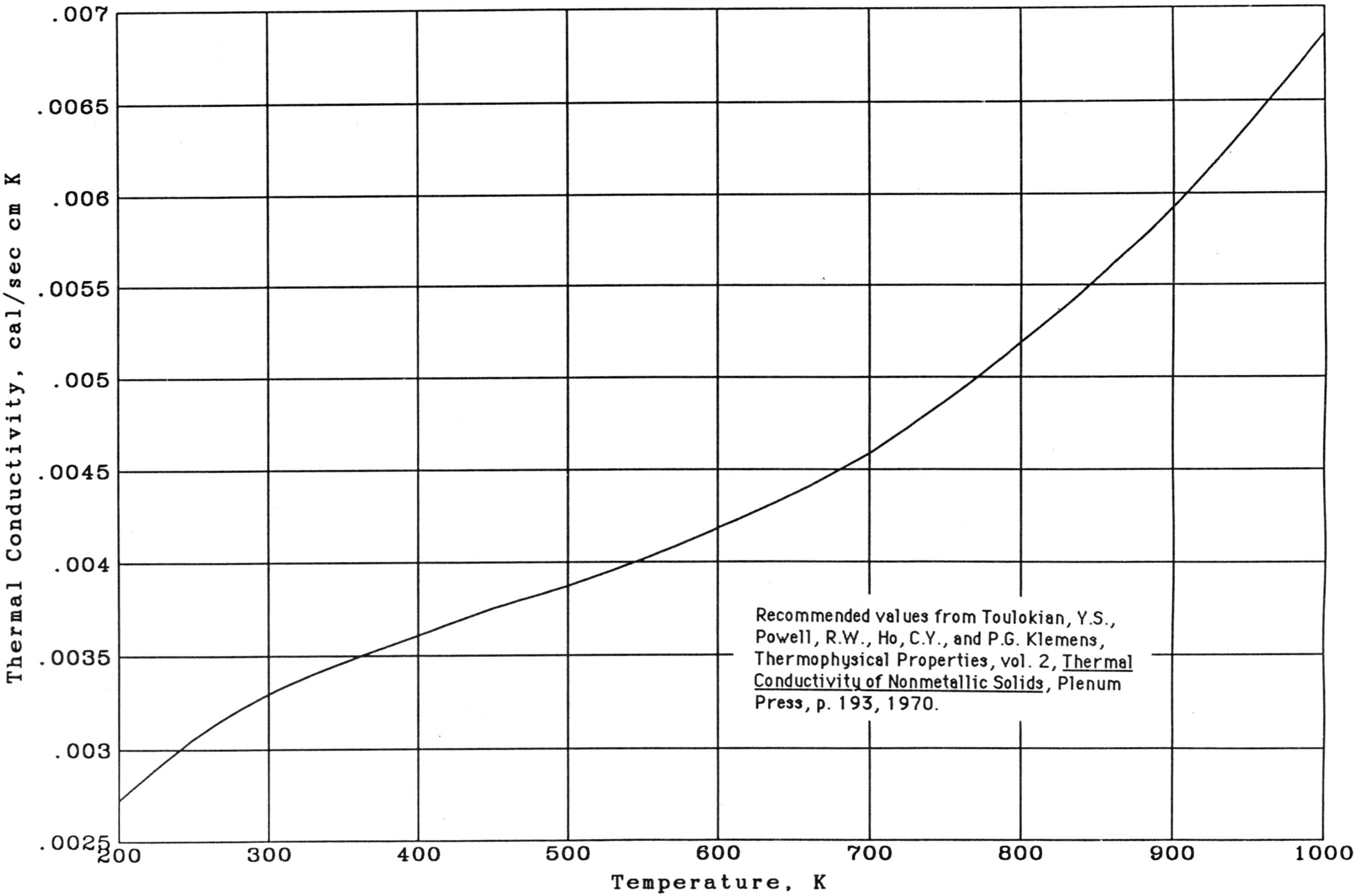

FUSED QUARTZ

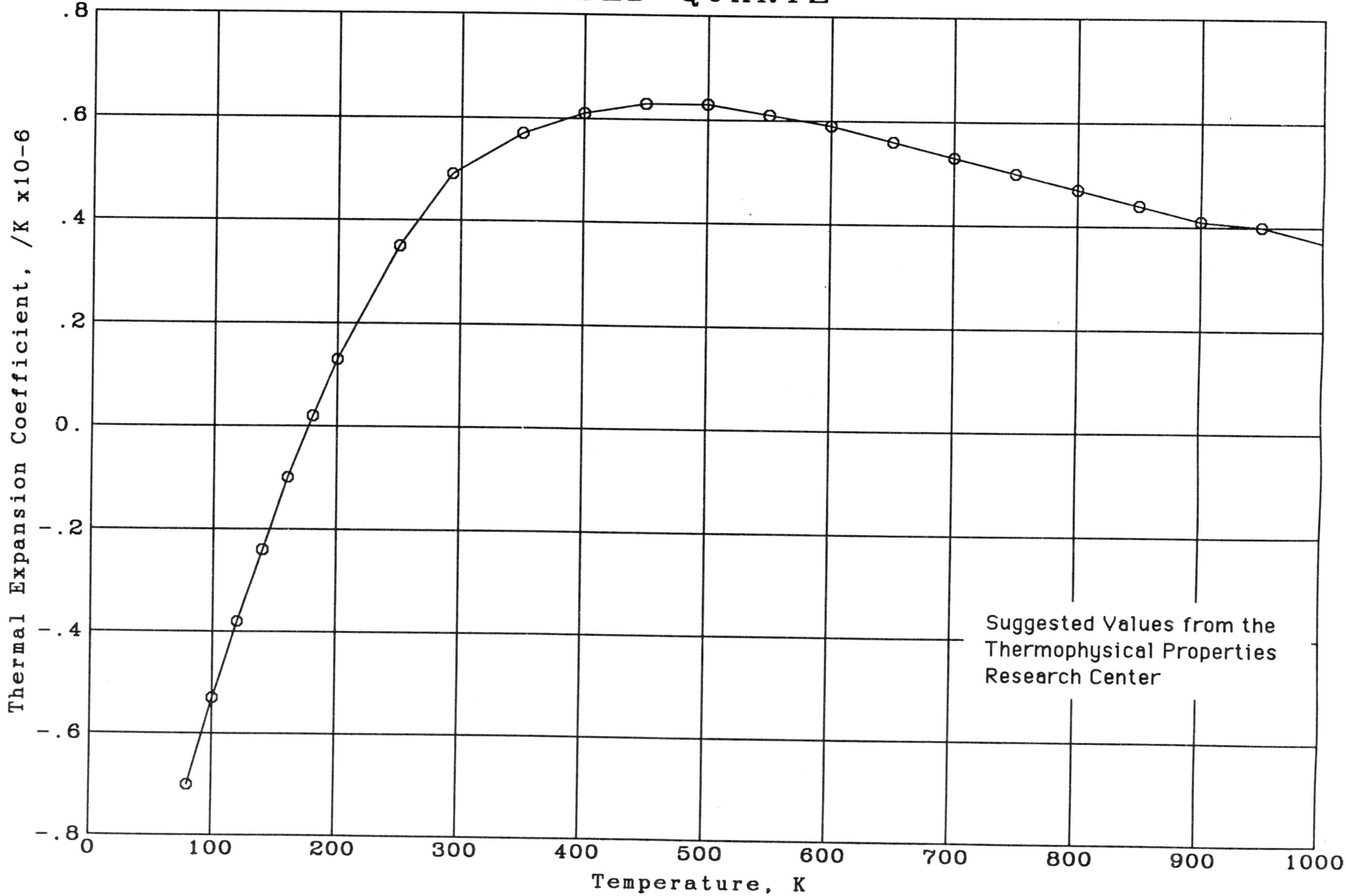

FUSED QUARTZ

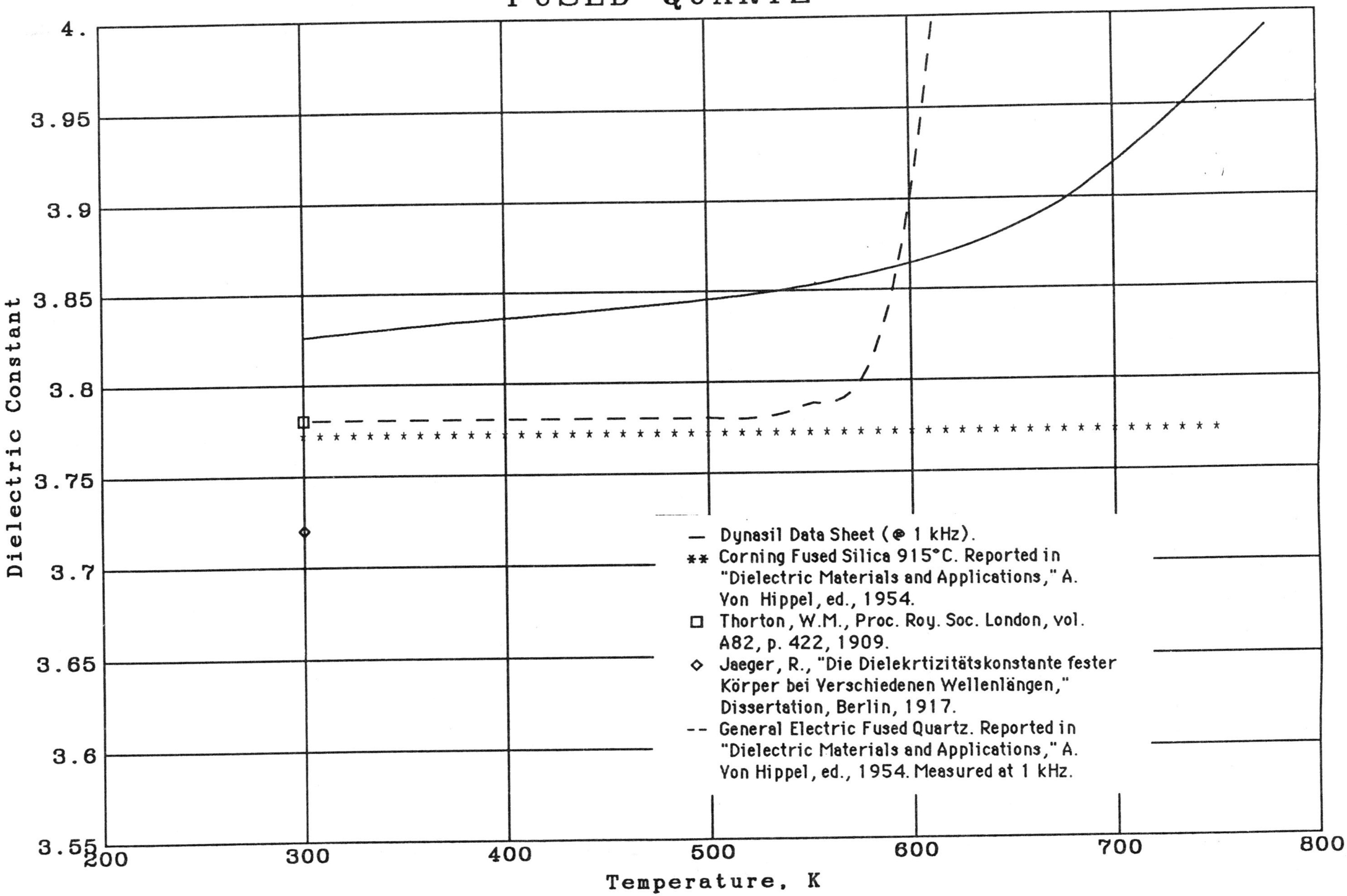

FUSED QUARTZ

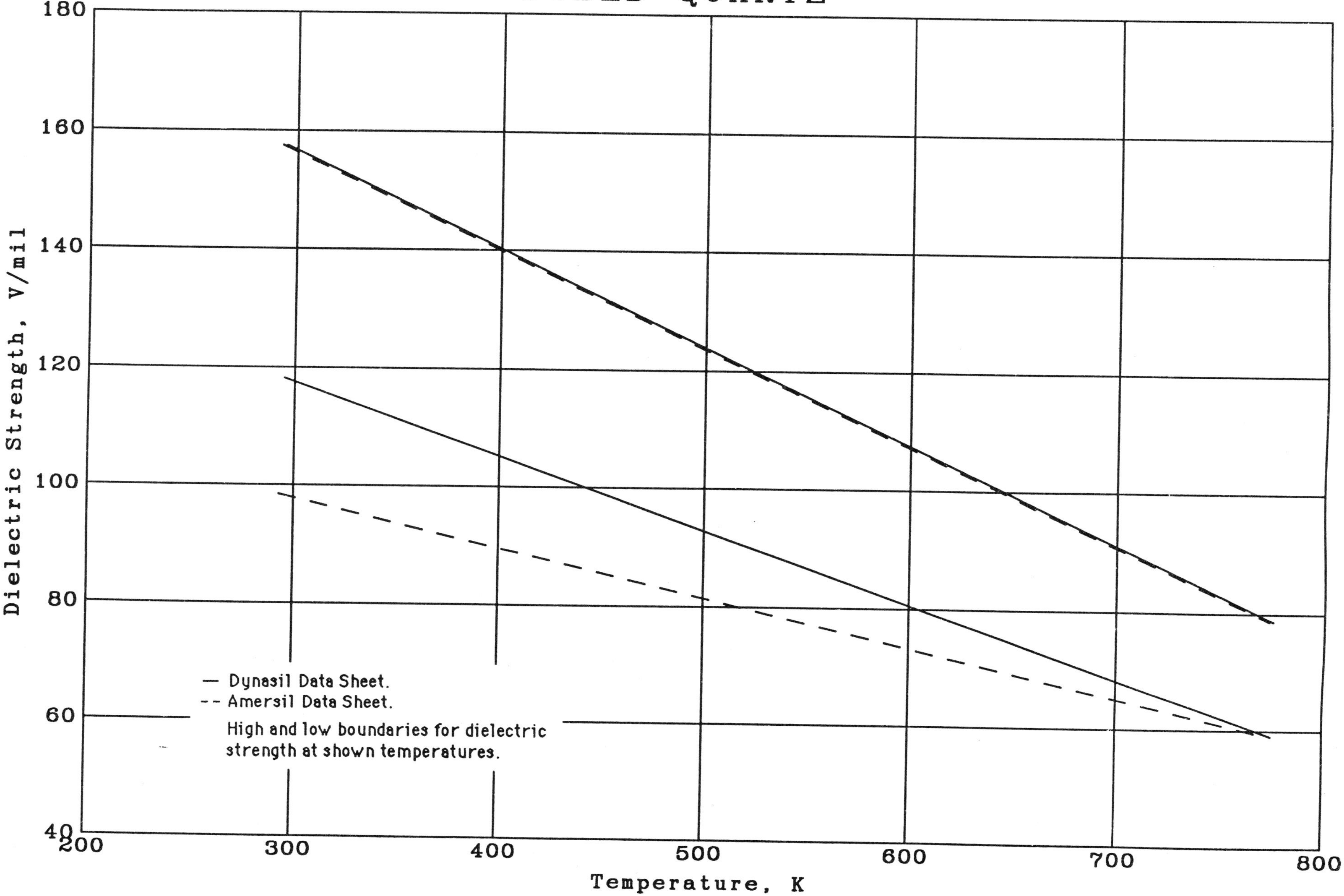

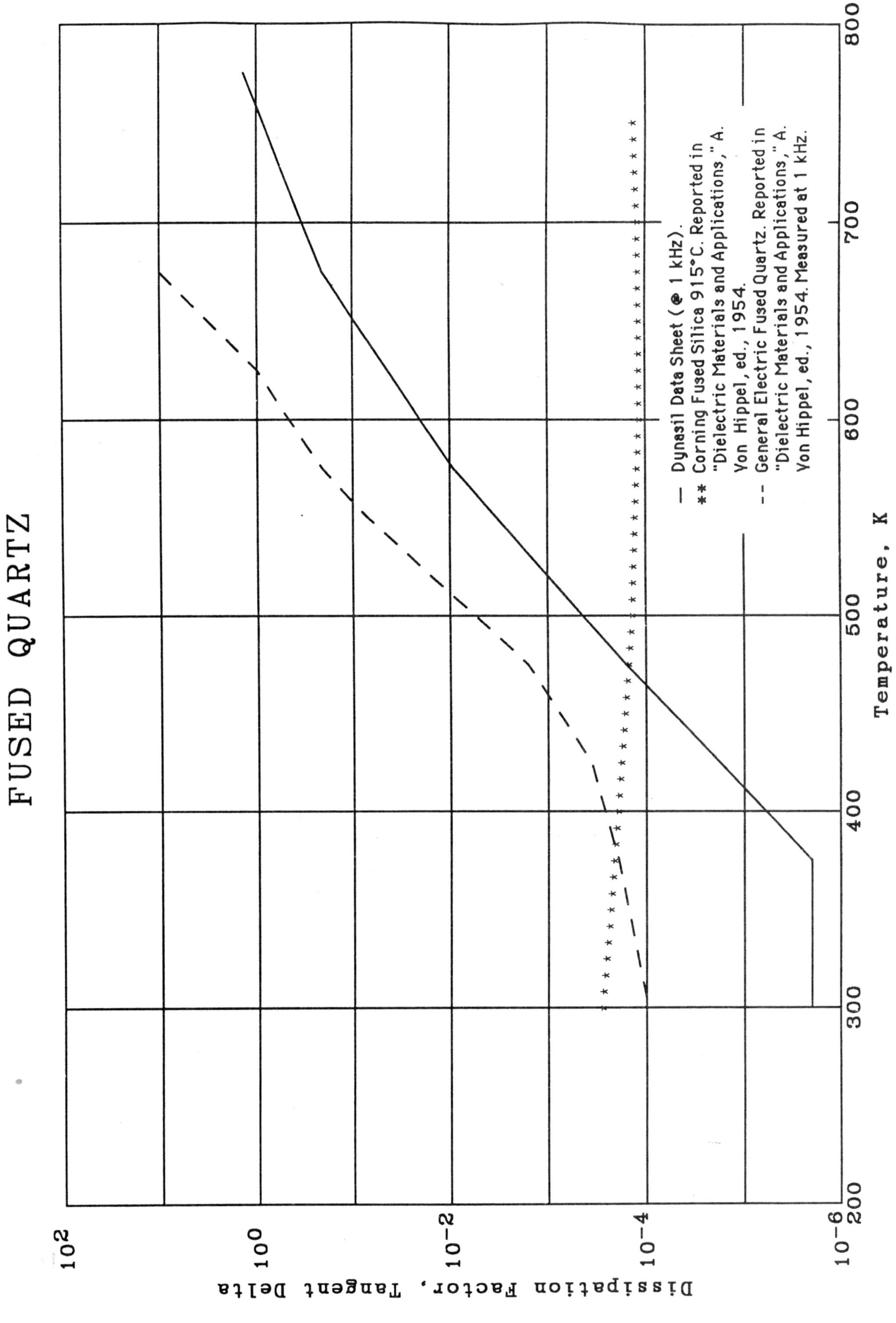
FUSED QUARTZ
Dissipation Factor, Tangent Delta
10^{2}
10^{0}
10^{-2}
10^{-4}
10^{-6}
200
300
400
500
600
700
800
Temperature, K
— Dynasil Data Sheet (@ 1 kHz).
** Corning Fused Silica 915°C. Reported in "Dielectric Materials and Applications," A. Von Hippel, ed., 1954.
-- General Electric Fused Quartz. Reported in "Dielectric Materials and Applications," A. Von Hippel, ed., 1954. Measured at 1 kHz.

FUSED QUARTZ

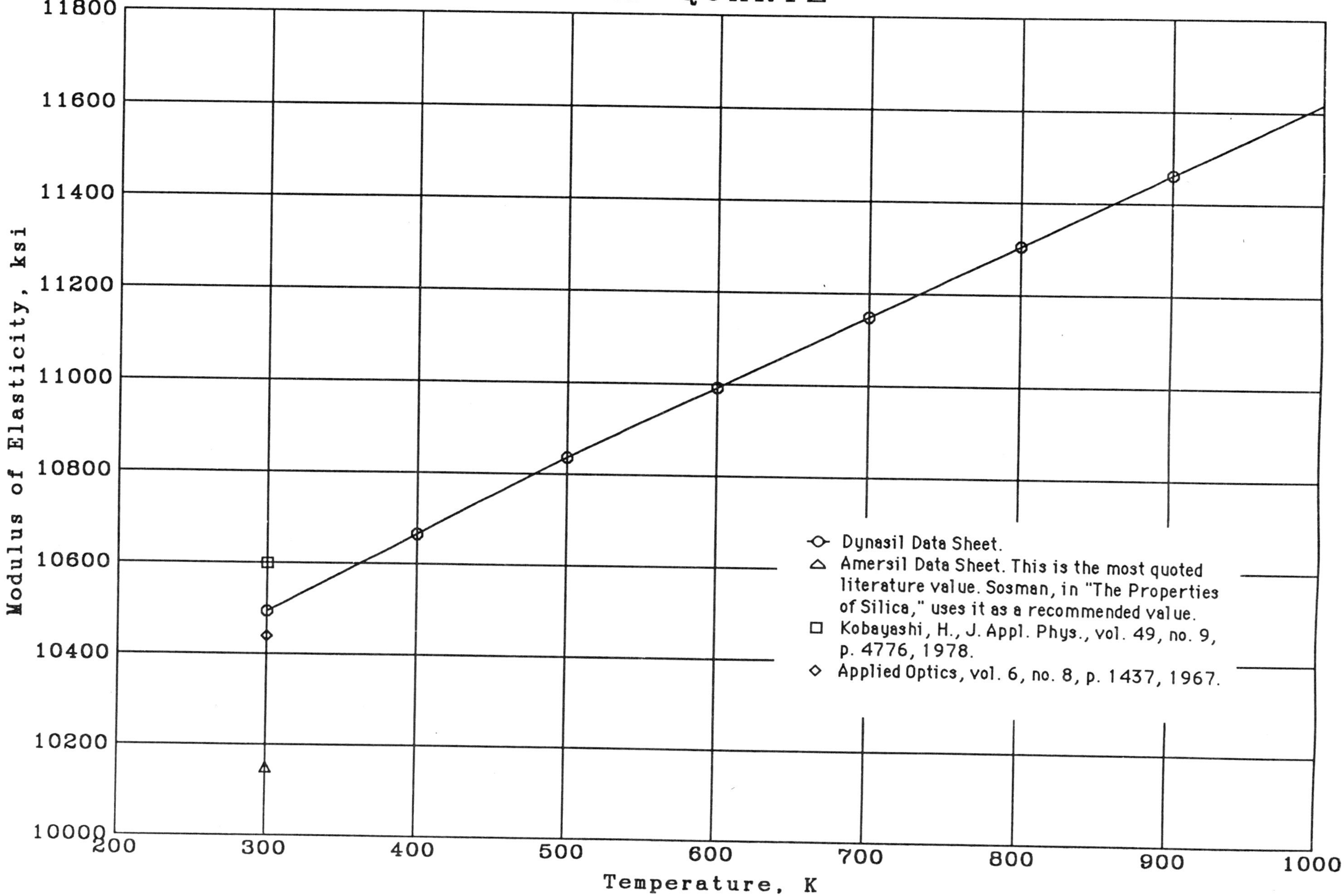

FUSED QUARTZ

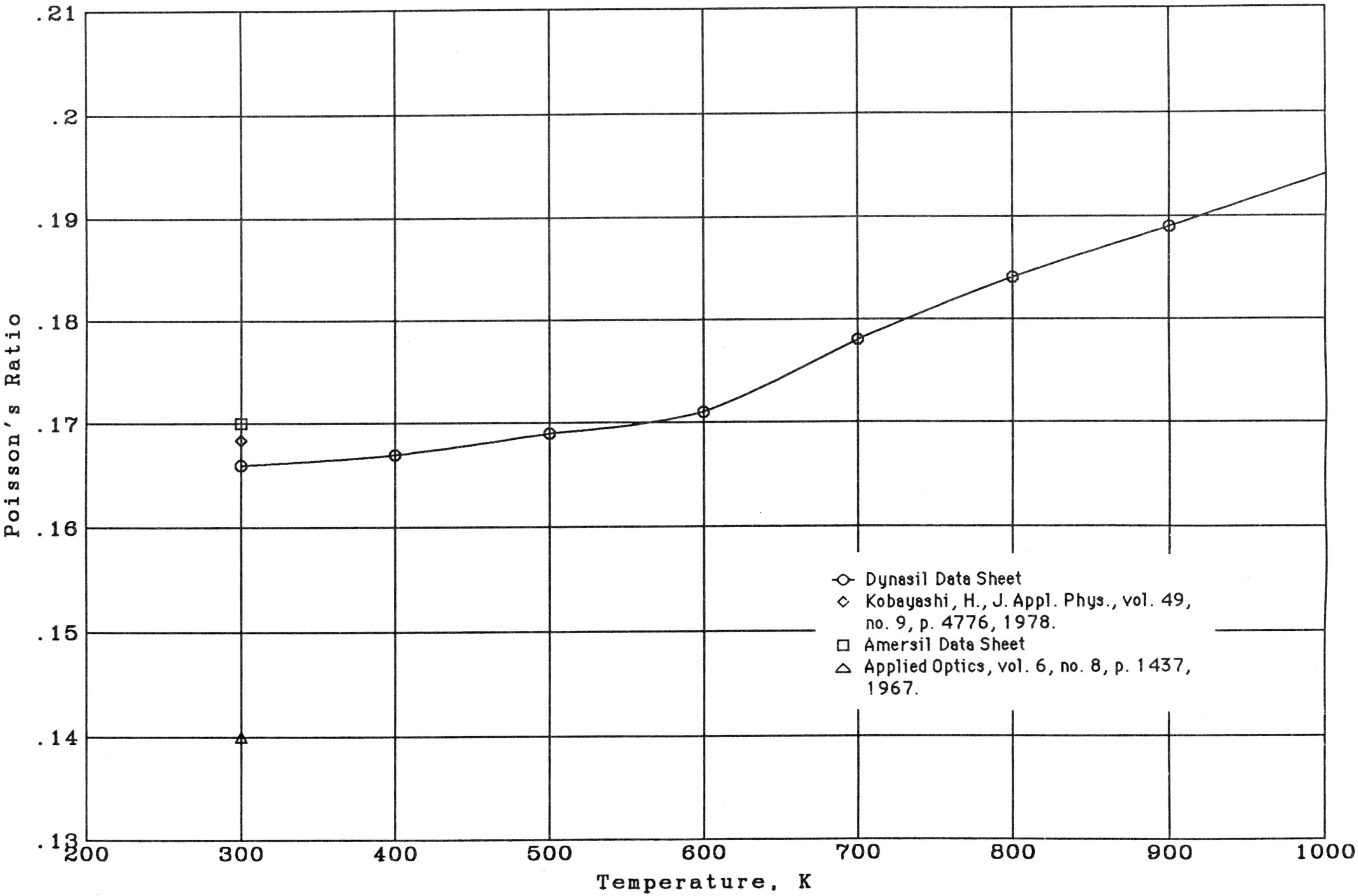

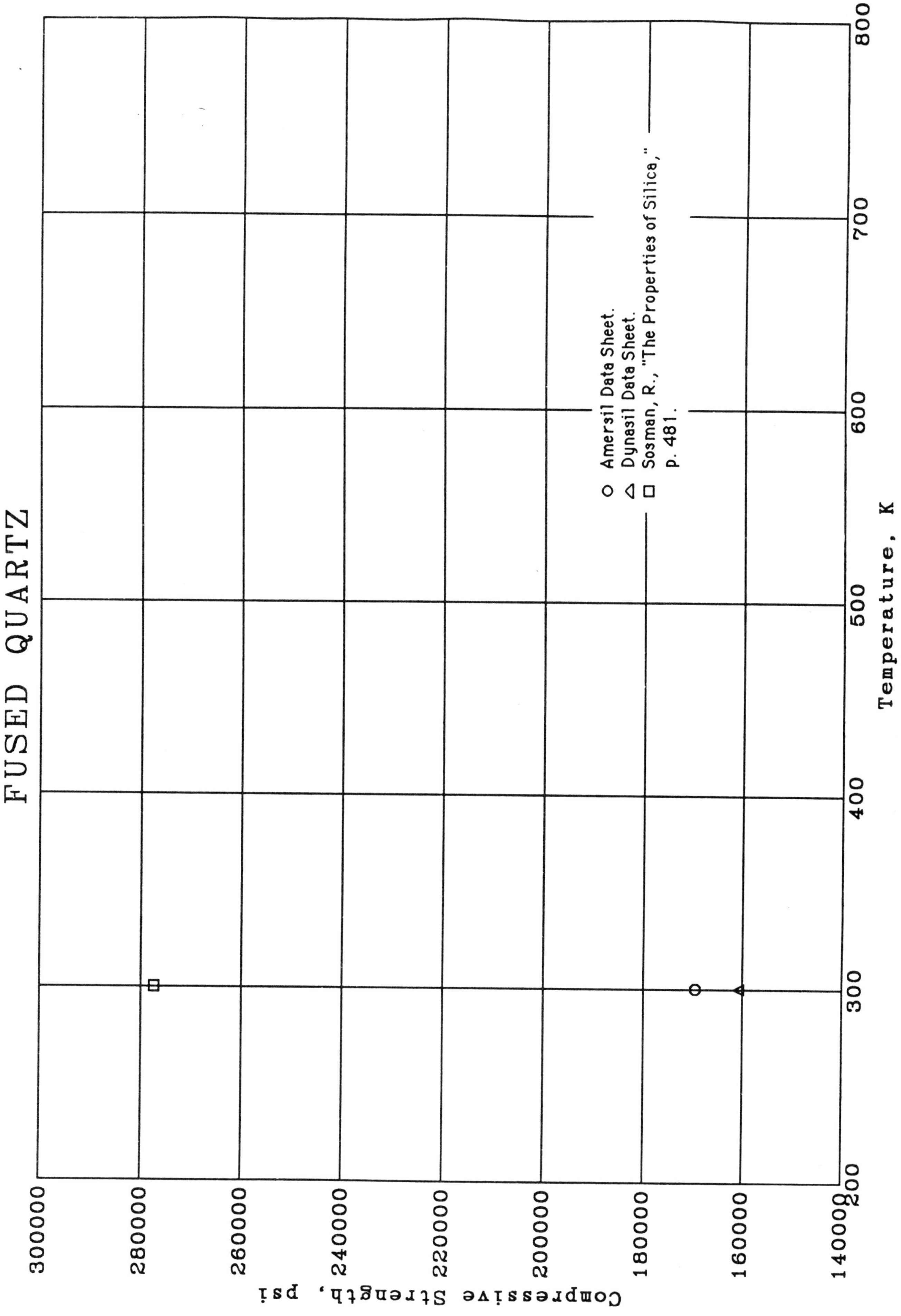

FUSED QUARTZ
Compressive Strength, psi
Temperature, K
300000
280000
260000
240000
220000
200000
180000
160000
140000
200
300
400
500
600
700
800
○ Amersil Data Sheet.
△ Dynasil Data Sheet.
□ Sosman, R., "The Properties of Silica,"
p. 481.

FUSED QUARTZ

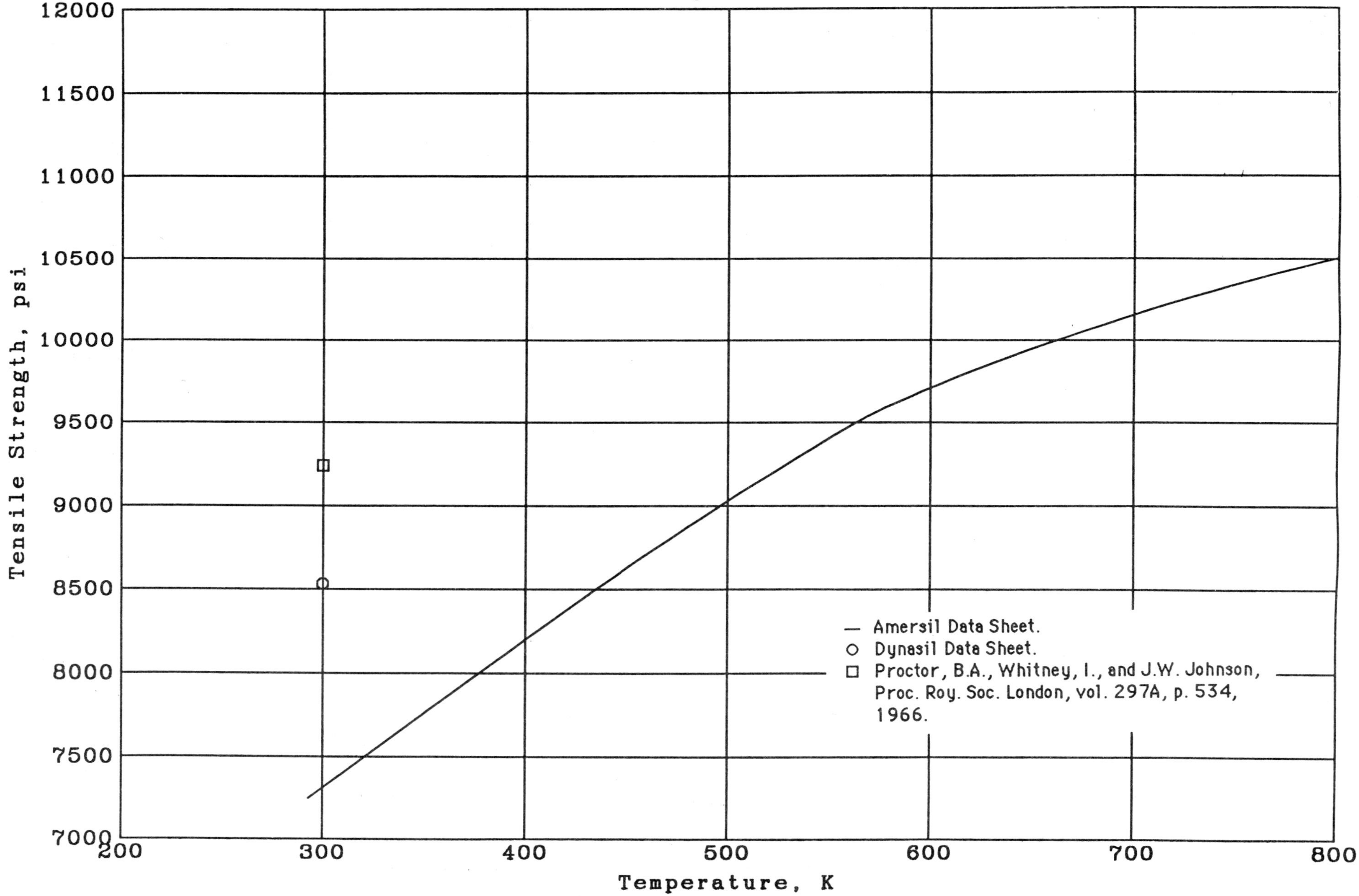

FUSED QUARTZ

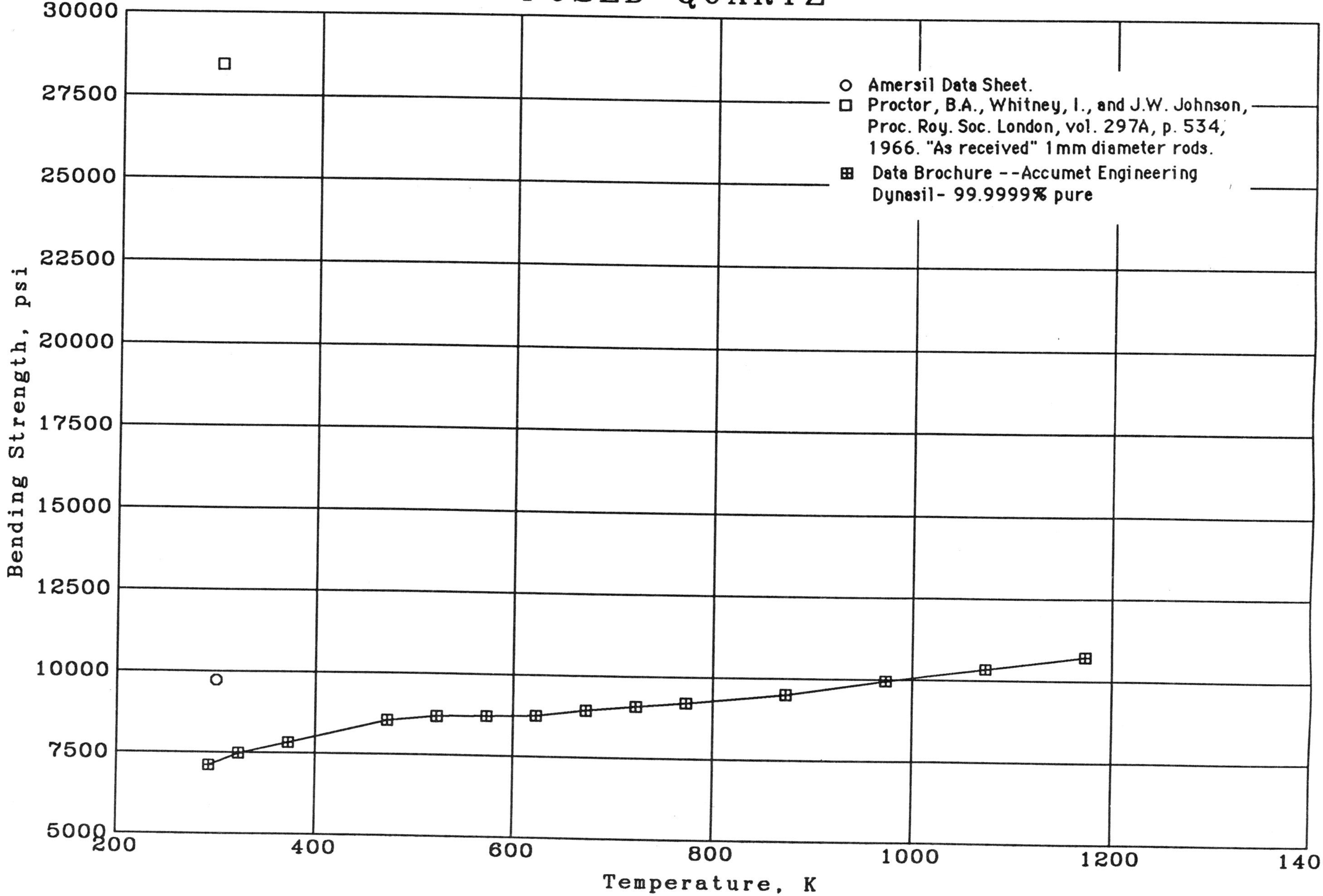

FUSED QUARTZ

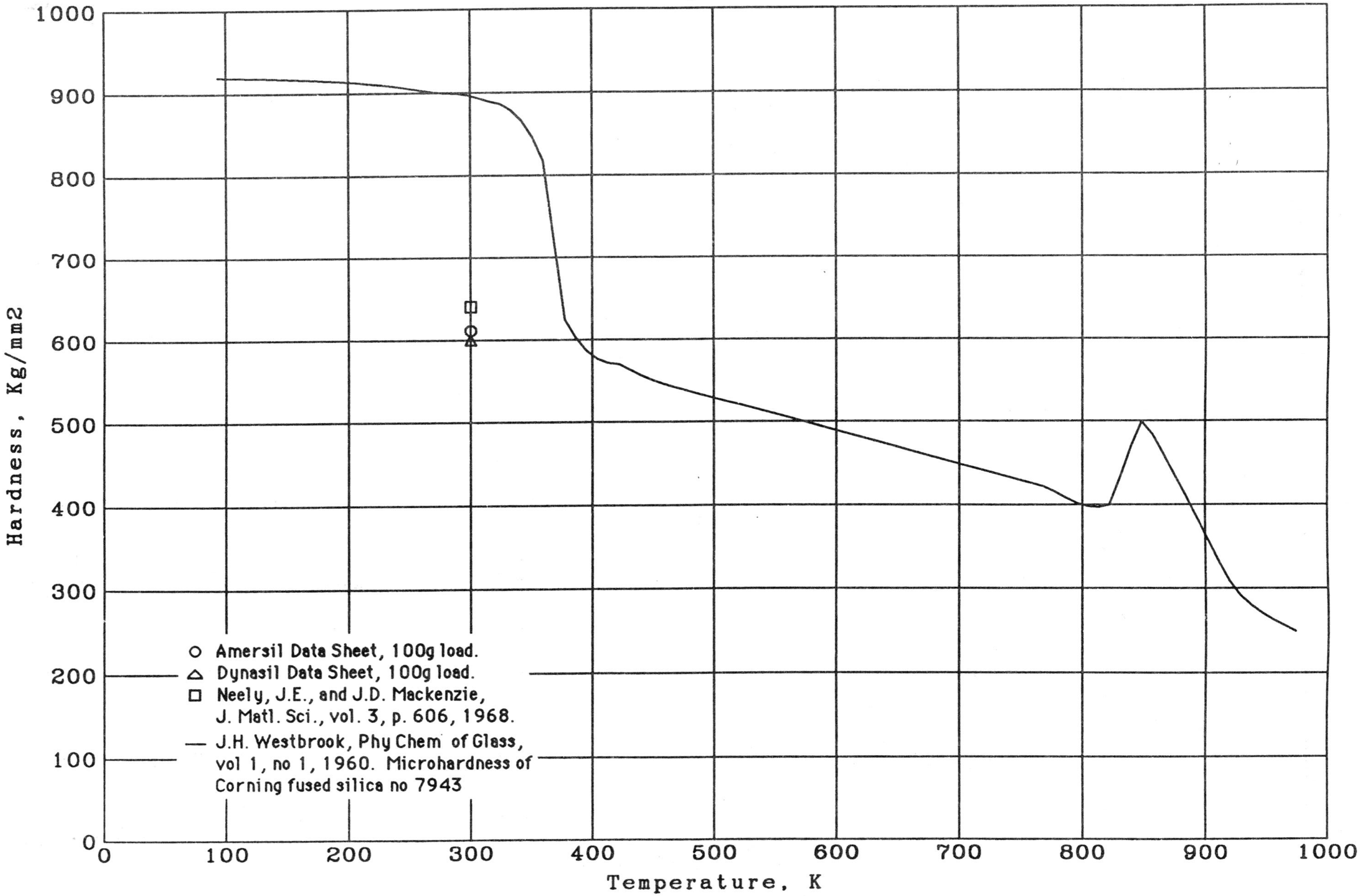

SILICON MONOXIDE

PROPERTY	@ 300K	UNITS	REFERENCES
Density [1,2]	2.125	g/cm^3	SO1 SO2 SO3
Specific Heat	0.2052	cal/g-K	SO4
Thermal Conductivity	0.00359	cal/cm-s-K	SO5 SO6
Thermal Exp Coef [3]		1/K	
Dielectric Constant	~ 5	------	SO7 SO8 SO9
Dissipation Factor [4]	0.001-0.04	------	SO7 SO10 SO11
Dielectric Strength [4]	10^2-10^4	V/mil	SO7 SO8 SO11
Young's Modulus [5]	72,000	ksi	SO12 SO13
Poisson's Ratio [5]		------	SO12 SO13
Yield Strength [6]		psi	SO12 SO13
Ult Tensile Strength	3 to 30	ksi	SO12 SO13
Elongation	3 - $30x10^{-2}$	%	SO12 SO13
Hardness	300	Vickers	SO14 SO15

PROPERTIES DEPEND ON DEPOSITION METHOD, RATE, PRESSURE, THICKNESS, SUBSTRATE, AND ELECTRODES.

[1] Electron beam evaporated material has a density 1.25 times and a dielectric constant 1.2 times that of resistance heat evaporated material. [SO16]

[2] Density varies with evaporation temperature. Material evaporated at 1050 C has a density of 1.97 g/cm^3, while material evaporated at 1350 C has a density of 2.3 g/cm^3. [SO2]

[3] No information is found regarding the coefficient of expansion, probably due to the unstable nature of the material. The thermal expansion of fused silica can be used as a first approximation.

[4] The dissipation factor and dielectric strength can reportedly be improved by exposure or heating in air to provide compositions closer to $SiO_{1.5}$. [SO17 SO18]

[5] Information regarding the elastic constants and mechanical properties is derived from the bulge test, which provides the reduced Young's modulus E/(1-). No independent determination of Poisson's ratio is found. As a first approximation, Poisson's

ratio of fused silica can be used.

[6] Silicon monoxide films are found to contain a residual compressive stress subsequent to deposition. Care must be used to insure that this stress does not cause separation of the film from the substrate. The presence of the residual compressive stress and the extremely low elongations make it undesirable to assign a yield strength.

SILICON MONOXIDE

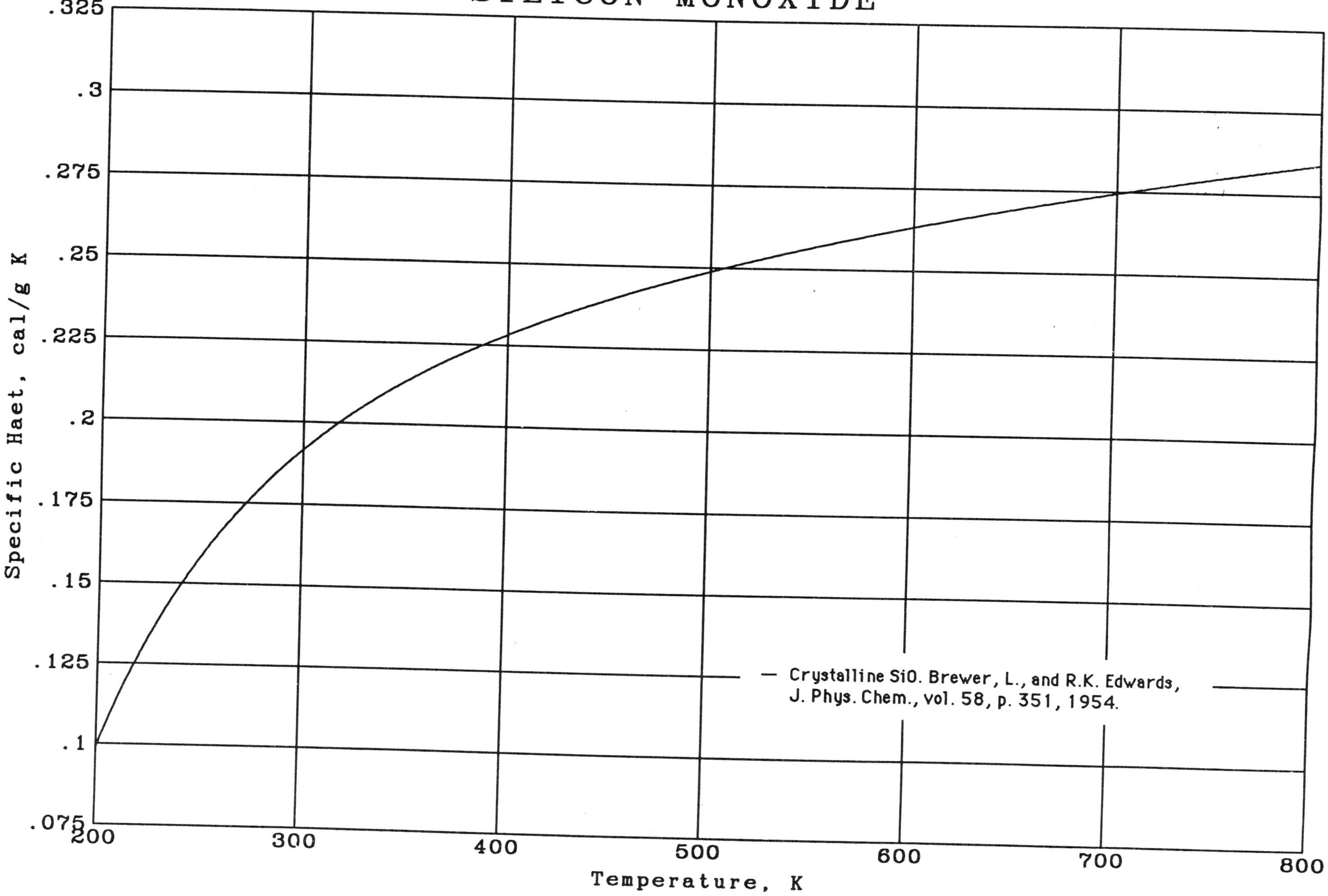

SILICON MONOXIDE

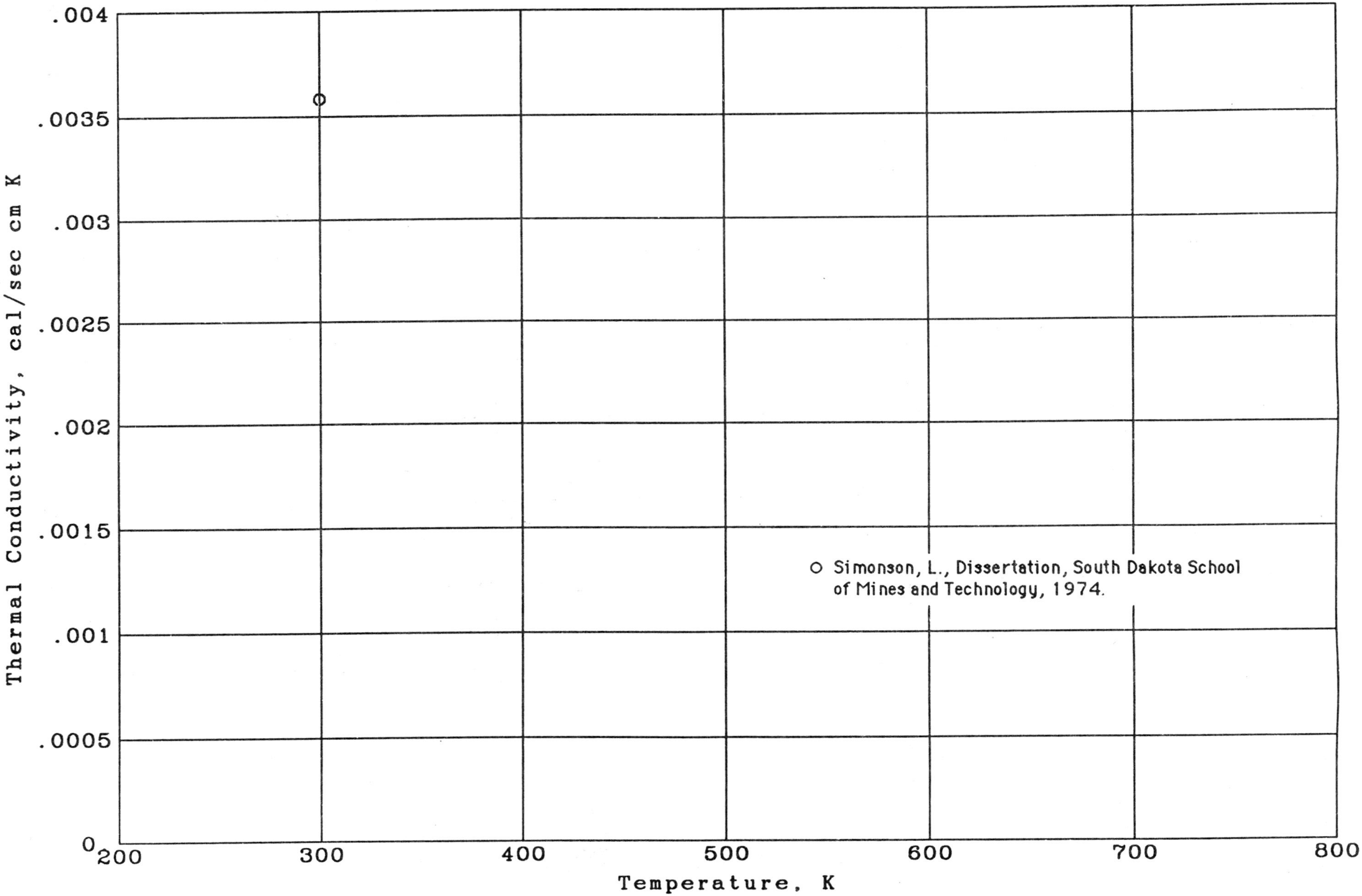

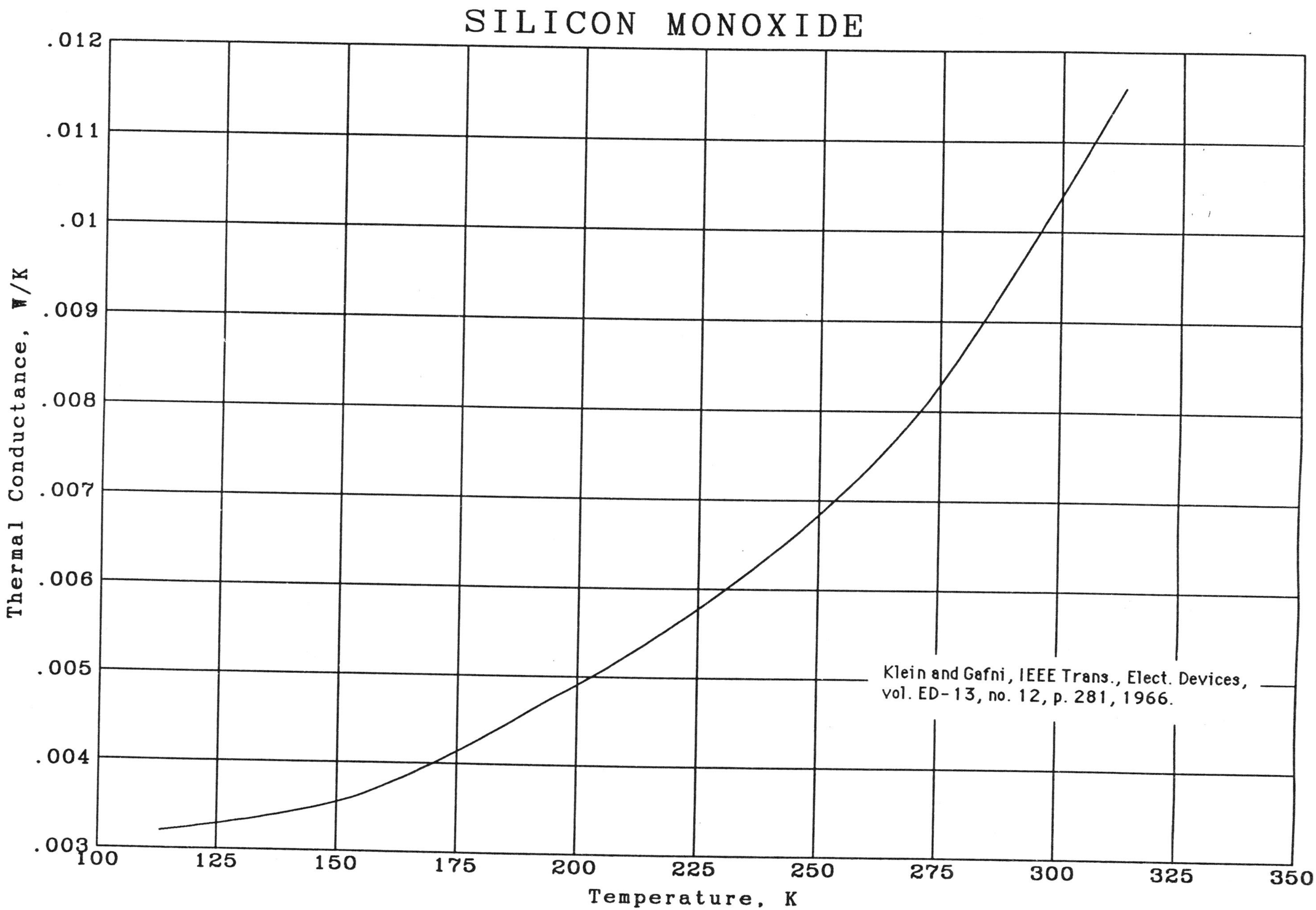

SILICON MONOXIDE
.012
.011
.01
.009
.008
.007
.006
.005
.004
.003
Thermal Conductance, W/K
100
125
150
175
200
225
250
275
300
325
350
Temperature, K
Klein and Gafni, IEEE Trans., Elect. Devices,
vol. ED-13, no. 12, p. 281, 1966.

SILICON MONOXIDE

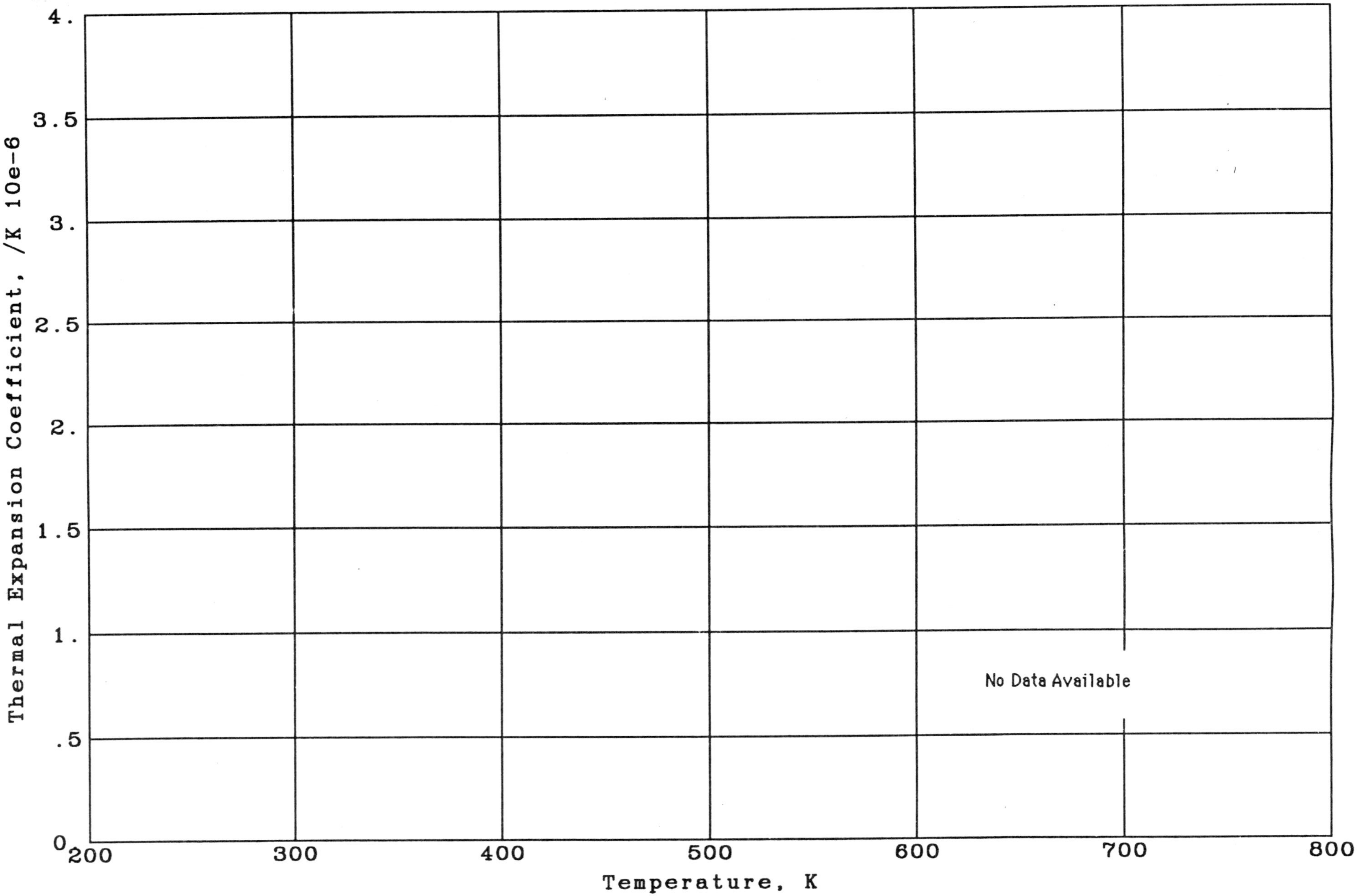

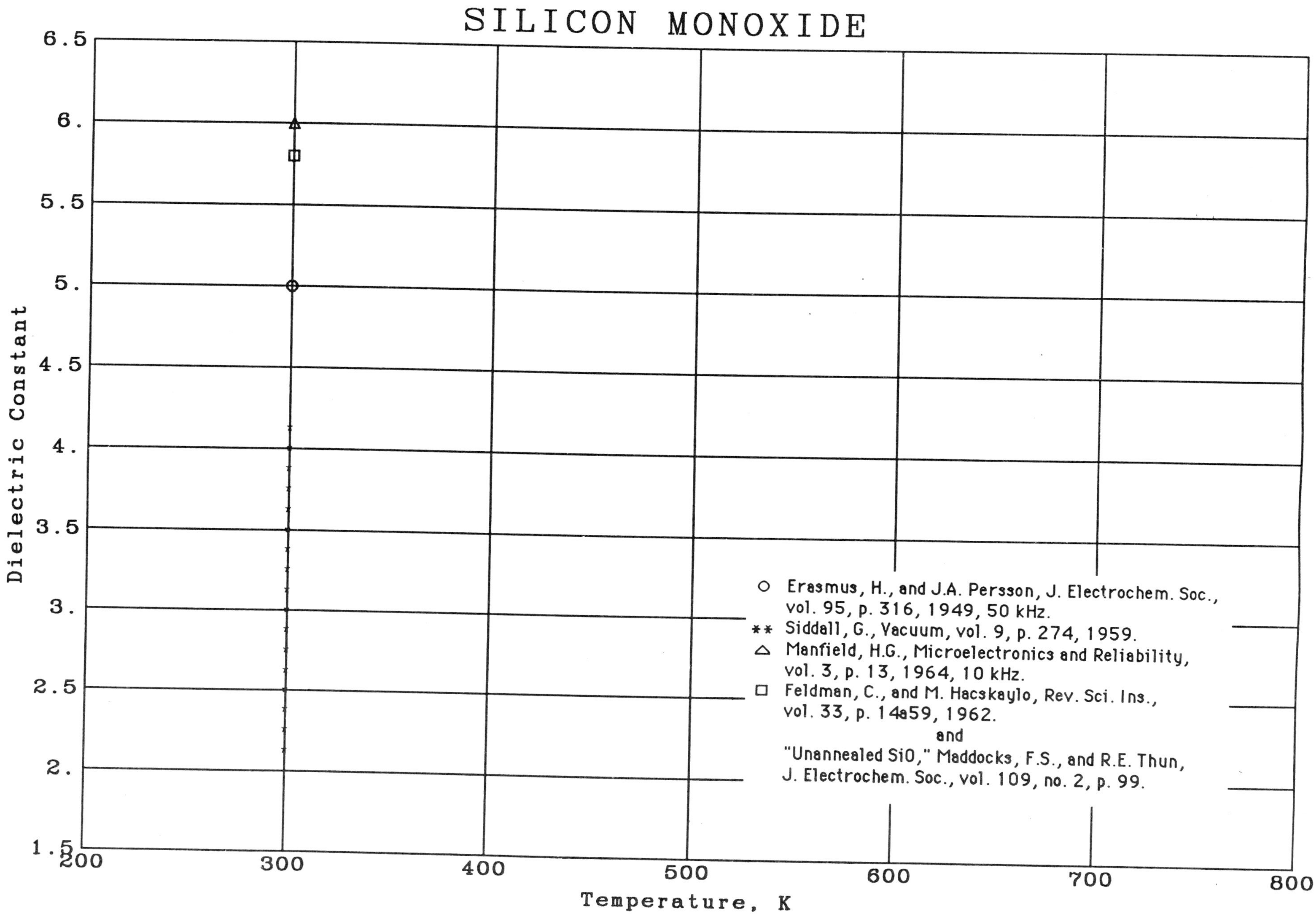
SILICON MONOXIDE
6.5
6.
5.5
5.
4.5
4.
3.5
3.
2.5
2.
1.5
Dielectric Constant
200
300
400
500
600
700
800
Temperature, K
○ Erasmus, H., and J.A. Persson, J. Electrochem. Soc., vol. 95, p. 316, 1949, 50 kHz.
** Siddall, G., Vacuum, vol. 9, p. 274, 1959.
△ Manfield, H.G., Microelectronics and Reliability, vol. 3, p. 13, 1964, 10 kHz.
□ Feldman, C., and M. Hacskaylo, Rev. Sci. Ins., vol. 33, p. 14a59, 1962.
and
"Unannealed SiO," Maddocks, F.S., and R.E. Thun, J. Electrochem. Soc., vol. 109, no. 2, p. 99.

SILICON MONOXIDE

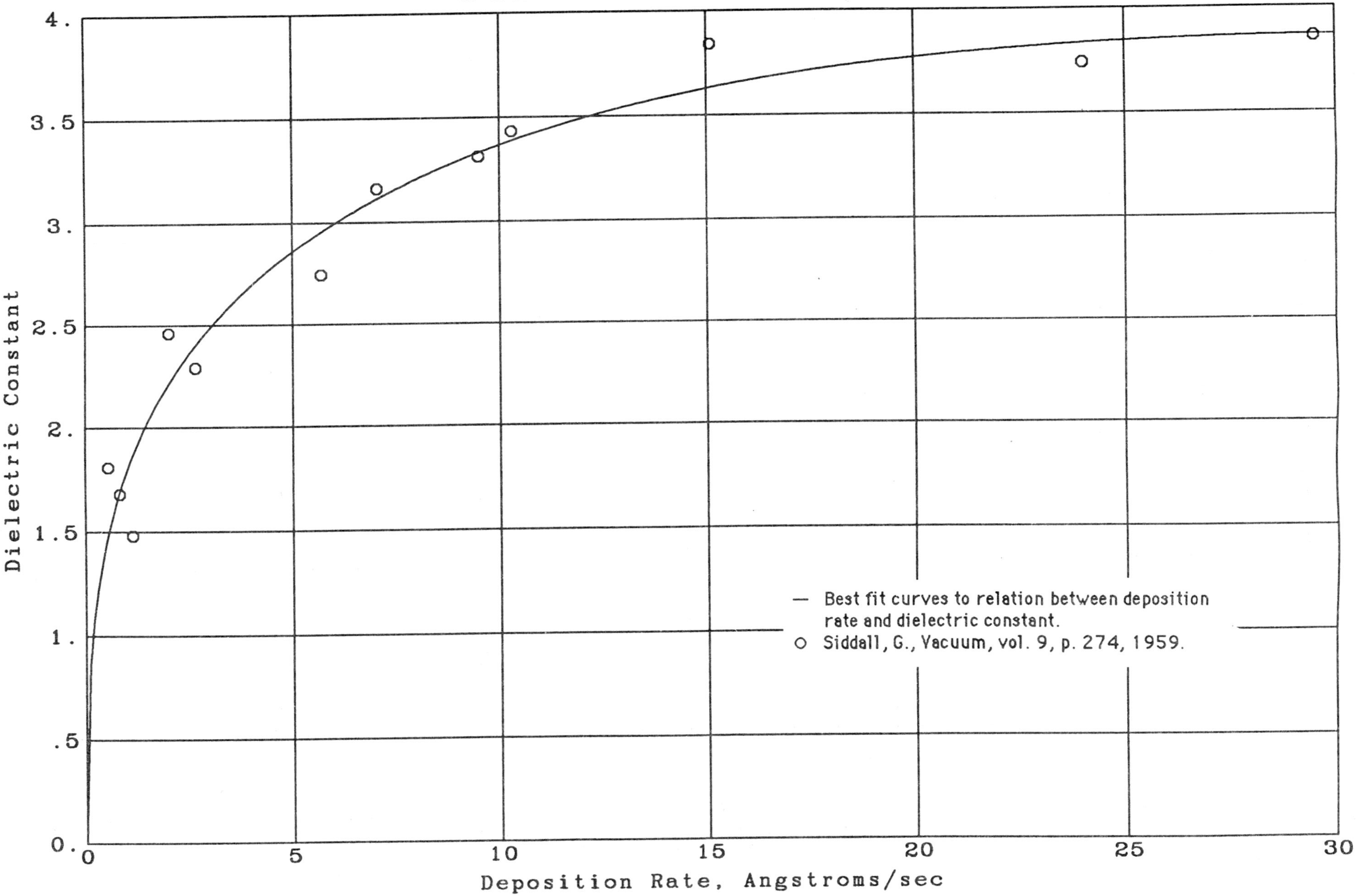

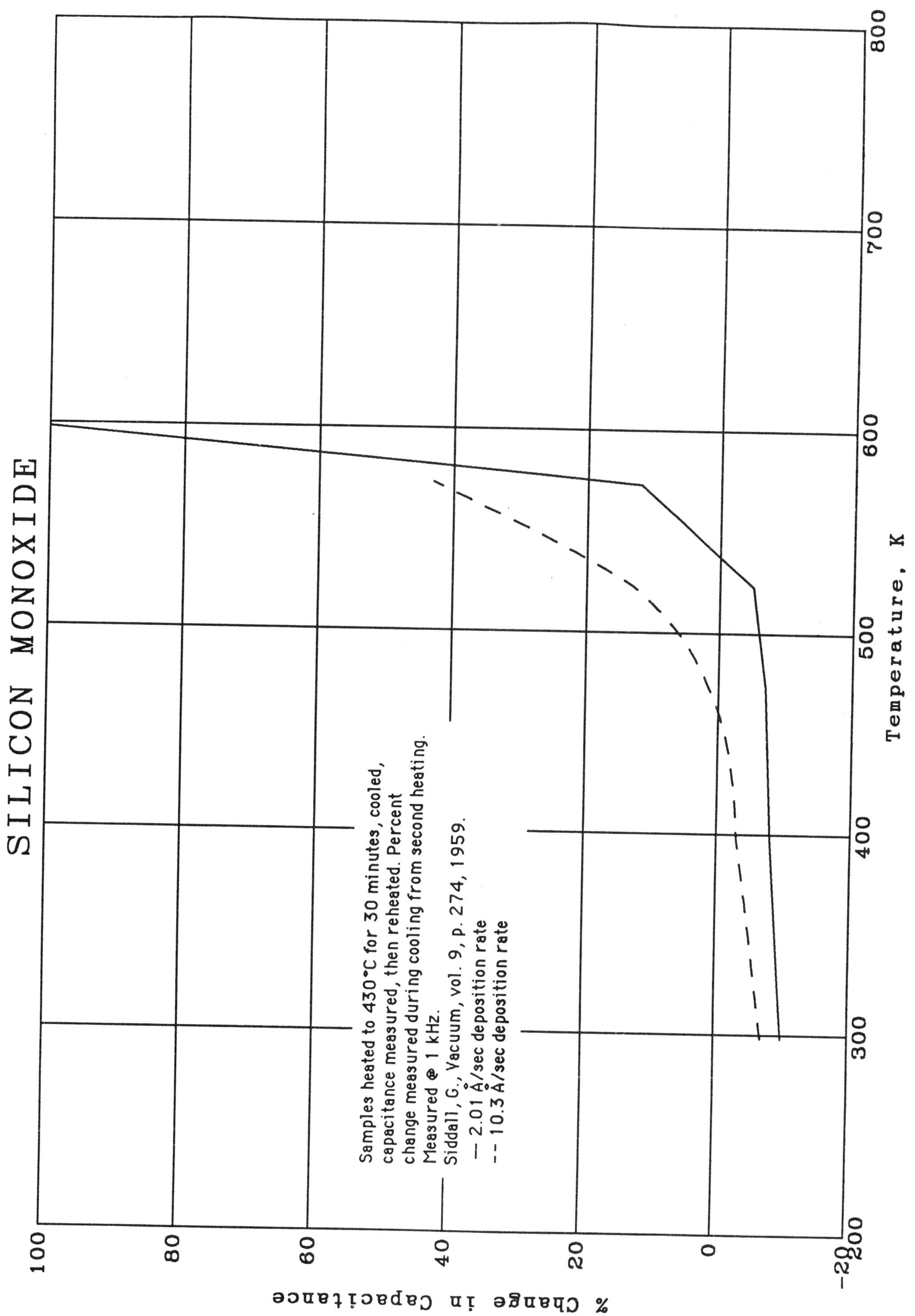
SILICON MONOXIDE
Samples heated to 430°C for 30 minutes, cooled, capacitance measured, then reheated. Percent change measured during cooling from second heating.
Measured @ 1 kHz.
Siddall, G., Vacuum, vol. 9, p. 274, 1959.
— 2.01 Å/sec deposition rate
-- 10.3 Å/sec deposition rate
% Change in Capacitance
Temperature, K
100
80
60
40
20
0
-20
200
300
400
500
600
700
800

SILICON MONOXIDE

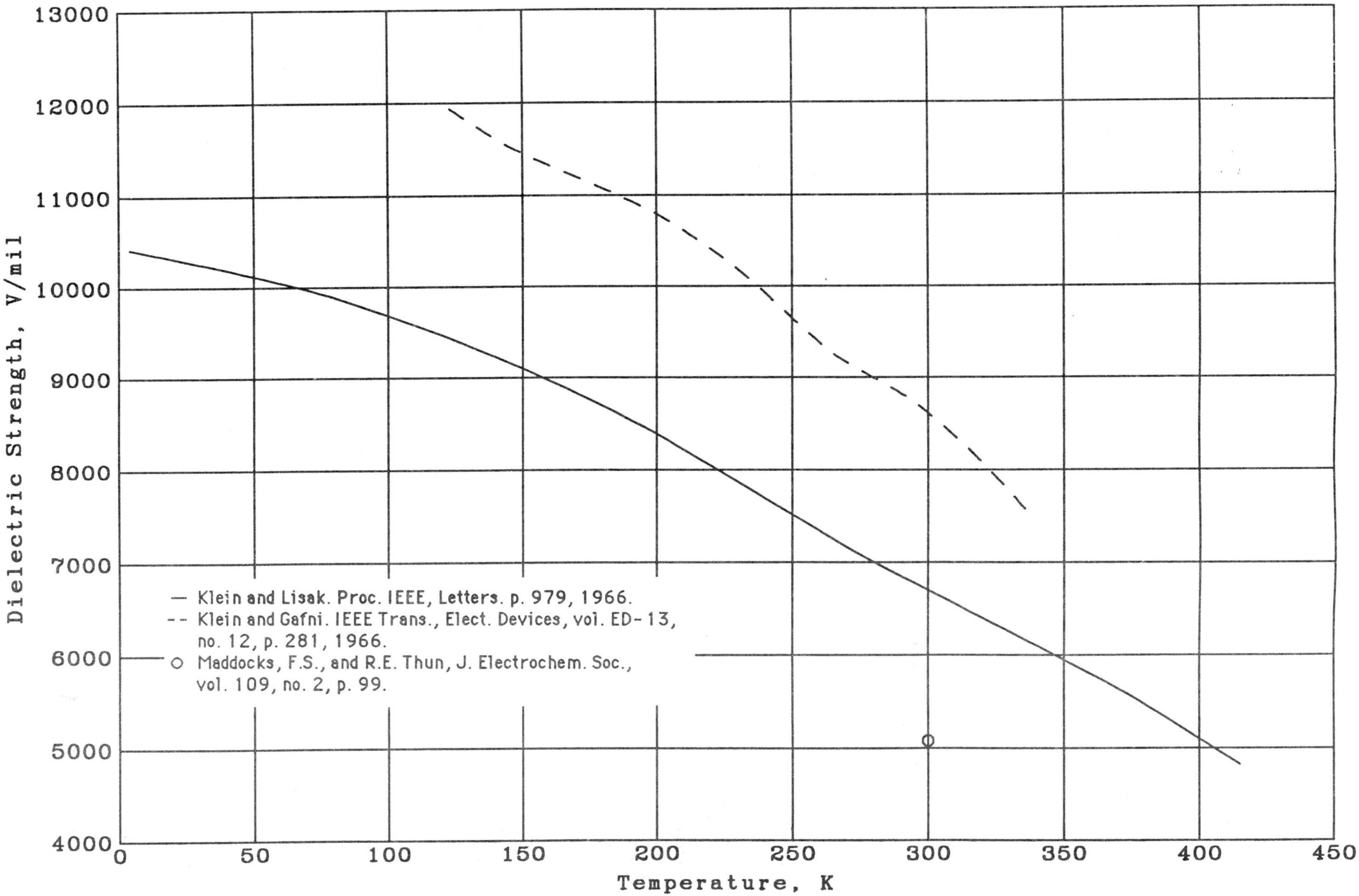

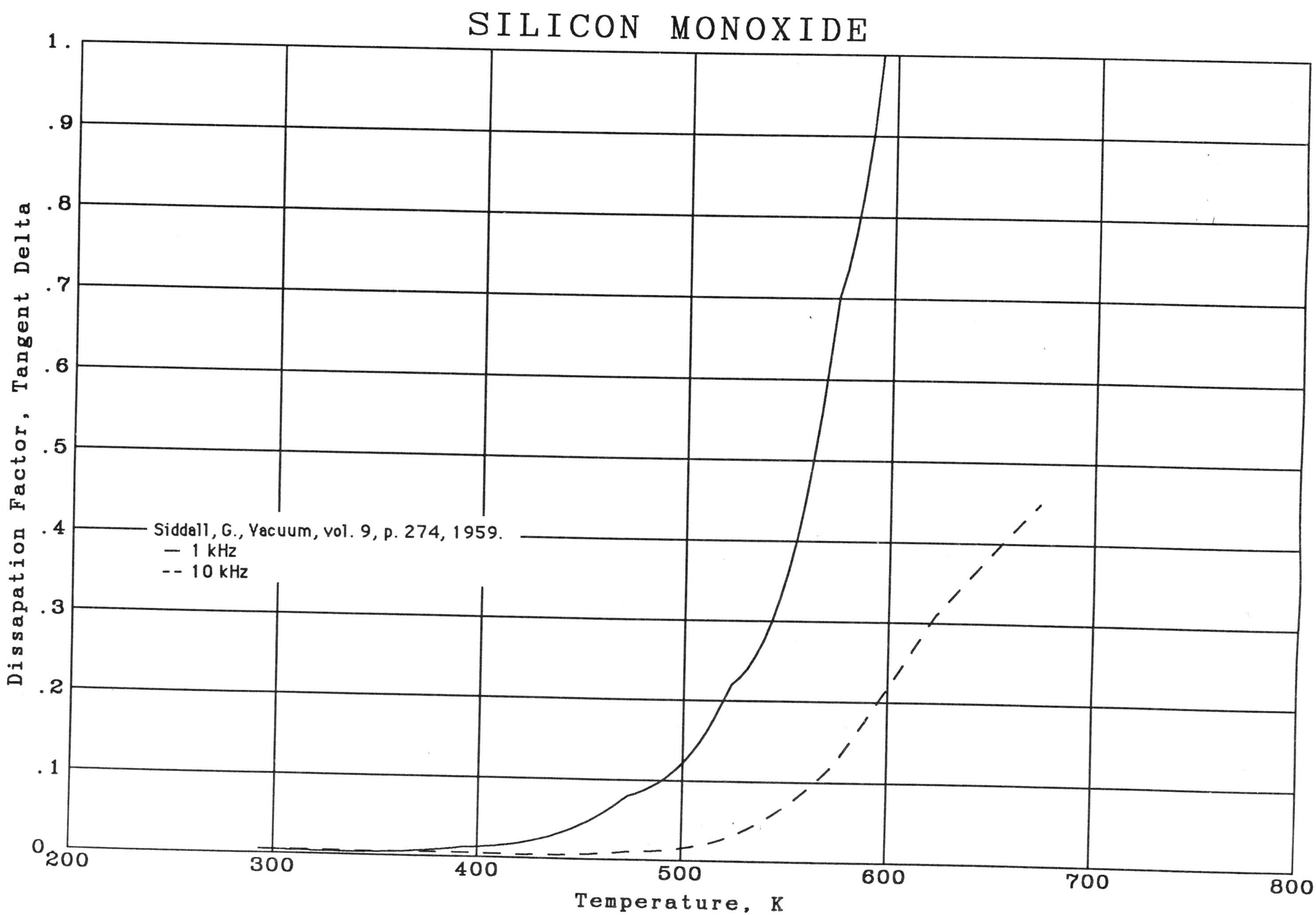
SILICON MONOXIDE
Dissapation Factor, Tangent Delta
Temperature, K
Siddall, G., Vacuum, vol. 9, p. 274, 1959.
— 1 kHz
-- 10 kHz
0
.1
.2
.3
.4
.5
.6
.7
.8
.9
1.
200
300
400
500
600
700
800

SILICON MONOXIDE

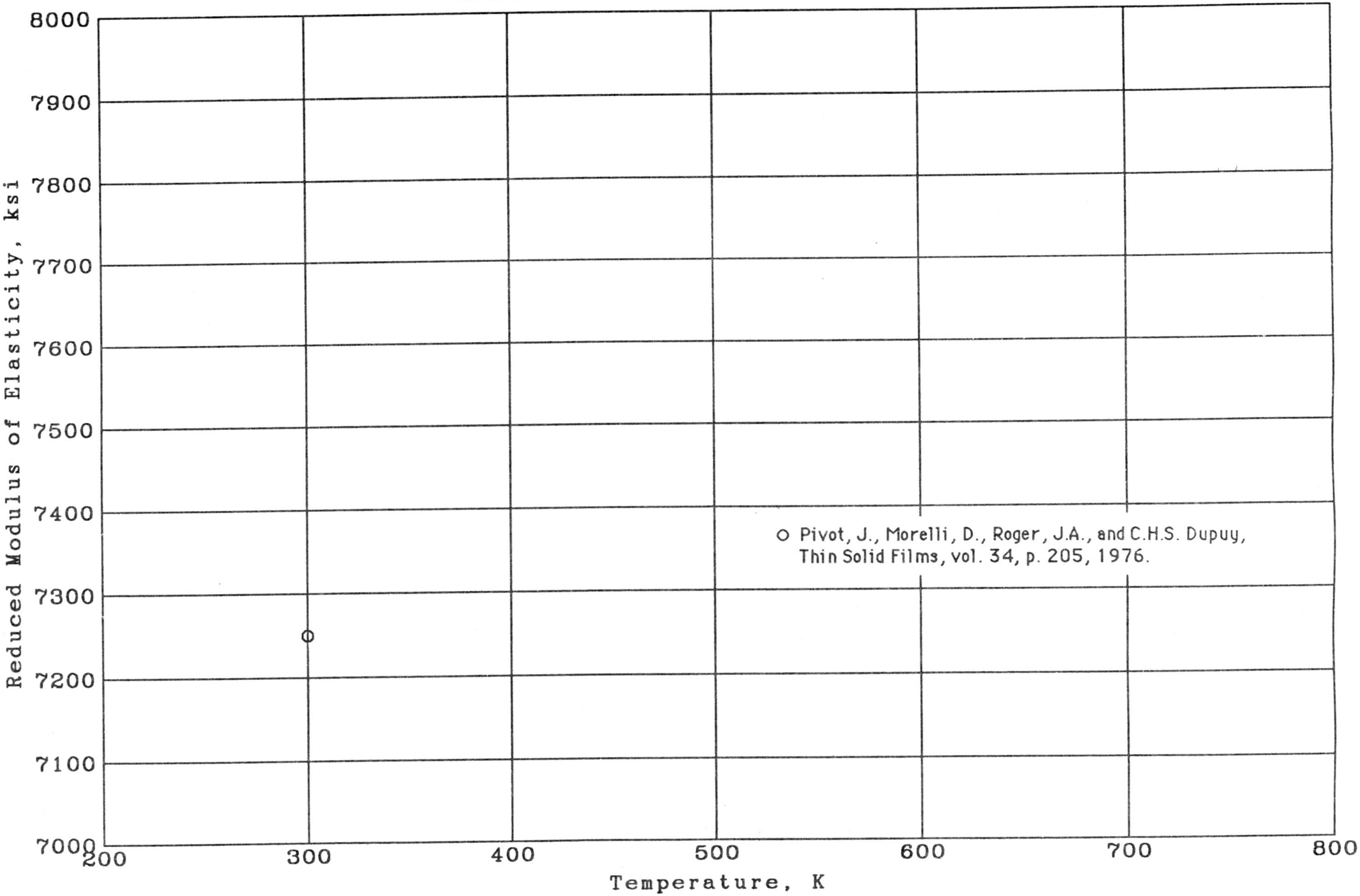

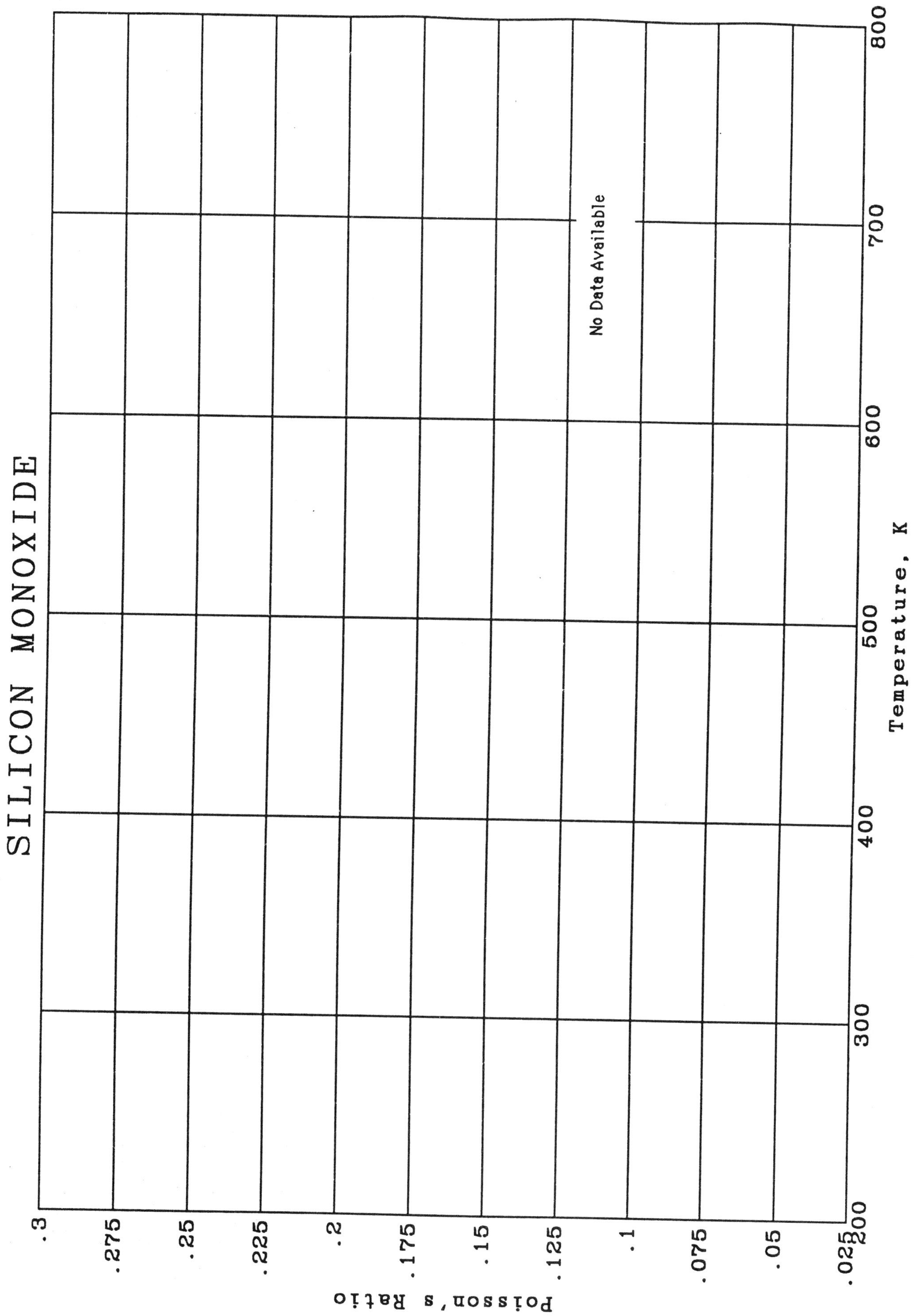
SILICON MONOXIDE
No Data Available
Poisson's Ratio
Temperature, K
.3
.275
.25
.225
.2
.175
.15
.125
.1
.075
.05
.025
200
300
400
500
600
700
800

SILICON MONOXIDE

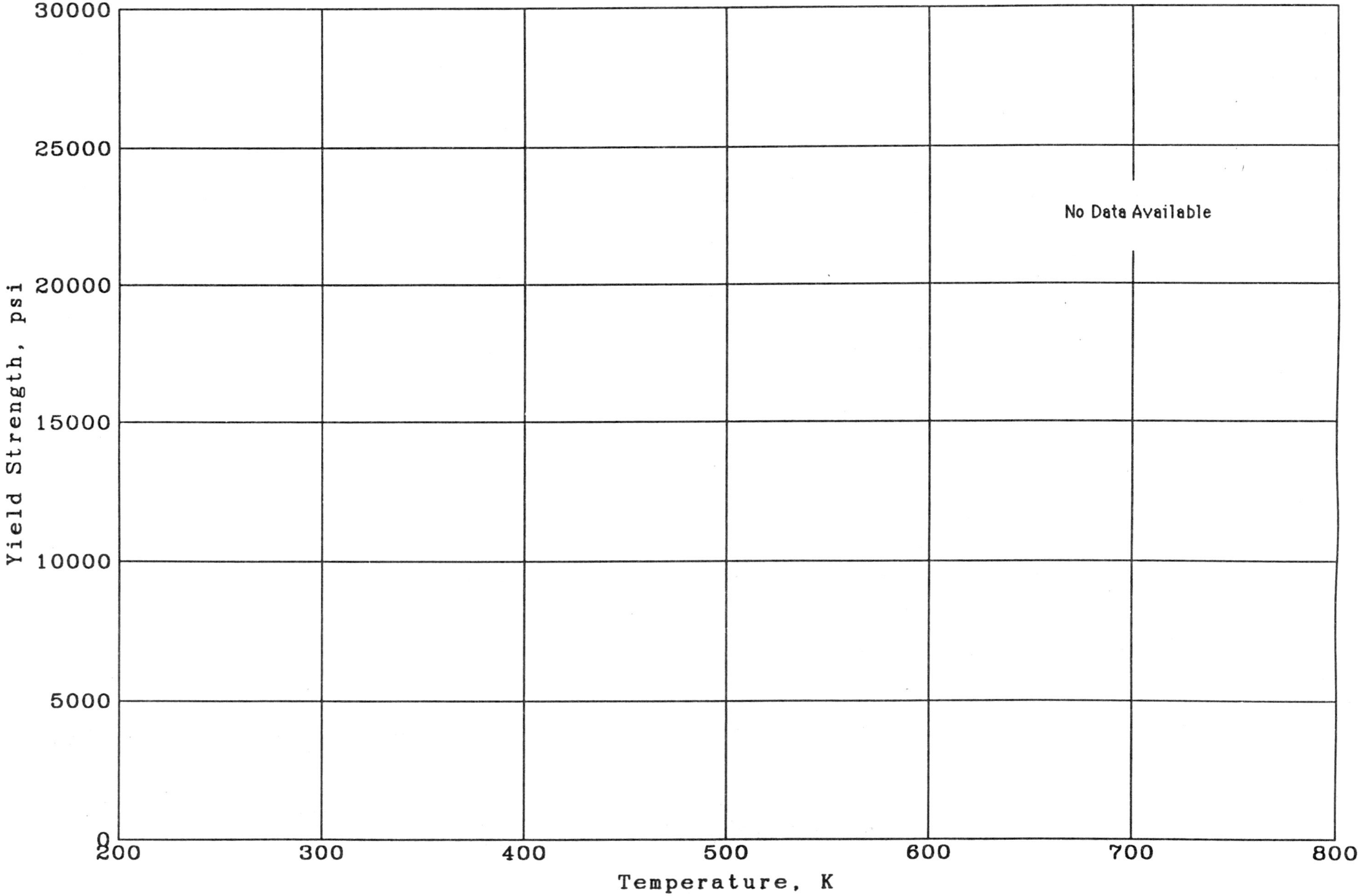

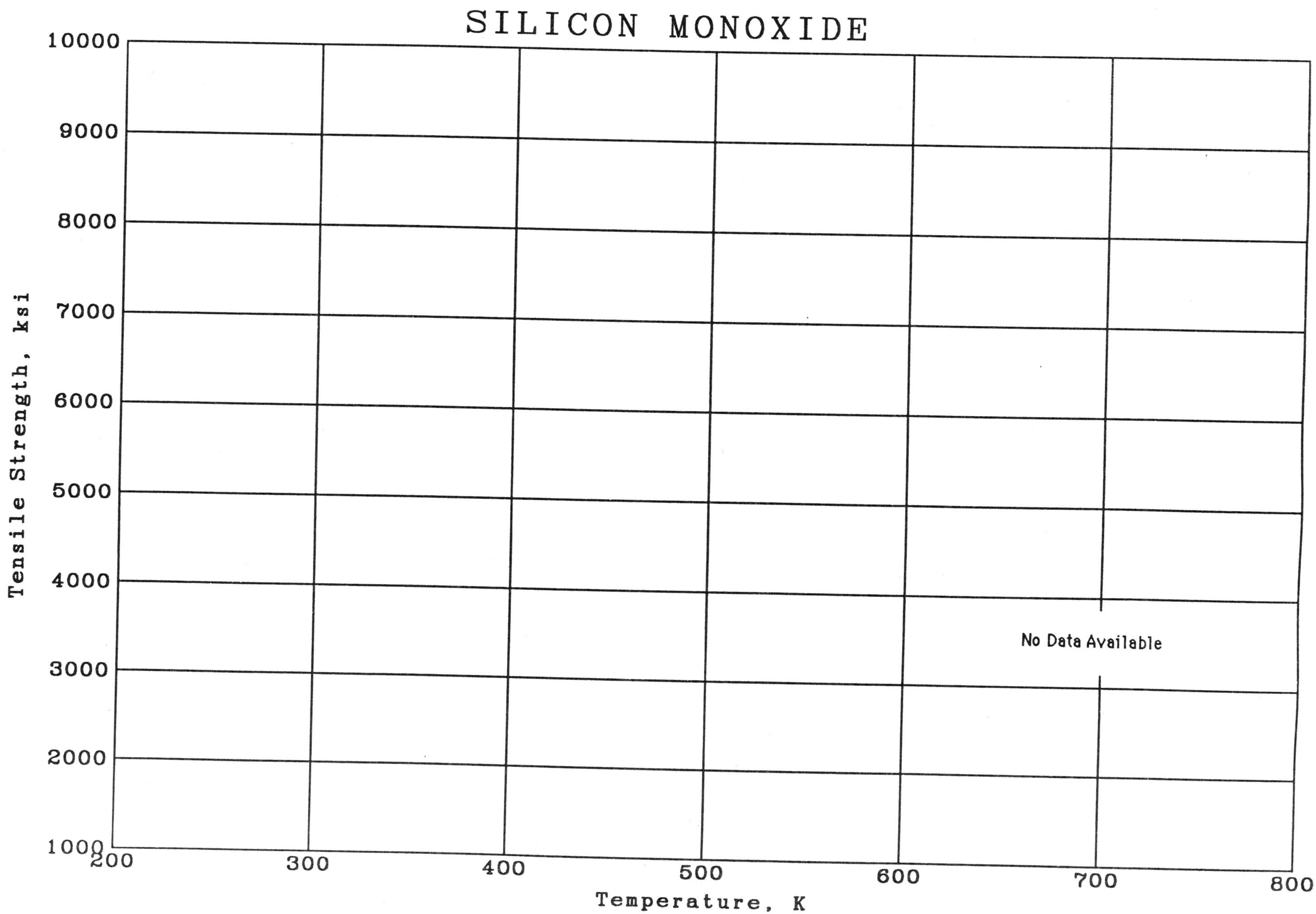
SILICON MONOXIDE
10000
9000
8000
7000
6000
5000
4000
3000
2000
1000
Tensile Strength, ksi
200
300
400
500
600
700
800
Temperature, K
No Data Available

SILICON MONOXIDE

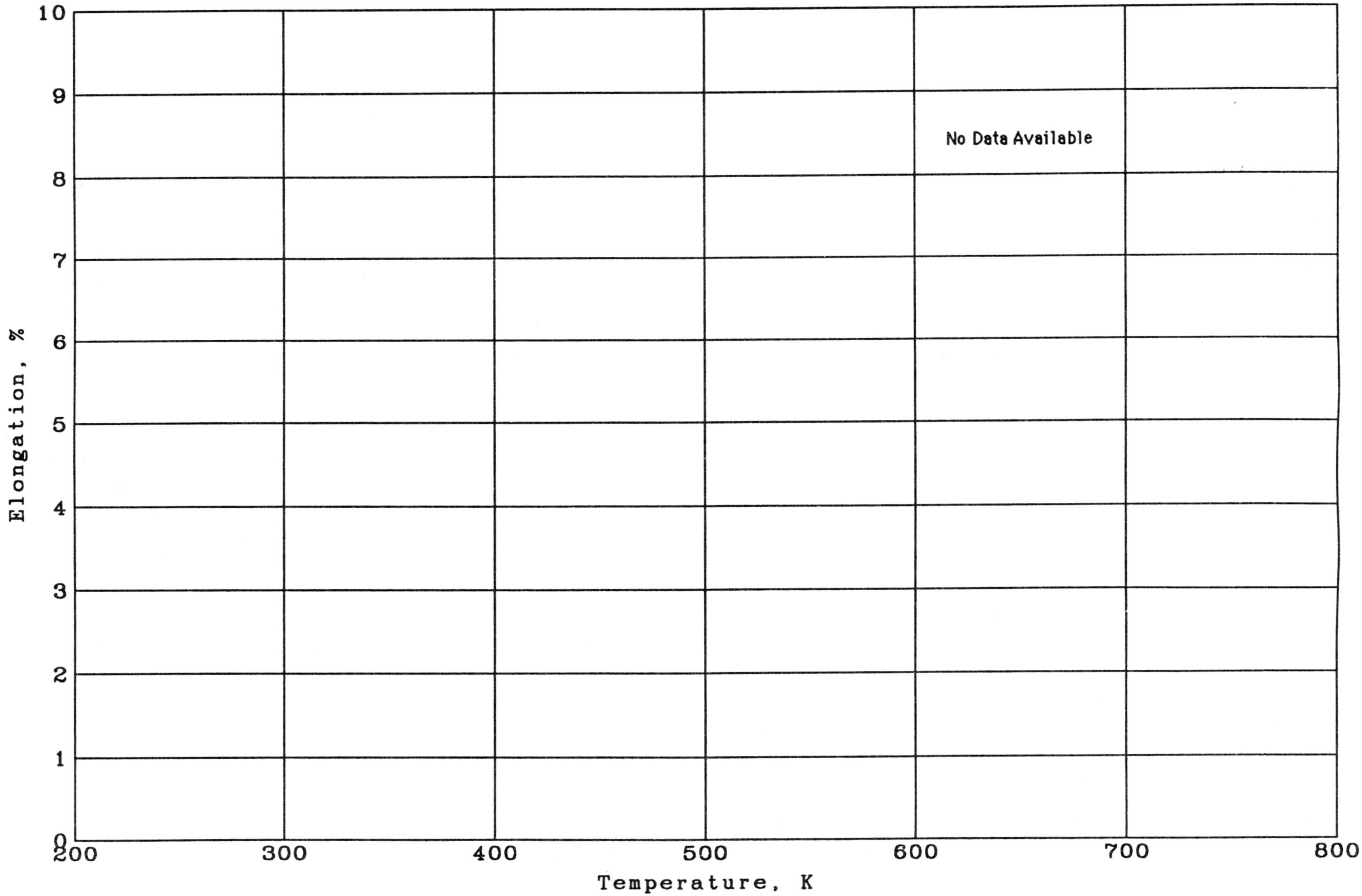

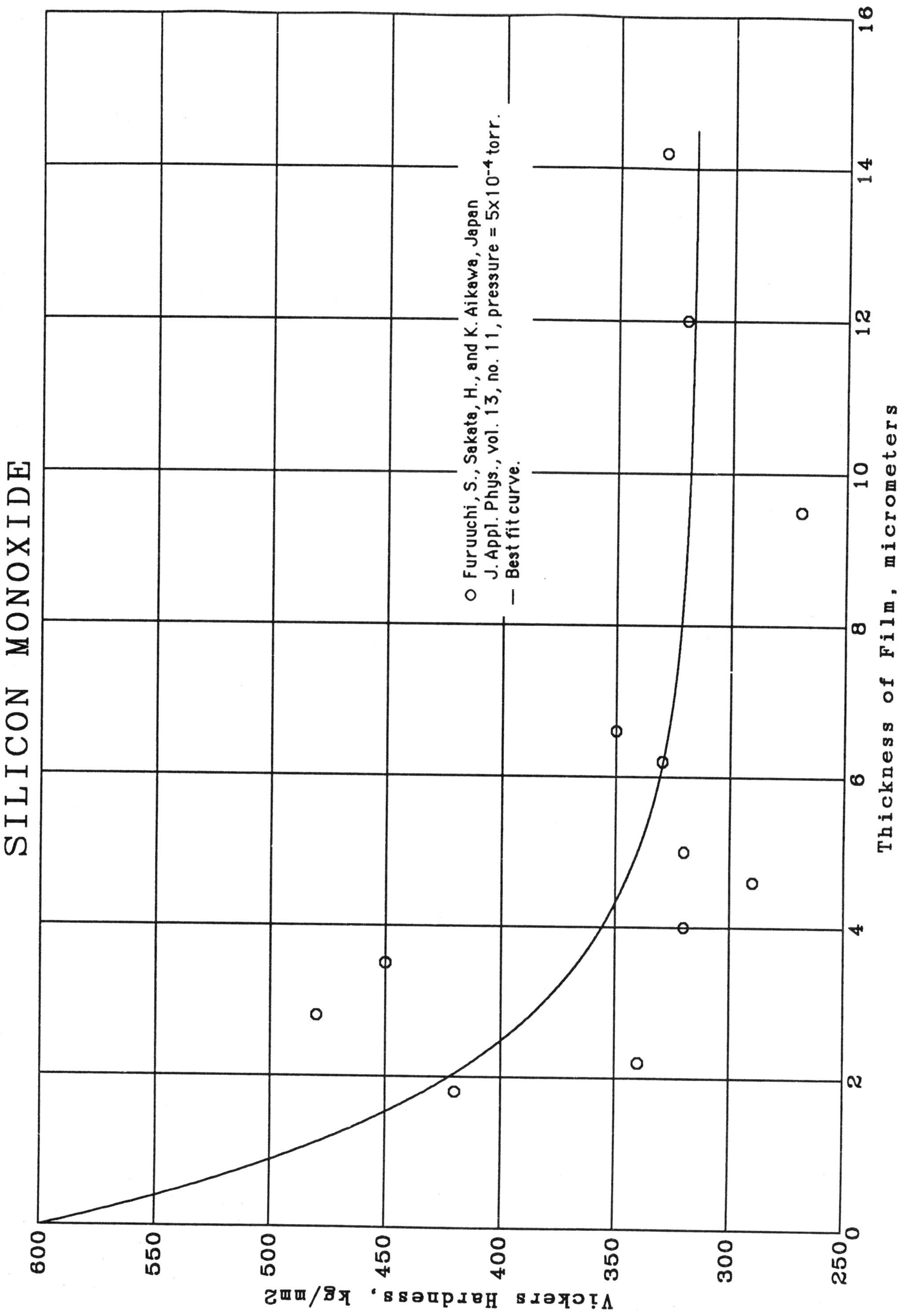
SILICON MONOXIDE
Furuuchi, S., Sakata, H., and K. Aikawa, Japan
J. Appl. Phys., vol. 13, no. 11, pressure = 5x10⁻⁴ torr.
Best fit curve.
Thickness of Film, micrometers
Vickers Hardness, kg/mm2
0
2
4
6
8
10
12
14
16
600
550
500
450
400
350
300
250

SILICON MONOXIDE

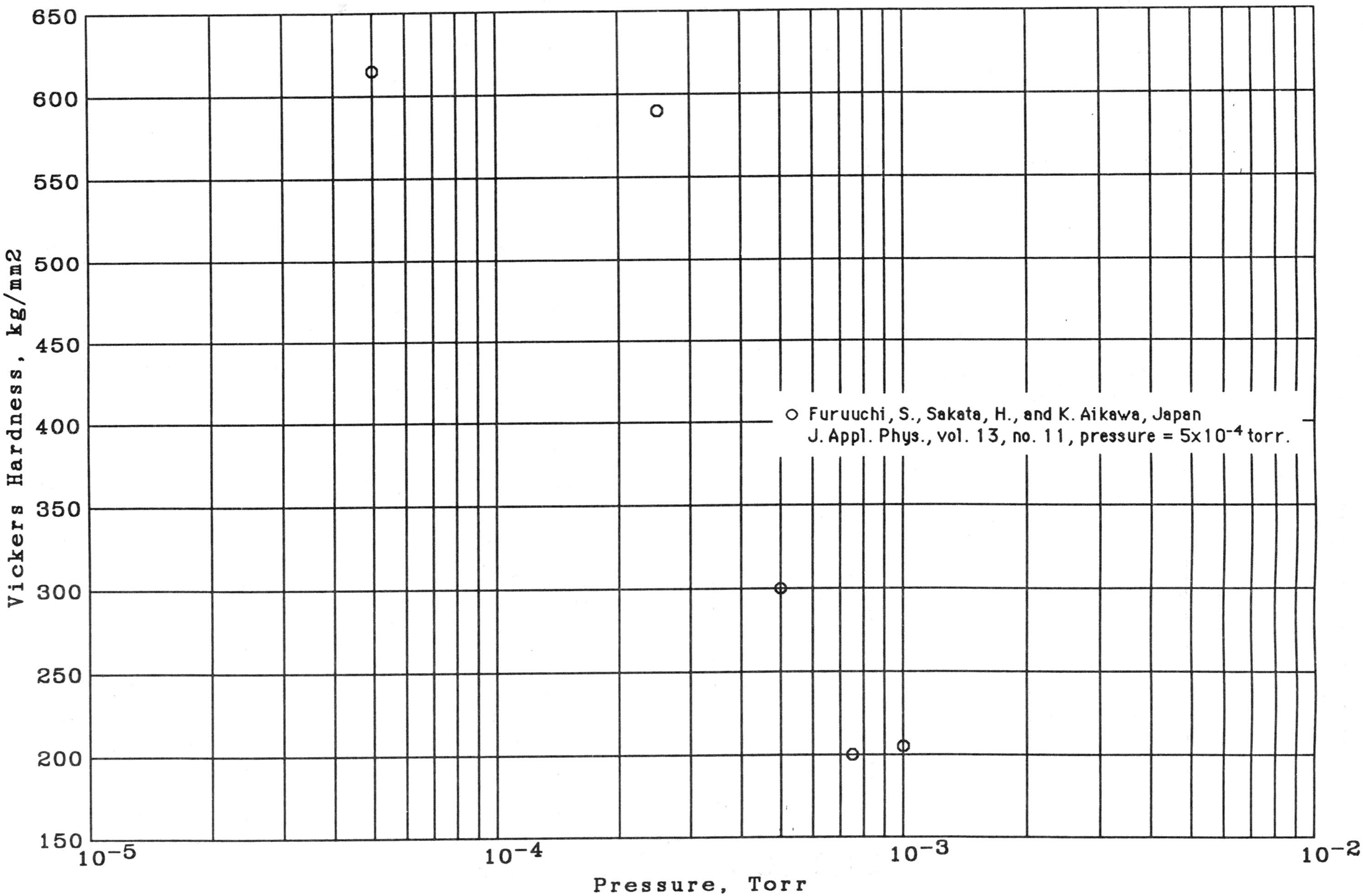

TANTALA[1]

PROPERTY	@ 300K	UNITS	REFERENCES
Density	8.0	g/cm^3	T1 T2 T3
Specific Heat	0.073	cal/g-K	T4 T5 T6
Thermal Conductivity		cal/cm-s-K	
Thermal Exp Coef	$1.2x10^{-6}$	1/K	T7 T8 T14
Dielectric Constant	25	------	T3 T9 T10
Dissipation Factor	.005	------	T9 T10 T11
Dielectric Strength	17,000	V/mil	T3 T9 T11
Young's Modulus 2	44,800	ksi	T12
Poisson's Ratio 2		------	T12
Yield Strength 3	13,000	psi	T12
Ult Tensile Strength[4]	50,000	psi	T12
Elongation	0.87	%	T12
Hardness	470	kg/mm^2	T13

1 Specific heat, thermal expansion coefficient and hardness are for crystalline material. All other properties are for anodically formed oxide.

2 The bulge test was used to determine the reduced Young's modulus and strengths of films of various thicknesses. No direct measurement of Poisson's ratio is found. The reduced modulus is found to be effectively constant for film thicknesses between 1500 and 2000 angstroms, decreasing at 3000 angstroms.

3 Ta_2O_5 films display anelastic behavior. No true yield point is observed. The stress at 0.2% biaxial strain is the value listed for a 2000 angstrom thick film.

4 The maximum fracture strength occurs at 2000 angstroms thickness, and drops abruptly at 3000 angstroms. The reduction in mechanical properties in thicker films is reportedly associated with a higher degree of crystallinity. Ta_2O_5 films are found to contain a residual tensile stress of 7000-15,000 psi. Anodic polarization in a 0.1M aqueous ammonium tartrate at a constant current density of 25 mA/cm^2 for 4 minutes provides the maximum strength and nearly the highest residual stress, compared to other time periods of polarization.

TANTALA

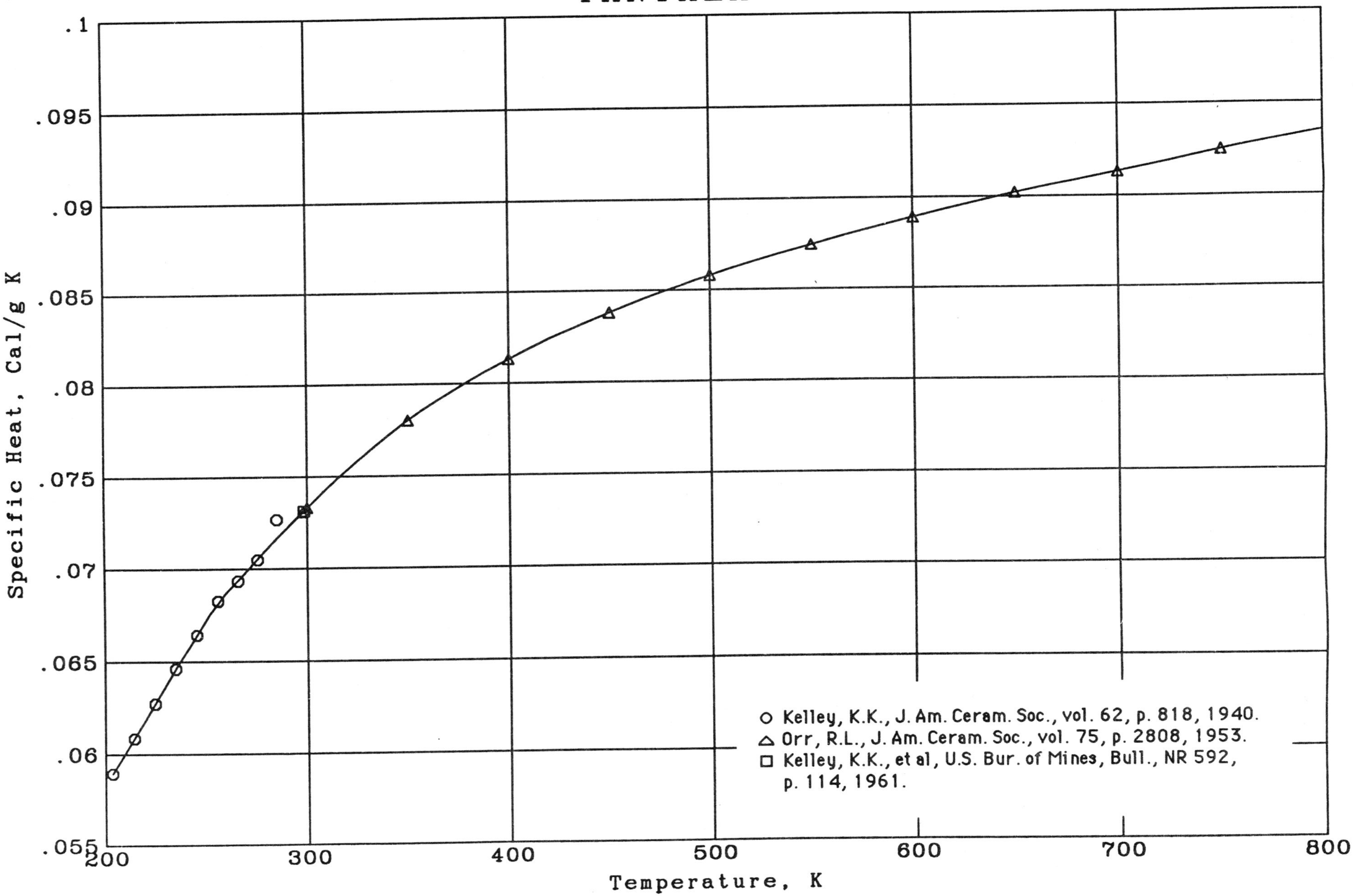
.1
.095
.09
.085
.08
.075
.07
.065
.06
.055
Specific Heat, Cal/g K
200
300
400
500
600
700
800
Temperature, K
○ Kelley, K.K., J. Am. Ceram. Soc., vol. 62, p. 818, 1940.
△ Orr, R.L., J. Am. Ceram. Soc., vol. 75, p. 2808, 1953.
□ Kelley, K.K., et al, U.S. Bur. of Mines, Bull., NR 592, p. 114, 1961.

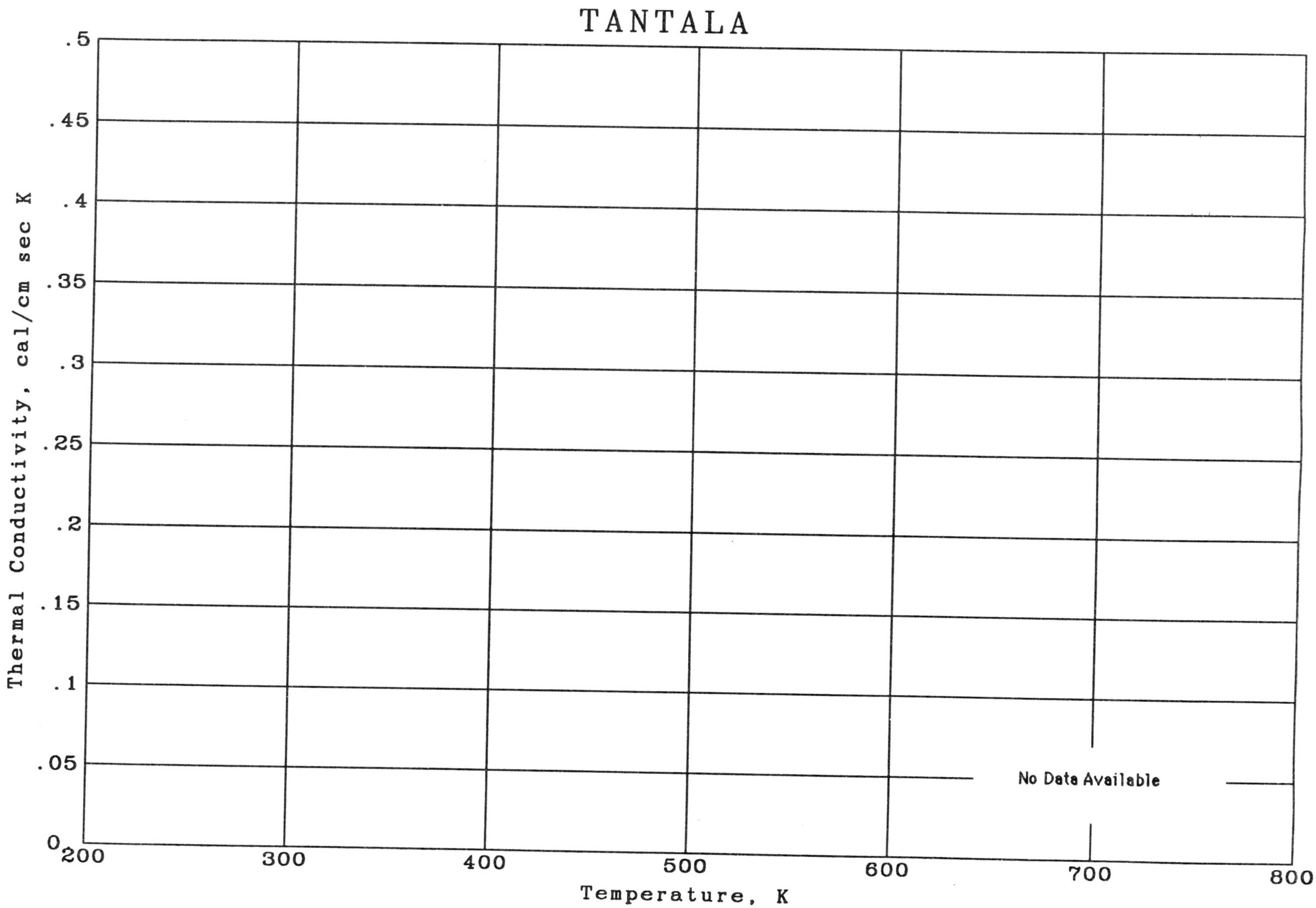

TANTALA
.5
.45
.4
.35
.3
.25
.2
.15
.1
.05
0
Thermal Conductivity, cal/cm sec K
200
300
400
500
600
700
800
Temperature, K
No Data Available

TANTALA

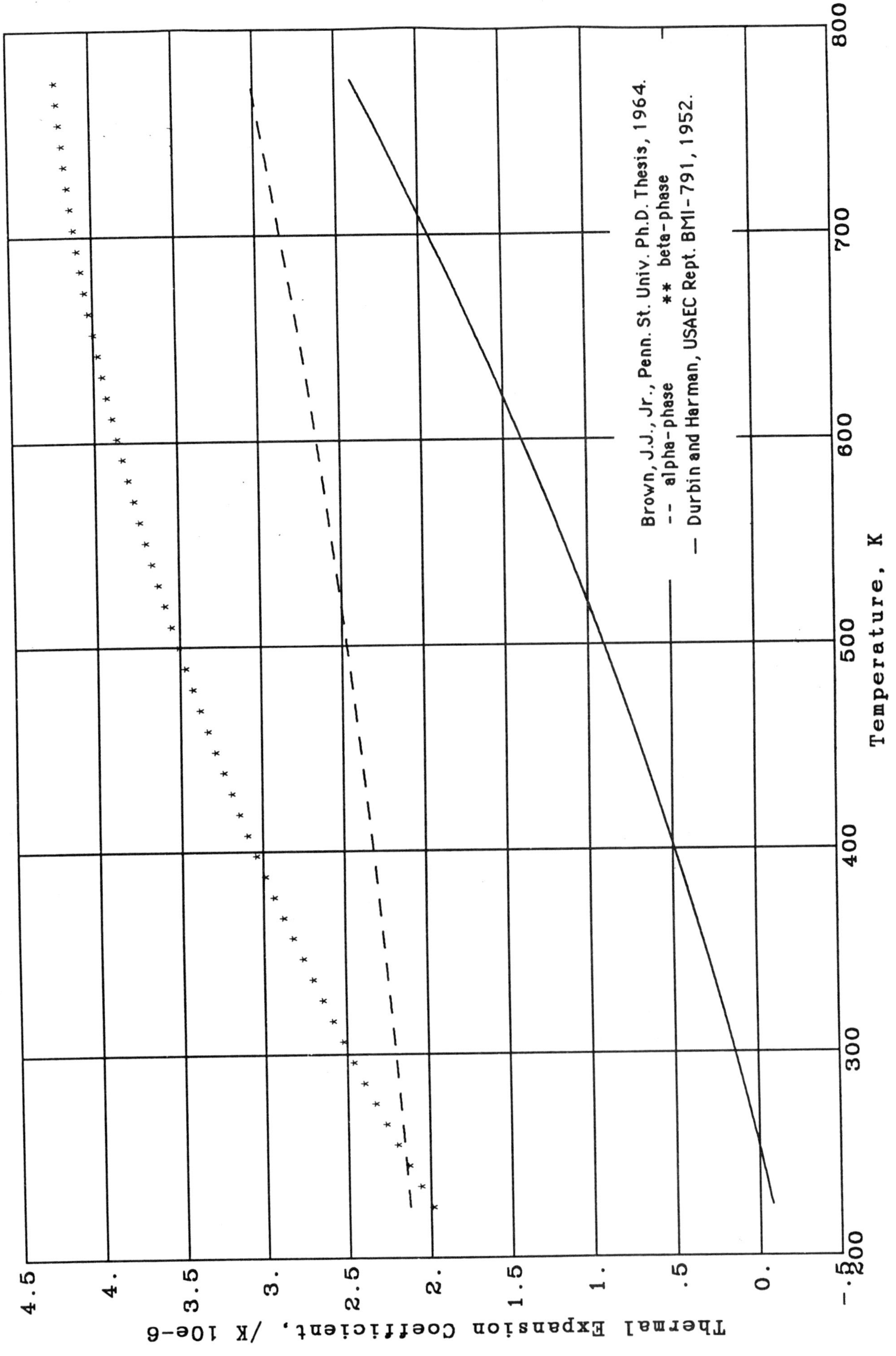

Brown, J.J., Jr., Penn. St. Univ. Ph.D. Thesis, 1964.
-- alpha-phase ** beta-phase
— Durbin and Harman, USAEC Rept. BMI-791, 1952.
Temperature, K
200
300
400
500
600
700
800
Thermal Expansion Coefficient, /K 10e-6
-.5
0.
.5
1.
1.5
2.
2.5
3.
3.5
4.
4.5

TANTALA

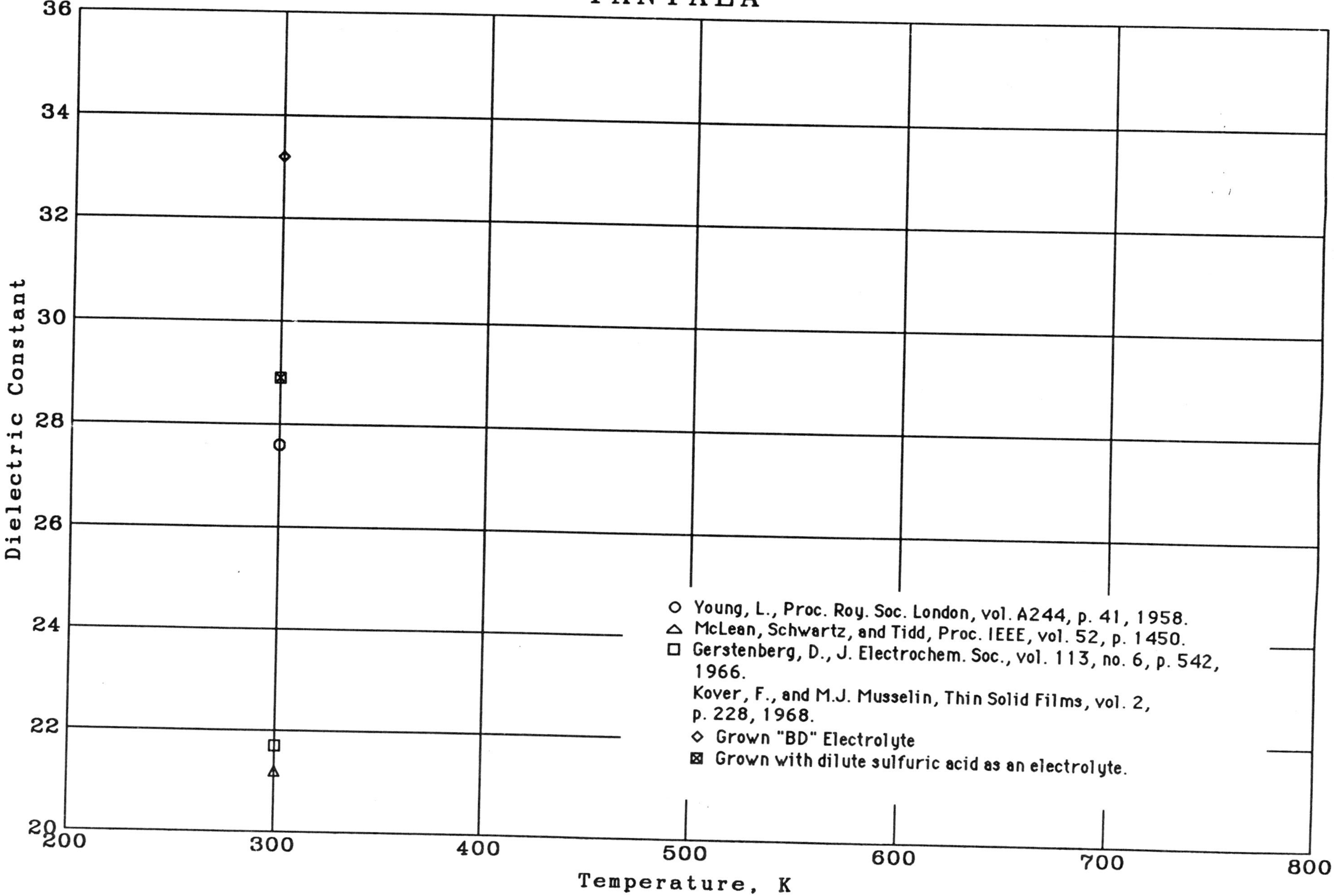

Dielectric Constant
20
22
24
26
28
30
32
34
36
200
300
400
500
600
700
800
Temperature, K
○ Young, L., Proc. Roy. Soc. London, vol. A244, p. 41, 1958.
△ McLean, Schwartz, and Tidd, Proc. IEEE, vol. 52, p. 1450.
□ Gerstenberg, D., J. Electrochem. Soc., vol. 113, no. 6, p. 542, 1966.
Kover, F., and M.J. Musselin, Thin Solid Films, vol. 2, p. 228, 1968.
◇ Grown "BD" Electrolyte
⊠ Grown with dilute sulfuric acid as an electrolyte.

TANTALA

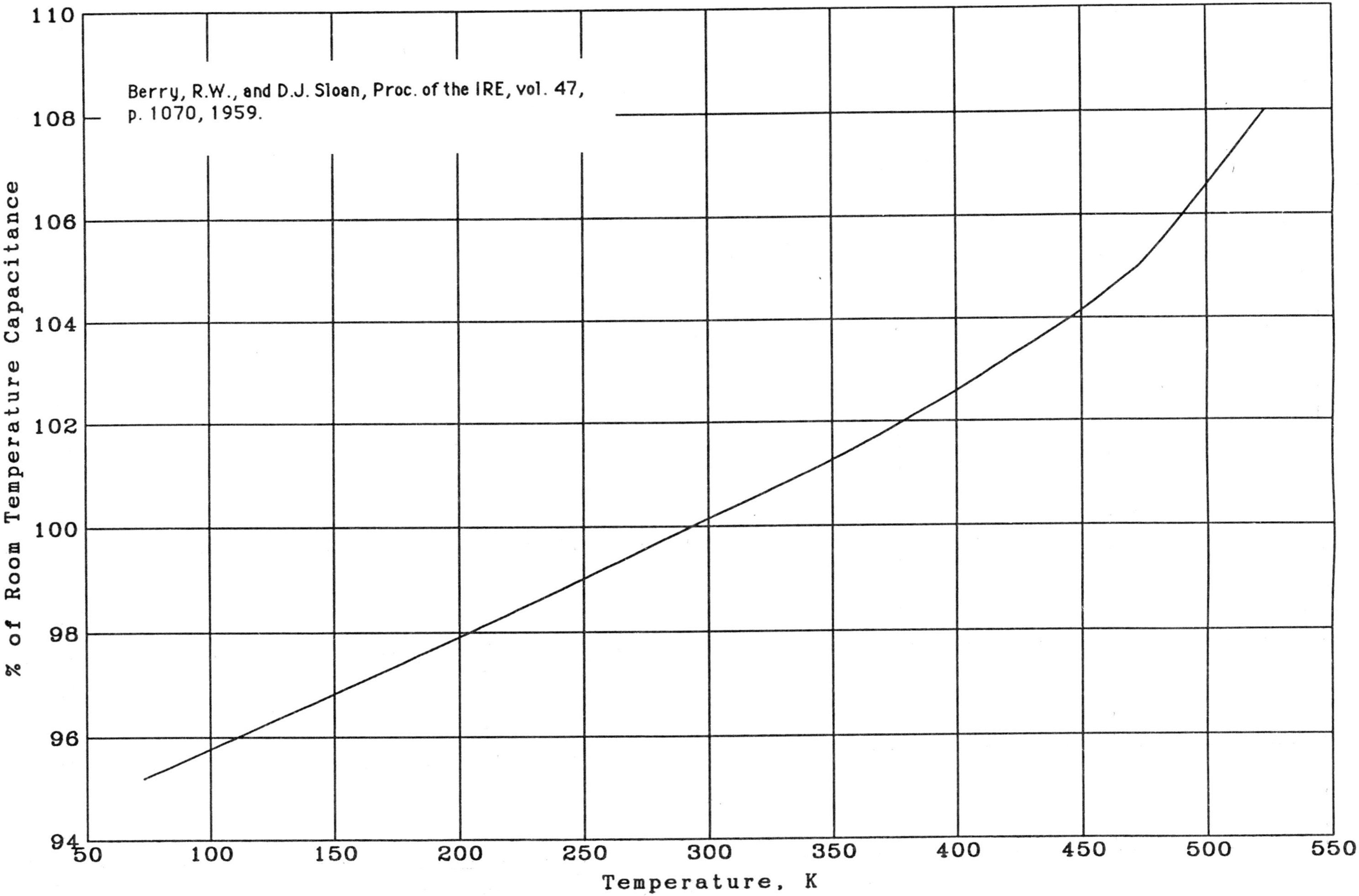

Berry, R.W., and D.J. Sloan, Proc. of the IRE, vol. 47, p. 1070, 1959.
110
108
106
104
102
100
98
96
94
% of Room Temperature Capacitance
50
100
150
200
250
300
350
400
450
500
550
Temperature, K

TANTALA

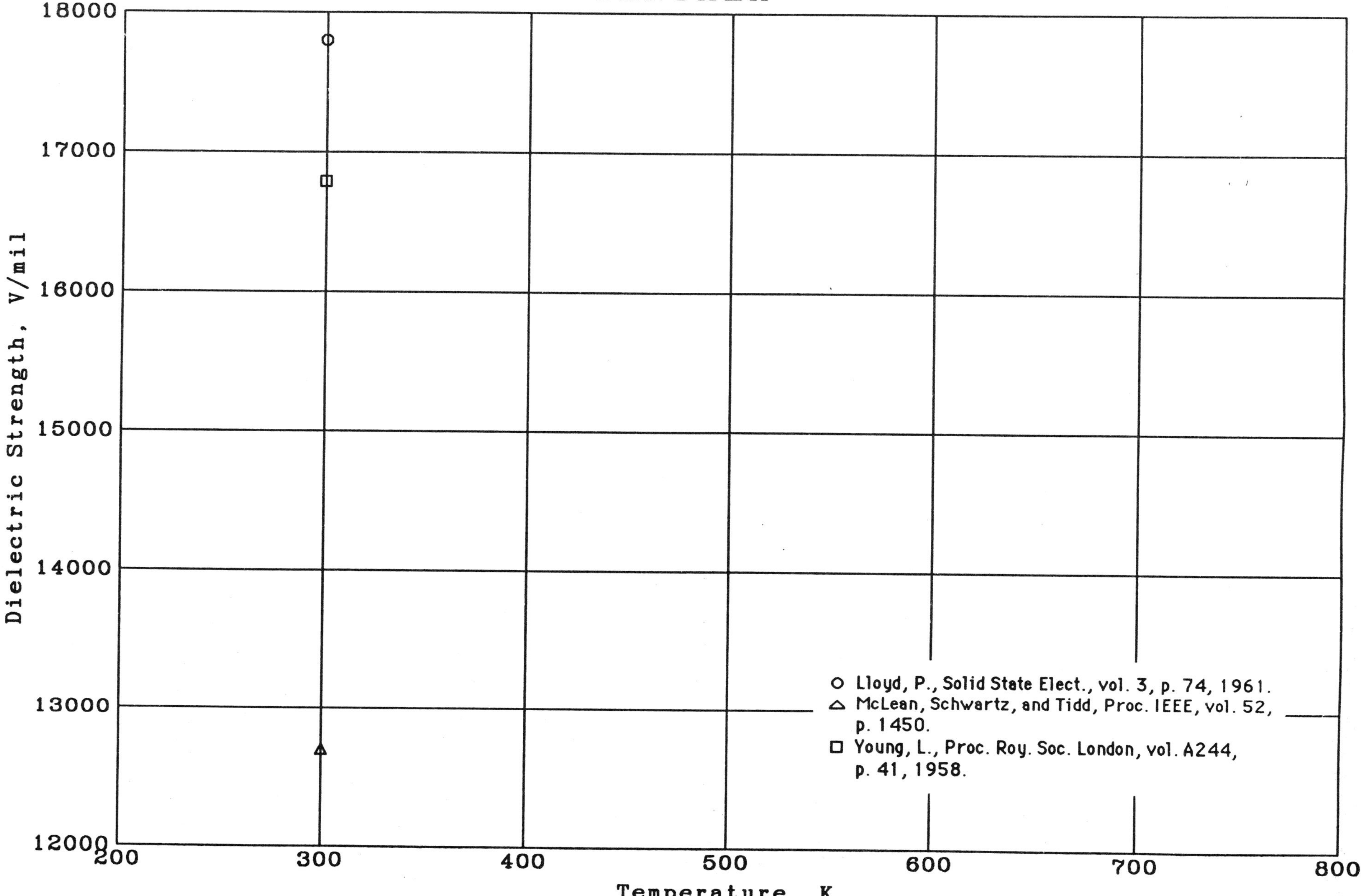
18000
17000
16000
15000
14000
13000
12000
Dielectric Strength, V/mil
200
300
400
500
600
700
800
Temperature, K
○ Lloyd, P., Solid State Elect., vol. 3, p. 74, 1961.
△ McLean, Schwartz, and Tidd, Proc. IEEE, vol. 52, p. 1450.
□ Young, L., Proc. Roy. Soc. London, vol. A244, p. 41, 1958.

TANTALA

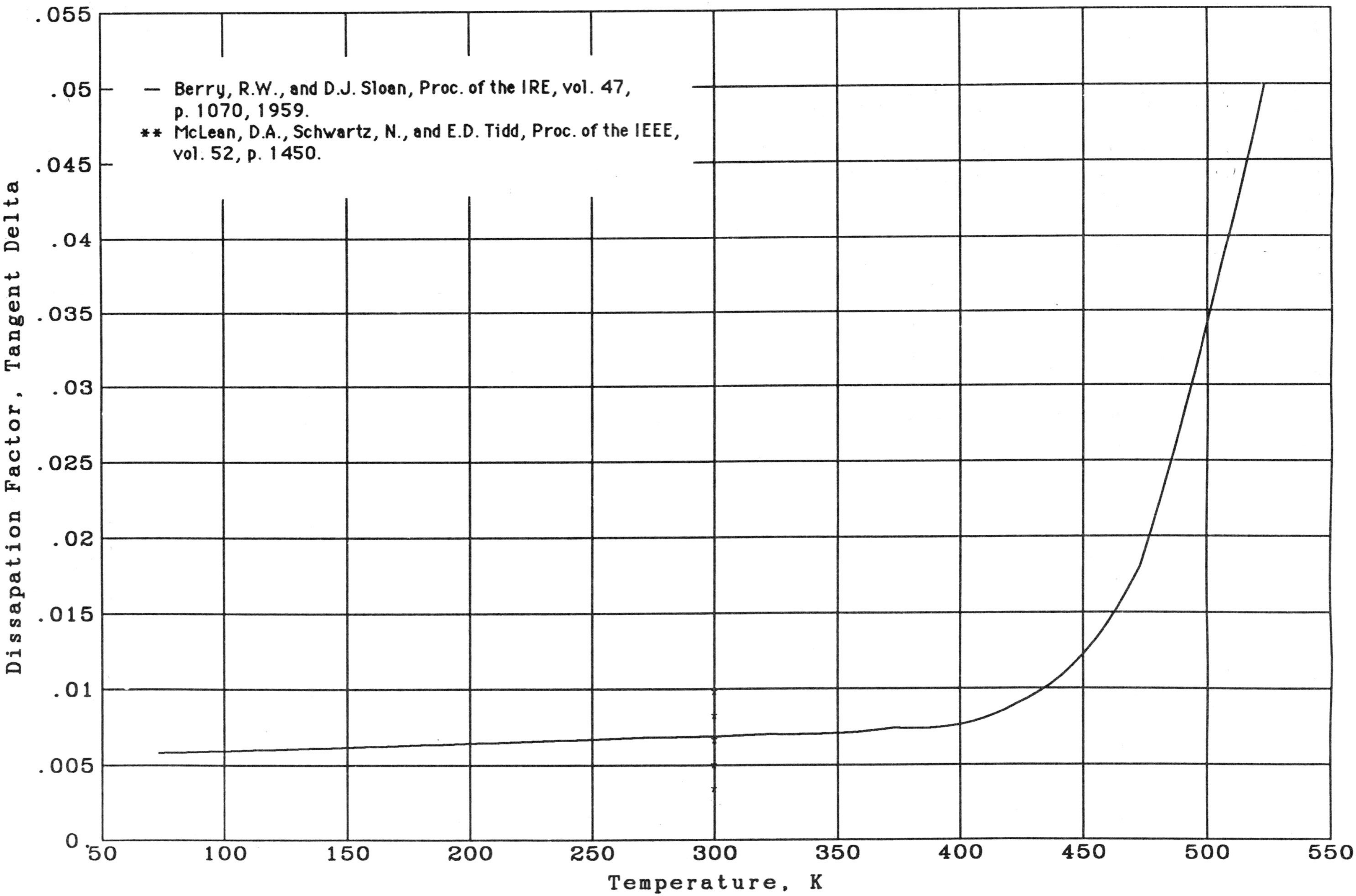
— Berry, R.W., and D.J. Sloan, Proc. of the IRE, vol. 47, p. 1070, 1959.
** McLean, D.A., Schwartz, N., and E.D. Tidd, Proc. of the IEEE, vol. 52, p. 1450.
Dissapation Factor, Tangent Delta
Temperature, K
.055
.05
.045
.04
.035
.03
.025
.02
.015
.01
.005
0
50
100
150
200
250
300
350
400
450
500
550

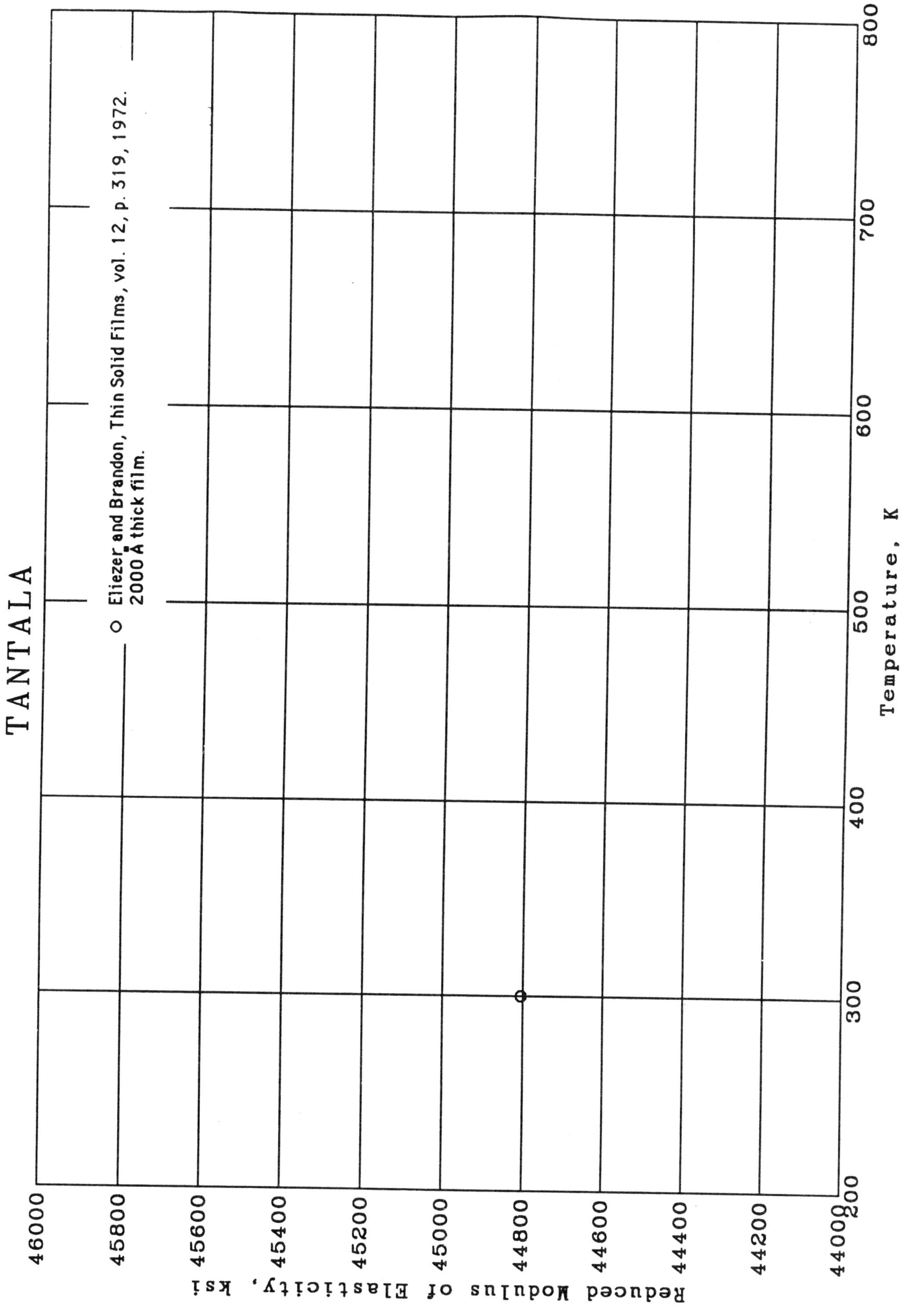
TANTALA
○ Eliezer and Brandon, Thin Solid Films, vol. 12, p. 319, 1972.
2000 Å thick film.
Temperature, K
200
300
400
500
600
700
800
Reduced Modulus of Elasticity, ksi
44000
44200
44400
44600
44800
45000
45200
45400
45600
45800
46000

TANTALA

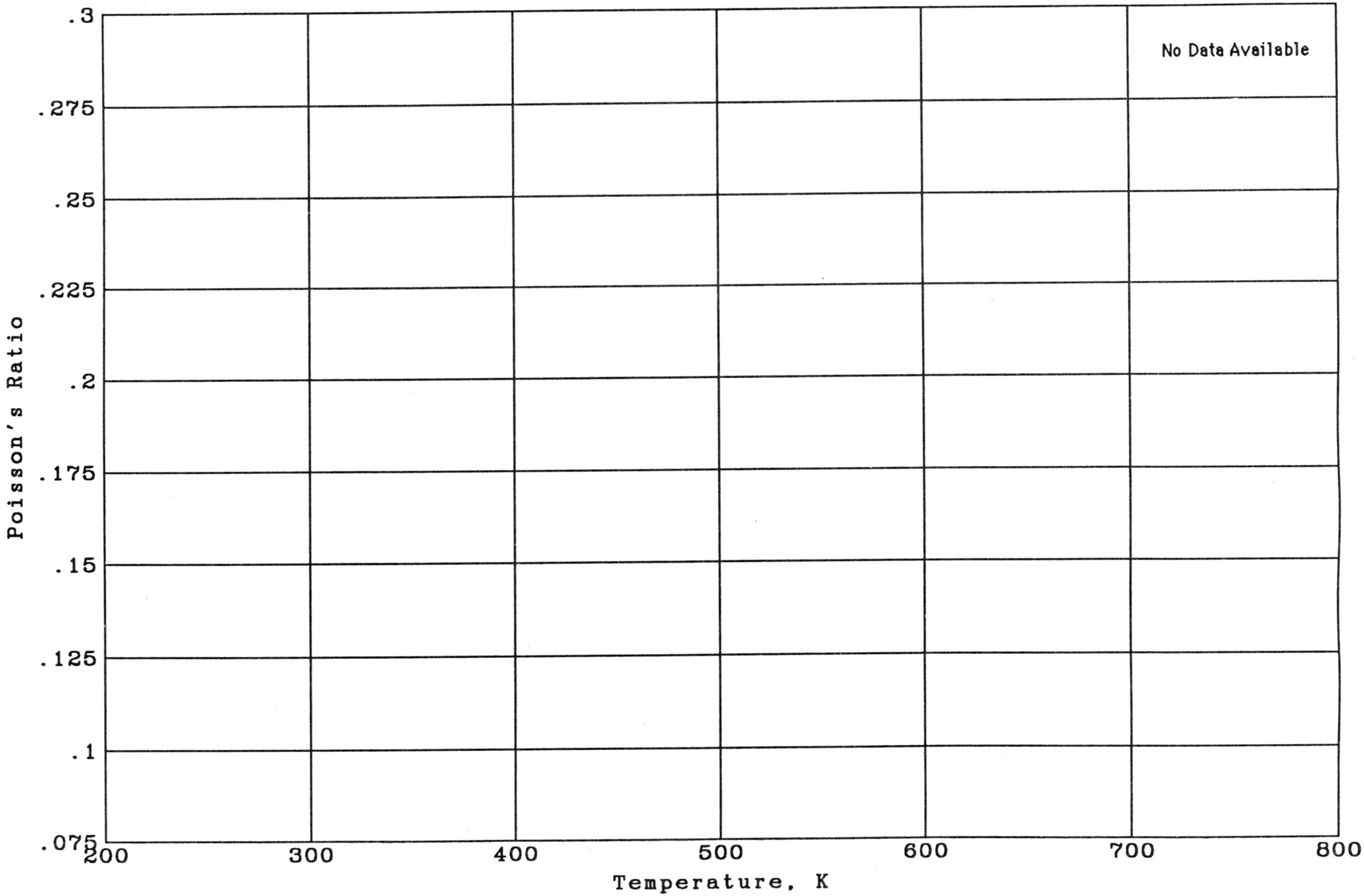

No Data Available
Poisson's Ratio
.3
.275
.25
.225
.2
.175
.15
.125
.1
.075
200
300
400
500
600
700
800
Temperature, K

TANTALA

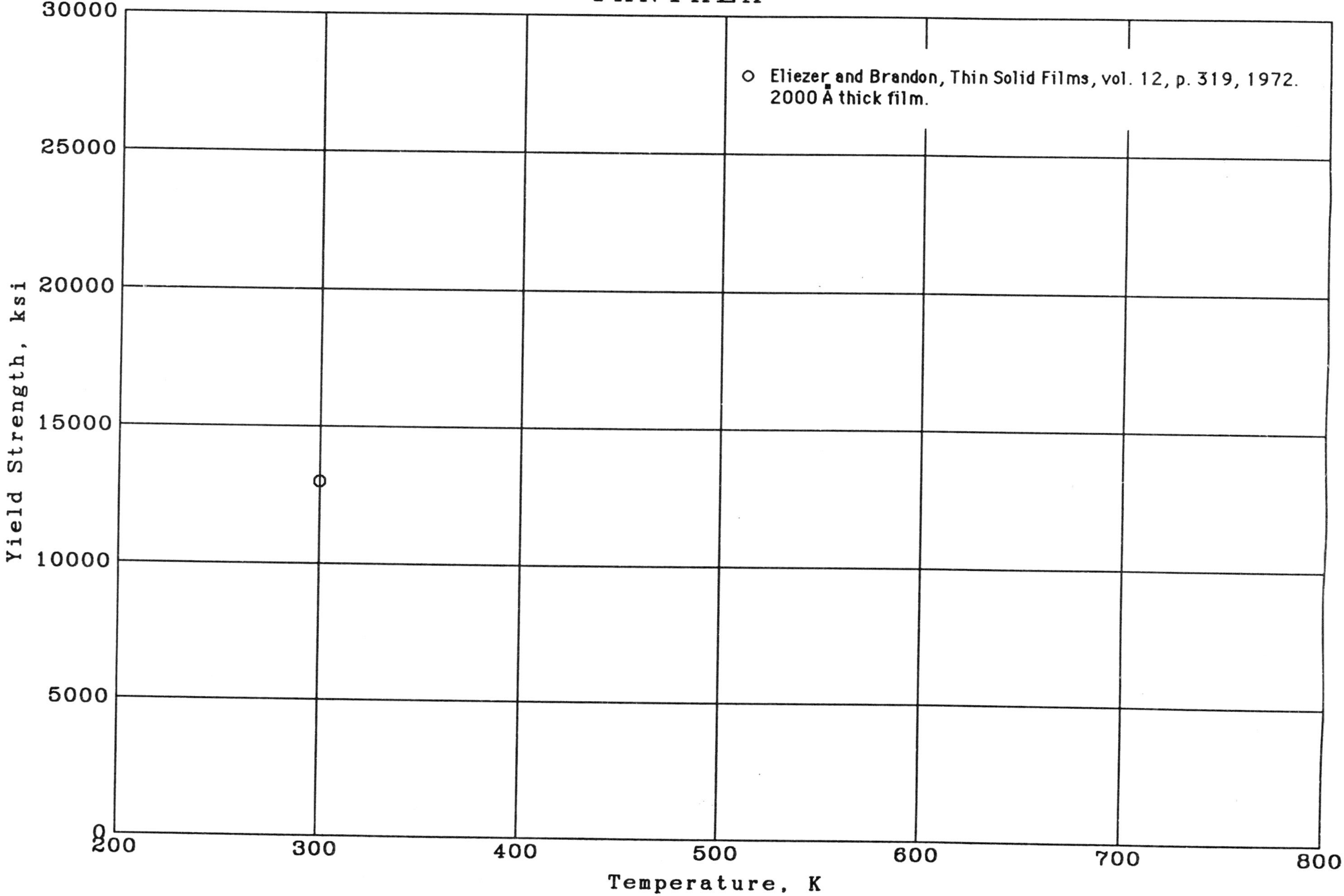
Eliezer and Brandon, Thin Solid Films, vol. 12, p. 319, 1972.
2000 Å thick film.
30000
25000
20000
15000
10000
5000
0
Yield Strength, ksi
200
300
400
500
600
700
800
Temperature, K

TANTALA

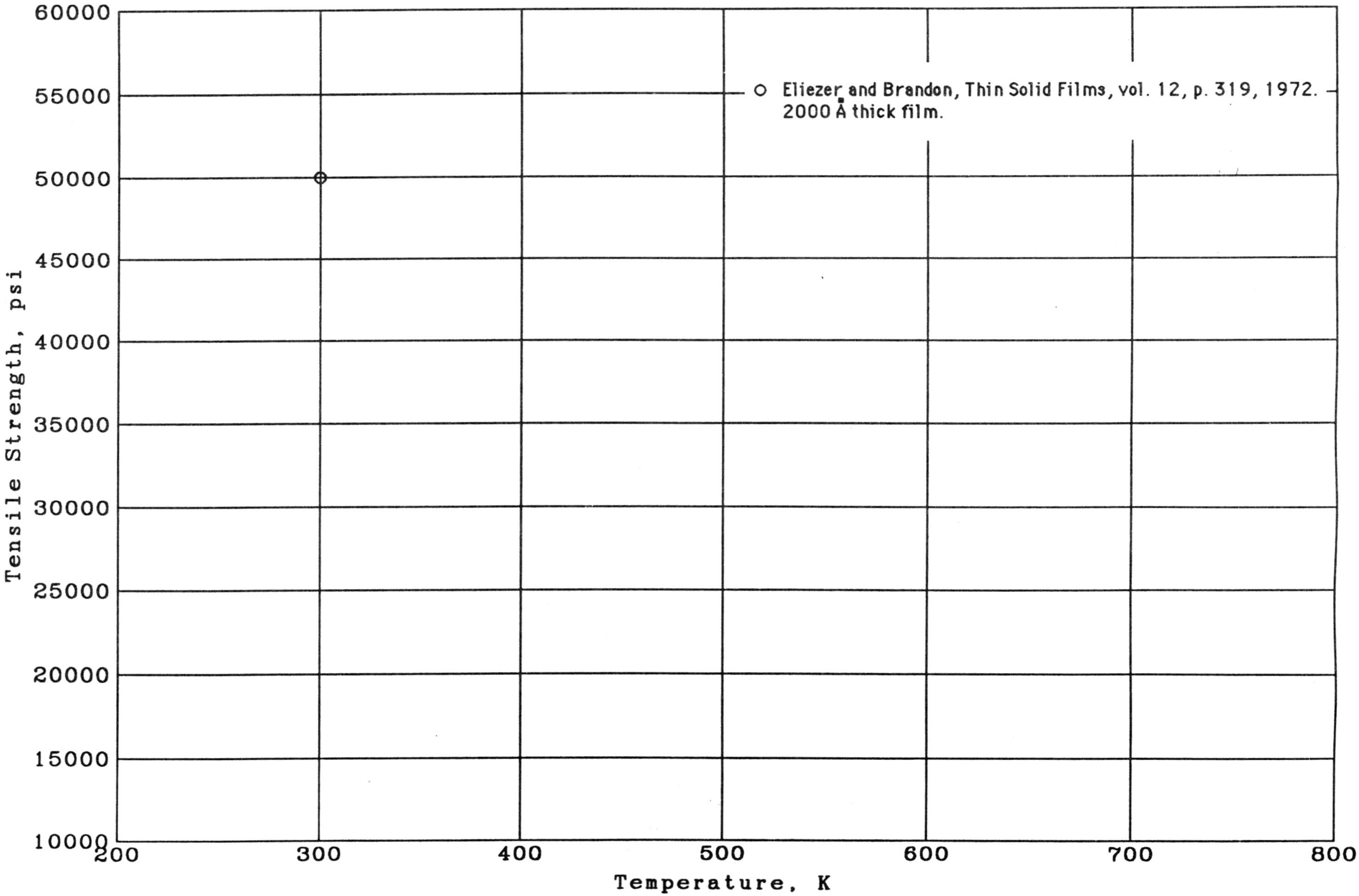

60000
55000
50000
45000
40000
35000
30000
25000
20000
15000
10000
Tensile Strength, psi
200
300
400
500
600
700
800
Temperature, K
○ Eliezer and Brandon, Thin Solid Films, vol. 12, p. 319, 1972.
2000 Å thick film.

TANTALA

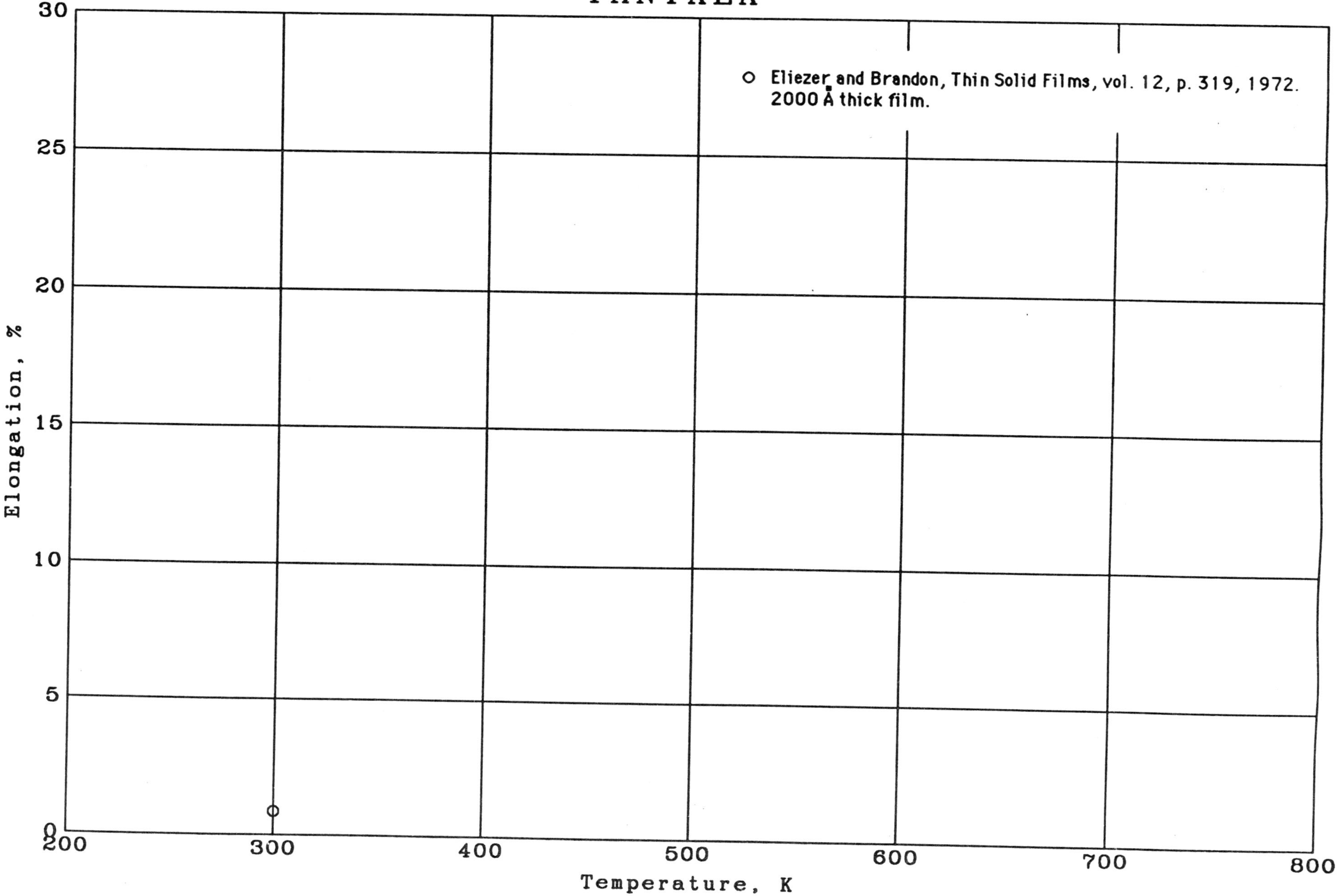

30
25
20
15
10
5
0
Elongation, %
200
300
400
500
600
700
800
Temperature, K
Eliezer and Brandon, Thin Solid Films, vol. 12, p. 319, 1972.
2000 Å thick film.

TANTALA

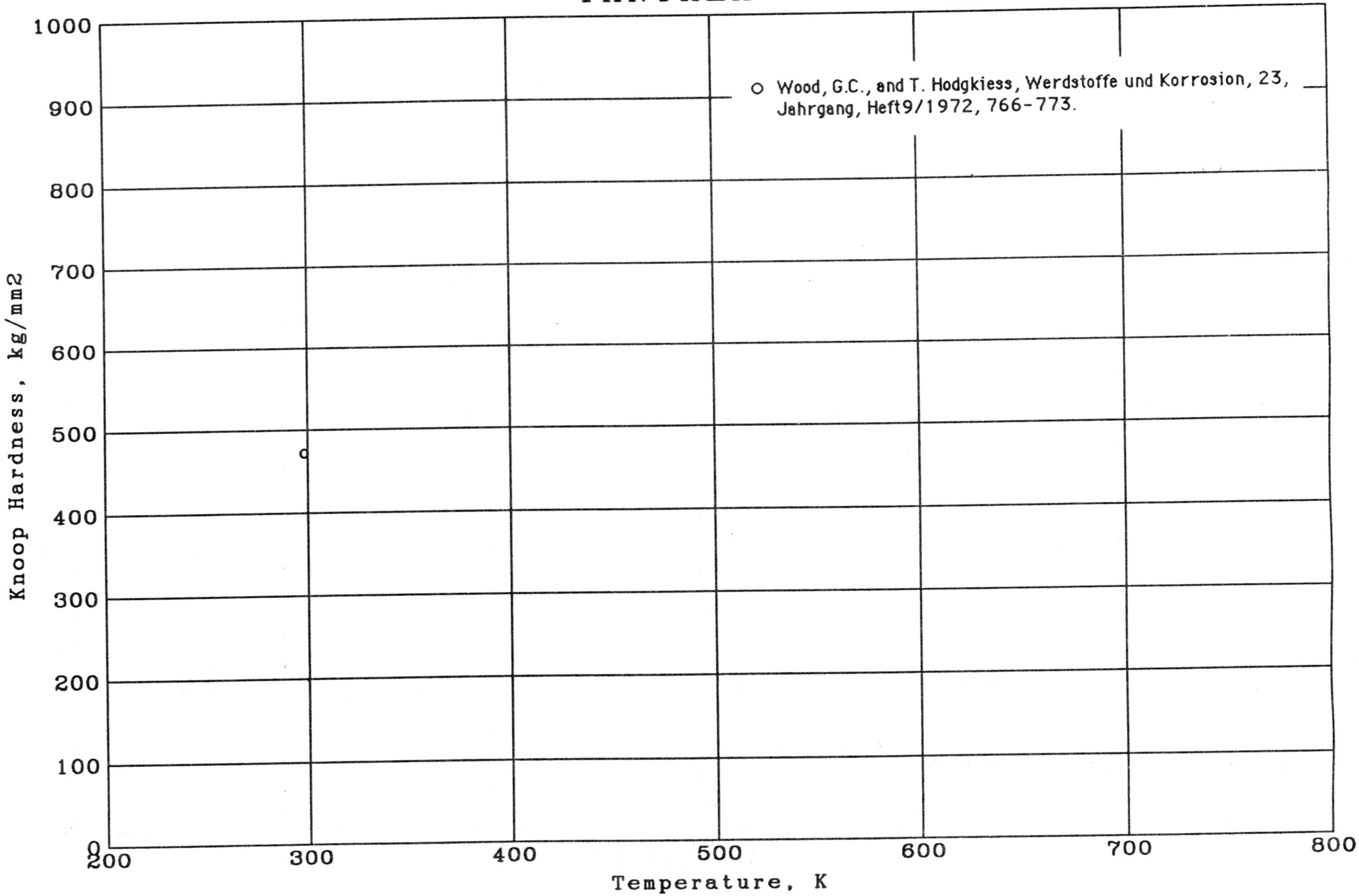
1000
900
800
700
600
500
400
300
200
100
0
Knoop Hardness, kg/mm2
200
300
400
500
600
700
800
Temperature, K
○ Wood, G.C., and T. Hodgkiess, Werdstoffe und Korrosion, 23, Jahrgang, Heft9/1972, 766-773.

REFERENCES

96% Alumina

96A1. Coors Data Sheet, 96% pure. Density = 3.72 g/cm^3. Thermal conductivity at 293K = 0.0590 cal/cm-s-K. Thermal expansion coefficient (mean) from 298K to 473K = 6.0x10^{-6} K^{-1}. Dielectric constant at 1kHz, 298K = 9.0. Dissipation factor at 1 kHz = 0.0011. Dielectric strength (0.25 in. thick) = 210 V/mil. Young's modulus = 44,000 ksi. Poisson's ratio = 0.21. Flexural strength (typical) at 298K = 52,000 psi. Tensile strength at 298K = 28,000 psi. Compressive strength at 298K = 300,000 psi. Hardness (Rockwell 45N) = 78.

96A2. GTE Wesco Data Sheet, 96% pure. Density at room temp. = 3.72 g/cm^3. Thermal conductivity at room temp. = 0.0612 cal/cm-s-K. Thermal expansion coefficient (mean) from 298K to 473K = 6.4x10^{-6} K^{-1}. Dielectric constant at 10 MHz, 298K = 9.30. Dissipation factor at 10 MHz, 298K = 0.0003. Dielectric strength (0.10-in. thick) at room temp. = 675 V/mil. Flexural strength at room temp. = 53,000 psi. Compressive strength (min.) at room temp. = 300,000 psi. Hardness (Rockwell 45N) = 79.

96A3. American Lava Corp. Chart No. 711. Density = 3.70 g/cm^3. Thermal expansion coefficient (mean) from 298K to 1173K = 7.9x10^{-6} K^{-1}. Dielectric constant at 1 MHz, 298K = 9.3. Dissipation factor at 1 MHz, 298K = 0.0003. Dielectric strength (0.25 in. thick) = 210 V/mil. Young's modulus = 47,000 ksi. Poisson's ratio = 0.22. Flexural strength = 46,000 psi. Tensile strength = 25,000 psi. Compressive strength = 375,000. Hardness (Rockwell 45N) = 78.

96A4. Shomate, C.H. and B.F. Naylor, J. Am. Chem. Soc., vol 67, 72, 1945. Specific heat at 298.15K = 0.1845 cal/g-K.

96A5. Ginnings, D.C. and R.J. Corruccini, RP 1797, J. Res. Nat. Bur. Std., vol 38, 593, 1947. Specific heat at 293K = 0.1830 cal/g-K.

96A6. Ginnings, D.C. and G.T. Furukawa, J. Am. Chem. Soc., vol 75, 522, 1953. Specific heat at 300K = 0.1861 cal/g-K.

96A7. Soxman, E.J., Study of Heat Transfer of Ceram. Matls., Alfred Univ., ONR, I, 1, 1955. Thermal conductivity at 300K = 0.0444 cal/cm-s-K, interpolated between 276K and 317K.

96A8. Product Data, Calcined Aluminas, CHE 920, Chemicals Division, Pittsburgh, PA, 15219. Young's Modulus = 39-40 x 10^6 psi. Tensile strength = 25,000 to 35,000 psi. Hardness = 9 Mhos or 1720 Knoop. All properties for 95%+ pure, vitreous bodied material at room temperature.

99% Alumina

99A1. Coors Data Sheet, 99% pure. Density = 3.83 g/cm^3. Thermal conductivity at 293K = 0.070 cal/cm-s-K. Thermal expansion coefficient (mean) from 298K to 473K = 6.3x10^{-6} K^{-1}. Dielectric constant at 1kHz, 298K = 9.4. Dissipation factor at 1 kHz = 0.0042. Dielectric strength (0.25 in. thick) = 215 V/mil. Young's modulus at 298K = 50,000 ksi. Poisson's ratio = 0.21. Flexural strength (typical) at 298K = 48,000 psi. Tensile strength at 298K = 31,000 psi. Compressive strength at 298K = 345,000 psi. Hardness (Rockwell 45N) = 80.

99A2. GTE Wesco Data Sheet, 99.5% pure. Density at room temp. = 3.86 g/cm^3. Thermal conductivity at room temp. = 0.0700 cal/cm-s-K. Thermal expansion coefficient (mean) from 298K to 473K = 6.9x10^{-6} K^{-1}. Dielectric constant at 10 MHz, 298K = 9.58. Dissipation factor at 10 MHz, 298K = 0.00003. Dielectric strength (0.10-in. thick) at room temp. = 800 V/mil. Flexural strength at room temp. = 45,000 psi. Compressive strength (min.) at room temp. = 300,000 psi. Hardness (Rockwell 45N) = 81.

99A3. Engineering Properties of Selected Ceramic Materials, Am. Ceram. Soc., 1966. Density = 3.98 g/cm^3. Young's modulus = 59,000 ksi. Poisson's ratio = 0.27. Flexural strength at 298K = 67,000 psi. Tensile strength at 298K = 38,000 psi. Compressive strength at 298K = 420,000.

99A4. Shomate, C.H. and B.F. Naylor, J. Am. Chem. Soc., vol 67, 72, 1945. Specific heat at 298.15K = 0.1845 cal/g-K.

99A5. Ginnings, D.C. and R.J. Corruccini, RP 1797, J. Res. Nat. Bur. Std., vol 38, 593, 1947. Specific heat at 293K = 0.1830 cal/g-K.

99A6. Ginnings, D.C. and G.T. Furukawa, J. Am. Chem. Soc., vol 75, 522, 1953. Specific heat at 300K = 0.1861 cal/g-K.

99A7. Truesdale, R.S., J.J. Swica, and J.R. Tinklepaugh, College of Ceramics, State Univ. N.Y. at Alfred Univ., Progr. Rept. 11, 1960. [AD 240278] 99.3% pure. Thermal conductivity at 300K = 0.0767 cal/cm-s-K, extrapolated parabolically from 374.2K.

99A8. Lang, S.M., ed., Nat. Bur. Std. Monograph 6, 1960. Material code 3, 99(+)% pure. Young's modulus = 52,000 ksi. Poisson's ratio = 0.239.

99A9. Product Data, Calcined Aluminas, CHE 920, Aluminum Co. of America, Pittsburgh, Pa 15219. Properties are for 99.5% Al_2O_3, vitreous body material at room temperature. UTS = 37,500-38,000 psi. Hardness = 9 Mhos. Dielectric Strength at 298K = 380 volts/mil.

99A10. Data Sheet, Millis Corp., Millis, MA. Properties are for 99.6% pure material. Dielectric constant at 298 K and 1 MHz =

9.90. Dissipation factor = 0.001 at 1 MHz. Thermal expansion coefficient (25-300C) = 6.2×10^{-6}.

99A11. Knudsen F. P., J. Am. Cer. Soc., vol 45, no 2, 94-5, 1962.

99A12. Spriggs, R. M. and L. A. Bissette, J. Am. Cer. Soc., vol 45, no 4, 198-9, 1962.

Alumina, Sapphire

S1. Precomp., Inc. Data Sheet. Density = 3.99 g/cm^3. Thermal expansion coefficient = 5.4x10^{-6} K^{-1}. Dielectric constant (parallel to c-axis) at 298K = 11.5. Dissipation factor at 1 kHz = 0.0001. Young's modulus = 62,000 ksi. Compressive strength = 300,000 psi. Hardness (Mohs) = 9.

S2. Clements, J.F. and J. Vyse, Trans. Brit. Ceram. Soc., vol 56, 296, 1957. Density at 298K = 3.983 g/cm^3.

S3. Hock, M. and D.A. Nitti, Univ. of Cincinnati, ASD-TR-61-528, 1, 1962. Density at 298K = 3.73 g/cm^3.

S4. Shomate, C.H. and B.F. Naylor, J. Am. Chem. Soc., vol 67, 72, 1945. Specific heat at 298.15K = 0.1845 cal/g-K.

S5. Ginnings, D.C. and R.J. Corruccini, RP 1797, J. Res. Nat. Bur. Std., vol 38, 593, 1947. Specific heat at 293K = 0.1830 cal/g-K.

S6. Ginnings, D.C. and G.T. Furukawa, J. Am. Chem. Soc., vol 75, 522, 1953. Specific heat at 300K = 0.1861 cal/g-K.

S7. Koenig, J.H., Rutgers Univ. N.J. Ceram. Res. Sta. Progr. Rept. 1, 1, 1953. Thermal conductivity at 300K = 0.0827 cal/cm-s-K, extrapolated linearly from 321.8K.

S8. Koenig, J.H., Rutgers Univ. N.J. Ceram. Res. Sta. Progr. Rept. 5, 1, 1954. Thermal conductivity at 300K = 0.0827 cal/cm-s-K, extrapolated linearly from 318.6K.

S9. McCarthy, K.A. and S.S. Ballard, J. Opt. Soc. Am., vol 41, 1062, 1951. Thermal conductivity at 299.2K = 0.0597 cal/cm-s-K.

S10. Nat. Bur. Std. Certificate, Std. Ref. Matl. 732, 1977. Thermal expansion coefficient at 300K = 5.48x10^{-6} K^{-1}.

S11. Adolf Metler Co. Data Sheet. Thermal expansion coefficient (mean) parallel to c-axis from 293K to 323K = 6.66x10^{-6} K^{-1}. Dissipation factor at 1 kHz, 298K = 0.000002. Dielectric strength = 1200 V/mil. Hardness (Mohs) = 9. Flexural strength varies from 65,000 to 100,000 when tensile axis varies from 30 to 60 degrees from the c-axis.

S12. Von Hippel, A., ed., Dielectric Materials and Applications, MIT Press, 1954. Dielectric constant (parallel to c-axis) at 298K = 10.55. Dissipation factor at 1 kHz, 298K < 0.0001.

S13. Fontanella, J., C. Andeen, and D. Schuele, J. Appl. Phys., vol 45, 2852, 1974. Dielectric constant (parallel to c-axis) at 1KHz, 300K = 11.5.

S16. Engineering Properties of Selected Ceramic Materials, Am.

Ceram. Soc., 1966. Young's modulus = 63,000 ksi. Flexural strength = 92,000 psi.

S17. Bernstein, B.T., J. Appl. Phys., vol 34, 169, 1963. Young's modulus at 300K = 57,000 ksi. Poisson's ratio at 300K = 0.236.

S18. Wachtman, J.B., Jr., et al., J. Res. Nat. Bur. Std., vol 64A, 213, 1960. Poisson's ratio at 298K = 0.232.

S19. Tefft, W.E., J. Res. Nat. Bur. Std., vol 70A, 277, 1966. Poisson's ratio at 300K = 0.234.

S20. Kotchick, D.M. and R.E. Tressler, J. Mat. Sci., vol 10, 608, 1975. Tensile strength (tested in water, strain rate 0.132/min.) = 370,000 psi.

S21. Lomer, P.D., Nature, vol 166, no 4213, 191, 1950. Dielectric strength of amorphous films varies from 28,000 volts/mil to 13,000 volts/mil as the film thickness varies from 3.4 micro-inches to 38.5 micro-inches at 290K.

S22. Nehring, V.W., and J.T. Jones, J. Am. Cer. Soc., vol 56, no1, 50, 1973. Average flexural strength = 75,000 psi.

S23. Yoshimura, M., and H.K. Vowen, J. Am. Cer. Soc., vol 64, no 7, 404-10. Dielectric strength > 2.54 x 10^3 volts/mil at room temperature.

S24. Hurley, G.F., Appl. Polym. Symp. 1973, No.21, 121-30. Compression strength of rods loaded along the <0001> direction was found to be 234 ksi. Tensile strength of 0.009 inch diameter filament loaded along the c-axis was found to be 370 ksi at room temperature. In a companion paper, Hurley reports the strength of polycrystals and machined crystals = 20 to 70 ksi.

S25. Coors Ceramics, Bulletin No. 301, Hardness of sapphire = 9 Mohs.

S26. Brenner, E.E., J. Appl. Phys., vol 33, 33, 1962. Strength of sapphire wisker = 370,000 at room temperature.

S27. Ryshkewitch, E., Ber. deut Keram. Ges., vol 22, 54-65, 1941. Compressive strength = 300,000 at room temperature.

Aluminum Nitride

AN1. Fichter, F. and A. Spengel, Z. Anorg. Chem., vol 82, 202, 1913. Density = 2.88 g/cm^3 (32.8% N) or 3.18 gcm^3 (31.3%N).

AN2. Taylor, K.M. and Lenie, C., J. Electrochem. Soc., vol 107, 308, 1960. Density = 3.26 g/cm^3. Thermal expansion coefficient (mean) from 298K to 473K = 4.03×10^{-6} K^{-1}. Dielectric constant at 298K = 8.5. Dissipation factor at 10 kHz = 0.004. Young's modulus at 298K = 50,000 ksi. Compressive strength = 300,000 psi. Hardness (Knoop, 100 g load) = 1225 kg/mm^2(ave.). Modulus of rupture (flexural strength) = 38,500 psi.

AN3. Ott, H., Z. Phys., vol 22, 208, 1924. Density = 3.25 g/cm^3.

AN4. Mah, A.D., et al., U.S. Bur. Mines, RI-5716, 1, 1961. Specific heat at 300K = 0.1767 cal/g-K.

AN5. Mezaki, R., et al., 3rd ASME Symp., Lafayette, Indiana, 138, 1965. Specific heat at 300K = 0.1696 cal/g-K.

AN6. Carborundum Data Sheet. Thermal expansion coefficient (mean) from 298K to 473K = 4.03×10^{-6}. Dielectric constant at 8.5 GHz, 298K = 8.15. Dissipation factor at 8.5 GHz, 298K = 0.0033. Young's modulus at 298K = 50,000 ksi. Specific heat (0 to 100C) = 0.1961 cal/g-K.

AN7. Borom, M.P., G.A. Slack, and J.W. Szymaszek, Ceram. Bull., vol 51, 852, 1972. Thermal conductivity at 300K = 0.16 cal/cm-s-K.

AN8. Long and Foster, J. Am. Ceram. Soc., vol 42, 53, 1959. Thermal conductivity at 298K = 0.0042 cal/cm-s-K.

AN9. Ceradyne, Inc., Costa Mesa, California, 92626. Thermal conductivity = 0.34. Dissipation factor = 0.001. Flexural strength = 64,000 psi. Compressive strength = 300,000 psi. Hardness = 1200 Knoop.

AN10. Slack, G.A. and S.F. Bartram, J. Appl. Phys., vol 46, 89, 1975. Thermal expansion coefficient at 300K = 2.56×10^{-6} K^{-1}.

AN11. Managalaraj, D., et al, Current Science, vol 51, no22, 1068, 1982. Dielectric strength = 5000 volts/mil for thin film. Dielectric strength of thin films decrease with increasing thickness.

AN12. Sharma, B.C., J. Solid. State Technol., vol 21, 48, 1978. Dielectric constant = 9.1

AN13. Duchene, J., Thin Solid Films, vol 8, 69-79, 1971. Dielectric strenghth for sputtered films varies from 2500 to 15000 volts/mil depending on deposition conditions.

AN14. Gibkov, V.N., et al, Porosh. Met. vol 10, no3, 84-88, 1970. Normal modulus of elasticity = 42.6 to 46.8 x 10^6 psi.

AN15. Zyatkevich, D.P., et al, Porosh. Met., vol 21, no 12, 65-69, 1982. Transverse strength = 45,000 psi for polycrystalline pressed from ultrafine powder.

AN16. Heard, H.C., and C.F. Cline, J. Matls. Sci., vol 15, 1889-1897, 1980. Compressive strength of polycrystalline material at 1 atm and room temperature = 450,000.

AN17. Cline, C.F. and J.S. Kahn, J. Elec Soc, vol 110, no 7, 773-775, 1963. Hardness = 1187 kg/mm^2.

AN18. Heraeus Corporation (personal communication with Katherine R. Kannak). Dielectric strength = 350 to 430 volts/mil for substrate material.

AN19. Norton Company. Technical data sheet DS101. 1987

Beryllia

B1. Brush-Wellman Data Sheet. Density = 3.02 g/cm^3. Thermal expansion coefficient = $8x10^{-6}$ K^{-1}. Dielectric constant = 6.7. Dissipation factor at 1 MHz = 0.0002. Dielectric strength (0.25 in. thick) = 270 V/mil. Young's modulus = 50,000 ksi. Poisson's ratio = 0.26. Flexural strength (average) for sintered samples = 49,000 psi. Tensile strength = 22,000 psi. Compressive strength = 225,000 psi. Hardness (Rockwell 45N) = 60 (minimum).

B2. Millis Corp. Data Sheet. Density (min.) = 2.85 g/cm^3. Thermal expansion coefficient (mean) from 298K to 1273K = $9.0x10^{-6}$ K^{-1}. Dielectric constant at 1 MHz = 6.5. Dissipation factor at 1MHz = 0.0004 max. Dielectric strength (0.25 in. thick) = 240 V/mil. Flexural strength (average) = 35,000 psi. Tensile strength = 22,000 psi. Compressive strength = 225,000 psi. Hardness (Rockwell 45N) = 60 (minimum).

B3. Coors Data Sheet: Density (typical) = 2.90 g/cm^3. Dielectric constant at 1 kHz, 298K = 6.8. Dissipation factor at 1 kHz, 298K = 0.001. Dielectric strength (0.25 in. thick) = 260 V/mil. Flexural strength (typical) at 298K = 40,000 psi. Compressive strength (typical) at 298K = 310,000 psi. Tensile strength (typical) at 298K = 20,000 psi. Hardness (Rockwell 45N) = 67.

B4. Kelley, K.K., J. Am. Chem. Soc., vol 61, 1217, 1939. Specific heat at 292.4K = 0.23773 cal/g-K.

B5. Walker, B.E., Jr., C.T. Ewign, and R.R. Miller, J. Chem. Eng. Data, vol 7, 595, 1962. Specific heat at 303.15K = 0.248 cal/g-K.

B6. Victor, A.C. and T.B. Douglas, J. Res. Nat. Bur. Std., vol 67A, 325, 1963. Specific heat at 298.15K = 0.2440 cal/g-K.

B7. Ditmars, D.A. and D.C. Ginnings, J. Res. Nat. Bur. Std., vol 59, 93, 1957. Thermal conductivity at 301.4K = 0.523 cal/cm-s-K.

B8. Sibley, L.B., et al., WADC TR 58-299, 1, 1958. [AD 203787] Thermal conductivity at 293.2K = 0.5198 cal/cm-s-K.

B9. Burk, M., J. Am. Ceram. Soc., vol 46, 150, 1963. Thermal conductivity at 308.2K = 0.674 cal/cm-s-K.

B10. Hedge, J.C., et al., U.S.A.F. Rept. ASD-TDR-63-597, 1963. [AD 424375] Thermal expansion coefficient at 300K = $7.41x10^{-6}$ (determined by differentiating L/L data).

B11. Bentle, G.G., J. Amer. Ceram. Soc., vol 49, 125, 1966. Young's modulus at room temperature = $5.705x10^7$ psi. Poisson's ratio at room temperature = 0.237.

B12. Cline, C.F., H.L. Dunegan, and G.W. Henderson, J. Appl.

Phys., vol 38, 1944, 1967. Young's modulus at 298K = 5.768x10^7 psi. Poisson's ratio at 298K = 0.204

Diamond

D1. Mykolajewycz, R. J. Kalnajs, and A. Smakula, J. Appl. Phys., vol 35, 1773, 1964. Density at 298K = 3.515 g/cm^3.

D2. Bochko, V.A. and Yu.L. Orlov, Sov. Phys. Dolkady, vol 15, 204, 1970. Density = 3.515 g/cm^3.

D3. Bearden, J.A., Phys. Rev., vol 54, 698, 1938. Density at 296.5K = 3.5154 g/cm^3.

D4. Victor, A.C., J. Am. Chem. Soc., vol 36, 1903, 1962. Specific heat at 298.15K = 0.1218 cal/g-K.

D5. DeSorbo, W., J. Chem. Phys., vol 21, 876, 1953. Specific heat at 300.57K = 0.1234 cal/g-K.

D6. Desnoyers, J.E. and J.A. Morrison, Phil. Mag., vol 3, 42, 1958. Specific heat at 300K = 0.1243 cal/g-K, extrapolated linearly from 278K.

D7. Berman, R. et al., Proc. Roy. Soc. London, vol A237, 344, 1956. Thermal conductivity (for type IIa) at 300K = 4.78 cal/cm-s-K.

D8. General Electric Data Sheet. Thermal conductivity at 298K = 4.78 cal/cm-s-K. Hardness (Knoop) = 7000 kg/mm^2.

D9. Diamond Research, 1976 (Suppl to Indian Diamond Review), 7. Thermal conductivity of type IIa is approximately equal to 5 cal/cm-sec-K.

D10. Novikova, S.I., Sov. Phys. Solid State, vol 2, 1464, 1961. Thermal expansion coefficient at 299K = 1.02×10^{-6} K^{-1}.

D11. Thewlis, J. and A.R. Davey, Phil. Mag., vol 1, 409, 1956. Thermal expansion coefficient at 293K = 0.78×10^{-6} K^{-1}.

D12. Sokhor, M.I. and V.D. Vitol, Sov. Phys. Cryst., vol 11, 632, 1970. Thermal expansion coefficient at 293K = $1.0\text{-}1.2 \times 10^{-6}$ K^{-1}.

D13. Gibbs, D.F. and G.J. Hill, Phil. Mag., vol 9, 367, 1964. Dielectric constant at 300K = 5.93.

D14. Bhagavantam and Narayana Rao, Nature, no 4097, May 8, 729, 1948. Dielectric constant at 1.6 MHz, 297K = 5.65.

D15. Goryunova, N.A., Chemistry of Diamond-like Semiconductors, Chapman and Hall, London, 1965. Dielectric constant = 5.7. Hardness (Knoop, 100 g load) = 8820 kg/mm^2.

D16. McSkimin, H.J. and W.L. Bond, Phys. Rev., vol 105, 116, 1957. Young's modulus = 152,000 ksi. Poisson's ratio = 0.104.

D17. McSkimin, H.J. and P. Andreatch, J. Appl. Phys., vol 43, 2944, 1972. Young's modulus = 152,000 ksi. Poisson's ratio = 0.103.

D18. Grimsditch, M.H. and A.K. Ramdas, Phys. Rev. B, vol 11, 3139, 1975. Young's modulus = 152,000 ksi. Poisson's ratio = 0.1042.

D19. Berman, R., Physical Properties of Diamond, Clarendon Press, Oxford, 1965. Tensile strength = 580,000 psi.

D20. Field, J.E., The Properties of Diamond, Academic Press, New York, 1979. Tensile strength = 412,000 psi. Hardness (Knoop) = 7750 kg/mm^2.

D21. Ratterman, E., Proc. Internat. Diamond Conf., 1967. Compressive strength = 1260 ksi.

D22. Heller, W.R. Phys. Rev., vol 84, no 6, 1130, 1951. Calculated theoretical dielectric strength = 190 to 760 volts/mil.

D23. Seitz, F., Phys. Rev., vol 76, no 9, 1376-1393, 1949. Maximum theoretical dielectric strength is calculated to be 3300 volts/mil.

D24. Hearn, A.J., Phys. Rev., vol 84, no 4, 798-802, 1951. "After a dc field of about 25.4 to 51 volts/mil is applied, the resistance (of some diamonds) drops from about 10^{13} ohms to 1 megaohm".

D25. Hull, E.H. and G.T. Malloy, J. Eng. Ind., Nov, 1966, 373-376. Compressive strength = 1,261,000 psi. Strength in torsion = 43,400 psi.

D26. Schwarzkopf, P., and R. Kieffer, Cemented Carbides, MacMillian Book Co., New York, New York, 236, 1960. Compressive strength = 284,000. Transverse bending strength = 42,600 psi.

Fused Quartz

FQ1. Stierlin, H., Viert, Natur. Ges. Zurich., vol 52, 382, 1907. Density at 273K = 2.203 g/cm^3.

FQ2. Stumpf, F., Ann. Physik., vol 40, 879, 1913. Density at 273K = 2.203 g/cm^3. Young's modulus = 10,100 ksi. Poisson's ratio = 0.133

FQ3. Chappius, P., Verh. Naturf. Ges. Basel, vol 16, 173, 1903. Density at 273K = 2.201 g/cm^3.

FQ4. Southard, J.C., J. Am. Chem. Soc., vol 63, 3142, 1941. Specific heat at 300K = 0.1775 cal/g-K.

FQ5. Lucks, C.F., H.W. Deem, and W.D. Wood, Am. Ceram. Soc. Bull., vol 9, 313, 1960. Specific heat at 293K = 0.164 cal/g-K.

FQ6. Westrum, E.F., Jr., IV Congress Int. du Verre, Paris, 396, 1956. Specific heat at 295K = 0.1762 cal/g-K.

FQ7. Devyatkova, E.D., et al., Sov. Phys. Solid State, vol 2, 681, 1960. Thermal conductivity at 300K = 0.00330 cal/cm-s-K.

FQ8. Hartunian, R.A. and R.L. Varwig, USAF TDR-594 (1217-01) TN-2, 1, 1961. [AD 606036] Thermal conductivity at 300K = 0.00330 cal/cm-s-K.

FQ9. Milnes, M.V., North Amer. Aviation Inc., SDL408, 1, 1963. Thermal conductivity at 298.2K = 0.0033 cal/cm-s-K.

FQ10. Asay, J.R., S.R. Urzendowski, and A.H. Guenther, U.S. Air Force Rept. AFWL-TR-67-91, 1, 1968. [AD 827596] Thermal expansion coefficient at 293K = 0.42×10^{-6} K^{-1}.

FQ11. Hallse, R.L., S.C. Colburn, and A.P. Beck, U.S. Air Force Rept. AFML-TR-64-398, 359, 1965. [AD 477414] Thermal expansion coefficient at 300K = 0.35×10^{-6} K^{-1}, determined by differentiating L/L data.

FQ12. Kirby, R.K. and T.A. Hahn, Nat. Bur. Std., Certificate of Analysis, Std. Ref. Matl. 739, 1971. Thermal expansion coefficient at 298K = 0.49×10^{-6} K^{-1}.

FQ13. Schulze, F.A., Ann. Physik., vol 14, 384, 1904. Dielectric constant = 3.20. Flexural strength = 9800 psi.

FQ14. Thornton, W.M., Proc. Roy. Soc. London, vol A82, 422, 1909. Dielectric constant = 3.78.

FQ15. Heraeus-Amersil Data Sheet. Properties at 293K. Dielectric constant up to 1 MHz = 3.70. Dissipation factor at 1 kHz <0.0005. Dielectric strength = 630-1020 V/mil. Young's modulus = 10,000 ksi. Poisson's ratio = 0.17. Flexural strength

= 9700 psi. Tensile strength = 7200 psi. Compressive strength = 169,000 psi. Hardness (Mohs) = 5.5-6.5.

FQ16. Monnier, R.C., Appl. Opt., vol 6, 1437, 1967. Material supplied by Owens-Illinois Co. Dissipation factor at 298K, 1 MHz = 0.0009. Young's modulus = 10,400 ksi. Poisson's ratio = 0.14. Hardness (Knoop, 200 g load) = 500 kg/mm^2.

FQ17. Dynasil Data Sheet. Dissipation factor at 298K, 1 kHz <0.000002. Dielectric strength at 293K = 760-1000 V/mil. Tensile strength at 298K = 8500 psi. Compressive strength at 298K = 161,000 psi. Hardness (Knoop, 100 g load) = 600 kg/mm^2. Flexural strength = 7100 psi.

FQ18. Sosman, R., The Properties of Silica, {? publ., date}. Tensile strength = 4300-11,600 psi.

FQ19. Berndt, G., Verh. Deut. Physik. Ges., vol 19, 314, 1917. Compressive strength = 280,000 psi.

FQ20. Austen, A.E.W. and S. Whitehead, Pr. Roy. Soc. A176, 33-50, 1940. Dielectric strength at room temperature = 1370 volts/mil.

Silicon Monoxide

SO1. Geld, V. and M.I. Kochnev, Zhur. Prikald. Khim., vol 21, 1249, 1948. Equimolar mixture of Si and SiO_2 heated and compressed in pellets at 1250-1350°C. Density at 300K = 2.146 g/cm^3.

SO2. Hass, G., J. Am. Ceram. Soc., vol 33, 353, 1950. SiO evaporated at 1200-1250°C at a pressure of about 0.01 microns Hg, deposition rate 6-12 angstroms/sec. Density at room temp. = 2.15 g/cm^3. Increasing pressure and lowering evaporation rate decreases density.

SO3. Wartenberg, H.V., Z. Elektroch., vol 53, 343, 1949. Density at 300K = 2.18 g/cm^3.

SO4. Potter, H.N., Trans. Am. Electrochem. Soc., vol 12, 191, 1907. Specific heat in the range of 303K to 372K = 0.2052 cal/g-K.

SO5. Simonson, L., Thermal Conductivity Measurements of Silicon Monoxide Using Thin Film Techniques, Dissertation, South Dakota School of Mines, Electrical Engineering, 1974. Thermal conductivity at room temp. = 0.00359 cal/cm-s-K.

SO6. Shifrin, G.A. and R. W. Gammon, Hughes Research Lab, Malibu, California. Thermal conductivity = 0.00073 cal/cm-sec-K.

SO7. Siddal, G., Vacuum, vol 9, 274, 1959. Electric resistance evaporated material, 1 to 10 kHz. Dielectric constant = 2 to 4. Dissipation factor at 293K = 0.01 to 0.04. Dielectric strength up to 250 V/mil.

SO8. Balmer, J.R. and J.H. Bruce, Inst. Electr. Engr., Paper No. 3762E, 1962. Resistance heated evaporated material, 1.5 to 1000 kHz. Dielectric constant at room temp. = 4.5 to 5.5. Dielectric strength at room temp. up to 2500 V/mil.

SO9. Erasmus, H. and J.A. Persson, J. Electrochem. Soc., vol 95, 316, 1949. Dielectric constant at 50 kHz = 5.0.

SO10. Gaffee, D.I., Inst. Electr. Engr., Paper No. 3633, 1961. Dissipation factor at 293K = 0.001 to 0.01.

SO11. Smith, E.E. and D.R. Kennedy, Inst. Electr. Engr., Paper No. 3637, 1962. Dissipation factor (for capacitors sputtered in oxygen, argon atmosphere) at 1 kHz = 0.005. Dielectric strength at 1 kHz = 7600 V/mil.

SO12. Pivot, J., et al., Thin Solid Films, vol 28, 357, 1975. Values taken from bulge tests on thin films. Reduced Young's modulus = 72,000 ksi. Mechanical properties determined from stress-strain plots.

SO13. Pivot, J., et al, Thin Solid Films, vol 34, 205-209, 1976. Mechanical properties taken from stress-strain plots.

SO14. Furuchi, S., et al., Japan J. Appl. Phys., vol 13, 1905, 1974. Thin films, about 6 microns thick. Vickers hardness = 300 kg/mm^2. Hardness depends on thickness and evaporation pressure.

SO15. McMillian, R.E., Trans. IEEE (Electrical Insulation), vol EI-5, 10, 1970. Hardness (Mohs) at room temp. = 7.

SO16. Shimoda, R.Y., et al., Vacuum, vol 18, 269, 1968.

SO17. Maissel and Glang, eds., Handbook of Thin Film Technology, McGraw-Hill, 12, 1970.

SO18. Inagaki, T. and Y. Nishimura, Fujitsu Sci. and Tech. J., 103, 1969.

Tantala

T1. Keeler, J., J. Electrochem. Soc., vol 112, 869, 1965. Density of anodic films at room temp. = 8.0 g/cm³.

T2. Schrijner, A.J., et al., J. ELectrochem. Soc., vol 111, 1167, 1964. Density of anodic films = 7.87 to 8.04 g/cm³, depending on the electrolytic solution.

T3. Young, L., Proc. Roy. Soc. London, vol A244, 41, 1958. Density at room temp. = 7.93 g/cm³. Dielectric constant = 27.6. Dielectric strength = 17,000 V/mil.

T4. Kelly, K.K., J. Am. Ceram. Soc., vol 62, 818, 1940. Specific heat at 294.2K = 0.07266 cal/g-K.

T5. Orr, R.L., J. Am. Ceram. Soc., vol 75, 2808, 1953. Specific heat at 298K = 0.07307 cal/g-K.

T6. Kelley, K.K., et al., U.S. Bur. Mines. Bull., NR 592, 114, 1961. Specific heat at 298.5K = 0.07312 cal/g-K.

T7. Durbin, E.A. and C.G. Harman, USAEC Rept. BMI-791, 1952. Thermal expansion coefficient at 300K = 0.21×10^{-6} K^{-1}, determined by differentiating L/L data.

T8. Brown, J.J., Jr., Penn. State Univ., PhD Thesis, 1964. Thermal expansion coefficient at 300K = 1.4×10^{-6} K^{-1} for α-Ta_2O_5 and 2.4×10^{-6} K^{-1} for β-Ta_2O_5, determined by differentiating L/L data.

T9. McLean, D.A., Proc. IEEE, vol 52, 1450, 1964. Dielectric constant at room temp., 1 kHz = 21.2. Dissipation factor = 0.002 to 0.01. Dielectric strength = 12,700 V/mil.

T10. Berry, R.W. and J. Sloan, Proc. IRE, vol 47, 1070, 1959. Dielectric constant at room temp. = 25. Dissipation factor at 293K = 0.008 to 0.01 over the range 100 to 100,000 Hz. Properties depend on the voltage of anodic formation.

T11. Lloyd, P., Solid State Electr., vol 3, 74, 1961. Dissipation factor at 293K <0.01 up to 30 kHz. Dielectric strength = 17,800 V/mil.

T12. Eliezer, D. and D.G. Brandon, Thin Solid Films, vol 12, 319, 1972. Mechanical properties of anodically formed film were determined by the bulge test. Young's modulus = 44,800 ksi. Ultimate tensile strength = 50,000 psi.

T13. Wood, G.C. and T. Hodgkiess, Werdstoffe und Korrosion, 23, Jahrgang, Heft9/1972, 766-773. Hardness of beta-Ta_2O_5 = 469 Kg/mm².

T14. Moody, W.E., Georgia Inst Tech, Final Report DA-AH01-67-

C1494. Thermnal expansion coefficient 0 to 100C for crystalline material = 1.0 ppm/K.

COMPOUND SEMICONDUCTORS

GaAs

PROPERTY	@ 300K	UNITS	REFERENCES
Density	5.3164	g/cm^3	GA1 GA2 GA3
Specific Heat	0.0770	cal/g-K	GA4 GA5 GA6
Thermal Conductivity	.07 to .14	cal/cm-s-K	GA7 GA8 GA9
Thermal Expansion	$6.478x10^{-6}$	1/K	GA10 GA11 GA12
Resistivity [1]	$3.3x10^8$	ohm-cm	GA13 GA14 GA15
TCR (218 to 398K)	-710 to 0.0	ppm/K	GA16 GA17 GA18
Band Gap	1.47	eV	GA19 GA20 GA21

1 Typical room temperature values of resistivity are in the range 10^7 to 10^9 ohm-cm, depending on doping levels and surface conditions. See Properties of Gallium Arsenide, EMIS Datareview Series No. 2, p. 2.1, pub. INSPEC, 1986.

GALLIUM ARSENIDE

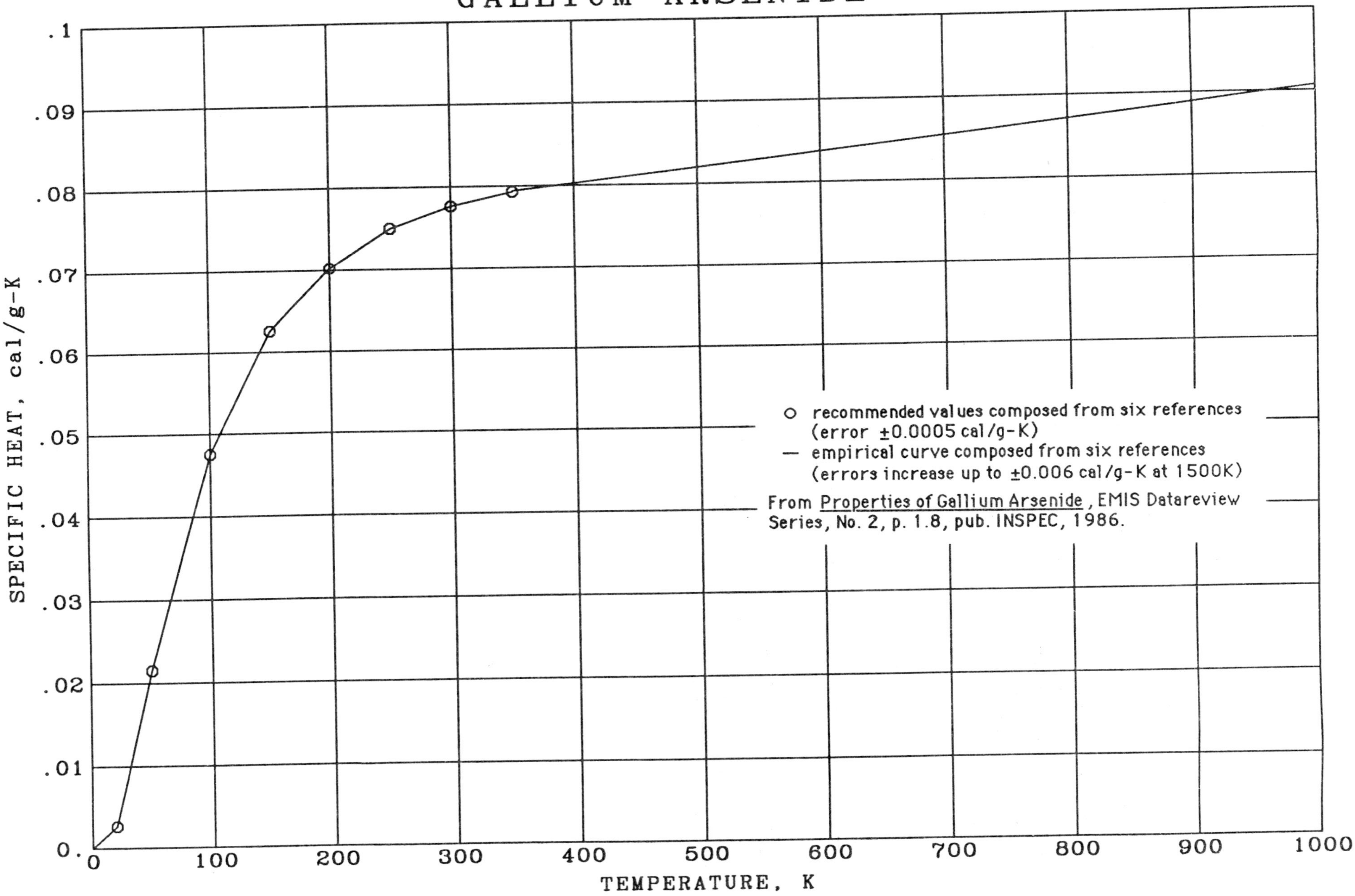

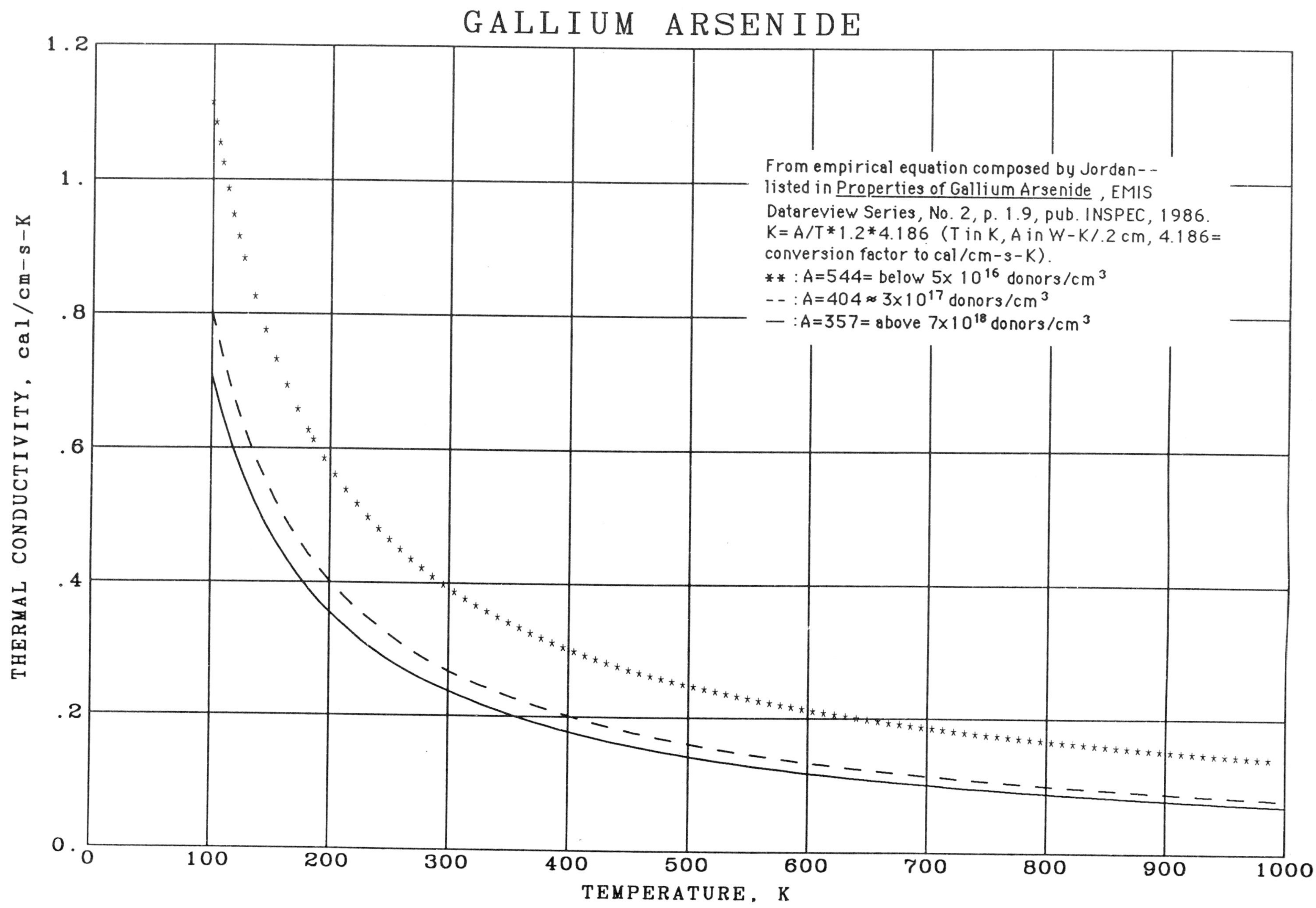
GALLIUM ARSENIDE
From empirical equation composed by Jordan--
listed in Properties of Gallium Arsenide , EMIS
Datareview Series, No. 2, p. 1.9, pub. INSPEC, 1986.
K= A/T*1.2*4.186 (T in K, A in W-K/.2 cm, 4.186=
conversion factor to cal/cm-s-K).
** : A=544= below 5x 10^16 donors/cm^3
-- : A=404 ≈ 3x10^17 donors/cm^3
— : A=357= above 7x10^18 donors/cm^3
THERMAL CONDUCTIVITY, cal/cm-s-K
TEMPERATURE, K
0.
.2
.4
.6
.8
1.
1.2
0
100
200
300
400
500
600
700
800
900
1000

GALLIUM ARSENIDE

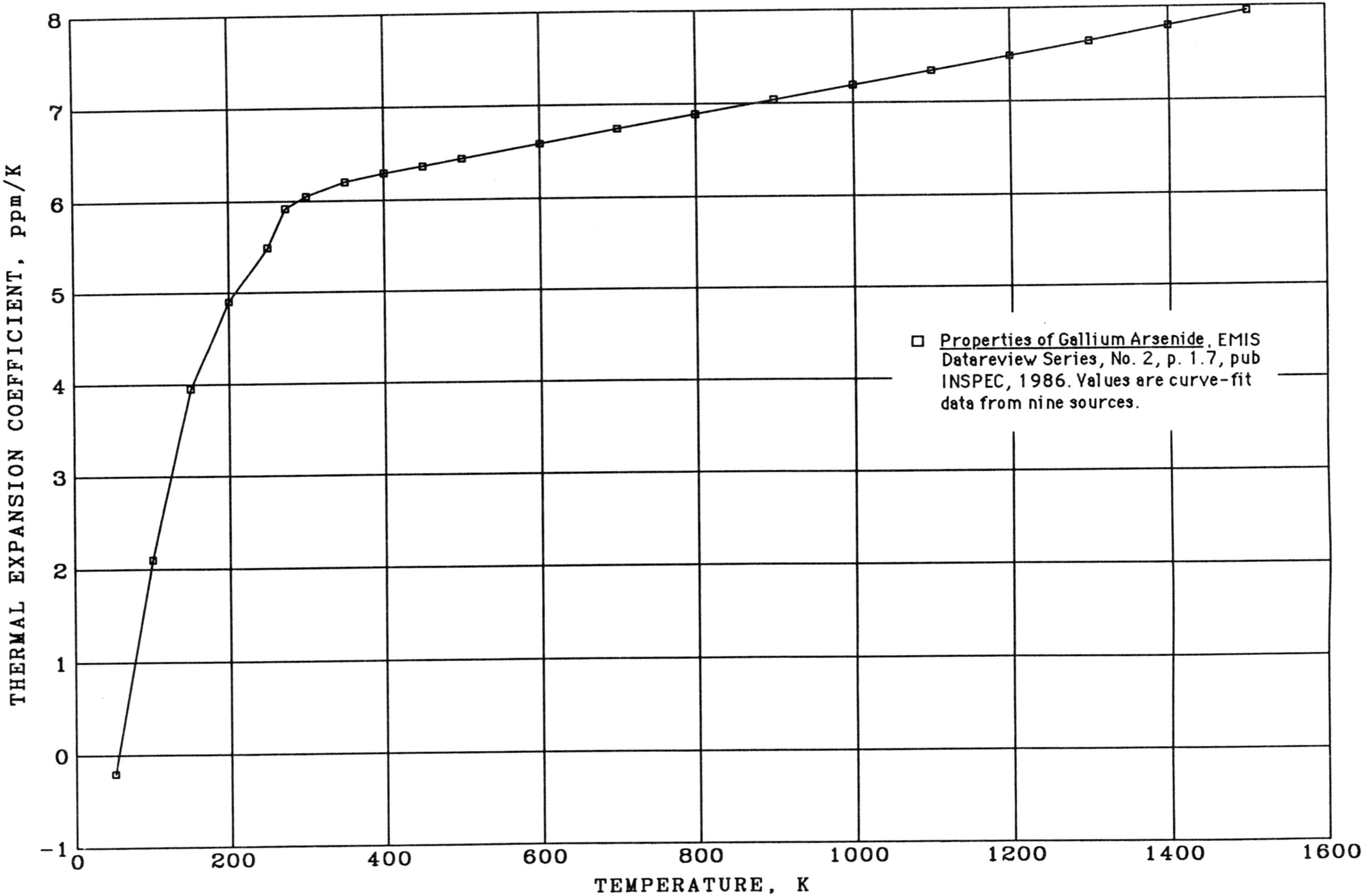

GALLIUM ARSENIDE

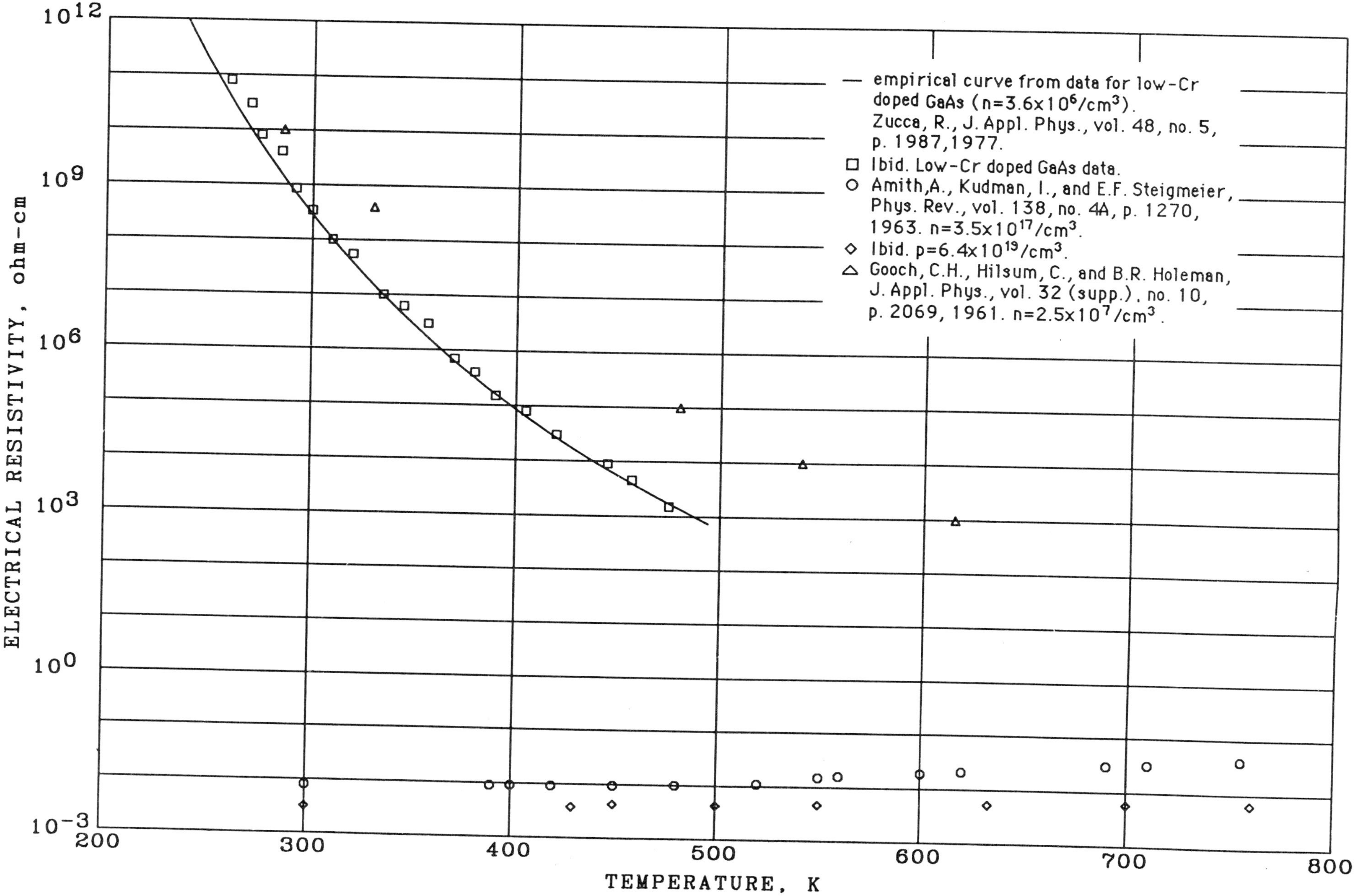

GALLIUM ARSENIDE

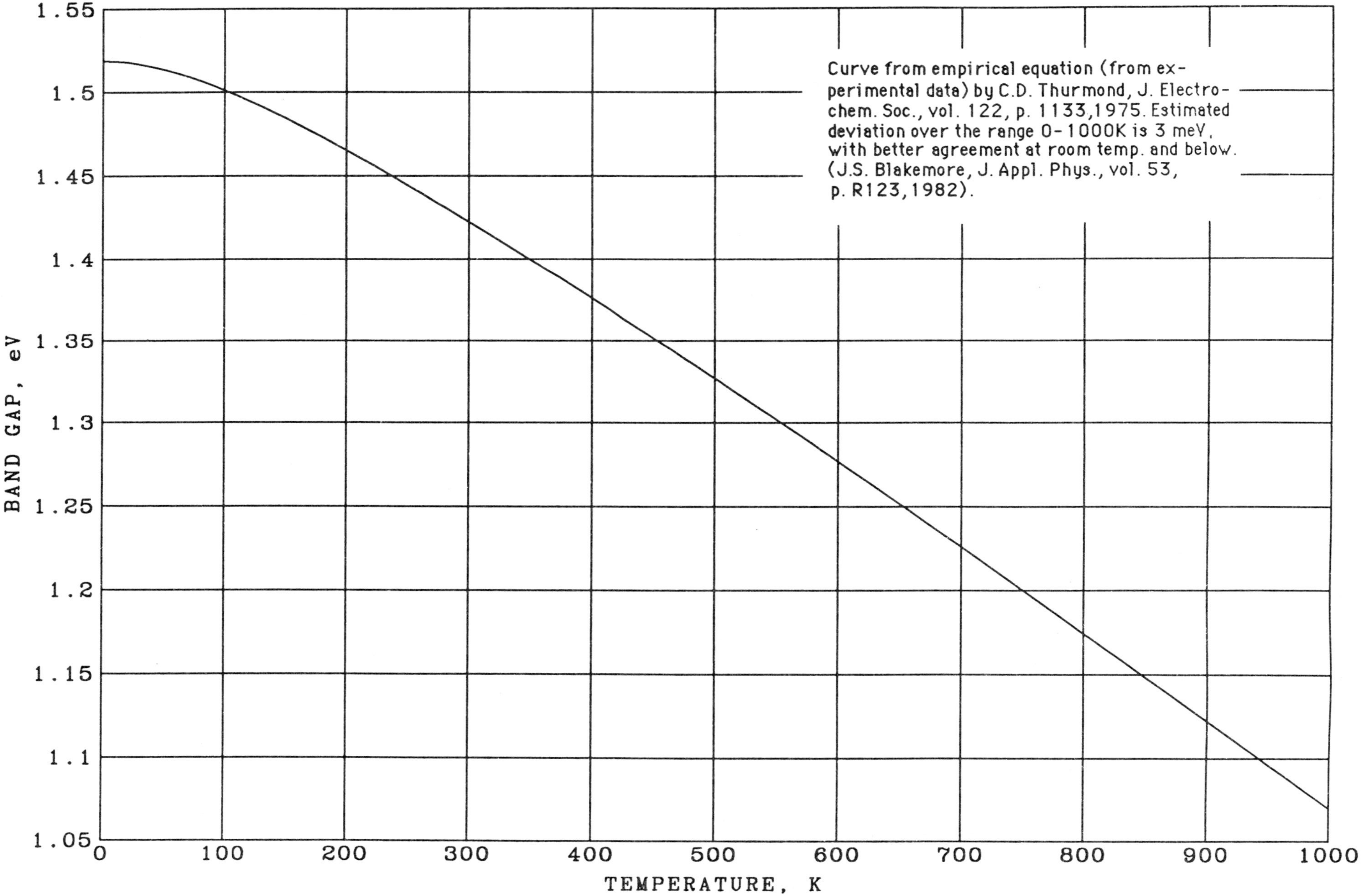

SiC

PROPERTY	@ 300K	UNITS	REFERENCES
Density	3.2	g/cm^3	SC1 SC2 SC3
Specific Heat	0.6732	cal/g-K	SC4 SC5 SC6
Thermal Conductivity	.143-.925	cal/cm-s-K	SC7 SC8 SC9
Thermal Expansion 1	3.4×10^{-6}	1/K	SC10 SC11 SC12
Resistivity	10^{-2}-10^{1}	ohm-cm	SC13 SC14 SC15
TCR (218 to 398K) 2	----	ppm/K	SC16 SC17 SC18
Band Gap	2.77	eV	SC19 SC20 SC21

1 Recommended value from TPRC Data Series, vol. 13, Thermal Expansion--Nonmetallic Solids, Y.S. Touloukian, ed., p. 873, 1976. Value is composed from four sources of polycrystalline alpha silicon carbide.

2 TCR varies from -.00068 to .0022 ppm/K over this range.

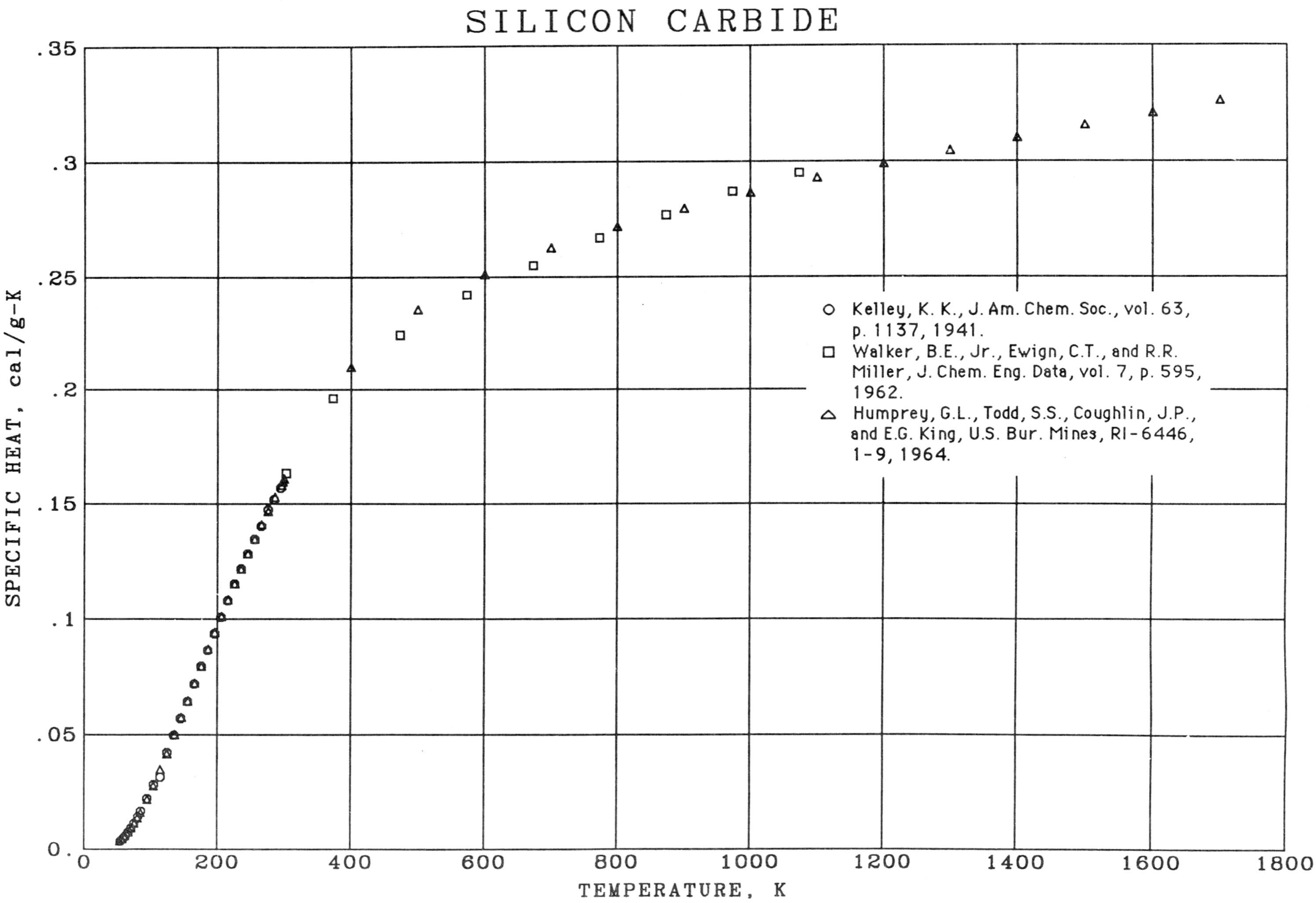
SILICON CARBIDE
SPECIFIC HEAT, cal/g-K
.35
.3
.25
.2
.15
.1
.05
0.
0
200
400
600
800
1000
1200
1400
1600
1800
TEMPERATURE, K
Kelley, K. K., J. Am. Chem. Soc., vol. 63, p. 1137, 1941.
Walker, B.E., Jr., Ewign, C.T., and R.R. Miller, J. Chem. Eng. Data, vol. 7, p. 595, 1962.
Humprey, G.L., Todd, S.S., Coughlin, J.P., and E.G. King, U.S. Bur. Mines, RI-6446, 1-9, 1964.

SILICON CARBIDE

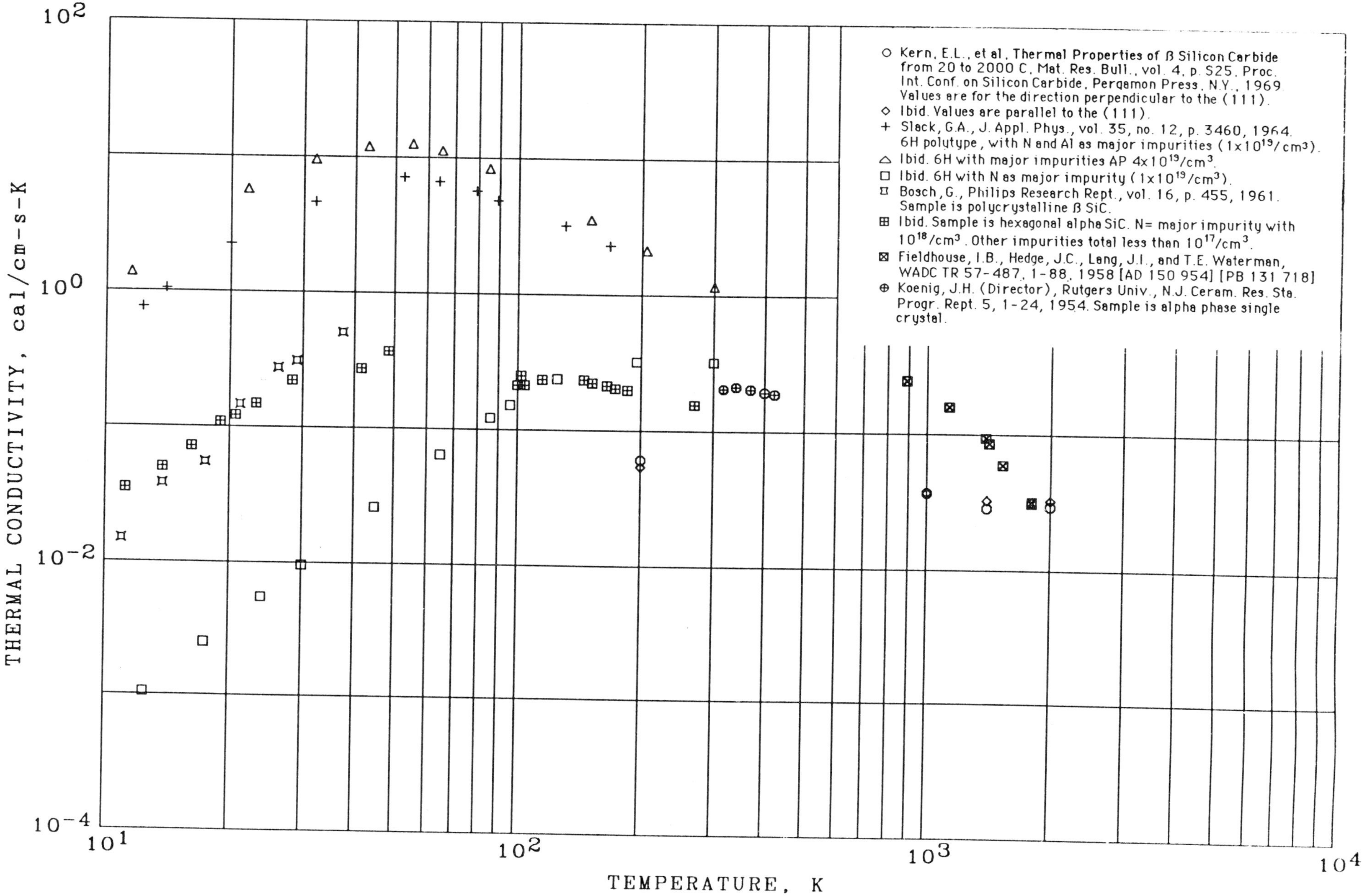

SILICON CARBIDE

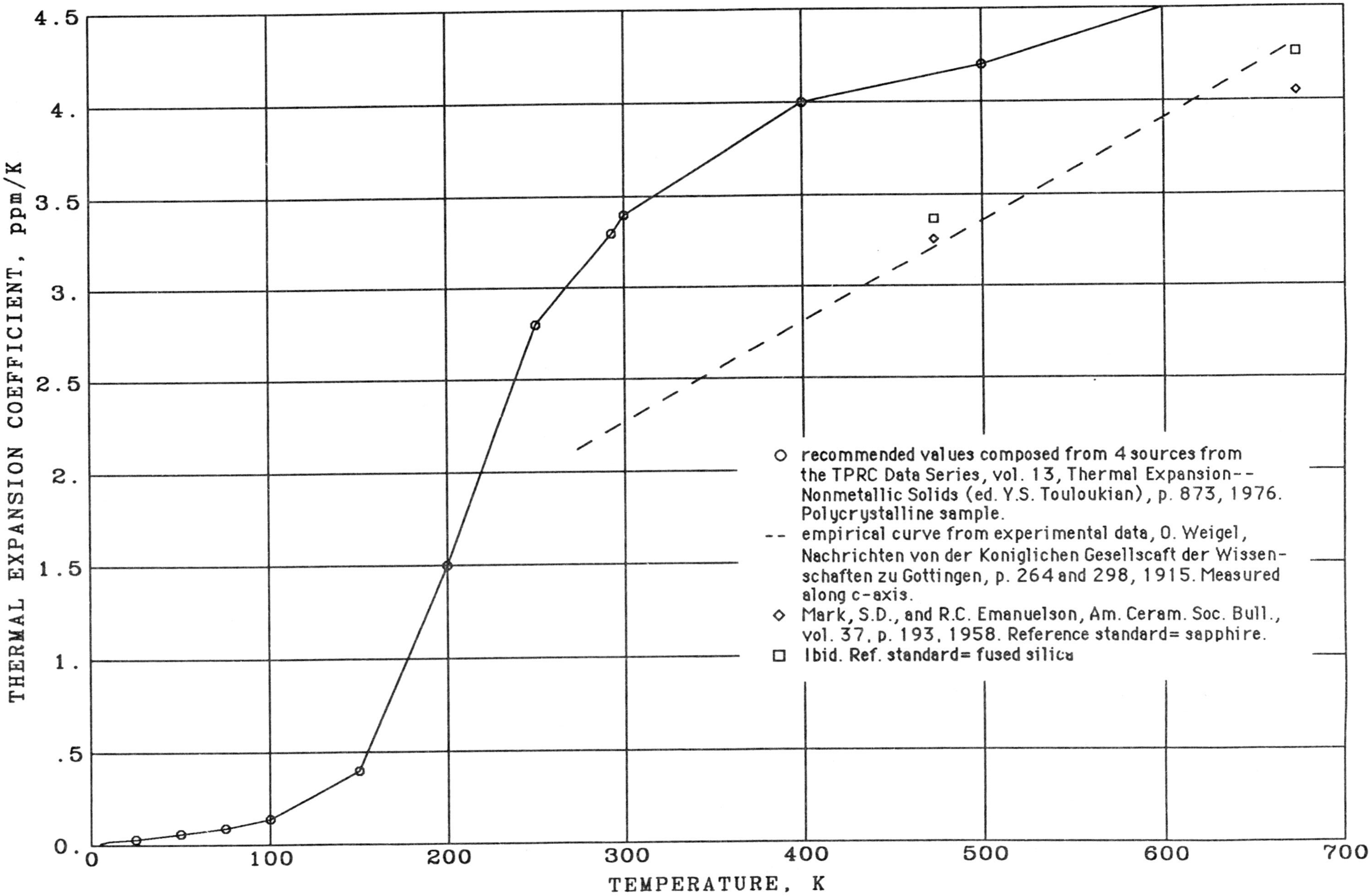

SILICON CARBIDE

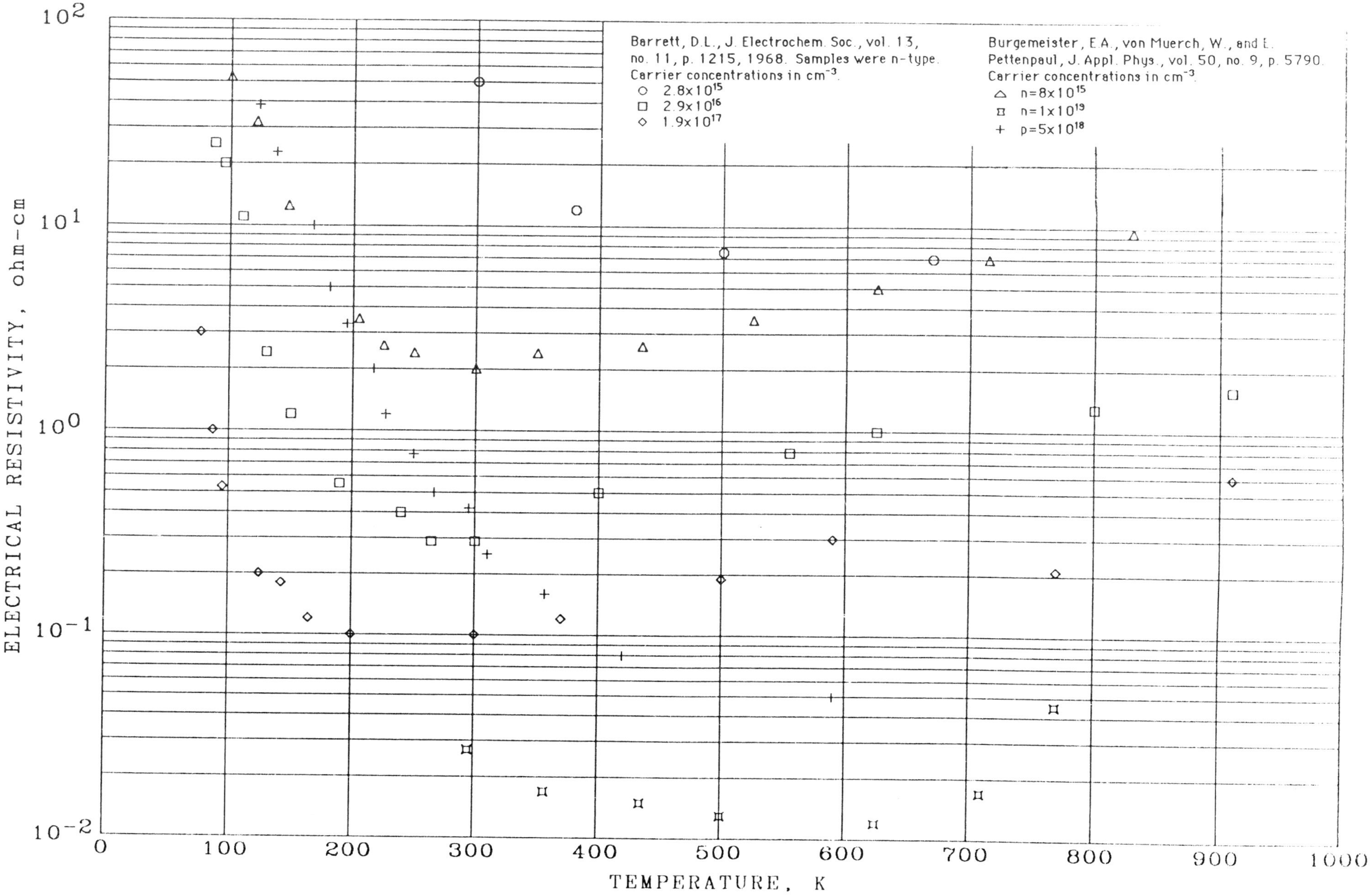

SILICON CARBIDE

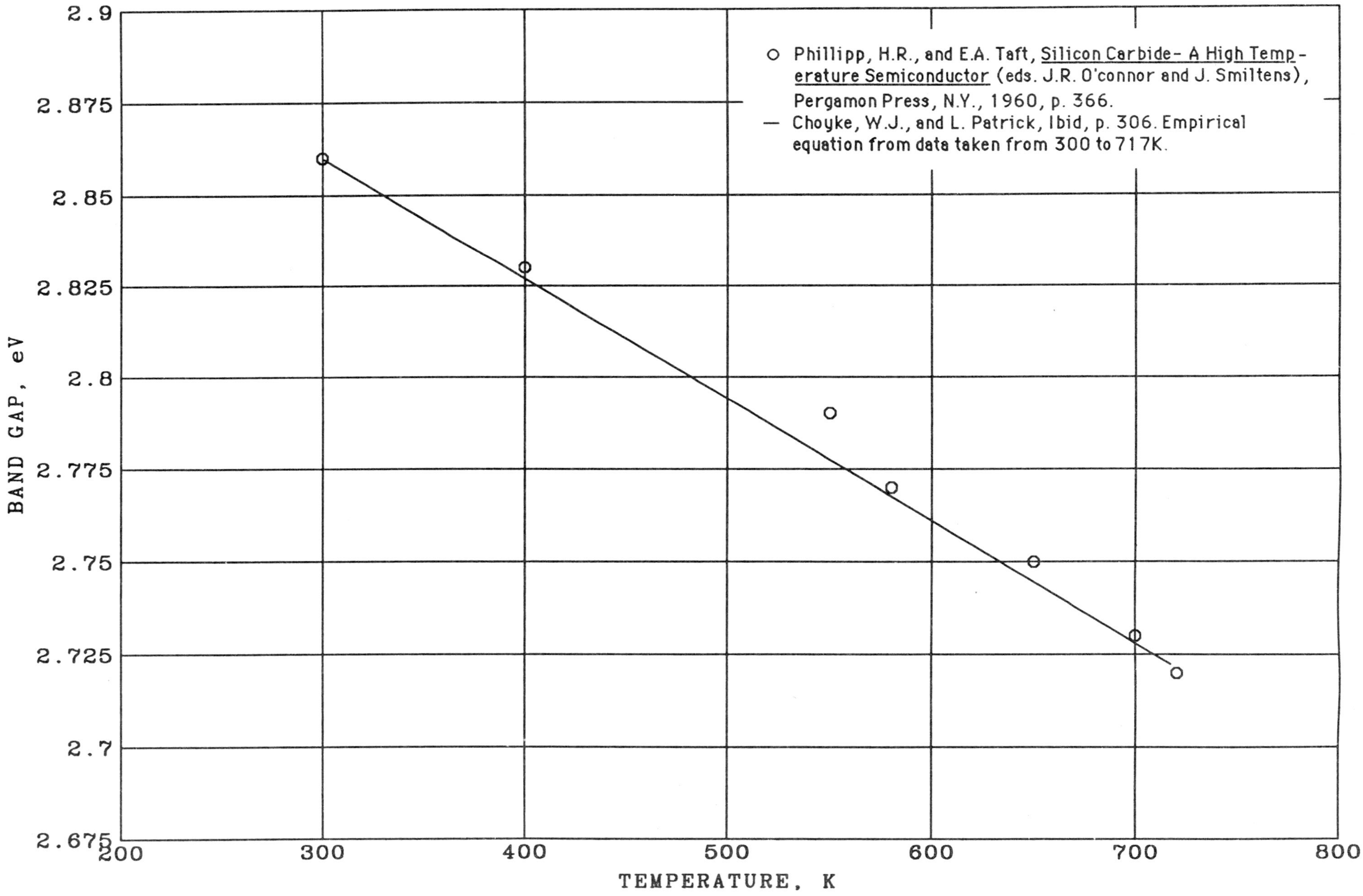

Ta_2N

PROPERTY		@ 300K	UNITS	REFERENCES
Density	1	15.6	g/cm^3	TN1 TN2 TN3
Specific Heat	2	0.0432	cal/g-K	TN4
Thermal Conductivity	2	0.024±0.005	cal/cm-s-K	TN5
Thermal Expansion	2	$5.2x10^{-6}$ 3	1/K	TN6
Resistivity	4	$1.75x10^{-4}$	ohm-cm	TN7 TN8 TN9
TCR (218 to 398K)	4	180	ppm/K	TN10 TN11 TN12
Band Gap	5	1.95-2.6	eV	TN13

1 Bulk density listed. Thin film density is reported to be 14.3 g/cm^3.[TN3]

2 Properties for bulk materials listed.

3 Value over the temperature range 293-1273K.

4 Bulk value listed. Thin film resistivities vary from 200-1500 micro-ohm-cm (see references). Thin film TCR varies from +2 to -400 ppm/K. See references.

5 Band gap range for $Ta_{1.15}N$. No data found for Ta_2N.

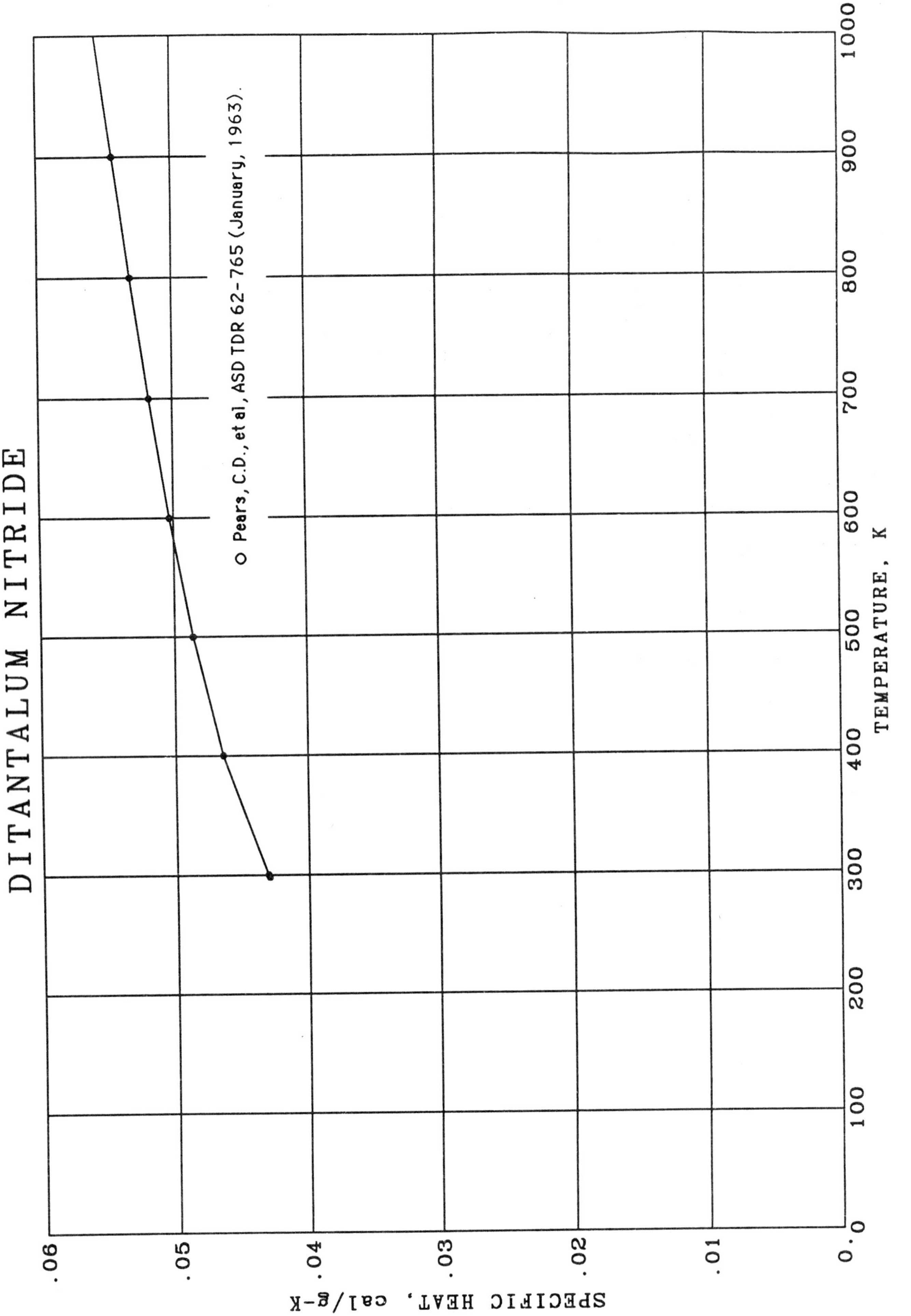
DITANTALUM NITRIDE
○ Pears, C.D., et al, ASD TDR 62-765 (January, 1963).
SPECIFIC HEAT, cal/g-K
.06
.05
.04
.03
.02
.01
0.0
TEMPERATURE, K
0
100
200
300
400
500
600
700
800
900
1000

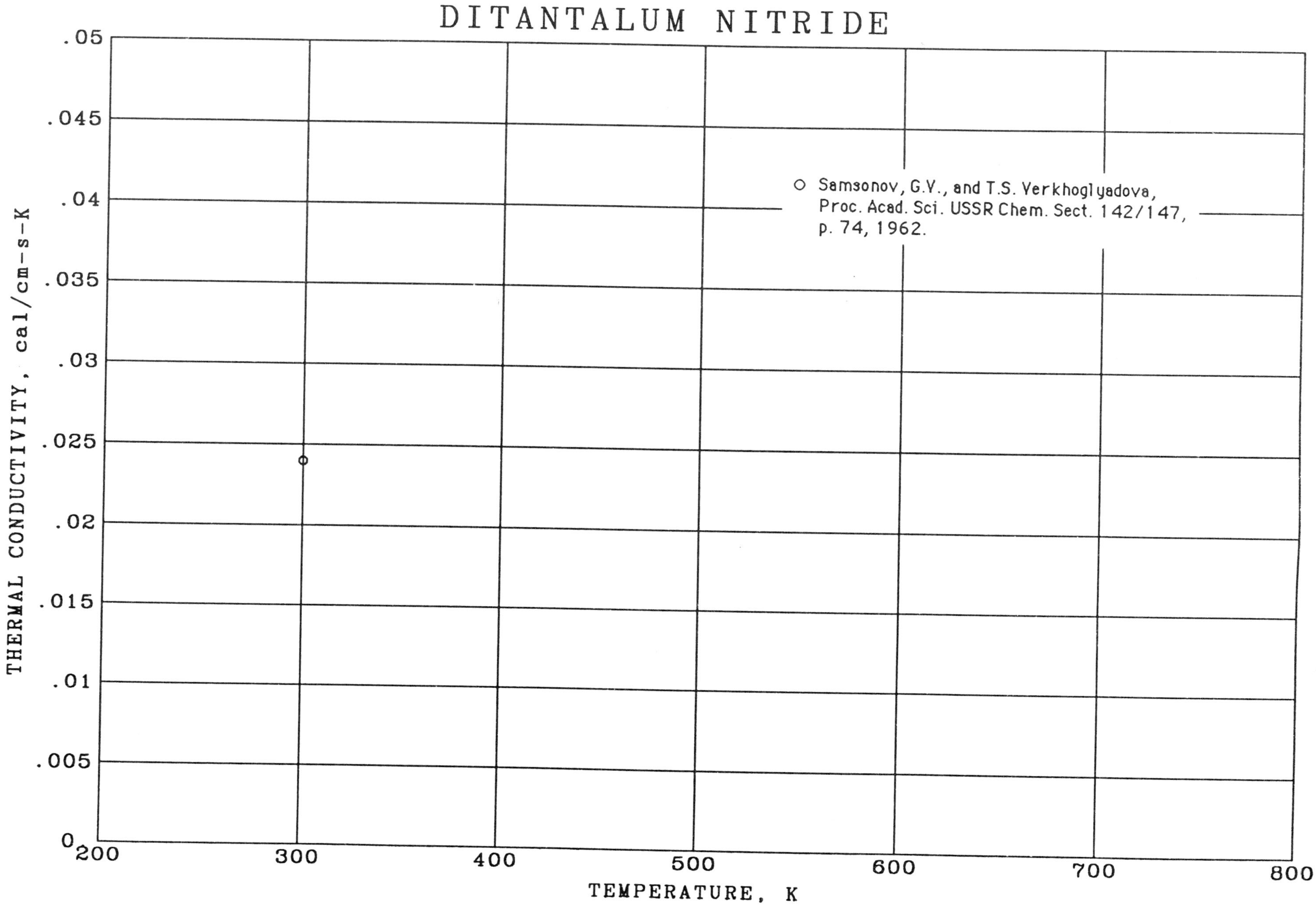
DITANTALUM NITRIDE
Samsonov, G.V., and T.S. Verkhoglyadova, Proc. Acad. Sci. USSR Chem. Sect. 142/147, p. 74, 1962.
THERMAL CONDUCTIVITY, cal/cm-s-K
TEMPERATURE, K
.05
.045
.04
.035
.03
.025
.02
.015
.01
.005
0
200
300
400
500
600
700
800

DITANTALUM NITRIDE

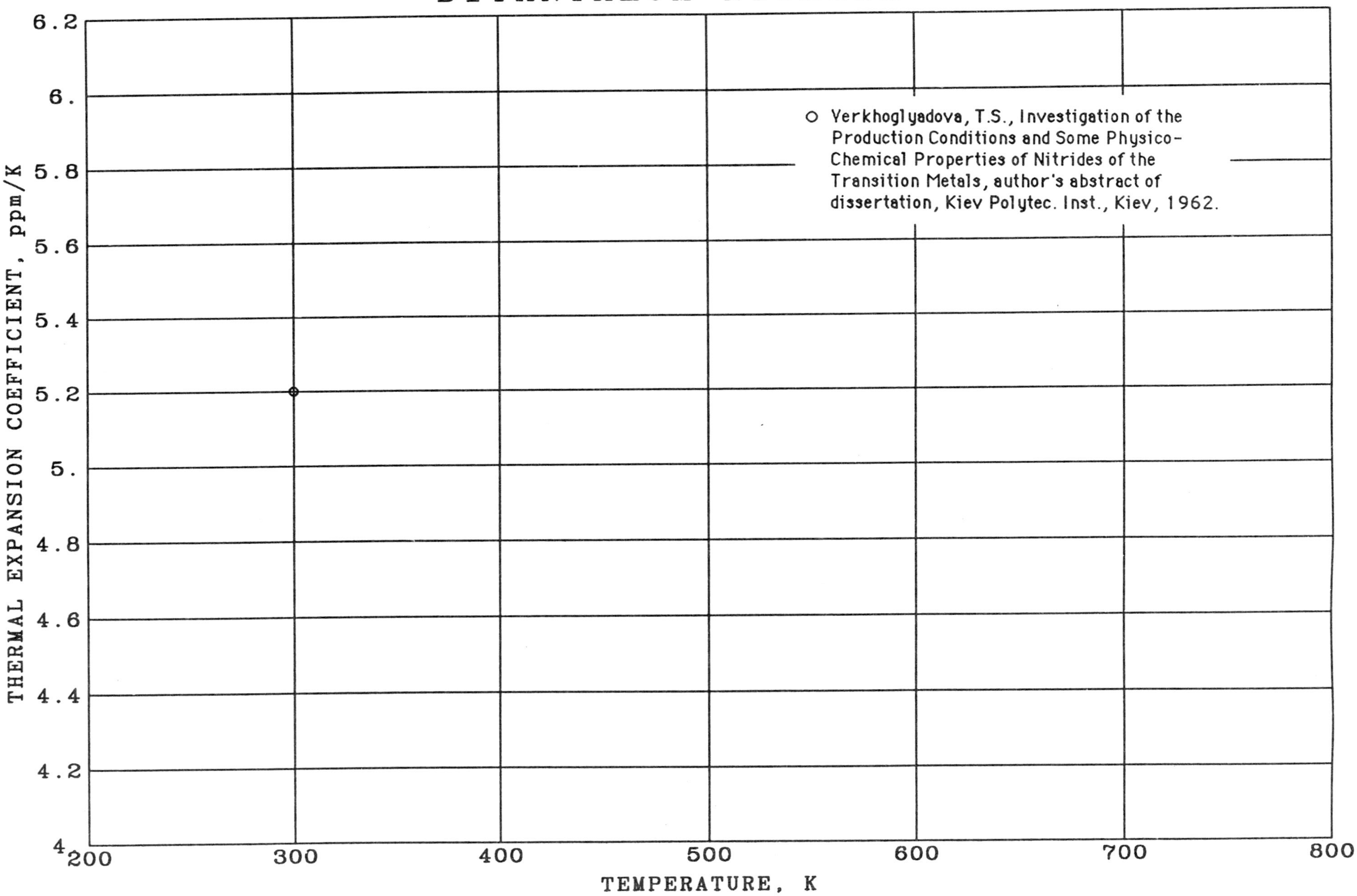

DITANTALUM NITRIDE

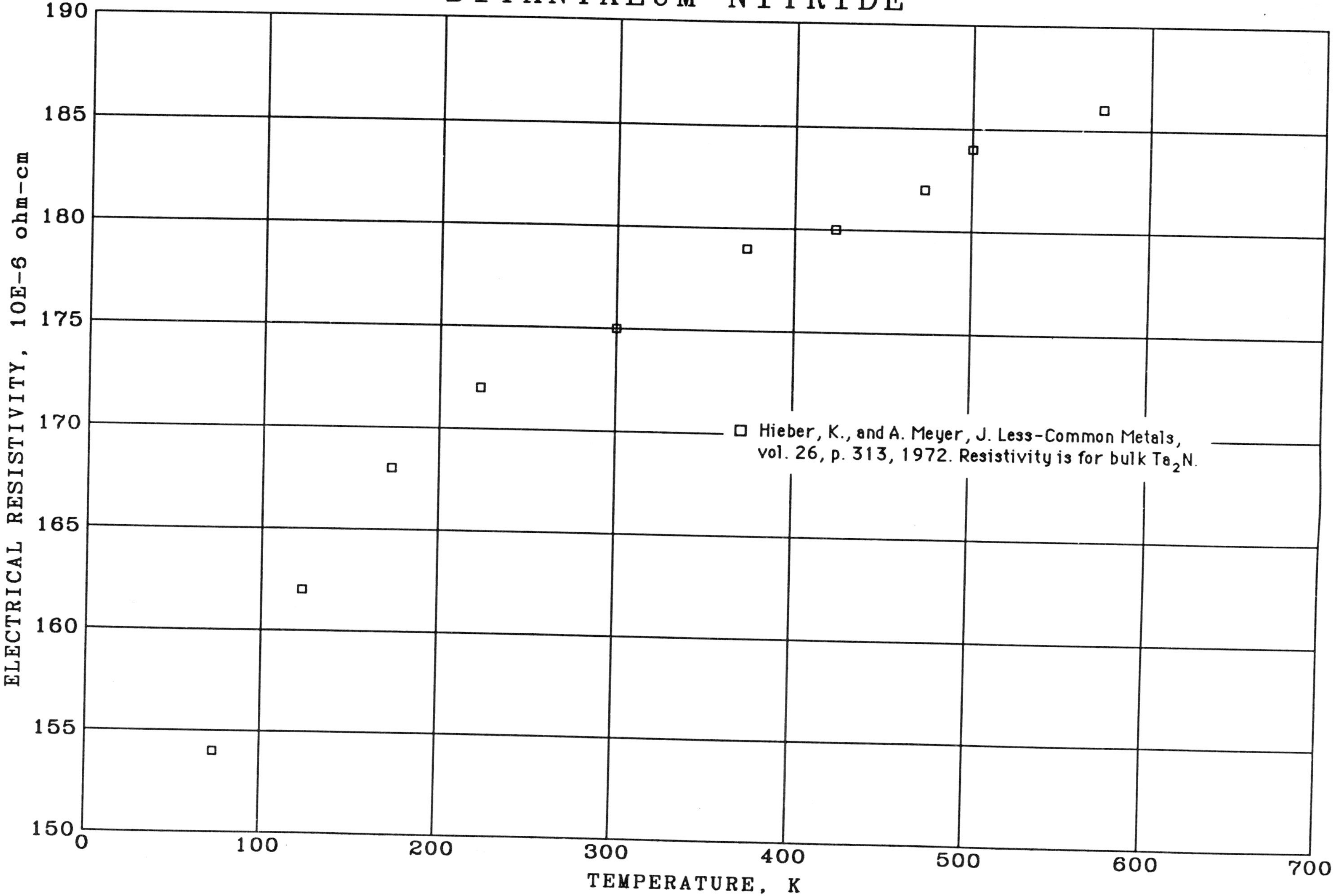

DITANTALUM NITRIDE

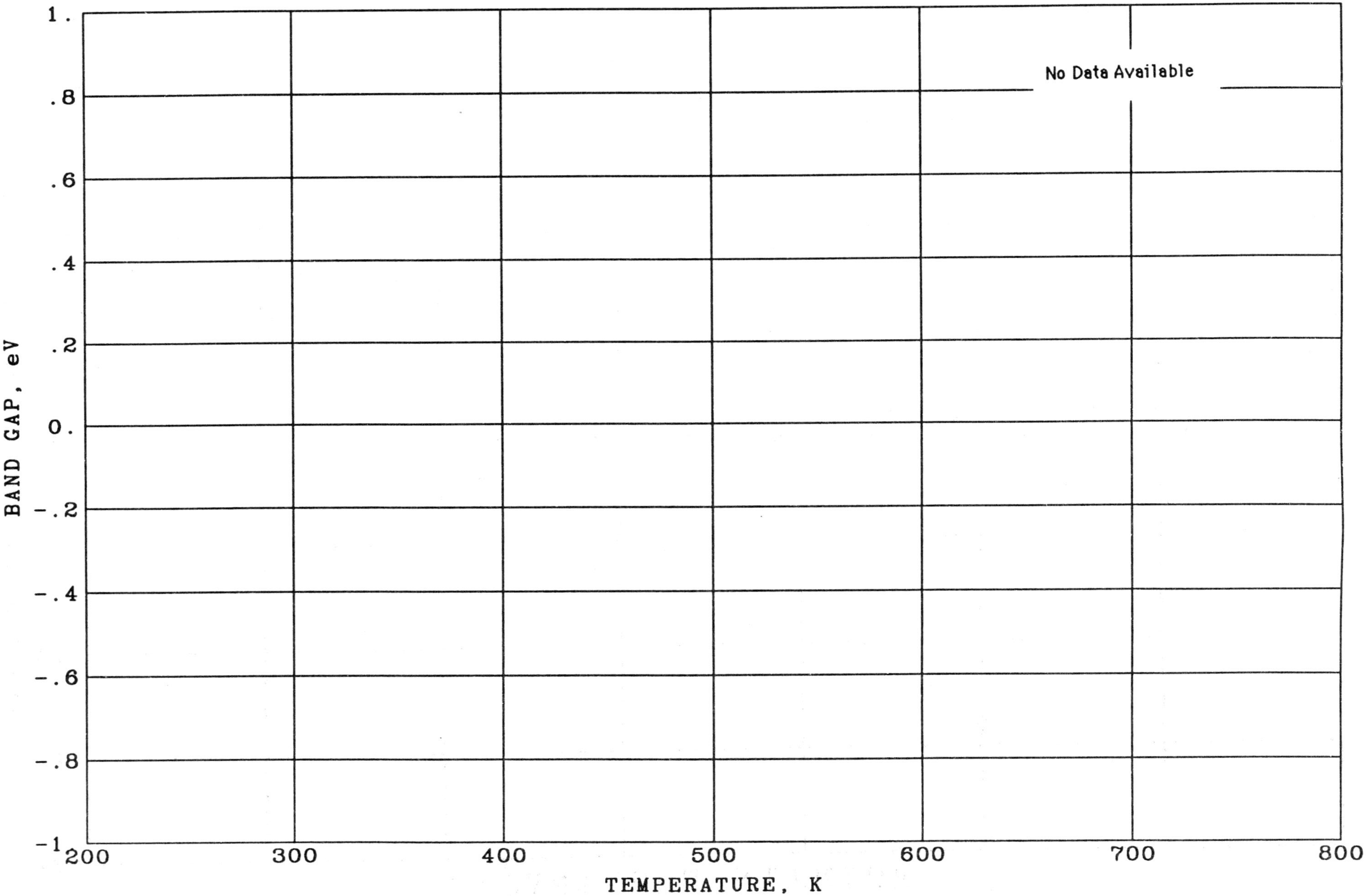

$CrSi_2$

PROPERTY	@ 300K	UNITS	REFERENCES
Density	4.95	g/cm^3	CS1 CS2 CS3
Specific Heat	0.13339	cal/g-K	CS4 CS5 CS6
Thermal Conductivity	$2.43x10^{-2}$	cal/cm-s-K	CS7 CS8 CS9
Thermal Expansion	$1.07x10^{-5}$	1/K	CS10 CS11
Resistivity	$8.25x10^{-4}$	ohm-cm	CS13 CS14 CS15
TCR (218 to 398K)	4139	ppm/K	CS16 CS17 CS18
Band Gap	0.34	eV	CS19 CS20 CS21

Properties listed are for bulk samples.

For the effect of doping level on transport properties, see graphs.

CHROMIUM DISILICIDE

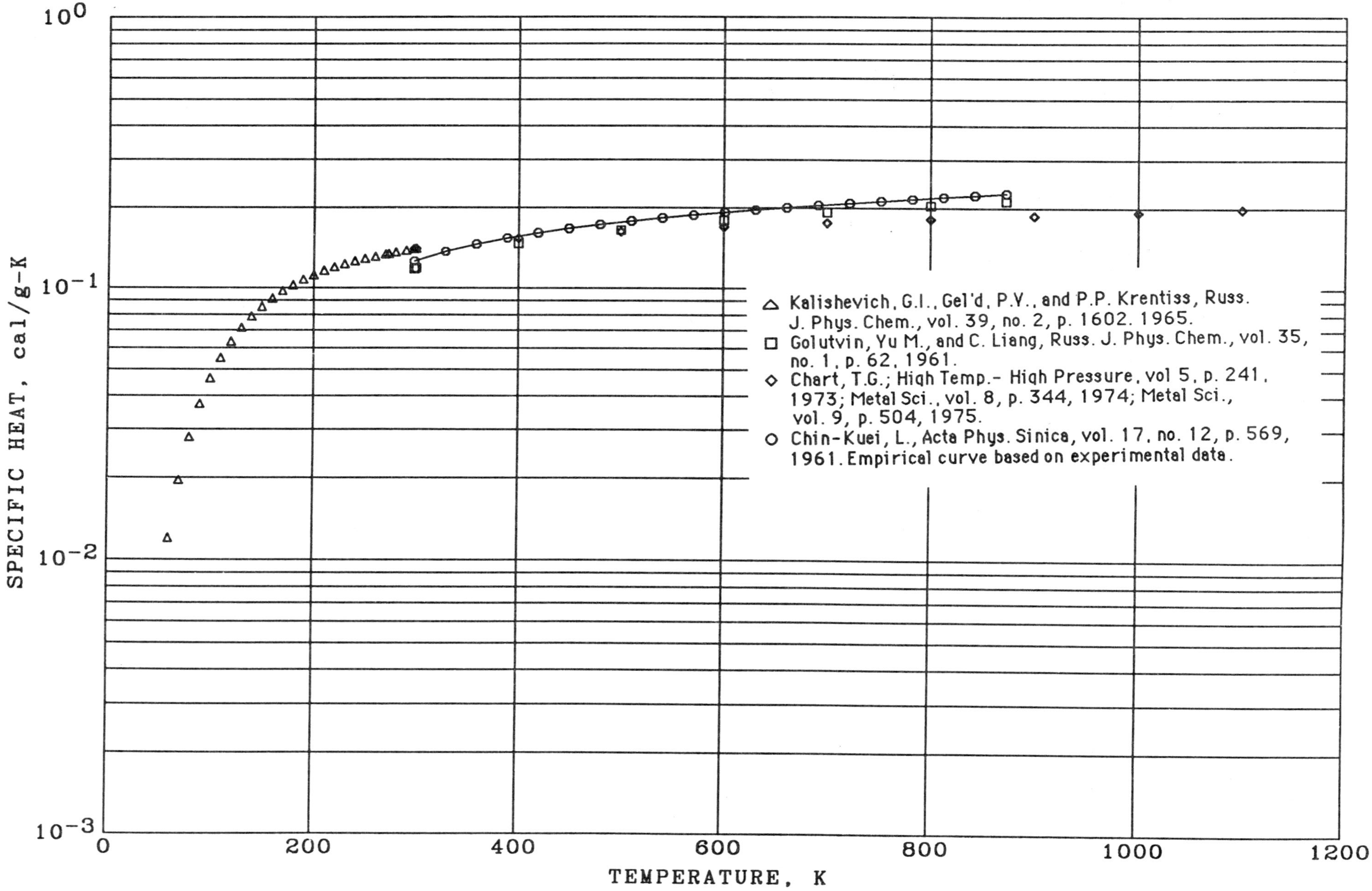

CHROMIUM DISILICIDE

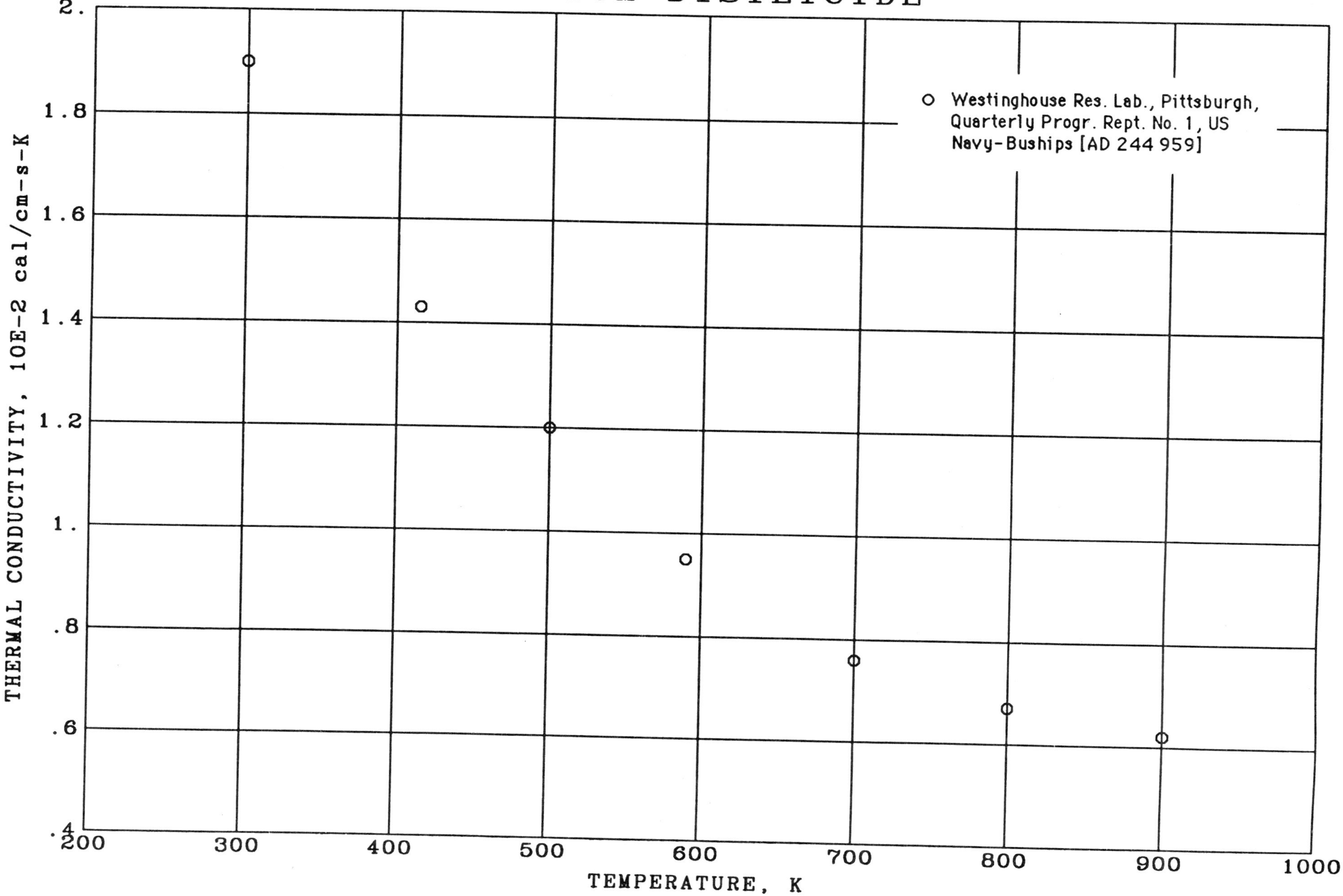

CHROMIUM DISILICIDE

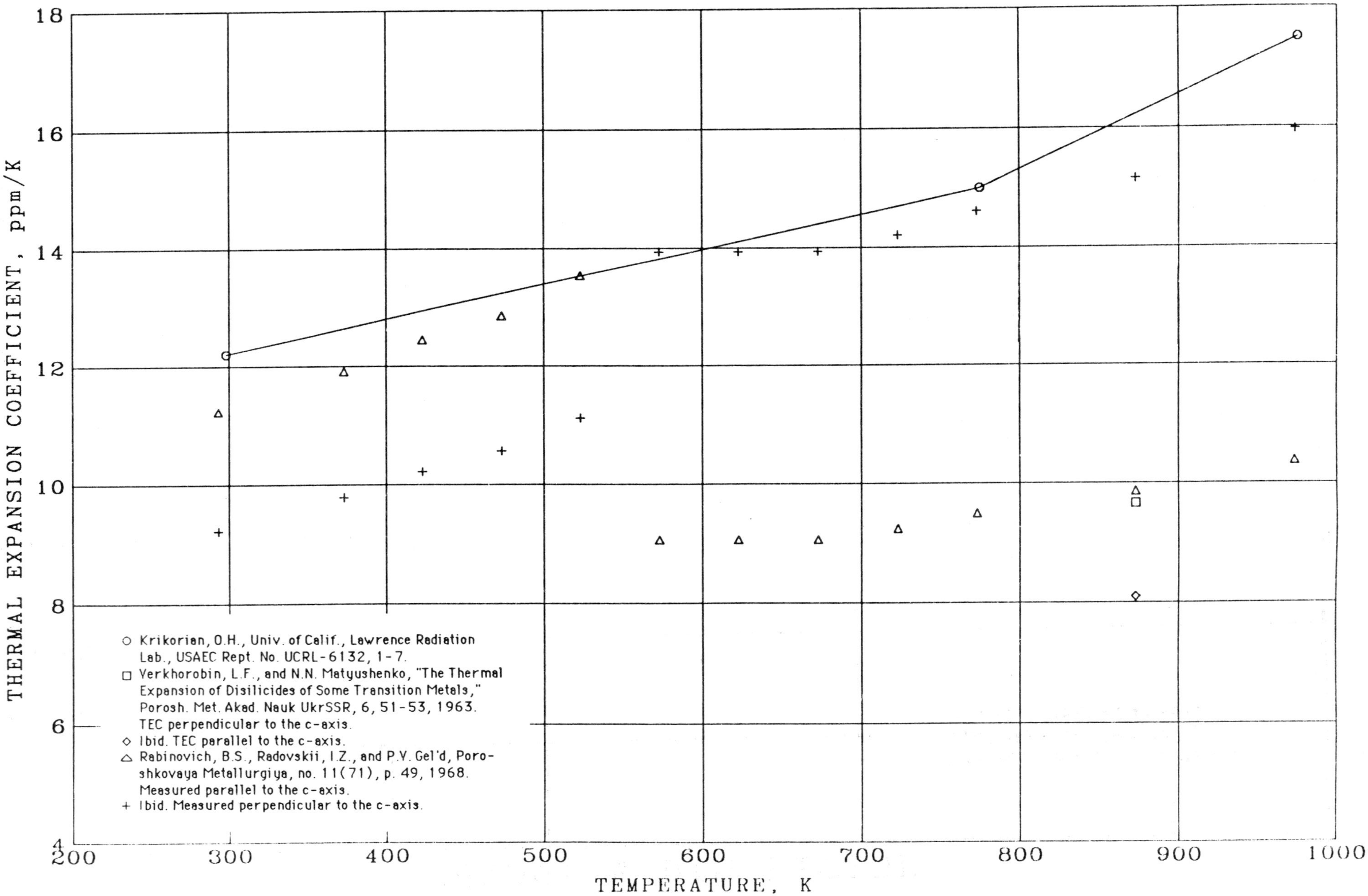

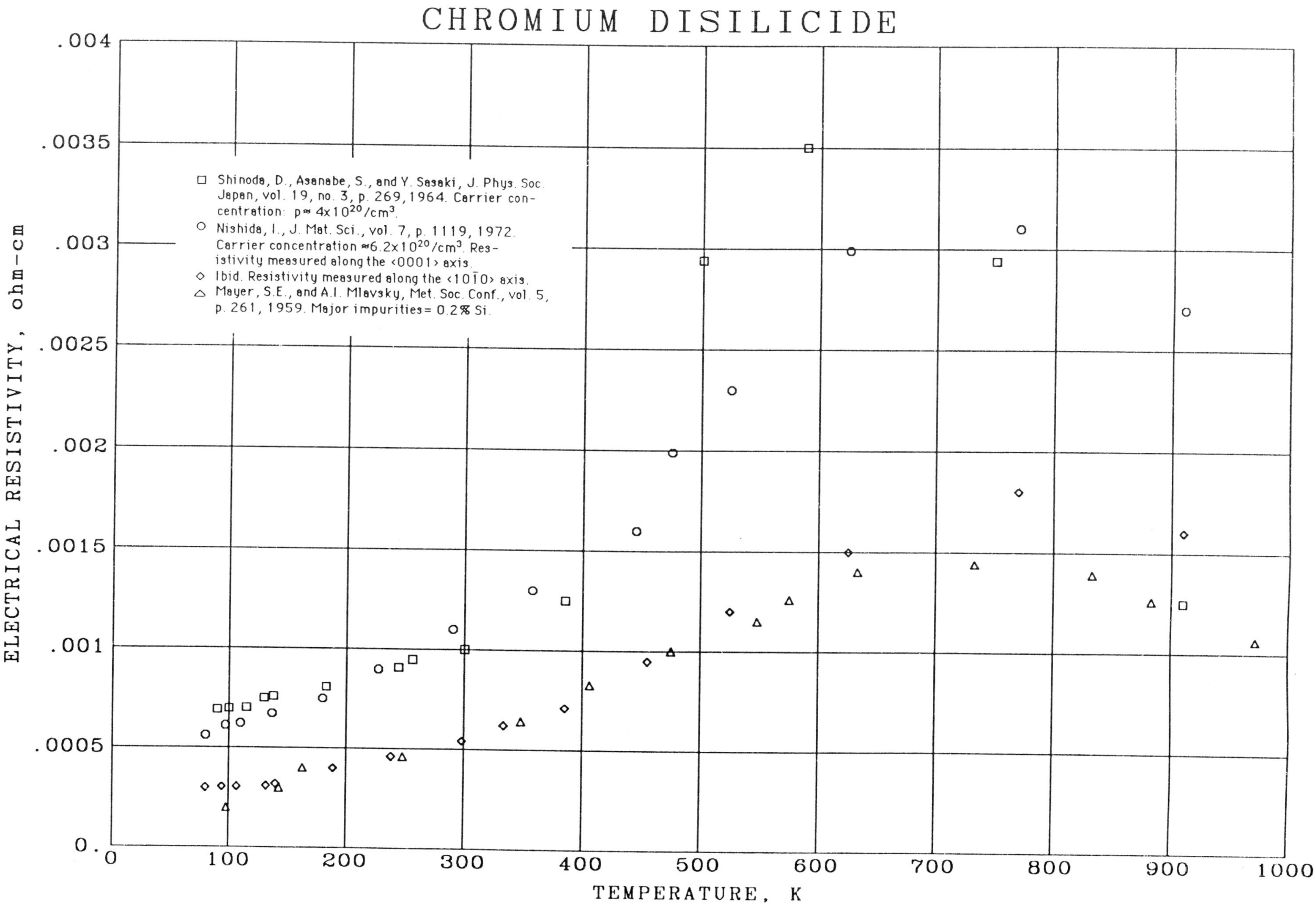
CHROMIUM DISILICIDE
ELECTRICAL RESISTIVITY, ohm-cm
.004
.0035
.003
.0025
.002
.0015
.001
.0005
0.
0
100
200
300
400
500
600
700
800
900
1000
TEMPERATURE, K
□ Shinoda, D., Asanabe, S., and Y. Sasaki, J. Phys. Soc. Japan, vol. 19, no. 3, p. 269, 1964. Carrier concentration: p≈ 4x10^{20}/cm^3.
○ Nishida, I., J. Mat. Sci., vol. 7, p. 1119, 1972. Carrier concentration ≈6.2x10^{20}/cm^3. Resistivity measured along the <0001> axis.
◇ Ibid. Resistivity measured along the <10$\bar{1}$0> axis.
△ Mayer, S.E., and A.I. Mlavsky, Met. Soc. Conf., vol. 5, p. 261, 1959. Major impurities= 0.2% Si.

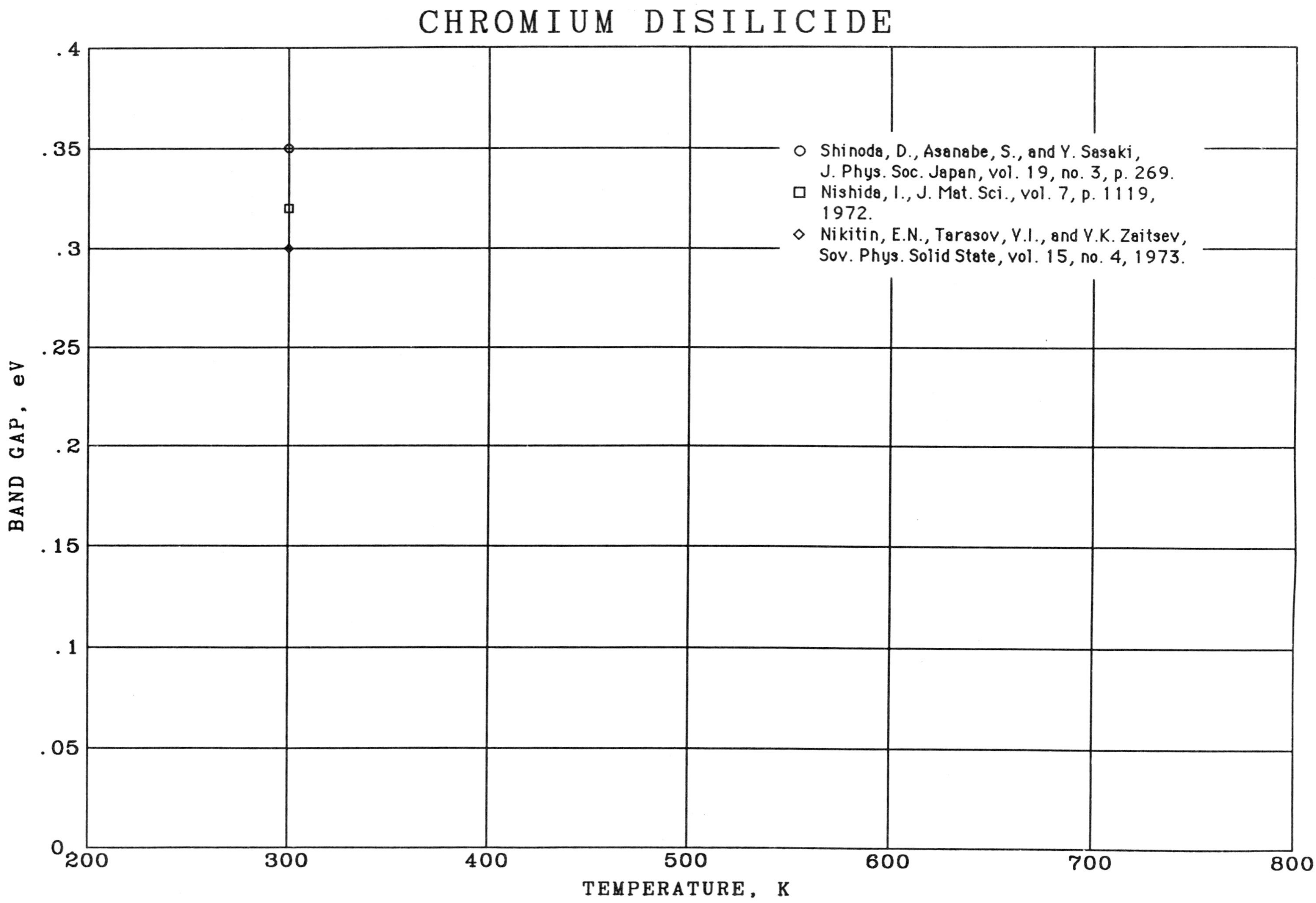
CHROMIUM DISILICIDE
BAND GAP, eV
.4
.35
.3
.25
.2
.15
.1
.05
0
200
300
400
500
600
700
800
TEMPERATURE, K
○ Shinoda, D., Asanabe, S., and Y. Sasaki, J. Phys. Soc. Japan, vol. 19, no. 3, p. 269.
□ Nishida, I., J. Mat. Sci., vol. 7, p. 1119, 1972.
◇ Nikitin, E.N., Tarasov, V.I., and V.K. Zaitsev, Sov. Phys. Solid State, vol. 15, no. 4, 1973.

Galium Arsenide

GA1. Weisberg, L.R., and J. Blanc, J. Appl. Phys., vol. 34, no. 4(I), p. 1002, 1963. Densities (in g/cm^3) @ 300K= 5.3162±0.0002, 5.3161±0.0001, and 5.3161±0.0002 (for three different samples).

GA2. Straumanis, M.E., and C.D. Kim, Acta Cryst., vol. 19, p. 256, 1965. Density @ 300K= 5.3175 g/cm^3.

GA3. Yarmolyu, N. I. et al, Sov. Phys. Semicond., vol. 14, p. 773, 1980. Density @ 300K=5.3160 g/cm^3. It was also shown that the density doesn't change with dopant concentrations up to an excess of $2x10^{18}/cm^3$.

GA4. Piesbergen, Von. U., Z. Naturforsch, vol. A18, no. 2, p.141, 1963. Specific heat @ 300K found from extrapolation from 273.2K and =0.0761 cal/g-K.

GA5. Lichter,B.D., and P. Sommelet, Trans. Metal. Soc. AIME, vol. 245, p. 1021, 1969. Specific heat @ 300K=0.0781 cal/g-K.

GA6. Dash, A.J., Finch, A., and P.J. Gardner, J. Chem. Eng. Data, vol. 19, no. 2, pp. 113, 1974. Specific heat @ 300K found from extrapolation back from 310K and =0.0767 cal/g-K.

GA7. Maycock, P.D., Solid State Elec., vol. 10, p. 161, 1967. Thermal conductivity (cal/cm-s-K) @ 300K= 0.0836 (p-type, $6.4x10^{19}/cm^3$), 0.0932 (n-type, $8x10^{18}/cm^3$), 0.1051 (high purity n-type, $5x10^{16}/cm^3$). Error in measurement ±10%.

GA8. Carlson, R.O., Slack, G.A., and S.J. Silverman, J. Appl. Phys., vol 36, no. 2, p. 505, 1965. Thermal conductivity (in cal/cm-s-K) @ 300K=0.1409 (high purity n-type, $10^{16}/cm^3$), and 0.1218 (n-type, $1.4x10^{16}/cm^3$).

GA9. Wright, T.R., Fackelmann, J.M., Kizer, D.E., and D.L. Keller, USAEC BMI-1644 (Del), U1-3, 1963. Thermal conductivity @ 300K=0.0788 cal/cm-s-K (p-type, $\simeq 10^{18}/cm^3$).

GA10. Pierron, E.D., Parker, D.L., and J.B. McNeely, Acta Cryst., vol. 21, p. 290, 1966. Thermal expansion coefficient @ 300K= $6.86x10^{-6}/K$.

GA11. Straumanis, M.E. and C.D. Kim, Acta Cryst., vol. 19, p. 256, 1965. Thermal expansion coefficient @ 300K= $6.4x10^{-6}/K$.

GA12. Nan, Shaw, and Liu Yi-Huan, Sci. Sinica, vol. 14, no. 11, p.1585, 1965. Thermal expansion coefficient @ 300K= $6.174x10^{-6}/K$.

GA13. Blakemore, J.S., J. Appl. Phys., vol. 53, no. 10, p. 520, 1982. Electrical resistivity @ 300K= $3.3x10^{14}$ $\mu\Omega$-cm (p-type, impurity concentration=$2.1x10^{-6}/cm^3$).

GA14. Sze, S.M., and J.C. Irvin, Solid State Elec., vol. 11, pp. 599, 1968. Electrical resistivity @ 300K= $11.0x10^6$ $\mu\Omega$-cm (n-type, impurity concentration=$10^{14}/cm^3$).

GA15. Zucca, R., J. Appl. Phys., vol. 48, no. 5, p. 1987, 1977. Electrical resistivity @ 300K= $3.4x10^{14}$ $\mu\Omega$-cm (low Cr doping, n=$3.6x10^6/cm^3$).

GA16. Zucca, R., J. Appl. Phys., vol. 48, no. 5, p. 1987, 1977. TCR over the range 218 to 398K= -710 ppm/K (from an empirical curve of GaAs with low Cr doping, n=3.6x $10^6/cm^3$).

GA17. Gooch, C.H., Hilsum, C., and B.R. Holeman, J. Appl. Phys., vol. 32 (supp.), no. 10, p. 2069, 1961. TCR over the range 218 to 398 K (extrapolated/curve-fit from data, 260 to 620K,n=$2.5x10^7$/ cm^3)= -6.9 ppm/K.

GA18. Amith, A., Kudman, I., and E.F. Steigmeier, Phys. Rev., vol. 138, no. 4A, p. 1270, 1963. TCR over the range 218 to 398K (extrapolated/curve-fit from data, 300 to 760K)= 0.00006 ppm/K (n=$3.5x10^{17}/cm^3$), 0.0008 ppm/K (p=$6.4x10^{19}/cm^3$).

GA19. Shay, J.L., Phys. Rev. B, vol. 2, p. 803, 1970. Band gap @ 300K= 1.42 eV.

GA20. Spitzer, W.G., and C.A. Mead, Phys. Rev., vol. 133, p.A872, 1964. Band gap @ 300K=1.37eV and 1.67eV.

GA21. Moss, T.S., J. Appl. Phys., vol. 32 (supp.), no. 10, p.2136, 1961. Band gap @ 300K=1.41eV.

Silicon Carbide

SC1. JCPDS X-ray Powder Data File, Card No. 22-173, Joint Committee on Powder Diffraction Standards, Swarthmore, Pa. Density @ 300K= 3.218 g/cm^3 (6H polytype).

SC2. De Mesquita, A.H.G., Acta Cryst., vol. 23, p. 610, 1967. Density @ 300K= 3.211 g/cm^3 (6H polytype).

SC3. Engineering Properties of Selected Ceramic Materials, American Ceramic Society, Columbus, Oh., 5.2.3-1, 1966. Density @ 300K= 3.208 g/cm^3 (hexagonal α-SiC).

SC4. Kelley, K.K., J. M., Chem. Soc, vol. 63, p. 1137, 1941. Specific heat @ 300K found from extrapolating from 294.6K to 300K and =0.1596 cal/g-K.

SC5. Humprey, G.L., Todd, S.S., Coughlin, J.P., and E.G. King, U.S. Bur. Mines, RI-6446,p. 1, 1964. Specific heat @ 300K (in cal/g-K)= 0.1607 and 0.1615.

SC6. Walker, B.E., Jr., Ewign, C.T., and R.R. Miller, J. Chem. Eng. Data, vol. 7, p. 595, 1962. Specific heat @ 300K found by extrapolating back from 303.15 K and =0.1615 cal/g-K.

SC7. Slack, G.A., J. Appl. Phys., vol. 35, no. 12, p. 3460, 1964. Thermal conductivity @300K (in cal/cm-s-K)=1.17 (n-type,major impurity, N, concentration= 1x10^{17} atoms/cm^3), 0.323 (p-type, major impurity AP 4x10^{19} atoms/cm^3),and 1.00 (n-type, major impurity, N, Al 1x10^{19} atoms/cm^3).

SC8. Bosch, G., Philips Res. Rept., vol 16, p. 455, 1961. Thermal conductivity @ 300K= .143 cal/cm-s-K (6H polytype, major impurity=N @ 10^{18} atoms/cm^3).

SC9. Burgemeister, E.A., and E. Pettenpaul, J. Appl. Phys., vol 50, no. 9, p. 5790, 1979. Thermal conductivity @ 293K (in cal/cm-s-K, measured normal to the c-axis, 6H-polytype)= 0.339 (p-type, carrier concentration$\simeq$10^{20}/cm^3, doped with Al_4C_3), 0.925 (n-type, carrier concentration=8x10^{15}/cm^3), and 0.705 (n-type, doped with N_2, carrier concentration=1x10^{19}/cm^3).

SC10. Whittemore, O.J., Jr., and N.N. Ault, "Thermal Expansion of Various Ceramic Materials to 1500°C," J. Amer. Ceram. Soc., vol. 39, no. 12, p. 443, 1956. Thermal expansion coefficient @ 300K= 4.000x10^{-6}/K, obtained by measuring the slope of the thermal expansion curve between 293 and 298K.

SC11. Schaake, H.F., "Study of Thermal and Mechanical Properties of Selected Solids from 4 Degrees to the Melting Point," USAF Rept. AFCRL-69-0538, 1969 (AD699 579). Thermal expansion coefficient @ 300K= 1.667x10^{-6}/K, obtained by measuring the slope of the thermal expansion curve between 273.2 and 299.6K.

SC12. Pojur, A.F., Yates, B., and Kelly, B.T., "Thermal Expansion at Elevated Temperatures, III. A Hemispherical Laminar Composite of Pyrolytic Graphite, Silicon Carbide and Its Constituents Between 300 and 800K.," J. Phys., vol. 5D, no. 7, p. 1321, 1972. Thermal expansion @ 300K=3.158x10^{-6}/K, obtained by measuring the slope of the thermal expansion curve between 331 and 350K.

SC13. Sasaki, W., J. Phys. Soc. Japan, vol. 7, p. 107, 1952. Electrical Resistivity @ 300K=6x10^{6} $\mu\Omega$-cm.

SC13. Nelson, W.E., Halden, F.A., and A. Rosengreen, J. Appl. Phys., vol. 37., no. 1, p. 333, 1966. Electrical resistivity @ 300K (in $\mu\Omega$-cm) of β-silicon carbide crystals= 7.0x10^{5} (carrier concentration=1.0x10^{16}/cm^{3}), 1.2x10^{5} (carrier concentration= 1.2x10^{17}/cm^{3}), and 3x10^{4} (carrier concentration=21.0x10^{19}/cm^{3}).

SC14. Barrett, D.L., J. Electrochem. Soc., vol. 113, no. 11, p. 1215, 1966. Electrical resistivity @ 300K (in $\mu\Omega$-cm)= 1.0x10^{5} (n=1.9x10^{17}/cm^{3}), 2.9x10^{5} (n= 6.5x10^{16}/cm^{3}), and 5.0x10^{7} (n=2.8x10^{15}/ cm^{3}).

SC15. Kang, H., and R.B. Hilborn, Jr., <u>Silicon Carbide</u>, Proc. Int. Conf., 3<u>rd</u>, 1973, p. 493. Electrical resistivity @ 300K (in $\mu\Omega$-cm, n-type 6H polytype)= 9.1x10^{4} (n= 5.2x10^{17}/cm^{3}), 5.0x10^{4} (n= 1.5x10^{18}/cm^{3}), 3.13x10^{4} (n= 2.5x10^{18}/cm^{3}).

SC16. Barrett, D.L., J. Electrochem. Soc., vol. 113, no. 11, p. 1215, 1966. TCR over the range 218 to 398K (in ppm/K) for β-SiC= .0022 (n=1.9x10^{17}/cm^{3}), and .0015 (n= 2.9x10^{16}/cm^{3}).

SC17. Burgemeister, E.A., von Muench, W., and E. Pettenpaul, J. Appl. Phys., vol. 50, no. 9, p. 5790, 1979. TCR over the range 218 to 398K (in ppm/K)= .0011 (n=8x10^{15}/cm^{3}) and -.0034 (p=5x10^{18}/cm^{3}).

SC18. Kang, H., and R.B. Hilborn, Jr., <u>Silicon Carbide</u>, Proc. Int. Conf., 3<u>rd</u>, 1973, p. 493. TCR over the range 218 to 398 K (in ppm/K)= -0.0038 (n=2.5x10^{18}/cm^{3}), -0.0075 (n=1.5x10^{18}/cm^{3}), and -0.00068 (n=5.2x10^{17}/cm^{3}).

SC19. Phillip, H.R. and E.A. Taft, Intrinsic Optical Absorption in Single Crystal Silicon Carbide, <u>Silicon Carbide, A High Temperature Semiconductor</u>, Proc. O'Connor, J.R. 3 Smiltens,J., p. 366, Pergamon Press, NY, 1960. Band gap @ 300K=2.86eV.

SC20. Choyke, W.J. and L. Patrick, Absorption of Light in Alpha Silicon Carbide Near the Band Edge, Ibid, p. 306. Band gap @ 300K=2.86 eV.

SC21. Ziomek, J.S. and P.B. Pickar, Phys. Status Solidi, vol. 21, p. 271, 1967. Band gap @ 300K=2.6 eV.

Ditantalum Nitride

TN1. Schonberg, N., Acta Chem. Scand., vol 8, p. 199, 1954. Density @ 300K= 15.42 g/cm^3.

TN2. Kamei, T., and E. Matsuzaki, Thin Solid Films, vol 77, p. 259, 1981. Density @ 300K= 14.3 g/cm^3.

TN3. Brauer, G, and K.H. Zapp, Z. Anorg. Allgem. Chem., 277 [1954] p. 129. Density @ 300K= 15.46 g/cm^3.

TN4. Pears, C.D., et al, ASD TDR 62-765 (January, 1963). Specific heat @ 300K= 0.0432 cal/g-K.

TN5. Samsonov, G.V., and T.S. Verkhoglyadova, Proc. Acad. Sci. USSR Chem. Sect. 142/147, p. 74, 1962. Thermal conductivity @ 300K= 0.024±0.005 cal/cm-s-K..

TN6. Verkhoglyadova, T.S., Investigation of the Production Conditions and Some Physico-Chemical Properties of Nitrides of the Transition Metals, author's abstract of dissertation, Kiev Polytec. Inst., Kiev, 1962. Thermal expansion coefficient @ 300K= 5.2×10^{-6}/K (range 293-1273K).

TN7. Verkhoglyadova, T.S., Investigation of the Production Conditions and Some Physico-Chemical Properties of Nitrides of the Transition Metals, author's abstract of dissertation, Kiev Polytec. Inst., Kiev, 1962. Electrical resistance @ 300K= 263.0±22μΩ-cm.

TN8. Hieber, K. and A.N. Meyer, J. Less-Common Metals, vol. 26, p. 313, 1972. Electrical resistivity @ 300K (bulk Ta_2N)= 175μΩ-cm.

TN9. Gerstenberg, D., and C.J. Calbick, J. Appl. Phys., vol. 35, p. 402, 1964. Electrical resistivity over the range 218 to 398K= 200-300 μΩ-cm (thin film).

TN10. Parisi, G.J., <u>Proc. Electron. Components, Conf., 1969</u>, Electrical and Electronic Engineers Inc., 1969, p. 367. TCR over the range 218 to 398K (thin film)= -50 to -400 ppm/K.

TN11. Gerstenberg, D., and D.J. Calbick, J. Appl. Phys., vol. 35, p. 402, 1964. TCR over the range 218 to 398K= -60 ppm/K (thin film).

TN12. Hieber, K., and A.N. Meyer, J. Less-Common Metals, vol. 26, p. 313, 1972. TCR over the range 218 to 398K (bulk)= 180 ppm/K.

TN13. Coyne, J.J., Jr., and R.N. Tauber, J. Appl. Phys., vol. 39, no. 12, p. 5585, 1968. Band gap @ 300K for $Ta_{1.15}N$ ranges from 1.95 to 2.6 eV.

Chromium Disilicide

CS1. Nishida, I., and T. Sakata, J. Phys. Chem. Solids, vol. 39, p. 499-505, 1978. Density= 4.94 g/cm^3.

CS2. Samsonov, G.V. Silicides and Their Application in Technology, Izd. AN UkrSSR, Kiev, 1959. Density @ 300K= 5.00 g/cm^3.

CS3. Boren, B., Ark. Kem. Min., 11 A Nr. 10 [1933/35] 1/28, 8/10. Density @ 300K= 4.91 g/cm^3.

CS4. Kalishevich, G.I., Gel'd, P.V., and R.P. Krentsis, Russ. J. Phys. Chem., vol. 39, no. 12, pp. 1602-3, 1965. Specific heat @ 300K=0.14086 cal/g-K.

CS5. Golutvin, Yu M., and C. Liang, Russ. J. Phys., Chem., vol. 35, no. 1, pp. 62-7, 1961. Specific heat @ 300K= 0.11849 cal/g-K.

CS6. Cart, T.G.: High Temp.-High Pressure vol. 5 (1973), p. 241; Metal Sci., vol 8, p. 344, 1974, vol. 9, p. 504, 1975. Specific heat @ 300K= 0.14082 cal/g-K.

CS7. Westinghouse, Res. Lab., Pittsburgh, Quarterly Progr. Rept., no. 1, US Navy-Buships. (AD244 959). Thermal conductivity @ 300K= $1.9x10^{-2}$ cal/cm-s-K.

CS8. Neshpor, V.S., Inzhenerno-Fizicheskii Zhurnal, vol. 15, no. 2, p. 321. 1968. Thermal conductivity @ 300K= $2.34x10^{-2}$ cal/cm-s-K (polycrystalline sample).

CS9. Voronov, B.K., Dudkin, L.D., and N.N. Trusova, Sov. Phys.--Cryst., vol. 12, no. 3., p. 448, 1967. Thermal conductivity @ 300K (in cal/cm-s-K)= $2.6x10^{-2}$, $2.78x10^{-2}$, and $2.55x10^{-2}$.

CS10. Krikorian, O.H., Univ. of Calif., Lawrence Radiation Lab., USAEC Rept. No. UCRL-6132, 1-7, 1960. Thermal expansion coefficient @ 300K= $12.2x10^{-6}$/K (over the range 298K to 973K).

CS11. Verkhorobin, L.F., and N.N. Matyushenko, "The Thermal Expansion of Disilicides of Some Transition Metals, " Porosh. Met. Akad. Nauk Ukr. SSR, 6, 51-3, 1963 [English translation in Sov. Powder Met. Metal. Ceram., 6, 468-9, 1963]. Thermal expansion coefficient (in 1/K, over the range 293 to 873K)= $9.2x10^{-6}$ (perpendicular to the c-axis) and $7.4x10^{-6}$ (parallel to the c-axis).

CS13. Nishida,I., J. Mat. Sci., vol. 7, p. 1119, 1972. Electrical resistivity @ 300K (in $\mu\Omega$-cm, carrier concentration= $6x2x10^{20}/cm^3$)= $1.3x10^3$ (along the <0001> axis) and $5.5x10^2$ (along the <10$\underline{1}$0> axis).

CS14. Mayer, S.E., and A.I. Mlavsky, Metals Society Conference, vol.5, pp 261, 1959. Electrical resistivity @ 300K=$5.5x10^2$ $\mu\Omega$-cm (p-type, impurities=0.2% Si).

CS15. Shinoda, D., Asanabe,S., and Y. Sasake, J. Phys. Soc. Japan, vol.19, no. 3, pp. 269-272, 1964. Electrical resistivity @ 300K= 1.0x10^3 $\mu\Omega$-cm (p=4x10^{20}/cm^3).

CS16. Shinoda, D., Asanabe, S., and Y. Sasaki, J. Phys. Soc. Japan, vol. 19, no. 3, p. 269. TCR over the range 218 to 398K= 5598 ppm/K (p-type, $\simeq$4x10^{20} carriers/cm^3).

CS17. Nishida, I., J. Mat. Sci., vol. 7, p. 1119, 1972. TCR over the range 218 to 398K (in ppm/K, hole concentration $\simeq$3.92x10^{20}/cm^3) = 2879 along the <0001> axis and 3282 along the <10$\underline{1}$0> axis.

CS18. Mayer, S.E., and A.I. Mlavsky, Metals Society Conference, vol. 5, p. 261, 1959. TCR over the range 218 to 398 K= 3737 ppm/K (p-type, impurities=0.2% Si).

CS19. Shinoda, D., Asanabe, S., and Y. Sasaki, J. Phys. Soc. Japan, vol. 19, no. 3, p. 269. Band gap @ 300K= 0.35 eV.

CS20. Nishida, I., J. Mat. Sci., vol. 7, pp. 1119-1124, 1972. Band gap @ 300K= 0.32 eV.

CS21. Nikitin, E.N., Tarasov, V.I., and V.K. Zaitsev, Sov. Phys. Solid State, vol. 15, no. 4, 1973. Band gap @ 300K= 0.30 eV.

GLOSSARY

Symbol or Abbreviation	*Definition*
α	true thermal expansion coefficient
α_m	mean thermal expansion coefficient
Btu	British thermal unit
°C	degrees centigrade
c_m	mean heat capacity
c_t	true heat capacity
cal	calories
cm	centimeter
δ	loss angle of a dielectric or path dependent derivative
E	Young's modulus
eV	electron volt
ϵ	dielectric constant or strain
ϵ_o	permittivity of free space
°F	degrees Fahrenheit
ft	foot
g	gram
in	inch
J	joule
j	heat flux
K	kelvin
k	thermal conductivity
ksi	thousand psi
kcal	kilocalories
kg	kilograms
kg_f	kilograms force
lb_m	pound mass
lb_f	pound force
m	meter
MPa	million pascals
N	newton
Pa	pascals or N/m^2
psi	lb_f/in^2
μ	micro or 10^{-6}
Ω	ohm
Q	quantity of heat
σ	stress
V	volt
W	watt (J/s)